INSECT NATURAL ENEMIES

INSECT NATURAL ENEMIES

Practical approaches to their study and evaluation

Edited by

Mark Jervis and Neil Kidd

School of Pure and Applied Biology
University of Wales, Cardiff
UK

CHAPMAN & HALL
London · Glasgow · Weinheim · New York · Tokyo · Melbourne · Madras

Published by Chapman and Hall, 2–6 Boundary Row, London SE1 8HN, UK

Chapman & Hall, 2–6 Boundary Row, London SE1 8HN, UK

Blackie Academic & Professional, Wester Cleddens Road, Bishopbriggs, Glasgow G64 2NZ, UK

Chapman & Hall GmbH, Pappelallee 3, 69469 Weinheim, Germany

Chapman & Hall USA, One Penn Plaza, 41st Floor, New York NY 10119, USA

Chapman & Hall Japan, ITP-Japan, Kyowa Building, 3F, 2-2-1 Hirakawacho, Chiyoda-ku, Tokyo 102, Japan

Chapman & Hall Australia, Thomas Nelson Australia, 102 Dodds Street, South Melbourne. Victoria 3205, Australia.

Chapman & Hall India, R Seshadri, 32 Second Main Road, CIT East, Madras 600 035, India

First edition 1996

Typeset in 10/12 Palatino by EXPO Holdings, Malaysia
Printed in Great Britain at the Alden Press, Osney Mead, Oxford

ISBN 0 412 39900 8

A catalogue record for this book is available from the British Library

Library of Congress Catalog Card Number: 94–68722

Printed on permanent acid-free text paper, manufactured in accordance with ANSI/NISO Z39 48-1992 and ANSI/NISO Z39 48-1984 (Permanence of Paper)

CONTENTS

CONTRIBUTORS

J.J.M. VAN ALPHEN
Institute of Evolutionary and Ecological Sciences,
Faculty of Mathematics and Natural Sciences,
University of Leiden,
Kaiserstraat 63,
P.O. Box 9516,
2300 RA Leiden,
The Netherlands

J. VAN DEN ASSEM
Van der Klaauw Laboratory,
Section Ethology,
University of Leiden,
Kaiserstraat 63,
P.O. Box 9516,
2300 RA Leiden,
The Netherlands

M.J.W. COPLAND
Department of Biological Sciences,
Wye College,
University of London,
Wye,
Ashford,
Kent TN25 5AH,
UK

M.A. JERVIS
School of Pure and Applied Biology,
University of Wales,
Cardiff,
P.O. Box 915,
Cardiff CF1 3TL,
Wales,
UK

N.A.C. KIDD
School of Pure and Applied Biology,
University of Wales,
Cardiff,
P.O. Box 915,
Cardiff CF1 3TL,
Wales,
UK

W. POWELL
Entomology and Nematology Department,
IACR Rothamsted
Harpenden AL5 2JQ,
UK

M.P. WALTON
School of Pure and Applied Biology
University of Wales,
Cardiff,
P.O. Box 915,
Cardiff CF1 3TL,
Wales,
UK

PREFACE

The past two decades have seen a dramatic increase in practical and theoretical studies on insect natural enemies. The importance and appeal of insect predators, and parasitoids in particular, as research animals derives from the relative ease with which some species may be cultured and experimented with in the laboratory, the simple life cycles of most parasitoids, and the increasing demand for biological control.

Unfortunately, despite a burgeoning of the literature on insect natural enemies, there has as yet been no general text available to guide enquiring students or research workers to those approaches and techniques that are most appropriate to the study and evaluation of such insects. Guidance on experimental design is particularly sought by newcomers to the subject, together with some idea of the pitfalls associated with various approaches.

The need for such a book as this was realised as a result of our experiences in supervising students in the large entomology post-graduate school at the University of Wales, Cardiff. We also took our inspiration from *Aphid Technology*, edited by H.F. van Emden (1972), which satisfied an important need among entomologists and ecologists by providing practical advice on how to study a particular group of insects. Our book is aimed at any student or professional interested in investigating the biology of predators and parasitoids, but will, we hope, be especially useful to post-graduates.

Our book is neither a practical manual nor a recipe book. Most chapters are accounts of major aspects of the biology of natural enemies, punctuated by practical information and advice on which experiments or observations to conduct and how, in broad terms, to carry them out. Detailed protocols are usually not given. Guidance is also provided, where necessary, on literature that may need to be consulted on particular topics. The coverage of the book is far from being exhaustive; some readers may be surprised at the omission of important topics such as dormancy, and others may be appalled that there is negligible treatment of systematics! For this, we apologise, but space was at a premium. Statistical aspects of sampling and experimental design are hardly gone into. This may be seen by some as an unforgivable lapse, but a wealth of advice on these essential aspects of scientific study is already contained in texts such as Southwood (1978), McDonald *et al.* (1989), Mead (1989), Hairston (1989) and Crawley (1993) (the latter text is concerned with a particularly valuable statistical software tool, GLIM). Ecologically-minded students of natural enemies will also find valuable advice in the special feature on statistics published in the September 1993 issue of the journal *Ecology*.

We hope that *Insect Natural Enemies* will successfully both encourage and assist the reader in his or her research into insect predators and parasitoids.

REFERENCES

Crawley, M.J. (1993) *GLIM for Ecologists*, Blackwell, Oxford.

van Emden, H.F. (1972) *Aphid Technology (with special reference to the study of aphids in the field)*, Academic Press, London.

Hairston, N.G. (1989) *Ecological Experiments: Purpose, Design and Execution,* Cambridge University Press, Cambridge.

McDonald, L.L., Manly, B., Lockwood, J. and Logan, J. (eds) (1989) *Estimation and Analysis of Insect Populations, Lecture Notes in Statistics 55,* Springer-Verlag, Berlin.

Mead, R. (1989) *The Design of Experiments,* Cambridge University Press, Cambridge.

Southwood, T.R.E. (1978) *Ecological Methods* (2nd edition), Chapman & Hall, London.

FORAGING BEHAVIOUR 1

J.J.M van Alphen and M.A. Jervis

1.1 BEHAVIOUR OF INSECT PARASITOIDS AND PREDATORS

1.1.1 INTRODUCTION

The study of insect predator and parasitoid behaviour is as old as ethology itself. Since Tinbergen's (1932, 1935), Salt's (1934), Ullyett's (1936) and Thorpe's (1939) pioneering studies our knowledge of predator and parasitoid behaviour in general has expanded considerably, for several reasons. Firstly, technological developments have improved our abilities to observe animals, to record our observations and to analyse the often quite complicated data gathered. Secondly, our knowledge of the stimuli that elicit behaviour has improved immensely, so enabling us to study with greater precision the causal aspects of behaviour. Thirdly, there has been a shift in emphasis from the almost strictly causal analysis to the study of the adaptive value (i.e. function) of behaviour. This move has been greatly facilitated by the use of mathematical models that generate testable quantitative predictions.

Students of parasitoid and predator biology are found among evolutionary biologists, ecologists, ethologists, taxonomists and agroentomologists. Although their research goals may be diverse, all these scientists require information on the behaviour of parasitoids. The study of parasitoid and predator behaviour is an important key to understanding how the insects live, how they influence the population dynamics of their hosts or prey and how they influence the structure of the insect communities in which they live. It is thus a necessary prerequisite for the selection of natural enemies for biological control programmes and for the evaluation of the performance of the insects after release (Luck, 1990). It can also contribute significantly to our knowledge of the taxonomy of parasitoids and of the co-evolution between parasitoids, their hosts and the plants on which the hosts feed.

This chapter is concerned with how to study the foraging behaviour of insect natural enemies. In behavioural ecology, the term 'foraging' is applied to all activities of animals concerned with the acquisition of food. It is easy to see how the term might be applied to insect predators, but describing parasitoid behaviour as foraging is perhaps stretching the meaning of the term somewhat, since it includes reproductive behaviour. It is, however, a common and accepted practice which we shall follow.

This chapter is not a review of the current state of our knowledge of foraging behaviour, although the reader is referred to published examples, most of them recent, throughout the text. Reference is made mainly to parasitoids rather than to predators, reflecting in part our greater familiarity with parasitoids. Despite this emphasis, the chapter will be of interest to students of any insect natural enemies, because there is a common body of theory and methodology. For example, predators, like parasitoids, often respond to olfactory cues emanating from the prey, they may react to chemicals produced by the plant on which the prey feeds, and they often show arrestment following contact with prey-derived substances such as honeydew. Like parasitoids, predators may adjust the time spent in a patch to the rate of encounters with prey, and they may show an aggregative response to prey distribution. Except for some typical parasitoid behavioural traits such as adjustment of progeny sex ratio in relation to local circumstances, superparasitism and host discrimination, few topics discussed in this chapter apply only to parasitoids.

The study of foraging behaviour requires a diverse array of skills, from how to design an experiment to how to analyse a complicated set of interdependent data. It is impossible to address all these specific skills within one chapter. We have chosen to refer to other literature, where possible, if such skills are required to solve a specific problem. We address ourselves mainly to questions of how to formulate testable hypotheses and of how to design experiments, stressing what, in our view, are the important questions. Not all aspects of behaviour are covered, and of those topics discussed we say more about some than others.

There is one important message we wish to convey which relates to every aspect of behaviour discussed: the particular experimental design adopted affects the outcome of the experiments. One should therefore always be very clear in one's mind as to why one wishes to carry out an experiment, and for what purpose the results will be used. A good approach to deciding on the correct experimental set-up is to ask the question: 'under which natural conditions has the particular behaviour evolved?' In laboratory experiments, host or prey densities, encounter rates, patch sizes or patch distributions should be chosen within the range of those found in nature. Unfortunately, for many agricultural and stored product pests, data on these aspects of parasitoid, predator, host and prey ecology are hard to obtain. A somewhat different approach may be required when considering the biological control potential of a parasitoid. In this case, it may be more appropriate to ask: 'under what conditions is the natural enemy expected to perform following its release in a biocontrol programme?'.

1.2 METHODOLOGY

1.2.1 THE CAUSAL APPROACH

Until the late 1970s, parasitoid foraging behaviour was mostly studied from a proximate (i.e. causal or mechanistic) standpoint, with a strong emphasis on identifying which stimuli parasitoids respond to both in finding and in recognising their hosts. Through this approach, fascinating insights into parasitoid foraging behaviour have been gained, and it has been demonstrated that often an intricate tri-trophic relationship exists between phytophagous insects, their host plants and parasitoids. We now know the identities of several of the chemical compounds eliciting certain behaviours in parasitoids. Some of the research in this field has been devoted to the application to crops of chemical substances, as a means of manipulating parasitoid behaviour in such a way that parasitism of crop pests is increased.

It has also been found that many parasitoid species display individual plasticity in their responses to different cues. Associative learning of odours (subsection 1.5.2), colours or shapes related to the host's environment has been described for many species (Vet and Dicke, 1992).

Often, causal questions do not involve elaborate theories. Questions of whether an organism responds to a particular chemical stimulus or not, or whether it reacts more strongly to one stimulus than to another, lead to straightforward experimental designs. It is in the technical aspects of the experiment rather than the underlying theory that the experimenter needs to be creative. However, the study of causation can be extended to ask how information is processed by the central nervous system. One can ask how a sequence of different stimuli influences the behavioural response of the animal, or how responses to the same cue may vary depending on previous experience and the internal state of the animal (Putters and van den Assem, 1990).

Two different causal approaches have been adopted in the study of the integrated action of a series of different stimuli on the behaviour of a foraging animal:

1. The formalisation of an hypothesis into a model of how both external information and the internal state of the animal result in behaviour, and the testing, through experiments, of the predictions of the model. Waage (1979) pioneered this approach for parasitoids. More recently, neural network models have been used to analyse sex allocation behaviour in parasitoids (Putters and Vonk, 1991; Vonk *et al.*, 1991);
2. The statistical analysis of time-series of behaviour to assess how the timing and sequence of events influences the behaviour of the organism. An example of this approach is the analysis of the factors influencing patch time allocation of a parasitoid, using the proportional hazards model (Haccou *et al.*, 1991).

Figure 1.1 Foraging decisions: In adopting either the functional or the causal approach to studying predator and parasitoid foraging behaviour, it is useful to consider that foraging natural enemies are faced with a number of consecutive or simultaneous decisions. Listed above are the decisions that may need to be made by a gregarious host-feeding parasitoid.

1.2.2 THE FUNCTIONAL APPROACH

The functional approach to the study of parasitoid behaviour is based on Darwinian ideas initially formalised by MacArthur and Pianka (1966) and Emlen (1966). Termed 'natural selection thinking' by Charnov (1982), it asks how natural selection may have moulded the behaviour under study.

Because foraging decisions (Figure 1.1) determine the number of offspring produced, foraging behaviour must be under strong selection pressures. Assuming that natural selection has shaped parasitoid searching and oviposition behaviour in such a way that it maximises the probability of leaving as many healthy offspring as possible, it is possible to predict the 'best' behaviour under given circumstances. In the real world, no Darwinian 'monsters' exist that can produce limitless numbers of offspring at zero cost. Because resources are often limiting and because reproduction incurs a cost (e.g. in materials and energy and foraging time) to an individual, increasing investment in reproduction must always be traded off against other factors decreasing fitness (e.g. more offspring often means smaller offspring having shorter lifespans). Thus, producing the maximum possible number of offspring may not always be the optimal strategy.

We refer to natural selection thinking as the **functional approach**, because its aim is to define the function of a particular behaviour. To achieve this goal, it is necessary to show that the behaviour contributes more to the animal's fitness than alternative behaviours in the same situation. The foraging behaviour of female parasitoids has a direct influence on both the number and the quality of their offspring, so it is particularly suited for testing optimisation hypotheses.

The functional approach can be applied not only to theoretical problems but also to problems such as the selection, the evaluation and the mass-rearing of natural enemies for biological control.

There are two ways of investigating functional problems in behavioural ecology. One is to predict quantitatively, using optimality models, the 'best' behaviour under given conditions. The other is to take account of the possibility that the optimal behavioural strategy will be dependent on what other individuals, attacking the same host or prey population, are doing.

Optimality Models

Optimality models are used to predict how an animal should behave so as to maximise its fitness in the long-term. They can be designed by determining:

1. **what decision assumptions apply,** i.e. which of the forager's choices (problems) are to be analysed. Some of the decisions faced by foraging natural enemies are shown in Figure 1.1. Sexually reproducing gregarious parasitoids need to make the simultaneous decision not only of what size of clutch to lay but also of what sex ratio of progeny to produce. The progeny and sex allocation of such parasitoids may be easier to model if the two components are assessed independently, i.e. it is assumed that only one decision need be made by the female. In a formal model, the decision studied must be expressed as one or more algebraic **decision variable(s)**. In some models of progeny (clutch size) allocation, the decision variable is the number of eggs laid per host, while in most models of patch exploitation the decision variable is patch residence time;
2. **what currency assumptions or optimality criteria apply,** i.e. how the various choices are to be evaluated. A model's currency is the criterion used to compare alternative values of the decision variable (in other words, it is what is taken to be maximised by the animal in the short term for long-term fitness gain). For example, some foraging models maximise the net rate of energy gain while foraging, whereas others maximise the fitness of offspring per host attacked;
3. **what constraint assumptions apply,** i.e. what factors limit the animal's choices, and what limits the 'pay-off' that may be obtained. There may be various types of constraint upon foragers; these range from the phylogenetic, through the developmental, physiological and behavioural, to the animal's time-budget. Taking as an example clutch size in parasitoids and the constraints there may be on a female's behavioural options, an obvious constraint is the female's lifetime pattern of egg production (fecundity). In a species which develops eggs continuously throughout its life, the optimal clutch size may be larger than the number of eggs a female can possibly produce at any one time. An example of both a behavioural and a time-budget constraint upon the behavioural options of both parasitoids

and predators is the inability of the forager to handle and search for prey simultaneously. Here, time spent handling the prey is at the cost of searching for further prey. For a detailed discussion of the elements of foraging models, see Stephens and Krebs (1986).

Sometimes the investigator knows, either from the existing literature or from personal experience, the best choices of decision assumption, currency assumption or constraint assumption. If it is impossible to decide on these on the basis of existing knowledge, one can build models for each alternative and compare the predictions of each model with the observed behaviour of the parasitoid or predator. In this way, it is possible to gain insight into the nature of the selective forces working on the insect under study (Waage and Godfray, 1985; Mangel, 1989a).

Early optimality models assumed a static world in which individual parasitoids search for hosts. While these models are useful research tools, they ignored the possibility that for a forager, today's decision may affect tomorrow's internal state which may in turn affect tomorrow's decision, and so on. The internal state of a searching parasitoid changes during adult life: its **egg load** (the number of mature eggs in the ovaries) and its energy reserves may decrease, and the probability that it will survive another day also decreases. The optimal behavioural strategy will depend on these changes. Likewise, the environment is not static. Bad weather, or the start of an unfavourable season can also influence the optimal strategy. Dynamic foraging models have been designed to take into account internal physiological changes and changes in the environment (Chan and Godfray, 1993; Weisser and Houston, 1993).

Implicit in some optimality models is the assumption that the forager is omniscient or capable of calculation, e.g. that a parasitoid wasp has some knowledge of the relative profitability of different patches without actually visiting them (Cook and Hubbard, 1977). Behavioural studies on parasitoids have shown, however, that insects can behave optimally by employing very simple quick 'rule' mechanisms such as the mechanism determining patch time allocation in *Venturia canescens* described in section 1.12 and the males-first mechanism, used by some species in progeny sex allocation (subsection 1.9.2 and Figure 1.14). These mechanisms approximate well the optimal solution in each case.

Evolutionarily Stable Strategies

Almost all parasitoids leave the host *in situ*, i.e. where they encountered it. Thus, there is always the possibility that other parasitoids may find the same host and also oviposit in it. The optimal behaviour of the first female thus depends on what other parasitoids may do. Likewise, the best time allocation strategy for a parasitoid leaving a patch in which it has parasitised a number of hosts depends both on the probability that other wasps will visit that patch and on the probability that other parasitoids may have already exploited the patches it visits next. For this reason, problems concerning the allocation of patch time, progeny and sex require models in which the evolutionarily stable strategy (ESS) (Maynard Smith, 1974) is calculated. The ESS approach, which is based on game theory, asks what will happen in a population of individuals that play all possible alternative strategies. A strategy is an ESS if, when adopted by most members of a population, it cannot be invaded by the spread of any rare alternative strategy (Maynard Smith, 1972). In seeking an ESS, theoreticians are looking for a strategy that is robust against mutants playing alternative strategies. The ESS, like the optimum in models for single individuals, is calculated using a cost-benefit analysis. We refer the reader to Maynard Smith (1982) and Parker (1984) for descriptions of the classical ESS models and for details of how to calculate the ESS in static models.

Why use Optimality and ESS Models?

Sometimes, experimental tests of optimality and ESS models will produce results not predicted by the models. At other times, only some of the predictions of the theoretical model are confirmed by empirical tests. Rarely is a perfect quantitative fit between model predictions and empirical test results obtained. Irrespective of whether a good fit is obtained, valuable insights are likely to be gained into the behaviour of the insect. Construction of models helps in the precise formulation of hypotheses and quantitative predictions, and allows us to formulate new hypotheses when the predictions of our model are not met. Thus, optimality and ESS models are nothing more or less than research tools.

Ideally, both causal and functional questions should be asked when studying the foraging behaviour of insect parasitoids and predators. In the subsection on superparasitism (1.8.4) and that on patch time allocation (1.12.2), we will show how, by ignoring functional questions, one may hamper the interpretation of data gathered to establish that a certain mechanism is responsible for some type of behaviour. Ignoring causal questions can likewise hamper research aimed at elucidating the function of a behaviour pattern, e.g. research into causal factors can demonstrate the existence of a constraint, not accounted for in a functional model, upon the behaviour of the parasitoid. Both causal and functional approaches are required for a thorough understanding of parasitoid behaviour.

1.2.3 THE COMPARATIVE METHOD

The second functional approach to studying animal behaviour is the Comparative Method, which is concerned with adaptive trends among organisms. Phylogenetics tells us that species have not evolved independently (see the example discussed below), and that they cannot be treated as independent points in statistical analyses – to do so may greatly over-estimate the true number of degrees of freedom (Harvey and Pagel, 1991). The Comparative Method provides the appropriate statistical tools to overcome this problem of phylogenetic bias when one is seeking to understand whether characters such as behavioural traits have evolved either together or separately, how they have evolved, and how they adapt organisms to their environments (Harvey and Pagel, 1991 and Brooks and McLennan, 1991 for details).

The Comparative Method can be used to make predictions concerning the ecology of a species. Hardy *et al.* (1992) used a phylogenetic tree, based solely on morphological characters (Nordlander, unpublished), of the six species of *Leptopilina* occurring in Europe to predict where to find *L. longipes*, a species whose hosts and host habitat were unknown (Figure 1.2). The five other species are all parasitoids of *Drosophila*. The tree divides initially into two branches. When we examine how the character host habitat choice is distributed

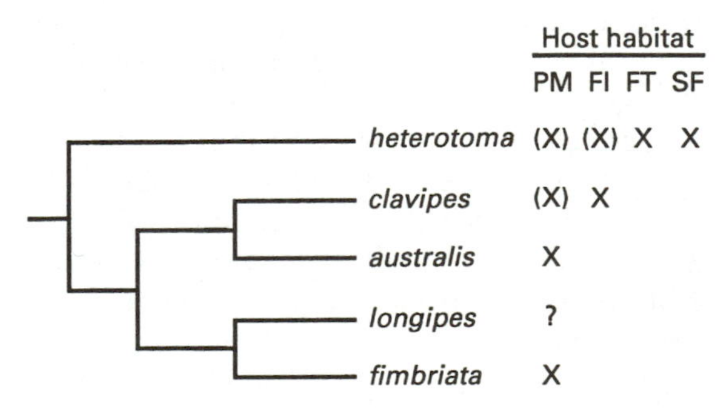

Figure 1.2 Cladogram, based on adult morphology (Nordlander, unpublished) of the *Leptopilina* species occurring in northwestern Europe. Microhabitat use is denoted by X for the principal microhabitat and (X) for microhabitats from which a species has occasionally been recovered. PM = decaying plant material; FI = fungi; FT = fermenting fruit; SF = sap fluxes. Microhabitat use by *L. longipes* was predicted from that species' position on the cladogram.

over the tree, it appears that the upper branch of the tree contains the species finding its hosts in fermenting fruits (*L. heterotoma*), while the other branch contains species finding their hosts in fungi and/or decaying plant matter (*L. clavipes*, *L. australis* and *L. fimbriata*). Because *L. longipes* is most closely related to *L. fimbriata*, it was predicted that it is attracted, like its close relative, to decaying plants and/or fungi. Subsequently, *L. longipes* was trapped with baits comprising rotting cucumber containing *Drosophila* larvae. During recent field work it was also found on decaying stalks of *Heracleum* and also on fungi.

1.3 THE TREATMENT OF PARASITOIDS PRIOR TO THEIR USE IN EXPERIMENTS

It has been shown for several parasitoid species that an individual's previous experience can modify its behaviour. This phenomenon has been observed in all phases of the foraging process and involves responses to chemical stimuli (Vet and Dicke, 1992). Previous ovipositions in hosts of a certain species can also influence host species selection in choice experiments (van Alphen and Vet, 1986), while the decision to oviposit into an already parasitised host (i.e. superparasitise) also depends on previous experience with unparasitised or parasitised hosts (Visser *et al.*, 1992b).

Thus, when designing experiments, one should always be aware that the previous history of an individual may influence its behaviour during an experiment. It can affect the results of experiments on patch time allocation, superparasitism and also the results of experiments in which interactions between adult parasitoids are studied. Even storing parasitoids in the absence of either hosts or host-related cues can have an effect. For example, among wasps so treated, older individuals may superparasitise more frequently than younger ones. Visser *et al.* (1990) showed that it even matters whether wasps are stored in a vial singly or with other females prior to conducting an experiment.

Having said that, we should point out that conditioning parasitoids, by allowing them to search and oviposit for some time before an experiment, can be a sensible practice. Inexperienced parasitoids often show lower encounter rates and are less successful in handling their hosts (Samson-Boshuizen *et al.*, 1974). By allowing parasitoids access to hosts before they are actually used in an experiment, one can often save many hours which would otherwise be wasted in observing parasitoids that are unwilling to search. Often, however, it is advisable to use freshly emerged, inexperienced females, for example in choice experiments, either where different host plants, host instars or host species are offered or where the olfactory responses of parasitoids to different chemicals are studied.

Often, one is interested in the performance of natural populations. These comprise individuals with different experiences and different degrees of experience, so using only inexperienced females in the laboratory gives a distorted view of what happens in nature. One approach is to collect adults from the field for study in the laboratory. A large enough sample should give a reasonable idea of how individuals in the population behave on average.

Because experience can influence subsequent behaviour, the results of experiments in which an insect encounters two situations in succession can depend on which situation is encountered first. In such cases, one should take care that in half of the replicates one situation is encountered first, while in the other half the sequence is reversed.

Phenotypic plasticity in behaviour of insect parasitoids is itself often a subject for investigation. A good summary of how experience can modify the behaviour of parasitoids is given by Vet and Groenewold (1990) and Vet and Dicke (1992), who mainly discuss the causation of changes in behaviour.

Alterations in behaviour due to experience are also the subject of optimality models which predict how parasitoids should adjust their behaviour to the conditions they meet in the environment.

1.4 RECORDING AND STATISTICAL ANALYSIS OF BEHAVIOURAL DATA

1.4.1 RECORDING BEHAVIOUR

A variety of techniques, from pencil and paper to 'intelligent' video systems, are available for recording foraging and other behaviour. The rapid development of portable computers is making specialised behavioural recorders redundant in many cases (subsection 3.2.2). There are now cheap 'notebook' computers on the market which can be used as behavioural recorders, even in the field.

Commercial software, such as *The Observer*, available from Noldus Information Technology, Business and Technology Center, Costerweg 5, 6702 AA, Wageningen, The Netherlands, can be used to record data on a personal computer. *Camera,* from IEC ProGAMMA, P.O. Box 841 9700 AV, Groningen, The Netherlands, is a hardware and software package for recording and analysing behaviour. Data are recorded with a special keyboard, and stored in the memory of a personal computer. The behaviour of different individuals is recorded directly from the video screen in different runs, but these are merged as encoded files in the computer. *Camera* has the advantage in that it allows the observer to analyse interactions between individuals. *Micromeasure Version 3*, developed by Copland and Varley, Wye College, University of London, Wye, Ashford, Kent TN25 5AH, UK, is another commercially available package, designed to analyse video images of animal behaviour. It runs on Commodore Amiga computers, but unfortunately not on IBM-compatible computers. It is particularly suitable for the analysis of search paths on irregularly-shaped structures such as host plant leaves. An alternative to purchasing software packages is to create one's own; many researchers have written their own software for recording behavioural data (subsection 3.2.2).

1.4.2 ANALYSING BEHAVIOURAL DATA

Because insects may change their behaviour in response to experiences gained while foraging, and because their internal state (e.g. egg load) changes during the foraging process, one cannot simply add all the events of a certain category occurring during an observation period. In a time-series of events, the different behavioural events are not independent.

The standard statistical methods described in numerous textbooks are in general inappropriate for the analysis of behavioural data, because the connection between the succession as well as the duration of acts cannot adequately be taken into account. Most students of parasitoid and predator behaviour cannot avail themselves of the methods developed in mathematical statistics for the analysis of such data, because many such methods have been formulated in mathematical jargon. Fortunately, this situation has changed with the publication of Haccou and Meelis' (1992) book on the statistical analysis of time-series of behavioural events. The book caters for the non-specialist. Moreover, a user-friendly software package for applying the statistical methods described in Haccou and Meelis (1992) is currently being developed.

1.4.3 BEHAVIOURAL RESEARCH IN THE FIELD

Whether behavioural research is aimed at answering fundamental questions or deals with the use of parasitoids and predators in biological control, the ultimate goal of interest is the performance of the insects in the field. The small size of many parasitoids makes observation of their behaviour in the field often difficult or impossible. This applies particularly to the monitoring of the movements of individuals, for example between **patches** (patches can be defined either as units of host/prey spatial distribution or as limited areas in which natural enemies search for hosts/prey; often there is a hierarchy of patches, e.g. tree, branch, leaf, leaf-mine). The movements of larger insects such as ichneumonids and sphecids can be more easily observed. Dispersal of small parasitoids in the field can be studied by placing patches with hosts (e.g. potted, host-bearing plants) and releasing marked adults. By checking the host plants at regular intervals for the presence of marked individuals, it is possible to obtain information on the speed at which the insects move between host plants, on the time they spend searching each patch and on the spatial distribution of parasitoids over the available patches. When hosts are later dissected, the aforementioned data can be related to the amount of parasitism in each patch (subsection 1.14.2). By using marked parasitoid individuals, one can distinguish between insects released for the experiment and those occurring naturally. Wasps larger than 1 millimetre in length can be marked with paint on the thorax, using a fine paint brush. By using different colours or colour combinations one can distinguish between different individuals, or groups. Smaller wasps can be marked with fluorescent dusts, but this has the disadvantage that one must remove wasps from the experimental plot to detect the dust mark under ultraviolet light. Recently, genetic markers have been used to monitor parasitoids in the field (Kazmer and Luck, 1995) (subsections 4.2.10 and 4.2.11 contain further discussion of marking techniques).

Many species, when observed in the field, continue foraging normally. Janssen (1989) used a stereomicroscope mounted on a tripod in the field, and used it to observe the foraging behaviour of parasitoids on patches (sap streams and fermenting fruits) containing *Drosophila* larvae. Casas (1990) also recorded the behaviour of *Sympiesis sericeicornis* while the parasitoid searched for its leaf-miner host on potted apple trees in the field.

Other natural enemy species are easily disturbed when approached, and disturbance can be avoided in some cases by using binoculars (Waage, 1983).

1.5 HOST/PREY FINDING AND SELECTION BEHAVIOUR

1.5.1 INTRODUCTION

In parasitoids, two phases in host finding behaviour can be recognised:

1. location of the host's habitat (habitat location);
2. location of the host within its habitat (host location).

At each level in this sequence two sorts of stimuli operate: **attractant stimuli** which elicit orientation to areas that either contain hosts or are likely to contain hosts, and **arrestant stimuli**, which elicit a reduction in the distance or area covered per unit time by parasitoids moving within such areas. Once arrested in a host area parasitoids may respond to additional attractant and arrestant stimuli which tend to localise their movements in even smaller units of host distribution (Waage, 1978). Predators similarly respond to a hierarchy of stimuli.

1.5.2 HOST HABITAT LOCATION BY PARASITOIDS

The literature concerning host habitat location consists largely of papers showing which stimuli (cues) attract parasitoids and predators to the host's habitat (see Vinson, 1985, for a review). Few papers deal with functional aspects of this step in the foraging sequence. The emphasis on causal aspects of host habitat finding reflects the fact that it is much easier to answer qualitative questions, such as which odour acts as an attractant, than it is to answer the question of why one odour should be attractive and another not in terms of the contribution to fitness of the parasitoid.

Parasitoids spend a significant proportion of their adult lives searching for places where hosts can potentially be found. They may use visual, acoustic or olfactory cues to locate potential host patches. Examples of the use of visual and acoustic cues are scarce (van Alphen and Vet, 1986), indicating either that these cues are less important than others in host location or that less time and effort has been devoted to their study. Certainly, for parasitoids olfactory cues are more important. Often, visual and acoustic cues can guide a parasitoid to its host over a short distance only, in contrast to olfactory cues which can act over much longer distances.

It is difficult to demonstrate the use, by parasitoids, of visual cues in host habitat location because the use of other, olfactory and acoustic, cues has to be excluded. Van Alphen and Vet (1986) investigated the searching behaviour of *Diaparsis truncatus*, an ichneumonid parasitoid of larvae of the twelve-spotted asparagus beetle, *Crioceris asparagi*. Larvae of the beetle feed inside the green berries of the *Asparagus* plant. It was shown, by placing green-painted wooden beads into *Asparagus* plants, that *D. truncatus* females respond to visual cues from the berries of *Asparagus*. The parasitoids landed more often on the slightly larger wooden beads than on the green *Asparagus* berries, which is consistent with the hypothesis that the parasitoids respond to visual cues. Such an approach may be adopted for parasitoids of other insects living in fruits. That parasitoids can recognise colours and shapes and use them in host finding has also been shown by Wardle (1990) and Wardle and Borden (1990). Herrebout (1969) showed that the tachinid fly *Eucarcelia rutilla* visually discriminates between twigs of *Pinus silvestris* and those of deciduous trees.

Some parasitoids respond to acoustic stimuli produced by the host, and so execute host habitat location and host location in one step. Cade (1975), whilst broadcasting the song of the male cricket *Gryllus integer* from a loudspeaker to study the mating behaviour of the crickets in the field, discovered that the tachinid fly parasitoid (*Euphasiopteryx ochracea*) of the cricket was attracted by the song. Burk (1982) similarly demonstrated this for the tachinid *Ormia lineifrons*. Soper *et al.* (1976), using tape recordings, showed that the sarcophagid parasitoid *Colcondamyia auditrix* finds male cicadas by this means, i.e. phonotaxis. Phonotaxis by the tachinids *Ormia depleta* and *O. ochracea* has been demonstrated using *synthesised* male calling songs (Fowler and Kochalka, 1985; Walker, 1993). Both Fowler (1987) and Walker (1993) carried out experiments in which the synthesised calls of a range of several host cricket species were simultaneously broadcast in the field. In Walker's experiments females of *O. ochracea* were attracted in the greatest numbers by the synthesised call of *Gryllus rubens*. The reader is referred to Walker's paper (1993) for details of equipment and methodology.

Chemical communication, both between insects and between plants and insects, plays a very important role in determining the behaviour of parasitoids and predators. Any chemical conveying information in an interaction between two individuals is termed an **infochemical** (Dicke and Sabelis, 1988). Infochemicals are divided into **pheromones**, which act intraspecifically, and **allelochemicals**, which act interspecifically. Allelochemicals are themselves subdivided into **synomones**, **kairomones** and **allomones**. A synomone is an allelochemical that evokes in the receiver a response that is adaptively favourable to both the receiver and the emitter; a kairomone is an allelochemical that evokes in the receiver a response that is adaptively favourable only to the receiver, not the emitter; an allomone is a allelochemical that evokes in the receiver a response that is adaptively favourable only to the emitter (Dicke and Sabelis, 1988). The majority of parasitoids respond to volatile kairomones or synomones in the long-distance location of their hosts. These chemicals may originate from:

1. the host itself, e.g. from frass, during moulting, during feeding, sex pheromones and aggregation pheromones, i.e. the chemicals involved are kairomones for the parasitoids;
2. from the host's food plant, i.e. the chemicals involved are synomones for the parasitoids; or
3. from some interaction between host and food plant, e.g. feeding damage, i.e. the chemicals involved are synomones for the parasitoids.

The attraction responses by parasitoids to odours from any source can be studied using various olfactometers and wind tunnels, or by observing the responses of parasitoids to odour sources following release of the insects in the field.

Two types of **air-flow olfactometer** are commonly used to study responses to olfactory cues. One is the glass or clear perpex Y-tube olfactometer (Figure 1.3). The insect can be given a choice either between odour-laden air (test) and odour-free but equally moist air (control) or between air laden with one odour and air laden with another odour. Although Y-tube olfactometers have been criticised on the grounds that odour plumes may mix where the two arms of the olfactometer meet due to turbulence, and that choice is no longer possible once the insect has passed the junction of the tube, impressive results have been obtained. Smoke can be passed through the apparatus to test for unwanted turbulence, but tobacco smoke must be avoided as

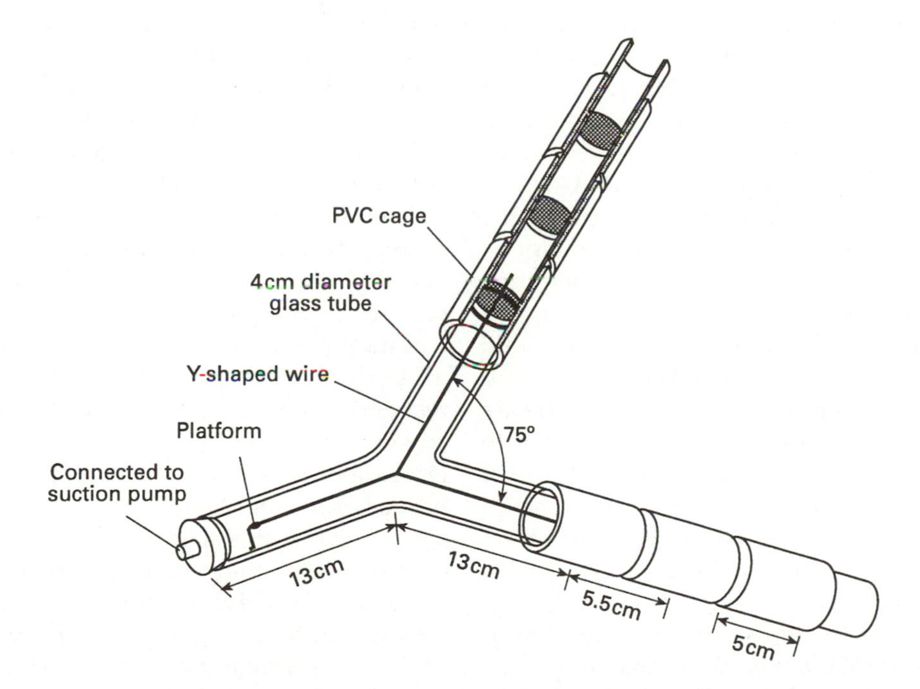

Figure 1.3 Host/prey habitat finding and host/prey finding behaviour in parasitoids and predators: Design of the Y-tube air-flow olfactometer used by Sabelis and van der Baan (1983). The Y-shaped wire within the tube cavity provides a walking surface for small predators and parasitoids. For details of operation, see text. Reproduced by permission of Kluwer Academic Publishers.

it is absorbed by the tubing and it can affect the outcome of future experiments. By passing NH_4OH vapour over HCl, a fine 'smoke' of NH_4Cl crystals can be created and the vapour channelled through the Y-tube. After testing, the crystals can easily be washed from the tubing. Turbulence, if detected, can often be reduced by adjusting the flow speed of the air.

With diurnally active insects, a light source is often required to illuminate the apparatus so as to encourage the insects to move towards the fork of the tube.

To eliminate the effects of any asymmetry in the apparatus, the chambers need to be alternated for each 'run'. It is recommended that parasitoids be tested individually, rather than in batches, because either interference or facilitation may occur between insects and bias the results. The apparatus should be washed, first with alcohol and then with distilled water, between runs to prevent any response of parasitoids to any trail left by previous individuals. Finally, consideration needs to be given to the possibility of left- and right-handedness in the insects. By analysing the number of left and right turns in the apparatus, it is possible to test, statistically, whether wasps tend to move more to the right or more to the left. The null hypothesis will be that the distribution of turns by parasitoids should be equal in both arms irrespective of the position of the chambers. An additional test of turn preference is to perform several runs when both chambers are empty, although insects may be unwilling to move through the apparatus in the absence of any odour. Some parasitoid species show 'handedness', i.e. a tendency to turn more in one direction than another (J. Pritchard, unpublished).

Y-tube olfactometers were used to investigate the host habitat location behaviour of the aphid parasitoids *Diaeretiella rapae* (Read *et al.*, 1970), *Aphidius uzbekistanicus* and *A. ervi* (Powell and Zhang, 1983). The insects were shown to be attracted by the odour of the host's food plant. Examples of other parasitoid species that have been investigated using Y-tube olfactometry are cited in Vinson (1976, 1985). Sabelis *et al.* (1984), using a Y-tube olfactometer containing a Y-shaped wire over which the insects could walk, showed that the predatory mite *Phytoseiulus persimilis* responds to odours of its prey *Tetranychus urticae*. A Y-tube olfactometer was used by Carton (1978) to show that the parasitoid *Leptopilina boulardi* (referred to in his paper as *Cothonaspis* sp.) is attracted by the odour of ethyl alcohol. Ethyl alcohol is emitted by fermenting fruits, in which the parasitoid's host, *Drosophila melanogaster*, develops as larvae.

Even when great care is taken in the design of olfactometer experiments and the analysis of data, the results of olfactometry may be difficult to interpret (Kennedy, 1978). This applies especially to Y-tube olfactometry. The Y-tube, when employing a light source, simultaneously presents test insects with two types of stimulus, light and air current, to which the insect might respond by phototaxis and anemotaxis, but presents the two odours (or odour and non-odour) separately at only one point in the apparatus – the fork, which represents the 'decision point'. Responding by phototaxis and anemotaxis to the common air current, insects might be entrained past the decision point and become behaviourally trapped in the wrong arm (Vet *et al.*, 1983). Another type of air-flow olfactometer, designed by Pettersson (1970) to study the responses of aphids, lacks this and other disadvantages. A modification (Figure 1.4) of the Pettersson apparatus by Vet *et al.* (1983), has proved quite popular among students of parasitoid behaviour, particularly as it lacks some of the disadvantages of Y-tube olfactometers. It is constructed mainly of transparent perspex, and has a central arena with four arms. Air is drawn out of the arena via a hole in the centre of the bottom plate. Air flows into the arena via four arms. Insects may therefore be exposed to as many as four different odours. Air speed in each arm can be

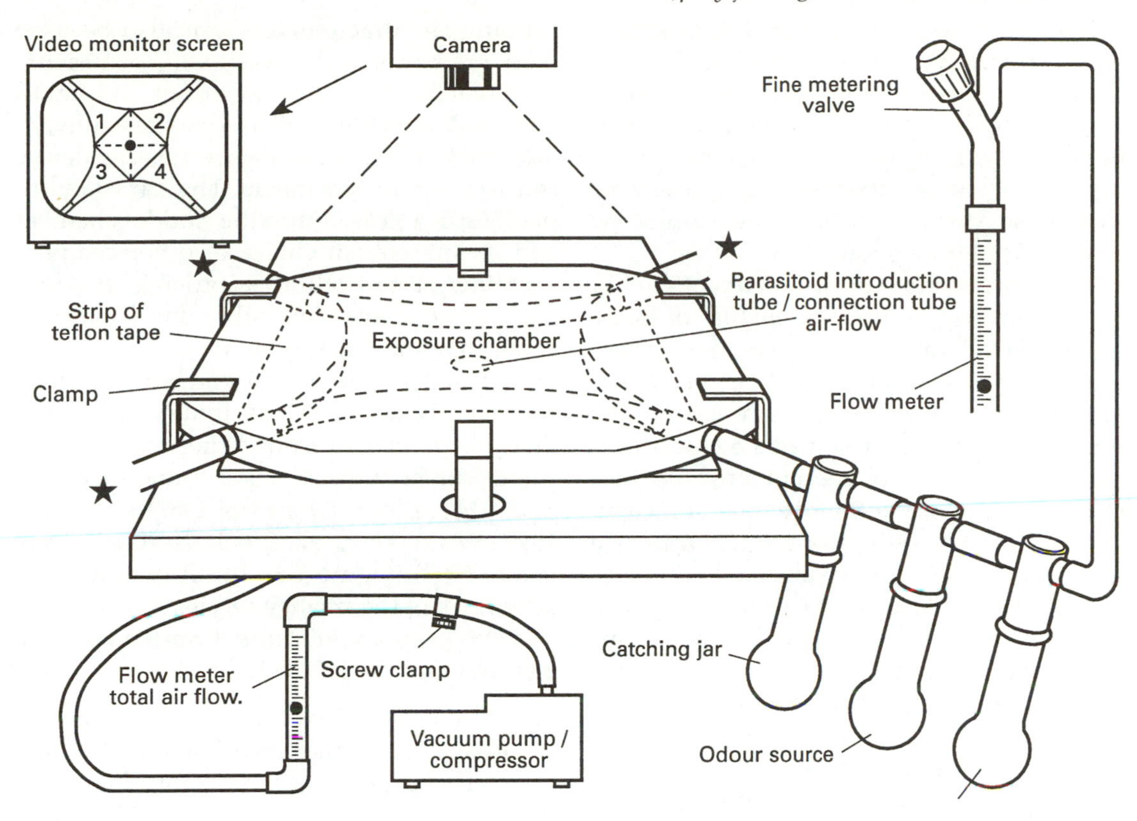

Figure 1.4 Host/prey habitat finding and host/prey finding behaviour in parasitoids and predators: Design of Pettersson olfactometer as used by Vet *et al.* (1983). The catching jar is used to collect any insects that move from the exposure chamber into an outflow tube. For details of operation, see text. Reproduced by permission of Blackwell Scientific Publications Ltd.

controlled with a valve and an anemometer. Care is taken that air speed is equal in all arms. Before an experiment is performed, an NH_4OH smoke test (see above) can be carried out to test for unwanted turbulence and and to show that a clear, straight boundary exists between odour fields. Diffuse light of equal intensity on all four sides of the arena prevents asymmetric attraction of insects to light. Insects are introduced through a hole in the bottom plate by temporarily disconnecting the tube from the air pump. Observations are best made using a video camera placed directly overhead, because a human observer may disturb the insects by his or her movements.

The Pettersson olfactometer thus allows an insect to choose between four different odour fields, and repeated choices by the insect are made possible. Vet (1983), studying the olfactory responses of *Leptopilina clavipes* with this design of olfactometer, recorded the following: (a) which odour field was chosen first by wasps; (b) which was chosen last (i.e. the wasp walked up one of the arms), and (c) the percentage of time allocated to each odour field.

In the Pettersson olfactometer, the final choice by an insect is made when it enters the narrow tube through which air laden with odour enters the arena. Both because air flow in this narrow region is strong and because many parasitoids have an aversion to entering narrow crevices, some insect species avoid this area and turn without entering.

Other parasitoids react to the odour stimulus by flying vertically upwards. Because flight is impossible in the narrow space between the base and the olfactometer cover, the insects will hit the top plate, and after a number of these aborted flight attempts become so disturbed that they cannot be expected to choose odour fields.

Often, one or more of the odours offered in an olfactometer comprises a mixture of many unidentifiable volatile substances, the concentrations of which in the odour fields are unknown. This does not pose a problem if the responses of an insect to a mixed odour source and a clean air control are compared, because only the test odour is the potential attractant. However, when testing for attraction to two odour sources (e.g. the odours of two different food plants of the host), there may be problems of interpretation. One of the odour sources may be more attractive than the other because the insect responds to one or more substances in that odour source that are lacking in the other. Alternatively, both odour sources may be qualitatively similar but the insect may be differentially attracted because of differences in the concentration of an attractant component of an odour. It also needs to be borne in mind that a combination of a qualitative difference and a quantitative difference may be responsible for differential attraction.

The ultimate solution to the above problem would be to isolate the attractants and test whether differential attraction to odour sources is due either to differences in chemical composition between the sources or differences in concentration of their chemical components.

With any air-flow olfactometer it is important to ensure, before carrying out any experiments, that air flows through the apparatus at a constant rate (usually the rate is low). With the Y-tube and Pettersson olfactometers, both of which are **hypobaric systems** (i.e. air is sucked out), a good quality vacuum pump should be used. Flow meters of the correct sensitivity i.e. neither over- or under-sensitive, should also be employed.

Static-air olfactometers can also be used with predators and parasitoids to measure chemotactic responses to odour gradients. One such olfactometer, used successfully by Vet (1983), is shown in Figure 1.5. The device consists of three chambers. The parasitoid or predator is released into the middle chamber and its subsequent choice of outer chamber containing a test odour recorded. Vet (1983) also recorded the time taken for females to reach an odour source chamber.

As noted above, not all parasitoids can be successfully tested in olfactometers, because they are prevented from flying. Flying parasitoids can be tested in wind tunnels (Noldus *et al.*, 1988; Elzen *et al.*, 1986, 1987; Kaas *et al.*, 1990; Wiskerke *et al.*, 1993; Grasswitz and Paine, 1993) (Figure 1.6), but it is difficult to keep track of the smaller species.

When even a wind tunnel cannot be used, one could try the following:

1. Place potentially attractive odour sources in an array either in the field, in a large field cage, or in a large controlled environment chamber;

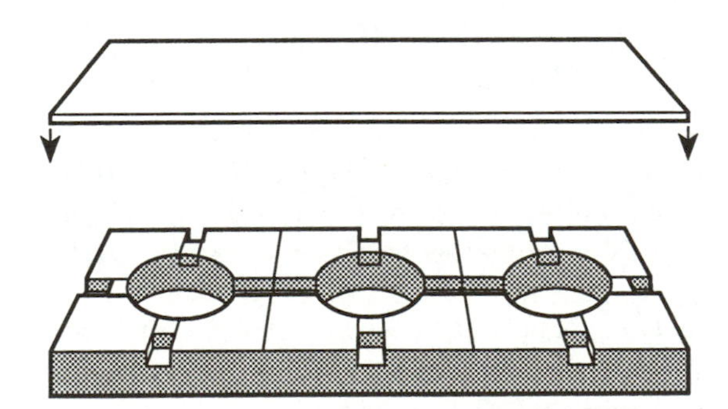

Figure 1.5 Host/prey habitat finding and host/prey finding behaviour in parasitoids and predators: Static air olfactometer of Vet (1983). The apparatus comprises three perspex boxes covered by a single glass lid. The excavations (chambers) in the boxes are connected by corridors. The chambers, measured internally, are 50 mm wide and 16 mm high; the corridors are 10 mm wide and 5 mm high. For details of operation, see text. Reproduced by permission of E.J. Brill (Publishers) Ltd.

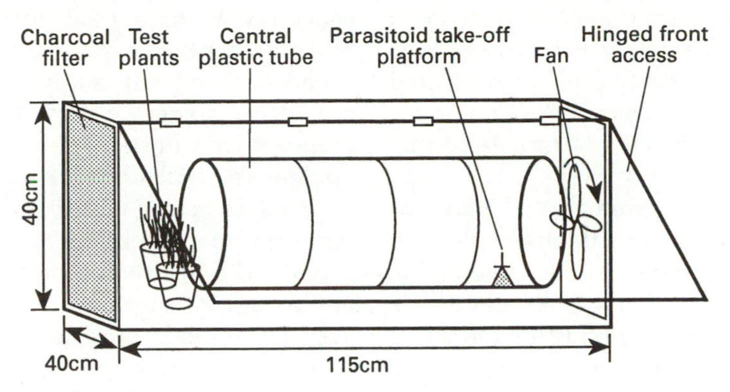

Figure 1.6 Host/prey habitat finding and host/prey finding behaviour in parasitoids: Design of wind tunnel used by Grasswitz and Paine (1993) to study the behaviour of *Lysiphlebus testaceipes* (Braconidae). The main (rectangular) chamber was constructed of Plexiglass, and the central (cylindrical) test section was constructed of Mylar. Reproduced by permission of Kluwer Academic Publishers.

2. Release a large number of adult females and examine the odour sources frequently;
3. Remove each insect that lands on the odour sources.

If more individuals than expected, on the basis of a random distribution, land on a particular odour source, this can be taken as evidence that that odour source is attractive. If different odour sources are offered, it may be possible to rank the different sources in terms of their attractiveness. The problem with this type of experiment is that the number of parasitoids trapped on a particular source is a function both of the number of parasitoids landing on the source and of the time they spend there. Ideally, parasitoids should be caught immediately after arrival on the source, but this is not always possible. Another problem is that the experimental design does not exclude the effect of interactions between individuals, e.g. some parasitoids repel conspecifics or chase them away. (subsections 1.5.4 and 1.14.3).

At Leiden, the field release method (involving counting of the numbers of females attracted to uninfested and infested cassava plants) has been used successfully in field experiments with *Epidinocarsis lopezi*. We have also used the method for comparing the attractiveness of different microhabitats containing *Drosophila* larvae to several species of *Leptopilina*. The method also allows a functional analysis of habitat choice. If one knows: (a) the encounter rates with hosts in the different microhabitats, (b) the species composition of the host larvae in each microhabitat, and (c) the survival rates of parasitoid eggs deposited in each of the host species, one can calculate the relative profitability of each microhabitat for the parasitoid and predict which ones the parasitoids should visit when given a choice. This approach was employed by Janssen *et al.* (1991). Another approach to studying functional aspects of host habitat location by parasitoids is to consider the reliability and detectability of a cue (Vet and Dicke, 1992). This approach contrasts cues having a high detectability but a low predictive value regarding the presence of hosts, with cues having a low detectability but a high predictive value. Cues with a high detectability are odours emitting from potential host plants. Cues with a high reliability are substances produced by the host plant in reaction to the presence of the host and substances emitted directly from the host. Vet and Dicke (1992) assume that high reliability cues are produced in smaller amounts than general host plant odours. It is unfortunately hard to see how reliability and detectability could be measured in a quantitative way. Therefore, it is unlikely that this concept could ever be translated into testable hypotheses.

As noted above, parasitoids may be attracted to the host's habitat in response to a stimulus arising from some interaction between the host and its food plant. An example of this phenomenon is the attraction of the parasitoid *Epidinocarsis lopezi* to cassava plants that have been infested by the cassava mealybug, *Phenacoccus manihoti*. The parasitoid is attracted not only to host-bearing but also to host-free parts of mealybug-infested plants, the stimulus being a volatile chemical that the plant produces when fed upon by the mealybugs, i.e. a synomone (Nadel and van Alphen, 1987, describe an experimental investigation, using a Petterson olfactometer).

The olfactory responses of foraging parasitoids and predators may vary with age, nutritional state and experience. It is important to take account of these factors when designing experiments. Ideally, preliminary experiments should be carried out to test for any effects. Synovigenic species (subsections 1.16.1 and 2.3.4) may spend the first few days of adult life searching, not for hosts, but for non-host foods such as nectar and honeydew which supply nutrients for egg development. Therefore, when young, they may be unresponsive to host plant and host odours. Some parasitoids may even be repelled very early in adult life by an odour which, later on in life, is used in host finding. *Exeristes ruficollis* responds in this manner to the odour of pine oil (Thorpe and Caudle, 1938). If females of *Leptopilina clavipes* are given oviposition experience with *Drosophila* larvae feeding on yeast, and are then tested in an air flow olfactometer offering the odours of yeast and decaying fungi (females inexperienced with hosts in yeast find the odour of decaying fungi attractive), most females preferentially choose the odour of yeast and females spend longest in a yeast odour field (Vet, 1983). In a mark-release-recapture experiment, Papaj and Vet (1990) released differently treated *Leptopilina heterotoma* (females previously allowed to oviposit in hosts contained within an apple–yeast substrate, females allowed to oviposit in hosts within a mushroom substrate and naïve females with no oviposition experience) in a woodland containing both apple and mushroom 'baits'. Wasps with experience of oviposition were recaptured more frequently than inexperienced wasps, and wasps tended to be recaptured on the bait they had been exposed to prior to release. The aforementioned behavioural changes in *Leptopilina* are the result of **associative learning**: the development of a response to a new stimulus after that stimulus has been experienced in combination with another stimulus to which the animal already shows an innate response. Associative learning, involving either stimuli associated with hosts or stimuli associated with host by-products, has been observed in several other parasitoids (Vet and Groenewold, 1990; Vet *et al.*, 1990 for reviews; Grasswitz and Paine, 1993 for a recent study). Note that associative learning is not confined either to olfactory responses or to host finding (Vet *et al.*, 1990).

1.5.3 HOST LOCATION BY PARASITOIDS

Having arrived in a potential host habitat, a parasitoid begins the next phase in the search for hosts. Often, insects show arrestment in response to contact with kairomones of low volatility deposited by their hosts on the substratum. Materials containing such kairomones (sometimes referred to as 'contact chemicals') have been shown to include host salivary gland or mandibular gland secretions, host frass, and homopteran honeydew and cuticular secretions. For example, the wax particles (cuticular secretions) left on cassava plants by the mealybug *Phenacoccus manihoti* arrest the parasitoid *Epidinocarsis lopezi* (Langenbach and van Alphen, 1985) and the coccinellid predator *Diomus* sp. (van den Meiracker *et al.*, 1990). Because mealybugs are never far away from such contaminations, arrestment behaviour increases the probability of encounter with hosts or prey. Contact with aphid and mealybug honeydew arrests both parasitoids and coccinellid predators (Ayal, 1987; Carter and Dixon, 1984; Evans and Dixon, 1986; van den Meiracker *et al.*, 1990; Budenberg, 1990; Hågvar and Hofsvang, 1991; Heidari and Copland, 1993).

Because stronger responses may be found to a kairomone after previous oviposition experience in the presence of the substance, an initial experiment ought to be performed using parasitoids with previous oviposition experience. The next series of experiments would involve comparing the response of parasitoids to patches of potential host habitat, within which hosts have never occurred (e.g. clean host plant leaves), with the response to patches within which hosts have previously occurred for some time. The following changes in behaviour might be observed in the searching insects: a decrease in walking speed, an increase in the rate of turning, a sharper angle of turn at the patch edge, an increase in the number and frequency of ovipositor stabs, an alteration in position of the antennae, an increase in the amount of drumming with the antennae and an increase in the amount of time spent standing still. Video recording equipment, together with the computer software discussed in subsection 1.4.1 can be used to record and analyse alterations in these behavioural components. If such equipment is not available, then the insect's path can be traced with a felt-tip pen on a Petri dish lid, and a map measurer then used to measure the distance travelled. If, during the experiment, the path trace is marked at regular, for example 3 second intervals, alterations in the speed of walking, over short time scales, can be measured. Path tortuosity can be evaluated by measuring the angle between tangents drawn at intervals along the path.

A useful additional analysis that can be carried out involves designating areas of an arena, e.g. the kairomone-treated area and the clean area, and measuring the proportion of the total time available that the insect spends searching each area. If the parasitoid or predator can be shown to have spent a greater proportion of its time in the treated area, then it has been arrested by the kairomone.

Once it has been demonstrated that patches within which hosts have occurred contain a stimulus to which parasitoids respond by arrestment, further experiments can be performed to elucidate the nature of the stimulus. To eliminate the possibility that the arrestment response is to some physical property of the patch (e.g. the texture of the wax secretions left by mealybugs, or depressions caused by feeding larvae), one can attempt to dissolve the putative kairomone either in distilled water, hexane or another suitable solvent, and then apply the solution to a surface, for example a leaf or a glass plate, which has never borne host larvae. If an arrestment response is still observed, it can be concluded that the soluble substance is a kairomone. For a detailed experimental study of the arrestment response in a parasitoid, conducted along these lines, see Waage (1978) (Figure 1.7a,b).

Kairomones provide quantitative, in addition to qualitative information. Several parasitoid species, when presented with several patches of kairomone in different concentrations, have been shown to spend longer periods searching those patches with the higher kairomone concentrations than the patches with the lower concentrations, at least over part of the range of concentrations (Waage, 1978; Galis and van Alphen, 1981; Budenberg, 1990) (Figure 1.7c shows Waage's experimental design). Because kairomone concentration varies with host density, parasitoids can obtain information concerning the profitability of a patch, even before they encounter hosts.

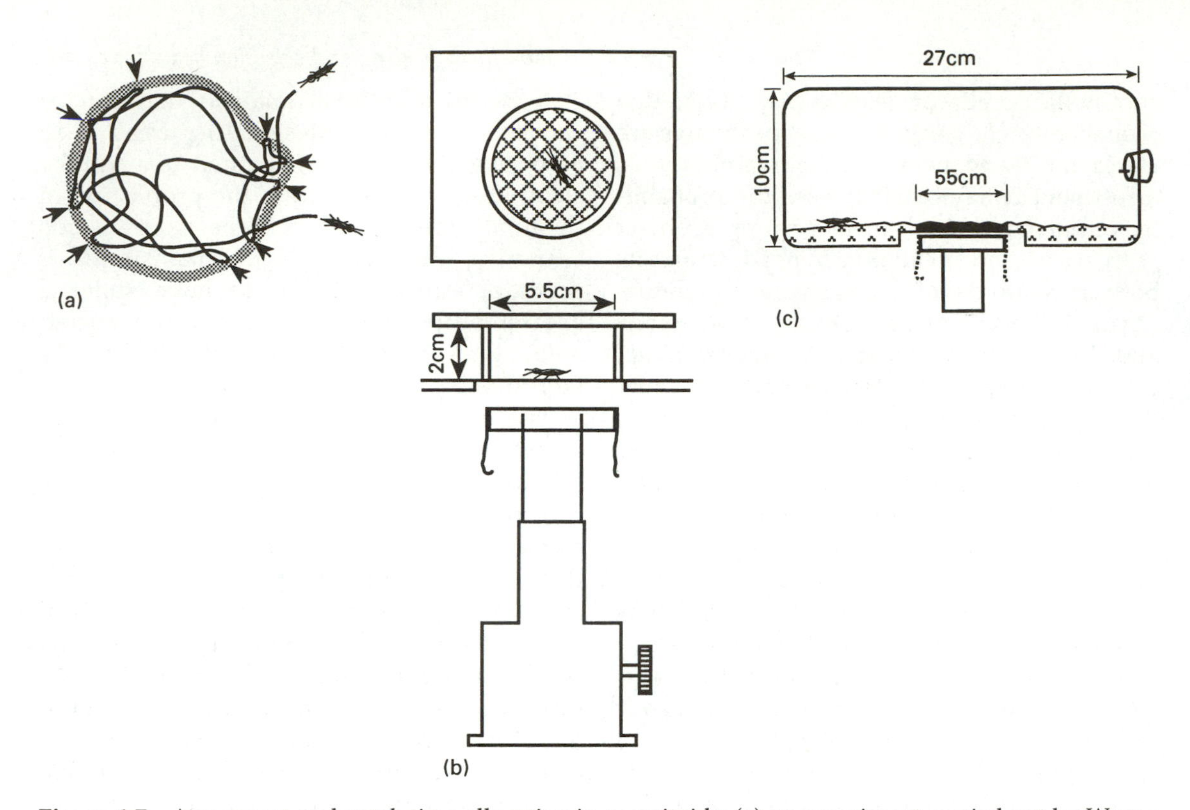

Figure 1.7 Arrestment and patch time allocation in parasitoids: (a) an experiment carried out by Waage (1978) to investigate arrestment behaviour shown by *Venturia canescens* in response to contact with a kairomone. The path of a walking female *V. canescens* was observed on a glass plate on which 1 ml of ether extract of ten pairs of host (*Plodia interpunctella*) mandibular glands had been placed and allowed to evaporate. Stippling denotes the edge of the patch. Upon encountering the patch edge from the outside, a female stops and begins to apply the tips of its antennae rapidly upon the the substratum. It then proceeds on to the patch at a reduced walking speed. Within the patch, the wasp occasionally stops walking and probes the substratum with its ovipositor. When the wasp encounters the patch edge, it turns sharply. Presumably due to waning of the arrestment response, e.g. through habituation or sensory adaptation to the chemical stimulus, the wasp eventually leaves the patch; (b) apparatus used by Waage (1978) to test the hypothesis that the patch edge response of *V. canescens* (see above) is to the removal of the chemical stimulus and not to the patch edge *per se*. The terylene gauze screen was impregnated with host mandibular secretion by confining ten fifth instar host larvae between two sheets of gauze. The lower sheet was then stretched over a Petri dish, as in the figure. By raising and rapidly lowering the contaminated screen, Waage could precisely control when a wasp (in the upper chamber) was 'on' and 'off' the patch. A wasp's movements were traced with a felt-tip pen on the glass plate roof of the chamber. Over the first centimetre travelled following stimulus removal, most wasps made a reverse turn, which may be considered to be a **klinotactic** (directed) response because the turn oriented the wasps back toward the point from where the stimulus was removed. Thus, Waage concluded that the patch edge response of *V. canescens* is to the removal of the chemical stimulus; (c) apparatus used by Waage (1978) to test the effect of kairomone concentration on patch residence time. Patches were made by confining different numbers of host larvae, together with food medium, between terylene gauze sheets for several hours. The larvae were then removed. For each kairomone concentration, the contaminated patch of food medium was stretched over the central part of the floor of the chamber (blackened area). An empty dish was raised beneath the patch (see next experiment). Two arbitrary intervals (14 s and 60 s continuously off a patch) were used as criteria for determining patch-leaving by wasps. Application of either of these criteria indicated that the duration of the first visit to a patch increased markedly with increasing kairomone concentration.

The apparatus was also used to test the effect of oviposition on patch residence time. A patch of host-contaminated food medium was stretched over the central part of the chamber floor, and at the onset of the experiment a dish containing thirty host larvae was raised beneath the patch. Each wasp was allowed to make one oviposition into a host as soon as it entered the patch. During the resting period following that oviposition, the dish containing host larvae was replaced with an empty dish. Oviposition was found to produce a marked increase in the duration of the first patch visit by a wasp. Another experiment was carried out by Waage (1978), which demonstrated that oviposition does not elicit a significant arrestment response in the absence of the kairomone. This experiment employed apparatus (b). A host-contaminated terylene screen with or without host larvae beneath it was raised beneath the chamber. A single wasp was exposed either to the chemical stimulus alone for the duration of one bout of probing or to this stimulus with hosts present for one oviposition of similar duration. The screen was then lowered, so removing the kairomone, and the time taken for the wasp to leave the chamber floor and climb one of the chamber sides recorded (this behaviour being interpreted as the cessation of any response elicited by the contact chemical). No significant difference in time taken to abandon the host area was observed between the treatment with and without oviposition. Reproduced by permission of Blackwell Scientific Publications Ltd.

1.5.4 RESPONSES TO PARASITOID ODOURS AND PATCH MARKS

Parasitoid Odours

Janssen *et al.* (1991) showed in olfactometer experiments that *Leptopilina heterotoma* is attracted to the odour of stinkhorn fungi containing larvae of *Drosophila phalerata*. When these patches are offered in an olfactometer together with similar patches on which searching females of *L. clavipes* are present, *L. heterotoma* avoids the odour fields of patches containing *L. clavipes* females. The conclusion from these observations is that *L. clavipes* produces an odour while searching, which repels its competitor *L. heterotoma*, at least when the latter is presented with the choice between host-containing patches emitting this odour and host-containing patches that lack the odour. Price (1981), citing Townes (1939), suggests that the function of the strong odour emitted by some female ichneumonids, noticed when these insects are handled, likewise signals the insects' presence to other parasitoids. Höller *et al.* (1991) found evidence that foraging primary parasitoids of aphids are repelled by odours produced by adult hyperparasitoids. In summary, adult parasitoids may produce odours which influence the subsequent arrival of conspecifics or even individuals of other species at a patch.

In cases where the odour of a parasitoid repels conspecifics, the substance is a pheromone, whereas in cases where heterospecific competitors are repelled, there is some justification in describing the substance as an allomone. However, because of similar problems to those mentioned when discussing patch marking (below), the use of the term allomone should be avoided here.

It is not known how widespread the use of repellent odours is among insect parasitoids, largely because it has not been studied in a systematic way. Like other infochemicals used by parasitoids, odours produced by adult parasitoids can potentially have a profound effect on patch choice and time allocation by individual wasps and thus on the distribution of parasitoids over a host population.

Patch Marking

Some parasitoid species are known to leave chemical marks on surfaces they have searched (Galis and van Alphen, 1981; Sheehan *et al.*, 1993). This marking behaviour can have a number of functions. By leaving a scent mark on the substratum, a parasitoid can avoid wasting time and energy in searching already visited areas. A female can also use the frequency with which she encounters marks to determine how well she has searched the patch, and so assist in the decision when to leave the patch. When encountered by conspecific or heterospecific competitors, marks sometimes induce the competitor to leave an area. *Pleolophus basizonus*, *Orgilus lepidus*, *Asobara tabida* and *Microplitis croceipes* (Price, 1970; Greany and Oatman, 1972; Galis and van Alphen, 1981;

Sheehan *et al.*, 1993) mark areas they search, and females spend less time in areas previously searched by conspecifics. In the case of a heterospecific competitor, the marker substance could be termed an allomone. However, leaving the patch may not always be in the interest of the competitor; the competitor may stay and superparasitise the hosts parasitised by the first female (subsection 1.8.4). Thus, the use of the term allomone should be avoided in this context.

The use of marker substances can be demonstrated by offering patches containing kairomone, but not hosts, to a parasitoid. After the parasitoid has left the patch, a second parasitoid is introduced on to the same patch. If the second insect always stays on the patch for a shorter time period than the first, the existence of a mark left by the first has been demonstrated.

One cannot easily demonstrate the reaction of a female to her own marks. When a female stays for a shorter period the second time she visits a patch, she may do so either because she recognises the patch from visual or tactile landmarks 'noted' during the first visit or because her motivation to search is always lower when she searches a second patch. The latter possibility can be ruled out by performing a control experiment in which the female is offered a new patch instead of the same one. It is harder to rule out the possibility that the female recognises the patch from landmarks.

1.5.5 SEARCH MODES WITHIN A PATCH

Whilst kairomones arrest parasitoids and predators in host patches and so increase the probability of encounter, host/prey location is itself likely to be in response to non-chemical, e.g. visual and tactile cues. For example, in coccinellid predators, prey honeydew acts as an arrestant stimulus for adults (van den Meiracker *et al.*, 1990 [*Diomus* sp., *Exochomus* sp.]; Heidari and Copland, 1993 [*Cryptolaemus montrouzieri*]), but the prey are located in response to visual cues (Stubbs, 1980 [*Coccinella septempunctata*]; Heidari and Copland, 1992 [*Cryptolaemus montrouzieri*]). Stubbs (1980) devised a method for calculating the distance over which prey are detected (Heidari and Copland (1992) describe a modification of the technique). Coccinellid larvae are arrested by honeydew (Carter and Dixon, 1984), but location of the prey occurs only upon physical contact (*Coccinella septempunctata*). Unlike the adults, the larvae do not use cues to orient themselves towards the prey.

It has been shown in a number of predators that arrestment occurs as a consequence of prey capture (Dixon, 1959; Marks, 1977; Nakamuta, 1982; Murakami and Tsubaki, 1984; Ettifouri and Ferran, 1993). In this way, the insect's searching activities are concentrated in the immediate vicinity of the previously captured prey, increasing the probability of locating a further prey individual. The adaptive value of such behaviour for predators of insects that have a clumped distribution, e.g. aphids, is obvious. Predators also show arrestment after capturing a prey individual but failing to feed on it – even a failed encounter with prey is an indication that a clump of prey has been found (Carter and Dixon, 1984). Carter and Dixon (1984) argue that the latter behaviour is particularly important for early instars of coccinellids, since the prey capture efficiency of these instars is relatively low. It has recently been discovered that in final instar larvae of the coccinellid *Harmonia axyridis* arrestment in response to prey capture only occurs if the predators are provided with the same prey species as they were reared upon, indicating a strong conditioning effect (Ettifouri and Ferran, 1993).

Arrestment in the above cases can be studied in the same way as arrestment of natural enemies in response to kairomones, i.e. by analysing the search paths of predators and parasitoids and by measuring the proportion of the total time available spent searching designated unit areas of an arena.

Species of parasitoids attacking the same hosts may differ in the way they search a patch. In parasitoids of concealed anthomyiid, calliphorid, drosophilid, muscid, phorid, sarcophagid and sepsid fly larvae, at least three different **search modes** exist (Vet and van Alphen, 1985). Wasps may either:

1. probe the microhabitat with their ovipositors until they contact a host larva (**ovipositor search**);
2. perceive vibrations in the microhabitat caused by movements of the host and use these cues to orient themselves to the host (**vibrotaxis**) which is then probed with the ovipositor; or
3. drum, with their antennae, the surface of the microhabitat until they contact a host (**antennal search**).

To determine which search mode a parasitoid species uses is easy in the case of ovipositor search or antennal search, where brief observation of a searching female suffices to classify her search mode. However, it can be difficult to prove that vibrotaxis occurs, because of the possibility that the parasitoid locates its hosts by reacting to a gradient in kairomone concentration or some other chemical cue, or to infrared radiation from the host.

Glas and Vet (1983) showed that *Diachasma alloeum*, a braconid parasitoid of apple maggot larvae, detects the host by vibrotaxis. They used the fact that parasitised hosts are paralysed by venom and do not move. With a mobile larva present in a hawthorn fruit or apple, ovipositor probes are concentrated in the immediate vicinity of the host larva. By contrast, when a paralysed larva is present in the fruit, ovipositor probes are distributed randomly over the whole surface of the fruit.

Lawrence (1981) showed that the braconid *Biosteres longicaudatus* finds moving hosts easily, but is unable to find dead hosts. Because the wasps probe with their ovipositors in response to scratches made from beneath the fruit surface with a dissecting needle, it is likely that they locate their hosts by vibrotaxis.

Sokolowski and Turlings (1980) showed that the braconid *Asobara tabida* finds it hosts by vibrotaxis. They used a special mutant of *Drosophila melanogaster* in which the larvae become immobile at a temperature equal to or greater than 29°C. Encounter rates with the mutant larvae were significantly lower than those with wild-type larvae at 29°C, while no difference in encounter rates was found at 20°C.

The reason why it is important to determine the search mode of a parasitoid or predator is that different search modes lead to different encounter rates with hosts in the same situation. Thus, a parasitoid using vibrotaxis as a search mode may be more successful in finding hosts when the hosts occur at low densities, while ovipositor search may be more profitable at high host densities. Antennal search results in encounters with larvae on the surface, while ovipositor search can also result in encounters with hosts buried in the host's food medium.

Often, the searching behaviour of a parasitoid comprises a combination of search modes, as the insect responds to different cues while locating a host. It is therefore not always possible to place the behaviour of a parasitoid into one category.

Predators may employ a combination of search modes. The larvae of the predatory water beetle *Dytiscus verticalis* may either behave as sit-and-wait predators when prey density is high, or hunt actively for prey when prey density is low (Formanowicz, 1982).

1.5.6 HOST RECOGNITION BY PARASITOIDS

Generally, specific (although not necessarily host *species* specific) host-associated stimuli need to be present for the release of oviposition behaviour by parasitoids following location of a prospective host. The role these stimuli play in host recognition has been investigated mainly by means of very simple experiments.

For many parasitoids, host size appears to be important for host recognition. Salt (1958) for example, presented female *Trichogramma* with a small globule of mercury – smaller than a host egg – and observed that the parasitoid did not respond to the globule. However, Salt then added minute quantities of mercury to the globule, whereupon a female would mount it, examine it and attempt to pierce it with her ovipositor. When Salt continued adding quantities of mercury to the globule, a globule size was reached where a wasp again did not recognise it as a prospective host.

Host shape can be important in host recognition. A number of workers have placed inanimate objects of various kinds inside either hosts or host cuticles from which the host's body contents have been removed, and have shown that some host shapes are more acceptable than others.

One needs to be cautious in interpreting the results of experiments where hosts or host dummies of various sizes and shapes are presented to parasitoids. If a parasitoid is found to attempt oviposition more often in large dummies than in small ones, or in rounded dummies than in flattened ones, the stimuli involved could be visual, tactile or both. Some authors have failed to determine precisely which of these stimuli are important. Similar caution needs to be applied to experiments in which dummies of different textures are presented to parasitoids.

As can often be inferred from direct observations on the behavioural interactions of parasitoids and hosts, movement by the host can be important in releasing oviposition behaviour. A simple experiment for investigating the role of host movement in host recognition involves killing hosts, attaching them to cotton or nylon threads, moving both these and similarly attached living hosts before parasitoids, and determining the relative extent to which the dead and living hosts are examined, stabbed, drilled or even oviposited in by the parasitoids.

Kairomones play a very important (although not necessarily exclusive) role in host recognition by parasitoids. Strand and Vinson (1982) for example showed in an elegant series of experiments how, if glass beads the size of host eggs are uniformly coated with material present in accessory glands of the female host (host eggs normally bear secretions from these glands), and are presented to females of *Telenomus heliothidis* (Scelionidae), the insects will readily attempt to drill the beads with their ovipositors. Female parasitoids, when presented with either clean glass beads or host eggs that had been washed in certain chemicals, were, on the whole, unresponsive. Strand and Vinson (1983) analysed the host accessory gland material and isolated from it (by electrophoresis of the material) proteins, two of which were shown to be more effective in eliciting drilling of glass beads. It cannot be assumed from these findings that *T. heliothidis* will recognise any object as a host provided it is coated in kairomone. Host size and shape are also important criteria for host acceptance by *T. heliothidis*.

A similar series of experiments, again employing glass beads, was carried out by Nordlund *et al.* (1987) on *Telenomus remus* and *Trichogramma pretiosum*, and showed that host accessory gland materials of the particular host moth species contained host recognition kairomones.

A useful investigation to carry out on the role of kairomones would be to take a polyphagous parasitoid species and determine whether the recognition kairomone is different or the same for each of its host species. Van Alphen and Vet (1986) showed that the braconid parasitoid *Asobara tabida* discriminates between the kairomone produced by *Drosophila melanogaster* and that produced by *D. subobscura*.

Acceptance of a prospective host for oviposition also depends upon whether the host is already parasitised. This very important aspect of parasitoid behaviour is dealt with later in this chapter (section 1.8).

1.5.7 HOST AND PREY SELECTION

Host Species Selection

Many parasitoid species are either polyphagous or oligophagous. Strictly monophagous species are relatively uncommon. When different potential host species occur in different habitats, a parasitoid 'decides' which host species is to be attacked by virtue of its choice of habitat in which to search. Sometimes, potential host species can be found coexisting in the same patch (e.g. two aphid species living on the same host plant, larvae of different fly species feeding in the same corpse). In these cases, experiments on host species selection are relevant, and can demonstrate whether the parasitoid has a **preference** for either of the species involved. Preference is defined as follows: a parasitoid or predator shows a preference for a particular host/prey type when the proportion of that type oviposited in or eaten is higher than the proportion available in the environment. This is the traditional 'black box' definition (Taylor, 1984), so-called because it does not specify the behavioural mechanism(s) responsible. For example, a parasitoid may encounter a host individual and accept it, but the host may then escape before the parasitoid has an opportunity to oviposit (likewise, prey may escape from a predator following acceptance). If host types differ in their ability to escape, they will be parasitised to differing extents even though they may be accepted at the same rate. Conversely, they may be accepted at different rates but be parasitised to the same extent. It could be argued that preference, to be more meaningful behaviourally, ought to be defined in terms of the proportion of hosts or prey accepted. However, it may not be possible in experiments to observe and score the number of acceptances (one reason being that the insects do not display obvious acceptance behaviour).

Often (section 1.11 describes a different approach), experiments designed to test for a preference score the number of hosts parasitised or prey fed upon after a certain period of exposure where equal numbers of each species have been offered. There is, however, a problem with this approach: the number of hosts oviposited in/prey eaten depends on the number of encounters with individuals of each species, and the decision to oviposit/feed on the less preferred species may be influenced by how often the female gets the opportunity to oviposit/feed on the preferred species. Encounter rates (section 1.15) may also be unequal for the two host/prey species, due to factors such as differences in size or activity. Therefore, species selection should preferably be investigated in such a way that encounter rates with both species are equal. This requires pilot experiments, with equal numbers of each species offered simultaneously, to calculate the ratio in which both types should be presented so as to equalise encounter rates.

Mathematical formulae used for quantifying preference (whether for species or for stages) are many and varied (Cock, 1978; Chesson, 1978, 1983; Settle and Wilson, 1990), but the most widely used measure of preference is the following (Sherratt and Harvey, 1993):

$$\frac{E_1}{E_2} = c\frac{N_1}{N_2} \tag{1.1}$$

where N_1 and N_2 represent the numbers of two host/prey types available in the environment, and E_1 and E_2 represent the numbers of the two host/prey types eaten or oviposited in. The parameter c is the preference index and can be viewed as a combined measure of preference and encounter probability (section 1.11). A value of c between zero and one indicates a preference for host/prey type 2, whereas a value of c between one and infinity indicates a preference for host/prey type 1. Mathematical formulae used in testing whether preference varies with the relative abundance of the different host/prey types are discussed in later in this chapter (section 1.11).

Optimal host selection models predict that the acceptance of a less profitable host species depends on the encounter rate with the more profitable host species. The less profitable species should always be ignored if the encounter rate with the more profitable species is above some threshold value, but should be attacked if the encounter rate with the more profitable species is below that threshold value (Charnov, 1976; Stephens and Krebs, 1986). Note that if recognition of prey is not instantaneous, then acceptance of the less profitable host species depends on the encounter rates with both host species and on the time it takes for recognition to take place. Often, for the convenience of the researcher, relatively high densities of hosts, resulting in high encounter rates, are offered in laboratory experiments. This will produce a bias towards more selective behaviour. For example, in laboratory experiments with high encounter rates, the *Drosophila* parasitoid *Asobara tabida* is selective when offered the choice between two host species differing in survival probability for its offspring (van Alphen and Janssen, 1982) and avoids superparasitism (van Alphen and Nell, 1982). However, in the field, when encounter rates are equal to or lower than one host per hour, wasps always generalise and superparasitise (Janssen, 1989). If one is interested in knowing the performance of a parasitoid species in the field, where host densities are often very low, one should use host densities equivalent to those occurring in the field. The high densities often offered in the laboratory may allow the researcher to obtain much data over a short period of observation, but the insect's behaviour in such experiments may not be representative of what happens in the field.

To understand the adaptive significance of host preferences the relative profitability of different host species can be assessed, in the first instance, by recording the survival rates of parasitoid progeny in the different hosts. If no differences in the probability of parasitoid offspring survival are recorded, one cannot assume that the host species concerned are equally profitable. Handling times may vary with host species, as may the fecundity and other components of the fitness of parasitoid progeny, and ideally, these ought to be measured.

Experiments on prey choice by predators are influenced by prey densities offered in a manner similar to that described above for parasitoids. Because searching activity is influenced by the amount of food in the gut, a predator's feeding history may determine the outcome of experiments on prey choice (Griffith, 1982; Sabelis, 1990).

So far, we have considered innate host and prey preferences. Preferences may alter with experience, as shown by Cornell and Pimentel's (1978) study of *Nasonia vitripennis*. In *N. vitripennis*, preference for a host species increased if females had previous experience of feeding on that species. Previous experience also influences the host species selection behaviour of *Asobara tabida*. Wasps conditioned on *Drosophila melanogaster* accept more *D. melanogaster* larvae for oviposition than wasps that have experienced mixtures of *D. melanogaster* and *D. subobscura* (van Alphen and van Harsel, in van Alphen and Vet, 1986).

Host species selection is further discussed in section 1.11.

Host Stage Selection

Parasitoids may encounter different developmental stages of the host within a patch. Those stages potentially vulnerable to attack may differ in their profitability. For **idiobionts** – parasitoids which do not permit the host to grow beyond the stage attacked and which therefore exploit a fixed 'parcel' of host resource (Figure 1.8a) – small host stages may provide inadequate amounts of resource to permit the successful development of offspring. Even where successful development of idiobiont progeny is possible in small hosts,

Figure 1.8 Idiobiont and koinobiont parasitoids (both gregarious) of the same host species, *Pieris brassicae* (Lepidoptera: Pieridae): (a) *Pteromalus puparum* (Pteromalidae) ovipositing into the pupa of the host. The pupa is a fixed parcel of resource, as it is a non-feeding stage; the parasitoid is therefore an idiobiont; (b) *Cotesia glomerata* (= *Apanteles glomeratus*) (Braconidae) ovipositing into newly hatched host larvae which will continue to feed, grow and develop during parasitoid development. The parasitoid is therefore a koinobiont. (Premaphotos Wildlife)

the offspring are small and therefore oviposition constitutes less of a fitness gain (in parasitoids, body size determines components of fitness such as fecundity, longevity and searching efficiency (2.7.3 and 2.8.3). Although they exploit a growing amount of host resource, **koinobionts** – those parasitoids which allow their hosts to continue to feed and develop (Figure 1.8b) – also may display a positive relationship between adult body size and host size, although the relationship may not be linear (Sequeira and Mackauer, 1992b, Harvey *et al.*, 1994) (subsection 2.9.2). For both idiobionts and koinobionts, smaller hosts may require less time for handling and represent less of a risk of injury resulting from the defensive behaviour of the host (Gross, 1993). For koinobionts (most of which are endoparasitoids), small hosts may present parasitoid progeny with less of a mortality risk from encapsulation (van Alphen and Drijver, 1982) (subsection 2.10.2). Females of both idiobionts and koinobionts may also gain in fitness from ovipositing in or on older larvae, owing to the fact that host mortality resulting from predation and/or intraspecific competition is more severe in early host stages than in late ones. It may even pay the parasitoid to oviposit preferentially in host larvae which have almost completed growth because of the mortality occurring in the younger host larvae (Reznik *et al.* 1992). Thus, it is often the case that parasitoids prefer to attack certain host stages and even avoid or reject some stages for oviposition.

The distribution of hosts of different size over the host plant may influence encounter rates and thus host stage selection behaviour. Later instars of mealybugs are often surrounded by earlier instars. Encounter rates with younger instars may be higher, and those with larger ones lower than predicted on the basis of their densities. Young maggots of calliphorid and drosophilid fly species feed near the surface of the substrate, while older larvae may burrow deeper, possibly out of the reach of parasitoids.

Often, hosts are not passive victims of their parasitoids. Maggots and caterpillars may wriggle, or otherwise defend themselves. Mealybugs may 'flip' the posterior end of their body, or throw droplets of honeydew on to parasitoids attempting to parasitise them. Aphids may drop from the host plant, while leafhoppers and planthoppers may jump away (Gross, 1993, gives a comprehensive review of behavioural defences of hosts against parasitoids). Such behavioural defences of hosts (which are often more effective in later host stages) can cause a problem of data interpretation. Should encounters which do not result in parasitism of the host be scored either as acceptances or as rejections? If the parasitoid clearly displays behaviour that is normally associated with host acceptance, e.g. turning and stinging shown by encyrtids (Figure 1.9), the encounter should be classified as an acceptance.

The ability of late stage hosts to defend themselves from attack better than early stage hosts may account for a host stage preference. The cost in terms of lost 'opportunity time' (time that could be spent in more profitable behaviour) when attacks on late stage hosts fail, may outweigh the fitness gain per egg laid (Kouame and Mackauer, 1991). This hypothesis requires testing.

Host size selection by parasitoids is not limited to the decision regarding whether to oviposit or reject the host. It also involves the decision regarding which sex the offspring ought to be, and, for gregarious parasitoids, how many eggs to lay (Figure 1.1). For practical reasons, we analyse those decisions as isolated steps, but one should bear in mind that they are interrelated, and that it is wise to study host size selection in combination with clutch size and sex allocation decisions. Host size selection in relation to clutch size is discussed further in section 1.6.

Predators are usually less specific in their choice of prey than parasitoids, although some predators show a preference for larger prey or certain instars (Griffiths, 1982;

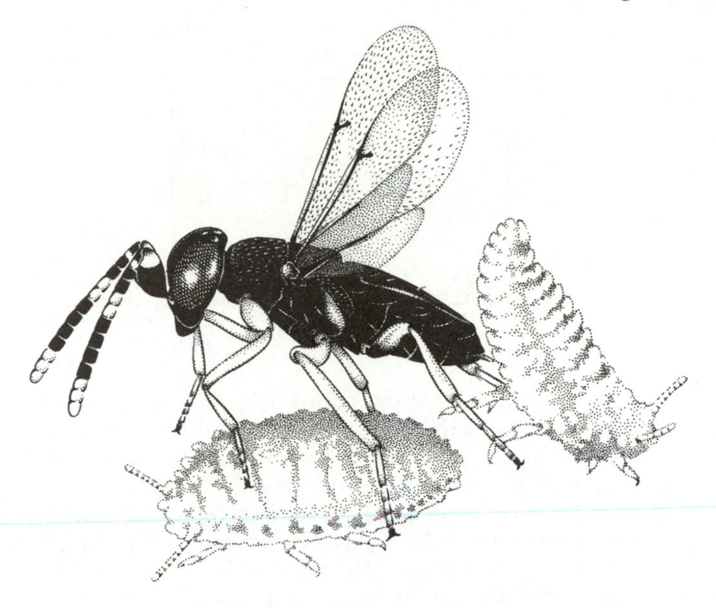

Figure 1.9 Host acceptance behaviour in the encyrtid parasitoid *Epidinocarsis lopezi*: The female examines the host with its antennae. Acceptance is indicated by the wasp turning towards the host, to insert its ovipositor. Sometimes, the host escapes whilst the wasp is turning. Acceptance therefore does not necessarily lead to oviposition.

Thompson, 1978; Cock, 1978). Prey size selection in predators may also change with the size of the predator (Griffiths, 1982).

Host selection decisions by one female may alter over time during an experiment because these decisions are affected by experience, egg load and stochastic variation in encounter rates. Such changes in decisions are one of the reasons why partial preferences are always found instead of the absolute, i.e. all-or-none preferences predicted by static prey choice models. If one is interested in questions such as how egg load should influence host selection, one should construct dynamic optimisation models as described by Mangel and Clark (1988) (e.g. see Chan and Godfray, 1993).

1.6 DECIDING WHAT CLUTCH SIZE TO LAY

Since a host represents a limited amount of resource and parasitoid offspring have the potential to compete for that resource (sections 2.9 and 2.10), gregarious parasitoids must make an additional decision after accepting a host for oviposition: how many eggs to lay in (or on) a host. This question has been addressed by Waage and Godfray (1985), Waage (1986) and Godfray (1987). Here we are mainly concerned with variation in the size of clutches allocated to hosts of a fixed size, although we shall eventually consider host size variation.

Given that the amount of resource a developing parasitoid obtains will determine its fitness, a fitness function $f(c)$ can be used to describe the fitness of each offspring in a clutch of size c allocated to hosts of a certain size. The fitness gain to the mother per host attacked is therefore the product of clutch size and the *per capita* fitness function, i.e. $cf(c)$. The value of c where $cf(c)$ is maximised is the parental optimum clutch size, known as the 'Lack clutch size', after David Lack (1947) who studied clutch size in birds. Predicted and observed fitness functions for three parasitoid species are shown in Figure 1.10. In

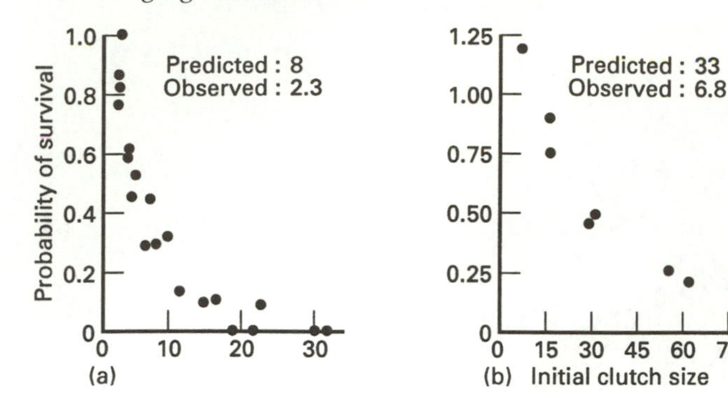

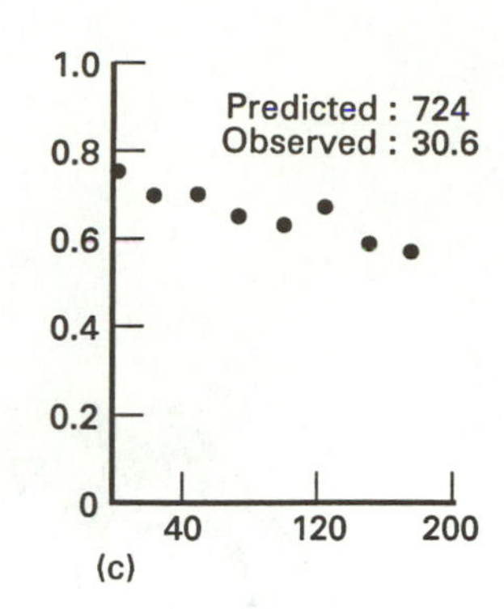

Figure 1.10 Optimal progeny allocation in gregarious parasitoids: The *per capita* fitness of offspring as a function of clutch size in gregarious parasitoids, estimated by the probability of survival in initial clutches of different sizes (observed = observed clutch size; predicted = predicted by calculation of *cf (c)*): (a) *Trichogramma evanescens* in eggs of the moth *Mamestra brassicae*; (b) *Telenomus farai* in eggs of the bug *Triatoma phyllosoma pallidipennis* (the overestimate of survival in this case is attributable to sampling error); (c) *Dahlbominus fuliginosus* on pupae of the sawfly *Neodiprion lecontei* (from Waage and Godfray, 1985 and Waage, 1986, who used data from Pallewata, 1986; Escalante and Rabinovich, 1979; Wilkes, 1963). In all three cases, there is a continuous decline *in per capita* fitness with increasing clutch size. For some other parasitoid species there is evidence of an Allee effect, that is, an initial rise then a fall in fitness. Such a dome-shaped fitness relationship may prove common among certain groups of gregarious endoparasitoids, due to the fact that in such parasitoids small larval broods often perish entirely either because of their inability to overcome host defences or to consume all the host tissues (a prerequisite in some species for successful pupation and emergence). Reproduced by permission of Blackwell Scientific Publications Ltd. and Academic Press Ltd.

each case, the probability of survival to the adult stage is used as the measure of fitness.

Fitness function curves can be constructed as follows:

1. by exposing hosts to individual parasitoids or and examining/dissecting some of these hosts immediately after oviposition to determine clutch size, and rearing parasitoids from the remainder to determine offspring survival (Figure 1.10(a), (b)). If larval mortality arising from resource competition occurs late in development, and dead larvae are not consumed by surviving larvae, one may simply record the numbers of emerged and unemerged offspring (Figure 1.10(c));
2. by manipulating parasitoid clutch sizes. This is relatively easy in the case of ectoparasitoids, as different clutch sizes can be obtained simply by adding or removing eggs, manually, from clutches present on the host's body surface (Hardy *et al.*, 1992). With this technique, however, there is a risk of damaging eggs during manipulation. With endoparasitoids, clutch sizes can be manipulated by interrupting oviposition, by allowing superparasitism to occur, or by exchanging the host for one of a different size after the wasp has examined it but immediately before it has the opportunity to begin ovipositing in it (Klomp and Teerink, 1962). However, a major problem with at least the latter technique is that the parasitoid may alter its sex allocation behaviour. The final egg clutch sex ratio could influence progeny fitness and therefore the optimum clutch size (Waage and Ng, 1984).

Other models predict that the best strategy for a parasitoid is to maximise fitness per unit time rather than per host attacked. If there is a cost in time to laying an egg, it may benefit a female to cease adding more eggs to a host and to allocate the time saved to locating a new host. The fitness gain from leaving hosts and searching for new ones will increase as the travel time between oviposition sites decreases, i.e. as host availability increases. As hosts become more abundant, females should leave each host sooner, i.e. produce smaller clutch sizes. Thus, with the maximisation of fitness per unit time models, females maximise fitness per host attacked (i.e. produce Lack clutch sizes) only when hosts are scarce. *Trichogramma minutum* appears to be a species whose strategy is to maximise fitness per unit time. Schmidt and Smith (1987a) presented females with nine host eggs attached by glue to a cardboard base and employed various egg spacing treatments: the eggs were situated with their centres either 2, 3, 4 or 5 millimetres apart on a grid. Clutch size was found to decrease with increased crowding of eggs, i.e. increasing host density per unit area.

There are also models that take into account egg limitation constraints, i.e. they assume that the parasitoid has a limited number of eggs to lay at any one time. Such a parasitoid is always in a position where available eggs are fewer than potential clutch sites. If eggs are severely limiting, i.e. egg load is much smaller than the number of hosts available (this could be due to the fact that the female has laid most of her eggs), a female should spread out her eggs between hosts so that the fitness gain *per egg*, rather than per clutch, is maximised. When *per capita* fitness of offspring decreases monotonically (as in Figure 1.10), the optimal clutch size under severe egg limitation is always one. Also, as the probability of a female surviving to lay all her eggs decreases, the optimal clutch size will increase, and be that which maximises fitness per host attacked (i.e. the Lack clutch size).

What if data and model do not match? As can be seen from Figure 1.10, clutch sizes predicted by optimality models tend to differ from the ones recorded in experiments. This discrepancy may occur because:

1. the wrong category of model has been used. For example, the parasitoid's strategy may be that of maximising fitness per unit time rather than per host attacked;
2. the model does not take account of stochastic variability in certain parameters (Godfray and Ives, 1988);
3. the measure of fitness (e.g. offspring survival) used may be inappropriate.

If fitness is measured instead by total offspring fecundity, a closer fit between model and data may be obtained (Waage and Ng, 1984; Waage and Godfray, 1985). Measuring offspring fecundity is likely to prove very time-consuming, so an alternative procedure is to measure offspring body size or weight; both of these factors are usually good predictors of fecundity in parasitoids (subsection 2.7.3). Le Masurier (1991) used a combined measure of fitness – the product of progeny survival and the calculated mean egg load at emergence (a measure of lifetime fecundity) of the surviving female progeny. The egg load for each emerging wasp was determined indirectly, from a regression equation relating egg load to head width. Interestingly, le Masurier found that the fitness function curve he constructed for a British population of *Cotesia glomerata* in larvae of *Pieris brassicae* showed no density-dependent effect of clutch size on fitness, and this therefore prevented him from calculating the optimum clutch size for that host. All he could predict was that females should lay at least the maximum number of eggs he recorded in a host.

An interesting point made by Hardy *et al.* (1992) in relation to progeny fitness measurements and mismatches between predicted and observed clutch sizes is that laboratory studies tend to underestimate the disadvantages of small body size.

A further factor to consider is variation in host size. If a gregarious parasitoid's strategy is that of maximising fitness per host attacked, then the optimal clutch size ought to increase with increasing host size. How do gregarious parasitoids (and solitary parasitoids for that matter) measure host size? Schmidt and Smith (1985) studied host size measurement in *Trichogramma minutum*. Females allocated fewer progeny to host eggs that were partially embedded in the substratum than into host eggs that were fully exposed. Since the eggs were of identical diameter and surface chemistry, it was concluded that the mechanism of host size determination is neither chemosensory nor visual, but is essentially mechanosensory, based on accessible surface area. Schmidt and Smith (1987b) subsequently observed the behaviour of individual *T. minutum*, during the host examination phase, on spherical host eggs of a set size and recorded: (a) the frequency of and intervals between contacts with the substratum bearing the eggs and turns made by the wasps, and (b) the number of eggs laid per host. In analysing the data, seven parameters were considered: the total number of substratum contacts, the mean interval between such contacts, the interval between the last contact and oviposition, the longest and shortest interval between contacts, the total interval between the first three contacts, and the interval between the first contact with the host and the first contact with the substratum (*initial transit*). Of these, only the duration of the initial transit across the host surface showed a significant linear relationship with the number of eggs deposited. As the duration of the wasp's initial transit increases, more eggs are laid. By interrupting the path of wasps during their initial transit, and thereby reducing their initial transit time, Schmidt and Smith (1987b) succeeded in reducing the number of progeny laid by a female. Schmidt and Smith concluded that wasps are able to alter progeny allocation by measuring short time intervals. Interestingly, the length duration of initial transit was found to be the same for both large and small wasps (Schmidt and Smith, 1987b, 1989).

Large-sized gregarious parasitoids (and solitary parasitoids, for that matter) are likely to measure host size in other ways, for example by determining whether the tips of the antennae reach certain points on the host's body (whether the stimuli the wasps respond to are tactile, visual or both, needs to be established), or by simple visual examination of the whole host.

1.7 HOST-FEEDING BEHAVIOUR

The females of many synovigenic parasitoids (subsections 1.16.2 and 2.3.4) not only parasitise hosts but also feed on them (Jervis and Kidd, 1986). **Host-feeding** supplies the females with proteinaceous materials for continued egg production (Bartlett, 1964; Jervis and Kidd, 1986). In some parasitoid species, host-feeding causes the host to die (so-called 'destructive' host-feeding), so rendering it unsuitable for oviposition. Even with those species that remove small quantities of host materials such that the host survives feeding ('non-destructive' host-feeding), the nutritional value of the host for parasitoid offspring may, as a result of feeding, be reduced and the female may lay fewer (gregarious species) or no eggs in it. Thus, while host-feeding potentially increases future fitness via subsequently increased egg production, the fitness gain is at the cost of current reproduction.

The decision either to host-feed or to oviposit, and the decision of what clutch size to lay upon a host that has been used for feeding, should depend on the immediate gain from oviposition weighed against the future gain from host-feeding. Since the future gain will depend on the probability of surviving and encountering hosts in the future, the decision whether to host-feed or oviposit will depend on the number of mature eggs that a parasitoid is carrying, i.e. its egg load. A female with a full supply of mature eggs can gain more from ovipositing than from host-feeding, while a female that has exhausted her supply of mature eggs could gain more from host-feeding. Clearly, the decision to host-feed will depend dynamically on the internal state of the parasitoid, and, because the probability of adult survival is likely to decrease with age, also on how old the female is. Host-feeding is thus a problem that can be studied using dynamical optimisation models (Chan and Godfray, 1993).

A general prediction of models of destructive host-feeding behaviour is that the fraction of hosts fed upon by females should increase with decreasing host availability, at least over the upper range of host densities (Figure 1.11) (Jervis and Kidd, 1986; Chan and Godfray, 1993) – a prediction borne out by the few empirical studies that have been carried out to date (DeBach, 1943; Bartlett, 1964; Collins *et al.*, 1981; Bai and Mackauer, 1990; Sahragard *et al.*, 1991).

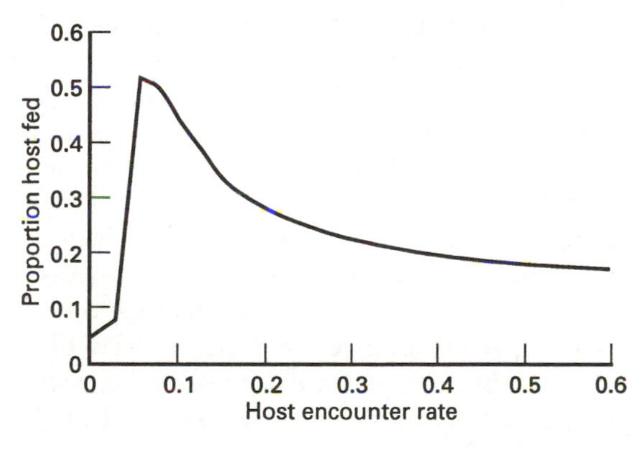

Figure 1.11 Host-feeding by parasitoids: Models by Chan and Godfray (1993) based on stochastic dynamic programming (a state variable approach) predict that with increasing host availability (i.e. increasing encounter rate) a decreasing proportion of host individuals will be fed upon. Jervis and Kidd (1986) found similar relationships with their models. As can be seen from the line, the reverse applies at very low levels of host availability. Shown here are the predictions of Chan and Godfray's (1993) 'resource pool' model (their paper gives details of this and other models).

Given that the fitness gains from ovipositing may vary in relation to host stage (subsection 1.5.6), it is likely that the decision either to host-feed or to oviposit also depends on host stage (Kidd and Jervis, 1991; Rosenheim and Rosen, 1992). Indeed, observational and experimental studies of destructively host-feeding parasitoids show in them a tendency to feed preferentially or exclusively on earlier host stages and to oviposit preferentially or exclusively on/in later ones (Kidd and Jervis, 1991; Rosenheim and Rosen, 1992). A similiar relationship is likely to apply to different-sized hosts of the same developmental stage.

Models developed by Chan and Godfray (1993) predict that the decision to host-feed versus oviposit depends on the parasitoid's egg load: host-feeding is more likely when egg load is low. Rosenheim and Rosen (1992) tested this prediction experimentally using the scale insect parasitoid *Aphytis lingnanensis*. Egg load was manipulated by using wasps of different sizes (egg load being a function of body size [subsection 2.7.3]) and also by holding parasitoids, prior to their exposure to hosts, at different temperatures (the rate of oöcyte maturation and therefore the rate of accumulation of mature eggs in the ovaries being a function of temperature (subsections 2.3.4 and 2.7.4). Manipulating egg load in this way ensured that previous history of host contact could be eliminated as a possible confounding variable. Alternative methods of manipulating egg load, e.g. depriving parasitoids of hosts or allowing them to oviposit, do not separate the effects of egg load and experience.

Rosenheim and Rosen (1992) found in their experiments that egg load did not significantly affect the decision to host-feed or oviposit on (small) hosts. It is unclear why the theoretical prediction for an egg load effect is not supported empirically.

1.8 HOST DISCRIMINATION

1.8.1 INTRODUCTION

Salt (1961) showed that the ability of parasitoids to recognise hosts containing eggs of conspecifics from those which are unparasitised occurs in the major families of parasitic Hymenoptera. This ability is known in the literature as **host discrimination**.

It is now understood that females of some parasitoid species are able to discriminate between:

1. parasitised hosts and unparasitised hosts (numerous published studies have shown this, although the conclusions drawn in some are questionable, see below);
2. parasitised hosts containing different numbers of eggs (Bakker *et al.*, 1990);
3. hosts containing an egg of a conspecific from one containing their own egg (van Dijken *et al.*, 1992; Hubbard *et al.*, 1987; McBrien and Mackauer, 1991).

Notwithstanding such sophisticated abilities, **superparasitism** – the laying of an egg in an already parasitised host – is a common phenomenon among insect parasitoids. The occurrence of superparasitism or, expressed statistically, the occurrence of a random egg distribution among hosts, has often led to the erroneous conclusion that a parasitoid is unable to discriminate between parasitised and unparasitised hosts.

1.8.2 INDIRECT METHODS

There are two approaches to determining whether parasitoids are able to discriminate between parasitised and unparasitised hosts. One is to dissect hosts (section 2.6, subsection 4.2.9) and calculate whether the recorded egg distribution deviates significantly from a Poisson (i.e. random) distribution (Salt, 1961). Van Lenteren *et al.* (1978) have shown that such a procedure is not without pitfalls. They point out that if the method is applied to egg distributions from hosts collected in the field, there is a risk that mixtures of samples with regular (i.e. non-random) egg distributions but different means may add up to produce a random distribution (Figure 1.12). This is one of the reasons why a random egg distribution does not constitute proof of the inability to discriminate. Another problem van Lenteren *et al.* (1978) identified concerning the analysis of egg distributions is that with gregarious parasitoids the distribution of eggs depends not only upon the number of ovipositions but also the number of eggs laid per oviposition.

There are further problems associated with the use of egg distributions. Van Alphen and Nell (1982) recorded random egg distributions when single females of *Asobara tabida* were placed with 32 hosts for 24 hours. Because not all replicates produced random distributions and because other experiments had unequivocally shown that females of this species are able to discriminate between parasitised and unparasitised hosts, the random egg distributions could not be explained by a lack of discriminative ability.

In van Alphen and Nell's experiments the replicates with a high mean number of eggs had random distributions, whereas replicates with lower means had regular ones. It was therefore concluded that *A. tabida* discriminates between unparasitised and parasitised hosts, but is unable to assess whether one or more eggs are present in a parasitised host. Egg distributions are a mixture of the regularly distributed first eggs laid in hosts and of the randomly distributed supernumerary eggs. At lower means, the contribution of the regular distribution of the first eggs is not masked by the random distribution of the supernumerary eggs, whereas at higher means it is.

Even when egg distributions more regular than a Poisson distribution are found, one cannot establish with certainty that a parasitoid is able to discriminate between parasitised and unparasitised hosts. The recorded egg distribution could result from parasitised hosts having a much lower probability of

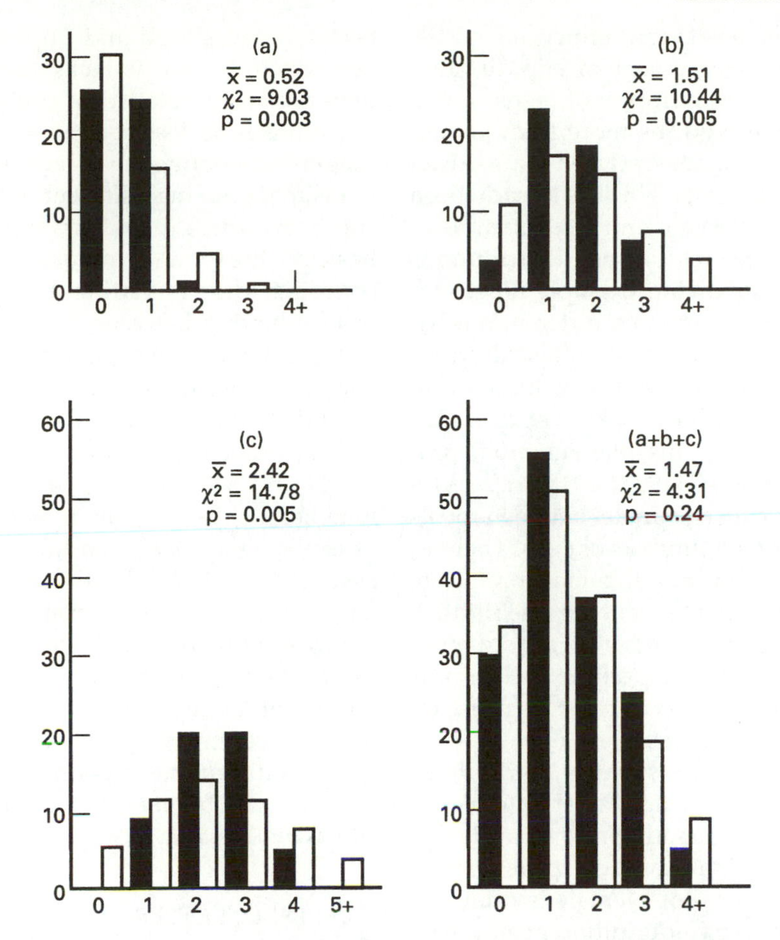

Figure 1.12 Host discrimination by parasitoids: Egg distributions for *Leptopilina heterotoma* (= *Pseudeucoila bochei*) parasitising *Drosophila melanogaster*. Three groups (A, B, C) of around 50 host larvae were presented to female wasps and the hosts subsequently dissected. The mean number of eggs recovered per host larva was different in each case. Although in all three cases superparasitism occurred, when the distribution was compared with the distribution that would have been obtained had the wasps oviposited at random (i.e. a Poisson distribution), the egg distributions are found to be more regular, indicating that the parasitoids discriminate. However, if data from the three distributions are pooled (A + B + C), a distribution is obtained that is indistinguishable from a Poisson distribution, a result that would lead to the erroneous conclusion that the parasitoids cannot discriminate. (Source: van Lenteren *et al.*, 1978.) Reproduced by permission of Blackwell Scientific Publications Ltd.

being encountered, either because they move less than healthy hosts or because they leave the host plant. It is also possible that encounter rates with parasitised hosts are lower because the parasitoid does not re-visit previously searched areas with the same probability, e.g. when it always walks upwards along branches or when it marks areas already visited and avoids re-searching such areas.

The previous examples show that there are major pitfalls associated with using egg distributions to determine whether a parasitoid can discriminate between parasitised and unparasitised hosts. Other components of the behaviour of the parasitoid, or of the behaviour of the hosts, can influence egg distributions. Moreover, a regular egg distribution with a mean number of eggs much greater than one requires more than just an ability to discriminate between parasitised and unparasitised hosts. This has already been illustrated in the above-mentioned example of *A. tabida* where no regular egg distributions are found. The following example illustrates how, in *Leptopilina heterotoma*, different mechanisms are responsible for egg distributions tending to be regular even at a high mean number of eggs per host (Bakker *et al.*, 1972). One explanation for this phenomenon, provided by Bakker *et al.* (1972), is that *L. heterotoma* is able to discriminate between hosts containing different numbers of eggs. There is, however, a second possible interpretation: when the parasitoid is able to distinguish hosts containing an egg of her own from those containing eggs of conspecifics and avoids ovipositing in the former, regular egg distributions would result. Therefore, it is impossible to decide, on the basis of egg distributions alone, whether a parasitoid is able to assess the number of eggs already present in a host.

Experiments therefore need to be carried out in which a parasitoid female is offered a choice of hosts containing different numbers of eggs, all laid by other (conspecific) females. Bakker *et al.* (1990) offered hosts containing two eggs and hosts containing one egg of other females to individual *L. heterotoma*. The wasps oviposited significantly more often in hosts containing a single egg, thus showing that *L. heterotoma* is indeed able to distinguish between hosts containing different egg numbers. Visser (1992) showed that *L. heterotoma* females are also able to recognise hosts containing their own eggs. Thus, both of the above mechanisms may have contributed to the regular egg distributions found by Bakker *et al.* (1972).

It is thus clear that by comparing observed egg distributions with those predicted by a Poisson distribution, one can neither conclude that a parasitoid is able to discriminate between parasitised and unparasitised hosts, nor conclude that it lacks this ability. This does not mean a statistical analysis of egg distributions is useless; it is possible to construct models predicting distributions of eggs for parasitoids having different abilities to avoid superparasitism (e.g. discriminating between healthy hosts and parasitised hosts and counting, discriminating but not counting, discriminating between hosts parasitised by self and hosts parasitised by others), and to compare the theoretical egg distributions with distributions recorded in experiments. This approach was adopted by Bakker *et al.* (1972) and Meelis (1982) when investigating whether wasps are able to assess the number of eggs already laid in a host. These authors assumed that parasitoids search randomly, and that there exists a certain probability that the wasp will lay an egg when it encounters a larva. This probability is 1.0 at the first encounter, but is lower at subsequent encounters. By keeping the probability of oviposition at the subsequent encounters constant, the model could be used to describe superparasitism by *A. tabida*.

1.8.3 DIRECT OBSERVATIONS OF BEHAVIOUR

The other method of determining whether parasitoids are able to discriminate between parasitised and unparasitised hosts involves observing the insects, and recording and comparing encounters resulting in oviposition and rejection of the different host categories. This method provides behavioural evidence that the parasitoid under study rejects parasitised hosts more often than unparasitised hosts. It is, however, wise to use other behavioural criteria in addition to acceptance/encounter ratios; differences in behaviour may also be found in patch time allocation, progeny (i.e. clutch size) and sex allocation (van Alphen *et al.*, 1987).

Because distributions of parasitoid eggs among hosts potentially have an important effect on parasitoid-host population dynamics, one requires a good statistical description of those distributions, for incorporation into population models. We prefer to use the observed behaviour as a basis for a model calculating egg distributions, instead of inferring the underlying behaviour from an analysis of the egg distributions.

An intriguing question is why it took so long before evidence was found of host discrimination by dipteran parasitoids. Host discrimination by hymenopteran parasitoids was discovered in 1926, but the phenomenon was described for tachinid flies only very recently (Lopez, Ferro and van Driesche, submitted). In field and laboratory experiments, Lopez and colleagues showed that *Myiopharus doryphorae* and *M. aberrans*, both parasitoids of Colorado Beetle (*Leptinotarsa decemlineata*) larvae, almost always reject parasitised larvae, whereas they readily oviposit in unparasitised larvae. Discrimination between parasitised and unparasitised fruits by tephritid fruit-flies is common (Prokopy, 1981), so host discrimination is probably a widespread phenomenon in Diptera.

1.8.4 SUPERPARASITISM

Many, if not all, parasitoids are able to discriminate between parasitised and unparasitised hosts, but superparasitism is a common feature in nature (van Alphen and Visser, 1990), posing the question: 'why and when should parasitoids superparasitise?'.

Van Lenteren (1976) addressed this question from the standpoint of causation. He assumed that superparasitism was caused by a failure to discriminate. Van Lenteren found that females of *L. heterotoma* inexperienced with unparasitised hosts readily oviposited in already parasitised hosts but avoid ovipositing in parasitised hosts after they have been able to oviposit in unparasitised ones. He concluded from this that parasitoids superparasitise because they are unable to discriminate between parasitised and unparasitised hosts until they have experienced oviposition in unparasitised hosts. A similar conclusion was drawn by Klomp *et al.* (1980) for *Trichogramma embryophagum*.

A functional approach to the problem is to ask whether it is adaptive for a parasitoid always to avoid superparasitism. Van Alphen *et al.* (1987) re-analysed the data of van Lenteren (1976) and Klomp *et al.* (1980), starting with the hypothesis that superparasitism can be adaptive under certain conditions. They reasoned that host discrimination is an ability which the parasitoid can use to decide either to reject a parasitised host or to superparasitise it, depending on the circumstances, i.e. superparasitism is not the result of an inability to discriminate. Van Alphen *et al.* (1987) argued that an inexperienced female arriving on a patch containing only parasitised hosts should superparasitise, because the probability of finding a better patch elsewhere is low.

Van Lenteren's (1976) inexperienced wasps not only rejected parasitised hosts more often than unparasitised ones but also encountered significantly fewer hosts in experiments involving patches containing only parasitised hosts, compared with similar experiments involving patches containing the same density of unparasitised hosts. It was known that *Leptopilina heterotoma* females search by stabbing with the ovipositor, twice per second, in the substrate, and it was possible to measure both the surface area of a host and that of the patches. It was possible therefore to calculate, from the numbers of encounters observed during a 30 minute observation period, that inexperienced wasps spent on average only 13.12 minutes searching and handling hosts when introduced on to a patch with parasitised hosts, whereas they spent on average 2.14 minutes when introduced on to patches with unparasitised hosts. It is unclear how such a difference could have escaped the attention of the observer! Van Alphen *et al.* (1987) interpreted the differences in behaviour between inexperienced wasps and experienced wasps as evidence that inexperienced wasps do recognise parasitised hosts. Thus, van Lenteren's (1976) data do not support the conclusion that host discrimination needs to be learnt.

Experiments by van Alphen *et al.* (1987), involving *L. heterotoma* and *Trichogramma evanescens*, confirmed that females inexperienced with unparasitised hosts are, like experienced wasps, already able to discriminate, although inexperienced females superparasitise more frequently. This example shows that alternative hypotheses can be overlooked if one asks only causal questions.

Static and dynamic optimality models as well as ESS models (subsection 1.2.2) have been published (Iwasa *et al.*, 1984; Parker and Courtney, 1984; Charnov and Skinner, 1985; Hubbard *et al.*, 1987; van der Hoeven and Hemerik, 1990; Visser *et al.*, 1990), showing that superparasitism is often adaptive. The models predict that oviposition in already parasitised hosts, though resulting in fewer offspring than ovipositions in unparasitised hosts, may still be the better option when either there is no time available to search for and locate unparasitised hosts or when unparasitised hosts are simply not available. By ovipositing into an already parasitised host under such conditions, a female may increase her fitness if there is a finite chance that her progeny will outcompete the other progeny. Experimental tests of some of these models have shown that parasitoids behave in such a way that models' predictions are at least met qualitatively (Hubbard *et al.*, 1987; Visser *et al.*, 1990; van Alphen *et al.*, 1992). Measurement of the pay-off from superparasitism is discussed in Chapter 2, subsection 2.10.2.

1.9 SEX ALLOCATION

1.9.1 INTRODUCTION

Because of the haplo-diploid genetic system of sex determination in Hymenoptera (subsection 2.5.2), sex ratio in parasitoid wasps can vary considerably. Students of parasitoid behaviour conventionally express sex ratios as the proportion of males. Male-biased sex ratios (i.e. ratios >0.5) in parasitoid mass-rearing units are undesirable, because males produced in excess of those required for ensuring that all females are inseminated, are useless for biological control. Thus, the study of sex allocation in insect parasitoids is an important area of study not only for those interested in evolutionary ecology, but also for practitioners of biological control (Luck, 1990).

Natural selection acts upon oviposition behaviour to produce sex ratios that are adaptive in relation to different conditions. There is therefore considerable interspecific and intraspecific variation in parasitoid sex ratios. Species such as *Melittobia acasta* always produce highly female-biased sex ratios (i.e. ≪0.5) (van den Assem, 1975); others may have sex ratios always close to 0.5 (van Dijken *et al.* 1989); whereas other species may have sex ratios that vary according to circumstances.

Several models predict optimal sex allocation for individual parasitoids under different conditions (Hamilton, 1967; Charnov, 1982; Waage and Godfray, 1985; Harvey, 1985; Waage, 1986; Nunney and Luck, 1988; Luck *et al.*, 1993). **The Theory of Local Mate Competition** (Hamilton, 1967) predicts that the evolutionarily stable sex ratio for a parasitoid population should decrease from 0.5 as mating becomes restricted to the offspring of fewer and fewer female wasps. If very few females colonise each host patch, for example due to the density of females in the habitat being low, due to females defending patches from colonisation by other females (subsection 1.5.4), or due to females tending to parasitise all the hosts in a patch and avoiding conspecific superparasitism, and there is also a high degree of (random) sibmating (due to a tendency of adults to mate before they disperse from their emergence site), the optimal sex ratio produced by a female should be low. When only one female colonises a patch, she should produce only enough sons to fertilise all her daughters. However, when several females colonise each patch, each female needs to produce sufficient sons to compete with the sons of other females, the evolutionarily stable sex ratio being 0.5 when a large number of females search together (Figure 1.13). In summary, Hamilton's model (which, although formulated prior to Maynard Smith's first papers on evolutionary stable strategies (subsection 1.2.2), was based upon 'ESS reasoning' (Parker, 1984)), predicts that the 'unbeatable' sex ratio is $(n - 1)/2n$, where n is the number of females colonising a patch of resource, on which their offspring mate at random.

The effects of modifying Hamilton's model have been explored by several authors (Nunney and Luck, 1988; King, 1993; Luck *et al.*, 1993).

Hardy and Godfray (1990) make the interesting point that while sex ratio theory assumes that females can produce offspring of both sexes, some females of arrhenotokous species are only able to produce sons, because they are virgins, i.e. they will show 'constrained' oviposition. Such females may in some cases be sufficiently common to select for an observable sex ratio bias by unconstrained females.

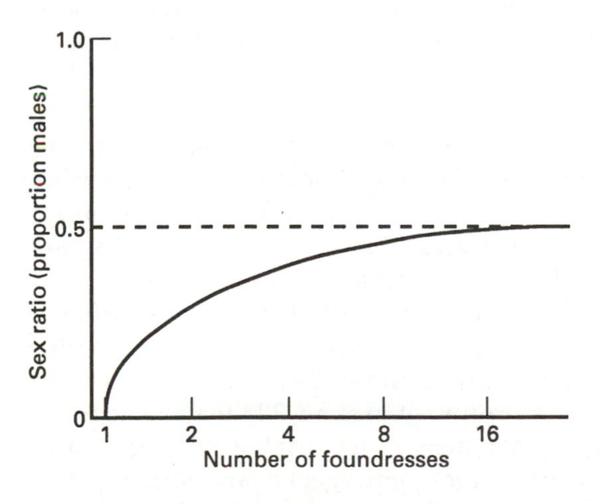

Figure 1.13 Sex allocation by parasitoids: The optimal sex ratio in relation to the number of females exploiting a host patch. (Based on information in Hamilton, 1967.)

Orzack (1993) should be consulted for very useful advice on the testing of sex ratio models and on data interpretation.

1.9.2 SUGGESTIONS FOR INVESTIGATIONS OF SEX ALLOCATION

Density Dependent Shifts in Sex Ratio

In a population of wasps adjusting sex allocation as predicted by Hamilton's (1967) model, the proportion of male offspring produced per female will be higher at high wasp densities than at low densities (this prediction is more likely to apply to idiobionts, since females of koinobionts will find it hard to estimate the final size of the host (see King, 1989)). Thus, individual optimisation does not go hand in hand with maximal female production in a population. This is one reason why the mass-rearing of parasitoids often does not result in desired female-biased sex ratios.

Models like Hamilton's (1967), which predict adaptive shifts in sex allocation in response to the presence of other females, raise the question of how these shifts can be achieved. Waage (1982) was the first to show that simple fixed mechanisms, such as always laying one or more male eggs first (Figure 1.14), can lead to variable sex ratios under different conditions, close to those predicted by the functional models (Waage and Ng, 1984; Waage and Lane, 1984). Rather than counting the number of hosts in a patch and calculating what fraction should be sons and daughters, the female can lay a son, then lay the number of daughters he can fertilise, then lay another son, and so on. Other mechanisms are also known where the stimulus to change the sequence of sex allocation comes from contacts with marks of conspecifics or

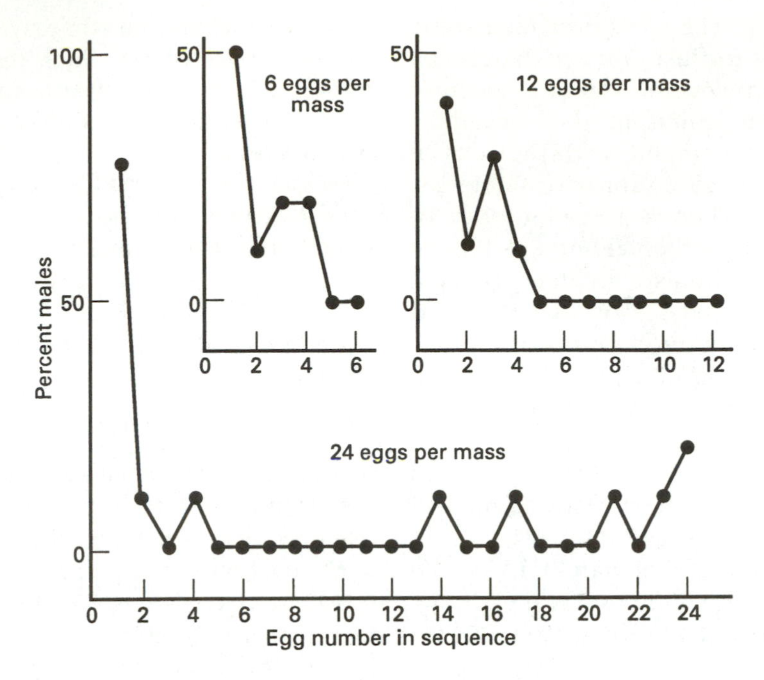

Figure 1.14 Sex allocation in parasitoids: The sequence in which the solitary scelionid parasitoid *Gryon pennsylvanicum* (= *atriscapus*) lays male and female eggs in host egg masses of 6, 12 and 24 eggs. Data based on observations of mated females and subsequent dissection of host eggs at the time of adult emergence. In all egg masses, males (usually one per mass) are placed in the first few host eggs. By this strategy, every host egg mass, independent of size, is ensured a male. In very large egg masses, e.g. of 24 eggs, a second male is sometimes produced towards the end of the sequence, suggesting that the ratio of males to females may be measured and kept constant for a particular size of egg mass by ovipositing females. The males-first strategy alone will also produce an adaptive increase in sex ratio with increased female crowding, since each wasp will lay fewer eggs per mass and therefore allocate proportionately more males. (Source: Waage, 1982.) Reproduced by permission of Blackwell Scientific Publications Ltd.

with parasitised hosts (Viktorov, 1968; Viktorov and Kochetova, 1971).

To show that females adjust sex ratio in response to the presence of other females, offer patches containing equal numbers of standardised hosts to different densities of parasitoid females (as in a mutual interference experiment, subsection 1.14.3). Primary sex ratio can be determined as described in section 1.9.3. If females use a simple 'males first' rule, a shift in sex ratio can be found with this set-up. Alternatively, one could offer a fixed number of standardised hosts per female in experiments with different numbers of females. In such an experiment, the number of hosts increases with the number of wasps. Where females use a 'males first' rule, no sex ratio adjustment should be found unless, that is, females react to parasitoid odour or marks.

Strand (1988) adopted a quite different approach in studying density-dependent shifts in sex ratio in *Telenomus heliothidis*. He kept groups of around 200 females together, without hosts, for variable periods of time shortly after emergence and mating, then allowed subsequently isolated females to oviposit in unparasitised hosts (Strand also tested for the effects of subsequent isolation by varying the isolation period). With this method, one can rule out the possibility that females alter sex ratio in response to encounters with already parasitised hosts.

Simple mechanisms such as 'males first' can easily be studied in solitary parasitoids by collecting a sequence of hosts parasitised by an individual female and rearing each of the hosts in separate vials. Investigating such behaviour in gregarious parasitoids is more difficult, but by interrupting oviposition at various points during the laying of egg clutches, and rearing the parasitoids, information on the sequence of male and female eggs can be obtained. Sometimes the sequence of male and female eggs can be inferred from the behaviour of the female (subsection 1.9.3).

Sex Ratio and Host Quality

Charnov (1979), using deductive modelling, argued that if hosts vary in the amount of resource they contain e.g. vary in size, and if the incremental gain in fitness per host is greater for one sex than for the other, then females should allocate that sex to the larger hosts (this prediction is more likely to apply to idiobionts). There are good reasons for assuming that females have more to gain from developing in a large host than do males, namely that major components of female fitness, in particular fecundity, are more strongly correlated with body size than major components of male fitness, in particular mating ability (van den Assem *et al.*, 1989). Empirical support for Charnov's hypothesis comes both from behavioural observations on parasitoids (Jones, 1982) and from parasitoid rearings from hosts, showing that females are mainly or entirely allocated to large hosts and males mainly or entirely to small hosts (Holdaway and Smith, 1932; Sandlan, 1979a). Note that the hypothesis predicts that a parasitoid should lay females in *relatively* larger hosts – a small host may become relatively large if no larger hosts are available to a female (Waage, 1986). Parasitoids offered only one size of host eventually allocate both sexes and achieve a more equitable lifetime sex ratio (Sandlan, 1979a; Avilla and Albajes, 1984), a theoretical prediction of a model that involves adjustment of sex ratio according to short- and long-term female experience (van den Assem *et al.*, 1984).

Host size selection experiments (subsection 1.5.7) can provide information on the influence of host quality on sex allocation, if they are combined with a technique for determining the primary sex ratio (subsection 1.9.3).

1.9.3 MEASURING PRIMARY SEX RATIOS

Theories of sex allocation deal with the oviposition decisions of female wasps. Tests of these theories require accurate measurement of the sex ratio of the oviposited eggs, i.e. the allocated or **primary sex ratio**. Often, however, due to differential mortality of male and female immatures the sex ratio of emerging parasitoids, the so-called **secondary sex ratio**, does not always reflect the primary sex ratio. To take account of this problem, the following methods can be used:

1. Wellings *et al.* (1986) designed a statistical method for estimating primary sex ratios from recorded secondary sex ratios. With this method, sex-specific larval mortality of immatures is corrected for;
2. Werren (1980), in laboratory experiments, determined the number of eggs laid in each host, and compared this number with the number of parasitoids emerging from each host in the remainder of a cohort. He found that progeny mortality was negligible and that secondary sex ratio could be taken as an acceptable measure of primary sex ratio, at least as far as his experiments were concerned. The problem with this method of determining how representative secondary sex ratio is of primary sex ratio, is that it requires large samples to minimise sampling error;
3. van Dijken (1991) dissected eggs laid by *Epidinocarsis lopezi* from the host mealy-

bugs and counted chromosome numbers of cells in metaphase. She was able to determine whether an unfertilised or a fertilised egg had been laid (subsection 2.5.2 describes details of the method);

4. Cole (1981), Suzuki *et al.* (1984) and Strand (1989) discovered for some species that one can actually determine whether a female parasitoid fertilises an egg or not, on the basis of differences in the insect's abdominal movements during oviposition – a feature common to these species is a pause during oviposition of a fertilised (female) egg;
5. Flanders (1950) and Luck *et al.* (1982) used a non-destructive method for *Aphytis* that involves determining the sex of an egg from the position in which it is laid: wasps lay male and female eggs on the host's dorsal surface and ventral surface, respectively.

It has been shown for a number of species that biased sex ratios, including all-female and all-male broods, are not the product of natural selection on oviposition decisions by the female (subsections 2.5.3 and 2.5.4 for discussion).

1.10 FUNCTIONAL RESPONSES

Solomon (1949) coined the term **functional response** when describing the response shown by *individual* natural enemies to varying host (prey) density; with increasing host availability, each enemy will attack more host individuals. Some authors (Young, 1980; Hopper and King, 1986) measure the functional response by recording changes in the attack rate of the natural enemy population as a whole. This is a questionable practice, since within a natural enemy population individuals respond not only to host/prey density but also to one another (subsection 1.14.3).

Several types of functional response are possible (Figure 1.15):

Type 1: where there is a rectilinear rise to a maximum (N_x) in the number of prey eaten per predator as prey density increases. The response is described by the following equation:

$$N_a = a'TN \tag{1.2}$$

where N_a is the number of hosts parasitised or prey eaten, N is the number of hosts or prey provided, T is the total time available for search, and a' is an acceleration constant, the instantaneous attack rate (equation 1.2 applies only when $N<N_x$).

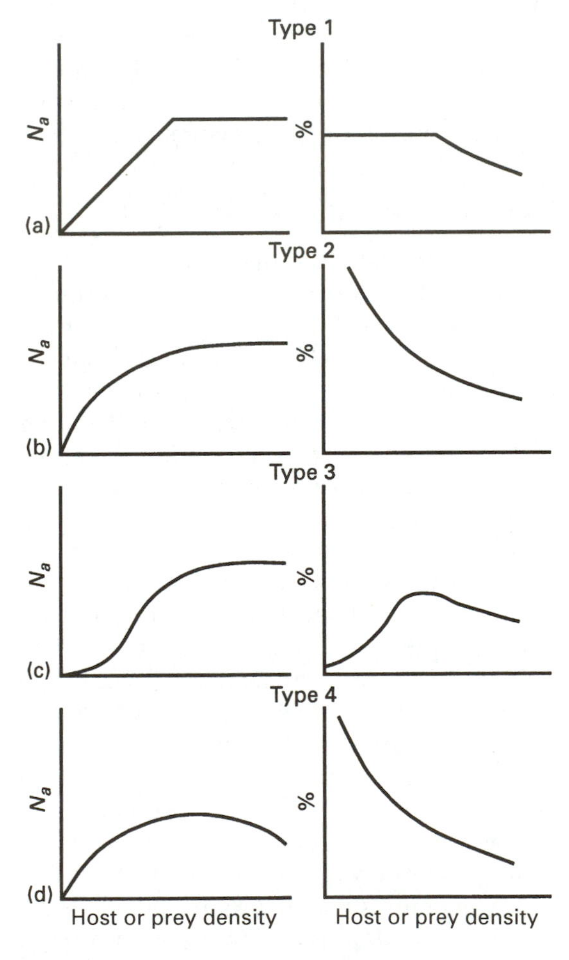

Figure 1.15 The four types of functional response observed in predators and parasitoids: (a) Type 1; (b) Type 2; (c) Type 3; (d) Type 4. N_a = number of hosts parasitised or number of prey eaten; % = percentage of hosts parasitised or percentage of prey eaten.

The Type 1 response is likely to be found when handling times (see below) are negligible and eggs are in limited supply.

Type 2: where the response rises at a constantly decreasing rate towards a maximum value, i.e. the response is curvilinear up to the asymptote, in contrast with the Type 1 response. Holling (1959a, b) predicted such a response, reasoning that the acts of quelling, killing, eating and digesting prey are time-consuming activities (collectively called the **handling time**) and reduce the time available for further search, and that with increasing prey density a predator will spend an increasing proportion of its total time available not searching:

$$T_s = T - T_h \cdot N_a \quad (1.3)$$

where T_s is the actual time spent searching, and T_h is the handling time.

Type 3: where the response resembles the Type 2 response except that at lower prey densities it accelerates. The response is thus sigmoid.

Type 4: where the response resembles the Type 2 response except that at at higher densities it declines, producing a dome-shape.

Sabelis (1992) also recognises a fifth type of response which is intermediate between the Type 1 and the Type 2. This response appears to be shown by some predatory mites, and will not be discussed further.

The functional response of a predator or parasitoid species is usually measured as follows: individual insects are confined in an arena (e.g. cage) with different numbers of prey or hosts, for a fixed period of time (Figure 1.16). At the end of the experiment, the natural enemies are removed and either the number of prey killed or the number of hosts parasitised (or both, in the case of some host-feeding parasitoids, see below) is counted. Hosts are either dissected or reared until emergence of the parasitoids. From the counts made, a graph can then be plotted relating the number of prey or hosts attacked to the number offered. The plots are then compared with mathematical models (Holling, 1959a,b, 1966; Rogers, 1972; Royama, 1972; Mills, 1982a; Arditi, 1983; Casas *et al.*, 1993). With predators, the functional responses of the different larval instars, as well as those of the adults, can be measured. With both predators and parasitoids, functional responses in relation to prey and hosts of different sizes can be measured.

There are two likely reasons why a Type 3 response may be recorded using the above-mentioned experimental set-up:

1. As host density decreases at the lower range of host densities, the parasitoid spends an increasing proportion of the total time available in non-searching activities. For example, at lower host densities, *Venturia canescens* spends a greater proportion of its time performing activities such as walking and resting on the sides of the experimental cage. Similar behaviour is probably responsible for the Type 3 response observed in parasitoids and predators that are offered unpreferred prey species. In *Aphidius uzbeckistanicus*, *Coccinella septempunctata* and *Notonecta glauca*, a Type 3 response was recorded when the unpreferred host and prey species was provided, compared with a Type 2 when the preferred species was provided (Dransfield, 1979; Hassell *et al.*, 1977).

   The parasitoid under investigation may be a host-feeder, mainly feeding upon hosts rather than ovipositing, at low host densities (section 1.7). N.B. in host-feeding parasitoids that feed and oviposit on different host individuals, we may distinguish between the following functional responses: that for parasitism alone, that for host-feeding alone, and that for parasitism and feeding combined (i.e. the 'total' functional response (Kidd and Jervis, 1989). If, as is likely, the handling time for feeding encounters is longer than for oviposition encounters, a Type 3

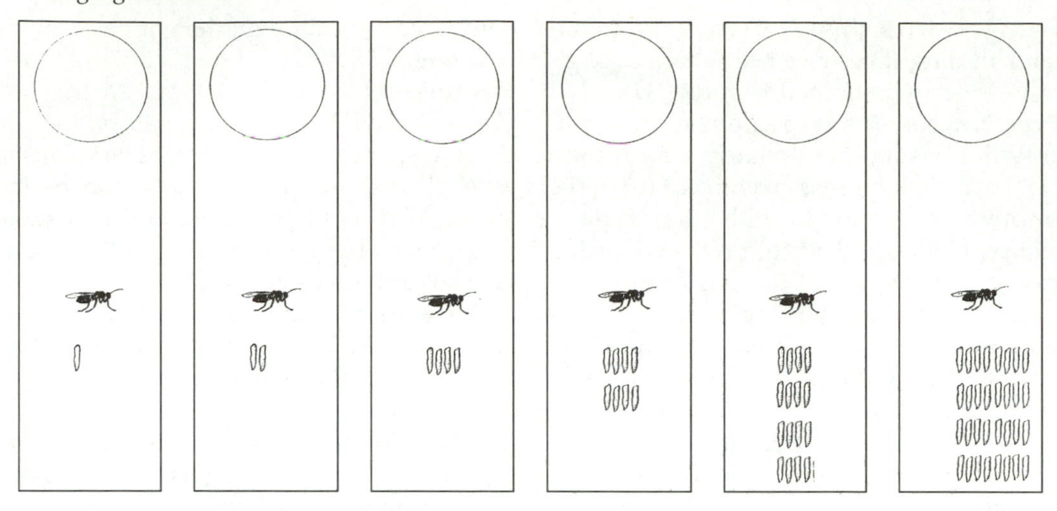

Figure 1.16 Functional responses of parasitoids and predators: Schematic representation of traditional design of a functional response experiment. Circles denote the experimental host patch and rectangles the experimental arena. See text for discussion.

response for parasitism may result (Collins *et al.*, 1981).

2. handling times may be shorter at higher host densities. In solitary parasitoids, this is an unlikely cause of a sigmoid functional response, but gregarious parasitoids may decrease clutch size at higher host densities (section 1.6), and so decrease handling time per host. Predators may ingest less food from each prey item at higher prey densities (Figure 1.17) and so reduce handling times. When extracting food from a prey item becomes increasingly difficult with the time spent feeding on it, predators may optimise the rate of food intake by consuming less of each individual prey item when the rate of encounters with prey is high (Charnov, 1976; Cook and Cockrell, 1978). This behaviour is predicted by optimal foraging models, while a similar prediction can be made on the basis of a causal model relating the amount of food in the gut to the amount eaten from each prey. Some authors have argued that the optimal foraging model can be refuted because there is a causal explanation for the observed behaviour. However, the reader should remember that causal and functional explanations are not mutually exclusive; indeed, they complement each other. When predictions of a causal and a func-

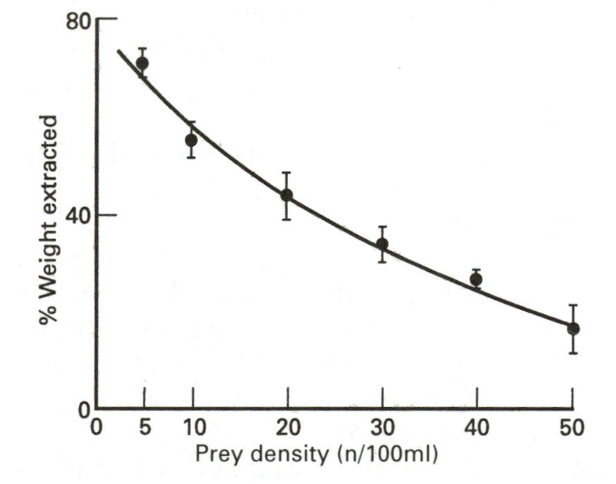

Figure 1.17 The relationship between prey availability and the percentage of the mass of each prey consumed by the belostomatid bug *Diplonychus rusticum*: The bug is more 'wasteful', eating less of each prey (*Chironomus plumosus*) as prey density increases. This effect is predicted by optimal foraging (i.e functional) and gut-filling (i.e. causal) models, and is shown by a wide variety of predators. (Source: Dudgeon, 1990.) Reproduced by permission of Kluwer Academic Publishers.

tional model are quantitatively similar, this can be taken as evidence that the mechanism does not constrain optimisation of a behavioural trait.

A Type 4 functional response will occur if:

1. when dealing with prey individuals, other prey individuals interfere with the predator and cause it to abort the attack more at high prey densities than at low densities;
2. the prey have a well-developed group defence reaction that is more effective at high prey densities than at low ones.

The classical functional response experiment assumes there is a homogeneous environment, or at least it does not consider the spatial distribution of prey and hosts. However, most insects are patchily distributed and the spatial distribution of hosts or prey within an experimental arena is likely to vary significantly with the density of the insects. Predators and parasitoids respond to differences in prey and host densities between patches by adjusting the amount of time spent in each patch (subsection 1.12.2). By allowing the parasitoid, rather than the experimenter, to determine the amount of time it spends in an experimental patch (in a so-called **variable-time experiment**), a different type of functional response may be obtained compared to experiments where the time spent is fixed by the experimenter (so-called **fixed-time experiments**) (van Alphen and Galis, 1983; Collins, *et al.*, 1981; Hertlein and Thorarinsson, 1987). Van Lenteren and Bakker (1978) suggest that in fixed-time experiments, some parasitoids are likely to show a Type 2 response, rather than a Type 3, because parasitoids are caused to revisit low density patches they would otherwise leave. Thus, a Type 2 response may be an artefact of the fixed-time experimental design.

Since the type of functional response found in an experiment depends very much on the experimental design adopted, one should first clearly define what sort of question one wishes to address before measuring a functional response. Often, a functional response is measured to provide insights into the suitability of a parasitoid as a biological control agent. The problem is then how one can use the information generated by the experiments to predict the performance of the parasitoid in the field. The context in which the data will be used is one of population dynamics, and thus relates to the response of the parasitoid *population* to host density. The spatial structure of natural host populations, and the interactions between individual parasitoids in the population, make it hard to relate the results of experiments on individuals in single-patch, single-parasitoid experiments to processes occurring at the population level. Thus, one cannot conclude from the type of functional response shown in a single-patch, single parasitoid experiment, whether or not a parasitoid species is potentially capable of regulating the host population (subsection 5.3.7).

Our personal opinion then, is that few meaningful insights into population dynamics can be gained by carrying out single-patch experiments. However, if such experiments are to be carried out, the minimum requirements for experimental design should be as follows: the foraging insect should be observed continuously (in most functional response experiments carried out to date, parasitoid and predator behaviour have not been examined directly at all, never mind continuously!), and a record made of how the parasitoid spends its time in the experimental arena. Parasitoids should be allowed to leave the arena when they decide to leave, so the experiment should be a variable-time one. It may prove difficult for the observer to decide when an experiment should be terminated. A parasitoid may leave the experimental patch for a short period, but then return and continue searching for hosts. Experiments may need to be terminated after the insect has spent an arbitrary period of time outside the patch (Waage, 1979; van Alphen and Galis, 1983), but of course the choice of the period is subjective and it acts as a **censor** in the data (a censor is a factor, other than a decision by the foraging insect, that terminates an experiment, e.g. a decision by the experimenter or an external disturbance (Haccou *et al.*, 1991)]). A solution to the problem of when to terminate an experiment is to use an arena containing two patches. Once the insect has left the first patch and arrived in the second one, the experiment can be terminated.

Ideally, functional response experiments should measure encounter rates with concurrently available patches containing different densities of hosts. To do this, a multi-patch experiment needs to be carried out. Such an experiment might show that high density patches are found more easily by the parasitoid, since such patches produce greater quantities of volatile attractants than low density patches (subsection 1.5.3).

Functional response experiments have so far not taken into account the possibility that the response of a parasitoid to patches of different densities depends on whether patches are scarce or common in the habitat. In 'poor' habitats when distances between patches are large and high density patches are scarce, parasitoids should, when exploiting low density patches, stay longer and parasitise more hosts. Finally, functional response experiments need to take account of the reaction of a parasitoid to the presence of conspecifics.

An alternative to measuring functional responses is to undertake an integrated analysis of all factors affecting patch time allocation in parasitoids (subsection 1.12.2).

Statistical methods for estimating the parameters of functional response models from experimental data are described in Livdahl and Stiven (1983), Williams and Juliano (1985), Juliano and Williams (1987), Houck and Strauss (1985) and Taylor (1988a). Trexler *et al.* (1988) review curve-fitting routines, and describe a method for identifying whether the functional response obtained in an experiment is a Type 2 or a Type 3 (Collins *et al.*, 1981 and Hughes *et al.*, 1992 describe a simpler method). Determining the type of response is an important step that needs to be taken by the investigator before he or she attempts to obtain parameter estimates from functional response models. Wrong estimates may be obtained if a model for a Type 2 response is used to estimate parameters from what is in reality a Type 3 response, and *vice versa*.

1.11 SWITCHING BEHAVIOUR

Species and host stage preference by natural enemies has been discussed in subsection 1.5.6. Preference (parameter c in equation 1.1) may not be constant but may vary with the relative abundance of two prey types, in which case if the predator or parasitoid eats or oviposits in disproportionately more of the more abundant type (c increasing as N_1 / N_2 increases) it is said to display **switching behaviour** (Murdoch, 1969) or **apostatic selection** (Clarke, 1962), the latter term being used by geneticists. Where disproportionately more of the rarer type is accepted (c increasing as N_1 / N_2 decreases) **negative switching** is said to occur (Chesson, 1984).

(Positive) switching behaviour has aroused the interest of students of population dynamics because it is associated with a Type 3 functional response (to prey type N_1) (subsection 5.3.7) (Murdoch, 1969; Lawton *et al.*, 1974).

Switching behaviour in parasitoids has been observed by Cornell and Pimentel (1978) in *Nasonia vitripennis*, van Alphen and Vet (1986) in *Asobara tabida*, and Chow and Mackauer (1991) in *Aphidius ervi* and *Praon pequodorum*, while switching in insect predators has been observed by Lawton *et al.* (1974) in the waterboatman *Notonecta glauca* and the damselfly *Ischnura*. Other examples are given in Sherratt and Harvey (1993).

Switching can be tested for by offering parasitoids mixtures of different host species in single-patch experiments. The combined density of the two host species should be kept constant, but the relative abundance of the two species should vary among treatments. If the mechanism causing the switching is to be determined, full records of parasitoid (and host) behaviour ought to be made. As in other host selection experiments (subsection 1.5.7) females should be observed continuously and the number of acceptances and ovipositions scored, to show whether females accept more or fewer individuals of a host type than they successfully parasitise (this possibility is usually ignored by authors).

One can either conduct fixed-time experiments in which depletion of the hosts is prevented by replacing each parasitised host by an unparasitised one of the same species, or allow depletion and terminate experiments when the parasitoid leaves the patch. Bear in mind that switching, like many other aspects of parasitoid and predator behaviour, is likely to be affected by previous experience of the natural enemy (see below).

The resulting data can be analysed using Murdoch's (1969) null or **no-switch** model:

$$P_1 = c.F_1 / (1 - F_1 + [c.F_1]) \quad (1.4)$$

where F_1 is the proportion of host species 1 in the environment, P_1 is the proportion of species 1 among all the hosts oviposited in, and c (a parameter we have already mentioned in subsection 1.5.7) is a combined measure of preference and encounter probability for species 1, given by:

$$c = \frac{N_2 \cdot E_1}{N_1 \cdot E_2} \quad (1.5)$$

which is a rearrangement of equation 1.1. In the absence of switching behaviour, c is a constant. c can be estimated in various ways, although it is convenient to estimate it when $N_1 = N_2$. The value of P_1 for any level of availability of species 1 can be estimated by substituting the estimated value of c in equation 1.4, and an expected no-switch curve plotted (Figure 1.18). If the parasitoid species' preference is not constant but alters with changing host availability (or encounter rate) the observed proportion of host species 1 among all the accepted or parasitised hosts will be higher than expected when species 1 is abundant and lower than expected when species 1 is rare.

Elton and Greenwood (Elton and Greenwood, 1970, 1987; Greenwood and Elton, 1979; see Sherratt and Harvey, 1993 for a recent discussion) provide a model which can be used for the detection of switching and other forms of frequency dependent selection, and which includes a measure of the deviation from constant preference as one of the parameters. Another model developed by Manly and colleagues (Manly *et al.*, 1972; Manly, 1972, 1973, 1974; see Sherratt and Harvey, 1993, for a recent discussion) takes account of prey depletion (this model requires modification before it can be used to take account of depletion of unparasitised hosts). The latter model can also be easily generalised for more than one prey or host type.

Tinbergen (1960) suggested, as a mechanism for switching, that predators form a **search image** of the most abundant prey species, i.e. they experience a perceptual change in the ability to detect a cryptic prey type, and this change does not occur when that type is rare (Lawrence and Allen, 1983; Guilford and Dawkins, 1987 give reappraisals of the evidence for search image formation). However, switching could well result from other behaviour such as active rejection of the less preferred host species as the preferred hosts becomes more abundant – a prediction of optimal prey selection models (subsection 1.5.7). Note that Murdoch's definition of switching is couched in terms of *relative* prey density, whereas optimal foraging models refer to *absolute* densities or encounter rates with prey.

Lawton *et al.* (1974) investigated whether experience with a particular prey species may be a contributory mechanism in the switching behaviour of *Notonecta* presented with *Asellus* and *Cloeon* (in this case, negative switching was recorded over days 2 to 4 of the experiment and positive switching over days 8 to 10 (Figure 1.18)). In a separate experiment, they measured the proportion of successful attacks on *Asellus* prey in relation to the proportion

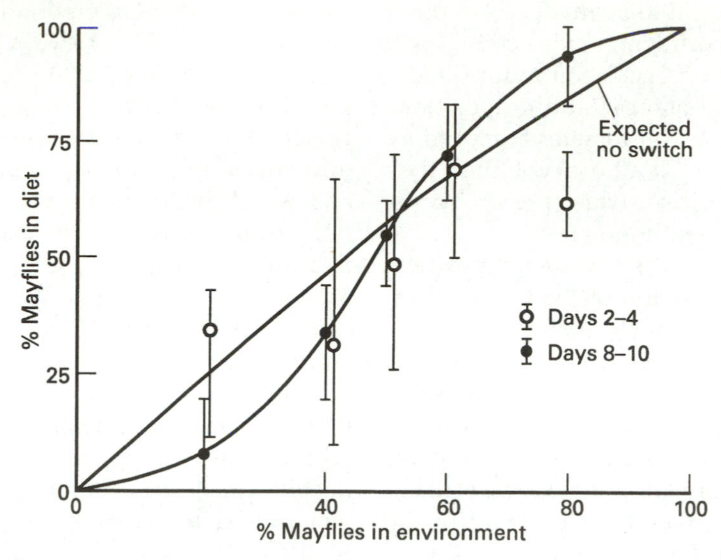

Figure 1.18 Switching in insect natural enemies: The percentage of mayfly larvae in the diet of *Notonecta* as a function of their relative abundance in the habitat. The almost straight line is the 'no-switch' curve. See text for further explanation. (Source: Lawton *et al.*, 1974.)

of this prey available in the environment during the previous seven days, and found that the more *Asellus* the predator was exposed to, the greater was the proportion of successful attacks recorded. While this strongly suggests that experience with *Asellus* affects the predator's prey capture efficiency, it does not prove conclusively that it does so, since no information was obtained on the encounter rates, and therefore the experience of the insects, during the pre-experimental period. The development of a search image could be ruled out as a mechanism for switching in this predator/prey system since: (a) in the switching test *Notonecta* took different prey species in a random sequence instead of attacking prey in 'runs' (Lawton *et al.*, 1974), and (b) the prey were unlikely to have been cryptic in the experimental tanks used.

Switching behaviour may not necessarily be adaptive. Chow and Mackauer (1991) found that *A. ervi* and *P. pequodorum* switched to the alfalfa aphid when pea aphids and alfalfa aphids were offered to wasps in a 1:3 ratio. However, Chow and Mackauer hypothesise that since alfalfa aphids are more likely than pea aphids to escape from an attacking wasp, a foraging wasp incurs a potentially higher cost in lost 'opportunity time' (subsection 1.5.6) when attacking alfalfa aphids. Furthermore, since it is possible that alfalfa aphids are poorer quality hosts in terms of offspring growth and development, wasps may not derive a fitness gain from switching to alfalfa aphids. This hypothesis requires testing.

For a comprehensive review of switching and frequency dependent selection in general, see Sherratt and Harvey (1993).

1.12 PATCH TIME ALLOCATION

1.12.1 INTRODUCTION

One aspect of parasitoid foraging behaviour where the causal approach and the functional approach have traditionally coexisted is patch time allocation. We will consider first which factors affect patch time allocation and second how one can analyse the interplay of the different factors.

1.12.2 FACTORS AFFECTING PATCH TIME ALLOCATION

Patch time allocation is likely to be affected by the following:

1. a parasitoid's previous experience;
2. its egg load;
3. patch kairomone concentration;
4. encounters with unparasitised hosts;
5. encounters with parasitised hosts;
6. the timing of encounters;
7. encounters with the marks of other parasitoids;
8. encounters with other parasitoid individuals;
9. superparasitism.

Figure 1.19 Patch time allocation by individual parasitoids and predators: Schematic representation of one suggested design for an experiment for studying patch time allocation. A randomised arrangement of patches (denoted by circles) of different host density is used. This experimental design can be used in the study of aggregative responses.

Some of these factors can be studied through experiments in which all the other factors are excluded. For example, the effect of kairomone concentration can be investigated without involving hosts at all (subsection 1.5.3). To eliminate the effects of encounters with other parasitoids and their marks, the experimental design shown in Figure 1.19 can be used. However, it may be impossible with some experiments to separate the effects of different factors. A notorious problem is the analysis of the factors that determine how long a parasitoid will stay on a patch which initially contains only unparasitised hosts. Because the parasitoid oviposits in the unparasitised hosts it encounters, the number of unparasitised hosts decreases while the number of parasitised hosts increases. Thus, with the passage of time, the parasitoid experiences a decreasing encounter rate with unparasitised hosts and an increasing encounter rate with parasitised hosts. Because both the temporal spacing and the sequence of encounters with parasitised and unparasitised hosts are stochastic in nature, encounter rates with both types of host do not alter in a monotonic, smooth fashion.

In some parasitoid species, encounters with unparasitised hosts have an incremental effect on the time spent in a patch (van Alphen and Galis, 1983; Haccou *et al.*, 1991). This poses the question: 'what effect do encounters with *parasitised* hosts have on patch time allocation, and how does the relative timing of encounters with parasitised and unparasitised hosts influence the period spent in individual patches?'.

1.12.3 ANALYSING THE INTERPLAY OF DIFFERENT FACTORS

Two distinct hypotheses can be formulated about the effect of encounters with parasitised hosts on patch residence times. The functional hypothesis is as follows: given that a parasitoid is able to discriminate between parasitised and unparasitised hosts, encounter rates with both host types provide the parasitoid with information on host density and the degree of exploitation of a patch. This information allows the wasp to determine when to leave the patch, e.g. high encounter rates with parasitised hosts in combination with low encounter rates with unparasitised hosts signal a high level of exploitation of the patch. Because it could be more profitable for the wasp to move on and search for a better quality patch, the insect might decide to leave. Van Lenteren (1976) recognised this as one of the functions of host discrimination, and showed through single-patch experiments that wasps continued to search on patches in which parasitised hosts were immediately replaced by unparasitised ones, whereas wasps allowed to search on similar unreplenished patches attempted to leave the experimental arena after most of the hosts had been parasitised. The functional hypothesis states that encounters with both unparasitised and parasitised hosts affect patch time, but it does not specify the mechanism involved.

The causal hypothesis formulates explicitly how encounters with parasitised hosts affect patch time. This hypothesis is an extension of a mechanistic model for patch time allocation proposed by Waage (1979) for the parasitoid *Venturia canescens*. Although this model was shown to be an incorrect description of the behaviour of *V. canescens* (Driessen *et al.*, 1994), it is still valuable as a conceptual model and it can be applied to many other parasitoid species. Waage assumed that a female parasitoid, when entering a patch containing hosts, has a certain motivation level for searching the patch, the level being set by previous experience and kairomone concentration on the patch. If the wasp does not locate and oviposit in hosts, the motivation level will decrease steadily over time to a threshold value whereupon the parasitoid leaves the patch. However, with each oviposition that occurs, an increment in motivation occurs. The initial level of motivation, combined with linear decreases of motivation during searching periods and increases in motivation following ovipositions, determine how long the parasitoid will stay in the patch (Figure 1.20a). The causal hypothesis assumes there is an additional effect of a rejection of a parasitised host causing a decrease in motivation level (Figure 1.20b). Like the functional hypothesis, the causal hypotheses predicts shorter patch residence times with increasing patch exploitation, all other things being equal.

A rigorous test of the causal hypothesis ought to demonstrate whether or not the mechanism by which shorter patch residence times come about is an increase in the tendency to leave the patch after a rejection of a parasitised host. Such a test implies that one is able to assess the relative effects on the motivation to search of ovipositions in unparasitised hosts (i.e. increments), of the time interval between encounters, and of rejections of parasitised hosts (i.e. decrements). To illustrate how difficult it is to determine whether the rejection of parasitised hosts causes a decrease in the motivation to search, we will discuss in some detail the experimental evidence given by van Lenteren (1991). In one experiment, individual females of *Leptopilina heterotoma* were allowed to search on a 1 centimetre diameter patch of yeast containing four unparasitised hosts and 16 parasitised hosts. Each host parasitised during the experiments was immediately replaced by an unparasitised one. As a control, single females of *L. heterotoma* searched a similar patch containing only four unparasitised hosts, and any unparasitised hosts parasitised during the experiment were replaced by unparasitised ones.

Wasps stay longer on patches with four unparasitised hosts than on patches with four unparasitised hosts and 16 parasitised hosts. Van Lenteren (1991) argued that because there were no significant differences in *average* time interval between ovipositions in unparasitised hosts in the two treatments, the differences in patch residence times between the treatments can be attributed only to a

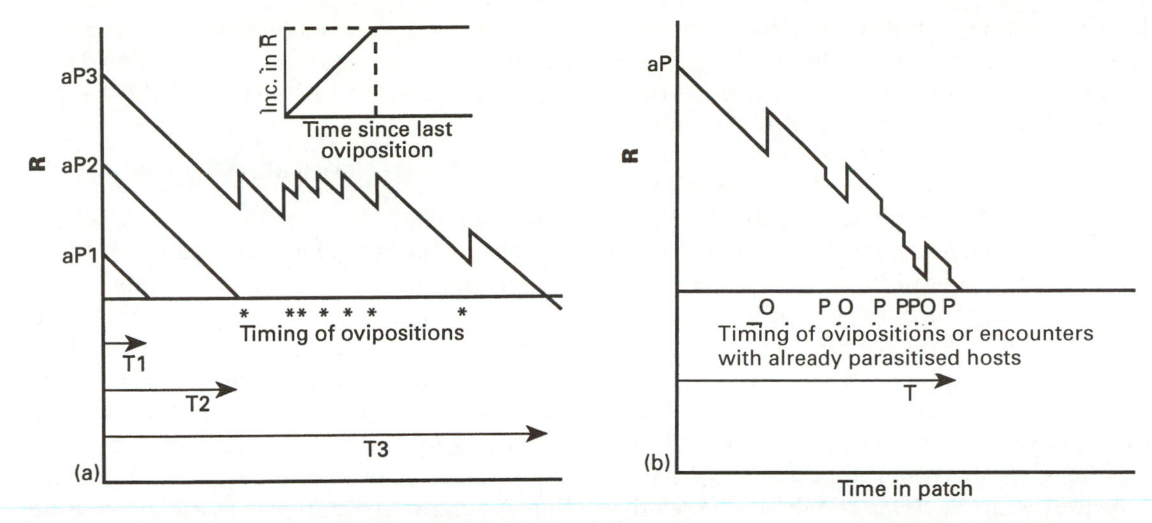

Figure 1.20 Patch time allocation in parasitoids: (a) Waage's (1979) causal model of patch residence time. R = responsiveness of the parasitoid to the patch edge (a function of the number of hosts in the patch (P1, P2, P3) and a constant a, the quantity of kairomone produced per host (asterisks denote ovipositions). An oviposition results in an increment of R (inset: the size of the increment depends linearly on the time elapsed since the previous oviposition, and the increment cannot exceed a maximum value); T1, T2 and T3 are the resulting patch residence times for three different cases; (b) Waage's (1979) model modified to incorporate a decremental effect of encounters with parasitised hosts. Symbols as in (a), except that O denotes an oviposition and P denotes an encounter with an already parasitised host. (a) reproduced by permission of Blackwell Scientific Publications Ltd.

decremental effect on patch residence time of encounters with parasitised hosts. In our opinion, however, this conclusion is not supported by van Lenteren's data.

First, consider whether it is at all valid to conclude from the observation that *average* time intervals between ovipositions did not differ between experiment and control, that there is no difference in the effect of ovipositions on patch residence time between the two treatments. This conclusion would be valid only if the parasitoid itself uses average intervals to assess patch profitability. As Haccou *et al.* (1991) have shown, the effect of an oviposition on the probability of a wasp leaving a patch depends on its timing; hence it is important *when* and *where* the longest intervals occur.

Despite a lack of statistical differences between the *average* values, important differences in interval times between ovipositions could occur between test and control treatments.

An alternative explanation for van Lenteren's (1991) results is that the differences in patch residence time are caused solely by the decrease in motivation over time that results from the extra time spent in rejecting parasitised hosts in the treatment with parasitised hosts. This time could otherwise be spent in ovipositing in unparasitised hosts. Rejection of a parasitised host takes between 2 and 6 seconds (Haccou *et al.* 1991), and with on average 33 rejections in the control treatment, this behaviour may account for an important part of patch residence time. If the decrease in the motivation to search (indicated by the sloping lines in Figures 1.20a and 1.20b) continues during the time spent in rejections, these small decrements may accumulate over time, causing the parasitoid to reach the threshold motivation rate for patch-leaving sooner than when no parasitised hosts are encountered. Intervals between encounters with unparasitised hosts would, on average, be slightly longer in experiments with parasitised hosts than in those without them, as indeed they were: 84 compared with 79 seconds. Although these differences are not significant, the time lost in rejection of parasitised hosts gradually accumulates, and so may be responsible for the ultimate differences in patch residence times.

Clearly, one cannot test the causal hypothesis simply by determining whether patch residence times and search times differ significantly between treatments. What is required is an analysis in which the relative weight of effects of the influencing factors and their timing are estimated from the data and tested statistically. For this reason, Haccou *et al.* (1991) adapted Cox's (1972) proportional hazards model to study the problem. They analysed a new set of experimental data using the model. No effect of encounters with parasitised hosts on the probability of patch-leaving was found. If such an effect exists at all, we expect it to be a small one. It might be detected in experiments in which there is a high proportion of encounters with parasitised hosts. Other attempts to determine whether rejections of parasitised hosts increase the probability of patch-leaving, all pre-dating Haccou *et al.* (1991), i.e. van Alphen and Galis (1983), van Alphen and Vet (1986), have, like van Lenteren's (1991) study, similar problems of interpretation. Hence the first evidence confirming the hypothesis first formulated by Waage (1979) comes from Hemerik *et al.* (1993). They used the proportional hazards model to analyse their experimental results and demonstrated in female *Leptopilina clavipes* that encounters with parasitised hosts decrease the tendency to search the patch.

Finally, the effect of encounters with parasitised hosts may depend on the previous experience of the parasitoid. It is thus possible that encounters with parasitised hosts could also *increase* the tendency to search on a patch, as is the case when they decide to superparasitise (van Alphen *et al.*, 1987).

1.13 PATCH DEFENCE BEHAVIOUR

Some parasitoids are known to defend patches of hosts against conspecific and heterospecific intruders. This behaviour serves both to prevent competing females from ovipositing in hosts not already attacked by the defending female (Waage, 1982) and to ensure that hosts already parasitised by the defending female are not superparasitised by competitors (van Alphen and Visser, 1990). Thus, defence of patches is an alternative competitive strategy to one of allowing conspecifics on the same patch and competing with them through superparasitism (subsection 1.8.4). Defence of patches is only advantageous under a limited set of conditions. The following factors favour defence of patches:

1. synchronous development of the hosts in the patch;
2. rapid development of the host to a stage which can no longer be attacked by the parasitoids, or rapid development of the parasitoid offspring to a stage at which they have a competitive advantage in cases of superparasitism;
3. short travel times between patches. When travel times are long, intruders are likely to be 'reluctant' to lose the contest for the patch. This would prolong fighting and increase the cost of defence;
4. species of parasitoid in which adult females have a very low probability of finding more than one host or host patch during adult life. This factor allows them to spend a long time guarding their brood, and it has undoubtedly played an important role in the evolution of brood-guarding (e.g. *Goniozus nephantidis*, Hardy and Blackburn, 1991). The females of brood-guarding parasitoids spend a

long period of post-reproductive life guarding one host or patch of hosts;
5. patches should be of a defensible size. Larger patches are harder to defend.

Patch defence was first described for scelionid egg parasitoids (Waage, 1982), which defend small and intermediate-sized host egg masses. However, it is also found in braconids (e.g. *Asobara citri*), ichneumonids (e.g. *Rhyssa persuasoria, Venturia canescens*), and bethylids (e.g. *Goniozus nephantidis*).

In some parasitoids, such as the abovementioned scelionid egg parasitoids, patch defence appears to be a fixed response to an intruder. However, in other species such as the braconid *Asobara citri*, patch defence and fighting behaviour decrease in frequency with increasing patch size and increasing numbers of intruders, and wasps may switch to competition through superparasitism.

Patch defence can have a pronounced effect on the distribution of adult parasitoids over host patches. It can lead to a regular distribution of parasitoids, and is thus one of the factors reducing aggregation.

Whether patch defence or competition by superparasitism is the better strategy depends on species-specific traits such as the encounter rate with hosts and the handling time. Thus, it is possible that one species attacking a certain host defends hosts or patches against intruders, while another parasitoid species attacking the same host does not. Patch defence is a poorly studied aspect of parasitoid biology and clearly deserves more attention from both pure and applied entomologists.

Patch defence should preferably be studied in multi-patch experiments. In single-patch experiments, one could easily underestimate the significance of defence behaviour. A classic example of this is the fighting and chasing which occurs when two females of *Venturia canescens* meet whilst searching the same patch. This aggressive behaviour is an important component of mutual interference in laboratory experiments with *V. canescens* (Hassell, 1978) (subsection 1.14.3). The function of the fighting and chasing is not easily understood from such experiments because the behaviour leads, on large patches, to a decrease in attack rate for both wasps but not to the permanent exclusion of the intruding wasp. However, field observations of *V. canescens* searching for *Anagasta* larvae feeding on fallen figs suggest that a fig containing a host larva can be successfully defended against intruding competitors; the latter move on to nearby figs following an aggressive encounter (Driessen *et al.*, 1994).

1.14 DISTRIBUTION OF PARASITOIDS OVER A HOST POPULATION

1.14.1 INTRODUCTION

The distribution of parasitoids over a spatially structured (i.e. heterogeneous) host population has attracted considerable attention from theoretical ecologists. Hassell and May (1973) and Murdoch and Oaten (1975), among others, have shown that this is one of the key features affecting stability of parasitoid-host population models (subsection 5.3.7). It is remarkable that the 'boom' in theoretical papers has not been paralleled by a similar surge in empirical studies on parasitoid behaviour.

1.14.2 AGGREGATION

The term **aggregation** is usually used to refer to the host-searching behaviour of parasitoids. Parasitoids may be more attracted to patches of high host density than to patches of low host density or they may show a stronger degree of arrestment in patches of high host density (subsection 1.5.3). Insect ecologists refer to an **aggregative response** of parasitoids and predators, because the aforementioned patch response behaviour leads to the concentration of parasitoids and predators in high density patches. Latterly, the term aggregation has also been applied to the concentration of parasitoids on patches of low host density or on certain patches irrespective of the number of hosts they contain; this can occur if parasitoids are attracted to some patches in response to stimuli that are either negatively correlated with or independent of host density. In studies of population dynamics the term aggregation has also been used in a statistical sense, in terms of both the variance in parasitoid distribution and the covariance between the distributions of host and parasitoid (Godfray and Pacala, 1992).

Aggregation of adult parasitoids is the result of two different processes:

1. differences among patches in the probability with which they are discovered by parasitoids. In a heterogeneous environment it is likely that not every patch has the same probability of being detected, even if all patches are otherwise similar. Patches may also differ in the probability of detection by parasitoids because of differences in host density or other aspects of quality of the patch;
2. the period of time that each parasitoid stays in a patch after discovering it. The number of parasitoids visiting a patch and the period of time they stay there determines the amount of 'search effort' devoted to a patch.

Aggregation can be measured in two main ways:

1. Individual parasitoids can be presented with several patches of different host density, as in studies of patch time allocation (Figure 1.19);
2. several parasitoids at one time can be presented with several patches of different host density (Figure 1.21).

When measuring aggregation using the second of these experimental designs, one should ideally monitor the behaviour of all parasitoids in all patches and record the time each parasitoid spends in each patch. In laboratory experiments with a modest number of host patches and with the insects continuously observed with video recording equipment, this is possible, but in field experiments such observations are very labour-intensive and often impossible to make. Published field studies on aggregation have therefore relied on periodic observations of the patches (e.g. Waage, 1983).

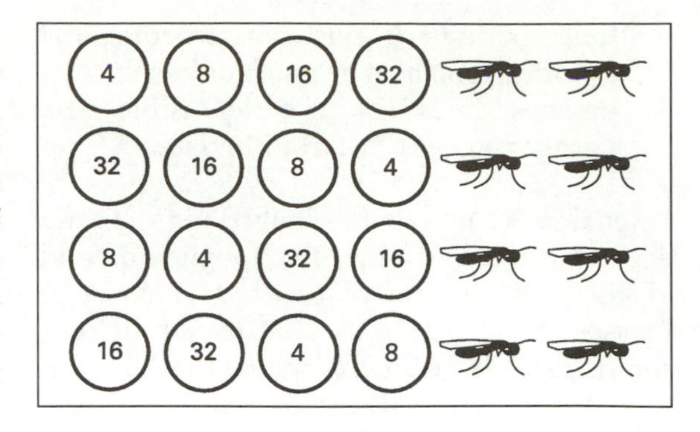

Figure 1.21 Aggregative responses of parasitoids and predators: Schematic representation of the design for an experiment which measures the aggregative response and which takes account of the effects of interactions between foragers.

One problem associated with studying aggregation in the field is deciding upon the spatial scale at which aggregation should be measured. Clumped distributions of hosts may occur at different levels of host distribution, and so may aggregation by parasitoids. It is often possible for practical purposes to define what a patch is. For example, when studying the distribution of parasitoids of the cassava mealybug within a cassava field, cassava plant tips infested with mealybugs are the most relevant foraging units, whereas if one wants to compare biological control between different fields, whole cassava fields can be considered as patches. Often, one can use the behaviour of a parasitoid to define patches. For example, the arrestment response of *Venturia canescens* to a contact kairomone produced by *Plodia* larvae defines the patch as an area containing the kairomone (Waage, 1978).

1.14.3 MUTUAL INTERFERENCE AND PSEUDO-INTERFERENCE

Before we describe these phenomena, we need to stress that mutual interference and pseudo-interference are concepts that can only be properly understood with reference to mathematical models, in particular those of searching efficiency (subsection 5.3.7). The reader is therefore recommended to consult the literature dealing with modelling, e.g. Hassell (1978).

The tendency for some parasitoids and predators to cease searching and to leave the immediate vicinity after an encounter with a conspecific (section 1.13) would account for the results of laboratory experiments designed to measure emigration rates in relation to parasitoid density. In these experiments, the proportion of female parasitoids leaving a single, fixed density host patch increased significantly with increasing numbers of parasitoids. It has also been observed that when females encounter either an already parasitised host or a parasitoid mark on the substratum, they move away from the area where the encounter occurred (subsection 1.5.4). Any of these various behavioural interactions are likely to cause the searching efficiency of a natural enemy in the single patch experiment to be reduced, a phenomenon known as **mutual interference** (Hassell and Varley, 1969; Visser and Driessen, 1991).

The study of mutual interference began with Hassell and Varley (1969) who noted an inverse relationship between parasitoid searching efficiency and the density of searching parasitoids (Figure 1.22):

$$\log_{10} a' = \log_{10} a - m \log_{10} P \qquad (1.6)$$

where P is the density of searching parasitoids; a' is the effective attack rate or area of discovery per generation, $a'P = \log_e$ (initial number of hosts/number of hosts surviving parasitism); a is the attack rate in the absence of interference (denoted by Q in Hassell and Varley, 1969). The gradient m is the measure of the extent of mutual interference.

Such a relationship is to be expected because, as parasitoid density increases, individual parasitoids will waste an increasing proportion of their searching time in encounters with other conspecifics. Rogers and Hassell (1974) and Beddington (1975) described this process mathematically. In their models of mutual interference (see their papers for details), they assume that following each encounter there is a period of time wasted during which searching does not occur.

The interactions studied by Hassell and Varley (1969) were of the type where different numbers of parasitoids were confined, for a fixed period of time, with a fixed number and *single patch* of hosts (Figure 1.23a). That is, interference was not considered in a multipatch context. The need to take account of host spatial heterogeneity can be explained with reference to experiments conducted by Hassell (1971a,b) using *Venturia canescens* (see also Hassell, 1978). Host larvae were confined in 15 small containers at densities ranging from 4 to 128 per container, and exposed to 1, 2, 4, 8, 16 or 32 parasitoids for 24 hours, at the end of which parasitism was scored. Hassell made continuous observations of the parasitoids, recording:

1. the total time spent per parasitoid on all containers;
2. the time spent on containers of different host density;
3. the frequency of encounters between parasitoids on a container leading either to a parasitoid departing or to an interruption of its probing behaviour without the insect departing.

The interference relationships revealed by the experiments are summarised in Figure 1.24a. Before discussing these relationships, it is

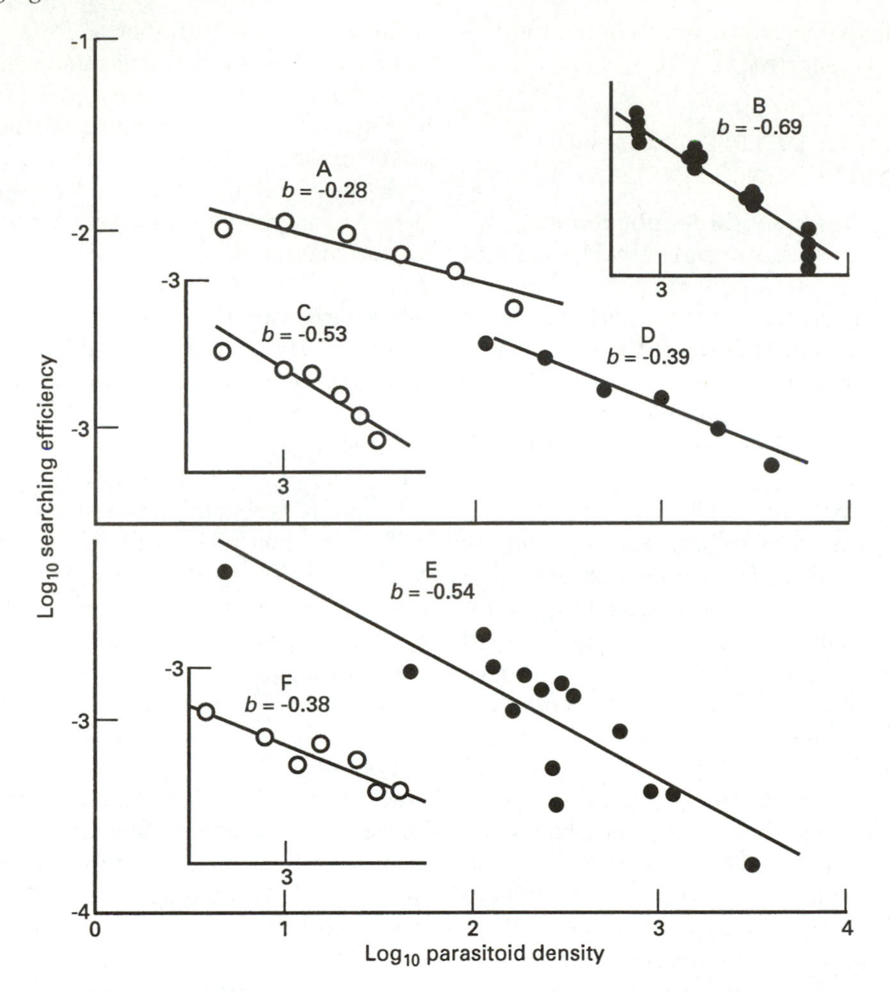

Figure 1.22 Mutual interference and pseudo-interference: relationship between searching efficiency (effective attack rate or area of discovery per generation, equation 1.6) and parasitoid density, for several parasitoid species (adapted from Varley *et al.*, 1973). Note that the response in each case is more likely to be curvilinear, tending to level off at very low predator densities when interference is negligible: A = *Dahlbominus fuscipennis*; B = *Leptopilina heterotoma*; C = *Chelonus texanus*; D = *Encarsia formosa*; E = *Venturia canescens*; F = *Itamoplex albitarsis*. Reproduced by permission of Blackwell Scientific Publications Ltd.

important to note that in obtaining lines A, B and C, Hassell (1978) used a model (see explanation, above, of parameter a' in equation 1.6) that treats all the patches as a single homogeneous area, that is, each of the three lines is based on the assumption of random search:

$$a' = \frac{1}{P_t T} \log_e \left[\frac{N_t}{N_t - N_a} \right] \qquad (1.7)$$

where a' is the *overall* i.e. average search rate, N_a is the total number of hosts parasitised in all patches, N_t is the total number of hosts in all patches, P_t is the number of parasitoids used in the experiment, and T is the duration of the experiment.

In obtaining line D, Hassell (1978) used a model that takes account of the difference in attack rates among patches and of the amount of time spent in the different patches. In other

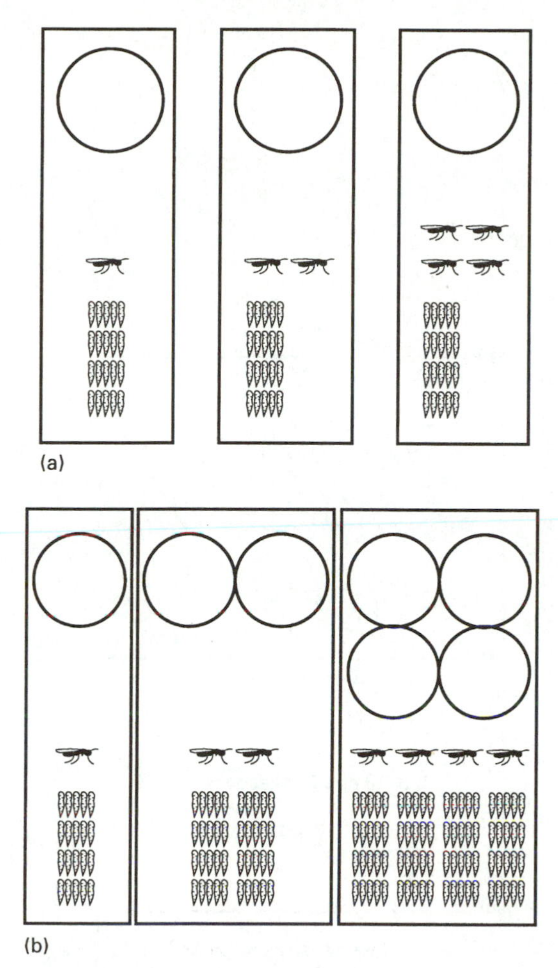

Figure 1.23 Mutual interference and pseudo-interference: Schematic representation of the design of two types of experiment for studying interference: (a) the design normally adopted for measuring interference, with one host patch (denoted by a circle); (b) the design used by Visser *et al.* (1990). Either a single parasitoid female searches a single unit patch containing 20 hosts, or two females search a double unit patch containing 40 hosts, or four females search a patch of four adjacent units of 80 hosts.

words, line D is based on the assumption of an aggregative response:

$$a' = \frac{1}{n}\sum_{i=1}^{n}\left[\frac{1}{P_t T_i}\log_e\left(\frac{N_i}{N_i - N_{ai}}\right)\right] \qquad (1.8)$$

where N_i, N_{ai}, and T_i are the number of hosts available, number of hosts parasitised, and the time spent respectively on the *i*th container, and a' is the overall search rate.

Line A in Figure 1.24a is the all-inclusive interference relationship obtained by counting the number of hosts parasitised and by assuming that the parasitoid searched for the duration of the experiment (i.e. 24 hours).

Line B is the relationship obtained if, instead of the apparent amount of time spent searching (24 hours), the real period of time spent searching is used. As Hassell found, the real searching time varied with parasitoid density (Figure 1.24b); each parasitoid generally spent a smaller proportion of the total time available on the host areas as parasitoid numbers increased. This is to be expected: at higher parasitioid densities the parasitoids will cease searching and disperse in response to contacts with parasitised hosts and parasitoid marks or in response to encounters with other parasitoids, long before the experimental period has elapsed. At lower parasitoid densities, a longer period of time will elapse before the parasitoids cease searching, because there are more hosts available per parasitoid and because parasitised hosts and parasitoid marks do not accumulate so fast.

Line C in Figure 1.24a is the relationship obtained if parasitoid–parasitoid encounters leading to interruption of probing for hosts are ignored (it was assumed that probing is resumed after a period of 1 minute). Line C therefore represents the relationship after all observed behavioural interference has been abstracted. Although parasitoids leaving host patches and parasitoids interrupting their probing accounted for much of the overall interference relationship, Line C represents a considerable amount of interference remaining. This poses the question: what mechanism is responsible for this residual interference? The answer to this question lies in the differential exploitation of patches of different host density (subsections 1.12.2 and 1.14.2).Free *et al.* (1977) argued, using deductive models, that marked parasitoid aggrega-

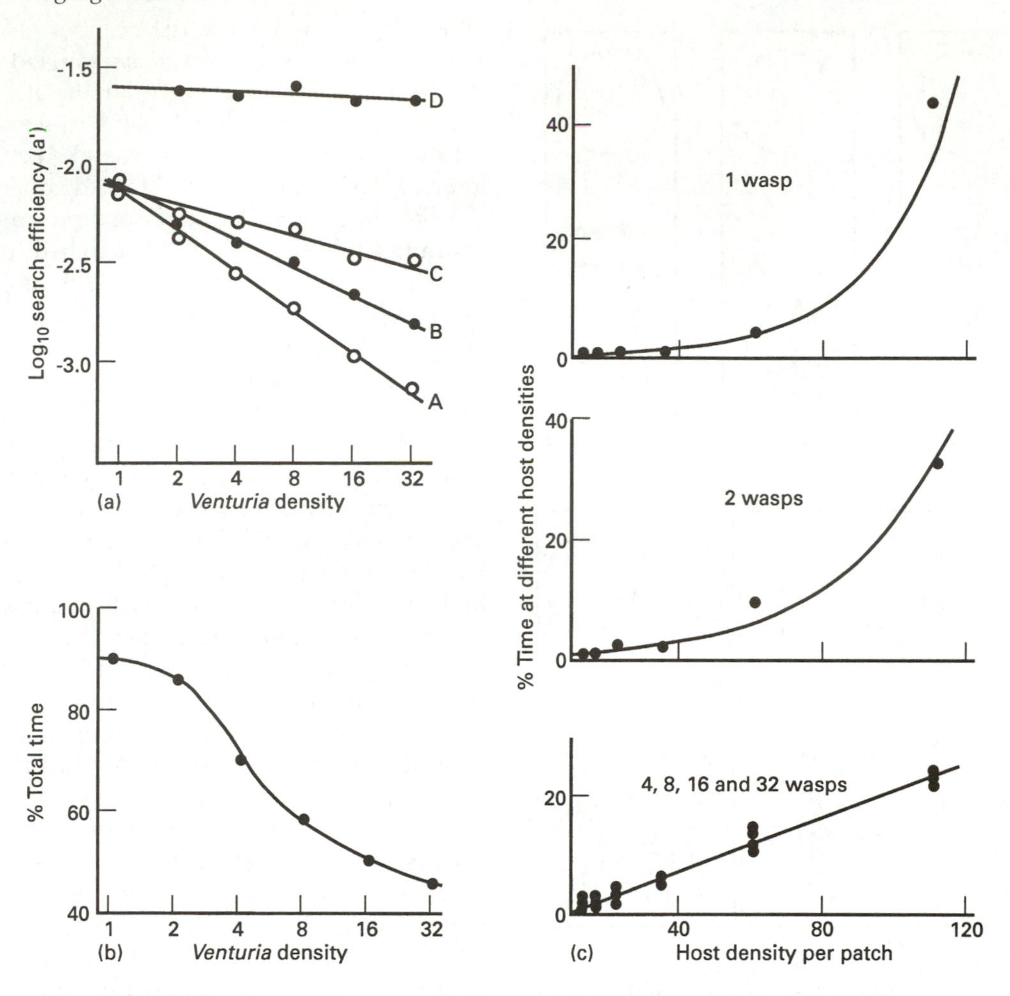

Figure 1.24 Mutual interference and pseudo-interference: (a) the relationships between searching efficiency and parasitoid density in *Venturia canescens*. See text for explanation; (b) the relationship between the proportion of time spent by individual wasps on host patches as a whole, and parasitoid density in *Venturia canescens*; (c) the relationship, for different parasitoid densities, between patch time allocation by individual *Venturia* and patch density. (Source: Hassell, 1978.) Reproduced by permission of Princeton University Press.

tion (e.g. resulting from a strong tendency of parasitoid individuals to spend longer periods of time in higher host density patches, and the consequent differential exploitation of patches) can lead to apparent interference, termed **pseudo-interference**, even if behavioural interference is lacking. As a consequence of parasitoids aggregating in high density regions (because these are initially the most profitable) a higher proportion of the hosts in the whole area (i.e. experimental cage) is parasitised than would be possible with random search. If parasitoids do not respond (i.e. by dispersal) rapidly to the declining profitability of the high host density (i.e. more heavily exploited patches), then overall searching efficiency will be lower at high parasitoid densities (line Y in Figure 1.25). Thus, pseudo-interference results from 'overaggregation' by the para-

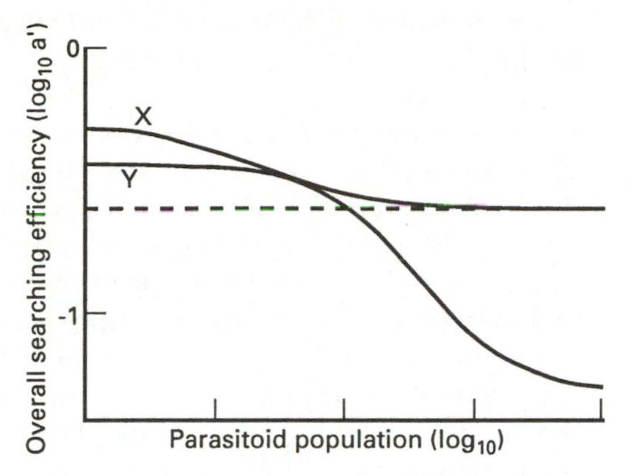

Figure 1.25 Mutual interference and pseudo-interference: theoretical relationship between overall searching efficiency and parasitoid density: Line X indicates a more profitable searching strategy (i.e. females maximise their oviposition rate) in which aggregation gives way to random search at high parasitoid densities; the parasitoids commence searching in the most profitable patches but then spread rapidly to other patches as exploitation proceeds until, eventually, parasitoids search randomly among all patches. Line Y indicates 'fixed aggregation' in which the distribution of parasitoids among patches is fixed; the parasitoids persist in searching the high density patches and pseudo-interference occurs through over-exploitation of those patches. The broken line indicates random search. (Source: Comins and Hassell, 1979.) Reproduced by permision of Blackwell Scientific Publications Ltd.

sitoids (Hassell, 1982a). In a population of optimally foraging parasitoids capable of responding rapidly to exploitation, overall searching efficiency would, at high parasitoid densities, be the same as for random search (line X in Figure 1.25).

Line D in Figure 1.24a differs from Line C in taking account of both the number of attacks and the actual searching time per host container. The fact that Line D is almost horizontal shows virtually all interference to have been been removed. The higher level of the line reflects how search rate is increased when changing from assumed random search (Lines A, B and C) to the observed aggregative search pattern shown in Figure 1.24c.

The fact that pseudo-interference is recorded in *V. canescens* by the above analysis, indicates that the parasitoids forage sub-optimally (Figure 1.25). Natural enemies are expected to forage optimally, on the assumption that all other things are equal. This is undoubtedly an unrealistic assumption. We might realistically expect the majority of parasitoids searching for patchily distributed prey to forage in a manner intermediate between the extremes of idealised optimal foraging and fixed aggregation (Comins and Hassell, 1979).

Traditionally, parasitoid attack rates (number of hosts parasitised per unit time) are used when considering interference relationships. However, if one is concerned with optimal behaviour, encounter rates should be considered. Visser *et al.* (1990) and van Dijken and van Alphen (1991) went further and calculated the mean number of realised offspring per female parasitoid per unit of patch time as a measure of individual efficiency. ESS models (subsection 1.2.2) developed by Visser *et al.* (1992a) predict that the presence of other females on a patch reduces this efficiency, even when the number of hosts per female is held constant (Figure 1.23b). This interference is not caused by behavioural encounters that decrease the encounter rate with hosts, but results from the parasitoids staying for longer on patches and superparasitising (subsection 1.8.4).

It needs to be stressed that when investigating mutual interference and pseudo-interference, the parasitoid densities used and the host spatial distribution pattern must reflect those found in the field. As pointed out by Free *et al.* (1977), few experimenters take account of this requirement.

Another factor to consider is the size of the experimental arena. Jones and Hassell (1988) found *per capita* searching efficiencies to be lower in field cages than in laboratory cages and interference to be more marked in the latter, the volume of which was relatively small. Jones and Hassell attributed the interference to an unnaturally high frequency of encounters between searching parasitoids (*Trybliographa rapae*). The much lower searching efficiency in the field cages was presumably due to the greater opportunities for parasitoids to spend time performing behaviour other than searching in close proximity to hosts.

1.15 MEASURING ENCOUNTER RATES

The encounter rate of individual parasitoids with hosts is an important parameter in many optimality models. Because not every encounter will be followed by oviposition, and because not every oviposition will be in an unparasitised host, encounter rate is not equal to the number of hosts parasitised per unit time.

Optimality models divide the time budget of a foraging animal into searching time, recognition time and handling time. Encounter rate is expressed and measured as the number of encounters per unit of searching time, thereby excluding recognition time and handling time. Because encounter rates are not always a linear function of host density, it is necessary to measure them at a range of host densities.

To measure encounter rates, observe a female parasitoid continuously during some time period and make a complete record of her behaviour. From this record, the number of encounters and the net period of time spent searching can be calculated. The encounter rate of a parasitoid searching a patch containing a number of hosts may not be constant over the foraging period for the following reasons:

1. parasitised hosts are encountered at a lower rate and the number of parasitised hosts increases during the observation period. A lowered encounter rate with parasitised hosts may occur either because hosts are paralysed by the wasp, and so move less (van Alphen and Galis, 1983) or because the hosts become aware of the presence of the parasitoid and alter their behaviour (e.g. aphids reacting to an alarm pheromone);
2. the search effort of the wasp decreases, either in response to contact with its own marker substance (subsection 1.5.3) or because its supply of ripe eggs dwindles. The same problem exists for prey choice by insect predators, for example aphids in a colony may alert one another to the presence of a syrphid larva by movements, so decreasing the probability of prey capture (Kidd, 1982).

One method of eliminating some of the causes of decreased encounter rate is to replace each parasitised host with an unparasitised one during the course of an experiment. This is not always possible, e.g. sessile hosts such as scale insects cannot easily be removed and replaced. Replacing parasitised hosts may also affect encounter rate: it may disturb the searching wasp and decrease encounter rate, or it may increase encounter rate when the parasitoid is of a species that reacts to host movements and the freshly introduced hosts move more than those already present. Finally, a parasitoid may learn, during the experiment, that the observer is introducing better quality hosts and simply walk towards the forceps or paint brush used to introduce the new host, as has often been observed with alysiine parasitoids.

Therefore, when measuring encounter rates, one should not replace parasitised hosts but instead keep the period of observation short, in order to avoid accumulation of parasitised hosts and marker substance.

Encounter rates can be used to calculate predicted rates of offspring deposited per unit time with a particular optimal foraging model.

Measuring encounter rates using a single-patch experimental design will overestimate the encounter rates that would be recorded in a multi-patch, i.e. natural environment, because the time spent in interpatch travel, i.e. transit time is not accounted for. Since it is often difficult or impossible to measure transit times, the simplest approach is to measure, over a fixed time period, the attack rate of a known number of parasitoids foraging in a spatially heterogeneous environment (Waage, 1979; Hassell, 1982a).

1.16 LIFE HISTORY CHARACTERISTICS AND SEARCHING BEHAVIOUR

1.16.1 INTRODUCTION

Insect parasitoids display an enormous diversity of life histories (Blackburn, 1991a). Some species have very short development times, can live only a few days as adults, and emerge with all their eggs ready to be laid, in contrast to other species which develop slowly, can live for several months as adults, and produce new eggs throughout adult life. Some species produce a large number of small eggs, whereas others produce a small number of large eggs. These different life-history traits are associated with differences in searching and host selection behaviour. When designing experiments on parasitoid behaviour, it is important to be aware that this is so.

1.16.2 EGG LIMITATION VERSUS TIME LIMITATION

Parasitoids can be divided into **pro-ovigenic** and **synovigenic** species (subsection 2.3.4). Pro-ovigenic parasitoids emerge with their full complement of mature eggs, whereas synovigenic parasitoids emerge with only part of their mature egg complement. Synovigenic parasitoids can themselves be divided into **anhydropic** species that carry few eggs at any one time, and **hydropic** species that carry many eggs (subsection 2.3.4 describes further differences between these two types). These different modes of reproduction, which undoubtedly represent ends of a continuum, can be understood as adaptations to differences in the spatial and temporal distribution patterns of hosts. Hosts in ephemeral resources (e.g. fly larvae in mushrooms) will be vulnerable to parasitism for a short period only, and so are likely to be attacked mainly by pro-ovigenic parasitoids. Hosts in longer-lived resources (e.g. diapausing pupae on the forest floor) will be vulnerable to parasitism for a longer period of time, and so are likely to be attacked mainly by synovigenic parasitoids.

Pro-ovigenic or hydropic species behave in most laboratory experiments in a time-limited manner, because even when large numbers of hosts are offered, these numbers do not exceed the number of mature eggs available to the parasitoid. On the other hand, synovigenic species behave in an egg-limited manner, often exhausting their daily egg supply in a few hours when the number of hosts offered to them exceeds the number of mature eggs in their ovaries. The question is whether such effects observed in the laboratory reflect the field situation or whether they are an artefact of unnaturally high host densities. To answer this question, some measure of oviposition rate under field conditions needs to be obtained and compared with the average rate of egg production in parasitoids.

The outcome of experiments on aspects of parasitoid biology as diverse as patch time allocation, functional responses, host selection, sex allocation, superparasitism or encounter rates, will all depend on whether the experimental conditions place the parasitoid under the constraint of time- or egg-limitation. Either experiments can be run under conditions representing both these constraints, or an experimental design can be chosen that is relevant to the particular question one is asking. For example, when asking about the performance of a parasitoid immediately following field release, present females in experiments with a superabundance of hosts so that they are egg-limited, but when asking about the performance of the parasitoid after the host population has been suppressed below a damage threshold (section 5.2), present females with low densities of hosts so that the parasitoids are time-limited. If one is asking evolutionary questions, it is advisable to choose a situation (e.g. range of host densities, host spatial distribution pattern) closest to what the wasps experience most often in nature.

1.16.3 THE COST OF REPRODUCTION

In many studies of time allocation, recognition time and handling time are taken to be the only time costs involved in oviposition. However, studies of the lifetime reproductive effort of organisms as diverse as rotifers, fruit-flies, water-fleas and birds have shown that reproductive effort has an effect on life expectancy, and that a trade-off exists between present and future reproduction (Stearns, 1992). There are good theoretical grounds for expecting such a relationship (Reznick, 1985; Bell and Kofoupanou, 1986). Recent work at Leiden has shown that the relationship exists in some insect parasitoids. Life expectancy in *Leptopilina heterotoma* decreases with the number of eggs laid (G. Driessen, unpublished). This means that rejection of a host not only saves handling time, but also prolongs the life of the parasitoid. The cost of reproduction therefore needs to be measured and included in the time costs of optimality models.

A simple estimate of the cost to life expectancy of ovipositions can be obtained from cohort fecundity-survival experiments in which females are provided with one of a range of host densities for their lifetimes (subsections 2.7.3 and 2.8.3; Bai and Smith, 1993). Bear in mind that in nature, females may have shorter life expectancies than indicated by laboratory experiments, because they spend a larger proportion of their time and energy in flight. Experiments with flight mills (section 6.5) indicate that in apple maggot flies (*Rhagoletis pomonella* (Tephritidae)) lifetime fecundity decreases with the time spent in active flight.

1.16.4 AGE-DEPENDENT FORAGING DECISIONS

Although parasitoids have a longer life expectancy when they lay fewer eggs, they do not live forever. The older they become, the less likely they are to survive to another day. Because of the diminishing probability of survival with increasing age, parasitoids should become less selective and accept more host types for oviposition (Iwasa *et al.*, 1984). This means that, all other things being equal, older wasps will superparasitise and accept less suitable hosts more readily than younger ones, a prediction that is supported empirically (Roitberg *et al.*, 1992, 1993). One can try to make use of this alteration in behaviour with age in experiments that require parasitoids to oviposit in non-preferred hosts, e.g. already parasitised individuals and unpreferred species.

1.17 FORAGING BEHAVIOUR AND TAXONOMY

Taxonomists describe species from preserved specimens and rely heavily on external morphological characters (Gauld and Bolton, 1988; Quicke, 1993). This is in most cases a satisfactory state of affairs, because differences in external morphology can often be found, even between closely related species. Sometimes, however, morphologically identical specimens can be collected from populations found in ecologically different situations, e.g. attacking a different host species, occurring on different host plants or in different geographical regions. The question then is whether these populations belong to one species or not – an important question, not only in deciding whether a parasitoid is a specific natural enemy of a target pest, but also because the scientific name of an organism is used in publications.

By comparing the host habitat-finding behaviour and host selection behaviour of different populations, one can establish whether important ecological differences exist between the two. Differences in host-habitat finding and/or host species selection can theoretically result in reproductive isolation between the two populations, which occupy different niches by virtue of the differences in their searching behaviour. When interpopulation differences in foraging behaviour are found, one should then determine whether cross-matings are possible. If such matings do not occur either in the laboratory or in the field, it is reasonable to conclude that the populations are 'good' biological species.

Vet *et al.* (1984) discovered *Asobara rufescens* by studying microhabitat location of wasps initially believed to be *A. tabida*. *A. rufescens* had until then gone unrecognised and been considered conspecific with *A. tabida*. Similarly, van Alphen (1980) discovered a new species of *Tetrastichus* which attacks the twelve-spotted asparagus beetle, *Crioceris duodecimpunctatum*, by showing that it rejected the eggs of *Crioceris asparagi*, the host of *Tetrastichus coeruleus*.

The classic example of the discovery, through a study of behaviour, of a 'cryptic' (sibling) species in predators, is the sand wasp *Ammophila adriaansei*. Marked differences in behaviour, including foraging behaviour, between individuals of this species and those of *A. campestris* were originally attributed to intraspecific variability (Tinbergen, 1974).

1.18 PARASITOID BEHAVIOUR AND COMMUNITY ECOLOGY

Like intraspecific competition (subsection 1.5.4), interspecific competition between adult parasitoids is in part mediated by infochemicals. One parasitoid species may avoid searching an area already visited by another species, in response to a chemical mark left on that area by the prior visitor (Price, 1970). A parasitoid species may search an area that has previously been visited by another species, but avoid oviposition in already parasitised hosts, in response to a mark left on or in the host by the first species to oviposit. Parasitoids may avoid a patch on which a competitor species is present, in response to an odour released by the competitor.

Interspecific competition between adult parasitoids may also take the form of physical 'fights' between adults. Under certain conditions, it can be advantageous for the first parasitoid to arrive in a patch to chase off later-arriving competitors.

The period of time spent in competitive interactions between adult parasitoids (be it in multiparasitism, the recognition and rejection of parasitised hosts or fighting, chasing and guarding) cannot be spent searching for hosts elsewhere in the habitat. Thus, interspecific competition between adult parasitoids will influence the distribution of eggs among hosts on the patch where the interaction occurs, and it will also affect the overall distribution of parasitoid eggs among the hosts in the habitat as a whole.

To appreciate how important competitive interspecific interactions are between adult parasitoids, behavioural studies, e.g. on host habitat location, host selection and competitive interactions between adults, as we have described them in the preceding sections, are necessary. Work at Leiden with *Drosophila* parasitoids has revealed that there is much more niche overlap between competing species than is suggested by parasitoid emergence data obtained from field-collected samples. The latter provide data on the **realised niche** while behavioural studies provide data on the **fundamental niche** of a species.

1.19 CONCLUDING REMARKS

The past two decades have seen rapid advances in the study of insect natural enemies, particularly as regards parasitoid behaviour. These developments have mainly been in response to advances in ecological theory. The availability of a whole suite of models on parasitoid-host population dynamics that incorporate important behavioural characteristics of parasitoids, combined with the rapid progress in behavioural ecology, has led to the formulation of more precise and more quantitative hypotheses. In addition, the tools for analysis of complex time-series of behaviour have become available, allowing us to address problems which previously could not be analysed properly. Behavioural studies of parasitoids have been conducted along two main lines: the functional analysis of behaviour, which has been guided by largely model-based theories on the evolution of animal behaviour, and the causal analysis of behaviour, which has been guided much less by models.

Developments in the causal and the functional analyses of behaviour have, for a large part, been independent. As we hope to have indicated in this chapter, research can benefit from an increased integration of the study of mechanisms and the study of function of behaviour.

Since biological control is part of applied ecology, the only way in which it can progress scientifically is by making full use of the potential of ecological theory. Behavioural and population models can point the way to selection criteria for natural enemies, such as knowing how a behavioural trait affects parasitoid-host population dynamics and how it affects the fitness of individual parasitoids.

There is a two-way interaction between theory and empirical research: behavioural and population models can guide us in the design of experiments, while the results of experiments stimulate new theory. Theories of population dynamics and behavioural ecology are, more often than not, concerned with the behaviour of parasitoids and predators in a spatially heterogeneous environment, with patchily distributed hosts or prey. For reasons of convenience, behavioural studies on natural enemies have often been conducted in single-patch environments. Empirical results from experimental studies on the behaviour of natural enemies in multi-patch environments, preferably natural environments, can provide the information needed both to test current theories and to develop new ones.

1.20 ACKNOWLEDGEMENTS

All students in the 'Ecology of Parasitoids and Predators' research group at the University of Leiden are acknowledged for the many stimulating discussions concerning experimental design and interpretation of results. J.J.M. van Alphen is grateful to Marianne van Dijken, Ian Hardy, Janine Pijls and Jan Sevenster for their comments on earlier drafts of the manuscript. Neil Kidd and Tom Sherratt are also kindly acknowledged for their comments.

THE LIFE CYCLE 2

M.A. Jervis and M.J.W. Copland

2.1 INTRODUCTION

This chapter is concerned with approaches and techniques used in studying those aspects of parasitoid and predator life cycles that have an important bearing on the major topics covered by other chapters in this book. To illustrate what we mean, consider the female reproductive system of parasitoids which we discuss in some detail (section 2.3). As pointed out by Donaldson and Walter (1988), at least some knowledge of its function, in particular of ovarian dynamics, is crucial to a proper understanding of foraging behaviour in parasitoids. The physiological status of the ovaries may determine: (a) the duration of any pre-oviposition period following eclosion; (b) the rate of oviposition, (c) the frequency and duration of non-ovipositional activities, e.g. host-feeding; and (d) the insect's response to external stimuli, e.g. odours, hosts (Collins and Dixon, 1986) (subsection 1.5.1). Note that egg load (defined in subsection 1.2.2) is now being incorporated into foraging models, as it is becoming increasingly clear that certain foraging decisions depend importantly upon the insect's reproductive state (Jervis and Kidd, 1986; Mangel, 1989a; Chan and Godfray, 1993). It also follows from the above that a female parasitoid's searching efficiency depends upon the functioning of its reproductive system (subsection 5.3.7), and this may in turn influence parasitoid and host population processes.

One particular aspect of natural enemy life cycles that is not addressed in this chapter is dormancy; this is discussed in depth by Tauber *et al.* (1986).

2.2 ANATOMICAL STUDIES ON NATURAL ENEMIES

2.2.1 INTRODUCTION

A general introduction to insect structure and function can be found in most standard entomological texts, e.g. Wigglesworth (1972); Chapman (1982); Richards and Davies (1977); Commonwealth Scientific and Industrial Research Organisation (1991). Individual topics are covered in texts such as Snodgrass (1935) on morphology; Davey (1965); Engelmann (1970); Kerkut and Gilbert (1985) on insect reproduction. There are also texts such as Hodek (1973); Canard *et al.* (1984); Gauld and Bolton (1988), that deal with some aspects of the morphology of particular taxonomic groups. This section is concerned with methods used for investigating the internal anatomy of predators and parasitoids, the emphasis being placed on the female reproductive system.

2.2.2 TECHNIQUES

Dissection

Many insect natural enemies, particularly parasitoids, are so small that routine investigations of their internal anatomy might at first sight seem impossible to undertake. One approach to anatomical investigation is to fix and embed insects in wax or resins and then to cut, using a microtome, serial sections of the body. This method is, however, technically difficult and there usually arise problems such as distortion (e.g. due to hardness of the cuticle), inadequate fixation and the difficulty of reconstructing sections into a three-dimensional model. A far easier approach is to dissect the insect.

In order to carry out dissection, the following equipment will be required: a binocular microscope with incident lighting (preferably fibre optics, see below), ordinary or cavity microscope slides, insect saline (7.5 g NaCl/litre) and some fine pins. The latter are best mounted, with sealing wax, either in glass tubes (4 mm diameter and approx. 50 mm long) or matchsticks.

For parasitoid Hymenoptera (Trichogrammatidae, Mymaridae and others to 25 mm long), place one droplet of insect saline on to a microscope slide and place the insect in the droplet. Use insects that have been recently killed with ether, carbon dioxide, some other suitable killing agent or by freezing. Individuals that have been dead for more than an hour at room temperature, and ones that have been preserved in alcohol are very difficult to dissect. Storing insects in a deep-freeze is highly recommended. When dissecting, ensure that the insect's body is dorsal side up, feet down. With one pin, restrain the insect from floating or otherwise moving in the saline by piercing its thorax, or hold the pin across the petiole. With the second pin, make small lateral incisions in the distal part of the gaster, preferably where there is an intersegmental membrane. Place the point of the second pin firmly upon the tip of the insect's gaster and pull the latter gently away from the remainder of the gaster. The abdominal wall should part in the region of the incisions and the abdominal contents spill out into the saline droplet. With a little practice, this technique will permit examination of the mid and hind gut, and of the entire reproductive system. By carefully noting the positions of all the various organs during dissection, it should be possible to reconstruct the spatial arrangement of the various organs and associated structures (Figure 2.1). More difficult manipulation may be required in the case of Hymenoptera with long ovipositors that are housed within the body as a spiral (e.g. Eurytomidae) or extended forward in a 'horn' above the thorax (e.g. *Inostemma* species (Platygasteridae)).

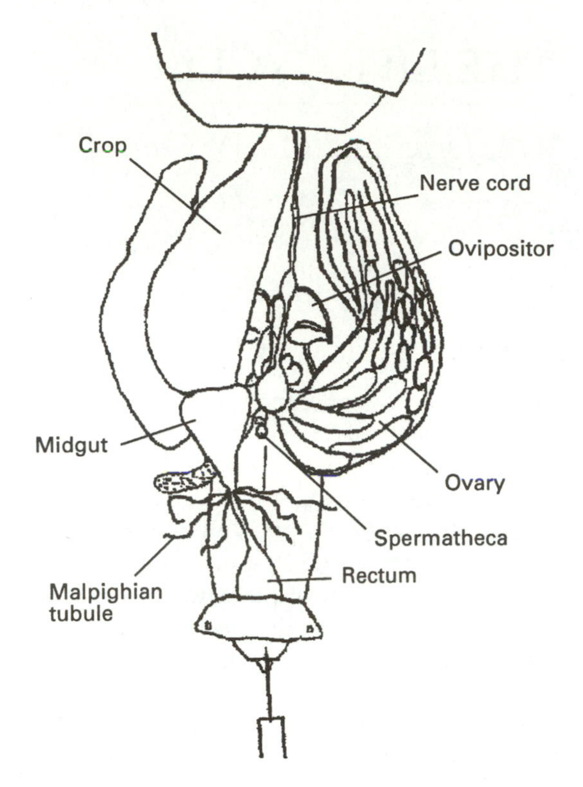

Figure 2.1 Dissection of gaster of female *Nasonia vitripennis* (Pteromalidae). The point of a micropin is used to pull away the tip of the gaster and reveal the internal organs.

Two points need to be borne in mind when using the above-mentioned technique. First, the insect must be kept covered in saline solution at all times. If it dries out, it cannot be satisfactorily reconstituted. Second, if water rather than saline is used, structures may expand and become seriously distorted. Unless a fibre optics system is being used, avoid using a below-stage light source (useful for assisting the examination of some structures) for periods longer than a few minutes only, as the specimen will dry out very quickly.

The above technique can be used for small predators and small dipteran parasitoids, but with large insects such as carabid beetles and hover-flies a small water filled wax-bottomed dish should be used instead of a microscope slide and saline droplet. Gilbert (1986) describes a technique for dissecting adult hover-flies (Syrphidae) (Figure 2.2a) that can also be applied to dipteran parasitoids and predatory beetles and bugs. The insect is placed on its back (on a slide or wax-bottomed dish, dry or under saline) and is secured with an entomological pin inserted through the thorax. Using a second entomological pin, a small tear is made in the intersegmental membrane at the junction of the thorax and abdomen. The end of one arm of a fine forceps is then inserted into this hole and the forceps are then used to grip the first abdominal sternite. Then, using a micropin (preferably one having a point that has been slightly bent), make lateral incisions in the abdomen, following the line of the pleura to the terminalia. Finally, peel back the abdominal sternites to reveal the internal organs (Figure 2.2b,c). The crop (very large in hover-flies) can be entirely removed using forceps, and its contents, e.g. pollen, examined and analysed. The reproductive system can be examined *in situ*, but to perform counts of ovarioles and mature eggs, the ovaries will need to be removed and the ovarioles teased apart. Carabid beetles are dissected in a similar fashion, except that the insect is placed on its front. Figure 2.3 shows the gut of a typical carabid beetle.

It is very difficult to interpret the structure of an insect's reproductive system or other organs if the organ has been fixed and preserved. If a permanent record of a dissection is needed, the insect's organs are best photographed or drawn as soon as possible. Semi-permanent mounts can be made with water soluble mountants such as polyvinyl pyrrolidone (Burstone, 1957) or glycerol, but anatomical features are best observed in freshly dissected insects. Anatomical features

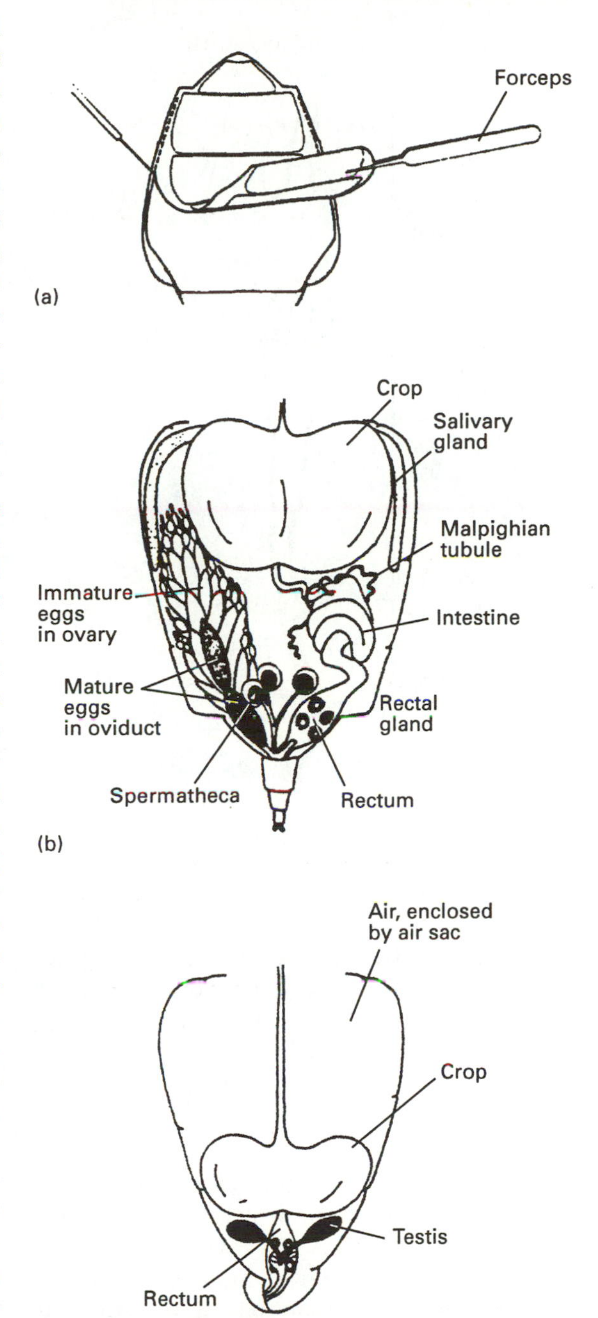

Figure 2.2 Dissections of hover-fly abdomen: (a) dissection procedure; (b) internal anatomy of female, (c) internal anatomy of male. (Source: Gilbert, 1986.) Reproduced by permission of Cambridge University Press

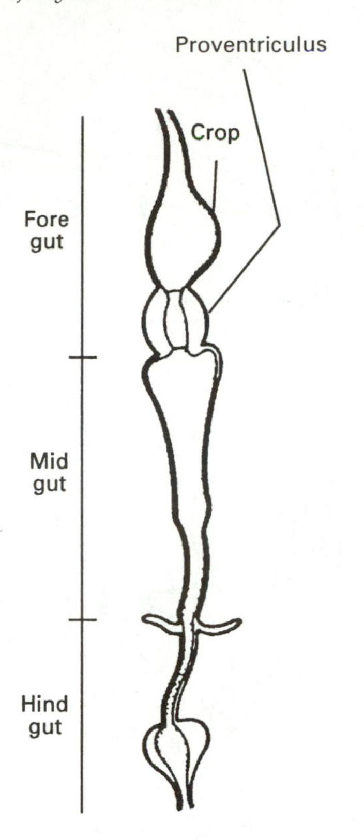

Figure 2.3 Gut of a typical carabid beetle. (Source: Forsythe, 1987.) Reproduced by permission of The Richmond Publishing Co. Ltd.

are enhanced by the use of specialist optics such as phase contrast, interference and dark ground illumination, with a transmission compound microscope.

Microscopy

There is a limit to the information that can be obtained from dissection. Histological and histochemical techniques will reveal the location of lipids, carbohydrates, nucleic acids and many more specific materials in, for example, the reproductive organs. Such techniques have been crucial to our understanding of oögenesis in parasitoids (King and Richards, 1969; King *et al.*, 1969b, 1971; Davies *et al.*, 1986). Combined with electron microscopy, they can reveal the detailed structure of secretory tissues, and can demonstrate the effects of diet and temperature on structures such as mitochondria and cell membranes. Davies (1974), for example, showed how in *Nasonia vitripennis* the ultrastructure of flight muscle changes with the age of the adult insect and with variations in adult diet.

2.2.3 OVIPOSITOR AND MALE GENITALIA

The ovipositor of female parasitoids may need to be examined in detail in order to understand the mechanics of oviposition, while the secondary genitalia of male dragonflies may need to be examined in order to study sperm competition (subsection 3.9.4). Light microscopy is usually employed to study these structures. For examination of whole mounts, they can be cleared and stained following standard protocols, whereas for examination of sections, e.g. of ovipositors, embedding, sectioning and staining needs to be carried out; standard protocols (embedding in Spurr's medium and staining with toluidine blue) were followed by Austin (1983) and Quicke *et al.* (1992). Greater detail of external morphology can be seen using scanning electron microscopy (SEM) (King and Fordy, 1970; Quicke *et al.*, 1992; Jervis, 1992). Specimens of small Hymenoptera and of Diptera are best prepared for SEM by critical-point drying them (Postek *et al.*, 1980), whereas specimens of larger and more hard bodied insects require only air drying.

Snodgrass (1935) described the basic structure of both male and female insect genitalia, while Scudder (1971) interpreted the structure of the ovipositor in Hymenoptera. Structural differences in the ovipositor of several families of Chalcidoidea were reviewed by Domenichini (1953). Studies by Copland and King (1971, 1972a,b,c,d), Copland *et al.* (1973) and Copland (1976) of cuticular structures and associated musculature have provided valuable insights into the functional significance of these differences. The mechanism, inferred from anatomical studies, for movement of eggs along the ovipositor of some Hymenoptera 'Parasitica' is described in detail by Austin and Browning (1981) and Austin (1983). Much less is known about the structure of the ovipositor and the mechanism of egg movement/release in dipteran parasitoids, e.g. Tachinidae, Conopidae, Phoridae, Pipunculidae.

SEM has been used on several parasitoid species to reveal the presence and diversity of sensilla on the female ovipositor (King and Fordy, 1970; Gutierrez, 1970; Weseloh, 1972; Hawke *et al.*, 1973; Greany *et al.*, 1977; van Veen, 1981; Jervis, 1992). The function (i.e. mechanoreception, chemoreception) of the sensilla can be provisionally inferred from their morphology but corroboration needs to be obtained both from a detailed examination, using transmission electron microscopy, and from a study of female oviposition behaviour. The role of ovipositor sensilla in host acceptance by parasitoids (section 1.5.5) has long been appreciated, but considerably more work needs to be done to establish their function.

Except for the studies by Domenichini (1953) and Sanger and King (1971), there has been little work on the functional morphology of male genitalia in either dipteran or hymenopteran parasitoids. The structure and function of the secondary genitalia of male dragonflies have been discussed by several authors, including Waage (1979, 1984).

2.3 FEMALE REPRODUCTIVE SYSTEM

2.3.1 OVARIES

The reproductive organs of hymenopteran (Figures 2.1, 2.4, 2.5) and dipteran parasitoids comprise a pair of ovaries which themselves comprise several ovarioles in which the eggs (oöcytes) develop. In hymenopteran (King and Richards, 1969) and dipteran parasitoids (Coe, 1966) the ovarioles are of the **polytrophic** type. Within each follicle, nurse cells or trophocyte cells (15 or more in hymenopteran parasitoids) surround the developing oöcyte, providing it with nutrients (Figure 2.6a). The oöcyte becomes increasingly prominent as it passes down the ovariole. Each oöcyte and its associated trophocyte cells originate from a single cell. It appears that in order to develop eggs as rapidly as possible, the protein production machinery of all the trophocyte cells passes materials into the oöcyte. The follicle cells, which may also pass materials from the haemolymph, secrete the egg shell, i.e. chorion. As the oöcyte matures, the trophocyte cells break down, leaving a small pore (the micropyle) in the chorion through which the sperm enters to penetrate the egg membrane and effect fertilisation.

To examine the ovarioles of a dissected insect, remove the ovaries (their attachment to the abdominal wall may need cutting), place them on a microscope slide in a drop of insect saline, and tease the ovarioles apart with micropins. Then gently place a cover-slip over the ovaries. The number of ovarioles can then be counted and their contents viewed.

In both hymenopteran and dipteran parasitoids, the number of ovarioles per ovary varies both interspecifically (Flanders, 1950; Price, 1975; Dowell, 1978; Jervis and Kidd, 1986) and intraspecifically (van Vianen and van Lenteren, 1986). Many chalcidoid wasps have an average of three ovarioles per ovary (*Encarsia formosa* has an average of eight to ten, depending on the population studied), whereas in ichneumonoid wasps the range of interspecific variation is much wider (Iwata, 1959, 1960, 1962; Cole, 1967. In some species of Ichneumonidae ovariole number alters according to whether the females are of the first or second field generation, female body size being taken into account, i.e. there is a seasonal dimorphism (Cole, 1967).

Predator ovaries (Figure 2.7) fall into several categories. Many, such as chrysopid lacewings

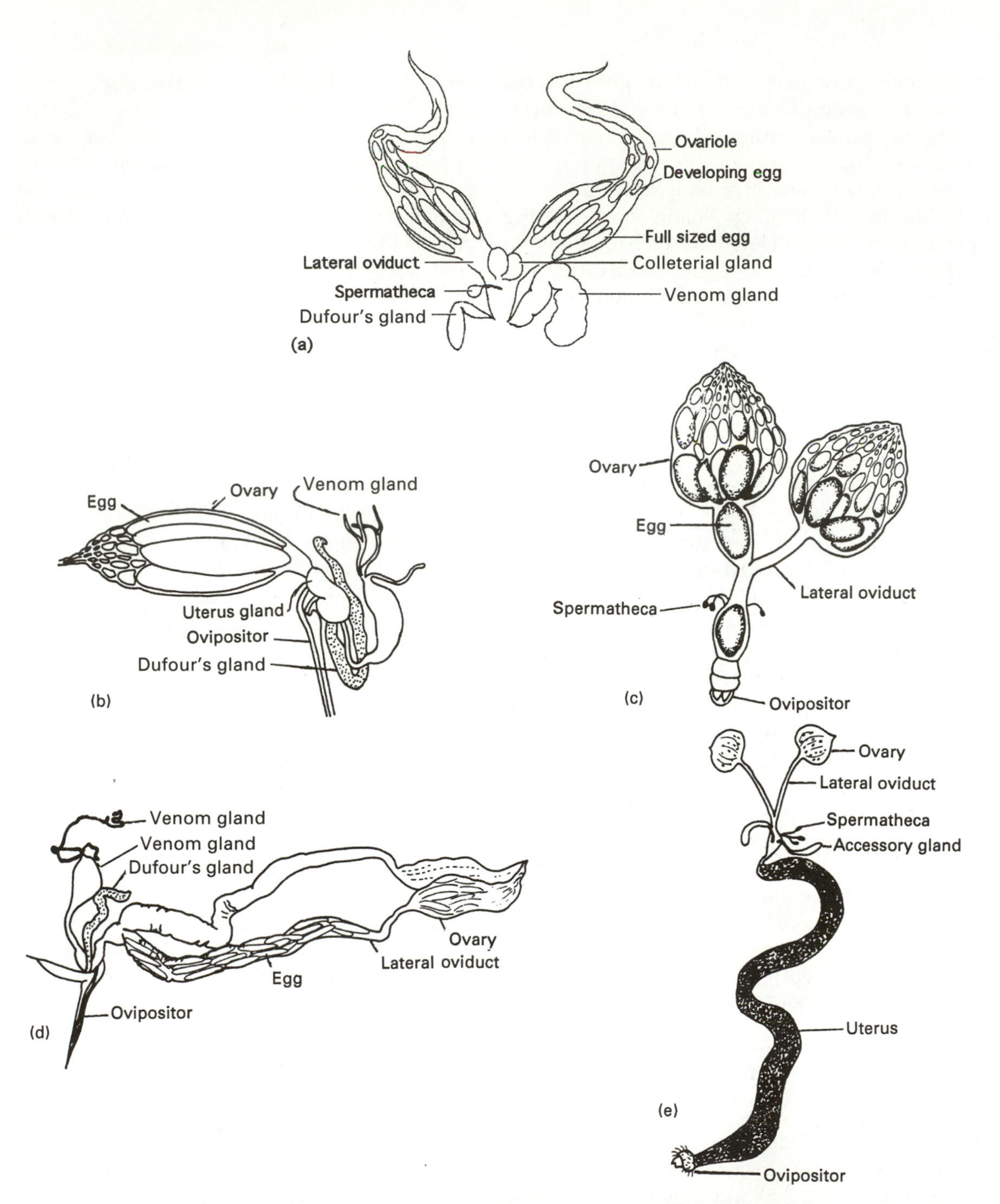

Figure 2.4 Reproductive systems of some hymenopteran and dipteran parasitoids: (a) gravid female *Coccophagus atratus* (Aphelinidae) (synovigenic-anhydropic) 24 h after emergence; (Source: Donaldson and Walter, 1988.) (b) *Trachysphyrus albatorius* (Ichneumonidae) (synovigenic-anhydropic); (Source: Pampel, 1914, in Price, 1975.) (c) *Hyperecteina cinerea* (Tachinidae); (Source: Clausen *et al.*, 1927, in Price, 1975.) (d) *Enicospilus americanus* (Ichneumonidae) (synovigenic-hydropic); (Source: Price, 1975.) (e) *Leschenaultia exul* (Tachinidae). (Source: Bess, 1936 in Price, 1975.) Note: acid gland = poison gland = venom gland. (a) Reproduced with permission from Blackwell Scientific Publications Ltd; (b), (c), (d) and (e) with permission from Plenum Publishing Corporation.

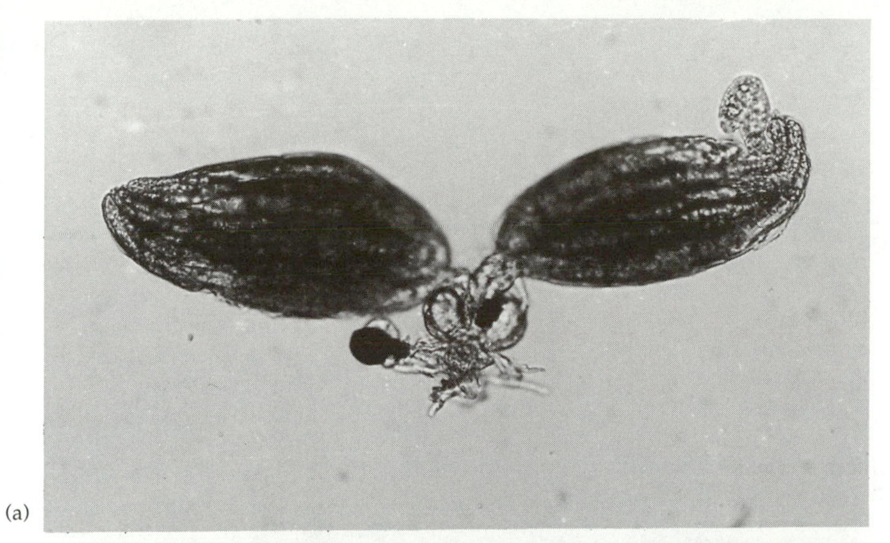
(a)

(b)

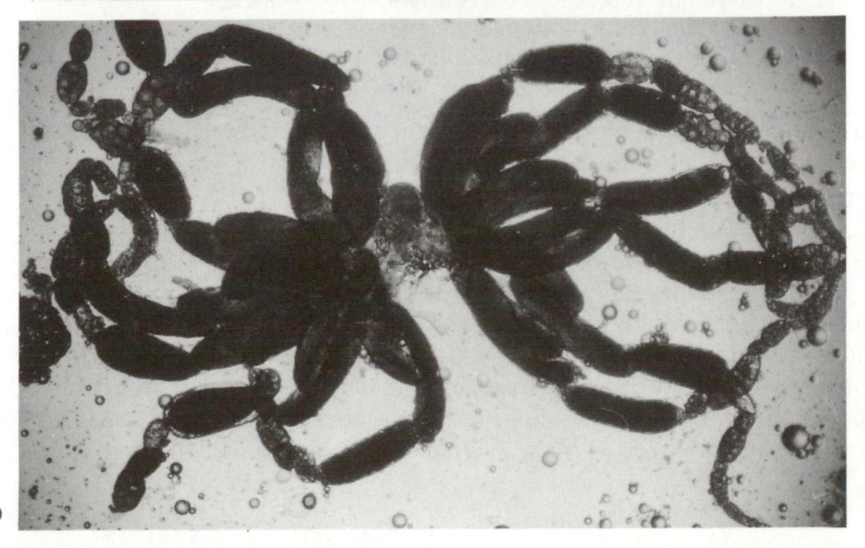
(c)

Figure 2.5
Reproductive systems of some hymenopteran parasitoids:
(a) *Gonatocerus* sp. (Mymaridae) (pro-ovigenic);
(b) *Cotesia* sp (Braconidae) (synovigenic-hydropic);
(c) unidentified Eulophidae (synovigenic-anhydropic).

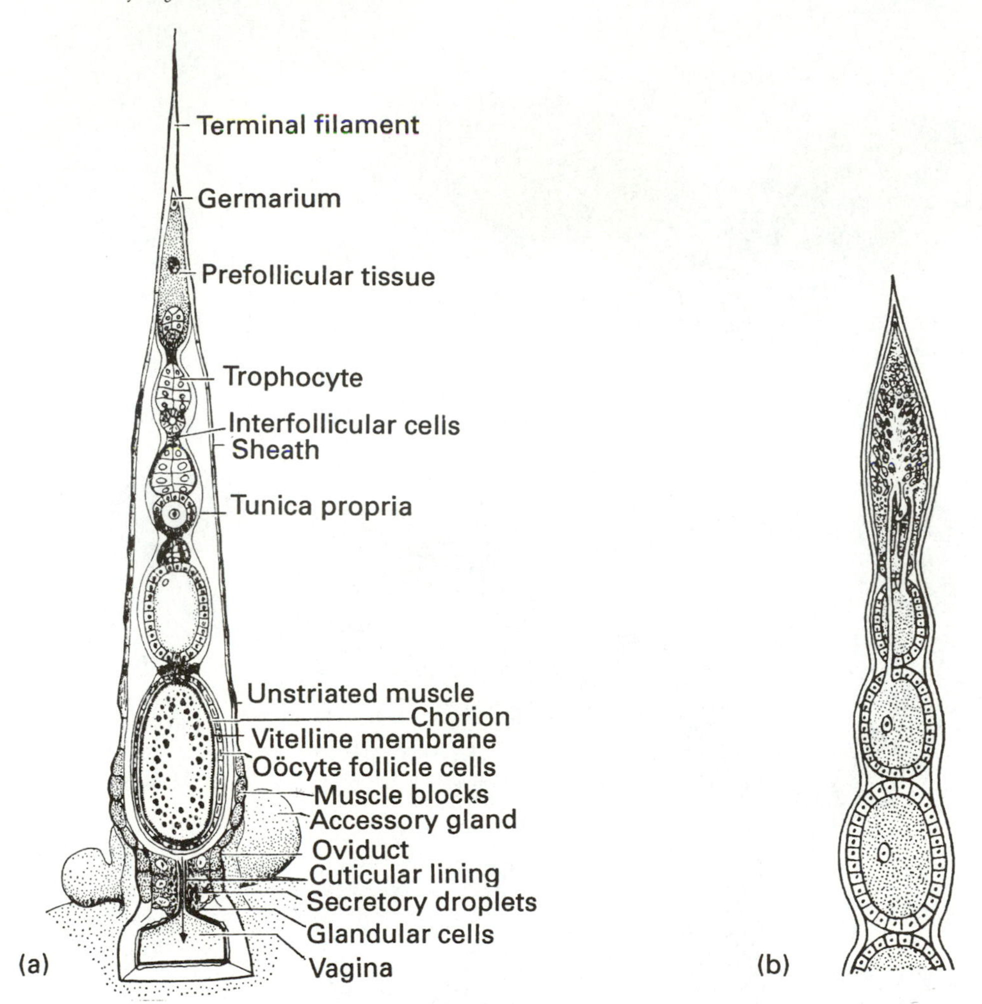

Figure 2.6 Ovariole structure: (a) polytrophic type, in *Nasonia vitripennis* (Pteromalidae) (Source: King and Ratcliffe, 1969.); (b) telotrophic type. (Source: de Wilde and de Loof, 1973.) (a) Reproduced with permission from the Zoological Society of London; (b) with permission from Academic Press.

and carabid and gyrinid beetles, have polytrophic ovarioles, but coccinellid beetles and predatory heteropteran bugs have **telotrophic** ovarioles (Figure 2.6b). In the latter, the trophocyte cells, instead of accompanying the oocyte as the it moves down the ovariole, remain in the swollen distal end of the ovariole and remain attached to the egg by a lengthening cytoplasmic strand that conveys the nutrients. Telotrophic ovarioles are therefore short, but they are often numerous. In predatory coccinellids, as in parasitoids, there is both intra- and interspecific variability in ovariole number (Iperti, 1966; Stewart *et al.*, 1991).

A measure of female reproductive potential can be obtained by counting the numbers of eggs within the ovaries. It is a fairly simple procedure to count mature eggs in species that possess enlarged lateral oviducts in which the eggs accumulate (subsection 2.3.2), but care is needed in the case of species that store (albeit for a brief period) some or all of their eggs

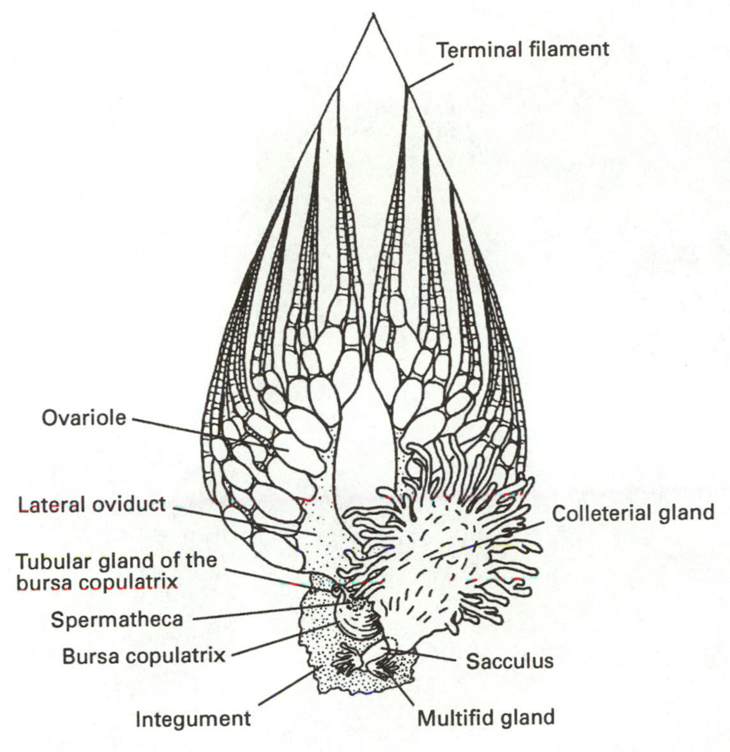

Figure 2.7 Schematic representation of reproductive system in female *Chrysopa septempuncata* (Neuroptera), dorsal aspect. (Source: Principi, 1949.) Reproduced with permission from W. Junk Publishers.

within the basal part of the ovariole. With practice, it is possible to recognise mature eggs by their slightly opaque appearance resulting from the presence of yolk within (i.e. in anhydropic species, subsection 2.3.4). Otherwise, a stain such as acetocarmine can be used. The stain is taken up by the immature oöcytes, because they lack a chorion; only the follicle surrounding the mature oöcyte stains (the follicle is eventually lost prior to the mature egg entering the oviduct).

A better measure of reproductive potential might be obtained by counting the number of mature eggs, the number of developing oöcytes and the number of egg primordia in the **germarium** (the germarium is the distal part of the ovariole from which the oocytes arise). Egg primordia can be counted in *Venturia canescens* (Ichneumonidae) (I. Harvey and J. Harvey, personal communication).

2.3.2 OVIDUCTS

The ovarioles empty into the lateral oviducts (Figures 2.4, 2.5, 2.8). In most Hymenoptera, each lateral oviduct includes an obvious glandular region – the calyx (Figure 2.8) – which secretes materials on to the egg as it is laid (Rotheram, 1973a,b). In some Braconidae and Ichneumonidae the calyx is the source of polydnaviruses (baculoviruses of the family Polydnaviridae) (Stolz and Vinson, 1979; Stolz, 1981; Cook and Stolz, 1983; Stoltz *et al.*, 1984; Strand *et al.*, 1988; Fleming, 1992). The latter, which replicate in the cells of the calyx, play a role in preventing encapsulation of the

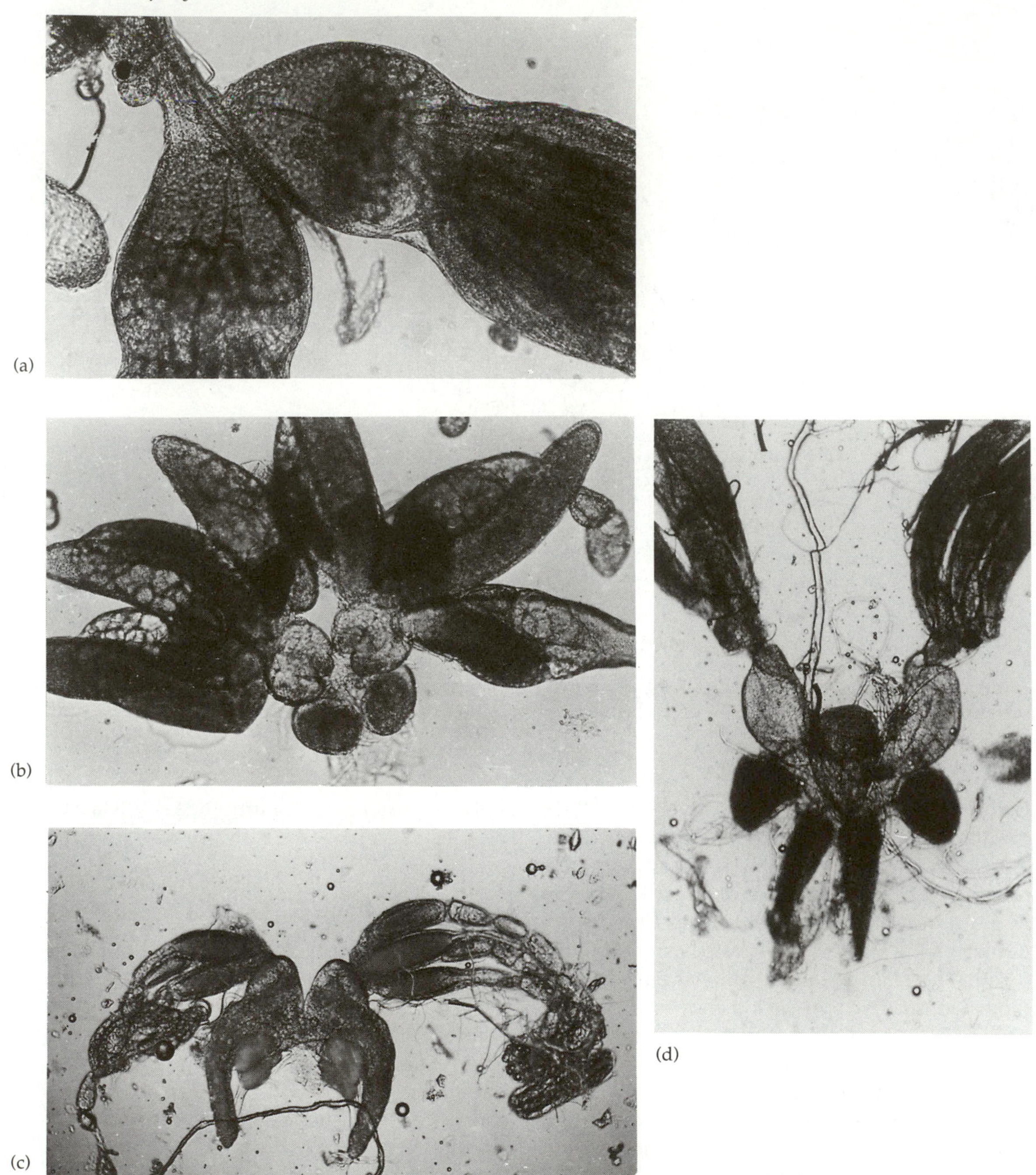

Figure 2.8 Calyx region of lateral oviduct in: (a) *Cotesia* sp. (Braconidae); (b) *Aprostocetus* sp. (Eulophidae) (also showing one pair colleterial glands); (c) *Torymus* sp. (Torymidae) (also showing two pairs of colleterial glands); (d) *Macroneura vesicularis* (Eupelmidae) (showing calyx lobes, i.e. the very long structures, and two pairs of colleterial glands).

parasitoid egg (subsection 2.10.2) and in modifying the host's growth, development, morphology and behaviour (Vinson and Barras, 1970; Vinson, 1972; Cloutier and Mackauer, 1980; Vinson and Iwantsch, 1980a; Stolz, 1986; Strand *et al.*, 1988). Cheillah and Jones (1990) raised an antibody against the extracted polydnaviral proteins of *Chelonus* sp. and then used it to reveal the location of polydnaviral proteins in the wasp's reproductive system.

In some synovigenic hymenopteran parasitoids the lateral oviducts can accommodate a small number of eggs, e.g. 9–12 per oviduct in *Coccophagus atratus* (Donaldson and Walter, 1988) (**Anhydropy**, subsection 2.3.4). In others the oviducts are greatly elongated, to form distinctive 'uteri', and can accommodate very large numbers of eggs (Figures 2.4d and 2.5b) (**Hydropy**, subsection 2.3.4).

The lateral oviducts join to form the common oviduct, a largely muscular structure that in turn becomes confluent with the vagina and (in wasps) the ovipositor stylets. In some tachinid parasitoids, egg storage (and incubation) occurs in the common oviduct, e.g. *Cyzenis albicans* (Hassell, 1968). In wasps, forward-pointing spines in the vagina push the egg into the ovipositor at or before oviposition (Austin and Browning, 1981). As it passes down the ovipositor, the egg is squeezed to a small diameter, a process that has been shown to trigger embryonic development (Went and Krause, 1973). Embryonic development of haploid (male) eggs of the ichneumonid parasitoid *Pimpla turionellae* can also be triggered by experimental injection, not involving egg deformation, of calcium ionophore A23187 (Wolf and Wolf, 1988). The chorion of the hymenopteran egg is remarkably flexible, so experiments on the initiation of embryogenesis can be carried out on mature eggs that have been removed from the ovarioles or lateral oviducts of a wasp. The eggs can be manipulated in various ways on a microscope slide, in saline solution, to show, for example, what degree of compression is required to trigger embryogenesis. In the tachinid *Cyzenis albicans* embryogenesis commences prior to oviposition; eggs, when laid, contain a fully-formed first-instar larva (Hassell, 1968).

2.2.3 SHAPE, SIZE AND NUMBER OF EGGS

The shape of eggs in parasitoid Hymenoptera and Diptera varies considerably between groups (Iwata, 1959, 1960, 1962; Hagen, 1964a). Egg types found among hymenopteran parasitoids include those with a simple ovoid shape, those that are greatly elongated (Figure 2.9a,b), those with a distinctive stalk at the micropyle end, and those with a double-bodied appearance (Figure 2.9c). For a review of the range of egg types found among parasitoids see Hagen (1964a).

Some eggs (hydropic-type eggs, subsection 2.3.4) characteristically increase greatly in size following deposition in the host's haemocoel. Among Braconidae for example, eggs of Euphorinae expand in volume a thousand times (Ogloblin, 1924; Jackson, 1928), and those of *Praon palitans* (Aphidiinae) over six hundred times (Schlinger and Hall, 1960).

Within a parasitoid wasp species the number and size of mature oöcytes in the ovaries is positively correlated to some extent with the size of the female (O'Neill and Skinner, 1990; Rosenheim and Rosen, 1992) (Table 2.1). This observation has important implications for foraging models, since larger females may obtain larger fitness returns per host and also, compared with smaller females, they can utilise a series of hosts in rapid succession (Skinner, 1985a; O'Neill and Skinner, 1990).

The number of mature oöcytes in the ovaries is a function of the number of ovarioles, which is also correlated with body size within a species. Data on oöcyte number, oöcyte size and ovariole number are based on

(a)

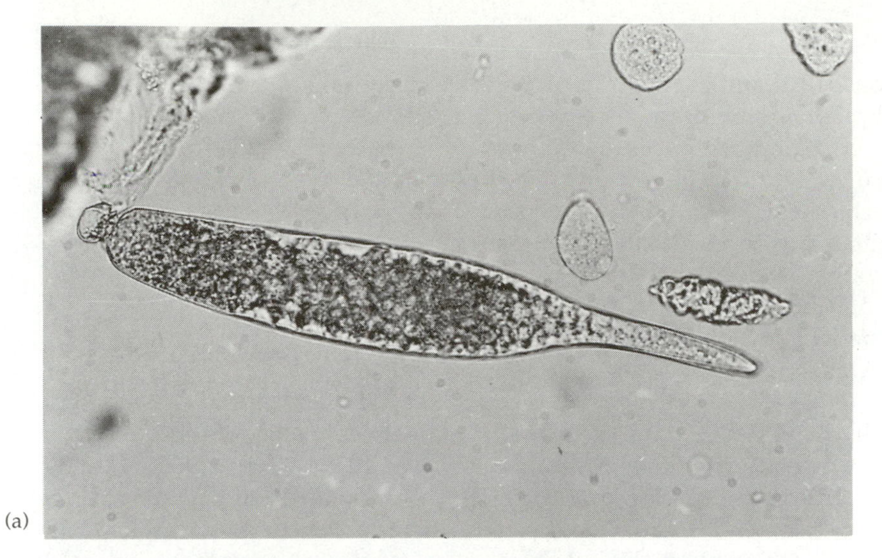

(b)

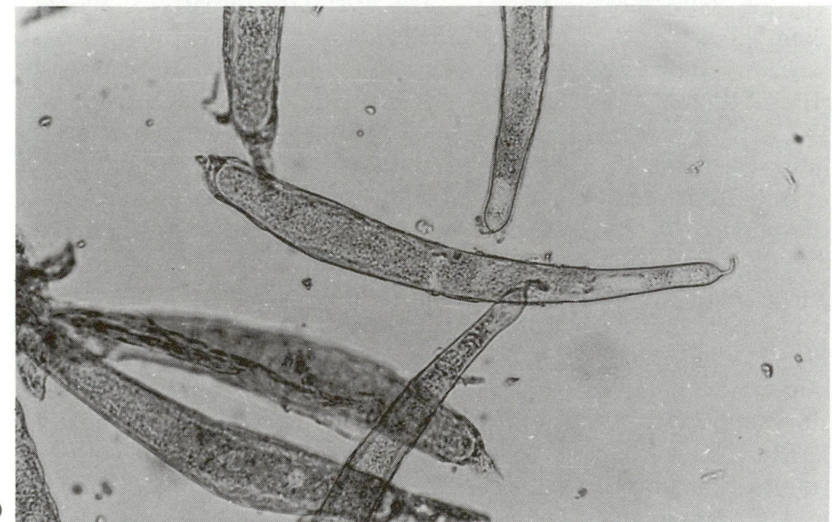

(c)

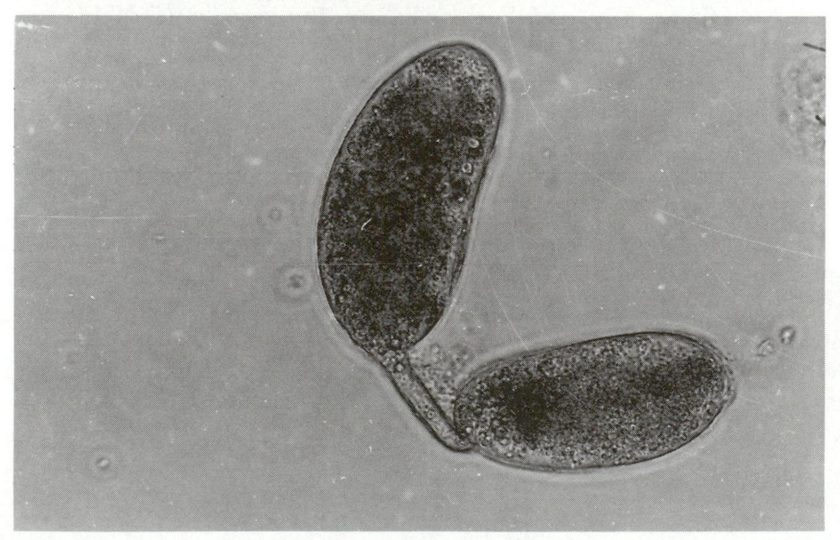

Figure 2.9 Eggs dissected out of the reproductive systems of parasitoid Hymenoptera: (a) unidentified Mymaridae; (b) *Cotesia* sp. (Braconidae); (c) unidentified Encyrtidae.

a limited number of parasitoid species, and for each species not all three variables have been measured. What is now required is a study of many more species, from a diverse range of taxa, to see how closely the three characteristics covary with body size.

In the damselfly *Coenagrion puella* and in the carabid beetle *Brachinus lateralis*, egg size is not correlated with female size (Banks and Thompson, 1987a; Juliano, 1985).

Between parasitoid species, ovariole number is a good predictor of fecundity, as Price (1975) has shown for Ichneumonidae and Tachinidae (Figure 2.10). It remains to be tested whether a correlation exists between body size and ovariole number, on a between-species basis.

Blackburn (1991b), showed through a comparative analysis (subsection 1.2.3) that among parasitoid Hymenoptera, there is a positive relationship between adult size and fecundity (fecundity is defined in subsection

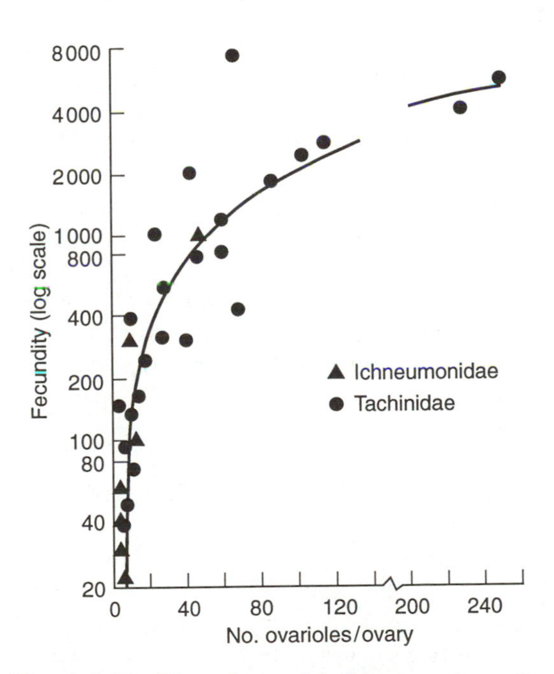

Figure 2.10 The relationship between fecundity (note log scale) and the number of ovarioles per ovary in Ichneumonidae and Tachinidae. Data points represent means for individual species. (Source: Price, 1975.) Reproduced by permission of Plenum Publishing Corporation.

2.7.1), but only when egg size is controlled for. When adult size is controlled for, species with a high fecundity (the maximum number of eggs reported to have been laid by an individual of a species) tend to have smaller eggs, indicating a trade-off between fecundity and egg size (small eggs need less of a material investment) (Blackburn's paper contains further discussion). The interspecific relationships among predatory Coccinellidae and Syrphidae with respect to ovariole number, mature oöcyte number, oöcyte size and female body size, and their biological significance, are discussed by Stewart *et al.* (1991) and Gilbert (1990).

2.3.4 PRO-OVIGENY AND SYNOVIGENY

In Lepidoptera (Boggs, 1977) and Hymenoptera (Flanders, 1950) species fall into either of two categories which Flanders (1950) termed **pro-ovigenic** and **synovigenic**. The females of pro-ovigenic species eclose with their full complement of mature or nearly mature eggs, whereas those of synovigenic species emerge with at most only part of their total mature egg complement but continue developing eggs during adult life. Pro-ovigenic insects thus complete oögenesis either before or very soon after eclosion, having used food reserves built up during larval life. Females may feed, but only for maintenance purposes (Jervis *et al.*, 1993). Synovigenic insects, on the other hand, need to feed (on hosts, honeydew or flowers) to achieve their full reproductive potential, although some species can mature a few eggs without first feeding (**autogenous** as opposed to **anautogenous** species). A few hymenopteran parasitoids (some Mymaridae; King and Copland, 1969, Eucharitidae and Trigonalyidae; Clausen, 1940) have been shown to be pro-ovigenic, while the majority appear to be synovigenic, being either autogenous or anautogenous (Jervis and Kidd, 1986). Hover-flies (Syrphidae) are either synovigenic-anautogenous or synovigenic-autogenous (Gilbert, 1991). The tachinid

Cyzenis albicans is synovigenic-anautogenous (Hassell, 1968). The green lacewing *Chrysoperla carnea* is synovigenic-autogenous; females can produce a few eggs without first feeding, provided the prey consumption rate during larval development has been sufficiently high. By contrast, some *Chrysopa* species need to feed on proteinaceous foods, e.g. aphid prey, to produce any eggs, i.e. they are synovigenic-anautogenous. Predatory coccinellids are also synovigenic-anautogenous.

To determine whether an insect species is either pro-ovigenic or synovigenic, dissect newly-emerged females and examine their ovarioles. If the ovarioles contain a majority of mature oöcytes, and any developing oöcytes present are all close to maturity, then the insect can be designated pro-ovigenic (Figure 2.5a). If the ovarioles contain a majority of immature oöcytes, then the insect may be designated synovigenic. Any such designation must be a provisional one, since there is good evidence that hymenopteran parasitoid species once thought to be pro-ovigenic may in reality be synovigenic-autogenous, having cycles of egg development during adult life (Walter, 1988). It is best to work on the assumption that pro-ovigeny and synovigeny are extremes of a continuum, and that among insects there are varying degrees of synovigeny.

To measure the rate of oögenesis in a synovigenic parasitoid in relation to different treatments, expose each of several large cohorts of standardised (e.g. newly emerged) females to a particular environmental condition, e.g. type of diet, temperature, and follow the cohorts through until the last females die. Each day, dissect part of each cohort and examine the condition of the ovaries in the females, recording the number of mature eggs. The age-specific and average daily rate of ovigenesis can be compared for different treatments. A detailed protocol for an investigation of this type, concerned with the effects of different temperatures, may be found in Kajita and van Lenteren (1982).

Flanders (1942, 1950, 1962) and Dowell (1978) recognised two divergent reproductive strategies among synovigenic hymenopteran parasitoids, based on the type of egg produced, the ability to store eggs and the ability to resorb them. Again, these strategies may represent extremes of a continuum.

1. **Anhydropy**: the eggs are **lecithal** (yolk-rich), large and cannot be stored for very long. Females have limited egg storage capacity (the lateral oviducts are very short) and the ovaries have few ovarioles (Figures 2.4a,b and 2.5c). A female can resorb unlaid eggs if deprived of hosts, so recycling nutrients for maintenance purposes and allowing her to survive until hosts again become available. Females of many species feed on hosts (Jervis and Kidd, 1986). Examples of such parasitoids are *Nasonia vitripennis*, *Bracon hebetor* and *Aphytis* species (Jervis and Kidd, 1986, give other examples);
2. **Hydropy**: the eggs are **alecithal** (yolk-poor), small and can be stored for long periods, because embryonic development relies upon immersion in the host's fluids and on uptake of host nutrients. Females generally have large numbers of ovarioles and store eggs in the lateral oviducts, which are enlarged (Figures 2.4d and 2.5b). Eggs, because they are not retained in the ovarioles and because they contain little energy and nutrients, are not resorbed. There is no host-feeding by females.

As pointed out by Jervis and Kidd (1986), this classification needs to be viewed with caution, as it was based on only a few species (this is still the case). Also, there are exceptions, e.g. *Coccophagus atratus* produces apparently yolk-rich eggs, but does not seem to have the ability to resorb them (Donaldson and Walter, 1988). However, it remains a useful scheme within which parasitoid reproductive biology can be investigated. The ecological significance of the anhydropy–hydropy dichotomy is not discussed here (Jervis and Kidd, 1986; Mackauer, 1990), but it is useful to note that anhydropic species would be expected to have a small maximum egg load and therefore be highly egg-limited compared with hydropic species (subsection 1.16.2).

In synovigenic-anhydropic parasitoids, oöcytes, when they become mature, are not immediately discharged into the oviduct. Usually a maximum of only a few (three in *Encarsia formosa*; van Lenteren *et al.*, 1987) mature eggs can be stored per ovariole at any moment in time. These eggs, however, can only be retained for a brief period, as they have limited storage life, and room has to be made for other oöcytes as they mature. If a female is deprived of hosts for a sufficiently long period (i.e. host are absent or else very scarce), she does not jettison such eggs but begins resorbing them, starting with the oldest (see below). In *Nasonia vitripennis* only the pycnotic residue of the follicle cell nuclei remains after resorption (King and Richards, 1968), although in a few species females may deposit partially resorbed eggs (Flanders, 1950). In some cases, even developing oöcytes may be resorbed (Jervis and Kidd, 1986; van Lenteren *et al.*, 1987 for reviews). By resorbing eggs, the female can use the energy and materials obtained from the eggs to maintain herself and to sustain oögenesis until hosts are again available. Through egg resorption, eggs are returned to the body of the wasps with only a partial loss of energy and materials, instead of total loss if the eggs were jettisoned.

Eggs that are undergoing resorption can be detected at the proximal ends of the ovarioles by their unusual shape compared with unaffected eggs (Figures 2.11a,b). Because of the partial removal of the chorion, eggs that have recently begun to be resorbed may, unlike unaffected eggs, increase in size when dissected out in water, and will certainly take up stains such as acetocarmine more readily (King and Richards, 1968).

As they are being resorbed, eggs shrink and finally disappear, leaving only remnants of the exochorion. The latter is probably voided through the egg canal at the next oviposition, although in some Encyrtidae part of the chorion (the aeroscopic plate) remains in the ovariole or is voided into the haemocoel (Flanders, 1942).

The time of onset of resorption in host-deprived wasps varies, depending on the availability of food. A female *Nasonia vitripennis* that is starved will begin resorbing eggs earlier than a female that is given honey (Edwards, 1954). In host-deprived, honey-fed females of *Nasonia vitripennis* oöcyte development continues, albeit slowly.

The rate of egg resorption can be measured using the chemical colchicine, which stops cell division by interfering with microtubule formation, and therefore halts production of further mature eggs. Rates measured for parasitoids vary from one to several days (Edwards, 1954; Benson, 1973; Bartlett, 1964; Anunciada and Voegele, 1982; van Lenteren *et al.*, 1987). In completely starved *Nasonia vitripennis*, when the terminal oöcyte of one ovariole has begun to be resorbed, it is followed by those of other ovarioles. With continued starvation, the penultimate oöcyte will also start being resorbed, first in one ovariole and then in the others, and so on (King and Richards, 1968).

If a female parasitoid is deprived of hosts for a long enough period for resorption to commence, the number of mature oöcytes in the ovaries (egg load), will depend on the rate of oögenesis (which will be much lower in starved females than in females that have access to non-host foods, subsection 2.7.3) and on the rate of resorption (which is not affected by diet) (King, 1963; van Lenteren *et al.*, 1987).

Egg resorption is another form of egg-limitation in synovigenic parasitoids, since whilst a female is in the process of resorbing eggs, she is temporarily incapable of ovipositing even if hosts become available (Jervis and Kidd, 1991).

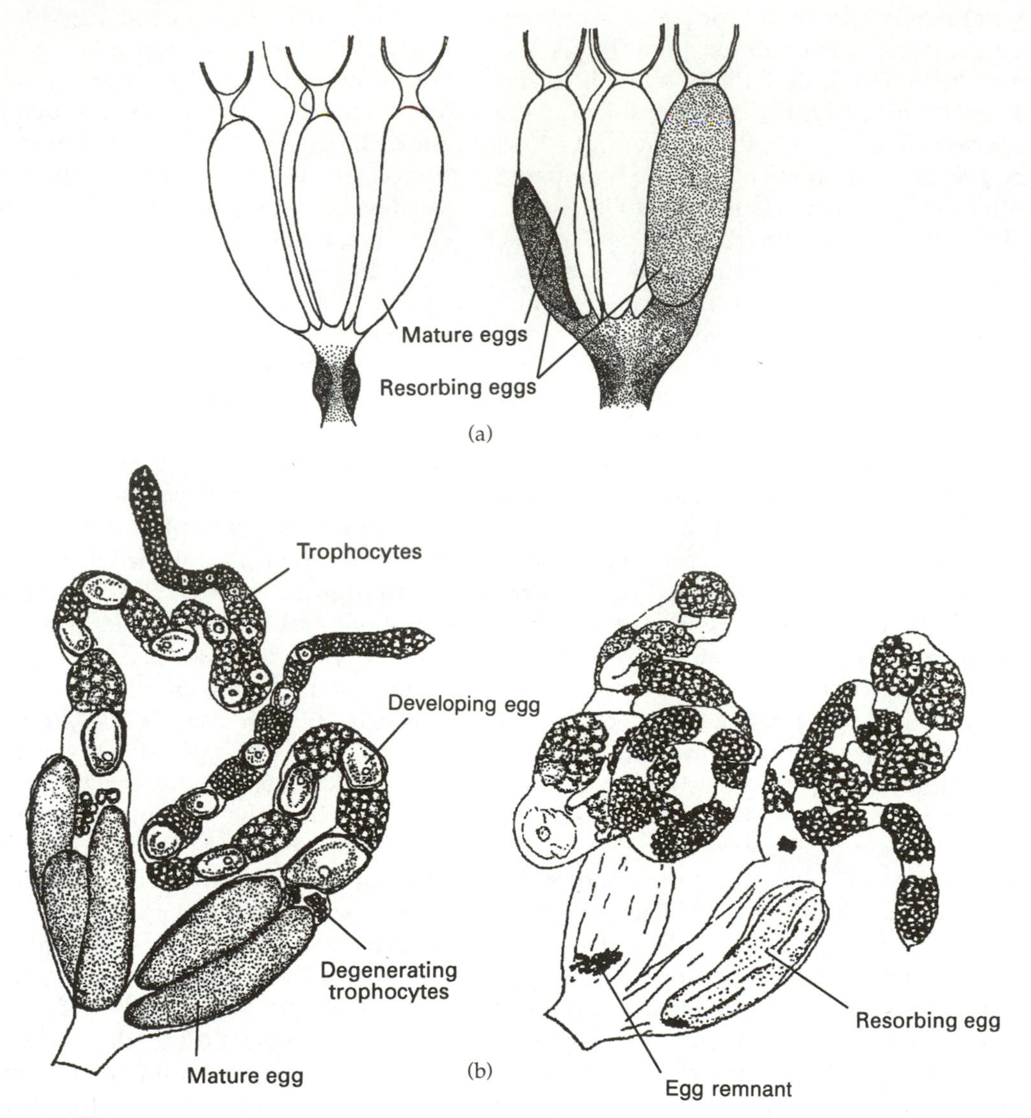

Figure 2.11 Egg resorption in synovigenic-anhydropic hymenopteran parasitoids: a) *Nasonia vitripennis* (Pteromalidae) (Source: King and Richards, 1968.); (b) *Bracon hebetor* (Braconidae). (Source: Grosch, 1950.) (In both cases, the ovarioles of a non-resorbing female are shown on left, and those of a resorbing female are shown on right.) (a) Reproduced by permission of the Zoological Society of London; (b) by permission of the Marine Biological Society, Woods Hole, Massachusetts.

Many predators fall into the synovigenic-anhydropic category; hover-flies are definitely known to resorb eggs if not allowed to oviposit. Many coccinellids, in the absence of prey, lay eggs but then eat them, so recycling nutrients (Heidari, 1989).

Techniques (involving light microscope histochemistry and transmission electron microscopy) for detecting yolk in eggs and measuring its quantity are described in King *et al*. (1971) and King and Richards (1969).

2.3.5 EGG LIMITATION

As discussed in Chapters 1 (subsection 1.16.2) and 5 (subsection 5.3.7), the degree to which a parasitoid is egg-limited is an important consideration when studying parasitoid foraging behaviour, from the standpoints of fitness gains and searching efficiency. The size of the parasitoid's egg load determines the number of eggs the female can lay at a given moment in time. What, then, sets the upper limit to egg load – is it the rate of ovigenesis or the storage capacity? If, in a species that is not currently resorbing eggs, not all the ovarioles are found to contain a mature egg at any instant in time when ovigenesis is at its maximum (the maximum can be found by exposing females to different host density and food regimes, over their lifetimes, subsections 2.7.2 and 2.7.3), i.e. there is asynchrony among ovarioles, then the ceiling to egg load is set by the rate of ovigenesis, not storage capacity. On the other hand, if at any instant in time all the ovarioles contain a full-sized egg and the lateral oviducts are also full of eggs, then the ceiling is likely to be set by storage capacity (in which case does ovigenesis cease when the maximum storage capacity is reached?). It is not clear whether there are any species in which the upper limit to egg load is set by the rate of ovigenesis; however, *Coccophagus atratus* is apparently a species of the other type. If females of this species are withheld from hosts but fed on honey following eclosion and are dissected after varying periods, the egg load is found to increase during the first 24 hours of adult life and thereafter remain constant (Figure 2.12). Since in this species there is no evidence for resorption, egg numbers are probably limited by the storage capacity of the ovarioles/lateral oviducts, with ovigenesis ceasing when there is no room for further eggs (Donaldson and Walter, 1988). This probably also applies to *Venturia canescens* (J. Harvey, personal communication). It would be interesting to know the frequency with which ovigenesis is switched on and off in females that are foraging under natural conditions.

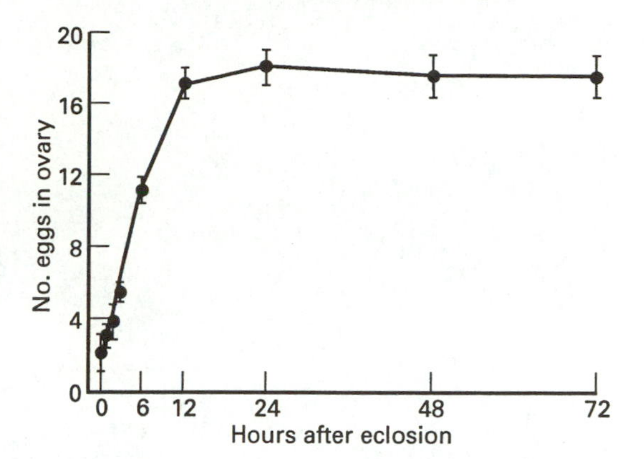

Figure 2.12 Number of full-sized eggs in the ovaries of *Coccophagus atratus* (Aphelinidae) at various intervals after female eclosion. (Source: Donaldson and Walter, 1988.) Reproduced by permission of Blackwell Scientific Publications Ltd.

2.3.6 MOTIVATION TO OVIPOSIT

A number of models indicate that the motivation to oviposit depends upon egg load (Mangel, 1989b; Jervis and Kidd, 1986; Chan and Godfray, 1993). How does a parasitoid know the size of its egg load? Donaldson and Walter (1988), in a detailed study on ovipositional activity and ovarian dynamics in *Coccophagus atratus*, showed that when females were exposed to an abundance of hosts they deposited eggs within defined bouts of ovipositional activity that were initiated only when the female had accumulated approx. 18 full-sized eggs (Figure 2.4a). This finding suggests that egg load, possibly perceived via stretch receptors in the lateral oviducts (Collins and Dixon, 1986), affects the motivation to oviposit.

2.3.7 SPERMATHECA

The spermatheca (Figures 2.13 and 2.14) is, as the name suggests, the sperm storage organ of

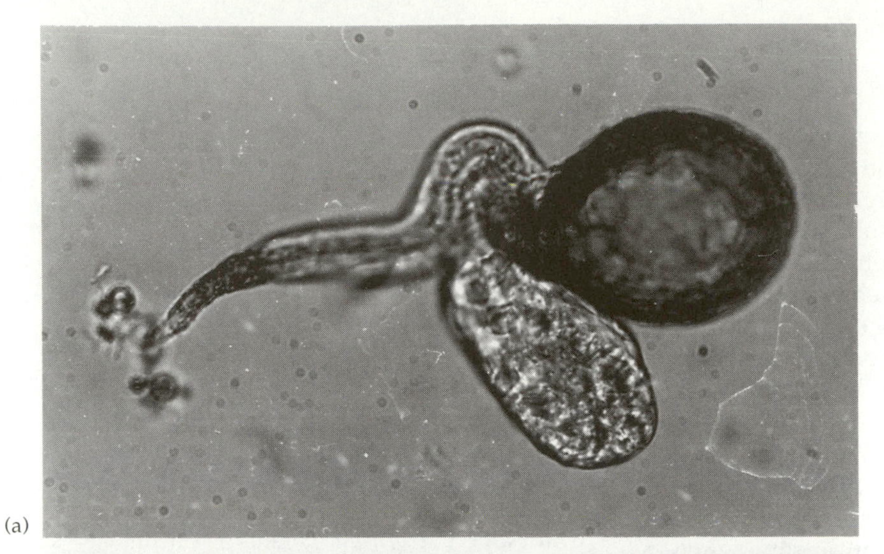
(a)

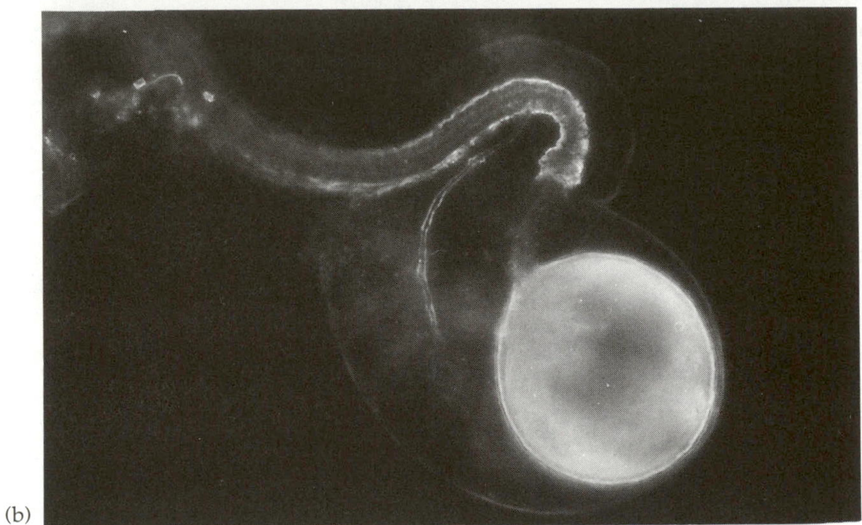
(b)

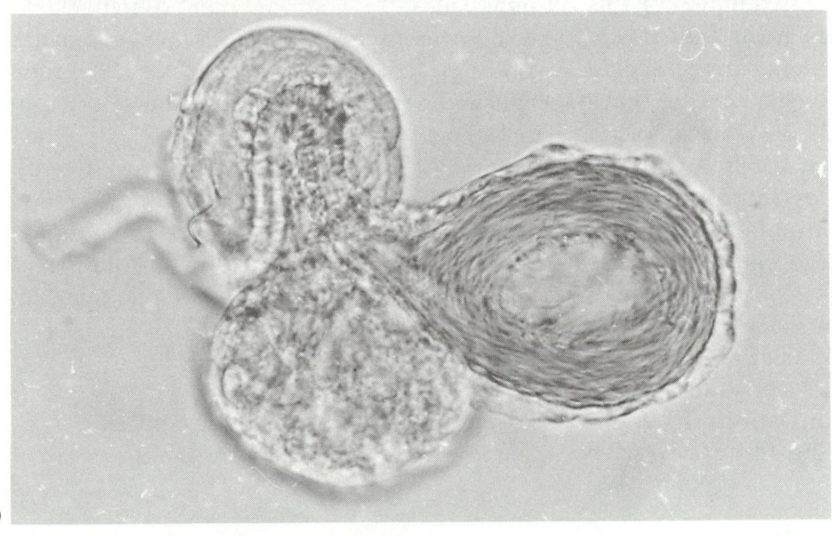
(c)

Figure 2.13 The spermatheca in hymenopteran parasitoids: (a) *Aprostocetus* sp. (Eulophidae) (showing pigmented capsule); (b) *Eurytoma* sp. (Eurytomidae) (capsule and gland united, showing gland collecting duct); (c) *Nasonia vitripennis* (Pteromalidae) (showing sperm).

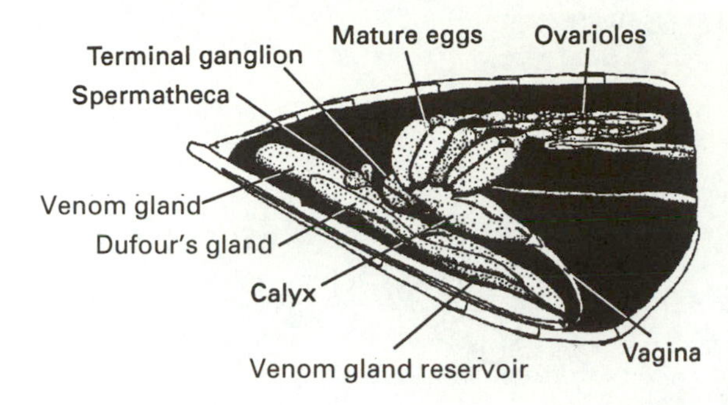

Figure 2.14 Reproductive system of female *Aphelinus* (Aphelinidae), showing position of spermatheca, Venom gland, Dufour's gland, Venom gland reservoir and other structures. (Source: Copland, 1976.) Reproduced by permission of Pergamon Press.

females. Syrphid flies and tachinid parasitoids have three (Figure 2.2c), whereasHymenoptera have only one. In Hymenoptera the spermatheca is situated at or near the confluence of the lateral oviducts. It comprises a capsule (the storage vessel), a gland which may help to attract, nourish and possibly activate sperm, and a muscular duct through which sperm are released (or not released, section 1.9) as an egg passes along the common oviduct. In many insects, the spermatheca is noticeably pigmented yellow or black (a possible adaptation for protecting sperm from the adverse affects of UV light), a useful feature to look out for when dissecting females. Using transmitted light it is usually possible to observe at high magnifications the movement of sperm, if sperm are present, within the capsule. To detect such movement, observations must be made within 5 minutes of dissecting the female. Hardy and Godfray (1990) determined whether or not field-caught parasitoids were virgins, by examining the spermatheca of dissected females. They were able to distinguish between empty spermathecae, those containing living sperm (present as a writhing mass) and those containing dead sperm (inadvertently killed by the dissection process).

Suggested studies on sperm use, depletion and competition are described in Chapter 3 (subsections 3.9.3 and 3.9.4).

2.3.8 ACCESSORY GLANDS

In many female insects there are obvious glands associated with the common oviduct which are termed **accessory** or **colleterial glands** (Figure 2.8). It is generally understood that they produce secretions which coat the egg as it is laid. These glands are present in nearly all chalcidoid parasitoids. Different families have different numbers and arrangements (King and Copland, 1969; Copland and King, 1971, 1972a,b,c; Copland *et al.*, 1973; Copland, 1976), but hardly anything is known about their function. They have been implicated in the formation of feeding-tubes by host-feeding Hymenoptera (Flanders, 1934), but they seem to be equally developed in species that do not host-feed. Some Torymidae have the largest glands, and *Eupelmus urozonus* has both large glands and enormous extensions from the calyx. *E. urozonus* is known for its ability to spin silk, but how it does so needs to be established. Noting the condition of the glands in dissected females in various experimental treatments may prove instructive as to their function.

2.3.9 DUFOUR'S GLAND

The Dufour's or alkaline gland (Figures 2.14 and 2.15) is well developed in Hymenoptera. It discharges into the proximal end of the common oviduct at the base of the ovipositor. In parasitoids it is the source of the parasitoid marker substances discussed in subsection 1.5.3 and section 1.18. The Dufour's gland is normally a thin-walled sac containing an oily secretion. It is a long tubular structure in most chalcids but may be extremely small in some braconid wasps, e.g. *Cotesia glomerata*, concealed among the bases of the ovipositor stylets.

2.3.10 VENOM GLAND

The venom gland = acid gland = poison gland, like the Dufour's gland, empties into the base of the ovipositor (Figure 2.16). It is either a simple structure as in

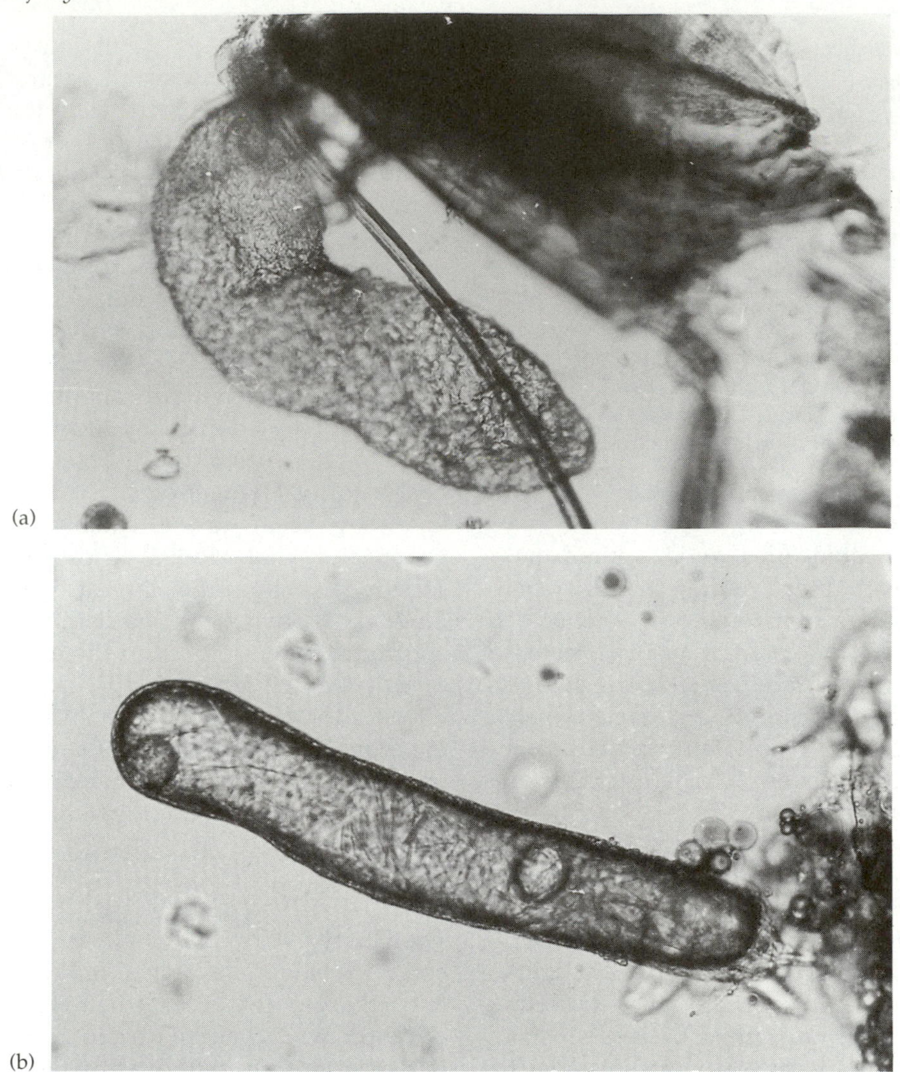

Figure 2.15 Dufour's or alkaline gland in hymenopteran parasitoids: (a) *Eurytoma* sp. (Eurytomidae); (b) *Colastes* sp. (Braconidae).

Chalcidoidea (Figure 2.16a), a convoluted tubular structure as in Ichneumonidae or a structure of intermediate complexity as in some Braconidae (Figure 2.16b). The venom of some idiobionts (subsection 1.5.6) induces permanent paralysis, arrested development or death in the host, whereas that of koinobionts induces temporary paralysis or no paralysis at all. Associated with the venom gland is a reservoir that has muscular walls; the reservoir may have additional secretory functions (Robertson, 1968; van Marle and Piek, 1986). The structure of the venom gland system of Hymenoptera has been comprehensively investigated by several workers (Ratcliffe and King, 1969; Piek, 1986), but there is considerable scope for further investigative work into gland function.

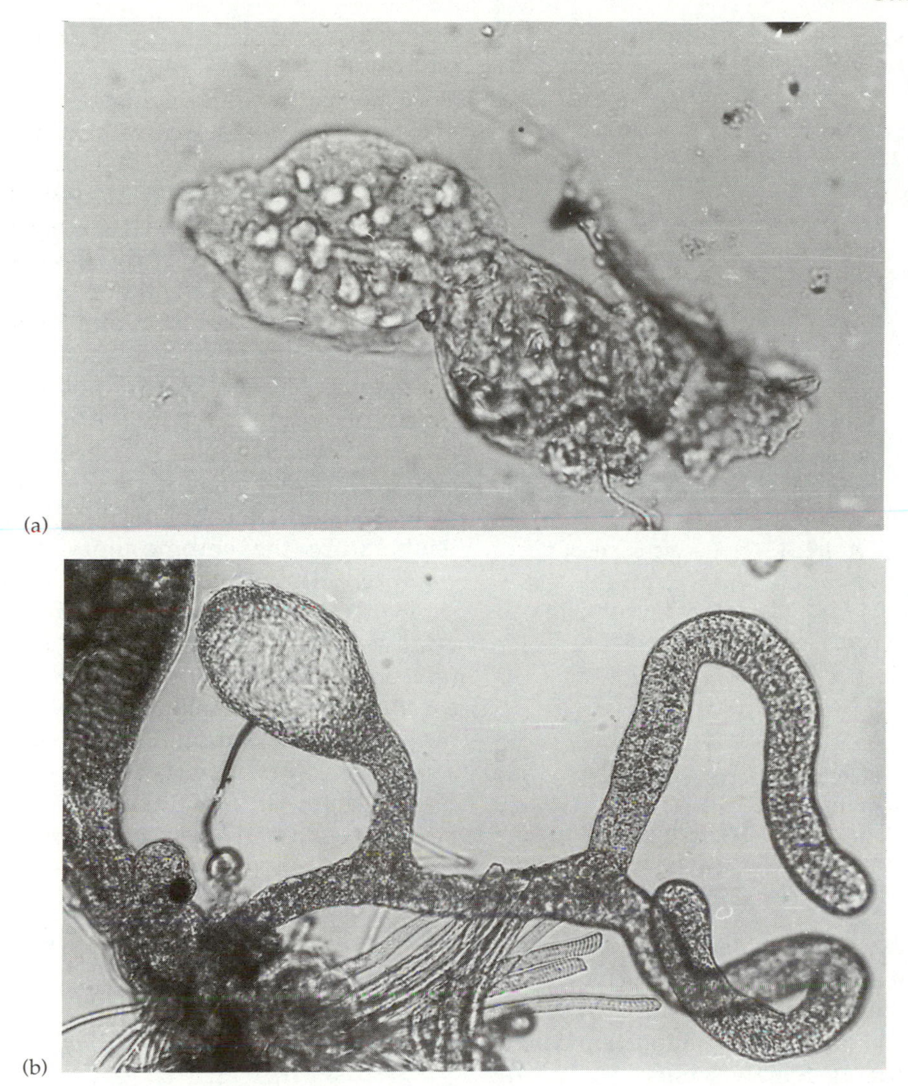

Figure 2.16 Venom gland in hymenopteran parasitoids: (a) unidentified Mymaridae (showing simple gland and reservoir); (b) *Cotesia* sp.(showing more complex, i.e. bifurcate, gland and reservoir).

2.4 MALE REPRODUCTIVE SYSTEM

An example of the reproductive system in male Hymenoptera is shown in Figure 2.17. The system comprises a pair of testes and usually a pair of accessory glands. The possible role of secretions from the latter in parasitoid mating behavior is discussed in subsection 3.10.3

2.5 SEX RATIO

2.5.1 INTRODUCTION

This section briefly discusses some aspects of sex ratio, mainly primary sex ratio, i.e. the sex ratio of progeny at the time of oviposition (section 1.9). It concentrates largely on hymenopteran parasitoids because of the

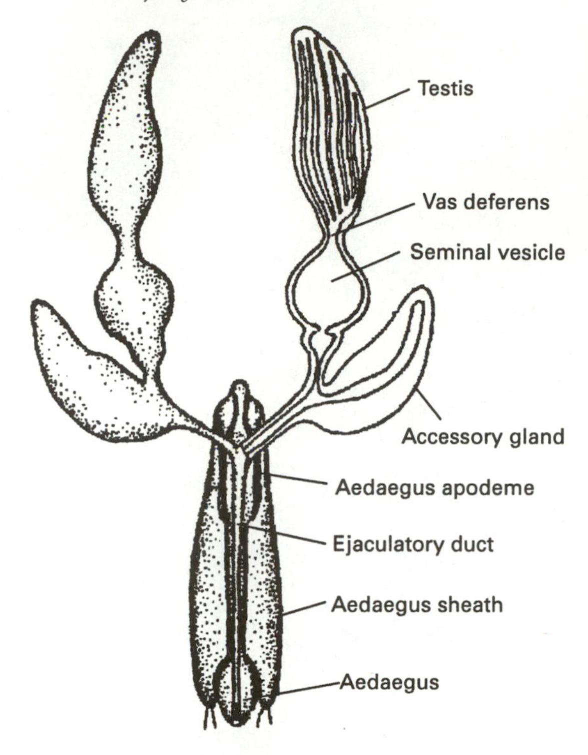

Figure 2.17 Schematic representation of reproductive system in male Chalcidoidea. (Source: Sanger and King, 1971.). Reproduced by permission of The Royal Entonological Society of London

considerable attention that is being paid, in behavioural and population studies on those insects, to sex ratio variation (section 1.9, subsections 3.2.4 and 5.3.7). Sex allocation behaviour by foraging female wasps has already been dealt with (section 1.9).

2.5.2 ARRHENOTOKY

The most common mode of reproduction in hymenopteran parasitoids is **arrhenotoky**, in which male progeny develop parthenogenetically from unfertilised (haploid) eggs and female progeny develop from fertilised (diploid) eggs. Females control the sex of their progeny when ovipositing, by selectively releasing sperm from the spermatheca (subsection 2.3.7).

In 'normally reproducing' arrhenotokous species, the primary sex ratio can be determined using a cytological technique with freshly laid eggs (van Dijken, 1991). Eggs are placed in a droplet of 2% lacto acetic orcein (a chromosome stain, made by dissolving 1.0 g of natural orceine in 10 ml of 85% lactic acid, 25 ml of glacial acetic acid and 15 ml of distilled water; this mixture is gently boiled for 1 h then cooled and filtered) and a coverslip gently placed over them. The eggs are then squashed to a monolayer after 25 min have elapsed and the cover-slip sealed (e.g. with nail varnish). After staining for 24 h at room temperature the chromosomes, which are 'captured' during the metaphase of cell division when they are maximally contracted, are examined and counted. Exact counts are unnecessary, as female eggs have two sets of chromosomes and males one set – an obvious difference. The optimum time to fix and squash the eggs may vary between species (van Dijken, 1991, found that this was 18–24 h post-oviposition in *Epidinocarsis lopezi*), so it is advisable to make a series of egg-squashes at different times following oviposition. It is also a good idea to examine first the eggs of virgin females so that the number of chromosomes in male eggs can be determined.

Kfir and Luck (1979) studied the effects of constant and variable temperature extremes on the sex ratio of F_1 progeny in two closely related arrhenotokous wasp species, *Aphytis melinus* and *A. lingnanensis*; the experimental design used is shown in Figure 2.18. Both species showed a significant reduction in the expected proportion of F_1 females when the parent was exposed to 32°C during its development, and the proportion of females was further reduced when mating and/or oviposition occurred at that temperature. These two effects appeared to be cumulative. Furthermore, brief preovipositional exposure of the females of both species to low and high

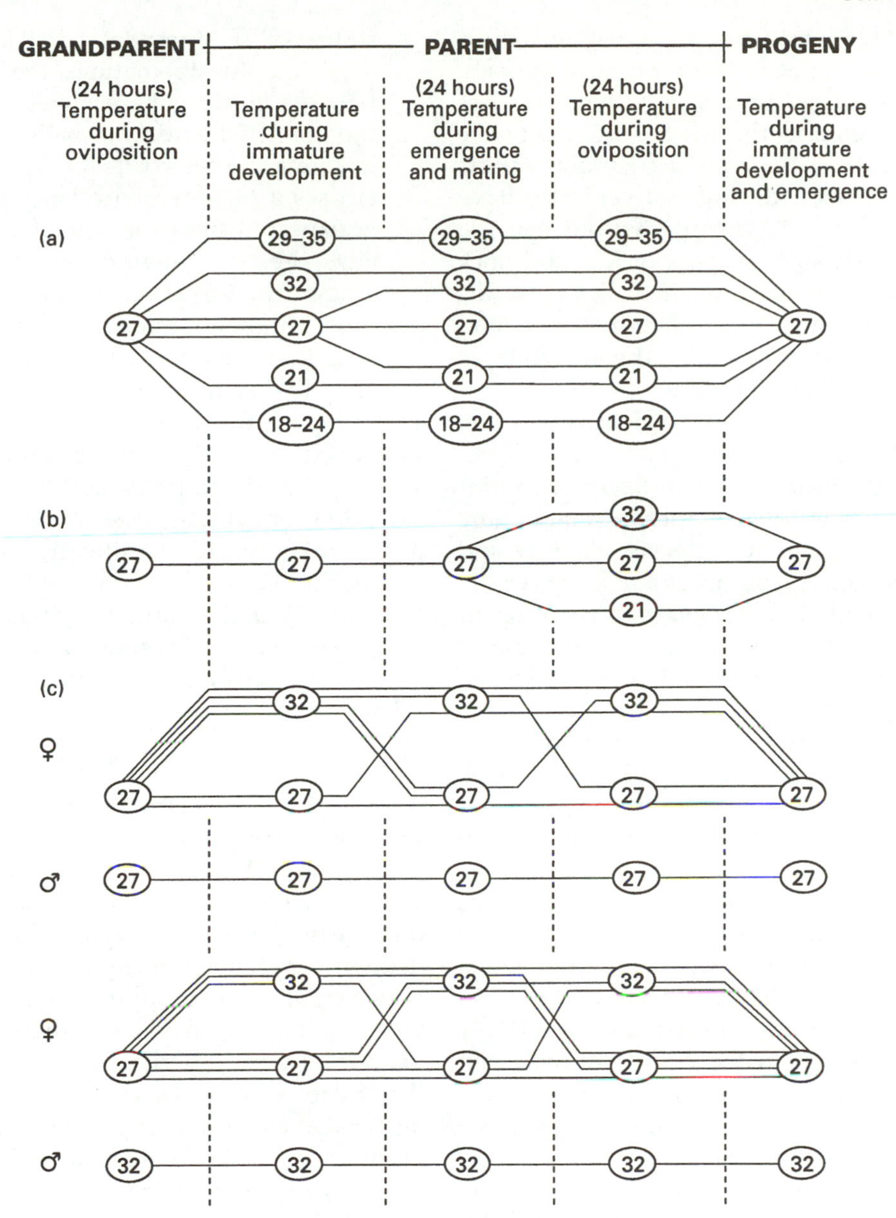

Figure 2.18 Schematic representation of experimental designs used to expose *Aphytis* (Aphelinidae) parasitoids to various constant and variable temperature regimes during their development and adult life (mating, preovipositional period, ovipositional period) and the effects of these exposures on subsequent F_1 progeny production and sex ratio: (a) male and female parents experience identical history of temperature exposure during development and mating; (b) female parents exposed to different temperatures only during ovipositional period; (c) male and female parents experience different histories of temperature exposure during development and mating; males reared at 27°C mated with females in design shown immediately above, males reared at 32°C mated with females in design shown immediately above. (Source: Kfir and Luck, 1979.) Reproduced by permission of Blackwell Scientific Publications Ltd.

extreme temperatures in most instances reduced the expected proportion of females. In general, the sex ratio of A. *melinus* was less severely altered by these exposures than were the ratios of A. *lingnanensis*. Similar effects of temperature extremes on sex ratio have been recorded in other parasitoid species. Investigations such as those of Kfir and Luck (1979) have helped to shed light on the species differences in climatic tolerance, and have provided criteria for the selection of biological control agents (5.4).

By what mechanisms might temperature affect progeny sex ratio in parasitoids? Does temperature in fact affect *primary* sex ratio? Kfir and Luck (1979) suggest that temperature extremes might reduce sperm viability and sperm mobility, and might also adversely affect mating behaviour and fertilisation (Kfir and Luck's 1979 paper gives a discussion of the particular mechanisms that might operate in the case of *Aphytis*).

Several hymenopteran parasitoids have been shown to produce diploid males as well as haploid males. In the Braconidae and Ichneumonidae that display this phenomenon (which results in a male-biased primary sex ratio), diploid male production is the result of a single-locus sex determination mechanism (Chalcidoidea have a different sex determination mechanism). Haploid individuals (arising from unfertilised eggs) have only one allele at the sex locus and are male, but diploid individuals (arising from fertilised eggs), have two alleles at the sex locus and can be either homozygous (with two identical copies of an allele, e.g. AA or BB) or heterozygous (with copies of two different alleles, e.g. AB or AC). Homozygous individuals are male, and heterozygous individuals are female.

Evidence for single-locus sex determination can be obtained by several methods (Stouthamer *et al.*, 1992):

1. examination of the distribution of offspring sex ratios from matings between siblings; a bimodal distribution is expected. In sib matings the male and female either share one sex allele (homoallelic) or do not have a sex allele in common (heteroallelic). In homoallelic crosses a male-biased offspring sex ratio is expected because half of the diploid eggs become males. In heteroallelic crosses the offspring sex ratio should be much higher. Note, however, that this detection method only works if the variance in sex ratio produced within each group of crosses is low;
2. genetic identification of diploid males (e.g. by electrophoresis (see subsection 4.3.11) or cytology (see above));
3. morphological identification of diploid males. Diploid males are larger than haploid males, and because they have larger cells, their bristle density is different (e.g. see Grosch, 1945).

2.5.3 THELYTOKY AND DEUTEROTOKY

A minority of species reproduce in other, entirely parthenogenetic, ways: **thelytoky** and **deuterotoky** (also known as **amphitoky**). With thelytoky there are no males, and unfertilised eggs give rise to diploid females. With deuterotoky, male progeny may be produced; however, as pointed out by Luck *et al.* (1993), the distinction between thelytoky and deuterotoky is ambiguous, the reason being that some parasitoid wasp species originally designated as thelytokous have been found to produce small numbers of males. These males are produced when the maternal females have been exposed to high temperatures. Whilst they have been considered to be non-functional, there is evidence that in some cases they are not only capable of mating but they are also able to pass on their genes to progeny, which reproduce thelytokously.

One cause of thelytoky is interspecific hybridisation (Nagarkatti, 1970; Pintureau and Babault, 1981; Legner, 1987). For example, Nagarkatti (1970) crossed a female of

Trichogramma perkinsii with a male of *T. californicum*, and the offspring comprised one thelytokous female, seven arrhenotokous females and ten males. Another cause is the presence of microbes. Arrhenotoky can be restored in several *Trichogramma* species either by treatment with antibiotics or exposure to a high temperature (Stouthamer *et al.*, 1990a,b). J.W.A.M Pijls (unpublished) showed that the arrhenotokous population of *Epidinocarsis diversicornis* attacking *Phenacoccus manihoti* (cassava mealybug) in Central South America is conspecific with morphologically identical but thelytokous populations in northern South America that attack *P. herreni*, by treating the thelytokous wasps with antibiotics. Males were obtained that interbred successfully with females of the arrhenotokous population. Both arrhenotokous and thelytokous forms are known in several parasitoid wasp species; these forms occur either allopatrically or sympatrically (Stouthamer, 1993).

2.5.4 OTHER FORMS OF SEVERE SEX RATIO DISTORTION

Other forms of severe sex ratio distortion are known in addition to thelytoky and diploid male production. Perhaps the best known is the **paternal sex ratio (psr)** element (Werren *et al.*, 1981, 1987; Werren and van den Assem 1986), which is chromosomal in origin and is effectively a piece of 'selfish' DNA (Nur *et al.*, 1988). Male *Nasonia vitripennis* (so far the only species in which it has been recorded) carrying the element cause the females they mate with to produce all-male progeny, even though their sperm fertilise the females's eggs (Figure 2.19). Each fertilised egg carries the maternal (i.e. haploid) set of chromosomes together with the psr element from the sperm, and it develops into a psr-carrying male i.e. only males develop from fertilised eggs and they inherit the psr factor from the paternal male.

Another sex ratio-distorting factor that has been recorded in *N. vitripennis* is the **maternal**

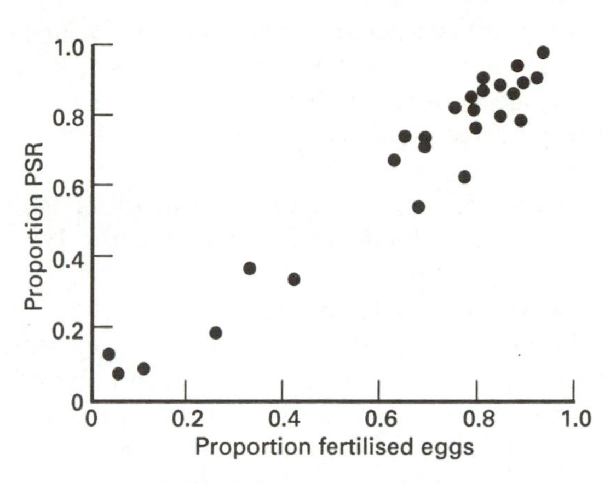

Figure 2.19 The relationship between the proportion of the male progeny that carry the paternal sex ratio factor (*psr*) and the proportion of eggs fertilised by mated females of *Nasonia vitripennis* (Pteromalidae) (Source: Werren and van den Assem, 1986). The latter proportion was determined for control crosses in which male and female wasps were of various *psr*-negative genetic 'strains' (the percentage of females in the progeny being the percentage of fertilised eggs). The former proportion was determined for test crosses involving male wasps taken from a known *psr* strain and females from the *psr*-negative strains. The *psr* trait was assayed as follows: females were mated with presumptive *psr* males and the sex ratio of the resulting progeny scored. If a greater than 90% male brood resulted, the male parent was taken to be a *psr* carrier. This phenotypic test involves a small bias since approximately 5% of control crosses also result in more than 90% male broods, as a result of inadequate mating. Where the assay proved ambiguous, at least five males from the F_1 brood were also tested, and if two or more produced all-male broods the parent was deemed to be a *psr* carrier. The data presented are not corrected for the small bias. There is a strong linear relationship between *psr* transmission and egg fertilisation, strongly suggesting that *psr* is a factor transmitted from sperm to eggs upon fertilisation. In fact, *psr* is an example of a parasitic *B*-chromosome (*B*-chromosomes are extra to the normal chromosome complement). *psr* destroys the other paternal chromosomes in the early fertilised egg. It disrupts the normal haplo-diploid sex determination system of *Nasonia* by converting diploid (female) eggs into haploid eggs that develop into *psr*-bearing males. (Beukeboom and Werren, 1993.) Reproduced by permission of the Genetics Society of America.

sex ratio (msr) factor, a possible extrachromosomal element. Virgin females carrying msr produce male offspring. After mating, their progeny are (practically) all female (Skinner, 1982).

A bacterium is the cause of **son-killer (sk)** in *Nasonia*. Female wasps carrying the bacterium produce female-biased broods; the majority of male progeny die as larvae (Skinner, 1985b). Bacterial transmission probably occurs via larval feeding on the blood of the parasitised host. Strictly speaking, son-killer affects secondary sex ratio.

Another cause of sex ratio distortion is the transmission of a *Rickettsia*-like bacterium in non-reciprocal crosses between strains (hence the label **non-reciprocal cross incompatibility** or **NRCI**). Eggs free of the bacterium can only be fertilised successfully by sperm from uninfected males, whereas eggs containing the bacterium can be fertilised by sperm from either infected or uninfected males (Breeuwer and Werren, 1990). In *Nasonia* this results in all-male broods being produced in crosses between uninfected females and infected males; it can be eliminated through treatment with antibiotics (Richardson *et al.*, 1987). The cause of NRCI in *Trichogramma deion* is not a bacterium but some maternally inherited factor associated with genes in the nucleus (Stouthamer, 1989, in Luck *et al.*, 1993).

Cases of sex ratio distortion that resemble msr, sk and NRCI have been recorded in a number of other parasitoid species, but the causal factor in each instance remains to be identified.

Every worker who consistently obtains only males among progeny should investigate whether the all-male broods occur simply because the females have not been inseminated. Females should be dissected and the spermatheca examined for the presence of sperm (subsection 2.2.5). Bear in mind that the mated females of some parasitoid species may run out of sperm (Dijkstra, 1986).

2.5.5 MORTALITY FACTORS DETERMINING SECONDARY SEX RATIO

Parasitoids

Son-killer has just been mentioned as a factor causing the secondary sex ratio to differ from the primary sex ratio in parasitoids. Later in this chapter (section 2.10), we discuss a variety of other factors, biotic and physical, that cause mortality of larvae. The possibility that some of these factors may affect the sexes to different extents ought to be investigated.

Predators

Predators and dipteran parasitoids are diploid insects; mating is usually essential to the production of viable offspring, and the sex ratio of offspring is close to unity. However, female-biased sex ratios have been recorded in two species of predatory ladybird (*Coccinella septempunctata, Adalia bipunctata*). Hurst *et al.* (1993) found that cytoplasmically inherited agents (possibly bacteria) are responsible in at least some populations, for killing male eggs (Ottenheim *et al.*, 1992). The factor is passed on through females and results in up to 100% of the male eggs of some broods failing to hatch, i.e. the secondary sex ratio is affected by the bacterium.

2.6 LOCATING EGGS IN HOSTS

Parasitoid eggs may have to be located in or on hosts for a variety of reasons, including the measurement of fecundity and parasitism (subsection 2.7.3, sections 5.2, 5.3), investigations of parasitoid behaviour (subsection 1.5.6, sections 1.6, 1.8, 1.10) and community studies (subsections 4.2.9, 4.3.5). The eggs of endoparasitoids are generally much more difficult to locate than those of ectoparasitoids; the degree of difficulty will depend upon factors such as the relative sizes of host and parasitoid eggs, the amount of fat body tissue, whether the eggs lie within organs or in the haemocoel, the size of other organs and the degree of sclerotisation of the host integument (Avilla and Copland, 1987). Preferably, hosts should be killed either by narcotising them (e.g. using CO_2 ethyl acetate) in which case they should be dissected shortly afterwards, or by placing them in a deep-freeze, in which case they can remain dissectable for several months. Attempting to locate eggs in hosts that have been preserved in alcohol is likely to prove very difficult indeed.

If endoparasitoid eggs prove difficult to locate, parasitised hosts should be kept alive long enough for the eggs to swell (i.e. in hydropic species) and/or the first instar larvae to form, the parasitoid immature stage in either case becoming more easily visible.

2.7 FECUNDITY

2.7.1 INTRODUCTION

The term **fecundity** refers to an animal's reproductive output, in terms of the total number of eggs produced or laid and should be distinguished from **fertility** which refers only to the number of viable progeny that ensue. From the standpoint of population dynamics, fertility is the more important parameter, as it is the number of individuals entering the next generation. However, it can be relatively difficult to measure (Barlow, 1961), so fecundity measurements are often used instead. We can also distinguish between **potential** and **realised** fecundity. A species' potential fecundity is usually taken to be the maximum number of eggs that can potentially be produced by females. For example, in the laboratory we might take a pro-ovigenic parasitoid (subsection 2.3.4), dissect its ovaries at eclosion and then count the number of eggs contained within. This number could be taken to represent the species' potential lifetime fecundity. This fecundity measurement could then be compared with the number of eggs actually laid over the life span when excess hosts are provided in the laboratory, i.e. realised fecundity. The figure for realised fecundity may fall slightly short of the estimate for potential fecundity, due to the insects dying before they can lay all their eggs. The differential is likely to be even greater if realised fecundity is measured under natural, i.e. field conditions (Leather, 1988).

Fecundity is a variable feature of a species, influenced by a range of intrinsic, and extrinsic (physical and biotic) factors. The evaluation of a natural enemy for biological control requires a study of the influence of these factors (and of possible interaction effects between certain factors) on potential and realised fecundity, and if possible, fertility. The data can be used in estimating a species' intrinsic rate of increase which is discussed later in this chapter (section 2.11). Fecundity is also used as a measure of individual fitness in insects.

When assessing the influence of a particular biotic factor on lifetime fecundity, it is important to determine to what extent variation in fecundity can be explained by variation in longevity. For example, take the positive relationship between female size and fecundity. The greater longevity of larger females compared with smaller females, rather than size itself, could be the main reason why larger females are more fecund. Females may have the same average daily egg production irrespective of body size, but by living longer, larger females lay more eggs over their life span (Sandlan, 1979b).

2.7.2 COHORT FECUNDITY SCHEDULES

A fecundity schedule for a parasitoid or predator species can be constructed by taking a cohort of standardised females (standardised in terms of physiological age, size, and sexual experience) and exposing them individually to some chosen set of constant environmental conditions from adult emergence until death. The number of eggs laid per female per day is then plotted, giving the **age-specific fecundity** of the species (Figure 2.20; see also Figure 2.67). The data obtained from the experiment

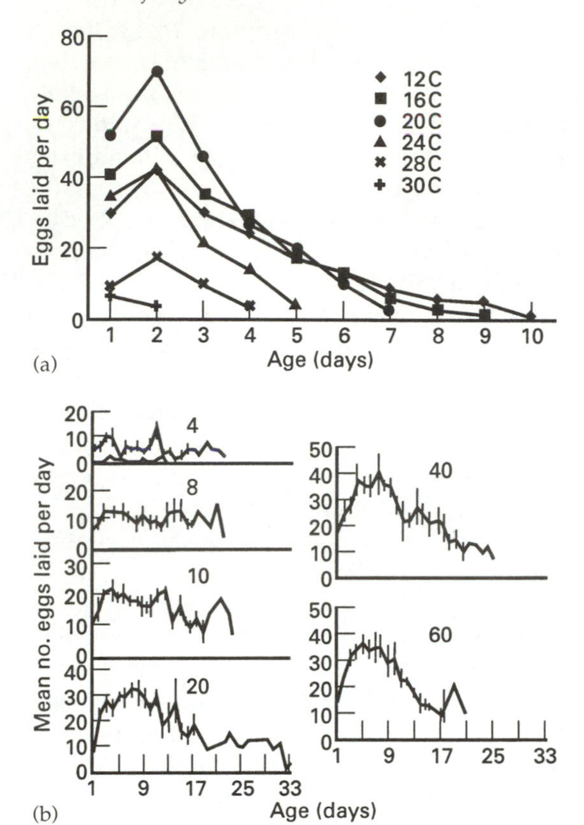

Figure 2.20 Fecundity schedule for two parasitoid species: (a) *Aphidius matricariae* maintained at different temperatures under constant host density conditions (Source: Hag Ahmed, 1989); (b) *Dicondylus indianus* (Dryinidae) maintained at different host densities (2–60) under constant temperature conditions. (Source: Sahragard *et al.*, 1991.) Reproduced by permission of Blackwell Wissenschafts–Verlag GmbH.

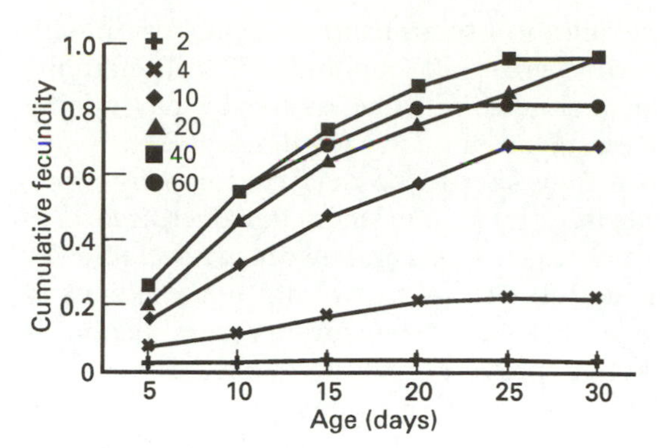

Figure 2.21 The cumulative realised fecundity of the dryinid wasp *Dicondylus indianus*, measured over the lifetime of females, at different levels of host availability. Fecundity is expressed as the proportion of the highest mean total number of eggs laid by females of any one treatment, this total representing the maximal fecundity/fitness that could be realised. (Source: Sahragard *et al.*, 1991.) Reproduced by permission of Blackwell Wissenschafts–Verlag GmbH.

can also be used to calculate both the lifetime fecundity of the species (used by evolutionary ecologists as a measure of fitness), and the **average daily oviposition rate** (lifetime fecundity divided by the average longevity). Using the same data, The **cumulative fecundity** of the parasitoids can also be plotted against either female age (Figure 2.21) or cumulative degree-days (Minkenberg 1989) (subsection 2.9.3). It is expressed as the proportion of the highest mean total number of eggs laid by females of any one treatment (e.g. temperature or host density treatment), this total representing the maximal fecundity/fitness realisable by females. The usefulness of the cumulative fecundity measure is that it tells us to what extent parasitoids achieve their maximum lifetime fecundity/fitness under particular conditions, and allows easier comparison of the effects of different treatments. Using the data from a fecundity schedule, the parameters m_x (age-specific fecundity) and l_x (age-specific survival) can be used in the calculation of the intrinsic rate of increase of the parasitoid population (section 2.11). If fecundity schedules are constructed for cohorts held under different host/prey availability regimes, the number of hosts or prey parasitised or eaten can be recorded and the data used to plot age-specific and lifetime functional responses (numbers parasitised or eaten vs. numbers available (section 1.10), as was done by Bellows (1985). Similarly, for predators, fecundity (age-specific or lifetime) can be plotted against prey availability (see **Food Consumption**, below).

An important consideration when using the above-mentioned experimental design is that as time goes on, the data are limited to progressively fewer females. To obtain fecundity data that are statistically meaningful, particularly data for the latter part of adult life, a very large starting density of females may be required. This, however, may increase the investigator's workload to an unacceptable level.

In parasitoids and predators, the fecundity schedule recorded in terms of oviposition (and, in synovigenic species, also in terms of ovigenesis) will show a rise in the number of eggs produced or laid per day until a maximum rate of productivity is reached. Thereafter a gradual decrease occurs until reproduction ceases altogether at or shortly before the time of death (Figure 2.20) (if there is a period of post-reproductive life, it is usually very short, see Jervis *et al.*, 1994, for exceptions). Fecundity schedules vary between species, depending on the reproductive strategies of the insects, e.g. pro-ovigeny and synovigeny (subsection 2.3.4). As we shall describe below, environmental factors (temperature, humidity, photoperiod, light quality, light intensity, host or prey availability) modify these patterns in a number of ways, and ideally the role of each factor in influencing the schedule ought to be investigated separately. This, however, may not be practicable, in which case the usual procedure is to expose a predator or parasitoid to an excess of prey or hosts (replenished/replaced daily), at a temperature, a relative humidity, and a light intensity similar to the average recorded in the field (Dransfield, 1979; Bellows, 1985).

To measure age-specific fecundity under cyclically varying conditions, e.g. of temperature, first establish the performance at a constant optimum. The likely effect of the alternative condition can be calculated using the programme shown in Appendix 2A. The exposure to the alternative can be of any duration as part of a 24 hour cycle.

2.7.3 EFFECTS OF BIOTIC FACTORS ON FECUNDITY

Host Density (Parasitoids)

If fecundity schedules are constructed for a parasitoid species over a range of host densities, it will be found that the parasitoids lay on average more eggs per day at higher host densities than at low ones, i.e. that the age-specific oviposition rate is positively density-related (Figure 2.22). Also, the lifetime pattern of oviposition, i.e. the shape of the curve, varies with host density. There may be a shift in the fecundity schedule, with wasps concentrating oviposition into the earlier part of adult life (Figure 2.20b). At high host densities, hosts are more readily available for the wasps to attack, whereas at low densities oviposition rates are of neccessity lower because the wasps have to search a greater area (and probably for longer), so expending energy that might otherwise be used in ovigenesis (Sahragard *et al.*, 1991).

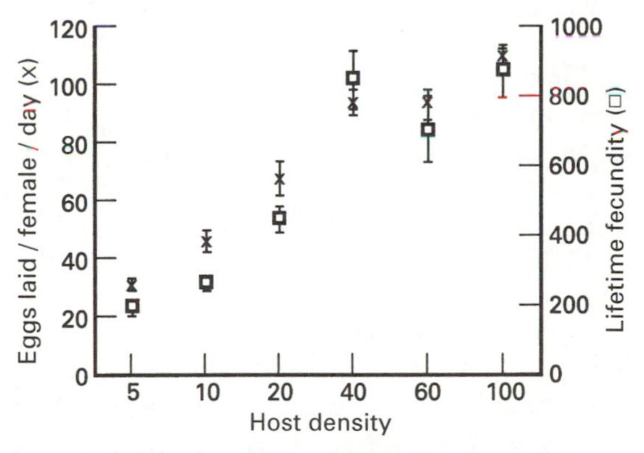

Figure 2.22 Relationship between fecundity (measured as the mean number of eggs laid per day and the total number of eggs laid over adult life) and host availability in the parasitoid *Aphidius smithi* (Braconidae). (Data from Mackauer, 1983.)

So far as lifetime fecundity is concerned, the relationship with host density is either a curvilinear one, resembling a Type 2 functional response (defined in section 1.10), or a sigmoid one, resembling a Type 3 functional response (Figure 2.22).

A difficulty that may arise when using low host densities is **ovicide**, i.e. the removal of eggs from parasitised hosts, although the number of (ecto)parasitoid species that practice ovicide is considerably smaller than the number of predator species that do so. Predaceous females of chrysopid lacewings are well-known for eating their own eggs in laboratory cultures (Principi and Canard, 1984). Where cannibalism is suspected, video recording techniques may help in determining the number of eggs lost in fecundity experiments.

Food Consumption

Non-predaceous Females

The females of many parasitoid and some predator species (e.g. *Chrysoperla carnea* (Chrysopidae) and adults of all aphidophagous Syrphidae) feed as adults on materials such as honeydew, nectar and pollen (Chapter 6), and consume substitute foods such as diluted honey in the laboratory. Females that are either deprived of such foods or experience a reduced intake of foods (but are given water) lay fewer eggs or no eggs at all. Some non-host foods have a more beneficial affect on fecundity than others (Leius, 1962; Wilbert and Lauenstein, 1974; Principi and Canard, 1984; Krishnamoorthy, 1984).

For an experimental investigation into the effects of adult nutrition on the fecundity schedule of a parasitoid to be ecologically meaningful, the effects of food provision need to be considered in the light of variations in host availability. This would be done by taking a cohort of standardised females and providing the insects with one of a range of host densities (see **Host Density**, above) and with a chosen diet for the duration of their lives, the hosts and food being replenished daily. If the effects on ovigenesis of combined host deprival/food provision are to be investigated, then, obviously, hosts are not provided to one set of females. One likely effect of providing food to females is that, at low host densities, females maintain a higher rate of oviposition than they can in the absence of foods. So far as the effects of food provision on lifetime fecundity are concerned, it will be necessary to carry out a statistical analysis to show whether or not any improvement in lifetime fecundity brought about by feeding is simply a result of an increase in longevity and not an increase in the daily rate of oögenesis (subsection 2.8.3).

Predaceous Females

We would expect the fecundity of predaceous females to be strongly influenced by prey availability. This relationship was modelled in a simple way by Beddington *et al.* (1976b) and Hassell (1978). If it is assumed firstly that some of the food assimilated by the female needs to be allocated to maintenance metabolism (and will therefore be unavailable for ovigenesis), and secondly that there is insufficient 'carry over' of food reserves from larval development for the laying of any eggs i.e. anautogeny, then there will be a threshold prey ingestion rate, c below which reproduction ceases, but above which there is some positive dependence between fecundity F and ingestion rate I. If it is assumed thirdly that this relationship is linear, then (Beddington *et al.*, 1976b):

$$F = \frac{\lambda}{e}(I - c) \tag{2.1}$$

where e, λ and c are constants. e is the average biomass per egg. There is empirical support for this model (Mukerji and LeRoux, 1969; Mills, 1981) (Figure 2.23). In Mills' (1981) experiment five feeding levels were used, the daily ration of individual females corresponding to between 1 × and 2 × the average female weight.

To express fecundity in terms of prey density, we first assume ingestion rate to be

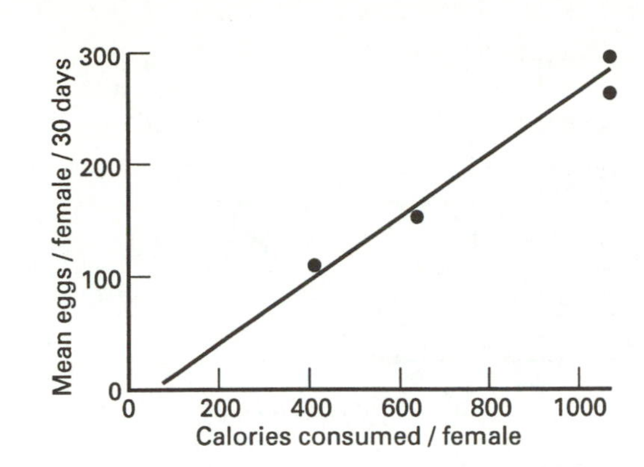

Figure 2.23 Fecundity as a function of ingestion rate in the pentatomid bug *Podisus maculiventris*. (Source: Beddington *et al*., 1976b, who used data from Mukerji and LeRoux, 1969.)

proportional to the number of prey eaten, N_a, so that:

$$I = kN_a \tag{2.2}$$

where k is a constant which depends upon the biomass (size) of each prey. Combining equations 2.1 and 2.2 with the simplest functional response model, Holling's (1966) disc equation (subsection 5.3.7), gives:

$$F = \frac{\lambda}{e}\left(\frac{kaN}{1+a'T_hN} - c\right) \tag{2.3}$$

This model predicts that fecundity will rise at a decreasing rate towards an upper asymptote as prey density increases, in the manner of the Type 2 functional response (section 1.10) but displaced along the prey axis . There is empirical support for this relationship, both from laboratory studies (Dixon, 1959; Ives, 1981a; Matsura and Morooka, 1983; (Figure 2.24a,b) and from field studies (Wratten, 1973; Mills, 1982) (Figure 2.25a,b (note that the data in (a) in this figure are expressed as logarithms). Anautogenous, obligate host-feeding parasitoids will have a similar fecundity/host density curve. In autogenous predators and host-feeding parasitoids however, reproduction can occur in the absence of prey, so the curve will not be displaced along the prey axis.

In the bug *Anthocoris confusus*, the viability (fertility) of eggs also varies with prey availability (Evans, in Beddington *et al*., 1976b). This relationship may be due to the female allocating less biomass per developing egg at

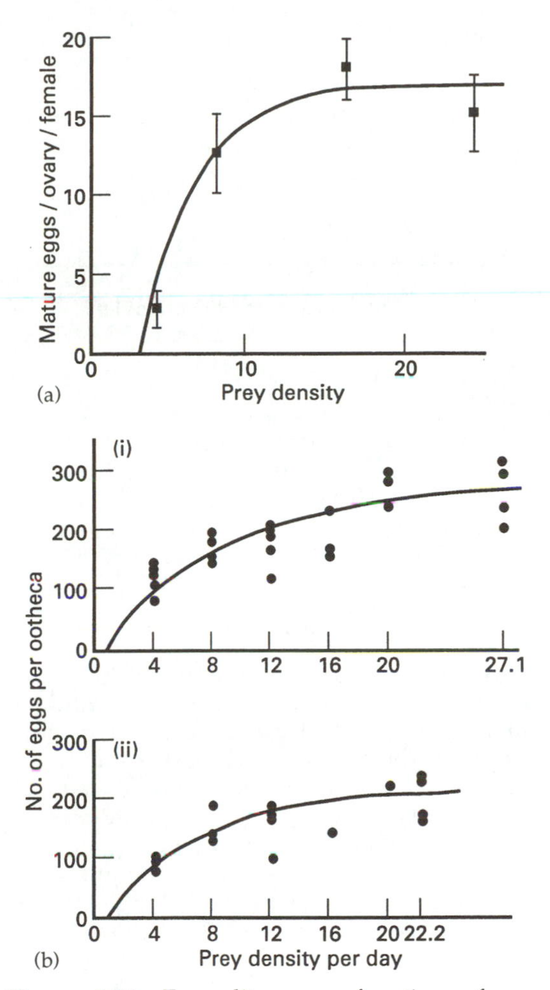

Figure 2.24 Fecundity as a function of prey density: (a) in the coccinellid beetle *Adalia decempunctata* (Source: Beddington *et al*., 1976b, who used data from Dixon, 1959); (b) in the mantid *Paratenodera angustipennis*: (i) first ovipositions, (ii) second ovipositions (ootheca = egg mass). Below the intersection of the curve (fitted by eye) with the prey axis, the insects allocate matter to maintenance processes only. (Source: Matsura and Morooka, 1983.) (a) Reproduced by permission of Blackwell Scientific Publications Ltd; (b) by permission of Springer–Verlag GmbH & Co. KG.

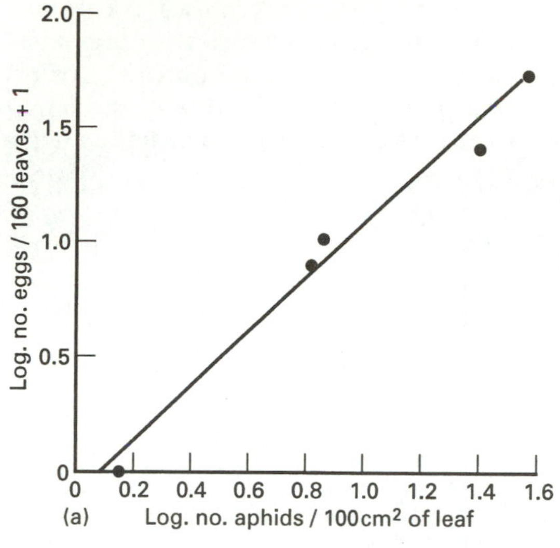

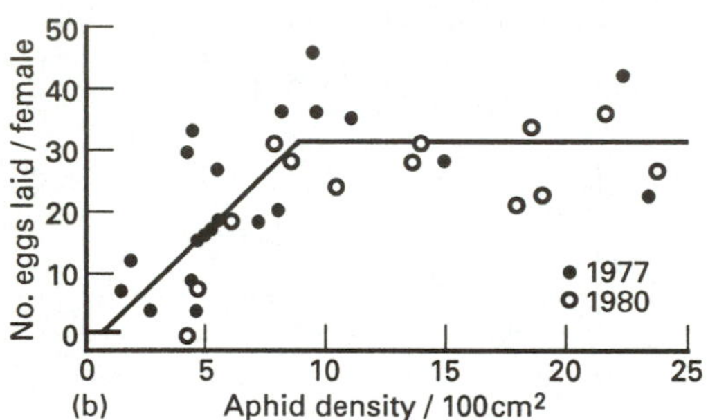

Figure 2.25 Fecundity as a function of prey density: (a) relationship between logarithm of number of eggs laid by the coccinellid *Adalia bipunctata*, and logarithm of density of aphids in the field, (Data from Wratten, 1973); (b) relationship between number of eggs laid per adult *Adalia bipunctata* and aphid density in the field. (Source: Mills, 1982b.) (a) Reproduced by permission of Blackwell Scientific Publishing Ltd; (b) by permission of The Association of Applied Biologists.

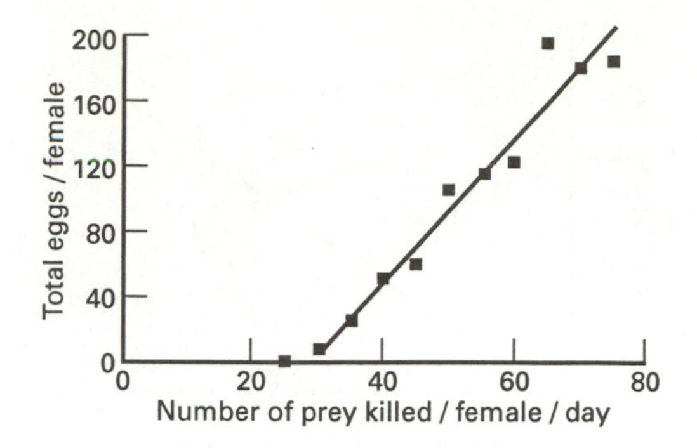

Figure 2.26 The relationship between fecundity and prey consumption rate in *Coccinella unidecimpunctata aegyptiaca*. (Source: Beddington *et al.*, 1976b, who used data from Hodek, 1973.)

lower prey densities, i.e. *e* in equation 2.1 is not a constant (Beddington *et al.*, 1976b).

There are also grounds for questioning the assumption that *k* in equation 2.2 is a constant (Beddington *et al.*, 1976b). If this assumption is correct, then the relationship between fecundity and the number of prey actually killed will be a straight line one, which is the case for *Coccinella undecimpunctata aegyptiaca* (Figure 2.26). However, as noted in Chapter 1 (section 1.10), when the rate of encounter with prey is high, some predators consume proportionately less of each prey item. This behaviour will alter the shape of the fecundity vs. prey killed curve, from rectilinear to curvilinear (Beddington *et al.*, 1976b). The shape of the fecundity vs. prey density curve will also be altered, having an earlier 'turnover' point and being more 'flat-topped' (Beddington *et al.*, 1976b).

Providing anautogenous predaceous females with non-prey foods together with prey might lower the ingestion rate threshold, since less of the prey biomass assimilated by the female needs to be allocated to maintenance metabolism. The fecundity–prey density curve of an anautogenous species will therefore be shifted along the prey axis, towards the origin. The shape of the curve is also likely to be altered. Experiments are needed to test this hypothesis.

Prey and Host Quality

Prey quality as well as availability is likely to affect fecundity, as has been shown for Coccinellidae and host-feeding Aphelinidae (Hariri, 1966; Blackman, 1967; Hodek, 1973;

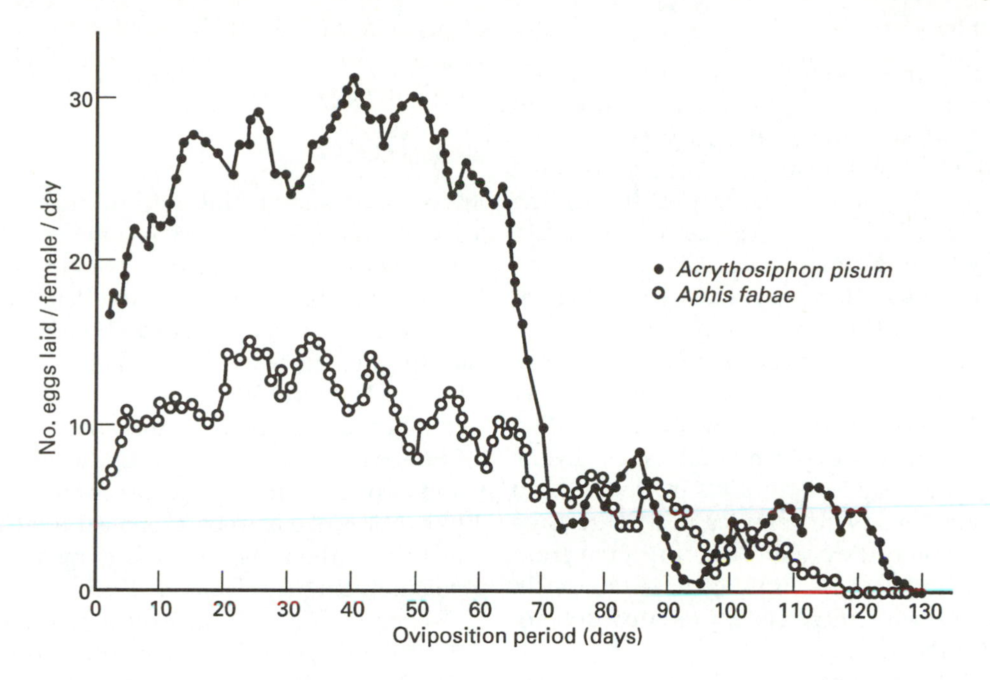

Figure 2.27 Fecundity of the coccinellid beetle *Adalia bipunctata* maintained on different prey species, *Acyrthosiphon pisum* and *Aphis fabae*. (Source: Hariri, 1966.) Reproduced by permission of W. Junk Publishers.

Wilbert and Lauenstein, 1974). Some coccinellids are unable to reproduce at all if confined to a diet of certain prey species (Hodek, 1973). Blackman (1967) found that adults of the coccinellid beetle *Adalia bipunctata* fed on *Aphis fabae* during both larval development and adult life were less than half as fecund as those fed on *Myzus persicae*. Their eggs were also smaller and less fertile. By carrying out another experiment in which adult beetles were fed on the opposite prey species to that fed upon by the larvae, Blackman (1967) tested whether the prey species given to larvae affected the fecundity of the adult. It did not; fecundity depended strongly upon the species fed upon by the adult. However, it is not clear from Blackman's experiments whether he controlled for the effects of prey availability. The results of a study by Hariri (1966) are shown in Figure 2.27.

In predators such as coccinellids the pre-oviposition period may be either lengthened or prolonged, depending on the prey species fed upon by the female (Hodek, 1973).

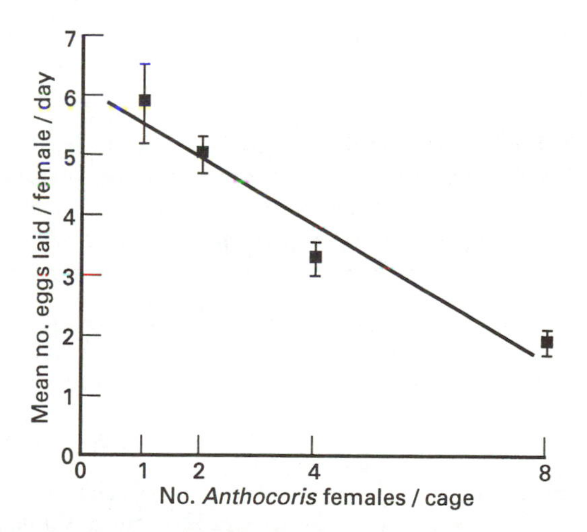

Figure 2.28 Relationship between fecundity and predator density in the aphid predator *Anthocoris confusus* (Anthocoridae). There was a decline in fecundity despite prey density being high at all times, i.e. the cause of the decline was mutual interference, not exploitation of prey. (Source: Evans, 1976.) Reproduced by permission of Blackwell Scientific Publications Ltd.

Mutual Interference

Mutual interference between female parasitoids results in a reduction in individual searching efficiency (subsection 1.14.3) which will result in a reduction in the rate of oviposition, i.e fecundity. In the predator *Anthocoris confusus* fecundity declined with increasing adult density, despite the fact that prey density was high at all times and was unlikely to limit egg production through prey exploitation (Evans, 1976) (Figure 2.28). To determine whether mutual interference was a result of confining predators in his experimental cages, Evans (1976) measured fecundity in relation to predator density in females in a large cage within which they were free to move from plant to plant. A significant decrease in fecundity with increasing predator density was still recorded.

Mutual interference, and therefore interference-mediated reductions in fecundity, cannot be assumed to occur in all predators. For example, Hattingh and Samways (1990) found no evidence for mutual interference in adults of three species of *Chilocorus* (Coccinellidae). Feeding rate did not decrease and dispersal did not increase with increasing beetle density.

Female Size

It has been shown that within many predators and parasitoids there is a positive correlation between body size or weight and one or more of the following variables: ovariole number, egg load (recorded either at or shortly after eclosion) and lifetime fecundity (Table 2.1, Figure 2.29), the first two of which are taken to be good comparative measures of potential fecundity. This is undoubtedly a valid assumption in most cases (Figure 2.30), but caution needs to be exercised in attempting to use them as 'short cut' measures of realised fecundity (Leather, 1988).

Takagi (1985) and Hardy *et al.* (1992) (Table 2.1) did not record directly the number of eggs laid during adult life; instead, they counted the number of adult offspring produced and took account of the intervening mortality processes, which were quantified.

In some species, the relationship between fecundity and body size correlated over only

Table 2.1 Parasitoid and predator species in which a positive correlation has been recorded between female body size or weight (or pupal weight) and reproductive variables in parasitoids and predators. Note that in pro-ovigenic species, egg load is potential fecundity, whereas in synovigenic species it is simply a part measure of potential fecundity

Body size vs.	*Species*	*Reference*
Ovariole number	*Metasyrphus corollae* (Syrphidae)	Scott and Barlow (1984)
	Encarsia formosa (Aphelinidae)	van Vianen and van Lenteren (1986)
Egg load	*Brachinus lateralis* (Carabidae)	Juliano (1985)
	Aphidius sonchi (Braconidae)	Liu (1985a)
	Dacnusa sibirica (Braconidae)	Croft and Copland (1993)
	Monoctonus pseudoplatani (Braconidae)	Collins and Dixon (1986)

Table 2.1 Continued

Body size vs.	*Species*	*Reference*
Egg load – *continued*		
	Cotesia glomerata (Braconidae)	le Masurier (1991)
	Cotesia rubecula (Braconidae)	Nealis *et al.* (1984)
	Venturia canescens (Ichneumonidae)	Harvey *et al.* (1994)
	Morodora armata (Pteromalidae)	O'Neill and Skinner (1990)
	Muscidifurax zaraptor (Pteromalidae)	O'Neill and Skinner (1990)
	Nasonia vitripennis (Pteromalidae)	O'Neill and Skinner (1990)
	Aphytis melinus (Aphelinidae)	Opp and Luck (1986)
	Aphytis lingnanensis (Aphelinidae)	Opp and Luck (1986) Rosenheim and Rosen (1992)
	Anagrus silwoodensis (Mymaridae)	Moratorio (1987)
	Anagrus mutans (Mymaridae)	Moratorio (1987)
	Trichogramma evanescens (Trichogrammatidae)	Waage and Ng (1984)
	Goniozus legneri (Bethylidae)	O'Neill and Skinner (1990)
	Myzinum quinquecinctum (Tiphiidae)	O'Neill and Skinner (1990)
Lifetime fecundity	*Coenagrion puella* (Coenagriidae)	Banks and Thompson (1987a)
	Chrysoperla carnea (Chrysopidae)	Zheng *et al.* (1993b)
	Eucelatoria bryani (Tachinidae)	Mani and Nagarkatti (1983)
	Notiophilus biguttatus (Carabidae)	Ernsting and Huyer (1984)
	Pimpla turionellae (Ichneumonidae)	Sandlan (1979b)
	Diglyphus begini (Eulophidae)	Heinz and Parrella (1990)
	Pteromalus puparum (Pteromalidae)	Takagi (1985)
	Lariophagus distinguendus (Pteromalidae)	Bellows (1985)
	Goniozus nephantidis (Bethylidae)	Hardy *et al.* (1992)

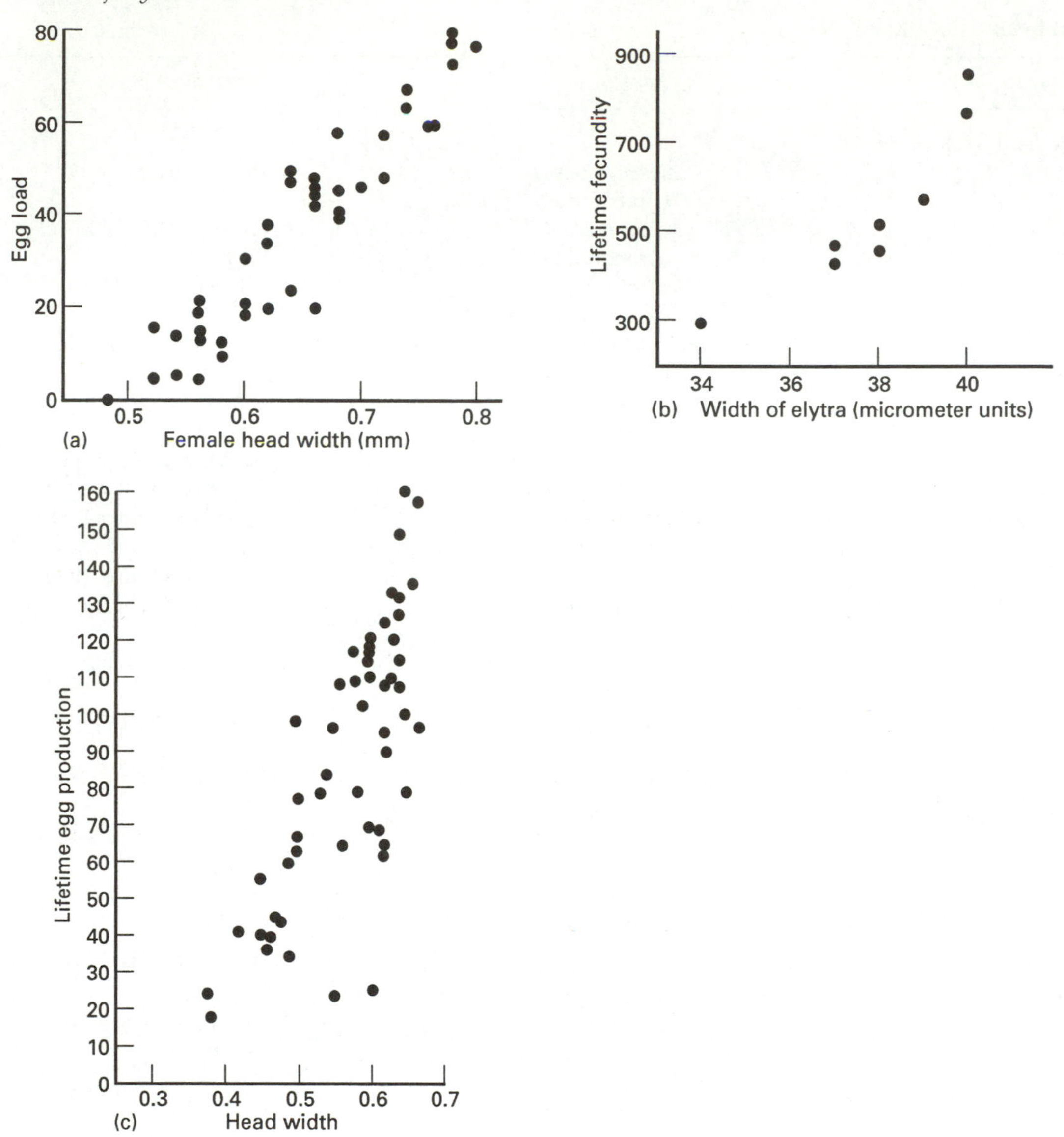

Figure 2.29 Positive correlation between fecundity measures (egg load, lifetime fecundity) and body size in females: (a) egg load in *Nasonia vitripennis* (Source: O'Neill and Skinner, 1990.); (b) lifetime fecundity in *Notiophilus biguttatus*; elytra width is expressed in micrometer units (100 units = 5.0 mm) (Source: Ernsting and Huyer, 1984.); (c) lifetime fecundity in *Lariophagus distinguendus* (Pteromalidae). (Source: van den Assem *et al.*, 1989.) (a) Reproduced by permission of The Zoological Society of London; (b) by permission of Springer-Verlag GmbH & Co. KG; (c) by permission of E.J. Brill (Publishers) Ltd.

part of the size range, with fecundity reaching a maximum in insects above a threshold size, e.g. *Aphidius ervi* (Sequeira and Mackauer, 1992b).

In predators, larger females have a shorter pre-oviposition period than smaller ones (Zheng *et al.*, 1993b), and this may contribute to their higher lifetime fecundity.

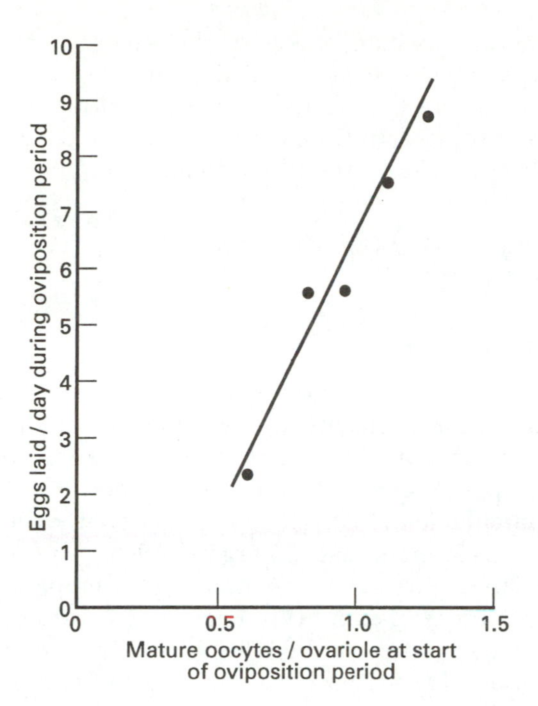

Figure 2.30 The relationship between the number of eggs laid per day and the number of mature oocytes per ovariole at the start of the oviposition period, in the parasitoid *Encarsia formosa* (Aphelinidae). It follows from this relationship that fecundity will be a positive function of ovariole number. (Source: van Lenteren *et al.*, 1987.) Reproduced by permission of Blackwell Wissenschafts-Verlag GmbH.

Body size is usually measured in terms of the width or length of some body part such as the head, thorax, or hind tibia. Some authors have measured dry body weight. Body size (or mass) is influenced by:

1. larval feeding history, i.e. prey availability, host size, host species during development, quality of host diet, clutch size, superparasitism (Dixon, 1959; Russel, 1970; Hodek, 1973; Dransfield, 1979; Sandlan, 1979b; Cornelius and Barlow, 1980; Beckage and Riddiford, 1983; Waage and Ng, 1984; Scott and Barlow, 1984; Liu, 1985a; Sato *et al.*, 1986; Eller *et al.*, 1990; Bai and Mackauer, 1992; Principi and Canard, 1984; Zheng *et al.*, 1993a,b; Harvey *et al.*, 1993, 1994) (Figure 2.31);
2. the temperature during larval development (Ernsting and Huyer, 1984; Nealis *et al.*, 1984) (Figure 2.32).

If an experiment, for whatever purpose, requires females to be of different sizes/fecundities, by far the simplest way of sorting insects according to size is to measure the parasitoids or predators when they are pupae, so avoiding any difficulties and/or harmful side effects associated with handling the adults.

Apparent exceptions to the body-size-fecundity generalisation are *Brachymeria intermedia* (Chalcididae) (Rotheray and Barbosa, 1984), *Edovum puttleri* (Eulophidae)

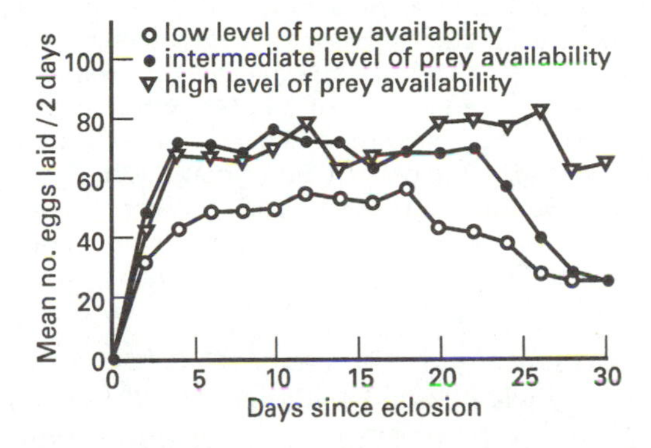

Figure 2.31 Effect of larval feeding history on fecundity in the lacewing *Chrysoperla carnea*. The data points indicate the average number of eggs laid, per 2-day period, of females provided with different levels of prey availability as larvae. Zheng *et al.* (1993b) showed that when lacewing larvae are fed fewer prey than they could potentially consume, they develop into smaller and less fecund adults than when they are given an overabundance of prey. Adults of *C. carnea* are non-predaceous, feeding on nectar, pollen and honeydew, and fecundity is also affected by consumption of these foods. Therefore female fecundity is determined by larval feeding history (through its effects on body size) and non-prey food consumption. (Source: Zheng *et al.*, 1993b.) Reproduced by permission of Kluwer Academic Publishers.

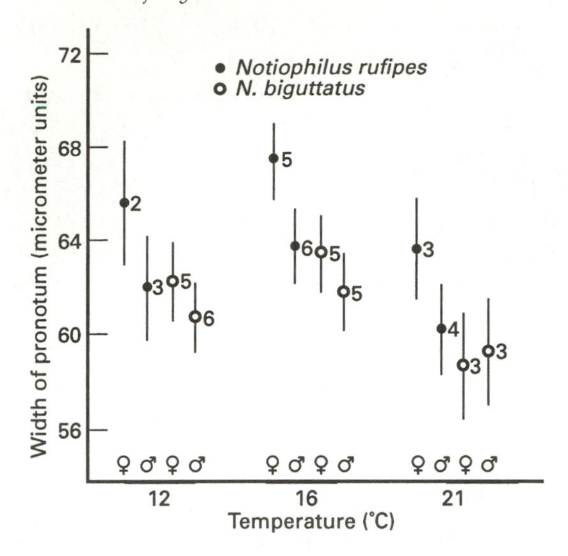

Figure 2.32 Effect of temperature during larval development upon adult size (as measured by pronotum width) in two species of carabid beetle. Means and the corresponding 95% confidence limits are shown, expressed in micrometer units (100 units = 2.5 mm). The data show a decline in adult size at either side of an optimum temperature for total biomass production. The effects upon size are translated directly into variations in fecundity. (Source: Ernsting and Huyer, 1984.) Reproduced by permission of Springer-Verlag GmbH & Co. Kg.

(Corrigan and Lashomb, 1990), and *Trichogramma maidis* (Bigler *et al.*, 1987). Bearing in mind that the relationship is non-linear for parasitoids, the lack of a correlation could in some cases simply be due to the adults measured lying within the upper range of body sizes.

Mating

Female predators and dipteran parasitoids, if they are either unmated or sperm-depleted, lay only a few eggs or (e.g. coccinellids) none at all. Eggs, if laid, are infertile. To achieve their full reproductive potential, females of some species may need to mate several times (Sem'yanov, 1970; Ridley, 1988). By contrast, if a female hymenopteran parasitoid lacks sperm for whatever reason, she can lay viable (male) eggs, so her fecundity and 'fertility' should not be affected by mating. However, as discussed in Chapter 3 (subsection 3.9.6), there is evidence to the contrary.

2.7.4 EFFECTS OF PHYSICAL FACTORS ON FECUNDITY

Temperature

The rate of egg production, and hence the age-specific and lifetime fecundity of predators and parasitoids will vary in relation to temperature (Force and Messenger, 1964; Hämäläinen *et al.*, 1975; van Lenteren *et al.*, 1987; Braman and Yeargan, 1988; Miura, 1990) (Figure 2.33). The influence of temperature on the fecundity schedule of a natural enemy species can be investigated by taking cohorts of standardised females, and expos-

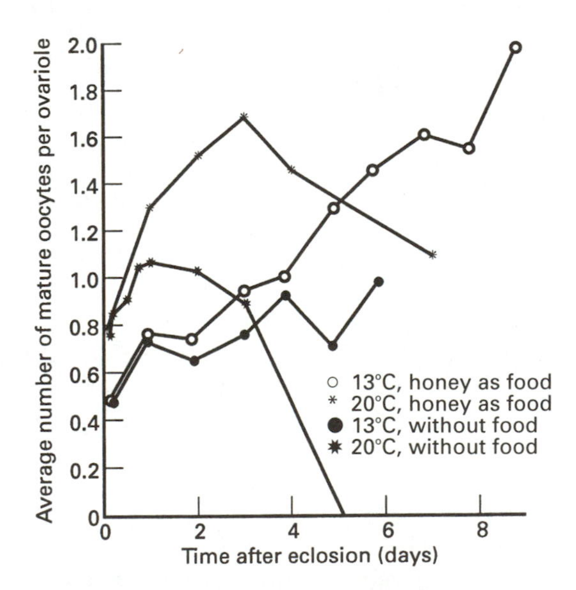

Figure 2.33 The number of mature oocytes per ovariole in the parasitoid *Encarsia formosa* (Aphelinidae) kept for several days after eclosion without hosts, either without food or on a diet of honey, at two different temperatures. (Source: van Lenteren *et al.*, 1987.) Reproduced by permission of Blackwell Scientific Wissenschafts–Verlag GmbH.

ing each of them to one of a range of temperatures for their lifetimes. Females of all the cohorts are exposed to the same conditions of host/prey, food and water availability (hosts and prey need to be replaced daily), humidity and photoperiod etc.. A constant humidity will probably prove to be the most difficult of all these factors to maintain. Temperature may influence the rate of prey consumption (Mills, 1981; Pickup and Thompson, 1990), so temperature-related variation in prey consumption should be looked for.

The effect of temperature upon egg load in a synovigenic insect can be investigated following the protocol, used for *Aphytis* parasitoids, of Rosenheim and Rosen (1992). Parasitoid pupae are isolated, and adults, when they emerge are kept with a supply of food (honey), at each of a range of temperatures for 24 hours. The adults are then dissected and the numbers of mature eggs they contain are counted. The results of Rosenheim and Rosen's (1992) study are shown in Figure 2.34, which also shows the influence of body size upon fecundity (subsection 2.7.3).

It is generally the case that there is an optimum temperature range outside of which the insect either cannot maintain ovigenesis and oviposition or is unable to do so for long (Force and Messenger, 1964; Greenfield and Karandinos, 1976) (Figures. 2.20a and 2.35). Although there is great variation from species to species, the limits to the favourable range for oviposition are often narrower than those for ovigenesis (Bursell, 1964).

Within the optimum range, one effect of higher temperature on the pattern of oviposition is to shift the fecundity schedule, with the ovigenesis/oviposition maximum occurring earlier in life (Siddiqui *et al.*, 1973; Ragusa, 1974; Browning and Oatman, 1981; Miura, 1990).

In the coccinellid *Adalia bipunctata* adult fecundity increases up to 20°C, correlating well with the increase in food consumption rate. However, above 20°C fecundity declines despite a continued increase in consumption.

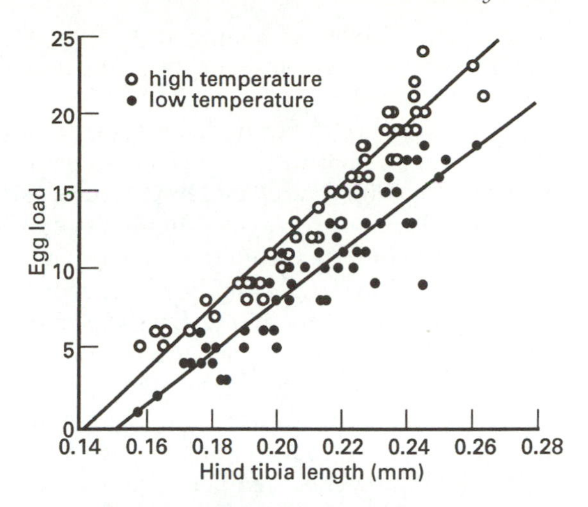

Figure 2.34 Influence on egg load of parasitoid size and the temperature at which females have previously been held from eclosion, in *Aphytis lingnanensis* (Aphelinidae). (Source: Rosenheim and Rosen, 1992.) Reproduced with permission from Blackwell Scientific Publishers Ltd.

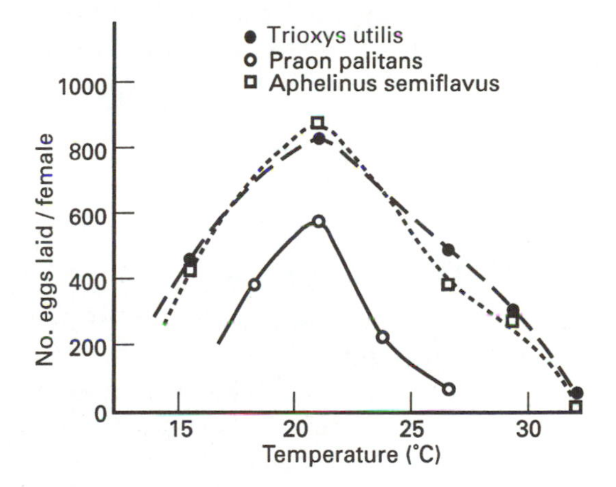

Figure 2.35 Comparison of mean lifetime fecundity of the aphid parasitoids *Praon palitans*, *Trioxys utilis* (Braconidae) and *Aphelinus semiflavus* (Aphelinidae), over a range of constant temperatures. (Source: Force and Messenger, 1964.) Reproduced by permission of The Ecological Society of America.

Higher temperatures may have a deleterious effect through increased costs in respiration, i.e. daily maintenance requirement (Mills, 1981), although Ives (1981a) found no significant influence of temperature on the maintenance requirement in the two *Coccinella* species he studied.

No attempts appear to have been made to describe mathematically the relationship between oviposition rate and temperature, as has been done with development. Several workers have found that alternating temperatures increase insect fecundity (Messenger, 1964a; Barfield *et al.*, 1977; Ernsting and Huyer, 1984), and thus it may be invalid to estimate oviposition rates in the field directly from constant temperature data. A similar approach to that used for estimating development based on cyclical temperature regimes might give more meaningful results but has not yet been attempted (subsection 2.9.3 and the computer program in Appendix 2A).

Some adult predators may be able to maintain maximal levels of oögenesis through thermoregulation achieved either by thermal preference behaviour (including basking behaviour), by employing physiological mechanisms and by employing physical adaptations such as melanisation of the integument (Dreisig, 1981; Brakefield, 1985; Miller, 1987; Stewart and Dixon, 1989).

Light Intensity and Photoperiod

The intensity, duration and quality of light has an important influence on the biology and behaviour of most insects. High light intensity seems to increase the general activity of both predators and parasitoids. For example, adults of the coccinellid beetle *Cryptolaemus montrouzieri* spend a greater proportion of their time walking and make more attempts to fly in bright light than under dim light conditions (Heidari, 1989). Light quality and intensity may also influence the close range perception of hosts. Care must therefore be taken in fecundity experiments to provide sufficient light for normal activity, but bear in mind that in the field, bright light conditions are normally associated with increased radiant heat. Laboratory experiments that involve varying light intensity alone will require the radiant heat component of light to be removed, using suitable glass and water filters. Even cold fibre optic lamps used in microscopy can raise the body temperature of dark-coloured insects by at least 2°C above ambient. A thermocouple (see Unwin and Corbet, 1991) placed into the body of a dead insect will enable the heat absorbed from a light source to be measured and suitable infra-red filters to be devised (Heidari and Copland, 1993).

Most natural enemy species will show strong diurnal peaks of behavioural activity, foraging being mainly confined to the photophase, as in many parasitoids and some carabid beetles (Ekbom, 1982; Ruberson *et al.*, 1988; Luff, 1978) (Figure 2.36).

Weseloh (1986) showed that the egg load of females of the egg parasitoid *Ooencyrtus kuvanae* (Encyrtidae) kept under long-day conditions increases more rapidly than that of females kept under short-day conditions; this is reflected in differences in progeny production.

Diurnal variation in activity obviously needs to be taken into account in fecundity

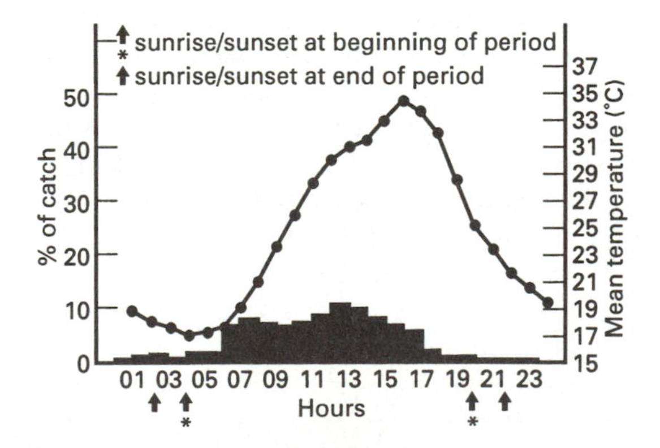

Figure 2.36 Diurnal flight activity patterns in *Encarsia formosa* (Aphelinidae) in the greenhouse. Data are for May – June. Percentage of the mean daily catch (by air suction trap) of wasps for each hour (histograms) and mean temperature (curve). (Source: Ekbom, 1982.) Reproduced by permission of Elsevier Science Publishers B.V.

experiments. Because of its effects on food consumption, photoperiod length may also influence larval growth and development rates in larval predators (which in turn will influence adult fecundity, subsection 2.7.3) and the rate of ovigenesis. A continuous light regime may help to avoid these complications, but it may result in a higher fecundity than would be achieved in the field (Lum and Flaherty, 1973).

Humidity

Decreasing humidity may increase fecundity in predators, through an increase in prey consumption (Heidari, 1989). In fecundity experiments care must therefore be taken to control humidity, so that it is around the field average.

2.8 ADULT LONGEVITY

2.8.1 INTRODUCTION

The life span of an individual insect can be divided into two phases: the development period from hatching of the egg until adult eclosion (section 2.9) and the period of adult life usually referred to as adult longevity. An obligatory or facultative period of dormancy may intervene during the lifetime of an individual to extend either development or adult longevity for a variable period of time. Dormancy is not dealt with in this chapter; for a comprehensive account, see Tauber *et al.* (1986).

Adult longevity may be studied from a variety of standpoints. For evolutionary biologists, it is a component of individual fitness (Waage and Ng, 1984; Hardy *et al.*, 1992), the assumption being that: (a) the longer a male can live, the more females he can inseminate and therefore the more eggs he can fertilise; (b) the longer a female can live the more eggs she will lay. In both cases, the proviso 'other things being equal' applies. Adult longevity is also studied from the point of view of population dynamics, because of its relationship to female fecundity, the prey death rate and the predator rate of increase. Most studies on natural enemies measure adult longevity in the laboratory; there is a dearth of studies that measure it under natural conditions. Individual marking techniques that can be used to measure adult survival in the field are discussed in Chapter 4 (subsections 4.2.10, 4.2.11).

Longevity, like fecundity, is a variable species characteristic, influenced by a range of physical and biotic factors. The commonest experiments into the effects of these factors involve taking a cohort of standardised females (subsection 2.7.2) and exposing them to some chosen set of constant environmental conditions from eclosion until death. The resulting data are then plotted, usually as a cohort survival curve. This also applies to data obtained by exposing different cohorts to one of a range of environmental conditions. Mean length of adult life can be plotted against variables such as body size, temperature, humidity, host or prey density, sugar concentration (in diet), and pesticide concentration, but this method of expressing longevity data has major drawbacks (see below).

2.8.2 SURVIVAL ANALYSIS

Frequently, in the literature, longevity data are presented as the mean length of adult life plus or minus its 95% confidence limit/standard deviation/standard error. However, when statistical comparisons between treatments are made, authors overlook the fact that individual longevity data are rarely normally distributed. For statistical comparisons between treatments to be biologically meaningful, the data are best presented in other ways such as cohort survival curves, which show the fraction of each cohort surviving at a particular moment in time (Figure 2.37). Such curves fall into 3 categories: **Type I**, in which the risk of death declines with age;

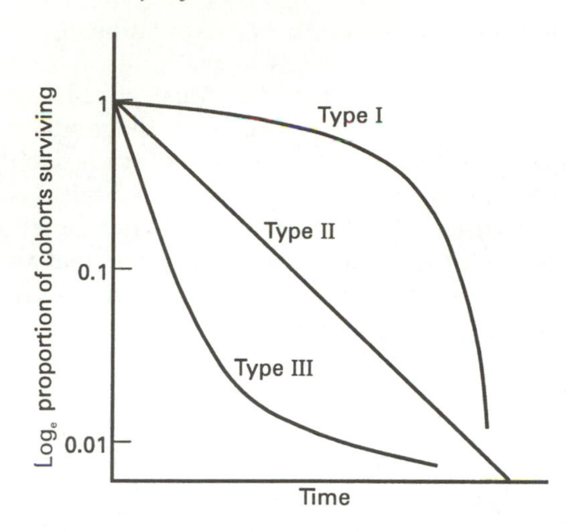

Figure 2.37 The three main types of survivorship curve: Type I – mortality concentrated in the oldest age classes; Type II – constant risk of death; Type III – mortality concentrated in in the youngest age classes. Note the logarithmic scale.

Type II, in which there is a constant risk of death; and **Type III**, in which the risk of death increases with age.

Survival data have been compared by plotting survival curves and calculating the time to 50% mortality (LT_{50}) for each treatment and testing the statistical significance of differences in this quantity. A major difficulty with this approach is that, at a particular point on the time axis, one or both of the curves might comprise few observations. Also, the 50% mortality level is subjective. As pointed out by Crawley (1993), generalised linear interactive modelling techniques, available in the form of the GLIM statistical package, offer the best means of analysing survival data. The data can be analysed statistically in terms of survivorship (proportion of individuals from the cohort still alive at a particular point in time), the age at death, and the instantaneous risk of death (also termed the 'age-specific instantaneous death rate' by biologists or 'hazard function' by statisticians). GLIM can be used to determine which of a variety of available models (exponential, log normal, Weibull) best describe the observed data. Having decided upon the most appropriate model, the effects of different experimental treatments can then be compared. For details of the procedure, see Crawley (1993).

The Weibull model has been used by Tingle and Copland (1989) and Hardy *et al.* (1992) to analyse survival data for parasitoids. The Weibull frequency distribution was originally considered as a model of human survivorship (Gehan and Siddiqui, 1973) and has commonly been used in engineering as a 'time to failure' model. The Weibull distribution is extremely flexible, possessing either positive or negative skewness, so allowing all three types of survival curve (I, II, III) to be analysed (Cox and Oakes, 1984). The advantage of using the Weibull model to describe survival curves is that it summarises the information contained in a curve as a rate parameter and a shape parameter. The fraction (F) of the cohort surviving at time t is given by:

$$F = 1 - \exp\left(-\left\{(t/b)^c\right\}\right) \qquad (2.4)$$

GLIM estimates the most appropriate value of the shape parameter c and allows the rate (or scale) parameter b to be a linear combination of explanatory variables. Hardy *et al.* (1992) examined survivorship females of the bethylid *Goniozus nephantidis*, and found that for each of two treatments a curve based on a Weibull distribution showed some systematic deviation from the observed curve. Having noted a relationship between longevity and body size in females (see below), they allowed the logarithm of the distribution's rate parameter to be a linear function of female size. Incorporation of female size significantly improved the fit, and therefore the explanatory power, of the model in the two treatments

If you are dealing with a particularly long-lived species, you do not have to wait until all the individuals have died in order to terminate an experiment. The experiment can be terminated earlier, and certain statistical analyses can be used to take account of individuals that die at an unknown time, i.e. insects that are 'censored', statistically speaking (Crawley, 1993). Such analyses can also be used to take account of individuals that are accidentally lost or killed during the experiment.

2.8.3 EFFECT OF BIOTIC FACTORS ON ADULT LONGEVITY

Host and Prey Density

Non-predaceous Females

In general, host/prey density appears to have little or no effect upon adult survival in these insects; at least this is the case if average longevity is used as the measure of survival (Liu, 1985b; Mackauer, 1983). Visual inspection of survival curves suggests the same, although a survival analysis of the type discussed above needs to be carried out on such data. With the exception of *Leptoplina heterotoma* there is no convincing evidence in parasitoid wasps of an expected trade-off between reproduction and life expectancy (section 1.12).

Predaceous Females

Most information on predaceous females relates to cases where the longevity of females deprived of hosts or prey (deprived for either the whole or part of an experimental period) is compared with that of females that were not deprived (Jervis and Kidd, 1986; van Lenteren *et al.*, 1987 have information on parasitoids). As one might expect, longevity was found to be shortest in the deprived females.

There are few published studies in which the longevity of predaceous females has been related to either availability of prey/hosts or consumption rate. Longevity is positively related to prey consumption rate in ovipositing *Coccinella undecimpunctata* over a wide range of prey densities (Ibrahim, 1955). Bai and Smith (1993) found a positive correlation between reproduction and survival in *Trichogramma minutum*, an occasional host-feeder, over a wide range of densities. In non-ovipositing *Thanasimus dubius* (Cleridae) longevity becomes a direct function of prey density only at low levels of prey availability (Turnbow *et al.*, 1978). A similar relationship was found for the ovipositing females of the host-feeding parasitoid *Dicondylus indianus* (Saharagard *et al.*, 1991). The probable reason for the lack of a relationship at higher levels of prey/host availability in these two cases is that at these levels maintenance requirements are fully satisfied.

Ernsting and Isaaks (1988) found, interestingly, that life expectancy *decreases* with increasing prey availability. They fed females of the carabid beetle, *Notiophilus biguttatus*, either two springtail prey per day or an excess of prey over their lifetimes. The beetles given two prey per day laid a smaller total of eggs than those beetles given an excess of prey per day, but they lived significantly longer. This is another apparent example of a trade-off between reproduction and life expectancy.

Prey and Host Quality

In predatory Coccinellidae adult longevity may be significantly affected by the prey species fed upon by the adult (Hodek, 1973). This also applies to host-feeding parasitoids (Wilbert and Lauenstein, 1974).

Non-host and Non-prey Foods

Many studies have shown that, in the absence of hosts or prey, parasitoids and predators given carbohydrate-rich foods, e.g. diluted honey solutions, live significantly longer than insects that are either starved or given only water (Hagen, 1986; Jervis and Kidd, 1986; van Lenteren *et al.*, 1987). A simple experiment involves providing predators or parasitoids with one of a range of different diets, e.g. different sugars or combinations of sugars in solution, or even different nectars or honeydews, and comparing the effects of these on survival. Whereas the monosaccharides glucose and fructose promote longevity in *Nasonia vitripennis*, only polysaccharides that can be broken down by the enzyme α-glucosidase are of value, suggesting that α-glucosidase is the only carbohydrate-hydrolysing enzyme in *N. vitripennis* (Copland, unpublished). For details of a more complex protocol for determining the effects of biochemical components of non-host foods on longevity, see Finch and Coaker (1969).

Body Size

A positive correlation between body size and longevity has been shown for the adults (in some cases males as well as females) of several parasitoid species (Table 2.2). Notable exceptions are *Encarsia formosa*, *Trichogramma maidis* and *Edovum puttleri* in which there is not a significant correlation (van Lenteren *et al*., 1987; Bigler *et al*., 1987; Corrigan and Lashomb, 1990), and host-deprived *Goniozus nephantidis*, in which larger females live longer than smaller ones if hosts are provided, but smaller females live slightly longer than larger ones if hosts are not available (Hardy *et al*., 1992).

Blackburn (1991a) showed through a comparative analysis (subsection 1.2.3) of 474 hymenopteran parasitoid species that across the species within a taxon, there is no correlation between body size and longevity.

Mating

As discussed in Chapter 3 (subsection 3.9.5) frequent mating may shorten life span in both females and males.

2.8.4 EFFECT OF PHYSICAL FACTORS ON ADULT LONGEVITY

Temperature

There will be an optimum range of temperatures outside of which survival is severely

Table 2.2 Parasitoid species in which there is a positive correlation between body size and adult longevity; F: female, M: male.

Species	*Reference*
Eucelatoria sp. nr. *armigera* (F) (Tachinidae)	Mani and Nagarkatti (1983)
Venturia canescens (F) (Ichneumonidae)	Harvey *et al*. (1994)
Pimpla turionellae (F) (Ichneumonidae)	Sandlan (1979b)
Pteromalus puparum (F) (Pteromalidae)	Takagi (1985)
Lariophagus distinguendus (F, M) (Pteromalidae)	Bellows (1985), van den Assem *et al*. (1989)
Pediobius foveolatus (F, M) (Eulophidae)	Hooker *et al*. (1987)
Diglyphus begini (F, M) (Eulophidae)	Heinz and Parella (1990)
Trichogramma evanescens (F, M) (Trichogrammatidae)	Waage and Ng (1984)
Trichogramma platneri (F) (Trichogrammatidae)	Hohmann *et al*. (1989)
Goniozus nephantidis (F) (Bethylidae)	Hardy *et al*. (1992)

reduced (Jackson, 1986; Krishnamoorthy, 1989).

In general, for both males and females, longevity decreases with increasing temperature within the optimum range (Abdelrahman, 1974; Hofsvang and Hågvar, 1975a,b; Sahad, 1982, 1984; Nealis and Fraser, 1988; (Figure 2.38), although for some species no more than a trend may be apparent (Barfield *et al.*, 1977a; Cave and Gaylor, 1989; Miura, 1990).

Most experiments designed to demonstrate the effect of temperature on adult longevity involve exposing insects to constant temperatures, which ignores the fact that in nature temperatures will fluctuate during each day, the lowest temperatures occurring at night. Ideally, longevity ought to be studied at temperature extremes that are part of a cyclical regime, but such an approach has rarely been adopted. Ernsting and Isaaks (1988) measured the survival of the carabid *Notiophilus biguttatus*, given excess prey, at a constant 10°C regime compared with a daily fluctuating (20°C day/10°C night) regime. The lower survival of beetles held under the fluctuating regime could simply be explained by the higher average daily temperature at that regime. Minkenberg (1989) incorporated a more realistic fluctuating temperature regime in his experimental design. He exposed the eulophid *Diglyphus isaea* to each of three constant temperatures (15, 20, 25°C) and to a fluctuating regime that involved the temperature increasing linearly from 0100 to 0300 h, decreasing from 1500 to 1700 h, and fixed at 22°C from 0300 to 1500 h and 18°C from 1700 to 0100 h. Survival of wasps held under the fluctuating regime (average daily temperature 20.3°C) was much lower than at the constant 20°C regime.

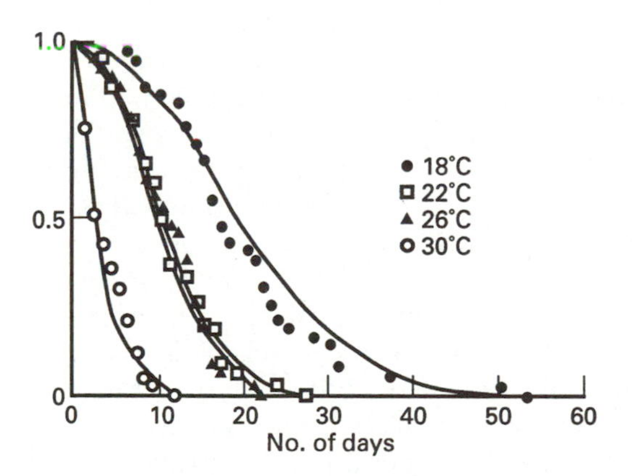

Figure 2.38 Longevity of *Anagyrus pseudococci* (Encyrtidae) at four different constant temperature regimes. (Source: Tingle and Copland, 1989.) Reproduced by permission of Lavoisier Abonnements.

The influence, on longevity, of an extreme (hot or cold) temperature forming part of a 24 hour cyclical regime and alternating with a tolerable temperature for which constant data are available, can be investigated using the QBASIC computer programme given in Appendix 2A.

Humidity

It is clear from many experimental studies that natural enemy adults have particular humidity requirements for survival (Kfir, 1981; Wysoki *et al.*, 1988; Herard *et al.*, 1988).

Although it would appear to be quite easy to carry out an experiment designed to measure survival at different humidities, there is the problem of maintaining the insects for a sufficiently long period for statistical comparisons to be made. Insects deprived of food are likely to die quite quickly. However, if they are provided with honey or sucrose solutions (see above), it may be difficult to separate the effects upon longevity of the water content of the air and that of the food. Similarly, it may be difficult to set up an experiment that incorporates some degree of biological realism in the form of a plant surface, since the latter will be actively transpiring.

Photoperiod

Little is known about the influence of photoperiod on longevity. Given that in predaceous females longevity may be affected by prey consumption rate (see below), and that some insects are active (i.e. consume prey) only during certain periods of the day, one might expect longevity to be influenced by photoperiod. Note that some entomophagous insects are nocturnal, e.g. certain Ichneumonidae (Gauld and Huddleston, 1976); Vespidae, Pompilidae and Rhopalosomatidae (Gauld and Bolton, 1988).

In the parasitoid *Ooencyrtus kuvanae* the photoperiod experienced upon adult eclosion influenced both longevity and the rate of progeny production. Short-day conditions resulted in females producing fewer progeny but living longer. Switching photoperiods after twelve days failed to alter this trait once it had been established (Weseloh, 1986).

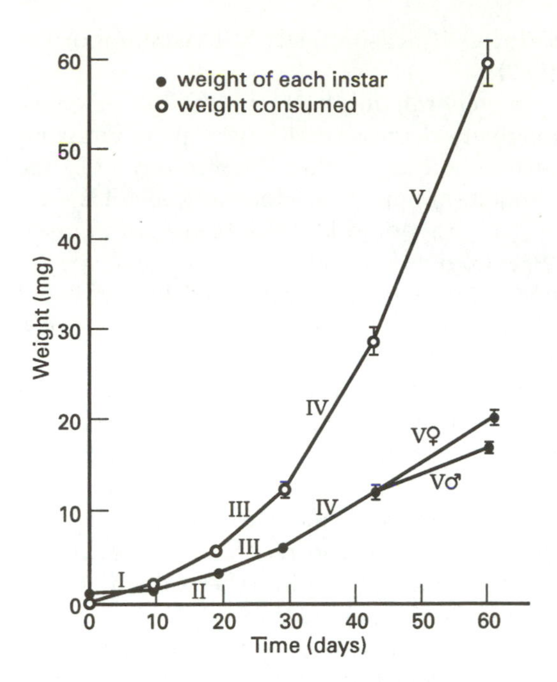

Figure 2.39 Development by the bug *Blepharidopterus angulatus* (Miridae) given excess food, showing the length of time spent in each larval instar, the body weight at the start of each instar, and the cumulative wet weight of lime aphids consumed up to the start of each instar. Roman numerals denote instars. (Source: Glen, 1973.) Reproduced by permission of Kluwer Academic Publishers.

2.9 GROWTH AND DEVELOPMENT OF IMMATURES

2.9.1 INTRODUCTION

Development refers to the morphological and anatomical changes shown by each individual insect from the time the egg is laid to the time the adult ecloses. **Growth** refers to the increase in biomass of the insect during the period between hatching from the egg and the end of the larval phase of the life cycle, as shown in Figure 2.39. The larval phase in predators comprises long periods of feeding and brief periods of moulting. Typically, biomass increases steadily throughout each instar. At the time of the moult, biomass falls slightly due to the loss of the exuvium and the loss of some water which is not immediately replaced as the insect is not feeding (Chapman, 1982). In some aquatic insects there is no decrease in biomass at the moult; instead there is an increase due to absorption of water through either the cuticle or the gut; in *Notonecta glauca* this increase is very large (Wigglesworth, 1972).

Field, as opposed to laboratory, measurements of growth and development in predators and parasitoids have been made on few species (Griffiths, 1980, on ant-lions; Banks and Thompson, 1987b, on damselflies).

For predators the protocols for measuring growth and development in relation to certain physical and biotic factors are relatively straightforward. For example, to study the influence of prey availability, take a series of cohorts of newly hatched larvae and present each insect with one of a range of chosen prey densities (prey of a fixed size), at a constant temperature, humidity, photoperiod, either for the duration of the insects' lives or for the duration of one or a few instars only. On each day of larval life, the prey are entirely replaced. Larval development is measured simply in terms of the period of time between moults. Larval growth can be measured as the dry or fresh weight gain, including exuvia weight (g/g/d) or body size increase (measured in terms of head width) between instars, and expressed as the ratio of weight gain or body size gain to instar duration. Some workers (Paradise and Stamp, 1990, 1991) have expressed growth rate differently, as the **relative growth rate** (RGR): (fresh weight gained/instar duration) × the average fresh weight of the predator during the instar.

For parasitoids, the protocols for measuring the influence of physical and biotic factors on growth and development may be rather more complicated than for predators. Endoparasitoids are a particular problem, since the sizes and weights of larvae cannot easily be measured and the larvae often cannot easily be assessed as to their stage of development (Mackauer, 1986; Sequeira and Mackauer, 1992b; Harvey *et al.*, 1994).

A necessary prerequisite for studying many aspects of larval development, particularly instar-related aspects of biology, in predators and parasitoids is the ability to distinguish between the different instars. In some cases, it is relatively easy to tell the instars apart, using features such as mouthpart structure the degree of wing development, the number and position of prominent setae, spines and other cuticular structures, structure of the tracheal system and associated spiracles, and body colour patterns. However, in other predators, obvious distinguishing features may be lacking. Morphometric techniques may therefore need to be used. Thompson (1975, 1978), for example, decided upon the instar of the damselflies larvae he studied (*Ischnura elegans*), using both a frequency distribution histogram of head widths of randomly field-collected larvae and a regression of modal head width against probable instar number. Even better discrimination between instars was obtained by plotting head width against body length (Thompson, 1978).

2.9.2 EFFECTS OF BIOTIC FACTORS ON GROWTH AND DEVELOPMENT

Food Consumption

Introduction

Predatory larvae need to consume several prey during development, and each successive instar will show a maximum rate of growth and development at different levels of prey availability. Generally with increasing prey density, at least across the low and medium ranges, larval predators consume more prey, develop faster, gain more weight and so attain a higher final size, compared with predators given low densities (Dixon, 1959; Lawton *et al.*, 1980; Scott and Barlow, 1984; Pickup and Thompson, 1990; Zheng *et al.*, 1993a,b). Where development rate increases non-linearly with prey consumption rate (see below), development rate stops increasing above a certain prey density while growth continues. Growth and development also vary in relation to prey quality.

Food consumption by insects is a subject in its own right, and the associated literature is very large (Waldbauer, 1968; Beddington *et al.*, 1976b; Kogan and Parra, 1981; Scriber and Slansky, 1981; Slanksy and Scriber, 1982, 1985; Slansky and Rodriguez, 1987; Karowe and Martin, 1989; Farrar *et al.*, 1989). The approach we are recommending here is that of Beddington *et al.* (1976b), as it provides one of the most useful bases for predicting predator–prey population dynamics (Chapter 5). The various problems inherent in measurements of food consumption and utilisation by insects and other arthropods are discussed in Waldbauer (1968), Lawton (1970), Ferran *et al.* (1981) and Pollard (1988).

The basic protocol for studying the effects of prey availability and prey consumption on growth and development in predators has already been outlined. Other measurements can also be taken in order that various **nutritional indices** can be calculated:

1. the biomass of the prey consumed and the biomass of prey captured but not consumed (biomass in each case is best measured in terms of dry weight, since prey remains are likely to lose water before retrieval), the predator's **efficiency of conversion of ingested food into body substance** (ECI) can be calculated as follows:

$$\text{conversion efficiency} = \frac{M}{C-D} \times 100 \quad (2.5)$$

where M is the increase in biomass of the predator, C is the biomass of the prey that are captured, and D is the biomass of the prey captured but not consumed ($C - D$ is therefore the biomass of prey actually ingested). According to Cohen (1984, 1989), predaceous insects with piercing, suctorial mouthparts (e.g. Heteroptera), because they obtain a larger proportion of highly digestible materials from their prey (a process assisted by pre-oral digestion), ought to have higher ECI values than predators with chewing mouthparts. So far as we know, no broad comparative studies have been carried out to test this hypothesis.

The ECI is a measure of *gross* growth efficiency, since biomass losses in the form of faeces and excreta are not considered.

2. C, D and the biomass which appears as faeces (F) and the products of nitrogenous excretion (U). The predator's **utilisation efficiency** – the efficiency with which the prey biomass *captured* is converted into predator biomass – can then be calculated:

$$\text{utilisation efficiency} = \frac{C-D-F-U}{C} \times 100 \quad (2.6)$$

3. C, D, F and U as in 1. and 2. The predator's **assimilation efficiency** – the efficiency with which the prey biomass *consumed* is converted into predator biomass – can then be calculated:

$$\text{assimilation efficiency} = \frac{C-D-F-U}{C-D} \times 100 \quad (2.7)$$

4. C, D, and F as in 1. and 2. The predator's **digestive efficiency** (also termed 'approximate digestibility') – the efficiency with which the prey biomass *ingested* is digested and absorbed – can be calculated:

$$\text{digestive efficiency} = \frac{C-D-F}{C-D} \times 100 \quad (2.8)$$

Other nutritional indices used in studies of food consumption by insects are discussed by Waldbauer (1968) and Slansky and Scriber (1982, 1985).

Growth Rate

At least some of the food a larva consumes needs to be allocated to maintenance metabolism. Because of this, growth will stop if consumption falls below a certain threshold (this threshold will become higher as the insect grows and its maintenance requirements increase). The energy allocated to growth can be assumed to be a linear function of food intake (Beddington *et al.*, 1976b):

$$G = \delta(I - B) \quad (2.9)$$

where G is the the growth rate (biomass accumulated per unit time, e.g. fresh weight gain [including exuvia weight] divided by the number of days spent in the instar) of each juvenile stage, I is the rate of ingestion of food (biomass of prey consumed per unit time, in comparable units to G) (equation (2.5)), and δ and B (the threshold ingestion rate, analogous with parameter c in equation 2.1) are constants. Mills (1981) gives an alternative model.

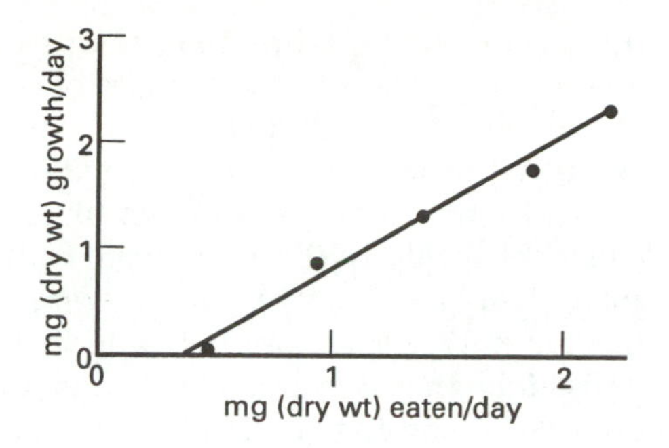

Figure 2.40 Growth rate as a function of ingestion rate in final instar of *Notonecta undulata*. (Source: Beddington *et al.*, 1976b, who used data from Toth and Chew, 1972.)

Figure 2.40 shows the relationship between growth rate and consumption rate in larval *Notonecta*; the relationship conforms to that predicted by equation (2.9). As can be seen from the intercept of the line with the abscissa, the predator needs to consume a minimum amount of food for any growth to occur.

Should the increase in respiratory rate be *non-linear*, then growth rate will be non-linear and conform to the following model (Beddington *et al.*, 1976b):

$$G = \delta\left(\log_e I - B\right) \qquad (2.10)$$

Development Rate

If W_i is the initial weight (biomass) of an instar (teneral weight), W_f is the final weight achieved, and W is the total weight gain, then $W = W_f - W_i$. The ratio W/G will define the duration, d, of the instar, and development rate, $1/d$, is given by the following linear model (Beddington *et al.*, 1976b):

$$\frac{1}{d} = \frac{\delta}{W}(I - B) \qquad (2.11)$$

If it is assumed, for simplicity's sake, that W remains a constant, then (Beddington *et al.*, 1976b):

$$\frac{1}{d} = \alpha\,(I - B) \qquad (2.12)$$

where α and B are constants. Equation 2.12 still predicts a simple, linear relationship between development rate and consumption rate.

As pointed out by Beddington *et al.* (1976b), equation (2.11) ignores the fact that in some predators larvae may, under conditions of food scarcity, moult to the next instar at significantly lower body weights than when food is abundant. W_i, W_f and W are therefore functions of consumption rate and thus of prey availability – weight gain in each instar cannot be assumed to be constant. Figure 2.41 shows how, in the damselfly *Ischnura elegans*, larvae fed at low prey densities moulted to smaller individuals, i.e. they moulted earlier than better fed larvae, after having gained less weight. Mills (1981) demonstrated, through a regression analysis of the dependence of W on consumption rate and on teneral weight in *Adalia bipunctata*, a significant dependence in both cases, with consumption rate explaining 47–75% of the variance.

Thus, the relationship in some predators is more complex than that described by

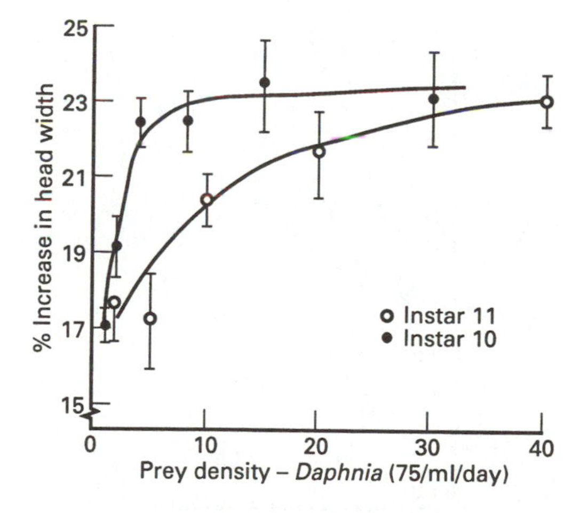

Figure 2.41 The effect of prey density on the percentage increase in head width at the moult in *Ischnura elegans*. (Source: Lawton *et al.*, 1980.) Reproduced by permission of Blackwell Scientific Publications Ltd.

equation 2.12. Lawton *et al.* (1980) provide the following non-linear model:

$$\frac{1}{d} = \alpha\left(\log_e N_a - B\right) \qquad (2.13)$$

where N_a is the number of hosts fed upon. An alternative non-linear model is provided by Mills (1981). Both models describe a negatively accelerating curve for the relationship between development rate and consumption rate, and curves of this type were obtained in the laboratory for both *Ischnura elegans* (Figure 2.42) and *Adalia bipunctata* (Figure 2.43).

Lawton *et al.* (1980) gave, as well as a dependence of *W* on consumption rate, three other reasons to account for non-linearity in the case of *Ischnura*:

1. variation in *k* (equation 2.2) with prey availability. In *Ischnura* this declined with prey availability (Figure 2.44), the predators wasting proportionately more of each of the prey they kill at higher densities (adaptive behaviour in many predators, section 1.10; Cook and Cockrell, 1978; Giller, 1980; Sih, 1980; Kruse, 1983; Bailey, 1986; Dudgeon, 1990). However, as Lawton *et al.* (1980) point out, a decline in utilisation efficiency in *Ischnura* cannot be the sole reason for the non-linear dependence of development rate upon prey consumption rate. If it is, daily growth rates plotted against prey biomass assimilated ($C - [D + F]$) ought to be linear (equation (2.9), which they are not (Figure 2.45).
2. a decrease in assimilation efficiency with increasing consumption rate. Lawton (1970) had suggested that this may occur with over-feeding at high levels of prey availability, causing defaecation to take

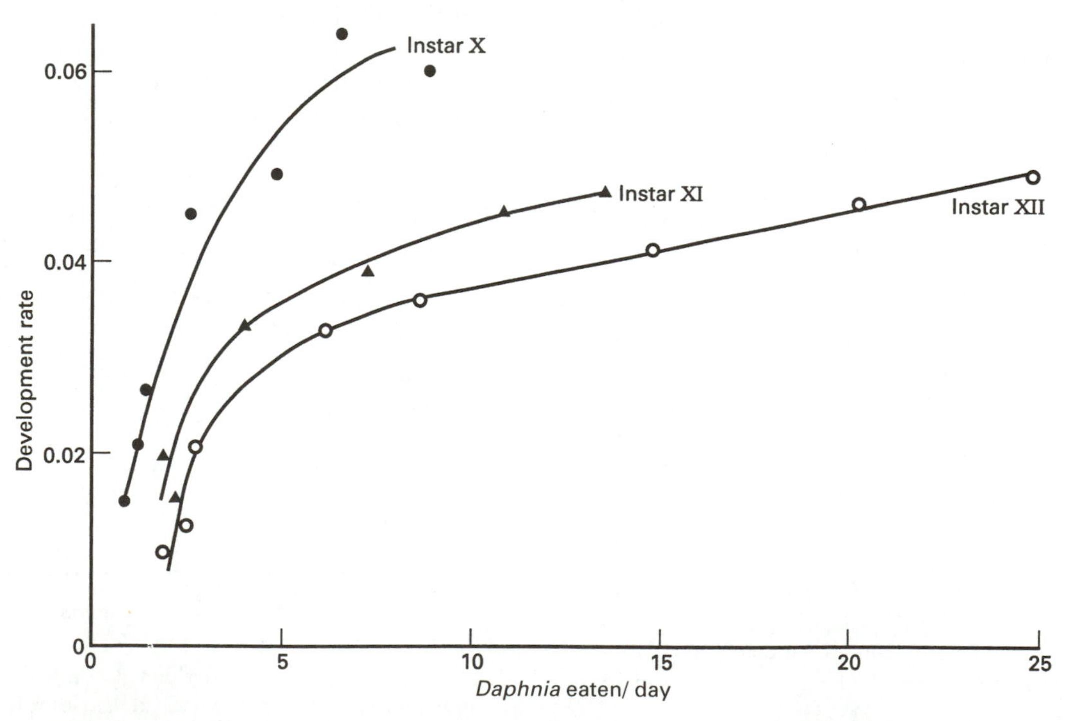

Figure 2.42 Development rates in *Ischnura elegans* (Odonata: Zygoptera) larvae as a function of the number of prey killed per day. Only larvae that succesfully completed their development in each stage were included in the calculations. (Source: Lawton *et al.*, 1980.) Reproduced by permission of Blackwell Scientific Publications Ltd.

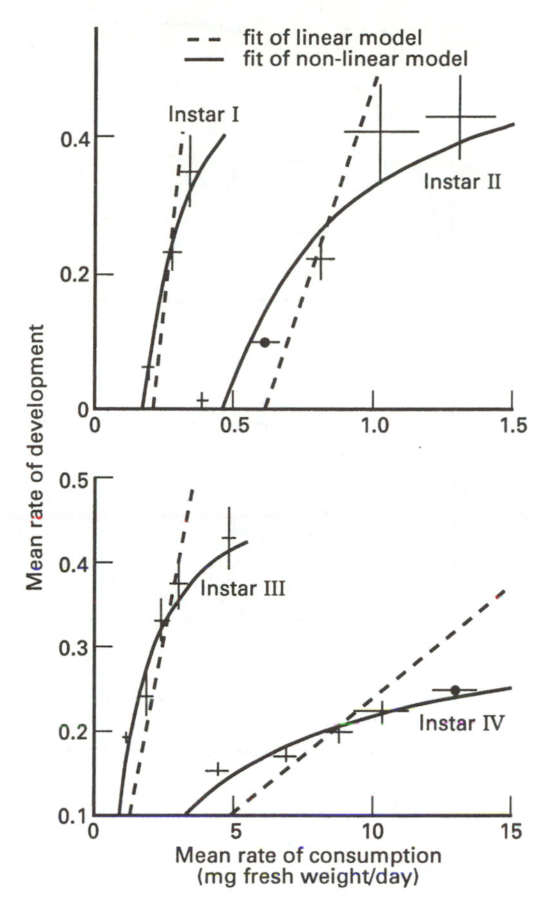

Figure 2.43 The relationship between mean development rate and rate of consumption for the four larval instars of *Adalia bipunctata* (Coccinellidae). Indicated is the fit of linear and non-linear models of development (see text). (Source: Mills, 1981.) Reproduced by permission of Blackwell Scientific Publications Ltd.

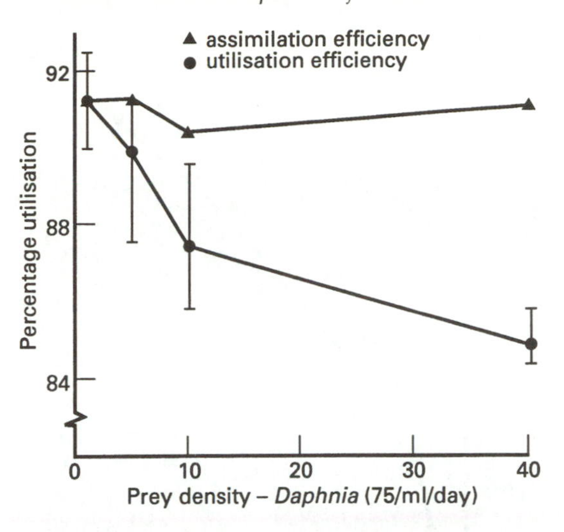

Figure 2.44 The relationships between assimilation and utilisation efficiencies and prey density in the eleventh instar of *Ischnura elegans* (Odonata: Zygoptera). Utilisation efficiency clearly declines with with increased prey density. (Source: Lawton *et al.*, 1980.) Reproduced by permission of Blackwell Scientific Publications Ltd.

place before digestion is complete. Lawton *et al.* (1980) investigated whether assimilation efficiency varied with decreasing prey availability. Since assimilation efficiency in *Ischnura* does not alter significantly with prey availability (Figure 2.44), it could not account for the non-linear dependence of development rate on consumption rate.

3. a non-linear increase in respiratory rates with increasing consumption rate (equation 2.10). Lawton *et al.* (1980) concluded that this effect, together with the variation in k and W, accounted for the observed relationship in Figure 2.42. Circumstantial evidence to support the conclusion regarding change in respiratory rates comes from Lawton *et al.*'s behavioural observations; larvae that have been held at high prey densities frequently engage in more waving of the gills than other larvae, suggesting that they are under oxygen stress. Respirometric methods would need to be used to confirm whether respiratory rates alter.

To obtain the relationship between development rate and prey availability, both equation (2.12) and equation (2.2) can be incorporated into the simple functional response model (Holling's disc equation, subsection 5.3.7) (Beddington *et al.*, 1976b):

$$\frac{1}{d} = \alpha\left(\frac{ka'NT}{1 + a'T_hN} - B\right) \qquad (2.14)$$

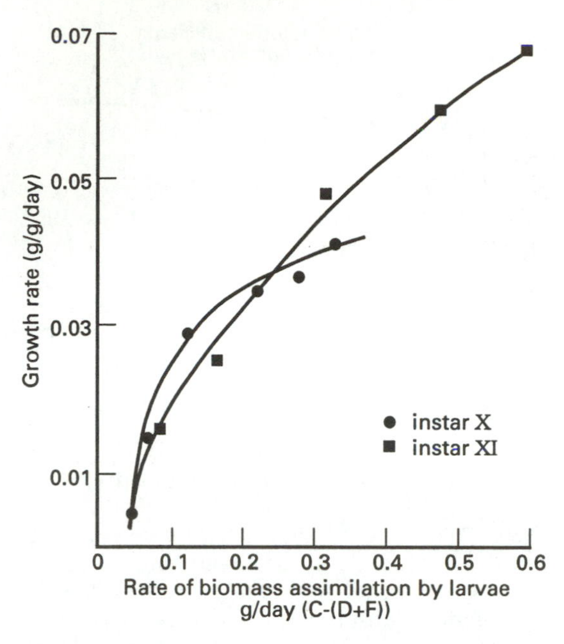

Figure 2.45 The effect of daily rate of biomass assimilation on growth rate for instars 10 and 11 of *Ischnura elegans* (Odonata: Zygoptera). Growth rate is measured as g/g/d increase in weight and is calculated by dividing weight gained during the instar by instar duration. These figures were corrected for the initial weight of the larvae. Wet weights were used for initial and final weights. Only larvae that successfuly completed their development in each instar were used in the calculations. (Source: Lawton *et al.*, 1980.) Reproduced by permission of Blackwell Scientific Publications Ltd.

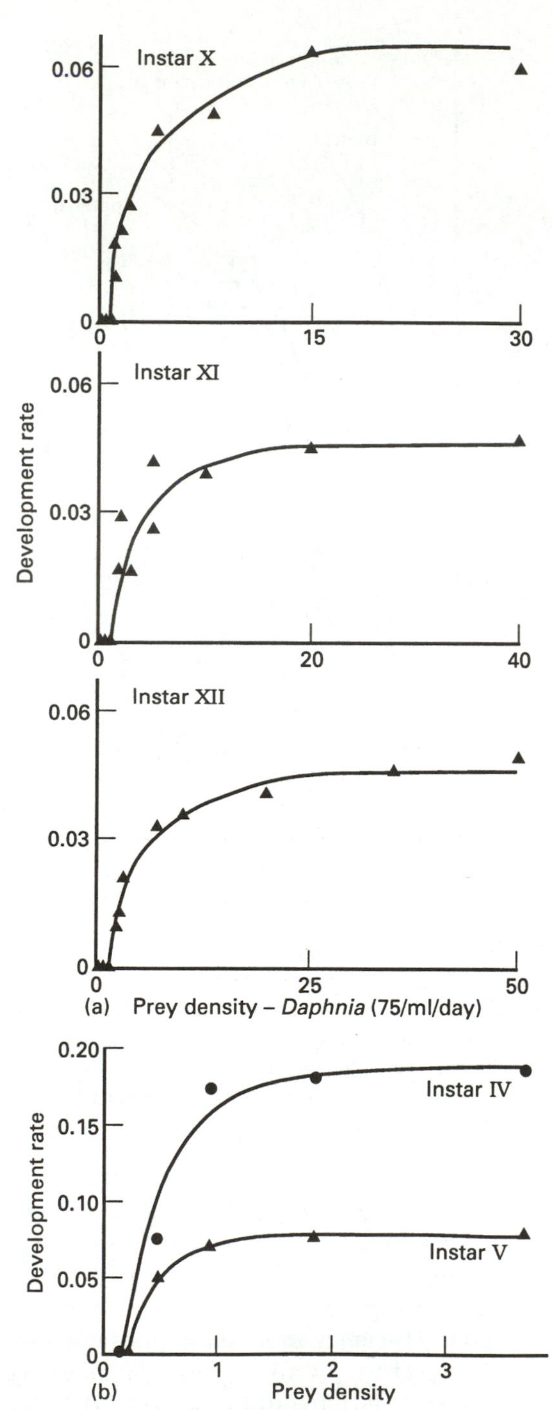

Figure 2.46 Development rates as a function of prey density in different instars of: (a) *Ischnura elegans* (Odonata: Zygoptera) (Source: Lawton *et al.*, 1980); (b) *Notonecta undulata*. (Source: Beddington *et al.*, 1976b, who used data from Toth and Chew, 1972.) (a) Reproduced by permission of Blackwell Scientific Publications Ltd.

Equation 2.14 describes a negatively accelerating curve, like a Type 2 functional response (section 1.10). As pointed out by Beddington *et al.* (1976b), the curve is unlikely to go through the origin. Unless the weight at which a species is able to moult to the next instar is very flexible, the effect of B will be to displace the curve along the prey axis. Put another way, there will be a threshold prey density (and therefore consumption rate) below which growth and development cannot take place. Examples of this are shown in Figure 2.46. In those species that consume proportionately less of each prey item when encounter rates, i.e. levels of prey availability are high (k declines) the curve will be somewhat different in shape: flatter-topped, with an earlier 'turn over' point (Beddington *et al.*, 1976b).

Variation in Growth and Development Between and Within Instars

Figure 2.39 shows both the cumulative increase in prey biomass consumed and the increase in weight of nymphs of the bug *Blepharidopterus angulatus* as they develop. Later instars account for most of the total consumption and growth that occurs. In the green lacewing *Chrysoperla carnea* the third (final) instar accounts for 80.5–82.8% of the total consumption and 80.45–85.6% of the total growth that occurs (data from Zheng *et al.*, 1993a).

Figure 2.46 shows that the development rate vs. prey availability curves differ between instars. As pointed out by Beddington *et al.* (1976b), this is to be expected from the instar differences that exist with respect to: (a) attack rate (a') and handling time (T_h) (parameters in the functional response model, subsection 5.3.7); (b) metabolic rate, which will increase with instar by a certain power of the body weight – this affects B in equation 2.13; and (c) the constants α and k (Beddington *et al.*, 1976b).

Examination of the growth rate vs. consumption rate plots for *Adalia bipunctata* (Figure 2.47) reveals that the slope (which represents conversion efficiency), decreases as the insects pass through the instars. This change in the slope is partly attributable to increased metabolic costs in later instars, as can be seen from the intercepts with the y-axis, representing basal respiratory rates. Howevever, the main cause is likely to be a decline in digestive efficiency, since compared with earlier instars, later instars of *Adalia* consume a greater proportion of each prey item, i.e. k increases with instar (Mills, 1981). To understand the relationship between the proportion of each prey consumed and digestive efficiency, consider the surface area/volume ratio difference between food boluses of different sizes. A larger bolus will have proportionately less of its surface area exposed to digestive fluids than a smaller bolus.

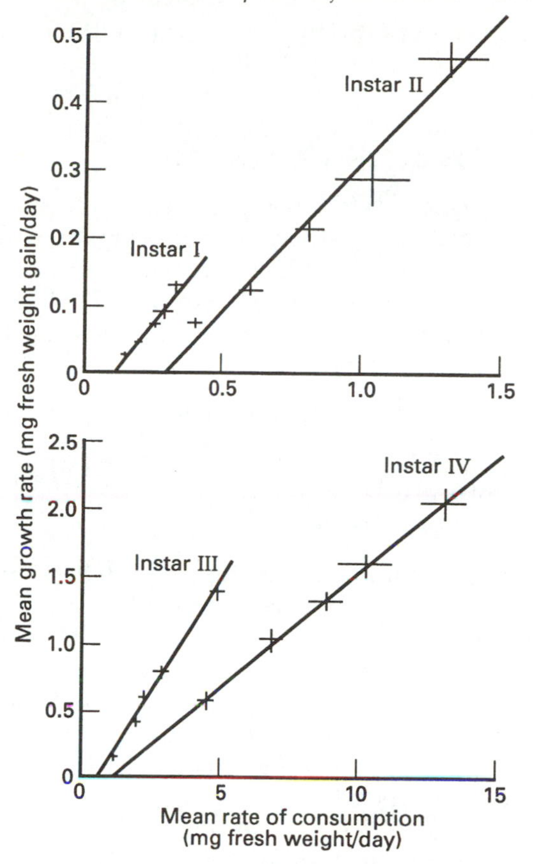

Figure 2.47 The linear dependence of average growth rates on food consumption rate for the four larval instars of the coccinellid beetle *Adalia bipunctata*. Note that the slope of the relationship, representing the gross food conversion efficiency, decreases as the insect develops. This is partly due to increased metabolic costs, as can be observed from the y-axis intercepts representing basal respiratory rates, but it is mainly due to a decline in digestive efficiency with instar, the larvae consuming a greater proportion of the total body of the prey as they increase in size with each instar. (Source: Mills, 1981.) Reproduced by permission of Blackwell Scientific Publications Ltd.

Conversion efficiency can also vary with consumption rate within an instar. Third instar larvae of *Chrysoperla carnea* provided with low prey densities have, as expected, a reduced consumption rate compared with third instar larvae given high prey densities, but they have a higher conversion efficiency (Zheng *et al.*, 1993a). A similar difference in conversion efficiency is shown by the early instars of the bug *Blepharidopterus angulatus* (Glen, 1973). Two possible reasons for this effect in the case of *C. carnea* were put forward by Zheng *et al*. (1993a):

1. digestive efficiency is increased, due to the smaller quantities of prey being ingested by larvae given low prey densities;
2. third instar larvae, like some spiders, reduce their metabolism in response to prey scarcity.

Respirometric methods would be required to test the latter idea.

Previous Feeding History

Can predators recover from the deleterious effects upon growth and development brought about in previous instars by prey scarcity? To answer this question, a cohort of larvae can be exposed to high levels of prey availability throughout two instars, e.g. the third and fourth in a coccinellid, and another cohort can be exposed to low levels in a prey availability regime that alters from low to high between these two instars. The fourth instar insects from the two regimes can then be compared with respect to weight gain and instar duration. In this experiment the consumption rate and the various nutritional efficiencies should be measured, to determine whether the compensatory effects shown by the test cohort are a result of changes in one or more of these factors within the later instar.

A similar experiment to the above was carried out by Paradise and Stamp (1991) on the mantid *Tenodera sinensis*. These authors showed that: (a) first and second instar mantids given a small quantity of prey attained a smaller size and spent more time in those instars than mantids provided with as much prey as they could eat, but that (b) in two out of three cohorts, mantids reared during the first instar on a poor diet recovered during the second instar when they were switched to a higher diet, gaining as much weight as and spending less time in that instar than those given a high diet throughout. The larvae of the later instar compensated for poor feeding in the earlier instar by having a higher consumption rate.

Zheng *et al*. (1993a) conducted a similar experiment with the green lacewing, *Chrysoperla carnea*, but over the whole of larval development. Larvae were either provided with a large quantity of prey over all three instars (HHH regime), or they were given a low quantity over the first two instars and a large quantity during the third (LLH regime). No significant difference in the duration of the third instar was found between larvae in the two regimes, but the overall duration of development from eclosion to pupation was significantly longer in the LLH larvae, i.e. recovery in development rate was partial. The dry weight gain of third instar larvae was not significantly different in the two treatments, and the same applied to the overall weight gain over the whole of larval development, i.e. recovery in growth was complete. Third instar larvae in the LLH regime consumed as many prey as those in the HHH regime, and the same applied to larvae over the whole of their development.

Limited recovery from suboptimal feeding conditions can, at least in the laboratory, be achieved in some Odonata (*Lestes sponsa*) by the larva passing through an additional instar. However, instar number is constrained and an increase in any linear dimension is limited to around 25–30% (D.J. Thompson, personal communication).

Can predators with higher growth rates in one instar maintain the advantage through subsequent instars? To answer this question, the above-mentioned experimental design can be reversed, so that in the test cohort the prey availability regime alters from high to low. Experiments carried out by Fox and Murdoch (1978) on the backswimmer *Notonecta hoffmani* show that larvae can maintain a growth advantage during larval development.

Non-Prey Foods

As with the fecundity–prey density relationship, two effects of providing non-prey foods together with prey might be to lower the prey ingestion rate threshold, so shifting the development rate vs. prey availability curve nearer to the origin, and to alter the shape of the curve. Predator larvae may require a lower minimum number of prey items in order to develop at all, and they may develop more rapidly at and above this minimum.

That development rate is increased by provision of non-prey foods is demonstrated by experiments conducted on larvae of the lacewing *Chrysoperla carnea* (McEwen *et al.*, 1993). At the three test prey densities offered to the predators during development, larvae given an artificial honeydew with prey required significantly fewer prey, developed significantly more rapidly, and attained a significantly higher adult weight than larvae given water with prey.

Some predators can complete larval development when prey are absent, provided certain non-prey foods are available, e.g. the bug *Orius insidiosus* (Anthocoridae) (Kiman and Yeargan, 1985), the coccinellid *Coleomegilla maculata* (Smith, 1961, 1965). Predators such as the bug *Blepharidopterus angulatus* cannot complete development on a diet of honeydew alone, but nymphs that are switched to a diet of aphids after the third instar can complete development (Glen, 1973).

Prey Species

Larval growth and development might be expected to vary in relation to prey species. Examples of studies demonstrating this effect in coccinellids are those of Blackman (1967) for *Adalia bipunctata* and *Coccinella septempunctata*, and Michels and Behle (1991) for *Hippodamia sinuata*. In *H. sinuata* the host species effect on development rate disappeared at temperatures exceeding 20°C.

Interference and Exploitation Competition and Other Interference Effects Among Conspecifics

Ecologists distinguish between competition through **interference** and competition through **exploitation**. In interference competition individuals respond to one another directly rather than to the level to which they have depleted the resource. In exploitation competition individuals respond, not to each other's presence, but to the level of resource depletion that each produces. With exploitation competition the intensity of competition is closely linked to the level of the resource that the competitors require, but with interference it is often only loosely linked (Begon *et al.*, 1990).

Larval predators show interference in the form of behavioural interactions. For example, larval dragonflies may interfere with one another's feeding through distraction (e.g. 'staring encounters' between dragonflies) and/or overt aggression (Baker, 1981; McPeek and Crowley, 1987; Crowley and Martin, 1989) (Figure 2.48). Such interactions are likely to result in reduced feeding or increased metabolic costs and therefore ultimately cause reduced growth, development and survival.

Van Buskirk (1987) carried out an experiment to test whether density dependent, interference-mediated reductions in growth, development and survival occurred in larvae of the dragonfly *Pachydiplax longipennis*. First instar larvae were raised in pools at initial densities of 38, 152 and 608 larvae/m^2, under two levels of prey availability (extra prey added daily to those already in pool; extra prey not added, i.e. food depletion likely to occur, pools in both cases containing the same initial density of prey), in a 3 × 2 factorial design. Van Buskirk found that with increasing predator density there was a decrease in growth and development rates, but he did not find any statistically significant interactions between prey addition and predator density, suggesting that some form of interference, rather than prey exploitation, was important. Within the prey-added treatment,

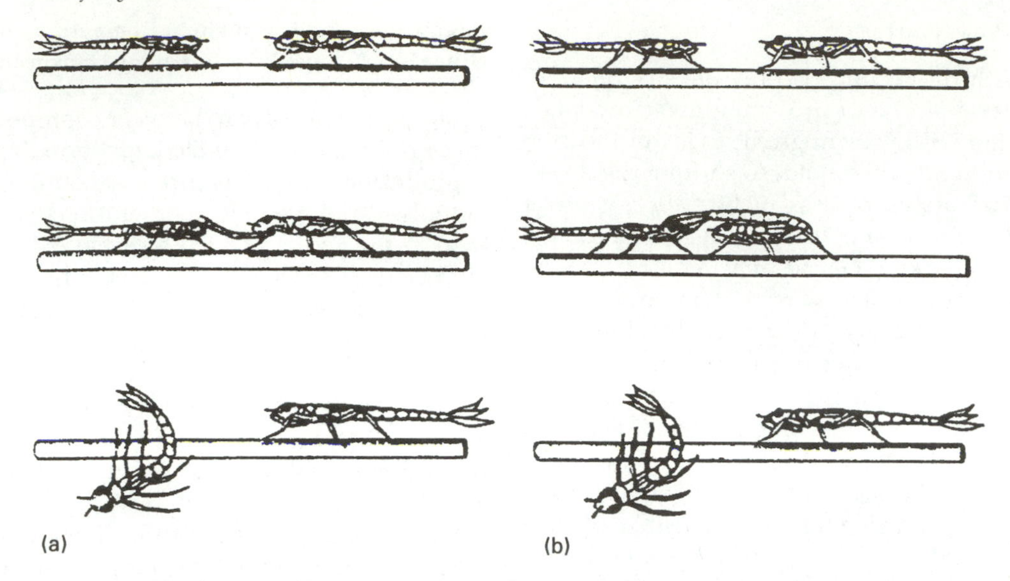

Figure 2.48 Aggressive interactions between damselfly larvae: (a) labial striking; (b) slashing with the gills. Larger larvae usually displace smaller ones, which may retreat by swimming off the perch. (Source: Williams and Feltmate, 1992.) Reproduced by permission of C.A.B. International.

the *per capita* amount of prey available was greatest at low predator densities (since identical amounts of prey were available at all predator densities). If larvae were competing by exploitation alone, prey availability would have had a greater *positive* effect at low predator densities than at high predator densities – this did not show up in the statistical analyses. Instead, prey availability increased survival by a similar amount at all predator densities. The positive effect of prey availability on survival suggests that food stress in the prey-absent larvae led to their becoming cannibalistic, the assumption being that larval dragonflies can survive long periods of time without prey and thus mortality could not be attributed to starvation (Lawton *et al.*, 1980) (section 2.10). Direct evidence of cannibalism was not, however, obtained.

Baker (1989), using the 'condition' (an index of the relative mass per unit head width of larvae) of larval dragonflies, related larval growth to larval density in a series of field sites. He found that for most of the year there was little evidence of food limitation (prey densities often exceeded known laboratory *ad libitum* levels) and, when larval 'conditions' were poor, there was no apparent change in prey densities, and larval predator populations were at their *lowest*. These results are in contrast to those obtained in the study by van Buskirk (1987) and in other studies (Pierce *et al.*, 1985; Johnson *et al.*, 1984; Banks and Thompson, 1987b) where the results suggest that aggressive interactions are important in limiting food intake of larval Odonata in the field. Among the reasons for this discrepancy given by Baker are that in his study larval densities were not high enough for either exploitation competition or interference to occur. Baker also points to differences in methodology and interpretation between his study and those of other workers (see discussion in his paper).

Anholt (1990) points to 'asymmetries in the burden of refutation' in several studies of competition in larval Odonata and other animals. Authors, when they have been unable to find evidence of prey depletion, have concluded by default that interference is the primary cause of density-dependent growth, development and survival. That is, they have made the assumption that if it is not competition through exploitation, then it must be competition through interference. Anholt's study represents a significant departure from previous work on Odonata in that he attempted to disentangle the effects of interference and exploitation by manipulating the rates of the two processes. Anholt manipulated the frequency of interactions between larval *Enallagma boreale* by altering perch availability at a *fixed density* of predators. Anholt argued that increasing the abundance of perches (i.e. increasing habitat complexity), will reduce the frequency of encounters and thereby reduce the intensity of interference competition without affecting the supply of planktonic prey, i.e. without depletion occurring. Anholt's experimental design was a fixed-effects analysis of variance: (a) with three factors (food availability, larval density, perch availability) completely crossed; (b) with two factors (larval density and food availability) crossed; and (c) with two factors (perch availability and starting instar) crossed. In Anholt's experiments, damselflies became more evenly distributed spatially among available perches as the predator density per perch increased, demonstrating that there were behavioural responses to the manipulation of habitat complexity (a prediction made by Crowley *et al.* (1987)). Food supply and predator density strongly affected survival, but the proportion of the variance in survival attributable to the habitat complexity manipulation, i.e. interference, was very small. Furthermore, whilst there were significant density-dependent alterations in growth or development, they were not attributable to food-related interference competition. Thus, despite the overt nature of the interactions between individuals, the costs appear to be minimal. Anholt (1990) suggested that the density-dependent reduction in larval growth and development observed in his experiments may have been due to both resource depletion, i.e. exploitation and resource depression. Resource depression is a term used to describe local reductions in prey availability that result from the prey minimising the risk of predation by becoming less active and/or altering their use of habitat space.

Gribbin and Thompson (1990) conducted laboratory experiments in which individuals of two instars (ones which commonly occur together in the field) of *Ischnura elegans* were maintained in small containers (transparent plastic cups) with a superabundance of prey (to avoid prey limitation) either: (a) in isolation, (b) with three larvae of the same instar, or (c) three larvae of different instars. Either one perch or four perches were provided to larvae in each treatment, and the experiment was treated as a two-way analysis of variance with perch availability as one factor and larval combination as the second factor potentially influencing development and growth. Small larvae showed increased development times and decreased growth (measured as percentage increase in head width) when kept with large larvae but similar effects were not shown when the smaller larvae were kept with other smaller larvae. Development time and size increases of large larvae were not significantly affected by the presence of small larvae, i.e. competition was asymmetric. Regardless of the instar combination used, reductions in growth and development (which were taken to be due to interference, since prey (approx. 200 *Daphnia magna*) was superabundant in all treatments) were lessened when there were more perches available, although only in a few cases was the lessening significant. Gribbin and Thompson found that in containers with only one perch, large larvae often occupied the perch, whilst the single small larva positioned itself on the side of the cup where the feeding efficiency was likely to have been reduced.

For a recent study of interference and exploitation competition in a species of carabid beetle, see Griffith and Poulson (1993). Interference competition has been shown by Griffiths (1992) to occur between larvae of the ant-lion *Macroleon quinquemacu-*

latus. Exploitation competition between two species of hover-fly has been studied by Hågvar (1972, 1973).

The deleterious effects of competition on larval growth (and fecundity) can be expressed by plotting *k*-values (defined in subsection 5.3.4) against $\log_{10}$ predator density. When describing such effects the terms 'scramble' and 'contest' competition are less appropriate than the terms 'exact compensation', 'over compensation' and 'under compensation' (Begon *et al.*, 1990; section 5.3).

Larvae may also show a reduction in feeding rates in the presence of predators (Murdoch and Sih, 1978; Sih, 1982; Heads, 1986). Such interference may reduce the rate of consumption of prey, even when the insects do not need to move in order to feed (Heads, 1986), with the potential result that growth, development and even survival are affected (Heads, 1986; Sih, 1982).

The early instar larvae of the waterboatman *Notonecta hoffmani* can suffer significant mortality due to predation from adult *conspecifics* (Murdoch and Sih, 1978; Sih, 1982), and the adult avoidance behaviour of larvae constitutes a form of interference. Sih (1982), in laboratory and field experiments, compared the behaviour of larvae when the adults were experimentally removed, with their behaviour in controls where adults were present. Early instar larvae avoided adults by altering their use of habitat space (spending less of the total time available in the central region of the pond or tub, where prey and adults occur at the highest densities), and some of the early instars also became less active. As a result of this behaviour, larvae of the first two instars experienced severely reduced feeding rates.

Host Size

Idiobionts

The concept of an individual host as a fixed 'parcel' of resource for a developing idiobiont parasitoid was introduced in Chapter 1 (subsection 1.5.6). For many idiobiont species host size determines the size (and/or mass) of the resultant parasitoid adult(s), as shown by data both on solitary and on gregarious species (Salt, 1940, 1941; Arthur and Wylie, 1959; Heaversedge, 1967; Charnov *et al.*, 1981; Greenblatt *et al.* 1982; Waage and Ng, 1984; van Bergeijk *et al.*, 1989; Corrigan and Lashomb, 1990). In parasitoids development rate is not necessarily positively correlated with the size of host oviposited in. For example, in *Trichogramma evanescens* development rate is highest in medium-sized eggs and lowest in small and large eggs (Salt, 1940), while in *Bracon hebetor* development time is unaffected by host larval size (Taylor, 1988a). The reasons for the lack of a clear relationship are complex, and the reader is referred to Mackauer and Sequeira (1993).

To investigate the influence of host size on growth and development in an idiobiont parasitoid species, present females (fertilised and, if necessary, unfertilised, to obtain data on both sexes) with hosts of different sizes and record the weight of the resultant adult progeny and the time taken from oviposition to adult eclosion (since adult eclosion is often influenced by light:dark cycles (Mackauer and Henkelman, 1975), observations should be carried out at the same time each day or under continous light conditions; video recording equipment can be used both to improve accuracy and to save time (Sequeira and Mackauer, 1992a)). If the parasitoid is a gregarious species, clutch size will have to be kept constant (section 1.6. describes clutch size manipulation techniques). One needs to bear in mind the possibly complicating effects of sex differences in food acquisition (and therefore growth and development) in broods of gregarious species. This problem can be partly circumvented by using uninseminated parent females, which will produce all-male egg clutches, but obtaining all-female clutches could prove very difficult (section 1.6).

For idiobiont parasitoids the *age* of the host may be a confounding factor. For example, some parasitoids that develop in host pupae may be able to utilise both very recently formed pupae and pupae within which the adult host is about to be formed. These different types of host pupa are likely to have the same external dimensions and similar mass but will represent very different amounts of resource.

To determine whether host stage i.e. instar and not host size *per se* mainly accounts for any variation in growth or rate of development, parasitoids e.g. idiobionts attacking larval Lepidoptera, can be presented with a range of host sizes within each host stage that overlaps with host sizes within the previous or subsequent stage.

Koinobionts

The hosts of koinobionts continue growing and feeding. One might not expect the same relationship between progeny size and host size at oviposition as exists for idiobionts, due to the variability in host growth that can occur during larval parasitoid development (Mackauer, 1986). Parasitism is known to influence a host's food consumption (amount of food or rate at which food is consumed), either decreasing it, as is often the case for solitary species (see Cloutier and Mackauer, 1979, for an exception), or increasing it, as is often the case for gregarious species (Slansky, 1986). Sequeira and Mackauer (1992a,b) and Harvey *et al.* (1994) have shown for *Aphidius ervi* and *Venturia canescens* respectively, that parasitoid adult size (mass) is not a linear function of host size (mass) at oviposition (Figure 2.49a). Although in both species, there is a linear increase in wasp size with increasing host instar up to the penultimate instar, beyond the latter, adult mass does not increase.

The relationship between development rate and the size of the host when oviposited in varies, being either linear throughout the whole range of available host sizes and highest in larger hosts (Fox *et al.*, 1967; Smilowitz and Iwantsch, 1973), or non-linear (Jones and Lewis, 1971; Avilla and Copland, 1987; de Jong and van Alphen, 1989; Harvey *et al.*, 1994) (Figure 2.49b).

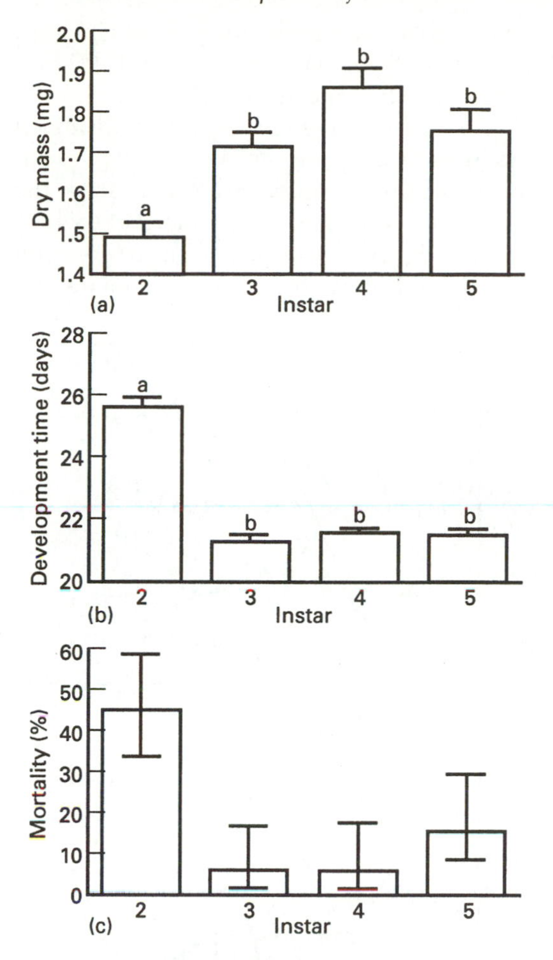

Figure 2.49 Growth, development and mortality of *Venturia canescens* (Ichneumonidae) reared on four instars of the moth *Plodia interpunctella*: (a) adult dry mass; (b) development time from oviposition to adult eclosion; (c) mortality. Treatments with the same letter do not differ significantly. (Source: Harvey *et al.*, 1994.) Reproduced by permission of The Ecological Society of America.

Valuable insights into the effects of host stage on growth and development can be obtained by plotting the 'growth trajectories' of both host and parasitoid (Sequeira and Mackauer, 1992b; Harvey *et al*., 1994) (Figure 2.50). Growth trajectories are studied by taking each host stage, dissecting parasitised hosts at various points in time after oviposition, separating the parasitoid larva from the host and measuring the dry weight of each. A trajectory is also plotted for unparasitised hosts. Using growth trajectories, Sequeira and Mackauer (1992b) showed that *A. ervi* responds to host-related constraints upon larval growth and does not regulate host growth for its own benefit, as some other parasitoids are known to do (Vinson and Iwantsch, 1980b). In *A. ervi* development time and adult dry mass covary positively (i.e. there is a trade-off between development *rate* and growth) with an increase in host size

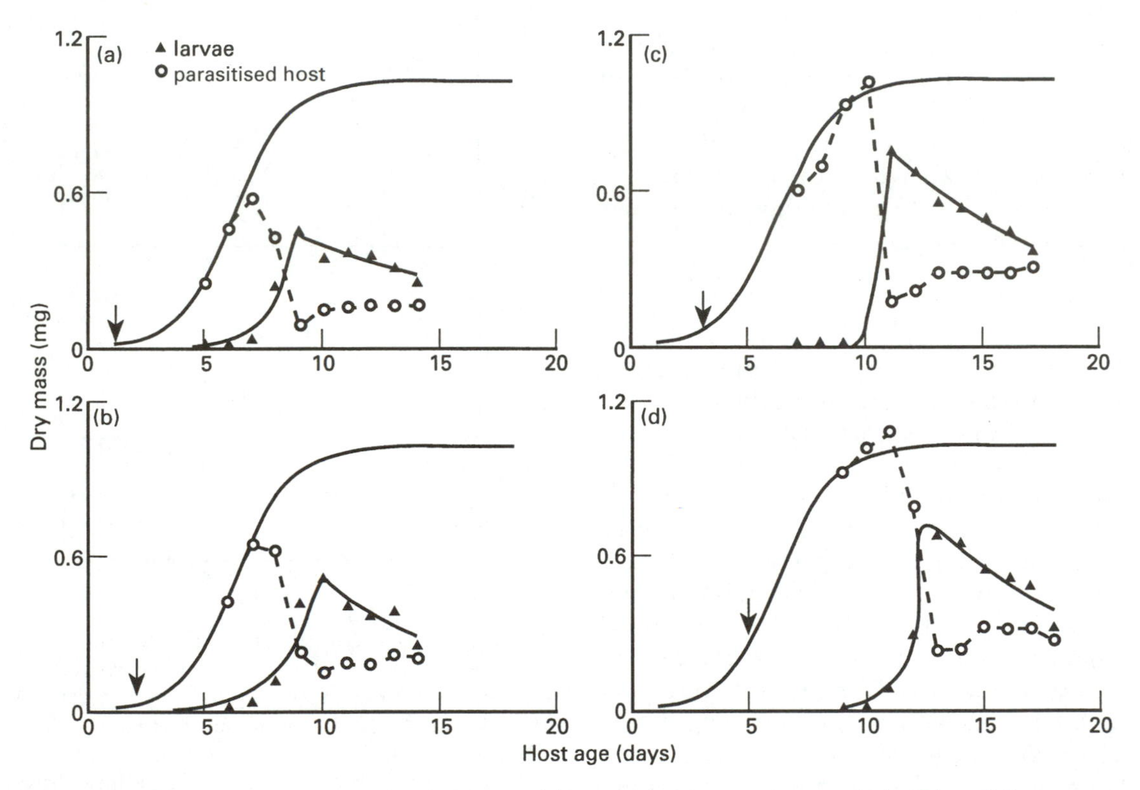

Figure 2.50 Growth trajectories of *Aphidius ervi* (▲) and of parasitised pea aphids (O) at different ages: (a) host nymphal instar one (24 h); (b) host nymphal instar two (48 h); (c) host nymphal instar three (72 h); (d) host nymphal instar four (120 h). The solid curve shows the corresponding trajectory of unparasitised aphids, samples of which were taken at various ages from birth to maturity. Arrows indicate the age of the host at oviposition. The turnover point of the parasitoid growth trajectory corresponds to parasitoid age of 8 days. The trajectory of parasitoid larval growth provides a direct measure of host 'quality' (Source: Sequeira and Mackauer, 1992b), reflecting as it does the nutritional relationship between the two insects during the course of parasitism, and its shape will be characteristic of the parasitoid species. All curves will, however, be 'J'-shaped, there being two functionally distinct phases in the development of holometabolous insects: first there is an exponential growth phase as the parasitoid larva feeds and converts host tissues into its own body mass, then there is a negative exponential decay phase between pupation and adult eclosion, when feeding has stopped and there is differential mass reduction due to respiration, water loss and voiding of the meconium (Harvey *et al*., 1994, and J. Harvey personal communication). Reproduced by permission of The Ecological Society of America.

from first to third instar, but they vary independently in parasitoids developing in fourth instar hosts. In the latter, adult mass does not increase further but development rate increases. The growth trajectories shown in Figure 2.50 indicate that in early instar hosts parasitoid growth and development rate are limited by the size and growth potential of the host (compare, in Figure 2.50, the average mass attained by parasitised aphids with that of unparasitised aphids of equivalent age). By contrast, in fourth instar hosts additional resources are available, and this allows an increase in development rate without an increase in adult mass.

As pointed out by Harvey *et al.* (1994), *A. ervi* may represent one end of a continuum, the other extreme of which is to delay parasitoid growth until the host has reached its maximum size (in which case we would expect wasp size to be unaffected by instar at oviposition but development rate to be reduced). *V.canescens* lies somewhere between the two extremes, as parasitoids developing in second instar hosts spend prolonged periods as first instars, ingesting host blood at a reduced rate until the host is well into its fifth instar (by contrast, parasitoids developing in third to fifth instar hosts do not need to arrest development because the host develops rapidly and provides sufficient food materials).

Graphical models of parasitoid growth and development have been constructed by Mackauer and Sequiera (1993).

The experimental protocol for studying the effects of host stage at oviposition upon growth and development is slightly more complex for koinobionts than for idiobionts inasmuch as the hosts need to be reared. Care must be taken to control for the effects of variations in host diet; Harvey *et al.* (1994) for example reared hosts with an excess of food.

Host Species

Given that hosts of different species are likely to constitute different resources in both a qualitative and a quantitative sense, we would expect parasitoid growth and development to vary in relation to the host species parasitised. This is indeed the case, as studies on idiobionts and a few koinobionts have shown, although few workers have measured both growth and development (Salt, 1940; New, 1969; Dransfield, 1979; Moratorio, 1987; Bigler *et al.*, 1987; Taylor, 1988a; Ruberson *et al.*, 1989; Corrigan and Lashomb, 1990). Salt (1940), for example, showed how the size of adult progeny of *Trichogramma evanescens* varied with the species of moth within the eggs of which larval development occurred (Figure 2.51). Moratorio (1987), working with *Anagrus mutans* and *A. silwoodensis*, showed that female progeny were larger when development occurred in the (large) eggs of *Cicadella viridis* but were smaller when development occurred in the (small) eggs of *Dicranotropis hamata*. But, whereas *A. silwoodensis* develops fastest in *C. viridis*, *A. mutans* develops fastest in D. *hamata,* i.e. development and growth countervary, not covary, in relation to host species in *A. mutans*. Development and growth also countervary in relation to species in *Telenomus lobatus* (Scelionidae); wasps develop more rapidly in eggs of *Chrysoperla* species (Chrysopidae) than in eggs of *Chrysopa* species, but the adults attain a larger size in eggs of the latter genus, the eggs being larger than those of *Chrysoperla* (Ruberson *et al.*, 1989).

In investigating the possible influence of host species, it would be useful to compare the performance of parasitoids in hosts of different species but of equivalent size (mass). If differences in performance are recorded, then this would suggest that the quality, rather than the quantity, of host resource affects growth and development.

Toxins in the Host's Food Plant

The species of plant fed upon by the host can be expected to influence parasitoid growth and development indirectly, through its

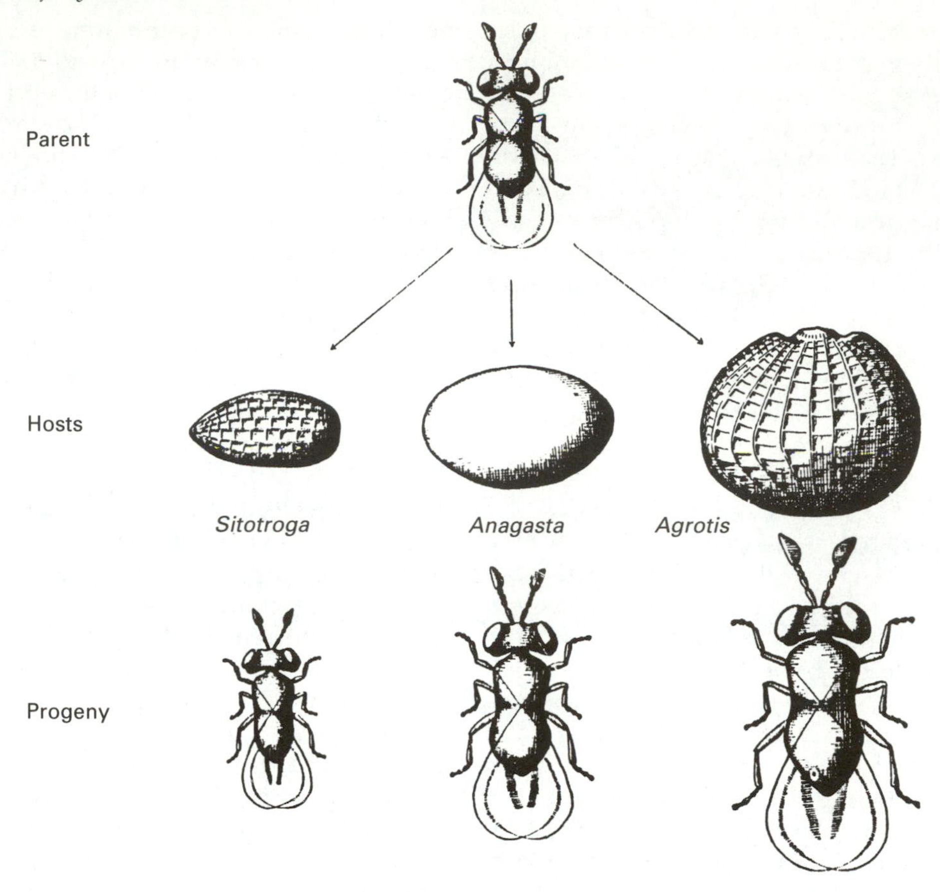

Figure 2.51 Relative sizes of female *Trichogramma evanescens* (Trichogrammatidae) and female progeny reared from different host species. (Source: Salt, 1940.) The reader should note that the confounding effects of progeny clutch size were not controlled for in Salt's experiments (development was solitary in *Sitotroga* and *Anagasta* but was either solitary or gregarious in *Agrotis*). The female shown here ex *Agrotis* developed solitarily; nevertheless, female developing gregariously in that host species were on average markedly larger than those developing in either of the other two host species. Reproduced by permission of Cambridge University Press.

effects on host size. Campbell and Duffey (1979) found evidence to suggest that the species of food plant can also affect parasitoid growth and development directly, due to toxic effects. They provided larvae of *Helicoverpa zea* parasitised by the ichneumonid *Hyposoter exiguae* with artificial diets containing varying amounts of the alkaloid α-tomatine, which occurs in tomatoes. Parasitoid larvae in hosts fed upon a medium containing the alkaloid developed slowly and attained a smaller adult weight than parasitoids in hosts given a tomatine-free diet. There were similar differences in the duration of the pupal stage.

The growth of the ichneumonid *Campoletis sonorensis* (and that of its host, *Helicoverpa virescens*) is improved when small amounts of gossypol are included in the host's artificial diet; high concentrations result in reduced growth (Williams *et al.*, 1988).

Superparasitism

Introduction

In Chapter 1 (section 1.8) superparasitism was defined as the laying of an egg (by a solitary parasitoid) or a number of eggs (by a gregarious parasitoid) in an already parasitised host. In the case of a solitary parasitoid species, only one larva survives in each superparasitised host. In a gregarious species the number of survivors per host will depend on: (a) the total number of eggs the host contains or bears, and (b) the size of the superparasitised host (Beckage and Riddiford, 1978; le Masurier, 1991). This section is concerned with ways of studying the fitness consequences for *surviving* larvae, and asks how larval growth development rate might be affected by superparasitism.

Solitary Parasitoids

Models of superparasitism as an adaptive strategy in solitary species (Visser *et al.*, 1990; van der Hoeven and Hemerik, 1990) are based on the assumption that superparasitism has no fitness consequences for the surviving larva, i.e. it does not increase larval development time or reduce adult size. This would seem to be a reasonable assumption, since **supernumerary larvae** (supernumeraries are larvae in excess of the number that can ultimately survive, i.e. complete development) are usually eliminated before they can utilise an appreciable amount of host resource. For example, Visser *et al.* (1992c) found no convincing evidence that *Leptopilina heterotoma* adults emerging from singly-parasitised hosts were larger than adults emerging from superparasitised hosts. However, as pointed out by Bai and Mackauer (1992) and Harvey *et al.* (1993), superparasitism *may* have fitness consequences for the larvae of some parasitoid species. Simmonds (1943) and Wylie (1983), for example, reported that in *Venturia canescens* (Ichneumonidae) and *Microctonus vittatae* (Braconidae) larvae take longer to develop in superparasitised hosts than in singly-parasitised hosts, although neither author recorded the number of eggs contained per host. Similarly, Vinson and Sroka (1978), subjected hosts of *Cardiochiles nigriceps* (Braconidae) to varying numbers of ovipositions, recorded the time taken from oviposition to larval emergence from the host, and showed that the higher the degree of superparasitism, the longer was the mean development time of the surviving larva (Table 2.3).

The fitness cost to *koinobionts* may be partly determined by the ability of the surviving larva to compensate for possibly reduced

Table 2.3 Percentage of hosts yielding larvae and time taken from oviposition to larval emergence from the host, in the solitary parasitoid *Cardiochiles nigriceps* (Braconidae) parasitising *Helicoverpa virescens* (Source: Vinson and Sroka, 1978.)

No. of ovipositions per host	*Percentage of hosts yielding a parasitoid*	*Mean time (days) to emergence*
1	92	12.3 + 1.6
2	58	12.2 + 1.9
3	63	14.7 + 2.7
4	29	15.6 + 2.5
5	27	15.9 + 3.0
>5	21	16.9 + 3.4

growth during embryonic and early larval development (when it must compete with the rival larva for host resources) by increasing growth later in development (Bai and Mackauer, 1992), and the same might apply to development rate. Bai and Mackauer (1992) carried out a simple experiment in which they subjected aphids to either one oviposition (single-parasitised) or several ovipositions (superparasitised) by *Aphidius ervi*. They used unmated females, in order to control for the possible bias resulting from differential development (and survival) between male and female larvae. They then compared the total development time and adult weights in the different treatments.

Interestingly, they found that *Aphidius ervi* gained 14% *more* dry mass in superparasitised hosts, i.e. growth was enhanced through superparasitism, and took no longer to develop, i.e. development rate was unaffected. The most likely explanation for this effect is that the superparasitised hosts ingested more food (Bai and Mackauer, 1992). As Bai and Mackauer point out, the fitness benefit, i.e. increased adult size gained by surviving larvae in superparasitised hosts, needs to be weighed against any costs in the form of reduced larval survival (subsection 2.10.2).

As we noted above, adult size in *Leptopilina heterotoma* is not affected by superparasitism; in this parasitoid either compensation is complete or there is no initial reduction in growth as a result of superparasitism. Studying the trajectory of parasitoid larval growth (see above) would shed light on this.

Larval development rate may also be influenced by **heterospecific superparasitism (= multiparasitism)** (McBrien and Mackauer, 1990).

Experiments aimed at investigating the effects of superparasitism on larval growth (as measured by adult size) and development would involve exposing a recently parasitised host to a standardised female and allowing the same or a different (conspecific or heterospecific) standardised female to deposit a specified number of eggs. The number of eggs laid in each case can be more easily monitored and controlled if the parasitoid is one of those species in which the female performs a characteristic movement during oviposition (subsection 1.5.6 and Harvey *et al.*, 1993). The time taken from oviposition to adult eclosion and the size or weight of emerging adults needs to measured and the different treatments compared with one another and with controls. The experiment could be expanded to take into account the possible effects of host size or host instar, as was done by Harvey *et al.*. The latter authors showed that superparasitism in *Venturia canescens* reduced development rate in parasitoids reared from both third instar and fifth instar larval hosts (the moth *Plodia interpunctella*), but that the reduction was greater in parasitoids reared from the later instar (Figure 2.52). The size of wasps reared from third instar hosts was unaffected by egg number, but adult wasps from both superparasitised fifth instar treatments (two eggs, four eggs) were significantly smaller than those reared from singly-parasitised hosts (Figure 2.53). Harvey *et al.* suggest that the reason superparasitism affected parasitoids from fifth instar hosts more than those from third instar hosts is that the fifth instar larvae were post-feeding, wandering larvae with a zero growth potential. Parasitism of such hosts would be more like idiobiosis than koinobiosis, and the surviving larva would be less able to compensate for any negative effects of superparasitism.

Gregarious Parasitoids

The fitness consequences of superparasitism for larvae of gregarious parasitoids have already been touched upon, from both a theoretical and an experimental standpoint, in Chapter 1 (section 1.8). Leaving aside Allee effects (defined in section 1.6), superparasitism will intensify competition among larvae for host resources, with the result that the *per*

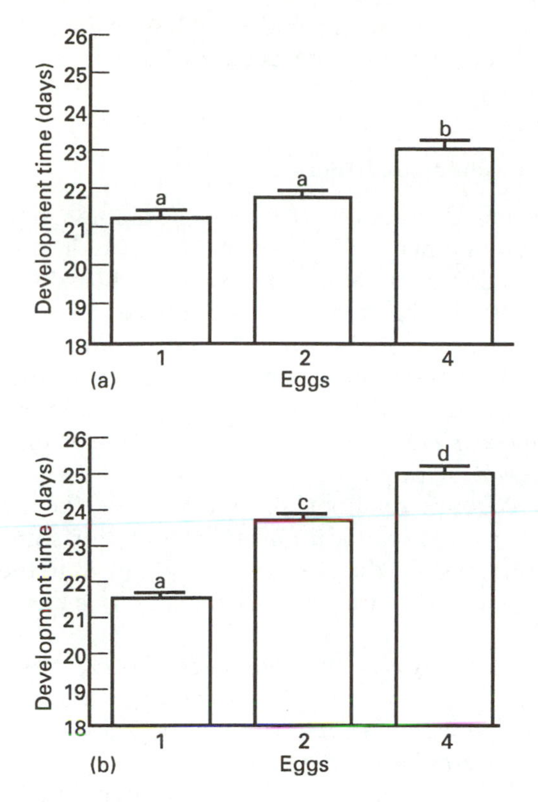

Figure 2.52 Effects of superparasitism on development in the ichneumonid parasitoid *Venturia canescens*. Development time (number of days taken from oviposition to adult eclosion) of wasps reared from from: (a) third instar; (b) fifth instar larvae of *Plodia interpunctella* containing one, two or four parasitoid eggs. Treatments with the same letter do not differ significantly. (Source: Harvey *et al.*, 1993.) Reproduced by permission of Blackwell Scientific Publications Ltd.

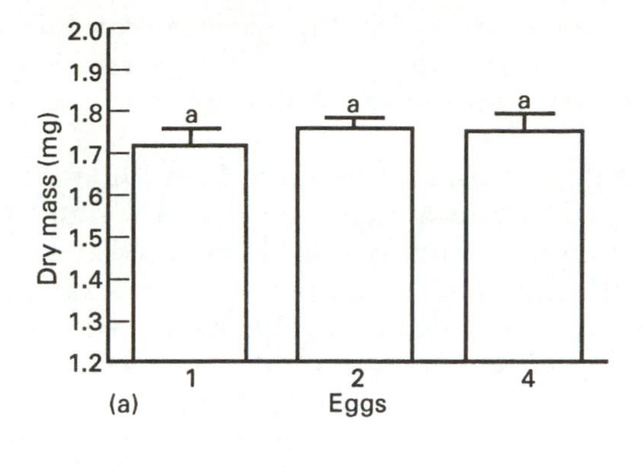

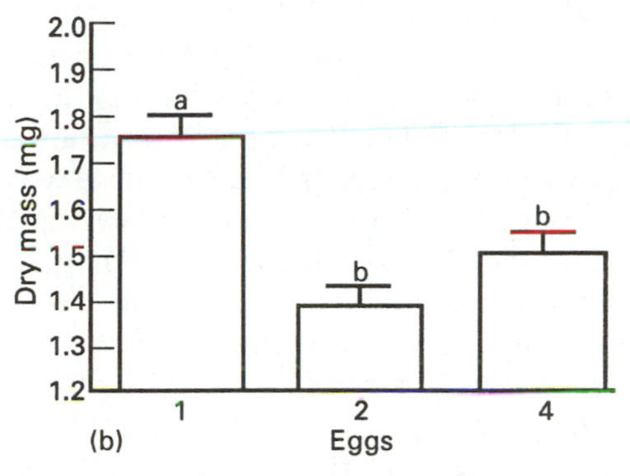

Figure 2.53 Effects of superparasitism on growth (as measured by adult dry mass) in the ichneumonid parasitoid *Venturia canescens*. The dry weight of wasps reared from: (a) third instar; (b) fifth instar larvae of *Plodia interpunctella* containing one, two or four eggs. Treatments with the same letter do not differ significantly. (Source: Harvey *et al.*, 1993.) Reproduced by permission of Blackwell Scientific Publications Ltd.

capita growth and development rate of the parasitoid immatures will be reduced. This is at least what one would expect, although Nealis *et al.* (1984) found that increased larval density per host slowed development of *Cotesia glomerata* only slightly (le Masurier, 1991, found no significant effect of clutch size on development time in this species) and tended to *increase* the rate of development in *Pteromalus puparum*. Le Masurier (1991) also found no significant decrease in body size with increasing clutch size in one population of *C. glomerata* parasitising *Pieris brassicae*, although he did find such an effect in another population parasitising *Pieris rapae*.

Experiments aimed at investigating the effects of superparasitism on larval growth (as measured by adult size) and development in gregarious parasitoids would involve:

1. in the case of intraspecific superparasitism, exposing a recently parasitised host to a standardised female and allowing the same or a different female to oviposit a further egg or clutch of eggs;

2. in the case of multiparasitism, exposing a host recently parasitised by a female of one species to a female of another species.

In both cases, the time taken from oviposition to adult eclosion and the size or weight of emerging adults need to be measured and the different treatments (i.e. initial and second clutches of different sizes) compared with one another and with controls. With ectoparasitoids, eggs can be artificially added to existing clutches of various sizes (section 1.6 and Strand and Godfray, 1989). Assuming competitive equivalence of clutches produced by different females, the effects upon parasitoid growth and development of simultaneous oviposition by two conspecific females would be analogous to the effects of increasing the primary clutch, as in Figure 1.10. That is, an increase in the number of eggs laid per host would have a negative effect irrespective of whether the eggs are laid by one or by two females, provided all the eggs are laid at the same time. However, the competitive disadvantage of a second clutch may be underestimated from a fitness function curve that is based solely on initial clutches, if there is a significant time interval between the laying of initial and subsequent clutches (Strand and Godfray, 1989). Measurement of any such disadvantage, in terms of growth and development, to a second clutch requires the progeny from the two clutches to be distinguishable by the investigator. This is only possible in those species in which there are mutant strains, e.g. the eye/body colour mutant 'cantelope-honey' in *Bracon hebetor*. To ensure, when using mutants, that competitive asymmetries do not bias the results of experiments, reciprocal experiments should be carried out for each clutch size and time interval combination (Strand and Godfray, 1989).

The possibly complicating effects of sex differences in larval food acquisition also need to be borne in mind in experiments on gregarious parasitoids; compared with the adding of a female egg, the adding of a male egg to an existing clutch could have less of an effect upon fitness of the progeny in the initial clutch.

Parasitoid Life History

Some differences between idiobionts and koinobionts with respect to growth and development have been discussed. Blackburn (1991a), through a comparative analysis (subsection 1.2.3) of the biological attributes of 474 parasitoid species, showed that several ecological differences among parasitoid species are correlated with aspects of egg, larval or pupal biology when the effects of taxonomic relationship and of body size are controlled for (adult size accounts for some interspecific differences in biology, but the proportion of the variance it explains is low):

1. Parasitoids of poorly concealed hosts have shorter egg incubation periods than parasitoids of better concealed hosts;
2. Koinobionts have longer pupal duration times and overall (i.e. oviposition to adult eclosion) development times than idiobionts;
3. Temperate species have longer overall development times than tropical species;
4. The overall development periods of egg parasitoids are longer than those of pupal parasitoids.

2.9.3 EFFECTS OF PHYSICAL FACTORS ON GROWTH AND DEVELOPMENT

Temperature

Development Rate

Figure 2.54 shows the typical relationship between an insect's rate of development and temperature. There is a threshold temperature below which there is no (measurable) development; this threshold is sometimes referred to as the **developmental zero**. There is also an upper threshold above which further increases in temperature result in only small increases in development rate. The overall relationship is non-linear (Mills, 1981), but over the intermediate range of temperatures normally experienced by an insect species in the field, the rate of development increases linearly with temperature. As noted by Gilbert *et al.* (1976) why this should be so is a mystery, since rates of enzyme action (which are presumably basic to development) usually increase exponentially, not linearly, with increasing temperature.

The deleterious effects of a high temperature extreme depends on how long the insect is exposed to it. As pointed out by Campbell *et al.* (1974) with reference to the development rate-temperature relationship shown in Figure 2.54, temperatures within the high range (i.e. the part of the relationship where the curve decelerates) only have a deleterious effect upon development if the temperature is either held constant in the range or fluctuates about an average value within the range. If the temperature fluctuates about a daily average within the medium range (i.e. the linear part of the the relationship) and the daily maximum reaches the high range, no deleterious temperature effect is observed. The influence, on development rate, of an extreme (hot or cold) temperature forming part of a 24 hour cyclical regime and alternating with a tolerable temperature (one within the linear part of the tolerable range) for which constant data are available, can be investigated using the QBASIC computer program given in Appendix 2A.

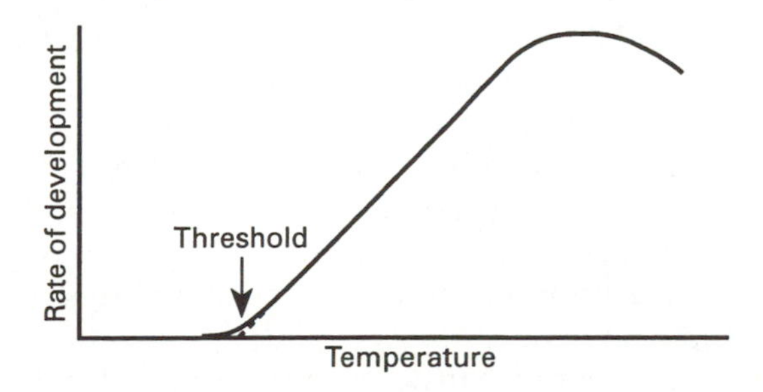

Figure 2.54 Rate of insect development as a function of temperature in insects. (Source: Gilbert *et al.*, 1976.) Reproduced by permission of W.H. Freeman & Co.

Given the fact that the development rate–temperature relationship is linear over the greater range of temperatures, the total amount of development that takes place during any given time period will be proportional to the length of time multiplied by the temperature above the threshold. With this physiological time-scale of **day-degrees** development proceeds at a constant rate, whatever the actual temperature. This concept is elaborated upon below.

To study the dependency of overall development rate on temperature in a parasitoid, expose hosts to female parasitoids at different constant temperatures (the range being chosen on the basis of field temperature records), and measure the time taken from oviposition to adult eclosion.

To demonstrate the effect of temperature on overall larval development in a predator, provide different cohorts of larval predators with a fixed daily ration of prey, at different temperatures, from egg hatch to adult eclosion. With both parasitoids and predators the thermal requirements for development can be determined for particular stages, i.e. the egg (Frazer and McGregor, 1992, on coccinellids), each larval instar and the pupa.

The data obtained from the above experiments can be described by a linear regression equation of the form:

$$y = a + bT \qquad (2.15)$$

where y is the rate of development at temperature T, and a and b are constants. Were the regression line to be extrapolated back, it would meet the abscissa at the developmental zero, t, which may be calculated from $t = -a/b$. The total quantity of thermal energy required to complete development, the **thermal constant** (K) can be calculated from the reciprocal of the slope of the regression line, $1/b$. Given in Appendix 2B is a QBASIC computer program listing for use in fitting either a linear or polynomial regression to data, and for calculating *t*, *K* and $1/b$.

Once *t* and *K* have been calculated from data obtained at *constant* temperatures, the rate of development under any *fluctuating* temperature regime can be determined by summation procedures. Unit time-degrees (day-degrees or hour-degrees) above *t* are accumulated until the value of *K* is reached where development is complete. This can be done by either accumulating the mean daily temperature minus the lower threshold or by accumulating the averages of the maximum and minimum daily temperature minus the threshold (i.e. $\sum [\{(T_{max} - T_{min})/2\} - \text{threshold}]$). However, both of these methods will result in great inaccuracies if a temperature contributing to the mean lies outside of the linear portion of the relationship. Means, by themselves, give no indication of the duration of a temperature extreme; an apparent tolerable mean temperature may actually comprise a cyclical regime of two extremes at which no development is possible. A much more accurate method is to use *hourly* mean temperatures (Tingle and Copland, 1988). Given in Appendix 2C is a QBASIC program, based on hourly temperature records, for either predicting the duration of development or validating the calculated thermal constants from field observations.

Summation has been used by Apple (1952), Morris and Fulton (1970), Campbell and Mackauer (1975), Lee *et al.* (1976), Hughes and Sands (1979), Morales and Hower (1981), Butts and McEwen (1981), Osborne (1982), Goodenough *et al.* (1983), Nealis *et al.* (1984) and Cave and Gaylor (1988). However, the method has been much criticised as it has two inherent faults. First, the assumed linear relationship is known to hold as an approximation for the median temperature range only (Figure 2.54) (Campbell *et al.*, 1974, on aphid parasitoids; Syrett and Penman, 1981, on lacewings). Second, the lower threshold on which summation is based is a purely theoretical point determined by extrapolation of the linear portion of the relationship into a region where the relationship is unlikely to be linear. The linear model is likely to underestimate development rates when average daily temperatures remain close to the threshold for long periods, although this can easily be corrected for (Nealis *et al.*, 1984). In an attempt to improve upon the thermal summation method, an algorithm was developed using a sigmoid function with the relationship inverted when the temperature is above the optimum (Stinner *et al.*, 1974). The assumed symmetry about the optimum is unrealistic, but Stinner *et al.* (1974) argue that the resultant errors are negligible. This algorithm has also been used in simulations for *Encarsia perniciosi* (McClain *et al.*, 1990a), fly parasitoids (Ables *et al.*, 1976) and other insects (Berry *et al.*, 1976; Whalon and Smilowitz, 1979; Allsopp, 1981). Ryoo *et al.* (1991) used a combination model involving upper thresholds to describe the development of the ectoparasitoid *Lariophagus distinguendus* (Pteromalidae).

In some cases the improvement in accuracy of simulations over the thermal summation method has been small or negligible, and it is questionable whether the use of complex models is necessary in relation to normal field conditions (Kitching, 1977; Whalon and Smilowitz, 1979; Allsopp, 1981). The method of matched asymptotic expansions was used to develop an analytical model describing a sigmoidal curve that lacks the symmetry about the optimum found in the algorithm of Stinner *et al.* (1974). Again, the authors concerned claimed excellent results (Logan *et al.*, 1976). However, comparisons of linear and non-linear methods to validate field data on *Encarsia perniciosi* showed no great differences (McClain *et al.*, 1990).

Other non-linear descriptions of the development rate-temperature relationship have also been developed. These include the logistic curve (Davidson, 1944) and polynomial regression analysis (Fletcher and Kapatos,

1983). Polynomial regression analysis can be used to select the best-fitting curve to a given set of data. Successively higher order polynomials can be fitted until no significant improvement in F-value results. This approach was found useful in describing data for *Diglyphus intermedius* (Patel and Schuster, 1983) and mealybug parasitoids (Tingle and Copland, 1988; Herrera *et al.*, 1989). The computer program in Appendix 2B calculates the best fit as well as any specified orders. Higher order polynomials may produce unlikely relationships between data points and fluctuate widely outside them. Before selecting a particular fit, it should be examined over the entire range of the data. It may be better to choose one that has a comparatively poor fit but is biologically more realistic. The computer program in Appendix 2D uses hourly temperature records for either predicting the duration of development or validating calculated coefficients of a polynomial fitted curve (Tingle and Copland, 1988).

Several authors have reported acceleration or retardation of development, when comparisons are made between development periods at cycling temperatures and at a constant temperature equivalent to the average of the cycling regime. The question of whether these effects are an artefact or are a real biological phenomenon is discussed by Tingle and Copland (1988).

Until recently, data on the development times of insects were almost always expressed in the form of means and standard deviations. This is not strictly justifiable as the distribution of eclosion times is not normal but shows a distinct skew towards longer developmental periods (Howe, 1967; Eubank *et al.*, 1973; Sharpe *et al.*, 1977). Several models have been developed which include a function to account for the asymmetrical distribution of development times (Stinner *et al.*, 1975; Sharpe *et al.*, 1977; Wagner *et al.*, 1984). Such models can be incorporated into population models (Barfield *et al.* (1977b), on *Bracon mellitor*). However, the poikilotherm model of Sharpe *et al.* (1977) did not give any great improvement in accuracy over day-degree models when predicting development of *Trichogramma pretiosum* (Goodenough *et al.*, 1983).

Biological control workers can use laboratory-obtained information on the effects of temperature on development in deciding which of several candidate species, 'strains' or 'biotypes' of parasitoids and predators to either introduce into an area or use in the glasshouse environment. In classical biological control programmes, the usual practice is to introduce natural enemies from areas having a climate as similar as possible to that in the proposed release area (Messenger, 1970; Messenger and van den Bosch, 1971; van Lenteren, 1986) (section 5.4). If there are several species, strains or biotypes to choose from, the one found to have a temperature optimum for development that is nearest to conditions in the introduction area should be favoured, all other things being equal. A classic example of a biological control failure resulting from the agent being poorly adapted to the climate of the introduction area is the introduction of a French strain of *Trioxys pallidus* into California to control the walnut aphid. This parasitoid was poorly adapted to conditions in northern and especially central California where it never became permanently established. The French strain was unable to reproduce and survive to a sufficient extent in areas of extreme summer heat and low humidity. A strain from Iran was subsequently introduced and proved far more effective (DeBach and Rosen, 1991).

As will be explained in Chapter 5 (5.3.8), data on development rate–temperature relationships are used in population models to predict both dynamics and phenologies. Morales and Hower (1981) showed that they could predict the emergence in the field of 50% of the first and second generations of the weevil parasitoid *Microctonus aethiopoides* (Braconidae) by using the day-degree method. McClain *et al.* (1990a) used the linear day-degree model and the sigmoid function model of Stinner *et al.* (1974) to predict the peaks of activity of parasitoids in orchards. The linear model predicted 8 of 13 peaks within ± 7 days, while the non-linear model was accurate for 7 of 13 peaks. Horne and Horne (1991) showed that simple day-degree models could account for the synchronisation of emergence of the encyrtid parasitoid *Copidosoma koehleri* and its lepidopteran host.

Growth Rate

Most studies on temperature relationships have dealt with development but have ignored growth. The relationship between growth rate and temperature in insects has been shown by direct measurement to increase linearly with temperature within the range of temperatures normally experienced by the insect in the field, in accordance with the following model:

$$\frac{1}{w}\frac{dw}{dt} = a(T - \theta) \qquad (2.16)$$

where T is the temperature, θ is the the threshold temperature below which no growth occurs, w is the larva's weight at time t, and a is a constant (Gilbert, 1984). Gilbert used this model to predict pupal weight, which determines fecundity, in the butterfly *Pieris rapae*. Tokeshi (1985) describes another method for estimating minimum threshold temperature and day-degrees required to complete growth, suitable for use with aquatic and terrestrial insects in either the laboratory or the field.

In some predator species the tendency is for successive larval instars to achieve a growth rate maximum at a higher temperature, e.g. in *Adalia bipunctata* the maxima recorded were 20°C, 22.5°C, 22.5°C and 25°C for the first, second, third and fourth instars respectively (Mills, 1981). Mills (1981) suggested these differing optima may reflect the increasing temperatures experienced by the coccinellid larvae as they progress through the life cycle in the field. However, in other predators, there is no such tendency, e.g. in the damselflies *Lestes sponsa*, *Coenagrion puella* and *Ischnura elegans* maximum development rates were recorded at the same temperature for the last five instars (Pickup and Thompson, 1990)

Interaction Between Temperature and Consumption Rate

Whilst temperature will affect growth and development rates of predators directly, one has to be aware that it can also exert an influence by changing the prey consumption rate (Mills, 1981; Gresens *et al.*, 1982; Sopp and Wratten, 1986; Pickup and Thompson, 1990) (Figure 2.55). The rate at which food passes through the gut will be positively temperature-dependent, and this will affect consumption rate by affecting hunger (insect hunger is directly related to the degree of emptiness of

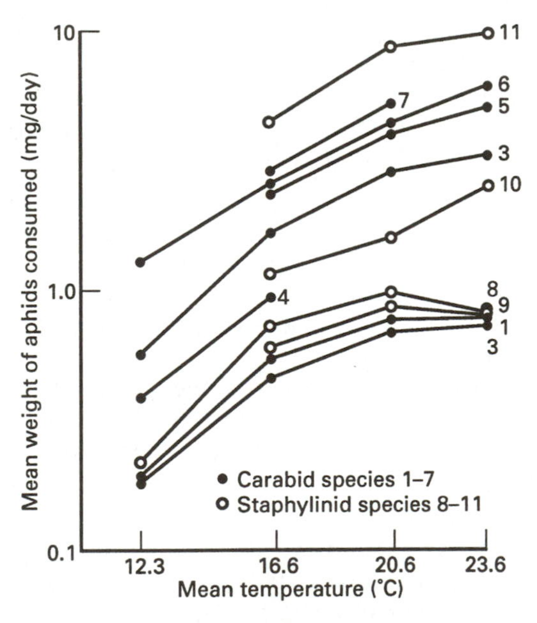

Figure 2.55 Effect of temperature on the mean weight of aphids (*Sitobion avenae*) consumed per day by eleven species of carabid and staphylinid beetles. In the experiments, individual beetles were given an excess of prey (first and second instar, in approximately equal proportions). (Source: Sopp and Wratten, 1986.) Reproduced by permission of Kluwer Academic Publishers.

the gut (Johnson *et al.*, 1975)). *B* in equations 2.9–2.14 (representing in part basal metabolic costs) will also be temperature-dependent (Pickup and Thompson, 1990), and consumption rate will increase to counteract an increase in *B*.

To take any confounding effects of varying consumption rate into account when assessing the influence of temperature on growth and development rates in *Adalia bipunctata*, Mills (1981) compared: (a) the mean growth and development rates recorded at the experimental range of temperatures (i.e. fixed daily ration of prey) with (b) those predicted from Figure 2.43 (i.e. constant temperature regime) for the appropriate rates of consumption (Figure 2.56). With this analysis, Mills (1981) recorded significant deviations from the predicted growth and development rates, indicating that temperature does have a direct influence on growth and development.

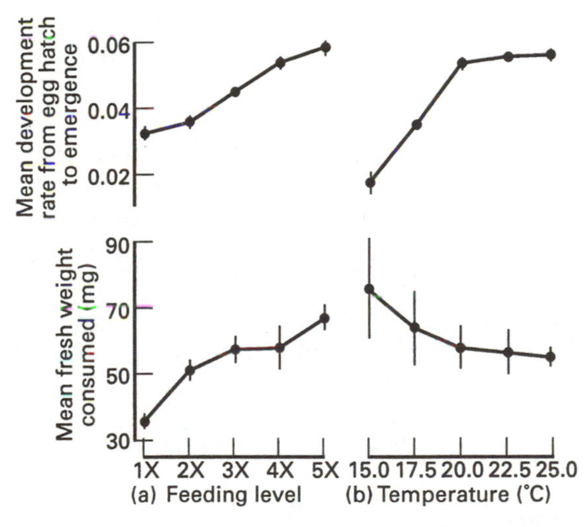

Figure 2.56 The mean rates of prey consumption and development of the immature stages of *Adalia bipunctata* (Coccinellidae) in relation to prey availability at: (a) a constant temperature (20°C) and various `feeding levels' (the weight of prey corresponding to 1× to 5× the average teneral weight of the instar); (b) a range of temperatures and one (4×) feeding level. (Source: Mills, 1981.) Reproduced by permission of Blackwell Scientific Publications Ltd.

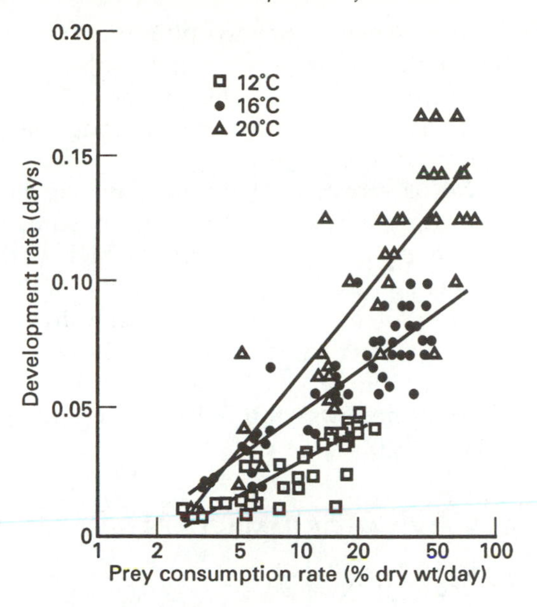

Figure 2.57 Development rate in relation to consumption rate (note log scale) in *Coenagrion puella* (Odonata: Zygoptera). Temperature affects development directly and indirectly by increasing the prey consumption rate. There is a clear interaction effect between temperature and consumption rate. (Source: Pickup and Thompson, 1990.) Reproduced by permission of Blackwell Scientific Publications Ltd.

A more straightforward approach to determining how consumption rate interacts with temperature to affect development rate involves plotting development rate against consumption rate, constructing regression lines for each temperature regime and then comparing the slopes of the lines. As can be seen from plots for the damselfly *Coenagrion puella* (Figure 2.57), higher consumption rates produce stronger developmental responses to increases in temperature.

Other Physical Factors

Diurnal predator larvae may, like the adults (subsection 2.7.4), show a reduction in daily consumption rate with decreasing photoperiod, and this will be reflected in a reduction in growth and development rates. Bear in mind, when varying photoperiod in experiments, that you may also be inadvertently varying the absorption of radiant energy by insects, thus altering their exposure to temperature.

Larvae of terrestrial predators may, like the adults, increase their rate of prey consumption with decreasing humidity, which will cause them to grow larger and more rapidly. Predator larvae may develop faster in an incubator than in a large environment chamber, even at the same temperature, because of the lower humidity in the former (Heidari, 1989).

2.10 SURVIVAL OF IMMATURES

2.10.1 INTRODUCTION

Below we discuss some factors that affect the survival of predator and parasitoid immatures. Parasitism and predation by heterospecifics are not considered (Chapter 5 gives a discussion of practical approaches), whereas predation by conspecifics, i.e. cannibalism (Sabelis, 1992) is. Mortality of parasitoid juveniles is largely dependent on that suffered by the hosts that support them. Hosts may be killed through predation, starvation and exposure to unfavourable weather conditions, and any parasitoids that are attached to or contained within the hosts will die. It follows that the probability of survival of parasitoid immature stages is strongly dependent on the probability of survival of the host. Price (1975) illustrated this relationship by reference to the host survival curves, which in insects are of either Type II or Type III (Figure 2.37), i.e. substantial mortality of hosts (very substantial in the latter case), and therefore of any parasitoid progeny they support, occurs by the mid-larval stage.

When investigating larval mortality, the possibility ought to be considered that some factors may cause higher mortality in one sex than in another. The same applies when studying parasitoids. For example, some parasitoids allocate male eggs to small host individuals, e.g. pupae, and female eggs to large individuals (subsection 1.9.1). If small hosts suffer a higher degree of mortality from a predator than larger ones, then the survival rates of male and female parasitoids will differ.

2.10.2 EFFECTS OF BIOTIC FACTORS ON SURVIVAL OF IMMATURES

Food Consumption by Predators

By recording deaths of individuals within each instar in the food consumption experiment outlined earlier, the relationship between food consumption and survival can be studied.

A model relating larval survival to prey availability was developed by Beddington *et al.* (1976b). If we assume that we are not dealing with a population of genetically identical individuals, then we would expect mortality through food shortage to take place at some characteristic mean ingestion rate μ_I with the population as a whole displaying variation about this mean value. Assuming that the proportion of the population experiencing 'food stress' is normally distributed about the mean, with standard deviation σ_I, the proportion (S) of the larval population surviving to complete development within any particular instar of duration d, at an ingestion rate I, will be given by (Beddington *et al.*, 1976b):

$$S = \frac{1}{\sqrt{2\pi}} \int_{-\infty}^{z} \exp\left(-\frac{z^2}{2}\right) dz \qquad (2.17)$$

where $$z = \frac{I - \mu_I}{\sigma_I}$$

Using equations 2.2 and 2.17, S may be expressed in terms of consumption rate or prey density (Figure 2.58). The relationship shown in Figure 2.58c is shown by predators in the laboratory (Figure 2.59). As pointed out by Beddington *et al.*, whether a survival curve rises extremely rapidly or slowly depends on the range of prey densities over which experi-

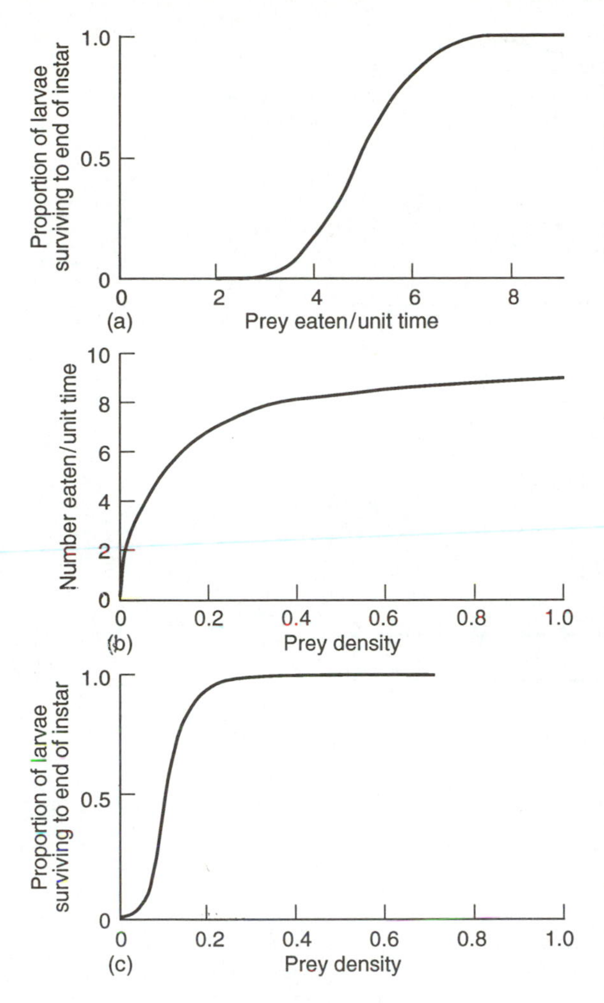

Figure 2.58 Hypothetical relationships between (a) the proportion of individual predators surviving to the end of an instar and their mean feeding rate during that instar; (b) predation rates and prey density; (c) the relationship obtained by combining (a) with (b). (Source: Beddington *et al.*, 1976b.) Reproduced by permission of Blackwell Scientific Publications Ltd.

ments are carried out and the graphical scales chosen for plotting the data.

Mortality due to nutritional stress apparently occurs at feeding rates very much higher than the minimum rate necessary for growth and development, so that individuals that are growing normally (albeit slowly) at low feeding rates are apparently highly likely to suffer high mortality at the moult (Beddington *et al.*, 1976b). There may be no relationship between the overall survival rate between entering and leaving an instar (S) and the duration of each instar (d), as in *Blepharidopterus angulatus*, or S may decline in a variety of ways with increasing d (examples in Beddington *et al.*, 1976b).

Survival rates vary between successive instars at comparable prey densities. Figure 2.60 summarises the relationship between instar and the feeding rate at which 50% of a larval cohort survive, in four predatory insects and a spider. The plots indicate a constant increase in feeding rate between instars to maintain survival rates at 50%. In the case of *Ischnura elegans* feeding rates necessary to ensure 50% survival approximately doubled between instars ten and eleven, and increased by a factor of 1.4 between instars eleven and twelve (Lawton *et al.*, 1980). The theoretical line in Figure 2.60 is for larvae that double their body weight between instars and in which minimum requirements increase by a 0.87 power of body weight (which they do for larvae of a damselfly closely related to *Ischnura*. The steeper slopes shown by *Adalia*, *Notonecta* and the spider perhaps reflect the higher exponents in their metabolic rate:weight relationships and/or larger increases in average body weights between instars (Lawton *et al.*, 1980).

Prey Species

The relationship between larval survival and prey species has been most thoroughly investigated in coccinellid beetles (Hodek, 1973). Coccinellid larvae have been presented with *acceptable* prey of various species, and larval mortality measured.

Consumption rate in relation to each prey species can be measured, and prey-related differences in survival correlated with differences in consumption rate. However, if there is a reduced rate of consumption on a prey

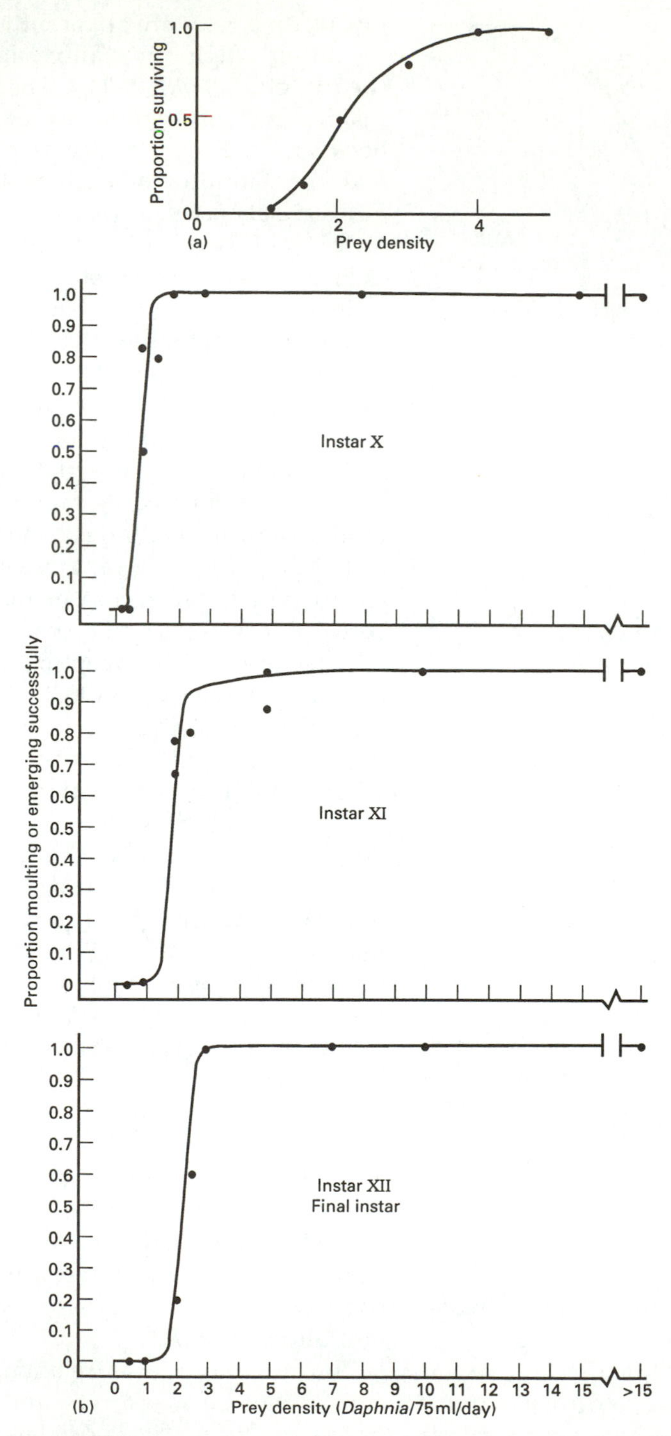

Figure 2.59 Relationship between the proportion of predators surviving to the end of an instar and the mean density of prey available during that instar. (a) First instars of the coccinellid beetle (*Adalia bipunctata*); (Data from Wratten, 1973) (b) tenth, eleventh and twelfth (final) instars of the damselfly *Ischnura elegans* (Source: Lawton *et. al.*, 1980). Reproduced by permission of Blackwell Scientific Publications Ltd.

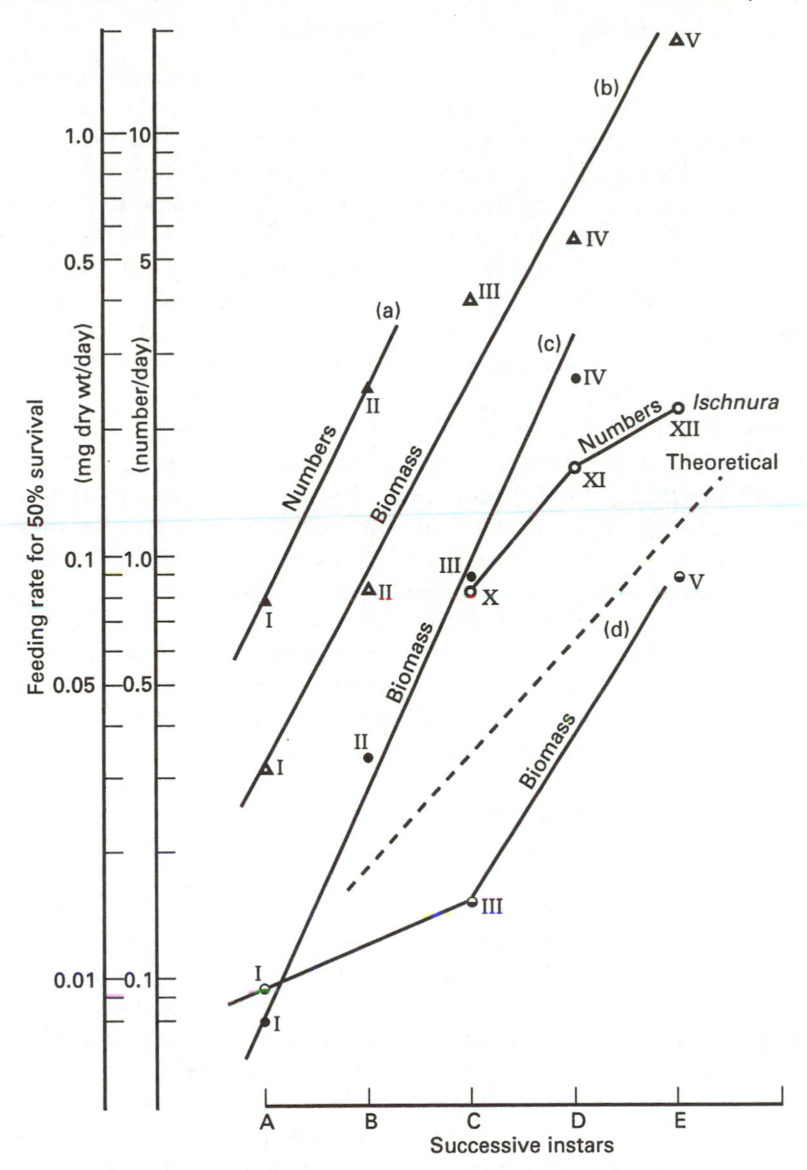

Figure 2.60 The relationship between instar number (in Roman numerals) and the consumption rate (feeding rate) at which 50% of a cohort of larvae in a particular instar successfully complete their development (LD_{50} for food stress): (a) *Adalia decempunctata* (Coccinellidae) (data from Dixon, 1959); (b) *Notonecta glauca* (Notonectidae) (Data from McArdle, 1977); (c) *Linyphia triangularis* (Arachnida: Linyphiidae) (Data from Turnbull, 1962); (d) *Blepharidopterus angulatus* (Miridae) (Data from Glen, 1973). The theoretical line is for larvae that double their body weight between instars, and where food-energy requirements increase by a 0.87 power of body weight. (Source: Lawton *et al*., 1980)

species and survival is also low on that species, one cannot necessarily conclude that poor survival is a direct result of reduced consumption rate. Survival on the 'better' prey species may still be higher than on the 'poorer' one at equivalent consumption rates (Hodek, 1973). If it is, then the prey-related difference in survival may be due to factors such as differences in the size or qualitative attributes of the prey species.

Interference and Exploitation Competition, Cannibalism

Interference from conspecifics and predators can, through its effects on feeding rates, potentially reduce survival (Heads, 1986; Sih, 1982) (subsection 2.9.2). Exploitation is an obvious potential cause of mortality among larval conspecifics, as prey may be depleted to a level at which larvae experience nutritional stress.

An additional cause of mortality in the immature stages of dragonflies, waterboatmen, coccinellid beetles, anthocorid bugs and ant-lions (Crowley *et al.*, 1987; Sih, 1987; Mills, 1982; Agarwala and Dixon, 1992; Naseer and Abdurahman, 1990; Griffiths, 1992) is cannibalistic behaviour. The adults and larvae of several Coccinellidae are known to be cannibalistic on eggs. In dragonflies, cannibalism may result in the death not only of one of the interacting pair (same or smaller instar larva) but also both participants, since it could attract the attention of predators (Crowley *et al.*, 1987) (this also applies to non-cannibalistic interference).

The effects of competition or cannibalism on survival in immature stages can be expressed as either percentage mortality plotted against predator density or as k-values for the mortality plotted against $\log_{10}$ predator density (Varley *et al.*, 1973) (subsection 5.3.4). If density-dependent mortality occurs, it will be shown within the upper range of densities only, i.e. there will be a threshold density of predators below which k is zero (Mills, 1982b) (or its value is slighty above zero, in which case one has to question whether the mortality recorded at low predator densities is entirely attributable to interference, exploitation or cannibalism). The manner in which k varies with $\log_{10}$ predator density indicates the nature of the density dependence, i.e. exact, over- or under-compensation (subsection 5.3.4) and whether competition is of the scramble-type or contest-type (for explanations of these terms, see Varley *et al.*, 1973; Begon *et al.*, 1990).

Host Size

As well as measuring growth and development of parasitoids, larval survival can also be recorded in relation to host size at oviposition. One might reasonably assume, for nutritional reasons, that generally for solitary idiobionts survival is highest in large hosts, although there could be cases where hosts above a certain size represent a resource in excess of the amount required by the larva to complete its development (in such cases larval survival may not be improved in the largest hosts, and it may even be reduced, e.g. due to putrefaction of the remaining host tissues).

By the same token, the relationship for koinobionts is likely to be more complex. For the solitary koinobionts *Lixophaga diatraeae* and *Encarsia formosa*, survival is highest in individuals that complete their development in medium-sized hosts (Miles and King, 1975; Nechols and Tauber, 1977). For the solitary koinobiont *Leptomastix dactylopii*, no significant differences were found between survival in different-sized hosts (de Jong and van Alphen, 1989). In *Venturia canescens* survival is highest in medium-sized hosts and lowest when the second instar host is oviposited in (Figure 2.49c). The probable reason for the lower survival in second instar hosts is injury to the host through insertion and removal of the ovipositor (this does not occur when later instars are attacked) (Harvey *et al.*, 1994).

A possible complicating factor in experiments is mortality from encapsulation, which may be higher in some stages than in others, so samples of hosts need to be taken and dissected during larval development to provide data on the frequency of encapsulation.

In those gregarious parasitoids in which progeny survival is 100% at the smallest clutch size (e.g. Figure 1.10a), 100% survival might also occur at larger clutches if larger-sized hosts are utilised. The slope of the relationship might also be less steep in the case of larger hosts.

Host Species

Given that different host species are likely to constitute different resources, both qualitatively and quantitatively, parasitoid larval survival may vary with host species, larvae (or eggs) dying through malnutrition, encapsulation or poisoning (if the host has sequestered toxins from its food plant) (Vinson and Iwantsch, 1980b). In *Telenomus lobatus* percentage eclosion, i.e. survival of progeny, was higher from the eggs of *Chrysoperla* species than from eggs of *Chrysopa* species (Ruberson *et al.*, 1989). In the gregarious idiobiont *Bracon hebetor*, survival within clutches was density-dependent both on a small moth species, *Plodia interpunctella*, and on a large moth species, *Anagasta kuehniella*, but the density-dependence in the latter case applied only to very high (artificially manipulated) clutch sizes (Taylor, 1988a).

To investigate the effect of host species on larval survival, present females with hosts of different species and, if possible, of equivalent size. Maintain the hosts until the parasitoids pupate, and maintain the parasitoid pupae until adults have ceased emerging. Any hosts that have received eggs but have not given rise to parasitoids should be examined (dissected in the case of endoparasitoids) for the remains of parasitoid eggs or larvae. Any parasitoid pupae that fail to produce adults should also be recorded. Sex differences in survival should also be looked for (Ruberson *et al.*, 1989).

Toxins in the Host's Food Plant

The effects of the alkaloid α-tomatine on the development and growth of *Hyposoter exiguae* has been discussed. Campbell and Duffey (1979) also showed that the percentage eclosion of adult parasitoids was significantly reduced in parasitoids that developed in hosts fed upon diets containing the alkaloid. Williams *et al.* (1988) recorded reduced survival of *Campoletis sonorensis* when high amounts of gossypol were added to the host's artificial diet.

Superparasitism

Solitary Parasitoids

In solitary endoparasitoids, supernumerary larvae are eliminated either through physiological suppression or, more usually, through combat, i.e. contest competition (Clausen, 1940; Starý, 1966; Fisher, 1971; Vinson and Iwantsch, 1980b). This applies not only to self- and conspecific superparasitism but also to heterospecific superparasitism (= multiparasitism).

Physiological suppression occurs either shortly after eclosion of the oldest embryo (apparently by a toxic secretion) or late in larval development (by starvation or asphyxiation). It may not be as common a process as was first thought, while evidence is increasing that the parent female may suppress rival offspring by injecting a toxin or viruses during oviposition (Mackauer, 1990).

The first instar larvae of almost all hymenopteran parasitoids are equipped with large, often sickle-shaped mandibles (Fisher, 1961; Salt, 1961) (Figure 2.61). Fighting often takes place between larvae that are of approximately the same age, although in some species first instar larvae will attack and kill later instars that either have reduced mandibles or lack mandibles altogether (Chow and Mackauer, 1984, 1986). Note that the possession by first instar larvae of large mandibles does not necessarily mean that fighting is the only mechanism employed in the elimination of rivals (Strand, 1986; Mackauer, 1990).

The mechanisms employed in the elimination of larval competitors in three solitary braconid parasitoids of aphids are summarised in Figure 2.62. As with other parasitoids, in cases of *intraspecific* larval competition the oldest larva generally survives in competition with a younger larva, although this may not apply where the larval

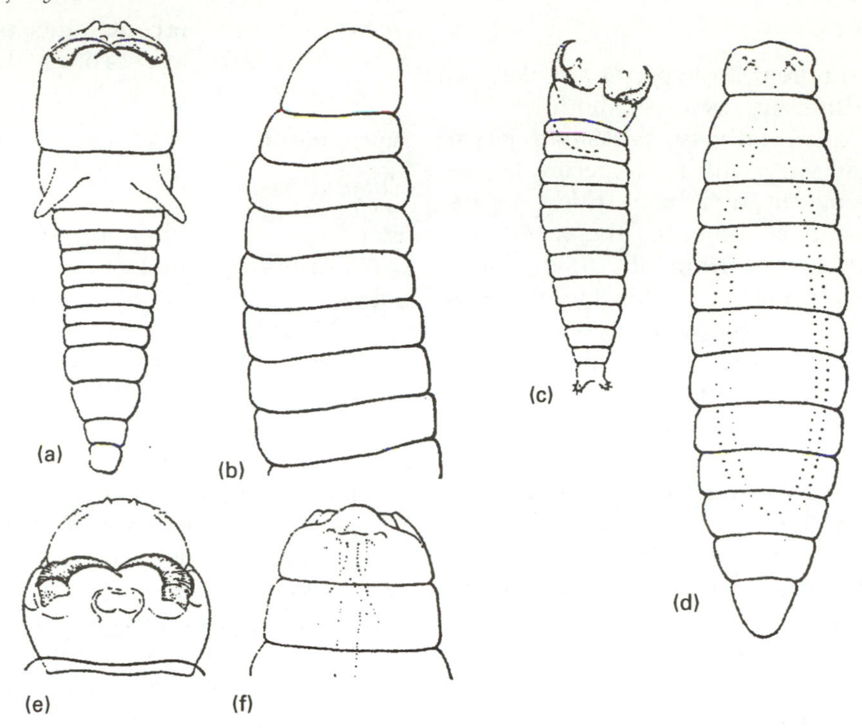

Figure 2.61 Larvae of parasitoid Hymenoptera that in the first instar have mandibles suitable for fighting (a,c,e) but do not have such mandibles in the second instar (b,d,f): (a,b) *Opius fletcheri* (Braconidae); (c,d) *Psilus silvestri* (Diapriidae), (e.f) *Diplazon fissorius* (Ichneumonidae). (Source: Salt, 1961.) Reproduced by permission of The Company of Biologists Ltd.

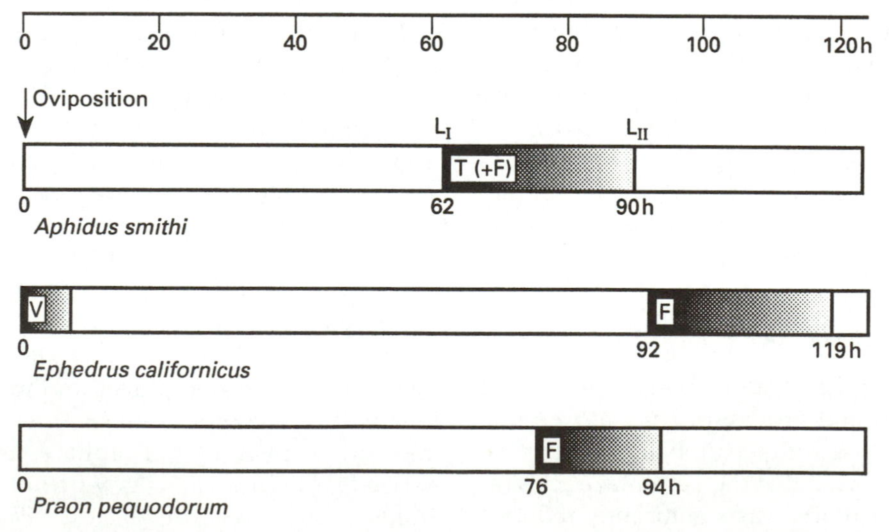

Figure 2.62 Mechanisms used in the elimination of rival competitors by the braconid parasitoids *Aphidius smithi, Ephedrus californicus* and *Praon pequodorum* in pea aphids. F = fighting among first instar larvae (L_I); T = toxin released at eclosion of L_I; V = venom injected by female at oviposition. Median times of eclosion to L_I and L_{II} refer to parasitoid larvae developing in second-instar pea aphids at 20°C. (Source: Mackauer, 1990). Published by permission of Intercept Ltd.

age difference is either very small or very large (in the latter case the older larva may have developed to the second, i.e. amandibulate instar by the time the second egg is either laid or hatches; Bakker *et al.*, 1985).

The 'oldest larva advantage' applies in some cases of *interspecific* larval competition among parasitoids but not to others (Tillman and Powell, 1992; Mackauer, 1990). Relative age differences can influence the outcome of an interaction, but the factors that appear to be more important in determining who survives are the particular competitive mechanism(s) and the developmental stage at which each comes into play. Bear in mind that the eggs of two species may be laid at the same time, but may hatch at different times and/or the development rate of the larva may be greater in one species than in another, and this may determine the 'window of interaction'. For example, the braconids *Aphidius smithi* and *Praon pequodorum* require about the same time to develop from oviposition to the second instar, but the embryonic period is much shorter in *Aphidius* than in *Praon*. This difference enables *Praon* to compete as a mandibulate first-instar larva with an older *Aphidius* larva. *Aphidius* larvae usually survive only if they have reached the end of the fourth (final) instar while *Praon* is still in the embryonic stage and thus unable to attack an older competitor (Chow and Mackauer, 1984, 1985).

A parasitoid species that wins under most conditions is described as **intrinsically superior** (Zwölfer, 1971; 1979). Ectoparasitoids tend to be intrinsically superior to endoparasitoids (Sullivan, 1971, gives one exception). The superiority of ectoparasitoids is a result of envenomation and/or more rapid destruction of the host, rather than a result of the endoparasitoid being attacked directly (Askew, 1971; Vinson and Iwantsch, 1980b).

An experiment designed to investigate the relative competitive superiority of solitary endoparasitoid larvae in instances of superparasitism would involve varying the time interval between ovipositions (from a few seconds to many hours), either by the same parasitoid species or females of different species. If heterospecific superparasitism is being studied, then the sequence of species ovipositions can be reversed. Whatever the type of interaction being investigated, by taking regular samples of the superparasitised hosts and singly-parasitised hosts at varying periods from the time of the second oviposition and dissecting them, the following can be recorded:

1. the stage of development (embryonic or larval) already reached by the older parasitoid at the time of the second oviposition (determine from dissection of singly-parasitised hosts);
2. the stage of development subsequently reached;
3. the stage of development of the younger parasitoid;
4. which, if any, of the eggs or larvae are dead or alive (exceptionally, both may be dead, as suggested by the data in Table 2.3);
5. any evidence of physical combat;
6. whether either of the parasitoids bear wounds (the latter may show signs of melanisation (Salt, 1961)).

Threshold time intervals for the different outcomes of competition (if there can be more than one outcome) can then be found. Note that for some interactions, the period of time between oviposition and the development of *the host* to a certain stage, indirectly determines which parasitoid species is the survivor. For example, in the case of *Trieces tricarinatus* and *Triclistus yponomeutae*, this interval determines the extent of development of the parasitoids after host pupation and the extent of development at the time of combat (irrespective of whether the host is singly- or multiparasitised, development of larvae beyond the first instar can only take place after host pupation) (Dijkerman and Koenders, 1988).

Instead of dissecting superparasitised hosts, the outcome of competition can be studied by rearing the parasitoids to adulthood. However, in studies of intraspecific superparasitism, this method requires that distinguishable (preferably morphologically distinguishable) strains be used. This method would also prove useful for studying intraspecific superparasitism when the interval between ovipositions is so short that it is not possible, through dissection, to distinguish between the progeny of the first female and the progeny of the second female. For example, Visser *et al.* (1992c) measured the pay-off from superparasitism in the solitary parasitoid *Leptopilina heterotoma*. They used two strains of this species: a wild type with black eyes and a mutant with yellow eyes. Hosts parasitised by females of one strain were exposed to females of the other strain and the interval between ovipositions was varied. The sequence of ovipositions was reversed to take account of any competitive asymmetry between strains. The probability of a second female realising an offspring from superparasitism, i.e. the pay-off, was then calculated for each strain.

Harvey *et al.* (1993) examined whether parasitoid mortality from superparasitism varies with host instar in cases of near-simultaneous oviposition by two conspecific females (of *Venturia canescens*). Parasitoids were reared from third and fifth instar hosts (the moth *Plodia interpunctella*) containing either one, two or four parasitoid eggs. Parasitoid mortality was found to be significantly higher in fifth instar hosts than in third instar hosts, but *within* instars did not vary with egg number (Figure 2.63). The probable reason for the higher mortality in fifth instar hosts is that there is some physiological incompatibility between the parasitoid and fifth instar hosts associated with pupation. (this same factor could have affected development and growth, subsection 2.9.2) (Harvey *et al.*, 1993).

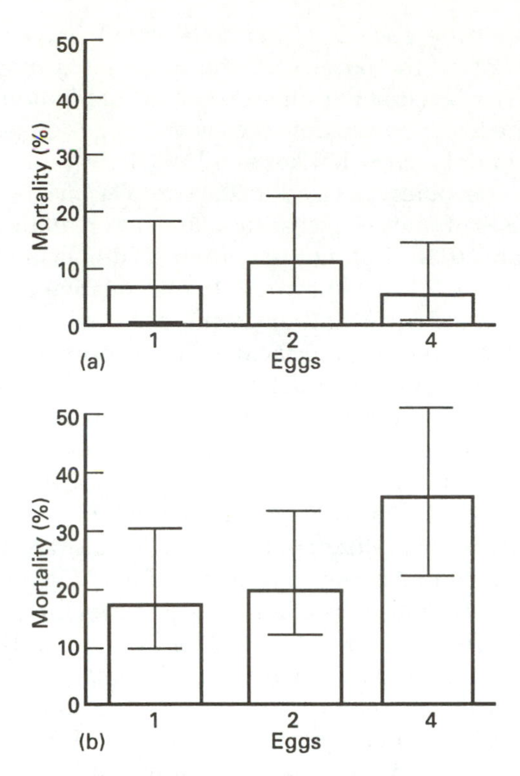

Figure 2.63 Effects of superparasitism on survival in the ichneumonid parasitoid *Venturia canescens*: Mortality of parasitoids reared from (a) third instar; (b) fifth instar hosts, containing one, two or four parasitoid eggs. Encapsulation was not a complicating factor in the experiments. (Source: Harvey *et al.*, 1993.) Reproduced by permission of Blackwell Scientific Publications Ltd.

Gregarious Parasitoids

In gregarious species where survival declines monotonically with increasing clutch size, the addition of an egg or clutch of eggs will (further) reduce percentage survival per host. The reduction will normally result from increased resource competition, since larvae of gregarious species do not engage in physical combat. In those species in which there is an Allee effect (1.6), there will be a threshold number of progeny per host below which all parasitoids die, so superparasitism of a host containing a clutch of eggs that is a number short of this threshold number is likely to *raise* the survival chances of the parasitoid immatures.

Assuming competitive equivalence of first and second clutches laid in or on a host, the effect on survival of simultaneous oviposition by two females would be analogous to the effects of increasing the initial clutch size (Strand and Godfray, 1989). However, in gregarious species mortality may vary not only with the number of eggs initially present but also with the time interval between ovipositions, i.e. it will depend on how soon superparasitism occurs after the laying of the initial clutch (Strand, 1986). Strand and Godfray (1989) demonstrated this for *Bracon hebetor*. In this species progeny survival within a second egg clutch, equal in size to the first, was approx. 42% (each clutch comprising 20 eggs), approx. 78% (each clutch comprising 10 eggs) and 83% (each clutch comprising four) when the first and second clutches were 'laid' simultaneously (they were placed on hosts by the experimenter, see below). However, when the time between 'ovipositions' was 12 hours or more, progeny survival within the second clutch was reduced to less than 10% for clutches comprising either 10 or 20 eggs (Figure 2.64).

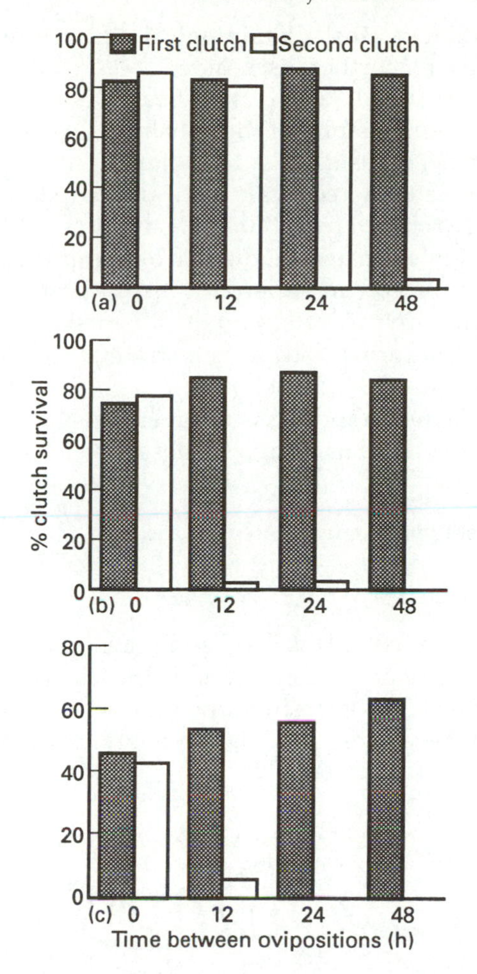

Figure 2.64 Relationship between *Bracon hebetor* (Braconidae) progeny survival within first and second clutches of eggs, and the time between 'ovipositions' for starting clutches of (a) 4; (b) 10; (c) 20 eggs. First and second clutches were equal in size for each experiment. (Source: Strand and Godfray, 1989.) Reproduced by permission of Springer-Verlag GmbH & Co. Kg.

Experiments aimed at investigating the mortality effects of superparasitism in a gregarious species can be conducted along the lines described in the section on superparasitism in relation to growth and development rates. Sex differences survival may be examined in such experiments; several studies (Vinson and Iwantsch, 1980b) have shown that with increased larval crowding there is a tendency for preferential survival of males.

Superparasitism in egg parasitoids can be investigated using *in vitro* techniques (Strand and Vinson, 1985; Strand *et al.*, 1986); parasitoid eggs and larvae can be added to a volume of culture medium equivalent in volume to a host egg.

As noted in Chapter 1 (section 1.8), the survival chances of parasitoid immatures can in some case be *improved* by superparasitism. For example, of eggs of *Asobara tabida* laid in larvae of *Drosophila melanogaster*, 1% survive in singly parasitised hosts whereas 7% survive in superparasitised hosts (van Alphen and Visser, 1990), encapsulation being the principal cause of mortality in both cases. Van Strien-van Liempt (1983) measured the survival of *Asobara tabida* and *Leptopilina heterotoma* in multiparasitised *Drosophila* hosts and compared it with survival in singly parasitised hosts. Percentage survival in instances of multiparasitism was not always lower than survival in singly parasitised hosts; in most cases, multiparasitism provided a mutual survival advantage. In cases such as these where parasitoid survival is increased through superparasitism, the mechanism is thought to be exhaustion of the host's supply of haemocytes (see **Host physiological defence reactions**, for further discussion).

Compared with an Israeli strain, a Californian strain of the aphelinid *Comperiella bifasciata* was subject to a higher encapsulation rate in red scale and also superparasitised more hosts. Blumberg and Luck (1990) suggested that since the risk of encapsulation is reduced in superparasitised hosts, the higher degree of superparasitism shown by the Californian strain is a strategy to avoid encapsulation.

The mortality effects of superparasitism can be expressed as *k*-values (subsection 5.3.4).

Host Physiological Defence Reactions

Introduction

Endoparasitoid larvae and eggs may die owing to a reaction of the the host's immune system. The term **immune system** is used in the loose sense that the hosts are capable of mounting a defensive response against foreign bodies. The response does not involve either a specific 'memory', with accelerated rejection of the second of two sets of an introduced foreign tissue, or a marked increase in the concentration of some specific humoral component, as has been shown for vertebrates. Thus, the probability of a parasitoid eliciting an immune response in an insect is independent of previous challenges (Bouletreau, 1986).

Host defence reactions are of several kinds (Ratcliffe, 1982; Ratcliffe and Rowley, 1987 for reviews), but the most commonly encountered type of reaction is encapsulation. Usually in encapsulation the foreign invader becomes surrounded by a multicellular sheath composed of the host's haemocytes (Figure 2.65). Successive layers of cells can often be discerned, and adjacent to the parasitoid there often develops a necrotic layer of melanised cells, representing the remnants of the blood cells that initiated the encapsulation reaction. The melanin deposits on the surfaces of encap-

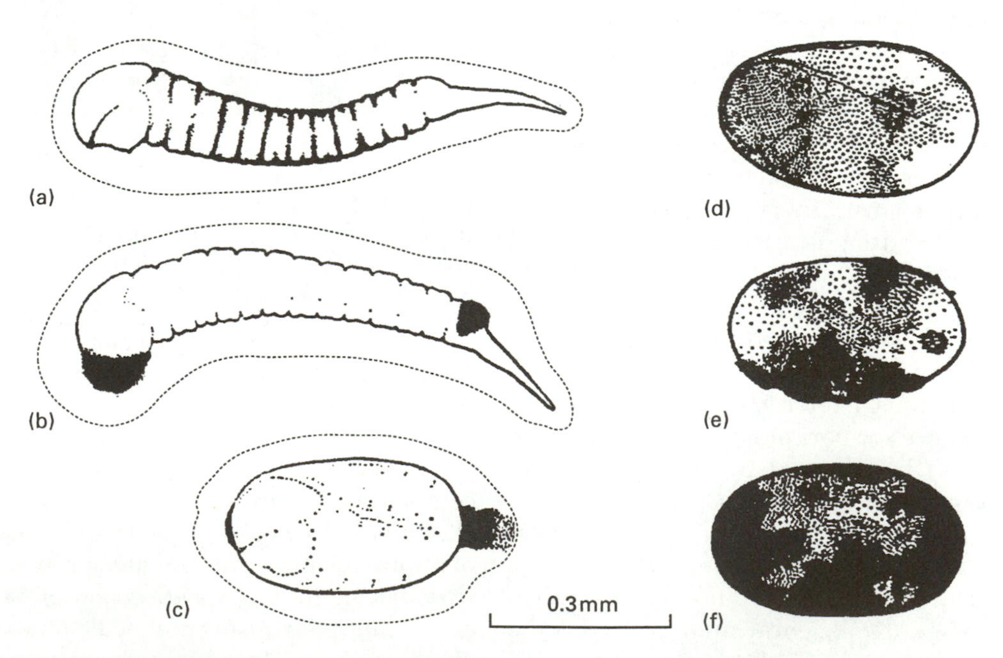

Figure 2.65 Encapsulation and melanisation: (a-c) encapsulated larvae and eggs of *Venturia canescens* implanted in a non-host insect (deposits of melanin can be seen); (d-f) deposits of melanin on eggs 24, 32 and 48 hours after implantation in a non-host insect. (Source: Salt, 1970.) Reproduced by permission of Cambridge University Press.

sulated parasitoid eggs and larvae often provide the first clue to the occurrence of encapsulation (Figure 2.65). Parasitoid immatures die probably from asphyxiation, although starvation may be the principal cause of death in some cases. Phagocytosis of parasitoid tissues gradually occurs, at least during the initial stages of encapsulation.

Parasitoids can resist, i.e. evade and/or supress the immune responses of their hosts. One means of evasion is the laying of eggs in refuges from encapsulation. Some parasitoids oviposit into specific host organs such as the nerve ganglia and salivary glands, where an egg cannot be reached by the host's haemocytes (Salt, 1970; Rotheray, 1979). In some parasitoids, the immature stages have surface properties that prevent encapsulation (Salt, 1968; Rotheram, 1967). The risk of encapsulation in a particular host species can be drastically increased by washing the surface of the parasitoid eggs using either solvents or water before they are artificially injected. Eggs of *Venturia canescens* removed from the ovarioles are encapsulated if they are artificially injected into the haemocoel of *Anagasta*, whereas eggs removed from the oviduct and calyx region do not become encapsulated.

Several parasitoids are known to inject materials into the host at oviposition which inhibit encapsulation (Rizki and Rizki, 1984; Vinson and Stoltz, 1986). In some species the materials responsible for inhibition are polydnaviral particles, although in some cases the parasitoid's venom is required to synergise the effect of the virus (Guzo and Stolz, 1985; Kitano, 1986). In *Cardiochiles nigriceps* (Braconidae) the larva is protected (presumably), by the virus, but the egg is protected by a resistant outer coating (Stoltz and Vinson, 1979).

Encapsulation is usually studied *in vivo* in either laboratory-cultured or field-collected hosts. However, some workers have successfully used an *in vitro* technique (Ratner and Vinson, 1983; Benson, 1989).

Population genetic and dynamic aspects of encapsulation are discussed by Bouletreau (1986) and Godfray and Hassell (1991).

Host Populations and Species

The ability of a host to encapsulate parasitoids is genetically determined and there may be considerable variability between populations of a host species in encapsulation rate (Bouletreau, 1986; Maund and Hsiao, 1991) (see also Dijkerman, 1990). For example, in *Drosophila melanogaster* there are clear differences between fly populations from different parts of the World with respect to the frequency with which *Leptopilina boulardi* is encapsulated (Bouletreau, 1986). Such effects are an important consideration when one is planning to release biological control agents (Maund and Hsiao, 1991).

The ability of a parasitoid species to avoid encapsulation may determine at least partly the range of host species that it parasitises in nature. Differential mortality in different host species due to encapsulation may also have played an important role in the evolution of host specificity, including preferences, of a parasitoid. To shed light on these aspects of the parasitoid–host interaction, female parasitoids can be presented in the laboratory with: (a) their natural host species, and (b) closely related species that are likely to be encountered in nature but either never produce parasitoid adults of that species in rearings or produce them only occasionally. If females can be induced to oviposit in the (b) hosts (this may prove impossible), the fate of their eggs can be monitored by dissection, and encapsulation frequency in a particular host species related to the frequency of parasitism in the field. If the frequency with which the parasitoid species is reared from different host species is determined by differential mortality of progeny, then we would expect an inverse correlation between the percentage of parasitised hosts in which encapsulation occurs and the percentage initial parasitism, i.e. percentage parasitism prior to the encapsulation taking effect.

Dijkerman (1990) observed that the abundance of *Diadegma armillata*, a solitary endoparasitic ichneumonid, in the parasitoid complexes associated with *Yponomeuta* moths varies among host species, being high in the complex associated with *Y. evonymellus* and very low in that associated with *Y. cagnagellus*. To determine whether this variation corresponds with the ability of each host species to encapsulate the parasitoid. Dijkerman used the following methods:

Parasitism experiments: Larvae of the different moth species were exposed to female *D. armillata*. Several days later, a sample of the hosts was taken and the insects dissected. The remaining hosts were maintained until the parasitoids emerged. By *dissecting* the hosts, the presence of parasitoid eggs or larvae was recorded, and the following noted:

1. the rate of infestation, i.e. the number of host larvae containing at least one egg of *D. armillata* as a percentage of the total number of larvae dissected;
2. percentage encapsulation, i.e. the number of encapsulated progeny as a percentage of the total number of eggs found at dissection (this measure of encapsulation efficiency might be less useful in cases where there is a high and variable degree of superparasitism among hosts, which was not the case in this study, see below).

By *rearing* hosts, the following were measured:

3. the rate of successful parasitism, i.e. the number of host individuals yielding *D. armillata* adults as a percentage of the total number of *Yponomeuta* yielding moths or parasitoids;
4. the percentage mortality of larvae, i.e. the number of larvae dying during their development as a percentage of the initial number of parasitoid larvae (if the mean number of parasitoid eggs per parasitised host significantly exceeds 1.0, a correction factor will need to be applied to the data to allow for the effects of parasitoid mortality through superparasitism).

Simultaneously, under the same conditions, host larvae that were not exposed to parasitoids were reared to moth emergence. This was done to establish whether the results could be biased by a higher mortality of parasitised hosts, compared with unparasitised hosts, in rearings.

Dissections of field-collected, late instar, hosts: The following were recorded:

5. percentage encapsulation (see above); percentage of successful attacks (successful at the time of dissection, notwithstanding encapsulation later on), calculated as: [the number of parasitoid eggs or larvae recorded at dissection, divided by the total number of hosts dissected] $\times$ 100. To exclude the potentially confounding effects of time and place, comparisons were made only for samples collected at the same locality and same time of day.

Except for one species, *Y. evonymellus*, infestation and successful parasitism rates recorded in the laboratory were markedly different. In *Y. mahalebellus* and *Y. plumbellus* no wasps were reared despite infestation rates of 30% and 95%, whereas in *Y. evonymellus* almost all infested larvae yielded adult parasitoids. Since mortality of parasitised hosts was not different from that of control larvae, and the mean number of parasitoid eggs per parasitised host was little more than 1.0, the differences between infestation and successful parasitism could be explained in part by encapsulation. The field dissections revealed that *Y. cagnagellus* suffers fewer successful attacks than *Y. evonymellus*, despite being the more abundant species at some localities. The low successful parasitism in *Y. cagnagellus* corresponds with the very low probability of survival of *D. armillata* in that species.

An interesting footnote to Dijkerman's findings is the observation that all the *Yponomeuta* species in which there was a high rate of encapsulation of *D. armillata* are considered to have diverged early in the evolution of the genus, whereas the more recently evolved moth species show either an intermediate rate of encapsulation or do not encapsulate eggs at all (Dijkerman, 1990).

An alternative approach was taken by Benson (1989), who used an *in vitro* technique. He tested the eggs of three aphidophagous ichneumonid species (Diplazontinae) against the haemolymph of a range of hover-fly species. The host ranges and preferences (including behavioural preferences) of each species were already well known, and this enabled rank orders of reaction among host species for a given parasitoid species, and among parasitoid species for a given host species, to be predicted. Haemolymph from a host species was mixed with insect tissue culture fluid and an egg of a diplazontine was added. After 24 h, the fluid was examined for changes in colour, the extent of the change, and the formation of a capsule. The predictions for each parasitoid species in different hosts and for each host species with different parasitoids were confirmed, strongly suggesting that differential host suitability has played a significant role in determining host specificity in diplazontine ichneumonids.

Host Plant

The rate of encapsulation of a parasitoid in a particular host species may vary with the species of plant that the insect grows on (Ben-Dov, 1972; Blumberg, 1991). For example, the scale insect *Protopulvinaria pyriformis* encapsulates a larger percentage of eggs of *Metaphycus stanleyi* when grown on *Hedera helix* or *Schefflera arboricola* than when grown on avocado plants (Blumberg, 1991). Similarly, the mealybug *Pseudococcus affinis* encapsulates a higher proportion of the eggs of the encyrtid *Anagyrus pseudococci* when reared on *Aeschynanthus ellipticus* than when reared on *Streptocarpus hybridus* (Perera, 1990).

Host Stage

With several parasitoids the probability of encapsulation occurring increases the later the host stage attacked (Berberet, 1982; van Alphen and Vet, 1986; van Driesche, 1988; Dijkerman, 1990). An explanation given by Salt (1968) for such a relationship is that earlier stages have fewer haemocytes available than later ones.

Host stage does not affect the probability of encapsulation of *Habrolepis rouxi* (Encyrtidae) in its red scale hosts (Blumberg and DeBach, 1979).

Insect eggs lack a cellular defence response to foreign bodies (Salt, 1968, 1970; Askew, 1971; Strand, 1986).

Superparasitism

The reduction in encapsulation ability of a host with superparasitism has already been discussed. Askew (1968) drew attention to this phenomenon. Explanations given in the literature are that the host is 'weakened' or that its supply of haemocytes becomes exhausted as a result of the increased parasitoid load.

Temperature

In some parasitoid species, the temperature at which the host is reared does not affect the frequency at which encapsulation occurs (e.g. *Habrolepis rouxi*: Blumberg and DeBach, 1979; *Aprostocetus ceroplastae*: Ben-Dov, 1972), whereas in others it does (e.g. *Epidinocarsis diversicornis*: van Driesche *et al.*, 1986; *Metaphycus stanleyi*: Blumberg, 1991). In *E. diversicornis* the rate of encapsulation is highest at the lower of two temperatures, whereas in *M. stanleyi* it is highest under high temperature regimes (Figure 2.66). It follows that in such species there may be seasonal or geographical variations in encapsulation rate.

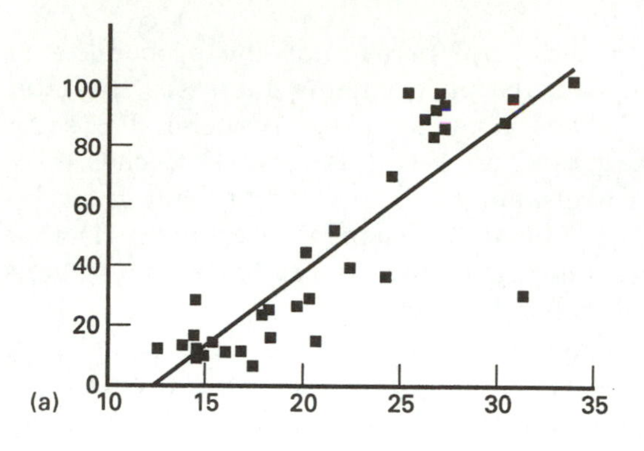

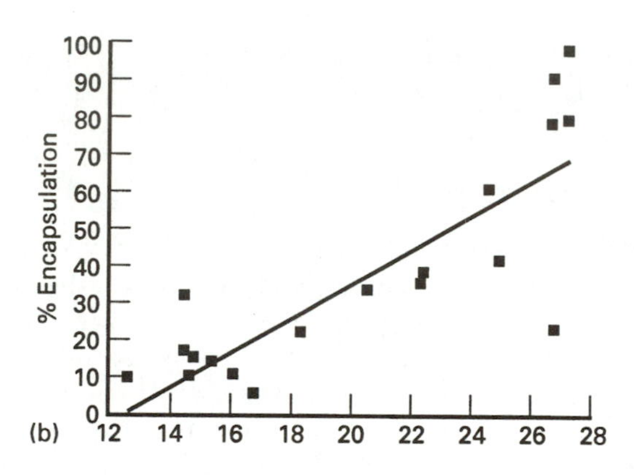

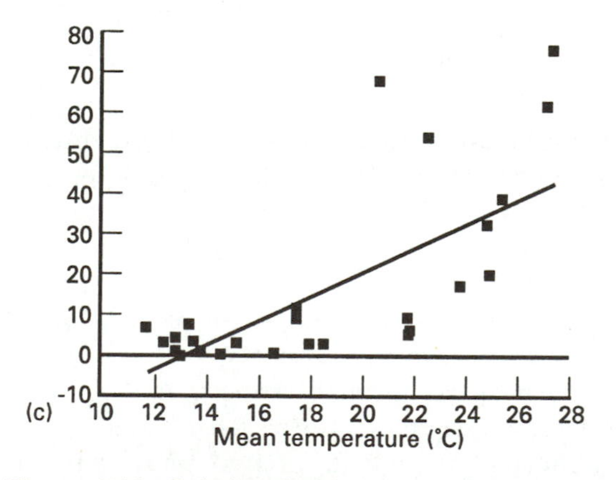

Figure 2.66 Relationship between rate of encapsulation of eggs and mean temperature in *Metaphycus stanleyi* (Encyrtidae) parasitising the pyriform scale on: (a) *Hedera helix*; (b) *Schefflera arboricola* under glasshouse conditions; (c) avocado in an orchard. (Source: Blumberg, 1991.) Reproduced by permission of Kluwer Academic Publishers.

2.10.3 EFFECTS OF PHYSICAL FACTORS ON SURVIVAL OF IMMATURES

Temperature

Parasitoids may be more hot/cold hardy than their hosts, in which case the lethal range of temperatures for the host will determine parasitoid survival. Prolonged exposure to extreme temperatures will kill the host first, and the parasitoid will then die as a result of starvation, anoxia or host putrefaction. Prior to death, a parasitoid's growth and development may be increased or decreased by the extreme temperature, (Tingle and Copland, 1988). On the other hand, parasitoids may be less hot/cold hardy than their hosts, such that they cannot tolerate the extremes of temperature that the host can tolerate, and so die as result of thermal stress.

Within the range of temperatures that are not *immediately* lethal to predator larvae, the lower the temperature, the longer totally starved larvae will be able to survive, and, in the case of larvae that have prey available, the less food will be required to stay alive (Lawton *et al.*, 1980).

Kfir and van Hamburg (1988) have shown that the outcome of heterospecific superparasitism can be influenced by temperature.

The influence of temperature on the host's ability to encapsulate parasitoids is discussed in the previous section.

Humidity

Low humidity can cause death of ectoparasitoid and predator larvae directly through desiccation, whereas high humidity can cause death indirectly by encouraging the growth of fungal pathogens.

2.11 INTRINSIC RATE OF NATURAL INCREASE

2.11.1 INTRODUCTION

The parameter known as the intrinsic rate of natural increase, describes the growth potential of a population under a given set of environmental conditions. It is often used, both by ecologists (Gaston, 1988) and by biological control workers (Messenger, 1964b), as a comparative statistic. In a biological control programme we may be faced with a choice of candidate parasitoid species; in the absence of other criteria we would select, for obvious reasons, the species with the greatest value for the intrinsic rate of natural increase (subsection 5.4.3).

This growth parameter is calculated, as described below, from age-specific survival and fecundity schedules. To understand first what it represents, we need to consider the most general of all population growth models, the exponential equation:

$$\frac{dN}{dt} = rN \qquad (2.18)$$

where N is the number of individuals in the population at any given time t, and r is the intrinsic rate of natural increase or the instantaneous *per capita* change in population size. Under conditions of an unlimited environment and with a stable age distribution, r is a constant.

For a given species, r can take a number of values. In theory at least, the species has an optimal natural environment in which its r will attain the maximum possible value, r_m, with a stable age distribution.

2.11.2 CALCULATING r_m FOR A PARASITOID WASP SPECIES

r_m is calculated by iteratively solving the following equation:

$$\sum_{x=0}^{n} e^{-r_m x} l_x m_x = 1 \qquad (2.19)$$

where x is the mid-point of age intervals in days, l_x is the fraction of the females surviving to the pivotal age x (or, put another way, the probability of a female surviving to age x), m_x is the mean number of female 'births' during age interval x per female aged x, and e is the base of natural logarithms. Trial r_m values are substituted into the above expression until the left hand side is (arbitrarily) close to 1. l_x and m_x are calculated by tabulating (Table 2.4) age-specific fecundity and age-specific survival data obtained from cohort fecundity and survival experiments (subsections 2.7.2 and 2.8.1 discuss the experiments; a graphical display of such data is given in Figure 2.67). If we find from examination of the **life table** that only 50% of wasps survive to the age of 5 days, then $l_5 = 0.5$. If we find that the average number of female offspring produced per individual alive during the age interval x is 25, then $m_x = 25$ (see legend to Table 2.4, for calculations based on another data set). The mean time taken from oviposition to adult eclosion, which can be measured in a separate

Table 2.4 Hypothetical life table for an experimental cohort of female parasitoids. x is the mid-point of age intervals (pivotal age) in days, l_x is the fraction of the females surviving to age x (in this example we assume no deaths occurred during development, so the proportion of females surviving to commence ovipositing is 1.0), and m_x is the mean number of female 'births' during age interval x per female aged x.

x	l_x	m_x	$l_x m_x$
12.5	1.0	12	12
13.5	0.9	14	12.6
14.5	0.8	18	14.4
15.5	0.7	22	15.4
16.5	0.5	25	12.5
17.5	0.3	13	3.9
18.5	0.1	4	0.4
			$\Sigma l_x m_x = R_o = 71.2$

experiment, is added to the pivotal age of each female. For example, this time period was 12.5 days for *Aphidius smithi* at 20.5°C (Mackauer, 1983). Parasitoid mortality during the immature stages also needs to be measured. In *A. smithi*, this mortality was negligible, so the probability of being alive at pivotal age 12.5 days + 1 day was set equal to 1.0 for all females (Mackauer, 1983). In *Aphidius sonchi* the time from oviposition to adult eclosion was 11.3 days and mortality of immatures was 8.0%, so the probability of being alive at pivotal age 11.3 days + 1 day was set equal to 0.92 (Liu, 1985b).

Once the values for l_x and m_x are calculated, then the following population statistics can also be calculated (Messenger, 1964b): the gross reproductive rate, GRR = Σm_x (the mean

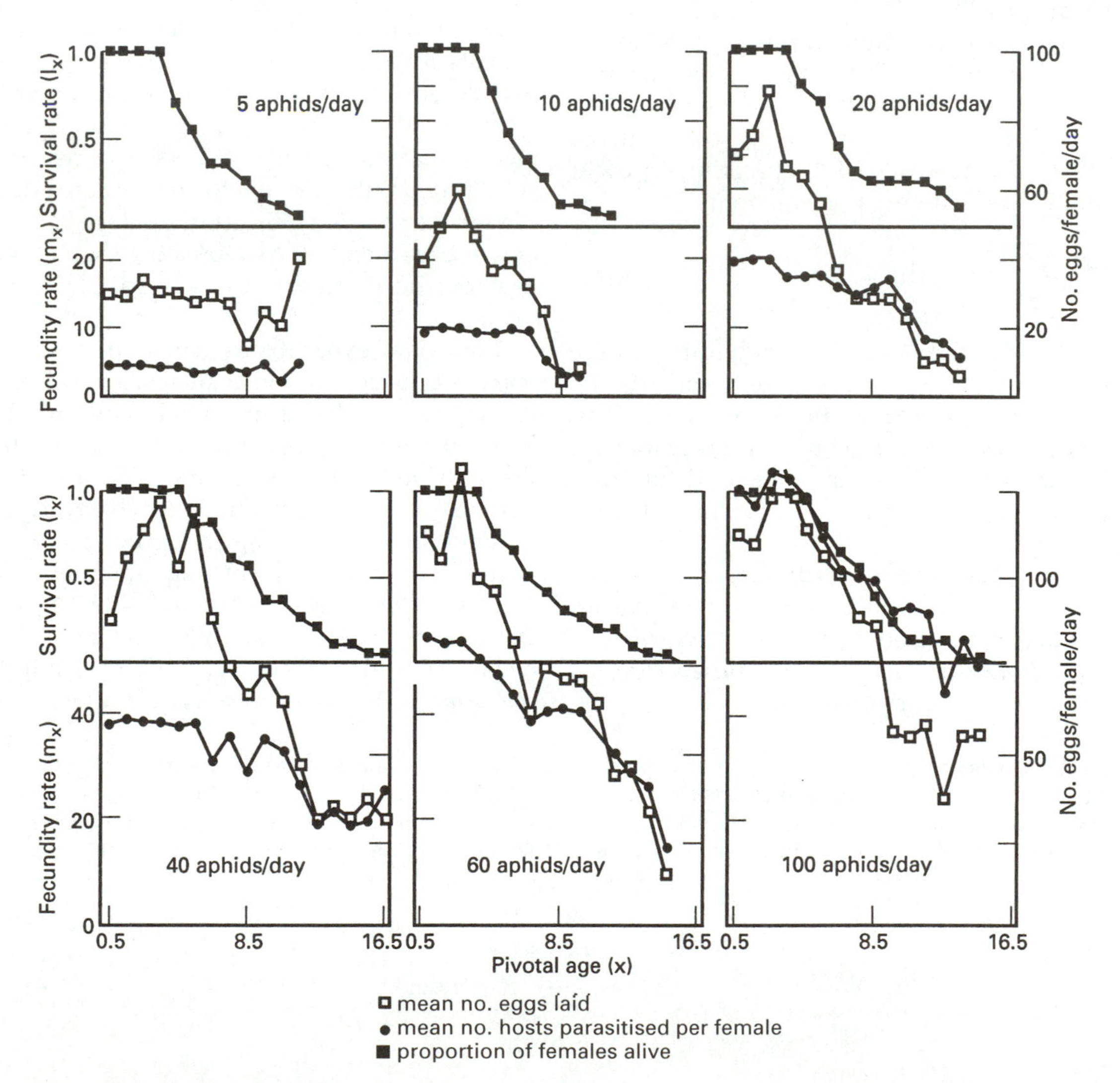

Figure 2.67 Age-specific fecundity and survival rates of *Aphidius smithi* provided with different densities of its host *Acyrthisphon pisum*. (Source: Mackauer, 1983.) Reproduced by permission of The Entomological Society of Canada.

total number of eggs produced by females over their lifetimes; measured in female/female/generation); the net reproductive rate, or 'basic reproductive rate' (the number of times a population will multiply per generation) $R_o = \Sigma l_x m_x$ (measured in females/female/generation); the finite capacity for increase, $\lambda = e^{r_m}$ (the number of times the population will multiply itself per unit of time; measured in female/female/day); the mean generation time, ($T = (\log_e R_o)/r_m$ (measured in days); and the doubling time (DT = $\log_e 2/r_m$ (the time required for a given population to double its numbers; measured in days).

Calculating the various parameters might appear to be a daunting task. However, all the calculations can be performed using a very simple program for which a QBASIC version is given in Appendix 2E. The only data required are age-specific survival and age-specific fecundity, which are entered for each day. The program comprises two stages: first there is the calculation of an approximate r_m (r_c, see below), and then there is iterative substitution to obtain an accurate estimate. Using the above data and the program in Appendix 2E, $r_c = 0.289$, $r_m = 0.296$, GRR= 108, $R_o = 71.2$, $\lambda = 1.344$, $T_c = 14.74$ (see below for explanation of this parameter), $T = 14.41$, DT= 2.34.

r_m can be measured (in female/female/day) at a range of host densities. It increases with increasing host density (Mackauer, 1983; Liu, 1985b). In *Aphidius smithi* this increase is also reflected in λ and in DT, which was less than half as long at the highest than at the lowest host density (Mackauer, 1983). Because in both *A. smithi* and *A. sonchi* the ovipositional pattern and the pattern of survival were similar to one another at the different densities (Figure 2.67), host density showed no significant effect on T.

To obtain a true measure of the influence of host density on the parasitoid's population statistics, some authors have based the m_x values on the number of hosts *actually parasitised* ('effective eggs' of Messenger, 1964b). This takes account of superparasitism; thus the number of hosts parasitised can be assumed to equal the number of progeny eventually produced (ignoring cases where no parasitoid progeny succeed in developing in a parasitised host)

Another factor that needs to be taken account of is the sex ratio of the progeny. This can be achieved by multiplying all m_x values in the life table by the overall sex population ratio, p, which is the proportion of females in all offspring produced. Regression of r_m on the natural logarithm of host density for different values of the sex ratio gives a series of parallel lines (Mackauer, 1983; Liu, 1985b; Tripathi and Singh, 1991) (Figure 2.68 shows regressions obtained for *Aphidius smithi*). The variation in r_m as a function of the parasitoid's sex ratio and host density can be shown as a response surface (Figure 2.69 shows the response surface for *Aphidius sonchi*) (Mackauer, 1983, gives details of the statistical procedure involved in obtaining the response surface). As can be seen from Figure 2.69, r_m increases as either host density or sex ratio increases, and at a given value of P the rate of increase in r_m slows at higher host densities. In *A. sonchi* the deceleration in r_m at high densities is such that the percentage increase in host density required to obtain a given percentage

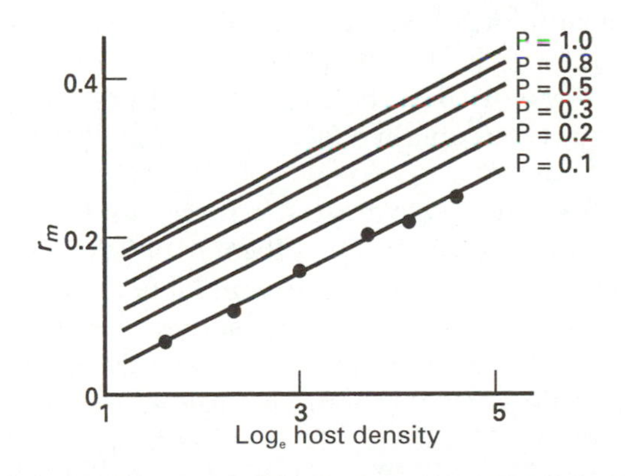

Figure 2.68 Relationship between the intrinsic rate of natural increase (r_m) of *Aphidius smithi* (Braconidae) and natural logarithm of host density, for different overall sex ratios. (Source: Mackauer, 1983.) Reproduced by permission of The Entomological Society of Canada.

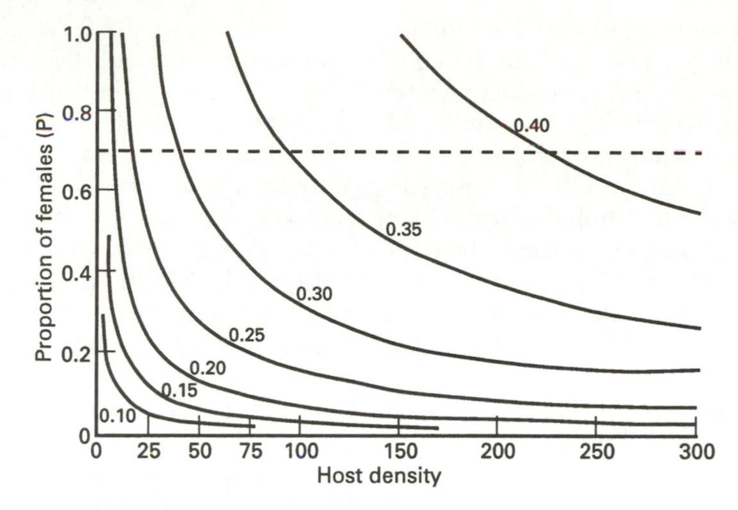

Figure 2.69 Response surface showing lines of equal r_m for *Aphidius sonchi* (Braconidae) for different host densities and parasitoid sex ratios (proportion of females). The broken line indicates a sex ratio of P = 0.7 observed in the laboratory. (Source: Liu, 1985b.) Reproduced by permission of Kluwer Academic Publishers.

increase in r_m is constant. For example, at P = 0.70, a 20% increase in r_m from 0.25 to 0.30 requires an increase of hosts from 15 to 39 per day, i.e. 24 hosts, while a 20% increase in r_m from 0.30 to 0.36 requires an increase of hosts from 39 to 101 per day, i.e. 62 hosts. This rule applies over the whole range of $0 \leq P \leq 1.0$, although the required increment in host density increases in absolute terms as the value of P declines. When P = 0.40, an increase of 48 hosts from 30 to 78 per day is required to obtain a 20% increase in r_m from 0.25 to 0.30.

The r_m of *Hyperomyzus lactucae*, the host of *A. sonchi*, is 0.3375. For a P value of 0.7, which is the sex ratio for *A. sonchi* in laboratory cultures, the parasitoid will achieve an r_m of 0.3378 at a host density of 74/day (preferably, the field sex ratio should be used in this computation, Mackauer, 1983). If the host density is increased to 200 per day, a sex ratio as low as 0.3 will yield an r_m of 0.3367, which is again close to that of the host.

In the absence of superparasitism (which will be higher at low host densities), the parasitoid's realised m_x will be equal to its oviposition rate, so yielding values of r_m higher than those computed. The minimum host density required to eliminate egg wastage through superparasitism can be determined. Theoretically, at that density the parasitoid's r_m will reach a maximum value that can be computed by setting m_x equal to the daily totals of eggs laid at the highest oviposition rate (Mackauer, 1983, gives details of statistical procedure involved).

Knowing how r_m varies in relation to factors such as host density (see above) and temperature (see below) can help us in deciding on the timing of introduction, for example in an inoculative release programme.

Equation 2.19 is not very 'transparent', that is, it is not particularly useful for any broad consideration of the relation between r_m and 'synoptic' life-history parameters such as generation times (Laughlin, 1965; May, 1976). A more useful statistic is r_c, the capacity for increase, which is an approximation for r_m. It is calculated as follows:

$$r_c = \left(\frac{\log_e R_o}{T_c}\right) \tag{2.20}$$

where T_c is the cohort generation time, defined as the mean age of maternal parents in the cohort at birth of female offspring (Laughlin, 1965) ($T_c = \sum_x l_x m_x / R_o$) (May, 1976) (for a discussion of the relationship between T and T_c, see May's paper; note that calculation of T involves first finding r_m from equation 2.19).

Equation 2.20 is based on the assumption that the reproductive period is brief relative to the total life cycle, which results in a small error in the estimation of generation time. r_c is a good approximation for r_m when R_o and thus population size remains approximately constant, or when there is little variation in generation length, or for some combination of these two factors (May, 1976).

Yu *et al.* (1990) used r_c as a measure of host quality for *Encarsia perniciosi*; r_c is largest for wasps provided with second instar California red scale and smallest for wasps given mature female scales.

A relatively simple method for calculating values for r_c was developed by Livdahl and Sugihara (1984). It dispenses with the need to construct detailed survivorship and fecundity schedules, and uses indirect estimates of R_o and T_c. It assumes the organisms being studied to have a Type III survivorship curve for the whole life cycle, i.e. high larval mortality and negligible adult mortality through the reproductive period; this assumption is only partly satisfied in the case of parasitoids, since in the laboratory there is likely to be low larval mortality while in the field there is likely to be high mortality of females during the reproductive period.

R_o is estimated from knowledge of the fraction of the original cohort maturing on day x and the future net fecundity of that fraction. In a cohort with an initial number of females, N_o, the fraction that survives to and matures on day x, may be denoted by A_x/N_o, where A_x is the number of new adult females produced at time x. This expression estimates the joint probability of surviving to age x and maturing on day x, and may be considered as the *per capita* production of adult females at time x. Because of the negligible adult mortality through the reproductive period, A_x/N_o contains all the important mortality information for the cohort. These data are then used to calculate R_o:

$$R_o = \sum_x \frac{A_x F_x}{N_o} \qquad (2.21)$$

where F_x is the future net fecundity of those individuals maturing on day x. The value of F_x can frequently be predicted from the mean body size of the new females emerging on day x:

$$F_x = f(\overline{w}_x) \qquad (2.22)$$

where $\overline{w}_x$ is mean female size, expressed in weight or linear dimension. The function $f(\overline{w}_x)$ can be determined empirically (**Female size,** in subsection 2.7.3). R_o may then be calculated as:

$$R_o = \frac{1}{N_o} \sum_x A_x \cdot f(\overline{w}_x) \qquad (2.23)$$

The terms in the equation (2.23) are analogous to the terms l_x and m_x. A_x contains survivorship information, and $f(\overline{w}_x)$ measures fecundity. A_x additionally contains information on maturation rate and $f(\overline{w}_x)$ is the future net fecundity of the average individual maturing on day x.

If a constant period, D, is required before adults can reproduce after eclosion (i.e. preoviposition period), T_c can be calculated as the weighted mean time to maturity plus D:

$$T_c = D + \left(\sum_x xA_x f(\overline{w}_x)\right) / \sum_x A_x f(\overline{w}_x) \qquad (2.24)$$

Using the estimated values of R_o and T_c, the following operational estimate of r_c (denoted r') can be obtained:

$$r' = \frac{\log_e \dfrac{1}{N_o} \sum_x A_x f(\overline{w}_x)}{D + \dfrac{\sum_x xA_x f(\overline{w}_x)}{\sum_x A_x f(\overline{w}_x)}} \qquad (2.25)$$

To use this equation, one only needs to observe cohorts during the maturation period in order to obtain measurements of the number of newly emerged adult females and their average size.

Livdahl and Sugihara also provide an equation for use as an index of population performance, when $f(\overline{w}_x)$ and D cannot be ascertained directly.

2.11.3 r_m AND TEMPERATURE

Since larval development rate, female survival and female fecundity vary with temperature (subsections 2.9.3, 2.8.4 and 2.7.3) we would expect r_m to vary also, which is the case. Figure 2.70 shows how r_m varies with temperature in three species of parasitoid and their aphid host (Force and Messenger, 1964). For examples of other studies, see Geusen-Pfister (1987) (*Episyrphus balteatus*), Cave and Gaylor (1989) (*Telenomus reynoldsi*), Lohr *et al.* (1989) (*Epidinocarsis lopezi*), Smith and Rutz (1987) (*Urolepis rufipes*), Mendel *et al.* (1987) (*Anastatus semiflavidis*); Miura (1990) (*Gonatocerus cinticipitis*). Siddiqui *et al.* (1973) provide a model to describe the relationship of $1/r_m$ to temperature. Using data for *Aphidius matricariae*, one of us (MJWC) found that the model gave a good fit to the data over part of the temperature range only. A simple polynomial model could express the relationship much more accurately.

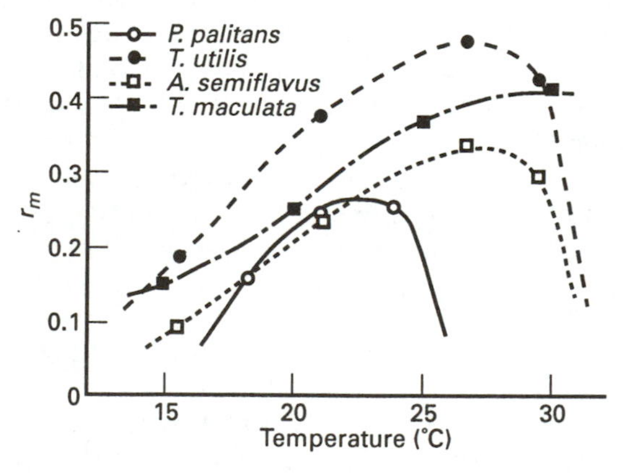

Figure 2.70 Comparison of intrinsic rate of natural increase (r_m) of the aphid parasitoids *Praon palitans*, *Trioxys utilis* (Braconidae), *Aphelinus semiflavus* (Aphelinidae) and their aphid host, *Therioaphis maculata*, over a range of constant temperatures. (Source: Force and Messenger, 1964.) Reproduced by permission of The Ecological Society of America.

2.12 ACKNOWLEDGEMENTS

We wish to thank Jacques van Alphen, Neil Kidd, Manfred Mackauer, Mark Shaw and David Thompson for commenting on parts of the chapter, Jeff Harvey for commenting on all of the chapter, Mike Majerus for kindly answering some queries, John Morgan and Kevin Munn for providing technical assistance, and Michael Benson for kindly allowing us to quote from his thesis.

APPENDIX 2A

Constant extreme (hot or cold) temperatures usually prove lethal to insects, and constant cold temperatures at best greatly increase the duration of experiments. This QBASIC program (Copland, unpublished) calculates the effect upon development, fecundity or longevity of an extreme temperature forming part of a 24 h cyclical regime alternating with a tolerable temperature (one within the linear part of the tolerable range) for which constant data are available. The program provides equivalent constant data for the extreme temperature which can be used in fitting regressions. The period of exposure to the extreme temperature can vary from 1 h to 23 h. If the effect of the extreme temperature is deleterious, then the program will show that temperature as being beyond the threshold. Simply type the data in lines 190 to 230.

PROGRAM 2A

```
10 REM  program to calculate the contribution made by an extreme
15 REM  temperature which is part of a 24 hour cyclical regime, with
20 REM  a tolerable temperature for which you have constant data.
25 REM  for development, longevity and fecundity studies (Copland, unpublished)
40 REM  -------------------------------------------------- read data
45 CLS
50 READ ct, cd, h, at, ad
60 REM  -------------------------------------------------- calculation - 2 dec points
70 et = INT(100 * (ad - (cd / 24 * (24 - h))) * (24 / h)) / 100
75 IF ad < cd THEN 110
80 et = 1 / ad - (1 / cd) * ((24 - h) / 24): IF et <= 0 THEN 110
90 et = INT(100 * (1 / (et * (24 / h)))) / 100
100 REM -------------------------------------------------- display results
110 PRINT "data for constant"; ct; "C is"; cd
120 PRINT "data for"; 24 - h; "hours at"; ct; "C and";
125 PRINT h; "hours at"; at; "C is"; ad
130 IF et <= 0 THEN 150
140 PRINT "calculated data for constant"; at; "C is"; et: GOTO 160
150 PRINT "calculated data for"; at; "C is beyond threshold"
160 END
170 REM
180 REM -------------------------------------------------- data
190 DATA 20 : REM  temperature of tolerable condition
200 DATA 40 : REM  data at constant tolerable temperature
210 DATA 12 : REM  hours at extreme temperature (in every 24)
220 DATA 36 : REM  temperature of extreme condition
230 DATA 30 : REM  data at 24 hour cycling temperature
```

APPENDIX 2B

Listing for a QBASIC program (Copland, unpublished, adapted from Lee and Lee, 1982) for fitting either a linear or a polynomial regression, calculating the reciprocal, thermal constant and threshold for development, and displaying results. The program can also be used for data on fecundity, longevity and other life-history parameters that vary with temperature over the integer range 0°–50°C. The number of data items must be typed in line 980 and the data themselves added in ascending temperature order from line 1000 onwards, with the data values separated by a comma. Use mean values for each temperature which assume an equal weighting for all data points. When running the program, 0 selects the best fit polynomial, while an order of 1 selects a linear fit. If required, the program calculates the reciprocal of y values, the thermal constant and the lower threshold. For these *hour*-degree calculations, data outside of the linear range can temporarily be hidden by using a REM in front of the DATA statement, and amending the value of the number of data items. The display shows data points (+), the fit over the temperature range (x), the point where they coincide (*), and regression coefficients and fit (r^2). The hour-degree constant (k) can be expressed in day-degrees by dividing by 24. Alternatively, alter the end of line 750 to: 750 PRINT TAB (59); "t1 =" -f (1) / f (2) : Print TAB (60) : "k ="; 1 / f (2).

PROGRAM 2B

```
10  REM program to fit a linear or polynomial regression, or calculate
20  REM reciprocal, threshold and thermal constant and display results
30  REM (Copland, unpublished) adapted from Lee and Lee (1982)
35  CLS
40  DIM p(50), r(50), t(50), u(50), v(50), w(50), x(50), y(50)
50  DIM z(50), a(10), b(10), c(10), d(11), f(10), g(10), k(10), q(10)
60  DIM s(10), a$(10), h$(3), q$(10), e(50), p$(50, 15)
70  REM  ------------------------------------------------  read data
80  FOR h = 1 TO 10: READ a$(h): NEXT h
90  READ n: FOR h = 1 TO n: READ x(h), y(h): w(h) = 1: NEXT h
100 n9 = n - 2: IF n9 > 9 THEN n9 = 9
110 PRINT "do you want reciprocal of y values (y or n)"; : INPUT a$
120 IF a$ = "y" THEN : FOR h = 1 TO n: y(h) = 1 / y(h): NEXT h
130 PRINT "type order required (0 for best, or 1 to"; n9; ")"; : INPUT po
140 IF po > n9 OR po < 0 THEN 130
150 REM  ------------------------------------------------  fit polynomial
160 m1 = po + 1: IF po <= 0 THEN m1 = 10
170 h = n - 1: IF m1 > h THEN m1 = h
180 m3 = m1 - 1: n2 = m3: a(1) = 1: s2 = w(1)
185 x9 = x(1): x1 = x(1): y9 = y(1): y1 = y(1)
190 FOR h = 2 TO n: IF x(h) <= x9 THEN 210
200 x9 = x(h)
210 IF x(h) >= x1 THEN 230
220 x1 = x(h)
230 IF y(h) <= y9 THEN 250
240 y9 = y(h)
250 IF y(h) >= y1 THEN 270
260 y1 = y(h)
270 s2 = s2 + w(h): NEXT h
280 y3 = (y9 + y1) / 2: y4 = (y9 - y1) / 2: IF y4 > 0 THEN 300
290 f(1) = y(1): n2 = 0: GOTO 640
300 FOR h = 1 TO n: v(h) = (y(h) - y3) / y4: d2 = d2 + w(h) * v(h) ^ 2
310 p(h) = 1: t(h) = 0: p1 = p1 + w(h) * v(h): s1 = s1 + w(h): NEXT h
320 s(1) = p1 / s1: c(1) = s(1): d2 = d2 - s(1) * p1
330 g(1) = ABS(d2 / (n - 1)): a1 = 4 / (x9 - x1): b1 = -2 - a1 * x1
340 FOR h = 1 TO n: u(h) = a1 * x(h) + b1: NEXT h
350 FOR h = 1 TO m3: d1 = 0
360 FOR j = 1 TO n: d1 = d1 + w(j) * u(j) * p(j) ^ 2: NEXT j
370 k(h + 1) = d1 / s1: w2 = s1: s1 = 0: p1 = 0
380 FOR j = 1 TO n: d1 = q(h) * t(j): t(j) = p(j)
385 p(j) = (u(j) - k(h + 1)) * p(j) - d1
390 s1 = s1 + w(j) * p(j) ^ 2: p1 = p1 + w(h) * v(j) * p(j): NEXT j
400 q(h + 1) = s1 / w2: s(h + 1) = p1 / s1
405 d2 = d2 - s(h + 1) * p1: g(h + 1) = ABS(d2 / (n - h - 1))
410 IF po > 0 THEN 470
```

PROGRAM 2B *continued*

```
420 IF h1 = 1 THEN 450
430 IF g(h + 1) < g(h) THEN 470
440 n2 = h - 1: h1 = 1: g1 = g(h)
445 FOR j = 1 TO m1: b(j) = c(j): NEXT j: GOTO 470
450 IF g(h + 1) >= .6 * g1 THEN 470
460 h1 = 0: n2 = m3
470 FOR j = 1 TO h
480 d1 = d(j + 1) * q(h): d(j + 1) = a(j)
485 a(j) = d(j) - k(h + 1) * a(j) - d1: c(j) = c(j) + s(h + 1) * a(j)
490 NEXT j: c(h + 1) = s(h + 1): a(h + 1) = 1: d(h + 2) = 0
500 IF h1 = 0 THEN 530
510 IF h <> m3 THEN 530
520 FOR j = 1 TO m1: c(j) = b(j): NEXT j
530 NEXT h
540 d(1) = 1: b(1) = 1: f(1) = c(1)
550 FOR h = 2 TO m1: d(h) = 1: b(h) = b1 * b(h - 1)
555 f(1) = f(1) + c(h) * b(h): NEXT h
560 FOR j = 2 TO m1: d(1) = d(1) * a1: f(j) = c(j) * d(1)
565 k1 = 2: j1 = j + 1: IF j1 > m1 THEN 590
570  FOR h = j1 TO m1: d(k1) = a1 * d(k1) + d(k1 - 1)
575 f(j) = f(j) + c(h) * d(k1) * b(k1): k1 = k1 + 1
580 NEXT h: NEXT j
590 FOR h = 1 TO n: j = n2 + 1: y5 = f(j): IF n2 = 0 THEN 610
600 FOR k = 1 TO n2: y5 = f(j - 1) + (x(h) * y5): j = j - 1
610 NEXT k: z(h) = y5 * y4 + y3: r(h) = (v(h) - y5) * y4: NEXT h
620 f(1) = (f(1) * y4) + y3: FOR h = 2 TO m1: f(h) = f(h) * y4: NEXT h
630 REM --------------------------------------------------- display results
640 m2 = m1 - 1
645 IF po > 0 THEN PRINT " order of polynomial specified ="; n2: GOTO 660
650 PRINT "maximum order of polynomial tested for ="; m2;
655 PRINT "  best found ="; n2
660 PRINT TAB(60); "coefficients"
670 n3 = n2 + 1: FOR h = 1 TO n3: PRINT TAB(60); a$(h); f(h): NEXT h
680 r2 = 0: FOR h = 1 TO n: r2 = r2 + r(h) ^ 2: NEXT h
690 a6 = 0: FOR h = 1 TO n: a6 = a6 + y(h): NEXT h: a7 = a6 / n
700 d6 = 0: FOR h = 1 TO n: d7 = y(h) - a7: d6 = d6 + (d7 * d7): NEXT h
710 r7 = (((d6 / (n - 1)) - (r2 / (n - 1 - n2))) / (d6 / (n - 1))) * 100
720 PRINT TAB(59); "r2="; r7 / 100
730 IF a$ = "y" AND n3 = 2 THEN 750: REM thermal constant, lower threshold
740 GOTO 760
750 PRINT TAB(59); "tl="; -f(1) / f(2): PRINT TAB(60); "k="; 24 / f(2)
760 LOCATE 5, 1
770 mn = 100: mx = -100: FOR h = INT(x(1)) - 1 TO INT(x(n)) + 1
780 FOR k = n3 TO 1 STEP -1: e(h) = e(h) + (f(k) * h ^ (k - 1)): NEXT k
790 IF e(h) < mn THEN mn = e(h)
```

PROGRAM 2B *continued*

```
800  IF e(h) > mx THEN mx = e(h)
810  NEXT h: IF y1 < mn THEN mn = y1
820  IF y9 > mx THEN mx = y9
830  sc = (mx - mn) / 15
840  FOR j = 15 TO 0 STEP -1: FOR h = 1 TO 50
845  p$(h, j) = " ": NEXT h: NEXT j
850  FOR h = 1 TO 50: IF e(h) = 0 THEN 870
860  p$(h, ABS(INT((e(h) - mn) / sc))) = "x"
870  NEXT h: FOR q = 1 TO n: p = INT((y(q) - mn) / sc)
880  IF p$(x(q), p) = "x" THEN p$(x(q), p) = "*": GOTO 900
890  p$(x(q), p) = "+"
900  NEXT q: FOR j = 15 TO 0 STEP -1: x = sc * j + mn
905  PRINT USING "###.###"; x; : PRINT TAB(8); "I";
910  FOR h = 1 TO 50: PRINT p$(h, j); : NEXT h: PRINT : NEXT j
920  PRINT TAB(8); " ------------------------------------------------------------------------------"
930  PRINT TAB(8); " 0    5    10   15   20   25   30   35   40   45   50"
940  PRINT TAB(8); "                t e m p e r a t u r e"
950  END
960  REM -------------------------------------------------- data
970  DATA "a=","b=","c=","d=","e=","f=","g=","h=","i=","j="
980  DATA 6                : REM number of data items
990  REM  more than 2 but less than 50 data points in ascending x order
1000 DATA 12, 33.12        : REM enter x values (temp) in integers
1010 DATA 16, 19.3         : REM then mean y values separated by comma
1020 DATA 20, 12.6
1030 DATA 24, 10.7
1040 DATA 28, 12.2         : REM omit for hour degree/thermal
1050 DATA 30, 14.3         : REM constant calculation and
1060 REM                           alter value in line 980
```

APPENDIX 2C

Listing for a QBASIC program (Copland, unpublished) that predicts the duration of development using accumulated **hour-degrees**. The thermal constant and lower threshold are typed in lines 110 and 120. When run, the program requests a temperature and then predicts the time required for completion of development. With a constant temperature regime, only a single figure is required. For controlled daily cyclical regimes, temperatures need to be given for each hour of the day. For field studies, hourly temperatures over the duration of the study are required. Where a data logger has been used and the data saved to disk, the data can be read by altering line 52 to read the disk file. The program can be used to validate the accuracy of the calculated values for the threshold for development and the thermal constant. The hour-degree constant can be derived from day-degrees by multiplying by 24. Alternatively, alter line 20 to read: 20 READ k: k = k * 24: REM thermal constant (hour degrees) and alter line 110 to: 110 DATA 183.04: REM day degrees thermal constant.

PROGRAM 2C

```
10   REM   a program to calculate accumulating hour degrees (Copland, unpublished)
15   CLS
20   READ k:              REM thermal constant
30   READ tl:             REM  lower threshold
40   h = 0:               REM set hour counter
45 REM ------------------------------------------------------ calculations
46   PRINT "enter one value for constant temperature"
47   PRINT "enter 24 hour values for daily cyclical controlled temperature"
48   PRINT "enter hourly data or read from data logger file for field data"
49   PRINT "exit with ctrl c": PRINT
50   h = h + 1: PRINT "input hour"; h; "temperature";
52   INPUT t:             REM input monitored temperature
55   dh = .00001: IF t <= tl THEN 70: REM no development below threshold
60   dh = (t - tl):       REM hour degrees accumulated per hour
70   dt = dt + dh:        REM cumulative hour degrees
80   PRINT "estimated life cycle"; (k / dt) * (h / 24); "days"
90   IF dt < k THEN 50
100  PRINT "life cycle complete in"; h / 24; "days": END
105  REM  ---------------------------------------------------- data
110  DATA :     REM hour degrees thermal constant k
120  DATA  6.3:           REM lower threshold tl
```

APPENDIX 2D

Listing for a QBASIC program (Copland, unpublished) that predicts the duration of development. The program is based on a polynomial curve fitted to non-reciprocal data (e.g. from Appendix 2B). The polynomial order is typed in line 160, the coefficients in line 170 onwards, and lower and upper temperature limits for which it can be used entered at the end. When run, the program requests a temperature and then predicts the time required for completion of development. With a constant temperature regime, only a single figure is required. With a controlled daily cyclical temperature regime, temperatures need to be given for each hour of the day. With field studies, temperatures need to be given hourly for the duration of the study. Where a data logger has been used and the data saved to disk, the data can be read by altering line 85 to read the disk file. The program can be used to validate the accuracy of fit of the polynomial curve.

PROGRAM 2D

```
10   REM program predicts life cycle from polynomial regression (Copland, unpublished)
15   CLS
20   DIM c(10):                            REM dimension polynomial coefficients
30   READ p:                               REM polynomial order
40   FOR k = 0 TO p: READ c(k): NEXT k:    REM coefficients
50   READ b:                               REM lower temperature limit
60   READ u:                               REM upper temperature limit
```

PROGRAM 2D *continued*

```
70  h = 0:                                  REM set hour counter
75 REM --------------------------------------------------- calculations
76  PRINT "enter one value for constant temperature"
77  PRINT "enter 24 hour values for daily cyclical controlled temperature"
78  PRINT "enter hourly data or read from data logger file for field data"
79  PRINT "exit with ctrl c": PRINT
80  h = h + 1: PRINT "hour"; h; "temperature";
85  INPUT t:                       REM input hourly monitored temp
90  dh = .000001: IF t < b OR t > u THEN 120: REM skip outside thresholds
100 FOR k = p TO 0 STEP -1: dh = dh + (c(k) * t ^ k): NEXT k
110 dh = (1 / dh) * .0416:         REM amount of life cycle in one hour
120 dt = dt + dh:                  REM accumulating life cycle
130 PRINT "estimated life cycle"; (1 / dt) * (h / 24); " days"
140 IF dt < 1 THEN 80
150 PRINT "life cycle complete in"; h / 24; " days": END
155 REM  ------------------------------------------------- data
160 DATA 4:   REM polynomial order (one less than no. coefficients)
170 DATA  160.74613          :      REM coefficient a
180 DATA -18.9775654         :      REM coefficient b
190 DATA  0.9183977          :      REM coefficient c
200 DATA - 0.0210875038      :      REM coefficient d
210 DATA  0.000204592138     :      REM coefficient e
220 DATA  6:                        REM lower limit to use polynomial
230 DATA 31:                        REM upper limit to use polynomial
```

APPENDIX 2E

Listing for a QBASIC program (Copland, unpublished) that calculates the intrinsic rate of natural increase (r_m) and associated life table parameters. The number of days for which data have been obtained should be typed in line 290. Lines 300 onwards should contain mean values (separated by commas) for day (x), age-specific survival (l_x), and age-specific fecundity (m_x). The dimension statement in line 20 assumes that there are data for a maximum of 100 days. The program calculates the innate capacity for increase (r_c, an approximation for r_m) and then uses iterative subtitution to calculate r_m. Accuracy and number of iterations can be changed by varying the value in line 130. Since the x interval is in days, the values of T and T_c are in days and r_c and r_m per day. Other time intervals may be used, and comparisions made of r_c and r_m for different populations or species using an appropriate multiplication factor (Laughlin, 1965).

PROGRAM 2E

```
10   REM program to calculate life table parameters (Copland, unpublished)
15   CLS
20   DIM x(100), k(100), m(100):                 REM dimension arrays (n<100)
30   READ n
35   FOR h = 1 TO n: READ x(h), k(h), m(h): NEXT h: REM  read data
40   PRINT "calculating data"
```

PROGRAM 2E *continued*

```
50  ct = 0: mx = 0: ro = 0: j = 1:                  REM set variables
60  REM ------------------------------------------- cumulative calculations
70  FOR h = 1 TO n
80  ct = ct + x(h) * k(h) * m(h): mx = mx + m(h): ro = ro + k(h) * m(h)
90  NEXT h
100 rc = LOG(ro) / (ct / ro): rm = rc:              REM calculate rc
110 crm = 0: FOR h = 1 TO n
115 crm = crm + k(h) * m(h) * EXP(-x(h) * rm): NEXT h
120 REM ------------------------------------------- iterative substitution for rm
130 IF ABS(1 - crm) < .00001 THEN GOTO 180: REM accurate to 4 dec points
140 IF crm > 1 THEN rm = rm + ABS(1 - crm) / x(1)
150 IF crm < 1 THEN rm = rm - ABS(1 - crm) / x(1)
160 PRINT j: j = j + 1: GOTO 110
170 REM ------------------------------------------- display results
180 PRINT "gross reproductive rate         (GRR)    ="; mx
190 PRINT "net reproductive rate           (Ro)     ="; ro
200 PRINT "capacity for increase           (rc)     ="; rc
210 PRINT "intrinsic rate of increase      (rm)     ="; rm
220 PRINT "cohort generation time          (Tc)     ="; ct / ro
230 PRINT "generation time                 (T)      ="; LOG(ro) / rm
240 PRINT "finite capacity for increase    (lambda) ="; EXP(rm)
250 PRINT "doubling time                   (DT)     ="; LOG(2) / rm
260 END
270 REM
280 REM ------------------------------------------- data
290 DATA 5:           REM number of days data
300 DATA 11.5, 1.0,  17.25: REM x (day), lx (survival), mx (fecundity) female offspring
310 DATA 12.5, .7,  20.7   : REM
320 DATA 13.5, .5,  10.6
330 DATA 14.5, .25,  7.05
340 DATA 15.5, .04,  2.15
```

MATING BEHAVIOUR 3

J. van den Assem

3.1 INTRODUCTION

3.1.1 WHAT IS MATING BEHAVIOUR?

The term **mating behaviour** refers to the events surrounding insemination. Alexander (1964) divides mating behaviour into: **insemination**, which may involve a copulatory act; **the events leading up to and responsible for bringing about insemination**, i.e. pair-formation and courtship; and **the events immediately following insemination**, i.e. specialised, temporary pair-maintenance. In this chapter competition for mating success will also be discussed, e.g. defence of mating territories and interactions with rivals, together with some aspects of fertilisation.

The taxonomic and biological diversity of insect natural enemies is enormous. Because of this, the choice of examples will be highly selective, concentrating mainly on my own experiences and approaches in studying chalcidoid wasps (Figure 3.1). This does not mean that this chapter is relevant only to a particular taxonomic group; parasitoids and predators need to solve the same problems: where to find a partner; how to recognise it; how then to behave so that insemination will result; and, in the case of males, what measures (if any) to take so as to uphold a sperm monopoly in fertilisation, i.e. how to counteract sperm competition. Males of those predatory insects that take large prey e.g. mantids, several robber-flies (Asilidae) and dance-flies (Empididae), may face a special problem: how to avoid being consumed by a female before a copulation has been secured (subsection 3.9.7).

Mating is a component of sexual reproduction, and thus it implies the existence of functional and morphological differences between partners (Parker *et al.*, 1972). Ignoring such questions as why organisms should reproduce sexually (Maynard Smith, 1978) and why there are only two sexes (Fisher, 1930), we are faced with two fundamental facts concerning insects. First, within most species there exist two sexes: females (the producers of macrogametes or eggs), and males (the producers of microgametes or sperm); second, a fusion of gametes is necessary for eventual reproductive success. Copulation allows the transmission of gametes from a donor (the male) to a receiver (the female) by way of the genitalia. Sperm are stored by the female in a spermatheca (subsection 2.3.7), and can be used, as and when necessary, for fertilisation at the time of oviposition. Insemination and fertilisation are separated in time; from minutes or hours, to years (as in honey bees).

Both male and female contribute to the production of each offspring: exactly one gamete each. A sperm constitutes, in terms of energy and material expenditure, much less of an investment than an egg. Although there seems to be an asymmetry in contributions from the two sexes, males produce ejaculates which contain many sperm, so the asymmetry is less extreme than it appears. Nevertheless, male investments are smaller than those needed for the production of egg masses plus the investments needed for the placement of eggs in sites where fitness returns are maximised (Dewsbury, 1982). Thus, one finds that males are able to mate

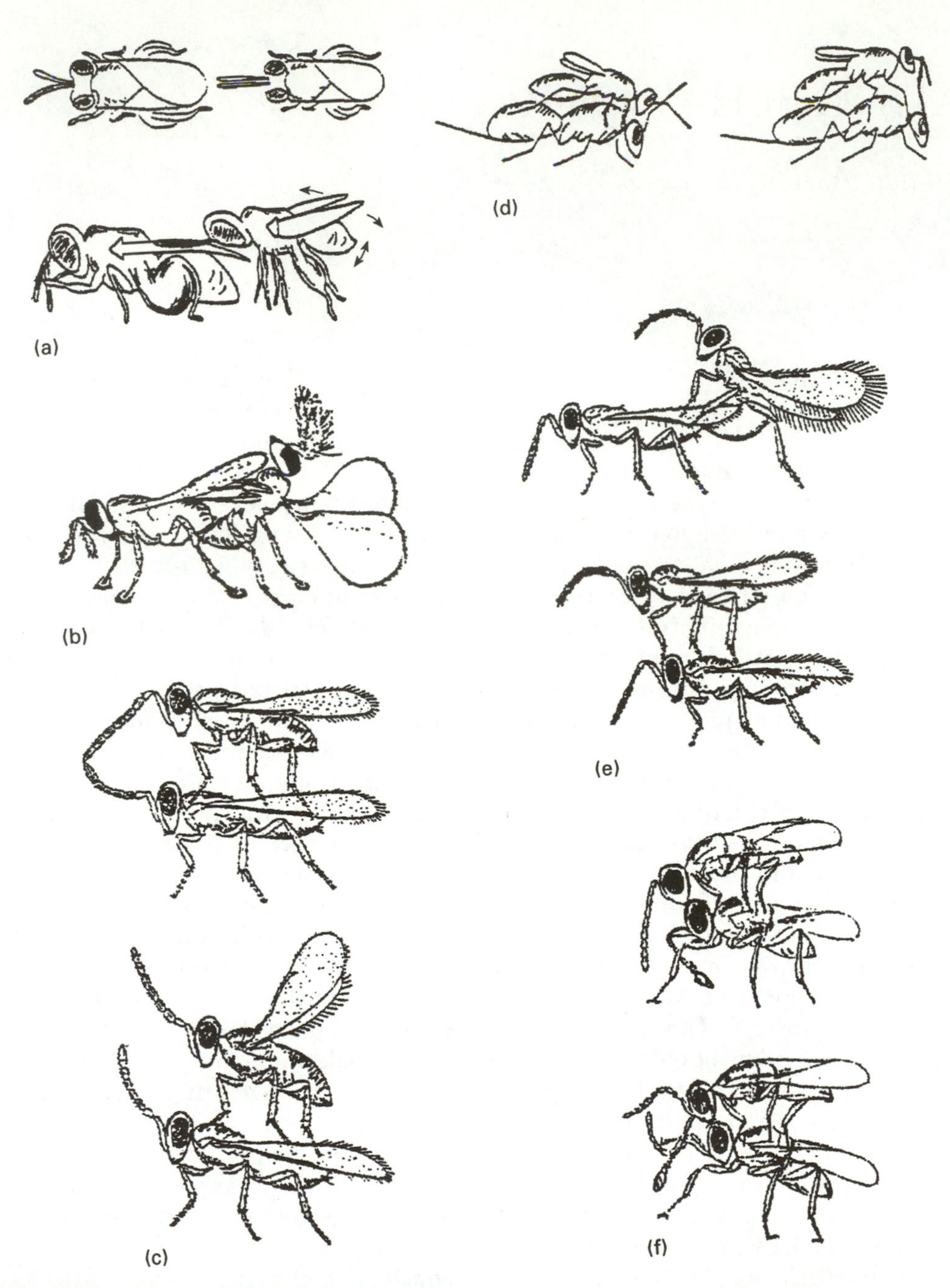

Figure 3.1 Various courtship and mating postures of parasitoid wasps: (a) *Brachymeria intermedia* (Chalcididae), male following female, and male performing courtship movements (wing pressing, and buzzing) (Source: Leonard and Ringo, 1978.); (b) *Trichogramma evanescens* (Trichogrammatidae), mating posture (Source: Hase, 1925.); (c) *Encarsia partenopea* (Aphelinidae), phases of post-copulatory courtship (Source: Viggiani and Battaglia, 1983.); (d) *Monodontomerus obscurus* (Torymidae), courting couple, male in low and in elevated position (Source: Goodpasture, 1975.); (e) *Archenomus longiclava* (Aphelinidae), mating, and post-copulatory courtship (Source: Viggiani and Battaglia, 1983.); (f) *Amitus vesuvianus* (Platygasteridae), phases of pre-copulatory display. (Source: Viggiani and Battaglia (1983.)

with more than one female, but that it is sometimes the case that females can (apparently) be made receptive only once in their lifetime – the basis both of male competition for mates and of female choice.

3.1.2 WHY INVESTIGATE MATING BEHAVIOUR?

Mating behaviour can be studied **for its own sake**, from the standpoint of:

1. its proximate **causation** and **organisation**, i.e. identifying and determining which mechanisms underlie the production of behaviour and how they operate, which factors (internal and external) release behaviour, how incoming information is processed, how motor activities eventually appear;
2. its **function**, i.e. measuring the consequences of behaviour, particularly in terms of reproductive success and fitness (its profitability);
3. its **evolution**, i.e. the steps by which behaviour patterns have changed over evolutionary time and have assumed their present form, and the directions any such developments have taken.

On the other hand, mating behaviour can be studied for **purely practical purposes**, in which case an investigation may be based upon:

1. the development of efficient techniques for the **mass-rearing** of insects that are required in large numbers at particular times, e.g. for field or glasshouse release in a biological control programme. A major problem with the mass-rearing of parasitoids is an overproduction of males (Waage *et al.*, 1985; Waage, 1986). A study of mating behaviour may uncover the reasons for this problem and so provide the key to the optimal production of males;
2. the precise **identification** of a species to be introduced as a biological control agent. It is essential, given the high degree of host specificity in many species, that biological control workers are absolutely certain about the taxonomic identity of the insects they intend to use. Sibling species (morphologically identical or near-identical species that are hard to identify) abound in the parasitoid Hymenoptera, and an analysis of mating behaviour may be the only means of distinguishing between them since, in general, mating displays tend to be species-characteristic (van den Assem and Povel, 1973; Matthews, 1975);
3. the demonstration, for **teaching** purposes, of the principles of animal behaviour. The mating behaviour of parasitoid wasps offers outstanding potential as teaching material. Many simple experiments that can be used in classwork are described below, using easily cultured parasitoid species (Barrass, 1976).

The approaches to any investigation of mating behaviour are determined by the kinds of questions that are being asked. The sort of quantitative analysis that is required for a study of the causation of behaviour requires a different data set to an investigation aimed at solving a taxonomic problem, and the two approaches will require different experimental designs. Yet, at the most fundamental level, these different approaches share many common features (section 3.2).

3.2 MAKING RECORDS

3.2.1 FROM EVENTS TO DESCRIPTION

Sequences of behaviour can never be stopped and observed at leisure. Nevertheless, to permit analysis, they have to be 'preserved' in some way, i.e. transcribed into a permanent record that can be referred to as often as desired. It is impossible to describe a sequence without partitioning it into separate events or 'acts' and assigning names to each.

Any description of behaviour needs to be as informative and unambiguous as possible. Vague and anthropomorphic characterisations, such as 'the male, using his antennae, caresses the female's antennae', must be avoided. Descriptions have to form the basis for further, usually quantitative, work, and this calls for unambiguous statements.

The records made should reflect the dynamics of behaviour as accurately as possible. This, however, is not as easy as it sounds, since there are so many aspects of the performance of even a simple sequence of an individual's behaviour that it is impossible to represent them all in one record. The motor patterns of the movements themselves may not be very variable, but the intensity and orientation of the movements will be, as will the timing of successive events. Furthermore, events may occur either in rapid succession or concurrently. Hence, a complete record of the stream of events, i.e. one that accounts for all shades and temporal characteristics of the performances, is practically unattainable even if one employs advanced recording techniques. This problem is compounded when one studies the behaviour of an interacting pair of animals. Therefore, the observer is faced with the dilemma of what to record and what not to record. The author's advice is not to start recording immediately, but first to acquaint oneself with the behaviour that one is seeking to investigate. Investigating mating behaviour takes time and requires patience.

3.2.2 EQUIPMENT USED FOR RECORDING MATING BEHAVIOUR

In attempting to record mating behaviour, it may help to observe events first and to make notes afterwards. In doing so, it will be necessary to repeat observations many times. A useful skill, which can be developed by training, is the ability to observe behaviour from an analytical standpoint. Usually during mating a number of body components (e.g. limbs, antennae, head, abdomen) take part in a posture or a sequence of behaviour such as a display. Do not attempt to memorise the details of movements of all the components but instead concentrate on one set of components, e.g. the front legs only, and memorise and subsequently describe their contribution to the entire sequence. Next, concentrate on another set of components, e.g. the middle legs, and repeat the same procedure. Then examine how the movements of the two sets of components are co-ordinated and ask whether any temporal relationships are recognisable. Then take a third set of components, and so on. Eventually, a synthesised description of the entire sequence can be made. A notepad and a pencil are sufficient for the above type of analysis. Further observations may be added to the description when previously unnoticed behaviour becomes apparent. One can also try to record the timing of separate events, or durations of intervals between events, in which case a stopwatch or other type of chronometer will be required.

Clearly, the above method can be applied only to rather simple, short and more or less stereotyped behavioural sequences. With more complex sequences that include either less stereotypy or more differentiated patterns of movements, memorising correctly the order of events – what came first, what followed next, or what coincided with what – will prove an impossible task. In such cases, it will be necessary to employ equipment for recording sequences of behaviour. An obvious item of such equipment is an audio tape recorder. With practice, one can achieve some measure of fluency using a tape recorder, mentioning 'essentials' and omitting the remainder. Signals can be included, e.g. a 'bleep' every few seconds, as temporal markers in the recording of the spoken reports. However, audio tape recording has severe disadvantages, particularly so far as movements of short duration are concerned. While speaking, one always lags behind the observed events. Also, transcribing spoken records is very time-consuming.

Alternatively, behaviour may be recorded on either cine film or video tape. A time signal can be included in the recorded sequences so as to indicate the exact points at which behavioural events occur. This method of recording is particularly helpful when adopting the 'analytical' approach because behavioural sequences can be repeatedly examined until all details have been noted. Filmed or video-taped records of courtship behaviour are indispensible for making precise descriptions of movements, and particularly for measuring the temporal relationships between the behavioural components involved. With high quality equipment, it is possible to make frame-by-frame analyses. Once a movement has been seen in detail in this way, it is much easier to recognise it at its normal speed and become aware of differences in performance on other occasions. Filmed or video-taped records are very useful when comparing the behaviour of related species (subsection 3.11.1). Excerpts from the respective displays may be viewed in rapid succession, and similarities and differences then recognised.

Filming and video-recording have disadvantages, however. It is usually not possible to view the entire field of action continuously – this becomes a problem when interactions between individuals are studied. There may be further technical problems, depending on the size of experimental animals. Many parasitoid wasps are too small for their behaviour to be viewed in sufficient detail with the naked eye, but they are also too large for viewing with a high-powered compound microscope. Thus, a dissecting microscope needs to be used, with artificial illumination. High-intensity illumination may, however, inhibit the performance of mating behaviour, and the heat produced by artificial illumination will certainly affect the behaviour of insects. The use of fibre optics is therefore highly recommended.

The alternative to a hand-written, filmed or video-taped record of behaviour is a machine-encoded one. Until relatively recently, the standard **event recorder** has been a totally mechanical device comprising a collection of keys with corresponding pens, ink and paper. Each key codes for a selected behavioural component; pressing a key causes its corresponding pen to deflect momentarily for as long as it is activated (the key is pressed for the duration of the behaviour in question). The pens are arranged in a row and the paper underneath is moved at a constant speed. For years, I used a 20-pen Esterline Angus recorder (manufactured by Esterline Angus, Indianapolis, USA) (Figure 3.2). Recording with such a device requires practice, but as soon as one has achieved a certain proficiency in using it, it is akin to playing an organ; sequences of behaviour can be recorded whilst 'playing' more or less automatically, and the operator can focus, relatively unhindered, on the behaviour itself. The method has two important advantages:

1. because of the constant passage of the paper underneath the pens, the timing of every deflection can be measured accurately and thus the timing of the observed events can be accurately estimated;
2. the record obtained is a visual representation of the temporal structure of the observed sequence; repetitions and other patterns will immediately catch the eye. This latter property should not be underrated as an aid in further analysis.

There are, however, many disadvantages associated with using the apparatus. Considerable time needs to be spent on analyses of records, even if one uses an automatic device for measuring the distance between any two deflections, i.e. measuring either the duration of one component of behaviour or the duration of the interval between two subsequent components. Practical disadvantages are the expensiveness of the high-quality paper, the inconvenience of filling the ink reservoir, and the fact that the pens are easily damaged or

Figure 3.2 (Left) Esterline Angus 20-pen event recorder (manufactured by Esterline Angus, Indianapolis. Ind. USA). (Right) Keyboard produced by the workshop of the author's laboratory. Each key corresponds to a separate pen. Also shown is part of an Esterline Angus-made record. A deflection of the pen corresponds to the onset of a particular 'act', and its return to the base line marks the end of the 'act'.

become clogged with dry ink when not used frequently.

A more advanced version of the event recorder is the electronic recorder which has a similar keyboard but does not utilise pens and paper. The *Soliprot* (Figure 3.3) has been developed by the electronics department of my laboratory. It has 25 keys, arranged in four rows. The keys code for behavioural events, and there is also a highly accurate, battery-powered time-marking facility. Pressing a key generates a signal that is recorded electronically. Like the entirely mechanical event recorder, the device records the length of time for which an individual key is pressed down. Activating more than one key at a time poses no problems, so events that occur simultaneously can be dealt with accurately. The electronic records can be read into a personal computer and processed with appropriate software. A print-out of the raw data can be obtained, i.e. a list indicating the points in time when certain keys were pressed and released, but this information is rather uninformative. With the appropriate software the data can be further analysed and, if so desired, be represented graphically. All necessary calculations made in data analysis can be done by the computer. I have always used my own software but those workers lacking appropriate knowledge or assistance may wish to purchase some commercially available software packages (subsection 1.4.1).

Currently, the most advanced method of recording is the on-line approach, using the computer's own keyboard as the recording

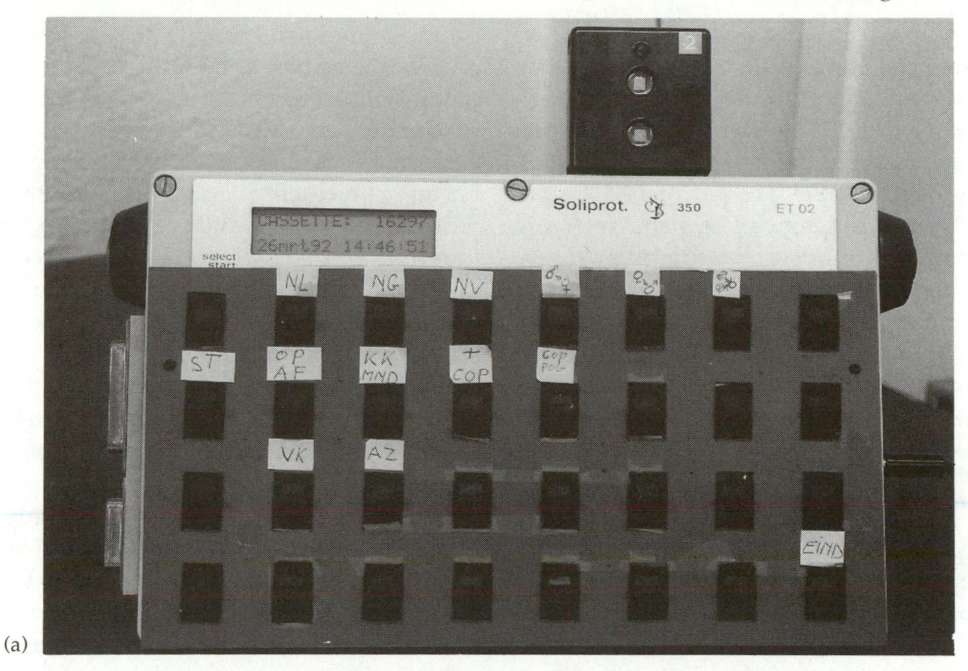

(a)

(b)

Figure 3.3 (a) *Soliprot* electronic recorder with keyboard and, above it, an 'eprom' (= erasable, programmable, read-only memory); (b) Detail of 'eprom' showing two windows. Data can be erased from the memory store by exposure, via the windows, to ultraviolet light.

device. The advantage of this is that, apart from using the computer's memory as a data store, one can actively interfere in an experiment by way of the on-line connection, and use the computer as an experiment-controller. Because computations are performed instantaneously and sequences of behaviour are monitored continuously, the computer can be programmed to produce a signal as soon as certain conditions have been met, and the experimenter may act upon the signal. These procedures allow for a kind of experimentation that was previously unattainable.

3.2.3 THE BIOLOGICAL CONTEXT OF MATING BEHAVIOUR

A necessary prerequisite for the fusion of micro- and macrogametes is a temporary coupling of male and female: the act of copulation. Clearly, a precondition for a copulation is that two individuals of different sex should meet (these are usually conspecifics). Where individuals have neither developed on or within the same host nor developed within the same patch, searching for mates becomes necessary. Male and female may come across one another by moving about randomly but the odds against them meeting in this way must be overwhelming. Non-random search, being more efficient, is likely to be selected for, and is thus likely to be the most common means of finding mates – hence, most parasitoids and predators produce signals. However, which sex should do the searching and which should do the signalling? Performance of either activity runs risks and incurs costs, albeit different ones (Greenfield, 1981). In ants, Hölldobler and Bartz (1985) recognise a female signalling/male searching syndrome and a male signalling/female searching syndrome. This dichotomy seems to be widespread, and parasitoid wasps are no exception: males either search for females or they swarm and presumably attract females by releasing, in concert, stimuli of some sort (subsection 3.5.1).

Once a male and female have met, a degree of 'readiness' has to be induced before copulation will occur: one partner has to be ready to donate and the other ready to accept. Generally, the male is the more active partner, producing a display of some sort to advertise its reproductive state, its species identity and possibly also (indirectly) the quality of genes on offer. The stimuli produced during displays of parasitoid wasps are discussed in section 3.6.

Frequently, a couple cannot go through the preliminaries of a copulation undisturbed. Competitors (usually rival males) may appear, which may complicate a situation considerably. For example, a female may become receptive following a certain amount of display by a male, but the copulation can be 'stolen' by a rival male (subsection 3.8.6). Males can also interact through sperm competition in those cases where a female can copulate more than once (Parker, 1970a,b,c,d; Waage, 1979, 1984). Functionally speaking, an earlier inseminator should take measures to improve the chances of his own sperm being used to fertilise a female's eggs; but as we shall see in subsection 3.9.4, counterploys have been evolved.

3.2.4 MATING IN PARASITOID WASPS

In arrhenotokously reproducing parasitoid Hymenoptera (subsection 2.5.2), skewed sex ratios are commonplace (Hamilton, 1967; also references in Luck *et al.*, 1993). In many species, because a female will mate either only once or at most only a few times, the number of females available for insemination is far less than the number of males ready to mate (males can inseminate many females). Put another way, the **operational sex ratio** (the ratio of males searching for mates to females available for mating) can be very male-biased even in those cases where the primary sex ratio (defined in subsection 1.9.3) is skewed in the opposite direction. For example, in *Nasonia vitripennis* the primary sex ratio may be as low as 0.1 (proportion of males) (Werren, 1980) whereas searching males outnumber available females at all times (van den Assem *et al.*, 1980a). A plausible prediction is that in almost any situation, because of a surplus of males, mating of parasitoid wasps is likely to occur in an intensely competitive context.

3.3 GENERAL METHODOLOGY

3.3.1 INTRODUCTION

This section deals with how mating behaviour can be investigated, what kinds of materials and equipment are required, how to design experiments, and what conclusions to draw from observations and manipulations. A few more special techniques will be described later (sections 3.5, 3.8) when discussing specific components of mating behaviour. Most examples given in this chapter come from my own research on two chalcidoid parasitoids of cyclorrhaphous fly pupae: *Nasonia vitripennis* (Pteromalidae) and *Melittobia acasta* (Eulophidae). His investigations were carried out in the laboratory because, like many other parasitoid wasps, *N. vitripennis* and *M. acasta* are very small insects, and observing them requires specialised equipment. Also, in the laboratory it is possible both to control environmental conditions and to standardise insect material. Important variables such as egg load or the period of time between successive matings are difficult to control in field experiments. This is not to say that field investigations are not worth pursuing; Parker's work on searching for mates in dung-flies (1970a,b,c,d, 1971, 1978) is proof to the contrary. The mating behaviour of parasitoid wasps can in some cases be observed in detail under field conditions (Tagawa and Kitano, 1981; Takahashi and Sugai, 1982), but often it is difficult to recognise individuals, and this prevents assessments of individual achievements within a group of interacting conspecifics. Insects such as digger wasps (Sphecidae, Pompilidae, Tiphiidae) or dragonflies can be marked either with bright, coloured paint or ink spots applied in convenient places on the body, or with tags glued to the dorsum. In principle, it is valid practice to use visually recognisable genetic mutants within parasitoid species, but a wingless mutant and an individual whose wings have been removed (e.g. in an experiment aimed at investigating the role of wings in display behaviour) must be seen as completely different entities. Even if the effect of a mutant trait on one aspect of behaviour (e.g. wing displays) can be taken into account, it is very difficult to determine whether it affects other aspects of behaviour (Hall *et al.*, 1980).

There is another, fundamental, point to consider. In the absence of field observations, it is questionable to what extent mating behaviour recorded under laboratory conditions constitutes natural behaviour. Conclusions regarding the function of behaviour will, when based solely on laboratory observations, be tentative, even more so than those regarding the causation of behaviour (Ewing, 1983).

3.3.2 STANDARDISING INSECT MATERIAL

Complete sequences of mating behaviour can only be observed in insects that are capable of performing. For the females of several parasitoid species mating seems to be a once-in-a-lifetime affair, so for these prior behaviour will have a major influence on subsequent behaviour. Even though males can mate repeatedly, their behaviour is also affected by what they have done previously (section 3.7).

Experimental animals should be of a standard quality. Inexperienced insects (those that have emerged in isolation, and as adults have never encountered conspecific individuals) are likely to be more uniform in quality than experienced ones. Inexperienced individuals of solitary species are easy to obtain: simply keep hosts (e.g. aphid 'mummies') or pupae, isolated from one another. All gregarious species whose behaviour has been observed will mate immediately following emergence from the host, so as a precaution such species need to be removed in the pupal stage and kept separate. Also, to avoid desiccation, the pupae need to be kept in an incubator at an appropriate temperature (usually 20–25°C) and relative humidity (usually 70–80%). The pupae of many species can be identified to sex using morphological features (van den Assem *et al.*, 1982a).

Age differences are an important source of variation in courtship behaviour (Barrass, 1960b; Tagawa *et al.*, 1985). If one has removed the pupae of species that pupate inside the host or host's covering (e.g. puparium) (see above) the adults that eclose may be 'premature'. Under natural conditions they would need to spend much more time emerging, during which changes in hormone titres or in the structure of the chitinous exoskeleton could occur, although for parasitoid wasps there is currently only circumstancial evidence for such changes (for other insects, see de Loof, 1987). Adults of the weevil parasitoid *Lariophagus* need to escape from the tunnel inside the grain of wheat that has housed the host. The gnawing of an exit-hole requires strong mandibles, and small wasps require much more time to escape than larger individuals (van den Assem, 1971). Gregarious species that pupate inside the host may face similar problems. Sometimes, an entire brood fails to escape because the wasps are too small and therefore weak, due to the effects of larval competition (see subsections 2.9.2 and 2.10.2), although this problem can be avoided in laboratory cultures by ensuring that not too many females are provided with too few hosts. Wasps within a clutch may vary in the period of time they spend inside a host before escaping, early-eclosing individuals probably doing most of the work involved in excavating an exit-hole. If this is the case, synchronous emergence of wasps from the host must not be taken to mean uniformity in age.

Freshly eclosed female wasps may refuse to mate. If they are receptive, males may have to display to them for a longer period of time than they would to older females. This phenomenon may have practical consequences for experiments which require unreceptive females of a known age. For example, in investigations aimed at measuring the maximum display production of males, females of uniform quality are required. Females collected from cultures are likely to have been inseminated and to have therefore become unreceptive, but usually their age will be unknown. The usual technique for providing standardised females (outlined in Figure 3.4) for experimental purposes is to expose virgin females of known age to two males in rapid succession. This will 'switch off' female receptivity for the time being. This manipulation should take place some hours prior to the start of an experiment, to eliminate, as far as possible, the side-effects of earlier displays (subsection 3.4.5). Note that switching off 'premature' *Nasonia* females takes approximately twice as long as for older ones.

3.3.3 COLLECTING INSECT MATERIAL

The kind of investigation planned will determine the nature of the insect material required, and *vice versa*. For a study aimed at understanding the temporal organisation of mating behaviour a species is required that can easily be mass-reared. For a comparative study, as diverse an array of species as possible is required. For applied work there may be no choice of insect at all. Several methods have been developed for collecting parasitoid wasps (Chapter 4) but the majority are designed for obtaining specimens for subsequent preservation, and are thus of no use for behavioural studies. A study of mating behaviour requires material in optimum condition that will perform normally. Therefore, collecting should never have adverse effects on the animals' condition, and the use of knock-down chemicals or anaesthetics, must be avoided at all times.

A diverse array of material can be obtained by rearing parasitoids from field-collected

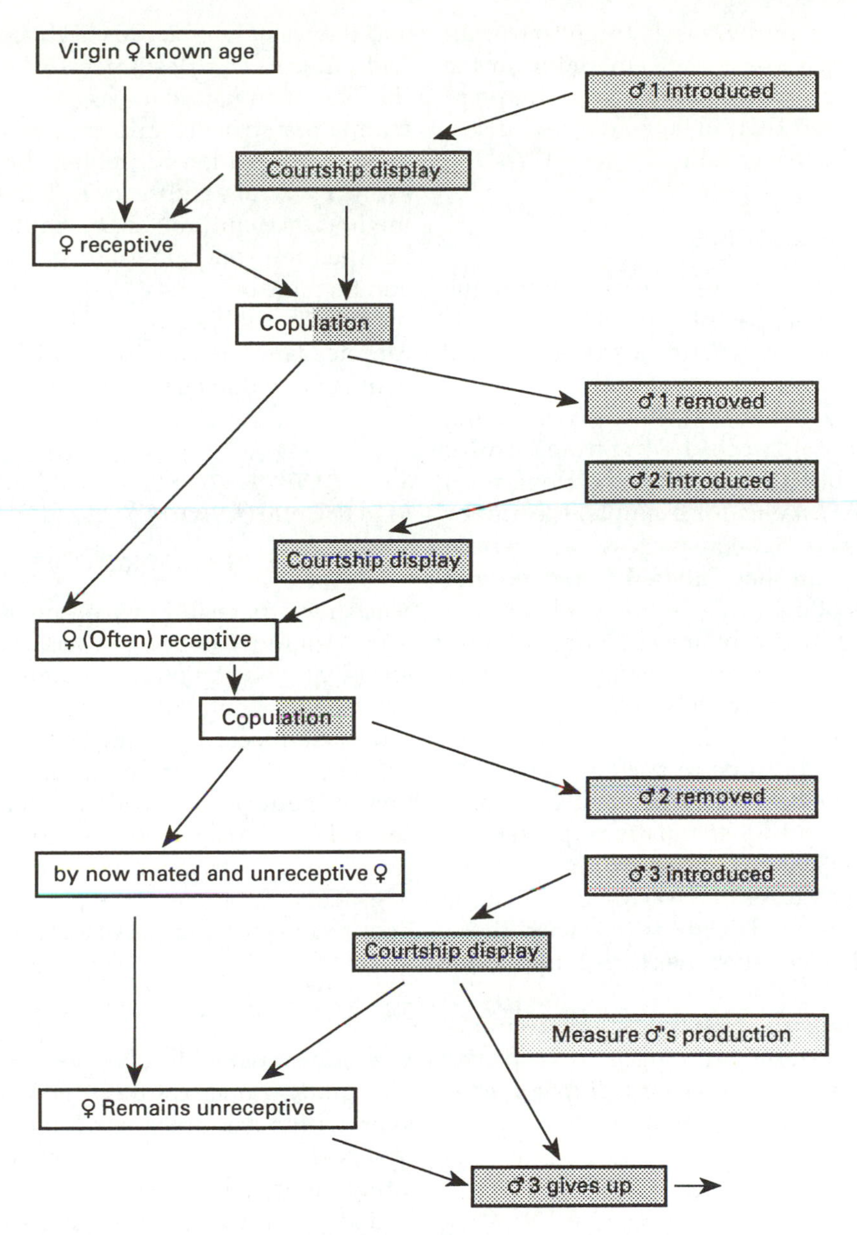

Figure 3.4 Investigating maximum courtship production by males. Unreceptive females are required. They can be made unreceptive by mating them once or twice (depending on the species observed) just prior to the experiment. The diagram outlines the standard technique for *Nasonia* females; males 1 and 2 make the female unreceptive, courtship production of male 3 can be measured. Unshaded and shaded parts of boxes denote contribution by female and male, respectively.

hosts (subsection 4.3.6). For example, parasitised aphids are easy to obtain and yield a variety of wasp species. The same is true of gall insects. For my own comparative work (van den Assem *et al.*, 1982a), galls have proved to be a rich source of material. Since many galls consist of hard tissues, parasitoid pupae are easily damaged when removed.

Therefore, it may be easier to collect adult parasitoids as soon as they emerge from the gall, although this requires very frequent sampling given that, once mated, the insect may never perform again.

3.3.4 INSECT CULTURES

Only well-kept cultures can ensure the constant availability of parasitoids that is required for many experiments (Waage *et al.*, 1985).

A problem associated with investigating the behaviour of insects derived from cultures is the possible adverse effects of long-term culturing. In *Nasonia* for example, the author found female responsiveness to certain display stimuli had altered after several generations of culturing, although changes in stimulus production by males were not found (subsection 3.8.4). Similar changes can occur in female receptivity thresholds and in the 'switching-off' mechanism (subsection 3.8.5), and are unlikely to be restricted to *Nasonia*. Gordh and DeBach (1976) compared sperm production and inseminative power of *Aphytis lingnanensis* (Aphelinidae) males drawn from cultures that were maintained for over 25 years in the laboratory with those of field-collected specimens, but found no differences.

One way of circumventing the undesirable effects of long-term culturing is to keep part of the parasitoid stock in diapause for extended periods.

3.3.5 HANDLING INSECTS

Handling of insects prior to carrying out experiments should be kept to a minimum and be performed as carefully as possible. Wasps can easily be coaxed into moving from one glass vial into another by holding the latter in the direction of a bright light source. To transfer a wasp from a glass vial into an observation cell (see below), simply tap the base of the tube once or twice with a finger, and the wasp will be caused to retract its legs and fall out. Females are easier to handle in this way than males; males seem to be able to maintain a stronger grip on a surface (probably an adaptation to prevent being pushed away by rivals (subsection 3.2.4)). Another method of handling is to capture a wasp between the soft hairs of a camel-hair brush, and then introduce it into the observation cell by a light rubbing of the brush. However, care needs to be exercised with this method, to avoid any damage to the insect.

3.4 EQUIPMENT USED IN OBSERVING MATING BEHAVIOUR

3.4.1 DISSECTING MICROSCOPE

Since many parasitoid wasps are at most only a few millimetres long, a dissecting microscope at recommended magnifications of between 6.5 and 40 ×, is usually needed. For most species, a magnification of 10 × is sufficient. With high magnifications, reduction in the field of vision and depth of focus severely constrains observation. Observing the mating behaviour of large parasitoids (e.g. Ichneumonidae) and predators usually does not require the use of a microscope.

3.4.2 OBSERVATION CELLS

It is only possible to observe mating behaviour under a microscope if the insects can be kept within certain spatial limits. A small observation cell is therefore often used, although insects for which swarming is a necessary part of their mating routine (subsection 3.5.1) need to be placed in large containers. The cells I used are obtained by cutting 7 mm-thick slices from a transparent perspex tube of internal diameter 25 mm. The particular diameter chosen fills the field of vision of a dissecting microscope used at a magnification of 10 ×. The cell is capped with a cover-slip (45 × 45 mm). Cells need to be placed on a strip of paper (150 ×

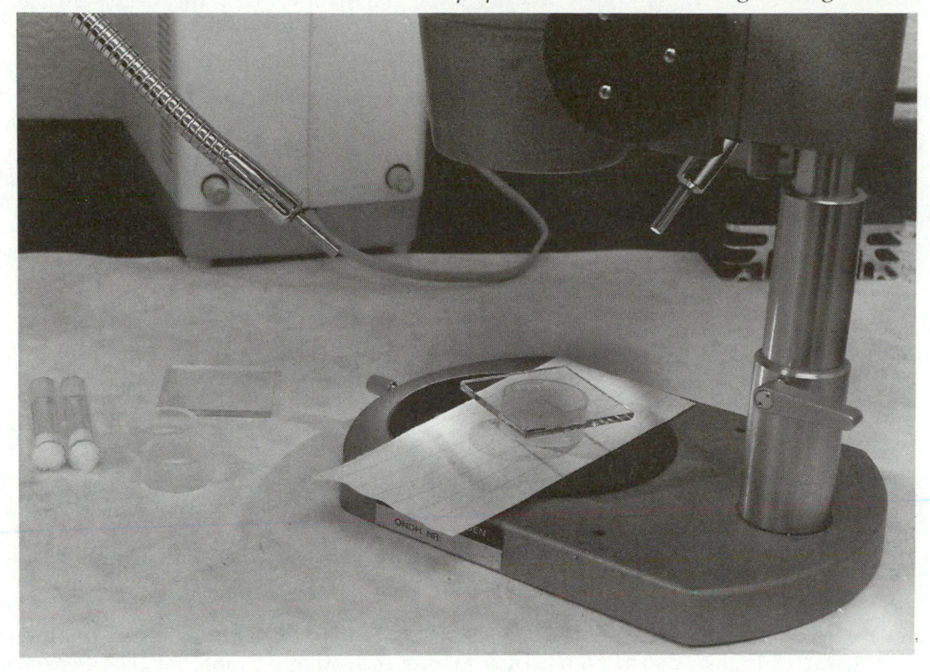

Figure 3.5 Perspex observation cell placed on a strip of paper under a dissection microscope and illuminated by fibre optics.

30 mm) during observations because it may become necessary to move the entire cell slightly to and fro (Figure 3.5). If, when observing insects, male and female apparently avoid one another for a long period of time, on no account attempt to encourage mating by tapping on the cell.

The following anecdote illustrates how specific the requirements for mating can be. Display and mating behaviour of *Tetrastichus coeruleus* were found to take place inside a large culture bottle with a sand-covered bottom, which also contained the remains of the hosts (the Asparagus Beetle, *Crioceris asparagi*) from which the wasps had emerged. Other specimens, which had eclosed in isolation, were introduced into an observation cell but males and females never reacted to one another. After several trials, a small quantity of sand was sprinkled on the bottom of the cell with the result that the insects mated successfully. The specific factor involved in eliciting courtship and mating was never identified; various kinds of sand were equally effective.

3.4.3 CONTROLLING AMBIENT TEMPERATURE

Good illumination is essential for carrying out observations, but heating of the observation cell must be avoided if meaningful quantitative data are to be obtained. To maintain a constant temperature, I use brass observation cells fitted with a Peltier unit (Figure 3.6), available commercially. With the Peltier unit heat is transported through a semiconductor by a reversible current. The temperature is raised at one end of the conductor and lowered at the other, depending on the direction of the current. The unit is placed in contact with the observation cell's undersurface. A control unit allows the choice of constant temperature; the actual temperature at the cell's undersurface can be read on a display. This set-up is accurate within 0.2°C.

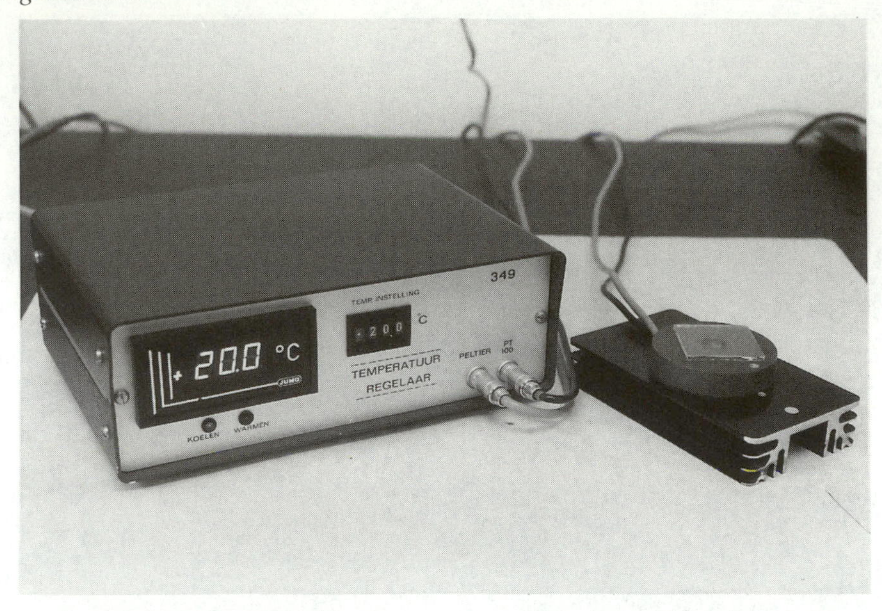

Figure 3.6 Peltier observation cell (right) with temperature display (left). The cell (diameter 14 mm, height 5 mm, made of brass) is surrounded by a mantle of insulating material. The Peltier unit and a sensor are situated underneath the bottom of the cell, and the ramified basal part of the device dissipates heat away from the cell.

If necessary, low temperatures can be obtained (down to approx. –10°C) but extra precautions are then required to dissipate heat by way of the cooling ribs situated underneath the cell. Extremes of temperature can be used to immobilise wasps and so allow precise experimental manipulations (e.g. sealing of mouthparts (subsection 3.6.3)) to be carried out. With less extreme low temperatures it is possible to slow down behaviour; all insect movements are performed at a slower rate, so enabling complex movements to be analysed more accurately (subsection 3.2.2). Display movements in *Nasonia vitripennis* can be slowed down two-fold in this way; the temperature inside the cell is set at 13°C, the minimum temperature for performing display behaviour.

3.4.4 NEGATIVE EFFECTS OF CONFINEMENT

If only easily observable events are to be recorded, e.g. the occurrence of a copulation or the duration of a display until copulation, it may be useful to observe wasps in the vial in which one of the wasps was kept after isolation. If this is done, care should be taken to do it in a standard manner, because different ways of bringing wasps together are likely to have different effects. The males of some parasitoids, e.g. *Nasonia* are known to apply substances (that may possibly function as chemical signals, subsection 3.10.3) to the wall of their vial. The various ways in which a pair of wasps can be brought together is illustrated in Figure 3.7.

Confining insects in a small space such as an observation chamber could have undesirable consequences. As a result of a performance, chemical substances could accumulate and affect the further progress of mating behaviour. In *Nasonia* there is evidence that courtship behaviour within the confines of a host puparium is inhibited, probably by chemical stimuli. The natural situation can be simulated by introducing a large number of

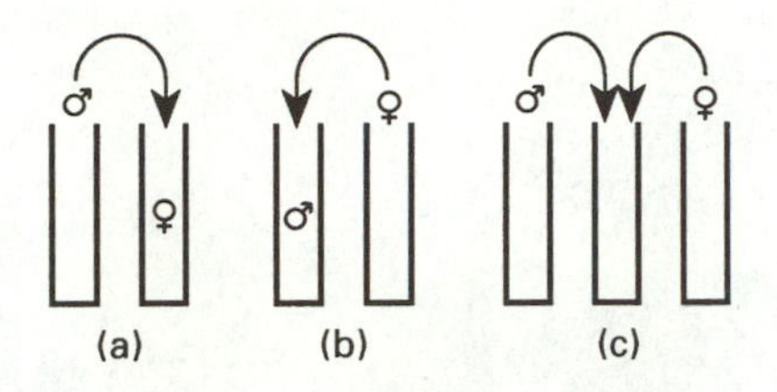

Figure 3.7 Three different ways of bringing a male and a female wasp together.

virgin females together with a male into a small vial or observation cell (van den Assem *et al.*, 1980b). The male will begin courting immediately and copulate as soon as a female is ready to do so, and will then copulate with another female, and so on. In due course the situation changes. First, females may spontaneously exhibit receptivity (opening the genital oritice, subsection 3.8.3) while the male is engaged elsewhere. They may approach the male before exhibiting, or they may become immobile and adopt the posture on the spot. Secondly, the male eventually stops walking around the vial and ceases displaying. When all the wasps within the vial are then transferred to a new vial, the male resumes his activities, only to become inactive some time later (Figure 3.8). The change in the male's behaviour suggests that a chemical stimulus is involved. The practical implications of the above experiment are clear: do

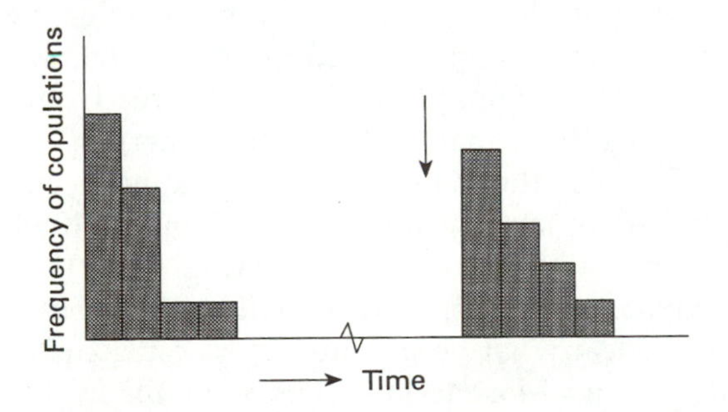

Figure 3.8 Typical temporal changes in frequency of copulations in a group of *Nasonia* (one male, many females) when kept in a small vial. Mating is resumed soon after the group is introduced into a new vial (vertical arrow).

not introduce more than a few specimens into a small, closed container vial for too long a period; mating behaviour may be negatively affected.

Several gregarious species (e.g. *Melittobia* spp.) are known to mate within the limited confines of a host puparium, and may occur in very large numbers. Inhibition of mating behaviour due to some effect of confinement does not occur either in the natural situation or in an artificial container. In such species volatile signals (as in *Nasonia*, subsection 3.6.3) would not operate efficiently and are therefore likely to be absent.

3.4.5 DUMMIES AS EXPERIMENTAL TOOLS

For a male wasp, confrontation with a real female is unnecessary for the release of courtship and mating behaviour. That the mere presence of certain substances may be sufficient can be demonstrated by killing females in liquid nitrogen, immersing and agitating them in a small quantity of a volatile solvent (e.g. hexane), then applying the solvent to filter paper. The residue remaining on the filter paper after evaporation of the solvent contains substances removed from the females. If a conspecific male is presented with the filter paper he will usually react immediately by performing the initial stages of display (subsection 3.5.3). In this way, I once observed a male *Pteromalus puparum* produce a complete display and copulate *in vacuo*, as it were. Seemingly intact females taken from the solvent and then dried fail to release any reaction, whereas dead females submerged only in water, and then dried, do. Clearly, a male need not 'recognise' the object to be courted as a female, but reacts in the first instance to one or more chemical substances that release mating behaviour. Other kinds of dummy will be discussed later (subsections 3.5.4, 3.8.3).

Not only external factors but also internal factors are involved in the production of mating behaviour. However, their relative importance alters continually. Investigations of these changes are possible if the external environment can be kept constant. Dummies would appear to be a valuable tool in this context, because they presumably present a set of constant external stimuli. But do they? The chemical stimuli (e.g. contact pheromones) that are initially present on the body surface of a dead female may gradually deteriorate or evaporate, and in time the dummy's quality may change appreciably. For example, in *Nasonia*, dummies deteriorate rather rapidly, whereas in *Lariophagus* the chemicals involved are less volatile and/or more stable and the dummies remain stimulatory for years. Completely artificial dummies may change in quality as a result of the behaviour that is directed at them. Males of many wasp species, when displaying, produce chemical substances (subsection 3.6.3) that may contaminate dummy females. With frequent use a dummy may no longer attract males, but repels them instead.

Figure 3.9 *Lariophagus distinguendus* wasps walking on millimetre graph paper, illustrating the large variation in body size.

3.4.6 BODY SIZE IN EXPERIMENTAL INSECTS

Parasitoid wasps, like most other insects, exhibit a high degree of intraspecific variation in body size (Figure 3.9). This variation is due to sexual dimorphism (males are usually [much] smaller than females), and intrasexual differences determined by food availability during larval development (subsection 2.7.3). As body size is likely to affect mating behaviour (smaller males may produce less stimulation than larger ones) and access to females (smaller males may be less successful in maintaining a dominant position, subsection 3.10.2), it should be controlled by the investigator. Wasps of similar size may be required for some experiments, while an array of different sizes may be necessary for others.

Solitary idiobiont species that parasitise hosts which vary in size allocate the sexes in a non-random manner, depositing most unfertilised (male) eggs on small hosts and most fertilised (female) eggs on large ones (subsection 1.9.2). Male progeny tend to be smaller than female progeny; the obvious explanation for this is that small hosts provide a smaller amount of resource (van den Assem, 1971; Charnov *et al.*, 1981). However, in solitary parasitoids that consume only a (small) part of their host (e.g. *Spalangia* species), males are still smaller on average than females. The explanation in this case is that despite there being an superabundance of food, the males still consume less (and thereby develop more rapidly). In parasitoid species such as *Spalangia* there is likely to be a selective advantage to males emerging on average earlier than females; males can be 'ready and waiting' for females as the latter emerge.

It is also the case that in gregarious species males are usually smaller than females. Male larvae appear to cease feeding and then pupate at an earlier stage than females, but it is unclear what proximate factors cause males to do this. More food is thus available for the females, and consequently females are larger in size and more fecund than they would be if male larvae continued feeding. Because in gregarious species the mating pool is usually strictly limited, and males will mate with the females with which they grew up (irrespective of whether they are sibs or non-sibs), males stand to gain in fitness by sacrificing large body size for early adult emergence, i.e. readiness to mate.

For experimental purposes, standardisation of body size in solitary parasitoids can easily be achieved by using standard-sized hosts. As a rule, measuring host size poses no special problems, but sometimes direct measurements cannot be made. For example, the larval grain-weevil hosts of the pteromalid wasp *Lariophagus distinguendus* have to be presented to female parasitoids *in situ* to be accepted for oviposition, and so only indirect measurements of hosts are possible. However, the largest diameter of the tunnel excavated by the weevil can be measured from X-ray photographs (Kirkpatrick and Wilbur, 1965). Maximum tunnel diameter and host body size correlate strongly (van den Assem, 1971).

The easiest way to proceed if only males of a different size are required is to present virgin females with hosts of various sizes. The hosts must be presented separately, otherwise the females may show a preference, avoiding smaller hosts.

3.5 THE EARLY STAGES OF MATING BEHAVIOUR

3.5.1 SEARCHING FOR MATES

Introduction

Figure 3.10 shows a typical sequence of mating behaviour, from the search for mates up to insemination and mate-guarding. In this and the following sections, investigations of the preliminaries of mating behaviour are discussed.

In parasitoids, the tendency is for males to search for females rather than *vice versa*, probably a consequence of male-biased operational sex ratios (subsection 3.2.4). Searching is likely to be concentrated in areas where the probability of finding mates is highest, i.e. sites where females either emerge, feed or oviposit (Figure 3.11). Males may attempt to monopolise these sites, i.e. show territorial behaviour, provided that the sites are in short supply and are sufficiently clumped. Where this is not the case, partners may meet as a result of signalling. Exactly what stimuli males and females respond to in finding one another cannot be stated in brief, since several different types of stimulus are involved – chemical, acoustic, tactile and visual. Chemical signals do not provide accurate information on the location of the signaller, at least not over relatively long distances. Acoustic signals can give this precise information but for purely physical reasons of size, small insects would need to employ very high frequency sounds in order to communicate through air over distances of more than a few times their body length (Michelsen *et al.*, 1982; Michelsen, 1983). Parasitoids undoubtedly communicate over short distances, but in structurally diverse environments such as they normally inhabit (e.g. vegetation), high frequency sounds would be unsuitable for long-range communication due to scatter, reflection and interference. For long-range communication then, parasitoids might therefore use substrate-borne vibrations. Some parasitoid wasps are known to produce such vibrations, but evidence that the latter function as signals, i.e. are used in communication, still needs to be sought.

Visual signals are probably only effective over longer distances during daylight hours where there is little cover (e.g. at the surface of water bodies or in open spaces; many dragonflies exhibit brilliant colours or conspicuous markings, which have a function in mutual recognition). Perception of visual cues over greater distances requires large eyes of sufficient acuity as those found in dragonflies and robber-flies (Asilidae). Parasitoid wasps such as chalcidoids, proctotrupoids and cynipoids are much too small to be able to perceive long-range visual signals; only short-range visual communication, in the order of 1 cm or less, is possible. Visual signals may be particularly effective at night, as light flashes. Fireflies emit species-characteristic, coded flashes (Lloyd, 1971). All signalling

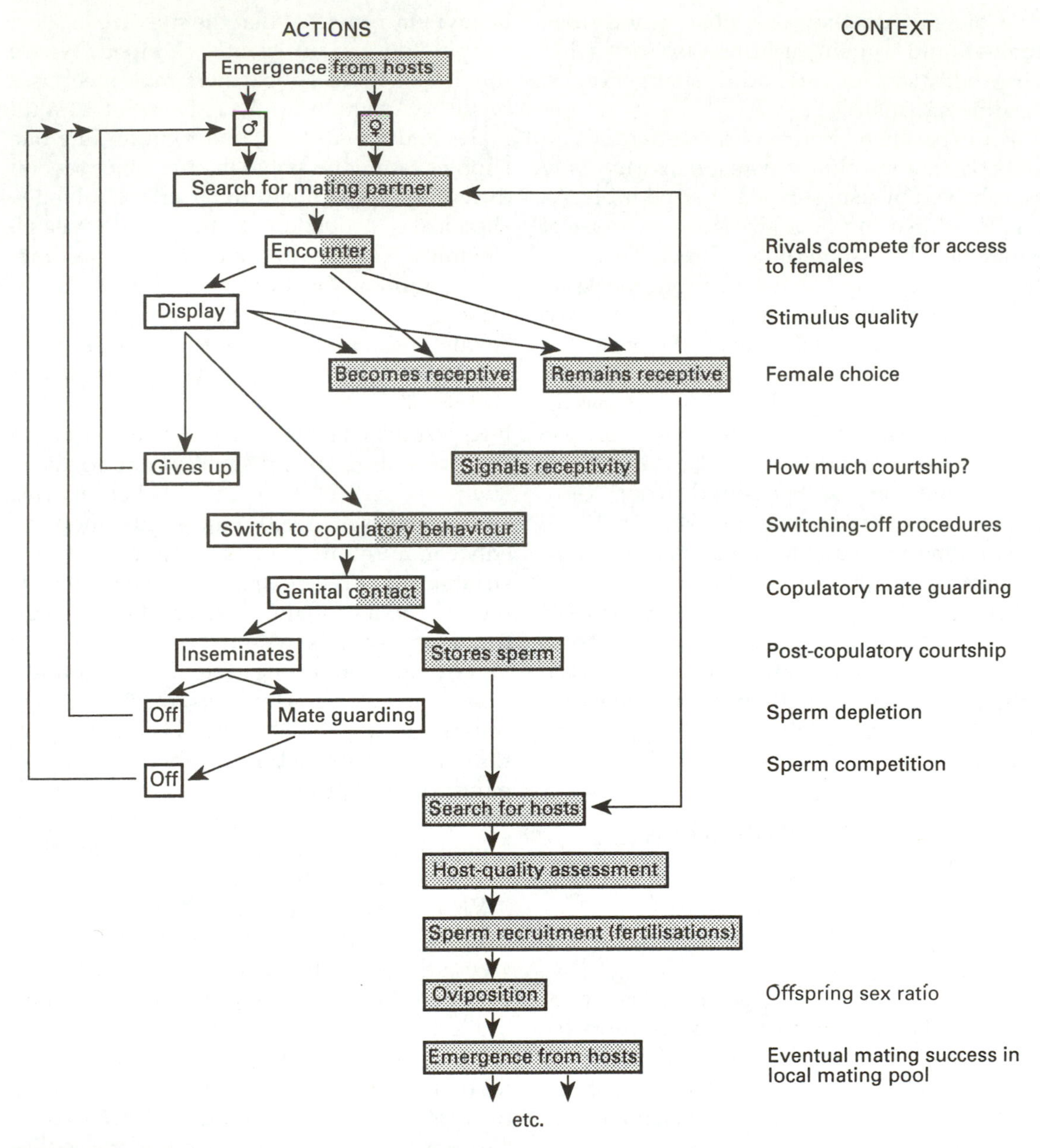

Figure 3.10 Schematic representation of the course of events in a typical sequence of mating behaviour. Unshaded and shaded parts of boxes denote contribution by female and male, respectively.

incurs risks from predation; for example the females of some firefly species attract heterospecific males for the purposes of feeding, not mating (Lloyd, 1975).

In parasitoid wasps, searching for mates has hardly been investigated in a systematic way. In gregarious species, males and females developing on the same host do not need to 'search' as such. In **genuinely solitary** species (solitary parasitoids whose hosts do not occur in masses, as distinct from **quasi-gregarious** parasitoids whose hosts do, e.g. scelionid para-

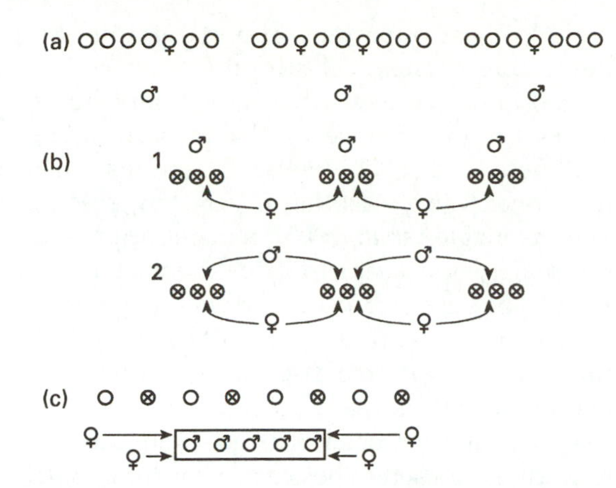

Figure 3.11 Schematic representation of different mate-searching strategies: (a) males monopolise sites where females emerge; (b1) males monopolise sites where females oviposit or search for food; (b2) males move continually between such sites; (c) sites cannot be monopolised, males aggregate and may signal in concert.

sitoids of heteropteran and lepidopteran eggs), the situation is different, and searching is necessary on the part of one or both partners.

Swarming Behaviour

The males of several solitary parasitoid species swarm, e.g. *Blacus* species, *Diachasmimorpha longicaudata* (Braconidae) (Donisthorpe, 1936; van Achterberg, 1977; Sivinski and Webb, 1989), *Diplazon pectoratorius* (Ichneumonidae) (Rotheray, 1981), *Aphelopus melaleucus* (Dryinidae) (Jervis, 1979) and some Encyrtidae, Pteromalidae and Eulophidae (Nadel, 1987; Graham, 1993). Nadel (1987) investigated mixed swarms of two species of encyrtid wasp (*Bothriothorax nigripes* and *Copidosoma* sp.) and a species of pteromalid wasp (*Pachyneuron* sp.). Males were observed either to fly or crawl over boulders. The swarms appeared at the same sites over a period of several weeks and contained several thousands of individuals of each species. In each species the vast majority were males. Swarms formed in the early morning and broke up during the night. Nadel established that females arrived for mating purposes; she dissected newly arriving females and found their spermathecae to be empty. Courtship and copulation occurred on boulders. Females were receptive upon arrival, were mated soon after, and became unreceptive.

Graham (1993) described the swarming behaviour of eulophids (in particular *Chrysocharis gemma*) and torymids (mostly *Torymus* spp.). Swarms comprising many thousands of individuals, the overwhelming majority being female, were observed year after year in exactly the same locations, i.e. on and around the same clump of trees. They appear to be present for a large part of the reproductive season. I collected a sample of *Chrysocharis* females on the spot, for dissection in the laboratory. The females' ovaries contained no mature eggs. However, it was not possible to ascertain whether the spermatheca in these females contained sperm; my guess is that all females were virgins. The function, if any, of this kind of swarming remains a mystery; they may not be mating swarms at all. The hosts of the observed species are not present locally.

Males in a genuine mating swarm may release chemical signals in concert and thus amplify the attractive effects. Species forming mixed swarms may use signals that are very similar to one another, but interspecific communication is not considered likely to occur. Most probably, swarms are not formed in response to active long-distance signalling by the participants; rather, certain peculiarities of the site itself attract both sexes when the insects are in mating condition. Nadel mentions that the sites she observed were on the only, low yet abrupt, peak within a 7.5 kilometre radius of a mountain ridge; the sites were no more profitable in terms of food or oviposition than the surrounding areas but they offered a distinct landmark and a certain degree of shelter from strong winds. 'Hilltopping' behaviour is known in many insects (Alcock, 1979) (Thornhill and Alcock, 1983 discuss how such a mating system may have evolved).

3.5.2 IS MATING RANDOM OR NON-RANDOM?

The question of to what degree mating is random in a more or less homogeneous environment has been addressed by Frank (1985). This problem is of theoretical importance and relates to the degree of female bias in progeny sex ratio. Sex ratio theory predicts female-bias in cases of non-random mating. Frank's investigations were on fig-wasps, but may inspire students to carry out similar work on parasitoids – the torymid parasitoids of these fig-wasps spring immediately to mind. Fig-wasps develop as larvae within the fig inflorescence (syconium) and mating takes place within the fig. Direct observations on mating are impossible to carry out because the figs have to remain sealed for mating to occur (Frank, 1984). Frank sampled figs randomly from trees. These figs were divided into pairs, and cut in half. Taking only one half of each fig, one male was taken (from among many others) and marked with enamel paint. These males were returned either to their own half or to the half of the other fig. The two halves were then joined for an hour using insect pins, and were then separated to enable Frank to record the location of each marked male. There was a further control: some males were returned to randomly selected fig halves in which they had not developed. The null hypothesis was that a male's searching is uncorrelated with the location of related females, related males or any cue that allows him to discriminate between the fig in which he developed and a foreign fig. Frank's (1985) results showed that males prefer their own fig half over a foreign half, although the proximate cues that they respond to were not identified.

3.5.3 PROXIMATE FACTORS AND THE RELEASE OF DISPLAY IN WASPS

In *Nasonia*, as in many gregarious species, newly emerged males remain on or near the host, and begin courting as soon as a female emerges. Males are able to locate a female within a host puparium, almost certainly by perceiving a chemical stimulus (subsection 3.10.3) and characteristically position themselves on the surface of the puparium, in readiness (King *et al.*, 1969a). By contrast, in a few species (e.g. *Melittobia*, van den Assem, 1976; Gonzalez *et al.*, 1985), females approach the males, and seem to offer themselves as mates. Contrary to other display sequences observed by me, those of *Melittobia* may last for quite some time (up to 30+ minutes), before *the male* is ready to copulate. Females that have not yet been courted often attempt to wedge themselves between a courting male and his partner. In *Melittobia,* males eclose before females even though they are larger than females (and would therefore be expected to take longer to develop). They remain inside the host's puparium or other covering. Often, the first male to emerge performs most if not all of the matings, after he has committed fratricide. The other males (which are usually sibs) are killed as pupae or as eclosing adults; upon encountering another (i.e. already emerged) male, a fierce fight ensues. For a mated female, a male's irresistibility quickly vanishes. Once copulated, females become positively phototactic and move away from the host. Because of the very skewed sex ratio (the sex ratio is around 0.02–0.05, but due to fratricide it becomes even more female-biased), surviving males have enormous mating opportunities.

In general, a female wasp's presence is first detected by olfactory means. Males begin moving their antennae up and down at a higher rate than previously, they vibrate their wings, and commence walking rapidly. Similar reactions are shown by males when they are presented with homogenates of females on filter paper (Obara and Kitano, 1974; Yoshida, 1978). *Anagrus* (Mymaridae) males will attempt to copulate with a fine brush that has previously been wiped over the bodies of virgin females (Ali, 1979). Attempts have been made to locate the source of the courtship-initiating pheromone in a number of parasitoids. In the braconid wasp *Diaeretiella rapae*, female gasters, when removed and presented alone, initiate courtship whereas other body parts fail to do so (Askari and Alishah, 1979). In *Cotesia glomerata* the pheromone emanates from a particular gastral segment (Obara and Kitano, 1974). Tagawa (1977, 1983) showed that in several *Cotesia* species a pair of glands in the female's reproductive system is the source of the pheromone. In the eulophid *Aprostocetus hagenowii*, extracts from the female thorax, but not from the head or gaster, released courtship behaviour (Takahashi and Sugai, 1982).

Yoshida (1978) designed a simple experiment for distinguishing between the effects of chemical and visual stimuli on the release of male courtship in the pteromalid *Anisopteromalus calandrae* (Figure 3.12). Using female pupae of different ages instead of adults, Yoshida found that pupae secrete the courtship eliciting pheromone, with a peak value around the *red-eye* stage; with older pupae there was a sharp decline in the effect. Full effects reappeared in adult females following emergence. A similar pattern has been reported for several Lepidoptera and Coleoptera.

Visual stimuli may work over very short distances: *Nasonia* males will follow a small black object that is moved to and fro a short distance away, from behind glass (Barrass, 1960a).

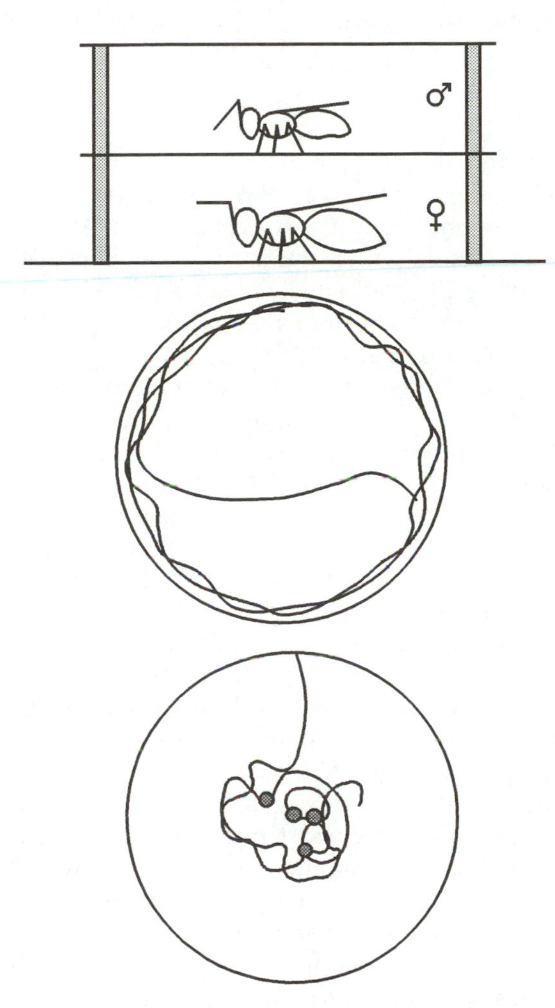

Figure 3.12 Experimental set-up for demonstrating the effect of chemical cues in the weevil parasitoid *Anisopteromalus calandrae*. Two glass rings (20 mm diameter, 4 mm high) are separated either by a thin glass cover or a millepore filter. With the glass there is no obvious reaction on the part of the male, which is walking along the wall of the cell; with the filter he restricts searching to the area with pores, together with antennations and wing vibrations (dots). (Source: Yoshida 1978.)

3.5.4 ORIENTATION WITH RESPECT TO THE FEMALE

Courting chalcidoid males either mount the female, as in *Nasonia* (Figure 3.13), or remain on the substratum positioned alongside or facing the female (subsection 3.11.2). In investigating which cues males respond to in courtship, dummy females can be used. A crude dummy made of cork, plywood or similar material may suffice to some extent,

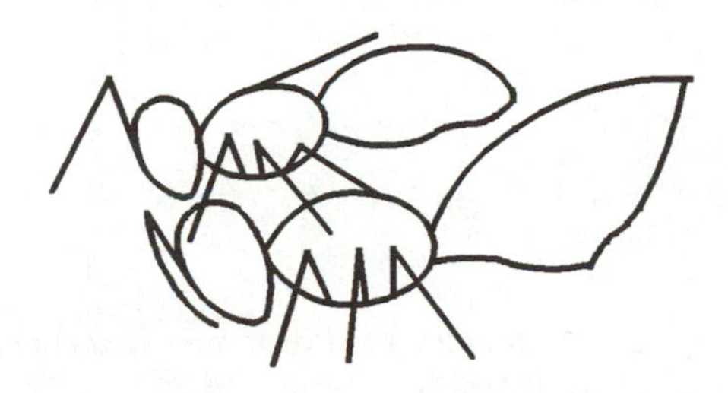

Figure 3.13 Stereotyped courtship position of a *Nasonia* male on top of the female.

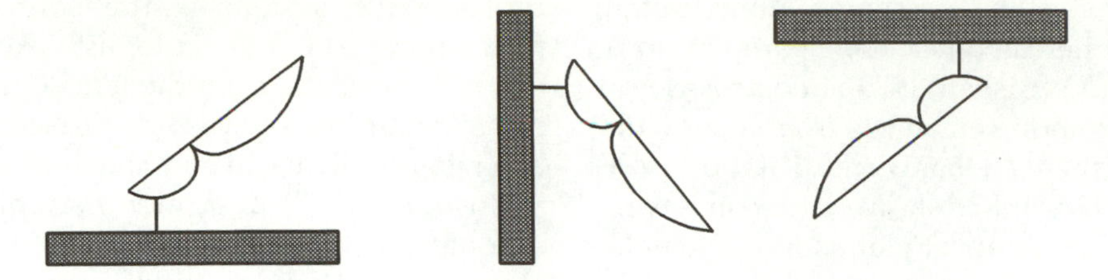

Figure 3.14 Dummies used by Yoshida and Hidaka (1979) for testing the influence of gravity on the male's orientation during courtship.

provided it carries the correct chemical cues (obtained from females that have been immersed and agitated in a volatile solvent (subsection 3.4.5)). Its actual size is unimportant, within certain limits. However, males presented with a very crude dummy will rarely proceed beyond moving about in an agitated manner while vibrating their wings, and soon dismount (Yoshida, 1978; Yoshida and Hidaka, 1979; Tagawa and Hidaka, 1982). Yoshida and Hidaka (1979) tested whether males orientate themselves in response to gravity. Dummies were attached to styrofoam at various angles (Figure 3.14). Yoshida and Hidaka found that courtship position was determined, not by the direction of gravity, but by the female's posture relative to the substratum.

The degree of specificity of the chemical stimuli that release male behaviour can be investigated using an olfactometer (subsection 1.5.2), although the following simple experiment might alternatively be carried out: using a set of standard dummies, e.g. identical pieces of plywood, apply a small amount of pheromone (obtained, as usual, by immersing and agitating female body parts in a solvent), each dummy receiving the pheromone from a different species. Males can be tested with a standard arrangement of dummies, e.g. all arranged on a circle or on the corners of a square, the male being placed at the centre (Figure 3.15).

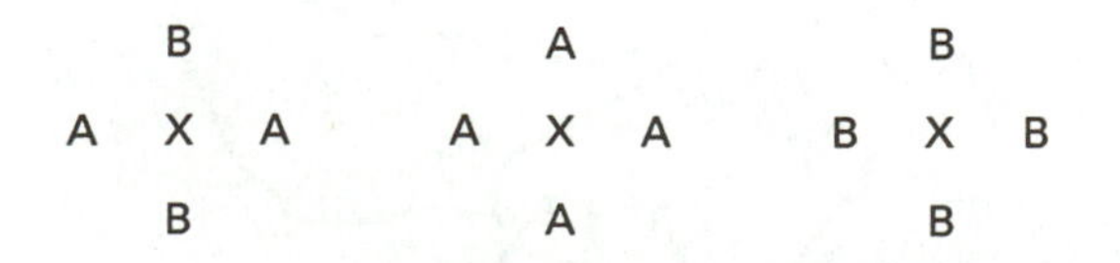

Figure 3.15 Arrangement of dummies in a choice experiment designed for testing the specificity of female sex pheromones.

A male's display position is stereotyped to a considerable extent and, because positions differ between groups, is thus of taxonomic value (subsection 3.11.1). *Nasonia* males mount the female and place the front tarsi on the female's head (placement of the other tarsi depends on the relative sizes of male and female). A stereotyped position requires specific stimuli for a correct orientation to be achieved, and these stimuli can be identified using dummies on which different real body parts can be placed, in various positions, using superglue. With such dummies, it is possible to determine what cues a male uses to arrive at the correct position, what cues cause him to remain in position and what cues release display behaviour. Examples of dummies used by the author are illustrated in Figure 3.16. In *Nasonia*, the position of the wings appears to provide directional cues (to the front versus to the rear), and the perception of something protruding at the end of a dummy (be it an antenna, an ovipositor, or some unnatural object) seems to be required to intitiate a display (Figure 3.17). Chemical stimuli are also involved: a display directed at an ovipositor lasts for only a short period of time, after which the male turns around

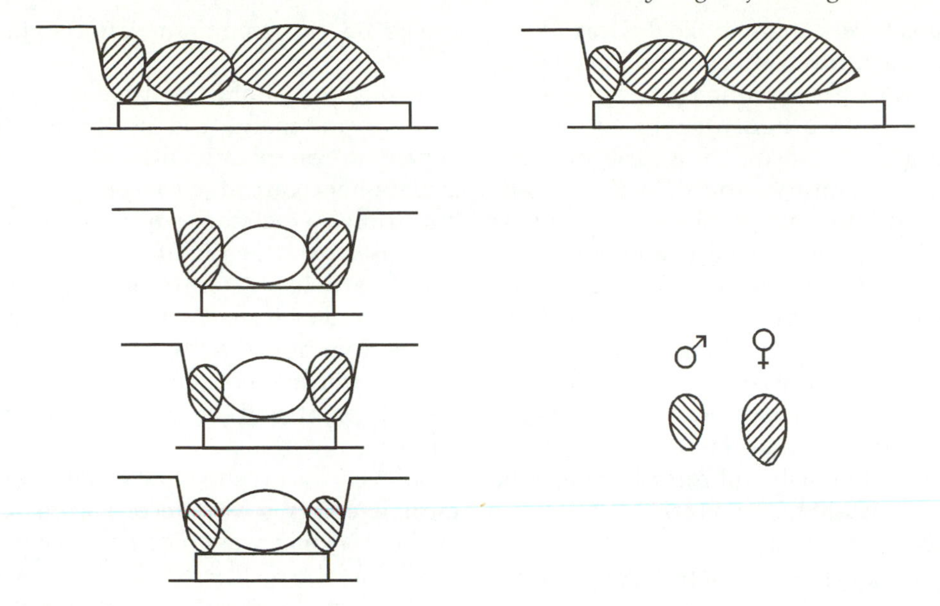

Figure 3.16 Examples of dummies that comprise male and female parts.

and moves to the front of the dummy. A male that courts a dummy comprising a female's body and a male's head will display for a shorter period of time than on a dummy that is comprised entirely of female body parts. Yoshida and Hidaka (1979) and Kitano (1975) used similar dummies to investigate the behaviour of other parasitoid species (*Cotesia glomerata* and *Anisopteromalus calandrae*).

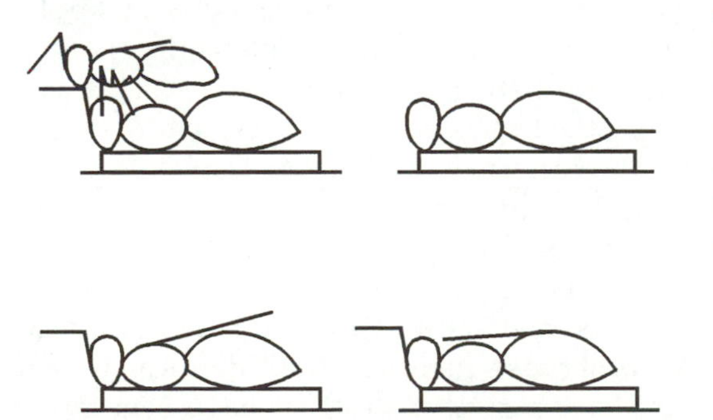

Figure 3.17 Examples of dummies that were used to determine what cues a male uses in orienting himself upon the female, what cues cause him to remain in position and what cues elicit display behaviour.

3.6 STIMULUS PRODUCTION

3.6.1 INTRODUCTION

Once a male has assumed the correct orientation relative to the female, and the female has become immobile (presumably in response to the male taking up the correct position), he will commence courtship. In certain parasitoid groups, courtship displays are very inconspicuous, i.e. the motor patterns involved are not eye-catching, whereas in others they are very striking. It should be stressed that of the various movements made during a display, some may not stimulate the female – they may simply be 'by-products' (subsections 3.6.2, 3.6.3, 3.11.1).

3.6.2 ACOUSTIC STIMULI

The males of many parasitoid wasp species, both upon approaching the female and during courtship *sensu stricto*, vibrate their wings, the vibrations coinciding with the production of sounds which can be recorded via the substratum using a variety of devices that act as transducers. Among the devices I have used is an electrodynamic microphone. Using the latter, direct observations are possible. A pair of wasps in a lightweight vial such as part of a gelatin capsule, or a glass cylinder (diameter ' 10 mm, capped with a half-capsule until the onset of the display), is placed directly on the microphone's membrane. The vial is removed if the couple happens to court upon the membrane itself. Signals from the microphone are recorded on tape, and behavioural sequences are observed from above. The records can be visualised and analysed with an oscilloscope and/or a sonagraph. Obviously, all records should be made at a constant temperature.

The above method has to be used with caution. In principle, it is inappropriate to rest objects on the membrane of an electrodynamic microphone which is then used for recording sounds: the membrane's mass will change and its response characteristics may be affected in such a way that the generated impulses no longer reflect the properties of the acoustic stimuli that one wants to record. There is always the dilemma of needing a vial to keep the wasps in place before any record can be made. Therefore, a few precautions and controls are necessary. A direct control that can be applied is to test the response characteristics of the microphone with a sound signal of a known frequency (I use 400 Hz, produced by a sine wave generator), with and without a vial in place. Also, a microphone is not designed to respond to contact vibrations; hence a test should be made with a vibrator in contact with the membrane (again with and without a vial) producing vibrations of appropriate frequencies. The signals derived from the vibrator can be displayed on a screen and compared with the simultaneous display of the signals from the membrane. The two types of signal should be identical. Most probably, the wasps' weight, being negligible in comparison to the membrane's mass, should not produce a noticeable distortion in the recorded vibration. It is also necessary to determine whether the precise location of the wasps on the membrane affects the signal recorded.

If the parasitoid or predator normally searches for mates and performs courtship behaviour on plants, then the stylus of a gramophone cartridge can be applied lightly but firmly to the plant surface close to where the insects occur (Ichikawa, 1976). Better results are likely to be obtained using an accelerometer instead of a gramophone cartridge. Experiments might be carried out in which recorded male signals are played back into the plant, and the the females' responses observed.

Sivinski and Webb (1989) give many technical details on how to record acoustic signals of the braconid wasp *Diachasmimorpha longicaudata*. Webb *et al.* (1983) describe an analysis of the acoustic aspects of behaviour of the Mediterranean fruit-fly *Ceratitis capitata*.

Nasonia males produce acoustic pulses of a constant quality throughout a display, the pulses coinciding with wing vibrations (Figure 3.18). However, the wings themselves are not the source of the acoustic emissions because neither immobilising them nor altering their surfaces, nor removing them altogether results in a noticeable alteration in signal. Similar observations were made by Leonard and Ringo (1978) on the chalcidid wasp *Brachymeria intermedia*. Male wasps can be made completely 'mute' simply by applying a tiny drop of superglue to the thoracic dorsum, between the wing bases. This prevents the thorax acting as a resonator.

The pattern of acoustic emissions produced during courtship is very species characteristic (Figure 3.18), suggesting a display function. This possibility can be investigated as follows: take a set of normal males and a set of otherwise identical (i.e. identical in size and experience) males that have been made mute by the above-mentioned method, and a set of males from which the wings have been removed. Compare the **courtship efficiency** (the amount of display necessary to induce receptivity in virgin females, subsection 3.8.2), and **mating success** (the proportion of

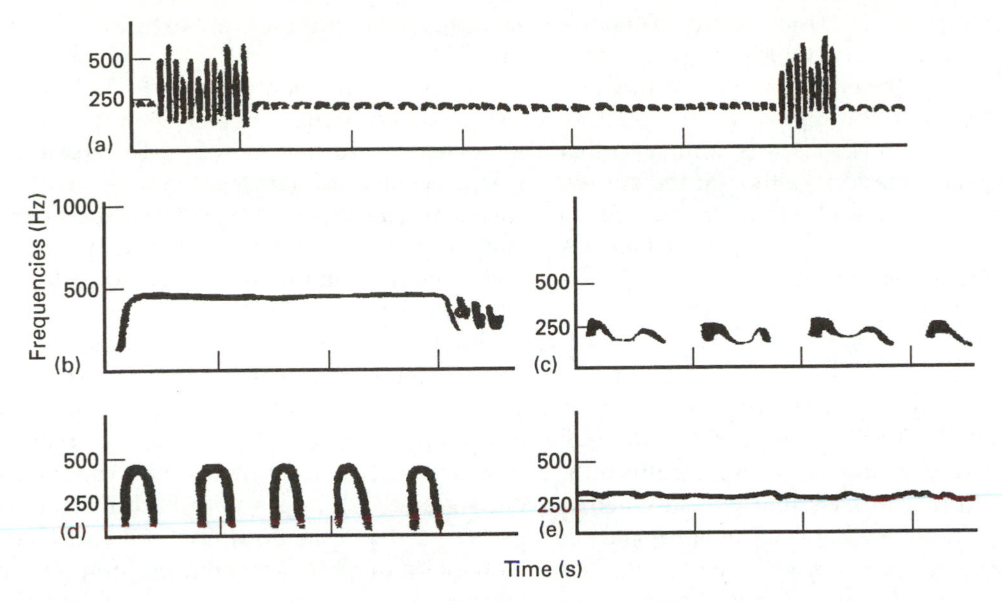

Figure 3.18 Sonagram tracings of the courtship sounds of: (a) *Nesolynx albiclavus*; (b) *Baryscapus daira*; (c) *Tetrastichus sesamiae* (all Eulophidae); (d) *Nasonia vitripennis*; (e) *Anisopteromalus calandrae* (both Pteromalidae).

mated females producing progeny of mixed sex) of wingless males, dumb males and normal males (van den Assem and Putters, 1980). In a study of *Nasonia vitripennis* I found no difference between the three treatments. That is, wing vibrations appeared to play no role in courtship with that particular experimental design. Mute males were similar to other males if young males were used, and young mute males were able to induce receptivity within exactly the same period as control males, and their mating success was similar. There was, however, one noticeable difference: females that were mounted by a mute male gave a 'startle response', as if they were mounted without being 'warned in advance'. This reaction resulted in a short delay in the onset of a display.

A positive effect of acoustic emissions was ascertained with old males: old, mute males and normal males of a similar age scored differently with respect to mating success. The following experiment was carried out: mute, old individuals were presented daily with a sequence of virgin females. *Nasonia* recordings and 'white noise' were alternately played back using a small loudspeaker at a barely audible sound level. The displays accompanied by courtship sounds were more successful than displays accompanied by white noise (van den Assem and Putters, 1980). The effectiveness of the playback indicated that the stimuli in question were normally airborne (i.e. were sounds) and not substrate-born vibrations. Acoustic stimuli probably play a secondary role in bringing about receptivity: younger males produce sufficient stimuli of primary importance (chemical stimuli in particular, subsection 3.6.3).

According to Barrass (1960b), old *Nasonia* males vibrate their wings less than young males and are less successful in courtship due to the deteriorating production of courtship 'sounds'. However, a deterioration in pheromone production could also have accounted for the poor success. I have not explored the species-recognition function of acoustic stimuli any further, except that I added the acoustic emissions of another species (*Mesopolobus mediterraneus*) to the display of 'dumb' old *Nasonia* males. This procedure proved effective, which suggests that at least in this case the emissions do not function in species recognition. However, the pitch of the *Mesopolobus* emissions was similar to that of the *Nasonia* emissions (although the pattern over time was different), so pitch may be an important cue. Clearly, there is room for further experimentation.

3.6.3 CHEMICAL STIMULI

Chemical stimuli have already been mentioned (subsection 3.5.3); they appear to play an important role in short-distance communication. As a rule, it is easy to establish whether chemical stimuli are involved in some part of display, and it is also easy to determine their site of origin. A simple experiment with *Nasonia* (van den Assem *et al.*, 1980b) demonstrated the importance of chemical stimuli in inducing receptivity in females (subsection 3.8.2). The onset of receptivity coincides with head-nodding by the male, a conspicuous component of his display (section 3.7). With head-nodding, the mouthparts are extruded with the upward component and retracted with the downward component of each nod. As a first step in determining whether chemical stimuli are released during mouthpart extrusion, it was necessary to manipulate males in such a way that head-nodding and mouthpart extrusion were prevented from occurring, separately and in combination, without affecting either of the other display components or the overall temporal patterning of the display. The experimental treatments were: (a) normal, i.e. control males (ones with no glue applied to any part of their body); (b) males with only the head immobilised; (c) males with the mouthparts sealed; (d) males with both the mouthparts sealed and the head immobilised. To this end, males in all treatments were immobilised by first exposing them to a low ambient temperature (4°C), then males in treatments (b), (c) and (d) had a minute drop of superglue applied to either the junction between the head and thorax (b), the maxillae (c), or to both (d). The glue was hardened by adding a small drop of water (Figures 3.19 and 3.20).

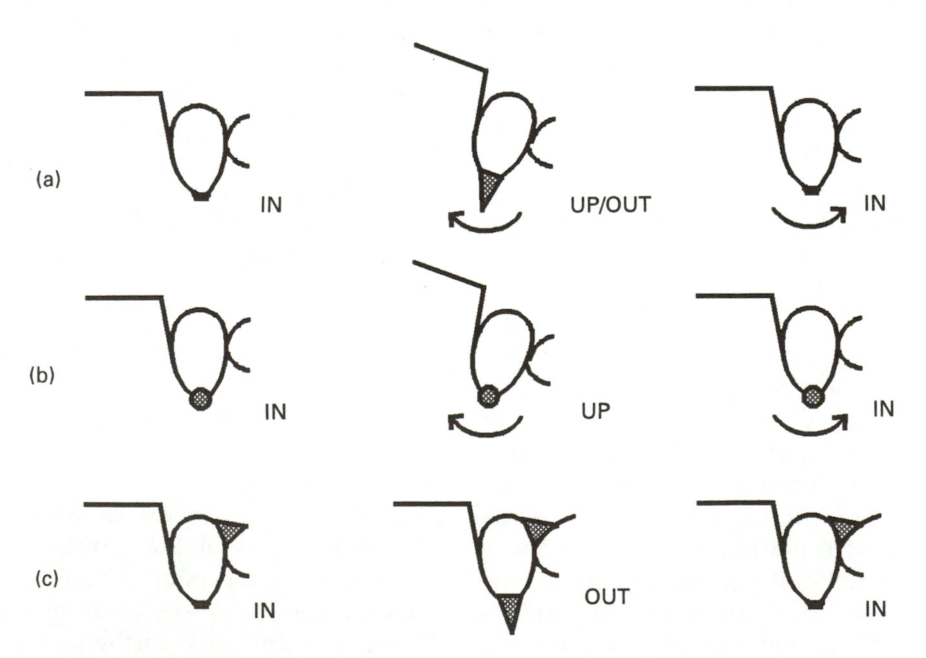

Figure 3.19 Head-nodding and mouthpart extrusion in: (a) normal *Nasonia* males; (b) mouthpart-sealed males; (c) males with immobilised head.

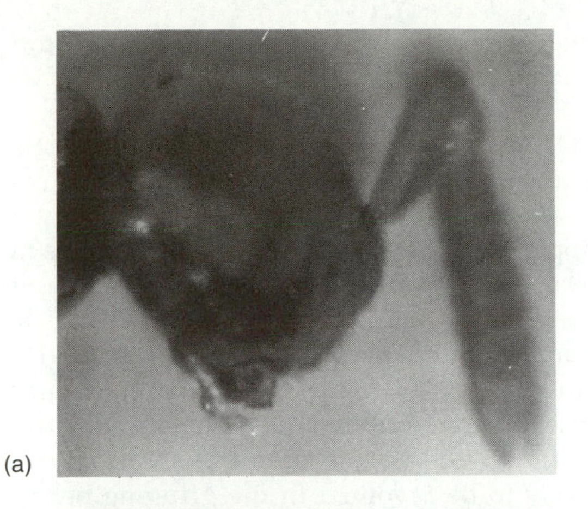

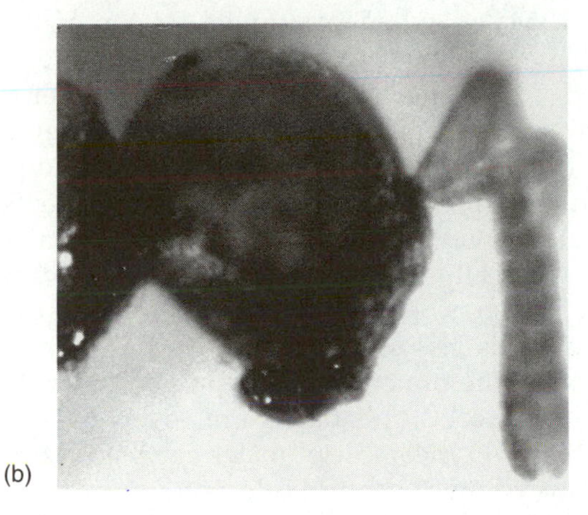

Figure 3.20 Heads of *Nasonia* males: (a) with intact mouthparts; (b) with sealed mouthparts.

The timing of courtship cycles was unchanged (e.g. antennal sweeps were performed on time) by glueing. Males with immobilised heads had to court for exactly the same period of time to induce female receptivity as did the normal-male controls. However, males with sealed mouthparts were totally unsuccessful in inducing receptivity.

The role of chemical stimuli in inducing female receptivity was demonstrated as follows (van den Assem *et al.*, 1980b): mouthpart-sealed males were presented with a virgin female; the males courted normally, but the females remained unreceptive. Using a 5 ml syringe, a quantity of air was removed from a vial containing many courting, normal, untreated males and females, and the air was slowly released over the female with the mouthpart-sealed partner whilst the latter was courting. Females immediately became receptive. None of the necessary controls (blowing air taken from an empty vial, from one containing males only or from one containing females only) were effective. Sealing a male's mouthparts appears to prevent the release of a pheromone that coincides with mouthpart extrusion. It is assumed that in related pteromalids mouthpart extrusions serve a similar function. Based on evidence from other Hymenoptera (e.g. honeybees and ants) it seems likely that the pheromones originate in the males' mandibular glands. This might be investigated by dissecting out the glands, agitating them in a solvent such as hexane and carrying out a behavioural assay, although the wasps' small size may be prohibitive. A solvent extract might also be used to study, by electroantennography, the receptors involved (the latter are probably located on the females' antennae), but again the small size of most parasitoid wasps is likely to make such an investigation difficult.

Contact pheromones, present on the cuticle, are probably very widespread. Because of their species-characteristic properties they play a role in the 'recognition' of mating partners (subsection 3.5.3). Once a display is underway, a constant input of stimuli originating from contact pheromones is necessary to keep the display going.

3.6.4 STIMULI OF OTHER KINDS

The displays of many species of wasp involve movements with the legs consisting of drumming, tapping, kicking and swinging. All of these types of movement might be used in tactile stimulation of the females. Males of some species have conspicuously decorated legs, for example males of *Mesopolobus* (Pteromalidae) bear coloured fringes or tufts. These decorations are moved in front of the female's compound eyes during courtship,

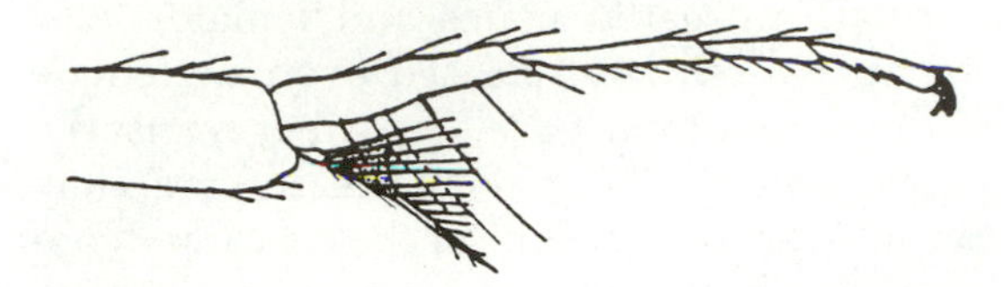

Figure 3.21 Distal part of tibia and tarsi of the middle leg of a male *Nesolynx glossinae*, showing a remarkable structure that probably provides specific stimuli during courtship. (Source: Graham, 1989.) Reproduced by permission of Gem Publishing Co..

suggesting that they have a visual function, but no-one has investigated this possibility, as far as we know.

In the eulophid *Nesolynx albiclavus*, the male flexes his middle legs during a display. The tibiae are inclined steeply inwards and the tarsi steeply outwards (a rather unusual posture), and the male also drums with these legs on the female's thorax. Graham (1989) noted an unusual modification in the middle legs of *Nesolynx* males; the tibial spur bears a basal triangular expansion from which long setae radiate like a fan, and the basitarsus bears remarkably long setae (Figure 3.21). The posture of the legs is such that these structures will make contact with the female during drumming, but it is not known whether they have any effect on the female's behaviour. The only way to investigate would be to remove either the setae or the legs and record whether there is any reduction in courtship success.

3.6.5 OVERALL DUMMIES

Courting males produce more than one kind of stimulus but not all kinds are of equal importance. The effects of separate stimuli are summated (Lorenz, 1939); moreover, one kind of stimulus may partly substitute another (Tinbergen, 1951). In order to ensure that the relevant stimuli have been identified, an *overall* dummy can be constructed and its effectiveness tested. Only when female receptivity is induced as it is by normal males, can one be sure that the stimuli have been properly identified.

Sometimes nature itself comes to the rescue of the experimenter, and presents the required kinds of dummy – males lacking one or more of what seem to be essential appendages. Males of *Melittobia acasta* are an example. Their antennae are peculiar in shape; they act as claspers for gripping the females' antennae during courtship. The scapes are swollen and bear small pores, and glandular tissue has been located within (Dahms, 1984a,b), so chemical stimulation is likely to be involved in the gripping behaviour. Males, when positioned on top of the female, also move their hind and middle legs, and they perform kicking movements towards her anterior, at the level of her eyes or beyond. Because the legs are provided with chitinous brushes, these movements could provide tactile stimuli. *Melittobia* males are mutually intolerant and engage in what are frequently fatal fights. Sometimes the damage is less severe than the severing of a head or the puncturing of a thorax, and only one or more appendages are severed (legs, antennae). 'Incomplete' males court as vigorously as intact ones and do not appear to be handicapped in any way – males lacking either antennae, hind legs, or middle legs are equally successful (van den Assem *et al.*, 1982a). Clearly, the question of what are and what are not essential attributes requires further investigation.

Burk (1981), in a study of acalypterate flies, addressed the question of why signalling in insects is simple in one species but complex in another. He suggested that the species' **mating system** plays a key role (the mating system of an animal is defined as the ensemble of behavioural and physiological adaptations specific to mating, as well as some of the social consequences (Vehrencamp and Bradbury, 1984). Burk recognised, in terms of behaviour only, the following systems: in species with resource-based male-controlled systems signalling is simple, males do not signal but merely jump on and attempt to copulate with any object. Where either encounter rates are low or mating decisions are made by the female, signalling is more highly developed. Extreme complexities are reached when mating takes place within groups of competing males, with control lying firmly in female hands. These proposals merit investigation, the more so because they do not hold for parasitoid wasps, since courtship in swarming species is not conspicuously complex (Nadel, 1987; Sivinski and Webb, 1989), whereas courtship in *Melittobia* (in which males are confined to the site of female emergence, there is a very high encounter rate, and the decision to mate is at least partly left to the female) seems to be very complex indeed (van den Assem *et al.*, 1982a). Investigating this contradiction may prove quite difficult. Categorising behaviour as either simple or complex will be fraught with difficulties as long as the relevant signals themselves have not been identified. Simple motor patterns may well coincide with the release of a complex set of chemical stimuli, and so on.

Figure 3.22 *Harpobittacus* hanging flies (Mecoptera) copulating. The female is feeding on the male's nuptial gift of a fly (Premaphotos Wildlife).

3.6.6 CHOICE OF MALES BY FEMALES

Although female choice is clearly an important driving force in sexual selection in animals, there is no evidence that female parasitoid wasps exert active choice. Usually, a virgin female will mate with the first conspecific male encountered. In fact, only a few examples of adaptive choice are known in insect natural enemies, the black-tipped hanging fly *Hylobittacus apicalis* (Mecoptera), investigated in the field by Thornhill (1980, 1988), being one. Both male and female hanging flies catch prey, usually Diptera, of various sizes and pierce it with the proboscis, injecting a paralysing venom together with enzymes that liquefy the prey's insides. Besides eating a prey, a male may offer one as a 'nuptial gift' which the female consumes during copulation (Figure 3.22 shows this for *Harpobittacus* sp., a related genus). Typically, large flies are offered. Whilst hanging by his fore legs from a perch, the male releases a pheromone originating from glands in the tip of his abdomen. Receptive females approach, take up a hanging position near the male and seize the prey. The male releases the prey as soon as genital contact is achieved. No conspicuous display movements are involved.

Females prefer males that carry larger prey, but this preference is not conclusive proof of adaptive mate choice. Thornhill obtained evidence for the latter as follows: prey size distribution and relative abundance were estimated from sweep net catches in the vegetation surrounding the mating sites. Females catch large and small prey in the same proportions as they occur in the vegetation. They do not discard small prey they catch themselves but do discard such prey if they are offered as nuptial gifts. Because courting males use large prey far in excess of the proportion present in the vegetation, some sort of active selection of prey by males must be involved. Females feed on the prey for the duration of the copulation. Males offering a large prey copulate for 20 minutes or longer and they, not the females, terminate genital contact. On the other hand, males with a small prey earn a contact of at most 5 minutes' duration (or no contacts at all) and genital contact is terminated by the females (Svensson *et al.*, 1990, demonstrated that in the empidid fly *Empis borealis* the time spent *in copula* is positively correlated with the volume of the nuptial gift). By interrupting copulations, Thornhill (1988) showed that a copulation lasting for up to 5 minutes is required for any sperm at all to be transmitted. Presentations of gifts of small prey therefore do not necessarily result in insemination. By contrast, the longer-duration copulations obtained by presenting larger prey will usually result in insemination (with copulations lasting between 5 and 20 minutes the amounts of sperm transmitted correlate positively with the time spent *in copula*). Following a full insemination, females become non-receptive for about 4 hours and produce an average of three eggs during this period.

Female *Hylobittacus* depend rather heavily on feeding by males during their oviposition phase. Thornhill (1980, 1988) suggests that nuptial feeding by *Hylobittacus* is adaptive because it reduces female mortality: the female is less exposed to predation than she would be if she had to hunt for herself, and so there is an increased probability that eggs fertilised by the male's sperm will actually be laid. An important point is that large prey are available in limited quantities and therefore only a small proportion of males will be involved in reproduction. Of the males offering nuptial gifts 90% carry large and 10% carry small prey; the latter will be discriminated against by the females. Thornhill's data indicate that males which capture large prey do so repeatedly, suggesting that males chosen by females may differ genetically from those that are spurned. This point requires verification.

Thornhill and Sauer (1991, 1992) investigated female choice and other aspects of reproductive success in another mecopteran, the scorpion fly *Panorpa vulgaris*, which is not a genuine predator but a scavenger, feeding primarily on dead arthropods. Females of *P. vulgaris* may not exhibit *active* mate choice but they take longer to mate with, and receive more sperm from, males that provide nuptial gifts compared with males that do not. Therefore, females adjust their behaviour during mating in a way that results in certain males siring more offspring than others. Thornhill and Sauer's study is interesting because it spans more than one generation of the insect, and genetic sire effects on offspring performance were brought to light. Male flies were collected from the field and tested immediately for ability to produce nuptial food (protein-rich saliva) during copulation. They were assigned to one of two categories: those that could produce saliva during mating – so called nutritionally *high* males and those that could not – so-called nutritionally *low* males. Because males have to feed in order to begin producing saliva, and because food is a limited resource for which there is strong competition, it was hypothesised that differences in saliva production may relate to differences in the ability to fight. Saliva production may be the basis for sexual selection in scorpion flies because of the positive correlation between amounts produced, duration of mating, and success in fertilisation.

Progeny of field-collected males were reared and paired with virgin full siblings (to minimise maternal genetic effects), and their offspring were reared in turn. Offspring ability to fight, and mating success of sons were then tested. In the former case, two males or females, the offspring of fathers of either category, were confronted in a small plastic box containing a piece of food (a segment of mealworm). In the latter case, the same pairs of contestants in the fighting tests were introduced into a similar box, a few days later, and presented with a virgin female. The male who initiated mating was designated the winner of the mating contest. Sons of high fathers won 75% of the food competition contests and 70% of the contests for mates. Likewise, daughters of high fathers won more fights over food than daughters of low fathers. These sire effects were not confounded by body size differences among contestants. However, there was no significant correlation between individual ability to win both a fight and a mating contest. Males seem to win matings not because of their better fighting ability, but probably because either they pursue females more vigorously or females actively prefer them as mates.

Thornhill and Sauer forward arguments for the latter possibility, turning to fluctuating asymmetry. Fluctuating asymmetry is defined as small, random deviations from perfect bilateral symmetry in a morphological trait (Thornhill, 1992). Fluctuating asymmetry is widely recognised as a measure of developmental stability, which is in turn an indicator of how good the genome is at controlling the processes of development to produce a perfectly symmetrical animal. Given that high fitness is likely to be associated with low fluctuating asymmetry, we would expect a negative relationship between fluctuating asymmetry and components of fitness (Palmer and Strobeck, 1986, give a review). Thornhill and Sauer measured, in sires and male and female offspring, the lengths of the left and right forewings and the lengths of the left and right basal tarsal segments. The measure of asymmetry they used was the ratio: [length of left wing (or tarsal segment) – length of right wing (or tarsal segment)/ length of right wing (or right tarsal segment) + length of right wing (or tarsal segment)] × 2.

Bilateral symmetry was shown to be heritable and to differ significantly both between the two sire categories and between the offspring of sire categories. Nutritionally low sires and their offspring were shown to be bilaterally more asymmetrical. The genetic paternal effects upon offspring fitness components may thus give rise to female choice based on genetic quality of potential sires. A female's mating preference could be based on some direct measure of a male's symmetry, but indirect correlations are more likely to be used.

A burning question remains unanswered from Thornhill and Sauer's study: if the ability of the male partner to produce a nuptial gift is so important to his reproductive success, then why was it that in the initial sample of males taken from the field a considerable percentage of individuals (44%) turned out to be *low* males, i.e. non-producers of gifts. Clearly, the whole story of mating success in *P. vulgaris* may not yet have been told.

Fluctuating asymmetry and mating success has also been studied in the damselfly *Coenagrion puella* (Harvey and Walsh, 1993). Males caught in the field were given an individual mark and the length of their wings measured before they were released. Every time a male was subsequently seen mating, its identity was recorded. Males with more symmetrical wings were found to obtain a higher number of matings.

In studies of fluctuating asymmetry, it is important to test for directional asymmetry (a consistent bias to one side) and antisymmetry (asymmetry the norm, but no one side consistently the larger) (Palmer and Strobeck, 1986). Antisymmetry is indicated by a platykurtic or bimodal distribution of left-right measurements.

3.7 SOME QUANTITATIVE ASPECTS OF MATING BEHAVIOUR

3.7.1 HOW MUCH COURTSHIP WILL A MALE CARRY OUT?

As a rule, males, when they meet a conspecific female, are immediately ready to court and are thus ready to make an investment – of time, energy and materials. This poses the question: how much of an investment is a male going to make? A standard amount per encounter perhaps?

The amount of time spent courting per female can easily be measured in relation to different external stimuli, e.g. different kinds of dummy females, before and after an event such as a copulation or an unsuccessful display (subsection 3.7.2). However, measured in these ways, it contains little information on courtship production.

A more thorough quantification of a male's investment would require separating the performance into its component acts (subsection 3.2.1). These components can then be quantified in two ways: as events which can be counted to provide totals and frequency measures, and as durations. Practical reasons will dictate the use of the one or the other. For a good general account of how to quantify behaviour, see Martin and Bateson (1993).

In terms of overall temporal patterning, displays are of two types (van den Assem, 1975):

1. **cyclical**: involving a repetition of similar units of movements that may include various body components (Figure 3.23a). Successive units, or cycles, may be identical, but more often they are slightly different, e.g. in *Nasonia* the number of head-nods per cycle changes according to a fixed pattern (Barrass, 1961);
2. **finite**: the display quality changing during the performance because movements that were included earlier are omitted, to be replaced by new components. Finite displays have a definite end, i.e. a finale, irrespective of whether female receptivity occurs or does not occur at that moment.

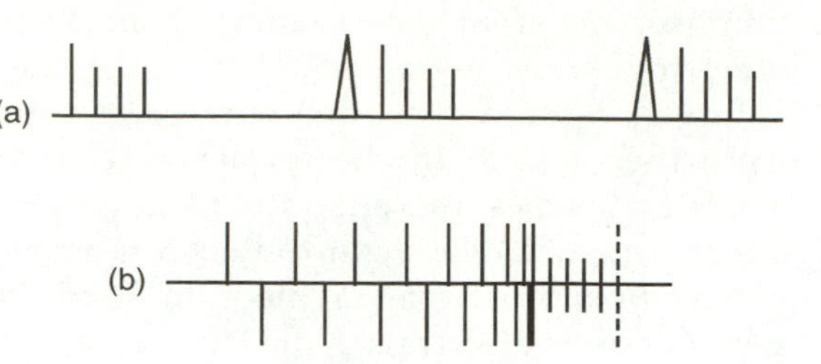

Figure 3.23. A schematic representation of: (a) 'cyclical'; (b) 'finite' display. Time runs from left to right. The pteromalid *Nasonia vitripennis* is an example of a species that produces a cyclical display. Triangles denote antennal sweeps, vertical lines denote head nods. Single nods are separated by short pauses and are clustered in time. First nods are denoted by longer lines than the following nods because they are more elaborate, last longer and coincide with the release of a pheromone. All series except the first are preceded by an antennal sweep. Series are separated by intervals. The period from the first nod of a series to the first nod of the subsequent series is termed a courtship cycle. A cyclical display consists of a repetition of identical or nearly identical cycles. The eulophid *Melittobia acasta* is an example of a species that produces a finite display. Long vertical bars above the horizontal denote antennal movements (flagellar vibrations concluded by grasping of the female's antennae), those below the horizontal denote series of swinging movements with the hind legs in combination with a loss of antennal contact between the male and female. Alternations accelerate until they coincide. At this point (denoted by a series of shorter vertical bars above and below the horizontal) the antennae are stretched downwards and the hind legs moved up and down, rubbing against the female's thorax, instead of swinging to and fro. After a few seconds, these movements are followed by a brushing of the middle legs against the female's eyes (denoted by the dashed vertical line at the far right). A finite display does not consist of distinct, more or less identical cycles, but changes markedly over time, leading to a succession of 'phases' such as 'introduction' and 'finale'.

Finite displays occur in *Melittobia* species (Figure 3.23b). Judging by Goodpasture's (1975) description of courtship in four species of *Monodontomerus* (Torymidae), a finite display seems to occur in those parasitoids also, although no mention is made of the timing of female receptivity (subsection 3.8.2).

As an example of a quantitative investigation of male production, the cyclical display of *Nasonia vitripennis* males will be discussed. Other species with cyclical displays can be investigated along similar lines. The head-nodding movements and mouthpart extrusions of *Nasonia* males have been mentioned already (subsection 3.6.3); consecutive series of nods, except for the first series, are preceded by an antennal sweep and are separated by intervals. Additionally, males drum on the female's compound eyes with their front tarsi, and vibrate their wings (Barrass, 1960a). Nodding is easy to quantify; separate nods and series of nods can be counted and recorded without using any specialised equipment. This is not the case with the between-series intervals. The succession of cycles is such that an automatic time-marker (subsection 3.2.2) is indispensable. Drumming movements are much too rapid and too irregular to be counted separately. They can be quantified per bout, on a presence or absence basis. Wing vibrations can be quantified precisely using the acoustic recording equipment discussed in subsection 3.6.2. The number of nods per series varies, as does the duration of intervals between series. General trends are illustrated in Figures 3.23, 3.24 and 3.25, and further data are contained in Barrass (1960b) and Jachmann and van den Assem (1993). The number of nods per series and the total number of series appear to be correlated; males exhibiting more nods per series exhibit fewer series up to the moment of dismounting.

In measuring a *Nasonia* male's display production, the first step would be to make external conditions constant through the use of either an unreceptive female or a dummy female. Males will differ in the duration of the display (the period between mounting and giving-up) that they are going to perform. This variability despite standardisation indicates that the performance is not exclusively under the control of external factors; internal factors also play a role, and these may alter periodically. This is an important point to bear in mind when observing a male on more than one occasion.

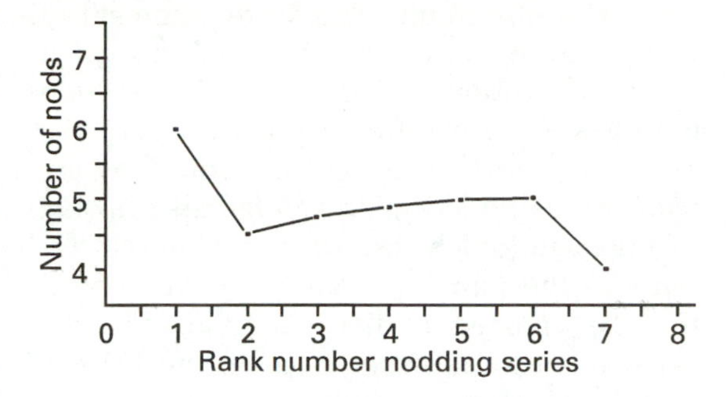

Figure 3.24 The number of nods in successive nodding series of courting *Nasonia* males.

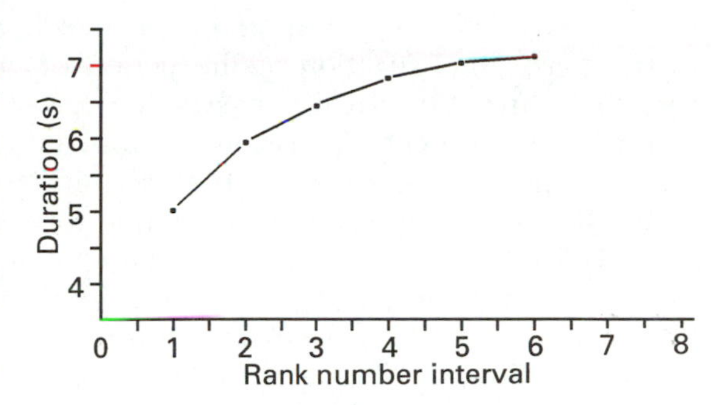

Figure 3.25 The duration of successive intervals between head-nod series in the courtship display of male *Nasonia*.

Some experiments require a more rigorous method of standardisation which can be achieved as follows: assume that because males give up and walk away, the complex of internal factors (readiness to court) has reached a minimum value at the moment of dismounting, and that all males are in a similar condition at this point. Dismounting is followed by a refractory period until the male is ready to mount a female again. During this period male and female can stay together, in which case the male himself terminates the pause by mounting the female that he courted earlier. Alternatively, the experimenter can separate male and female, preventing courtship for a certain period, and then either the partners can be re-united or the male can be presented with another female. In fact, treatments do not differ in their effects on the duration of the next display; it is the duration of the period of non-courtship (the pause) that is the key factor in this respect. The second display is recorded in the same way as the first one. With pauses lasting less than 24 hours, the second display is always shorter than the first; how much shorter depends on the duration of the pause. However, time is not the only controlling factor: males that have produced a longer first display (i.e. one with more nodding series) will also produce a longer second display than males with a shorter first display, following a pause of equal duration (Figure 3.26). This is probably an effect of inherent differences between individuals, which are resistant to further standardisation procedures.

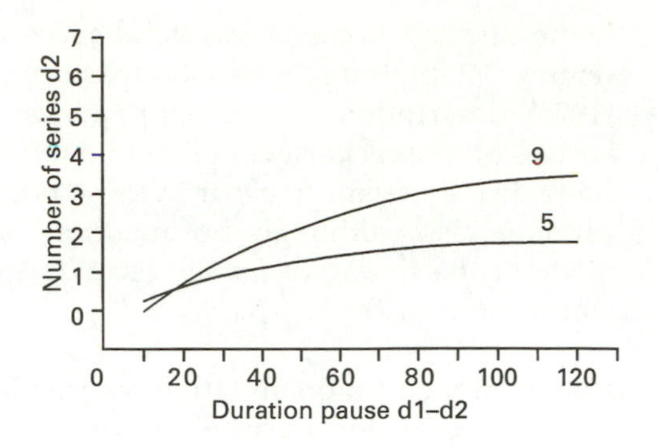

Figure 3.26 With pauses of equal duration, *Nasonia* males that produce more (9) nodding series in the first display 'recover' more rapidly than those that produce fewer (5) series.

The results suggest that during the performance of a display something is 'used up', and that during the periods of non-display a 'recovery' takes place. The extent of the recovery is time-dependent. With longer pauses

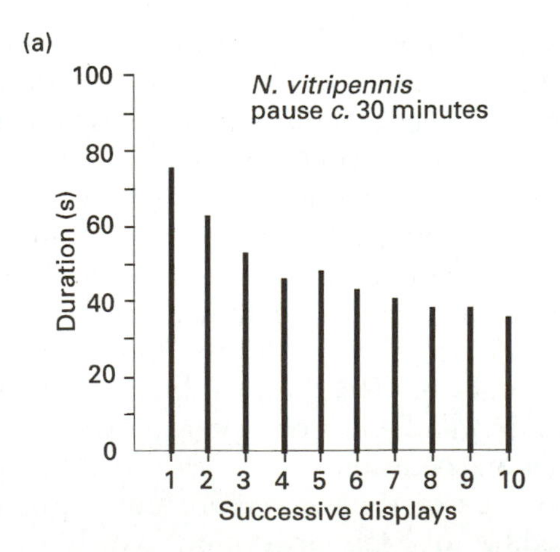

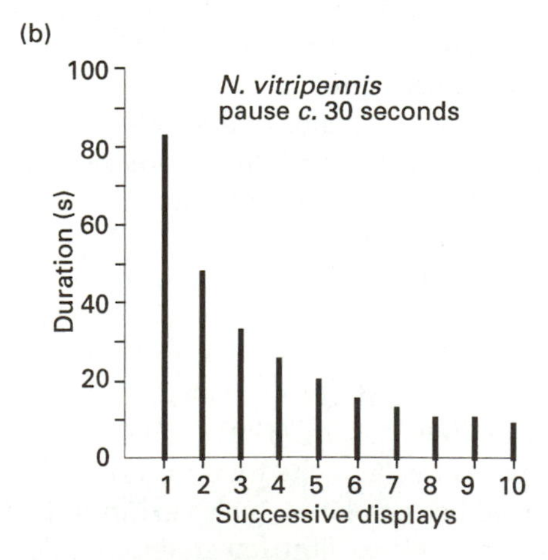

Figure 3.27 Durations of successive displays in *N. vitripennis*; inter-display intervals: (a) c. 30 min; (b) c. 30 s. *Nasonia* males cannot be described as simple courtship 'automata', as they do not produce unitary displays at any time.

there is more recovery, but the processes do not follow a linear course; they proceed rapidly at first but slow down later on.

Other conclusions that can be drawn are that males will court a female long before they are able to display maximally, and that their current display production is partly determined by what display was produced earlier. *Nasonia* males cannot be described as simple courtship automata; they do not produce unitary displays at any time. Clearly, for an understanding of the dynamics of display behaviour, records of the observed motor patterns do not provide sufficient data; measurements of between-display intervals are also required (Figure 3.27). Provisos, however, have to be given; the performance of finite displays as produced by *Melittobia* males seems to accord much more with the idea of an automaton's behaviour. Males produce a sequence of displays (on receptive and unreceptive females) of roughly similar duration (Figure 3.28), independent of the duration of intervening pauses (van den Assem *et al.*, 1982a).

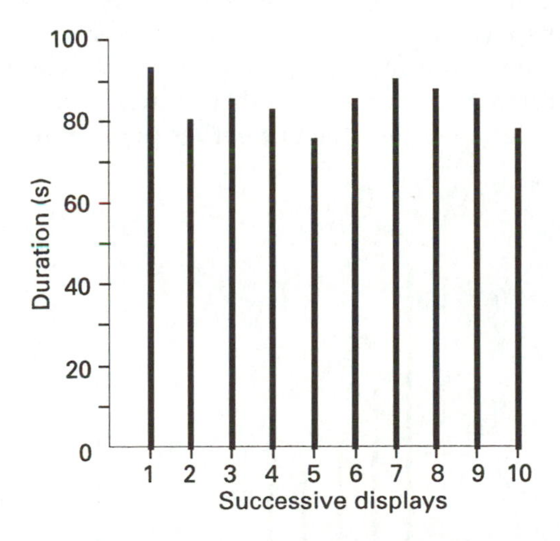

Figure 3.28 Duration of successive displays in *Melittobia acasta*, an illustration of the 'automaton' principle.

3.7.2 WHY DOES A MALE GIVE UP WHEN HE DOES?

Displays come to an end, but how? The male may respond to external agents, e.g. the onset of female receptivity (subsection 3.8.3). However, males that court either a dummy or an unreceptive female (subsection 3.7.1) give up for reasons other than being physically exhausted. Observations on such males show that the cessation of a display either coincides with or occurs immediately after head-nodding, suggesting that the performance itself brings about dismounting at a particular moment. I began with the working hypothesis that one and the same factor is in control of both the timing of dismounting and the occurrence of head-nods. I assumed that with the performance of each nod the readinesss to continue courting is reduced by an increment, so bringing the moment of dismounting one step closer. This hypothesis, however, needed to be modified because the assumptions appeared too simple. A simulation that reproduced almost all of the observed behaviour accounted for the contributions of individual head-nods and also allowed for some recovery of the tendency to court during the intervals between nodding series, more or less similar to the process that is assumed to occur during the refractory periods in pauses.

The modified hypothesis was tested in various ways, including interrupting a display in a subtle way. A male was made to halt his performance for some time without causing him to give up the courtship position. By doing this, the increase in the probability to dismount was supposed to be temporarily arrested and should have resulted in some additional recovery such that the display would be prolonged. The simulation also predicted that the precise timing and the duration of the interruption were of primary importance. The desired kind of interruption could be achieved with a dummy female (Figure 3.29) – a modified version of the type

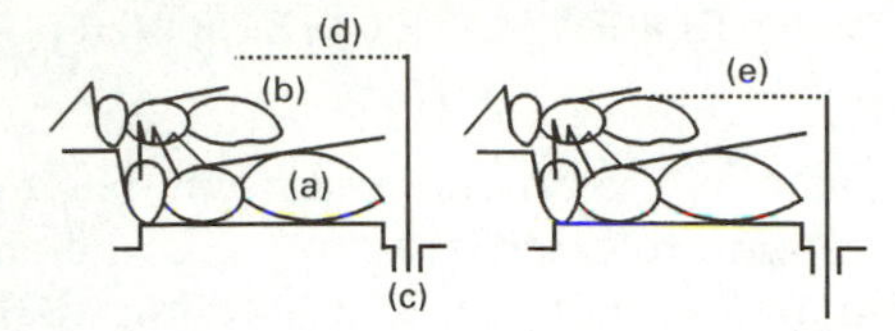

Figure 3.29 A construction for interrupting a male's display at a previously chosen point in time without causing him to dismount: (a) dummy female; (b) courting male; (c) opening in bottom of observation cell and vertical rod; (d) hair; (e) hair in 'strike' position.

used to investigate the effects of the onset of female receptivity (subsection 3.8.3). A stiff, horizontally protruding hair was fixed to the end of a vertical rod that could be moved up and down. The rod was positioned at the rear end of the dummy's 'gaster'. The hair was struck against the courting male's dorsum, causing him to cease displaying for some time. As predicted, both the timing and the duration of the interruption had a strong effect on how many extra series and extra nods were going to be produced. It was concluded that the display behaviour of *Nasonia* (and presumably that of other species) is under the control of rather simple rules. Going into further detail is beyond the scope of this chapter; for further discussion, see Jachmann and van den Assem (1993, 1995).

One more investigation of the recovery process should be mentioned because I happened to come across a peculiar phenomenon entirely by accident (van den Assem *et al.*, 1984). *Nasonia* males that have courted a number of unreceptive females in succession will produce only a short display on subsequent females, the duration of the display depending on the recency of the previous dismounting. Males that had courted five females in rapid succession were subjected to one of two treatments during a half-hour pause between courting the fifth and sixth female: they were either kept at room temperature (20°C) or placed in a freezer (–30°C) (males given the latter treatment were, because they were in a tube inside a styrofoam container, not actually exposed to 30°C all the time). After half an hour, the males from both treatments were presented with their sixth female. Although the freezer-treated males were completely immobile (appearing dead) when introduced to the females, they 'regained consciousness' surprisingly rapidly (only then were they presented with the sixth female). The room temperature-treated males produced a short display as predicted, but the freezer-treated males produced a display not significantly different in duration to that of inexperienced males, as if the effect of earlier performances

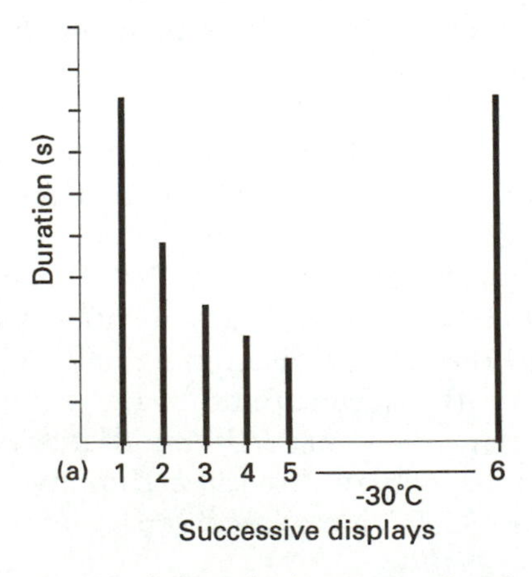

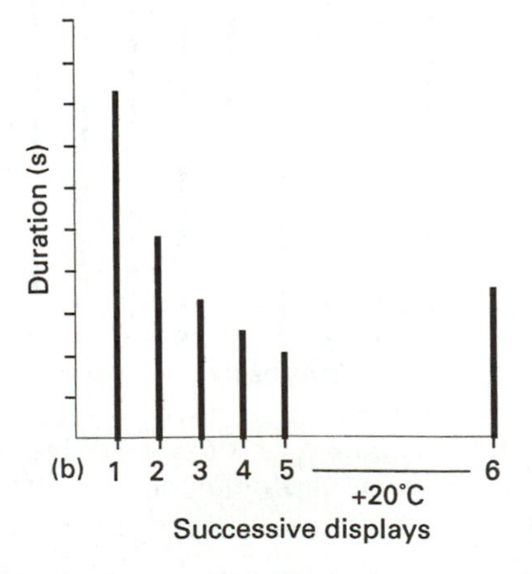

Figure 3.30 *N. vitripennis*: durations of each of five successive displays, and, following a pause of half an hour at either +20°C or –30°C, the duration of the sixth display.

had been completely lost (Figure 3.30). These observations rule out the possibility that during the pause something that is 'used up' during displaying has to be either replaced or requires processing. If, after the half hour in the freezer, the freezer-treated males were kept alone at room temperature for a period of around 5 minutes before being exposed to the sixth female, the effect of freezer treatment upon the duration of display was no longer demonstrable and the insects performed as if the entire pause had been spent at room temperature. This is experimental evidence that genuinely inhibitory processes, and not 'consumption' processes of some sort, play a role in bringing a display to an end. There is at present no satistactory explanation regarding the physiological mechanisms involved. The phenomenon is mentioned because sometimes what seems to be rather unorthodox procedure may yield unexpected results that, hopefully, will provide new insights into the causal processes underlying mating behaviour.

3.8 FEMALE RECEPTIVITY AND ITS CONSEQUENCES

3.8.1 INTRODUCTION

The mere presence of a male never results in overt female receptivity. The types of stimuli males need to produce to bring about the transition from latent to overt receptivity in the female have been discussed earlier in this chapter (section 3.6). Here, more specific questions are asked: are the relevant stimuli produced continuously or are they administered at intervals? If the latter applies, then what is the temporal pattern of the procedures? Is the onset of receptivity a random event during a display, does it relate to the performance of a particular display movement, or is it induced as soon as a certain amount of stimulation has accumulated, i.e. a threshold operates? Also, are courtship sequences 'chain reactions', with both participants reacting to one another in a step-by-step manner until copulation is achieved? This is often how they have been portrayed in the literature.

The last of these questions has already been answered. Males of many species court a dummy or a dead female in the same way as they do a living female, and they may attempt to copulate. Simple chain reactions are thus ruled out. The timing of the successive display cycles appears at first sight to be a matter of internal control. This is not to say that external stimuli are of minor importance in maintaining the male's display. On the contrary, courting *Nasonia* males monitor the position of the females' antennae continuously, as can be demonstrated with the appropriate dummy females (subsection 3.8.3).

3.8.2 STIMULUS BOMBARDMENT OR DOSED ADMINISTRATION OF STIMULI?

Direct observations are required to establish whether the onset of female receptivity coincides with the performance by the male of a particular display movement. However, coincidences may not be obvious. For example, in cyclical displays that show little motor differentiation, males repeat identical movements and may do so at a high rate. Displays of at least some ichneumonids fall into this category: males vibrate their wings vigorously and from the onset of courtship repeatedly attempt to grasp the female's genitalia with their claspers; at a certain point in the courtship sequence this 'wriggling' develops into genuine copulation. A similar sequence of behaviour is seen in *Trichogramma evanescens* (Hase, 1925, and my own observations). Probably, pheromones (the release of which need not coincide with conspicuous motor patterns) play a major role; females could well be bombarded with stimuli all the time, but this requires further investigation.

Coincidences between display components and the onset of female receptivity are easier to discern in cyclical displays that have more differentiated motor patterns. Earlier (subsection 3.6.3), experiments were mentioned involving mouthpart-sealed *Nasonia* males, that demonstrate the role of a chemical stimulus delivered with head-nodding. The onset of receptivity coincides with the first nod of a series (Figure 3.31a) and (almost) never with second or later nods. This alone is strong evidence for a periodic production of the essential stimuli that induce overt receptivity, but there are two additional pieces of evidence: (a) first nods differ from the additional nods in the character of the movement (subsection 3.7.1), and (b) females are able to perceive the receptivity-inducing stimuli continuously. Females may signal receptivity at any time once a certain concentration of the mouthpart pheromone is reached in the environment (subsection 3.4.4).

With finite displays, the production of a set of releaser stimuli seems to occur once per display (subsection 3.7.1), which makes the timing of the switch from latent to overt receptivity a very predictable event (Figure 3.31b). This is even stronger evidence of periodic production of stimuli.

3.8.3 RECEPTIVITY AND ASSOCIATED POSTURE

Generally, the switch from latent to overt receptivity in the female coincides with a change in her posture: she raises her gaster, so opening the genital orifice. There may be variations on this posture, depending on the taxonomic group. In eulophids for example, females seem to 'sag in the middle' while raising the abdomen and they direct the head and antennae upwards resulting in a posture that resembles the *lordosis* posture of copulating rodents. In *Nasonia* and related Pteromalinae, females lower the head and draw the antennal flagellae tightly towards the frontal region of the head (Figure 3.32).

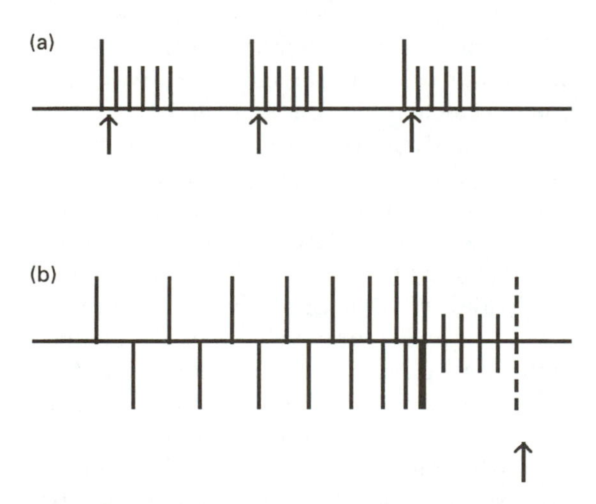

Figure 3.31 Onset of receptivity (denoted by upward-pointing arrows) tied to the performance of a particular event: (a) in the cyclical display of *Nasonia vitripennis*, where receptivity may occur at several points in a sequence, but always immediately following the performance of a first nod; (b) in the finite display of *Melittobia acasta*, where receptivity occurs at a unique point in time, at the end of a finale. For other symbols, see Figure 3.23.

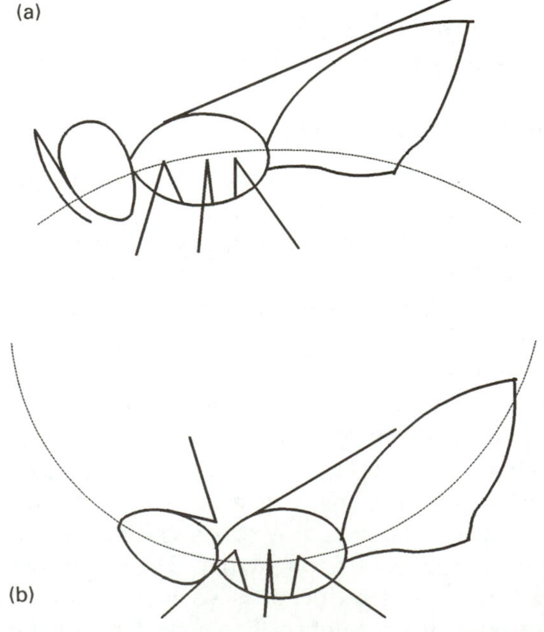

Figure 3.32 Female receptivity posture of: (a) pteromalids; (b) eulophids.

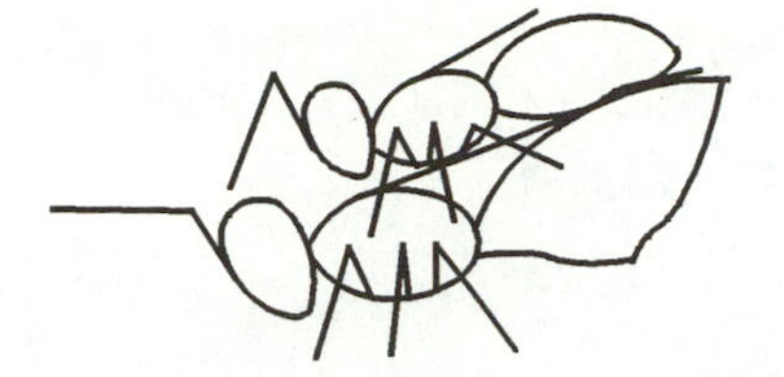
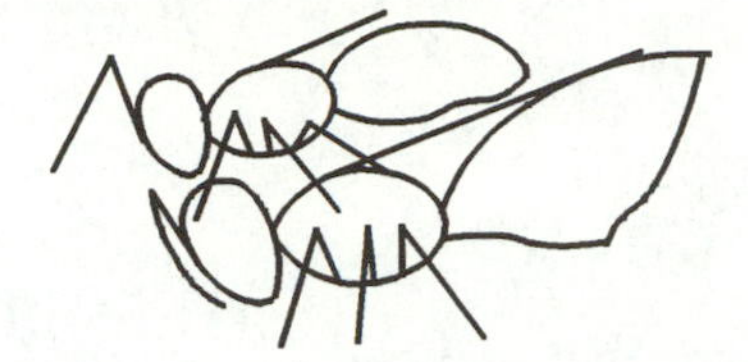

Figure 3.33 Position of the courting male in different species of Pteromalidae, and the possibility of the male perceiving directly the female's raising of the abdomen.

As soon as receptivity appears, the male ceases courting and switches to copulatory behaviour. Which factors are responsible for this change? I investigated this question using a dummy female. The following considerations were given to the dummy's construction: in many species, the raising of the abdomen by the female may provide a tactile stimulus to a mounted courting male, but this is not the case in *Nasonia* and related Pteromalinae (Figure 3.33). Because of the anterior position he adopts upon the female, this kind of stimulus is unlikely to have an effect. Another conspicuous movement the female performs at the onset of receptivity is the downwards folding of her antennae, which would be very easily perceived by the male. A dummy that had movable antennae was therefore tried (Figures 3.34 and 3.35). The dummy partly consisted of a freshly killed, antennectomised female fastened to the bottom of an observation cell with a human hair. The hair, running transversely over the female's dorsum, and kept in place by two small clips, was used instead of some more permanent way of fastening because frequent alteration to the wasp-part of the dummy was necessary (subsection 3.4.5). The mechanical part of the dummy consisted of a vertical rod bearing two tiny pieces of wire that represented antennae. The rod passed through a hole in the bottom of the cell, allowing the pseudoantennae to be moved up and down (van den Assem and Jachmann, 1982, give precise details of the mechanism). In the control treatment, males courted the dummy in exactly the same way as they did a living female; when the dummy was not activated, they would give up and dismount after a similar period of time to that recorded in males presented with unreceptive, living females. Downward movements of the pseudoantennae caused the males to cease courting and attempt copulation. This experiment demonstrates that antennal movements act as a genuine signal. It is assumed that similar movements in other pteromalids have the same function.

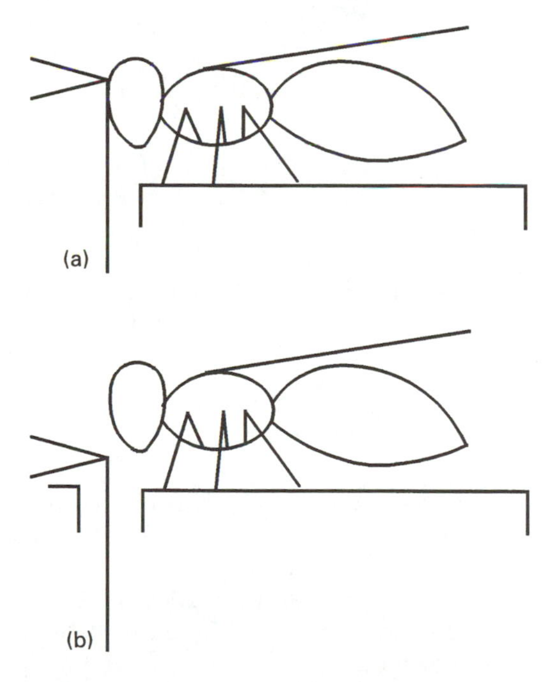

Figure 3.34 The dummy *Nasonia* with movable antennae; antennae in: (a) courtship-eliciting position; (b) receptive position.

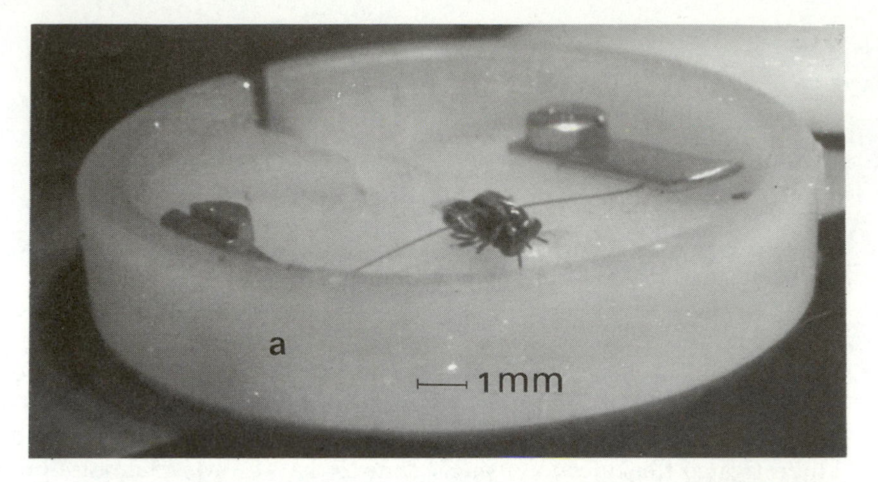

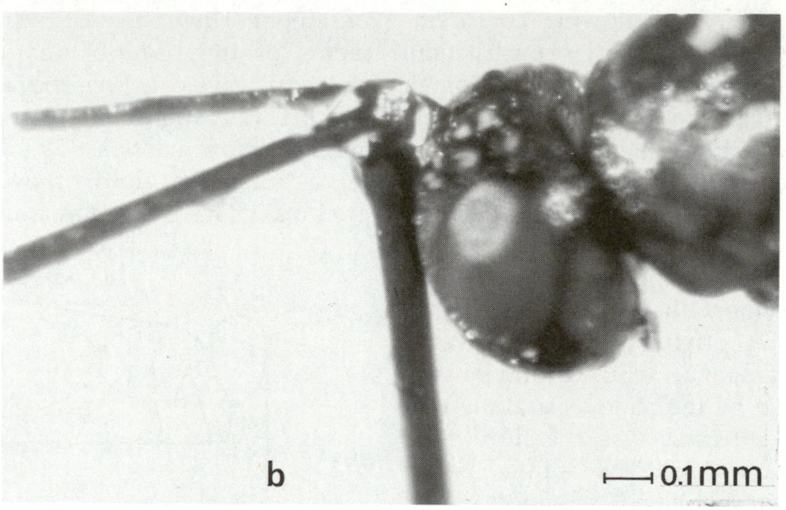

Figure 3.35 (a) A male *Nasonia* performing courtship behaviour on the dummy; (b) a close-up of the dummy female's head in contact with the pseudoantennae.

3.8.4 INTRASPECIFIC VARIABILITY IN THE ONSET OF RECEPTIVITY

Virgin females will usually become receptive when courted, sometimes even after the barest minimum of display by a male. However, inseminated females require a longer display. Inherent differences between individual females and/or individual males may be involved. Males may differ in their stimulus value, while females may differ in their sensitivity to certain stimuli. In *Nasonia*, for example, the amount of pheromone released by a male with a head nod may vary among individuals, but we have no evidence of this. In females, however, there is some experimental evidence of intrasexual variation in sensitivity to stimuli, as the following experiment demonstrates: virgin females were made receptive by a courting male and allowed to copulate, but the male was removed with a camel-hair brush before he could regain the anterior position for a bout of post-copulatory courtship (females will usually signal receptivity again during post-copulatory courtship (subsection 3.8.6)). Females were assigned to two groups: those that had signalled receptivity with a male's first head-nod series, and those that did so after three or more series. After a pause of half an hour the females were presented with another male, and all signalled receptivity again. A longer display was required to release this last response but the initial differences remained: females of the first group required significantly less stimulation than those of the second group. Thus, individual females differ in sensitivity to courtship stimuli. On the other hand, males that had induced receptivity with the first nod of the first series were not consistently more efficient compared with males that induced receptivity after several series, when presented with a second series of females half an hour later.

3.8.5 SWITCHING OFF OF FEMALE RECEPTIVITY

Normally, sexual receptivity of female wasps can be induced a limited number of times only, in some species apparently only once, e.g. *Melittobia* spp. Females will at some point refuse further matings. Both internal and external factors are probably involved in the **switching off process** that makes females permanently or temporarily unresponsive to courtship stimuli. Those related to insemination *sensu stricto* are obvious candidates; stored sperm could provide direct, 'external' stimuli (e.g. as a measure of the fullness of the spermatheca). However, an ejaculate contains materials other than sperm which might indirectly affect a female's receptivity (Leopold, 1976). The effect of spermatheca fullness would seem rather easy, and the effect of ejaculate substances difficult, to verify.

An investigation into the stimuli that switch off receptivity might be conducted along the following lines: virgin females are presented with an inexperienced male (to guarantee a maximum sperm donation (subsection 3.9.3). The first step is to measure how long a copulation should continue for, in order to bring about full insemination (subsections 3.6.6, 3.9.4). Next, copulations need to be interrupted at various times between the onset of insemination and its completion, in order to provide fully filled and partially filled females. Next, females are presented with a few hosts per day, so that they use up their sperm supply. They are then tested for receptivity once or several times daily. Repeated copulations should be avoided, but sometimes cannot be prevented, so it is advisable to use, as a precaution, normal males together with recognisable mutants (e.g. colour morphs). This will enable the investigator to distinguish between the earlier and later inseminator (daughters sired by each type of male can be distinguished). The most simple outcome that would be recorded is the reappearance of receptivity immediately after the moment of sperm depletion, i.e. a change from mixed sex progeny to all-male progeny.

Several complications may arise, however. One is a possible cumulative effect (positive or negative) of the daily receptivity tests. Hence, a control group of wasps should be used that are tested only once, at various points in time, i.e. after various degrees of sperm use. The control wasps will also provide data on the effect of sperm use versus the effect of time, i.e. the duration of the period following insemination. It might be the case that receptivity reappears after a certain interval, even if some sperm remain in the spermatheca, or because additional ejaculate factors may become ineffective with time. As a further check, another group of controls should be used: females that are never presented with hosts following insemination (at least part of this group should have an opportunity to perform host-feeding but not oviposition, so as to make them similar in this respect to females with hosts). If the case is a simple one, and if the spermatheca of host-deprived females remains full, receptivity will not reappear. However, various amounts of sperm may become absorbed, so direct checks on the amount of sperm in the spermatheca (subsection 3.9.3) are necessary. A much later reappearance of receptivity in the non-ovipositing group would point to the role of additional ejaculate factors but to investigate this possibility would be difficult. A further set of control females makes use of partially or completely sperm-depleted males (subsection 3.9.3) as 'inseminators', assuming that the depletion is paralleled by a depletion of ejaculate substances, in a repetition of the experiments described above. A deeper analysis of the effects of insemination requires sophisticated physiological techniques which go beyond the scope of this chapter.

In many other groups of insects, the reappearance of receptivity in females seems to be under the direct control of hormones which are involved either in egg maturation or oviposition *per se*. Such females may copulate either immediately after or immediately before the laying of a clutch of eggs whether or not there are sperm present in the spermatheca, e.g. dung-flies (Parker, 1970a) and dragonflies (Waage, 1973).

In several groups of parasitoid wasps there is yet another factor involved in the switching-off process. The occurrence of overt receptivity may itself initiate a process that inhibits its early reappearance. For example, in *Nasonia* (and also a number of species belonging to other groups) females that were made receptive but were never inseminated (because the male was brushed off before he could make genital contact) appeared unreceptive following the event. Brushing the male off at times other than after the onset of overt receptivity had no such effect.

It was pointed out earlier (subsection 3.3.4) that the receptivity threshold of females is one of the first characteristics that is affected by prolonged culturing in the laboratory. In some of my laboratory 'strains' females seem almost continuously ready to signal receptivity, with or without prior copulation occurring.

3.8.6 MALE READINESS TO COPULATE

In general, males can copulate without first courting. If a male encounters a female in the copulation posture (either a living female or a dummy that has been obtained by killing a copulating pair of wasps in liquid nitrogen), he will proceed immediately with copulatory behaviour, dispensing with courtship. This ability forms the basis of the **sneaky male strategy** in which a male, when he happens to come across a courting couple, immediately grasps the female's abdomen (Figure 3.36) and, at the onset of receptivity, copulates with her (usually in the first position, i.e. ahead of the male who did the courting and made the female receptive).

Immediately following a copulation, a *Nasonia* male re-assumes the anterior position and courts afresh, and the female usually signals receptivity for a second time. However, males never respond to a second signal; they remain at the front and continue displaying for a short period before dismounting. Hence, under identical external conditions, i.e. first and second signal, a male may or may not react to the female's signal (by backing up to copulate). Whether he does react depends on his previous behaviour which has changed his internal state. It is the performance of a copulation that causes the change in his reactivity: neither brushing off a male prior to or following a receptivity signal nor removal just prior to a genital contact makes a difference; in all these cases there is an immediate attempt at copulation by the

Figure 3.36 A courting *Nasonia* couple, and a 'sneaky' male clasping the female's abdomen.

male when he is again confronted with the signal. The refractory period following a copulation is rather short. A male that is prevented from performing a post-copulatory display (by being brushed off at the termination of a copulation) and is quick to mount and court another female, will not back up if she, in turn, is quick to signal (she must do so at the first nod of the first head-nod series). Apparently, the refractory period has yet to come to an end. If a longer display is necessary to induce receptivity in this second female, the male will back up immediately and copulate. The conclusion drawn from this is that the refractory state is not a *female-specific* phenomenon, coming to an end as soon as the male is off the female, but a *time-dependent* one.

The immediate readiness to copulate that is apparent in searching males is lost once display commences. Short-term, periodic changes in readiness become apparent. The dummy female with movable antennae (subsection 3.8.2) (Figure 3.34) was used to investigate the underlying processes. Live females signal receptivity at the first nod of a series of head-nods (subsection 3.8.2), and males back up (to copulate) immediately. When the dummy is activated at a 'natural' point in the courtship sequence there is a similar reaction. However, the dummy can do what living females almost never do, that is signal at 'unnatural' times in the sequence. An unnaturally timed signal has an instant effect: the male ceases displaying, but the time taken to respond with backing up depends on the recency of the previous first nod (Figure 3.37). Only signals occurring immediately prior to the onset of the next courtship cycle (onsets can be predicted accurately, subsection 3.7.1) induce an early reaction. An analysis of the changes in duration of refractory periods, i.e. the postponements of the backing up reaction, carried out by Jachmann and van den Assem (1993) led to the conclusion that the timing of the backing-up reaction results from the interaction of at least two different processes: (a) the effects of signalling (a signal represents a strong external stimulus to the courting male); and (b) the endogenous, periodic changes underlying courtship behaviour. The latter are produced by an oscillator-like system that is responsible for the timing of the succession of display cycles. A simulation that reproduced the observed behaviour most closely was mentioned earlier (subsection 3.7.2). Further details are to be found in Jachmann and van den Assem (1995).

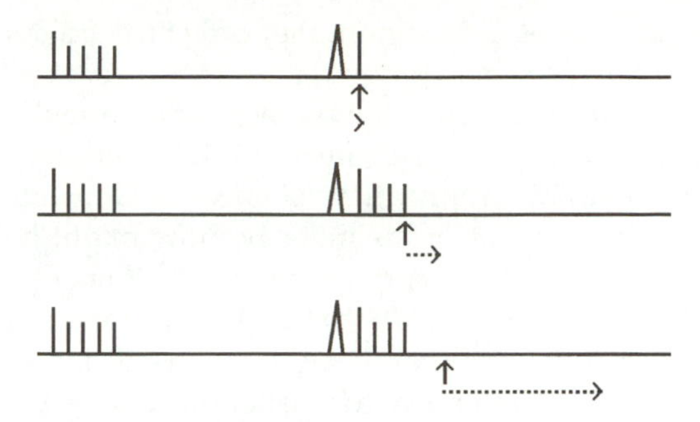

Figure 3.37 Signalling at various points of a display (vertical arrows), and duration of corresponding delays until backing-up (length of horizontal arrows). For symbols, see Figs. 3.31 and 3.23.

Observations made on several *Melittobia* species (*acasta*, *digitata*, *evansi*) suggest that fluctuating male readiness is a general phenomenon. *Melittobia* females signal receptivity at the display's finale, as if it is only by that time that the male has produced either a sufficient amount of stimulation or a necessary set of stimuli. Very rarely do females signal at some stage before the finale, in which case males do not back up but instead either dismount or start courting afresh. To all appearances, they are unable to make the switch to copulatory behaviour at the 'wrong' moment (van den Assem *et al.*, 1982).

3.9. ASPECTS OF MATING *SENSU STRICTO*

3.9.1 CHOICE OF FEMALES BY MALES

Female parasitoid wasps separate from their gametes at oviposition. By their choice of time and place of oviposition, they affect the fitness of their progeny. A copulating male separates from his gametes at the moment of an ejaculation and is thereafter faced with uncertainty because his reproductive success depends entirely on that of his mate. So how might he improve the prospects of his donated sperm? The only way in which a male could exert a direct influence would be by choosing larger females in preference to smaller ones, since the former have, on average, better prospects (they live longer and are more fecund) than the latter (van den Assem *et al.*, 1989). I have so far not found any evidence of mate size preferences by males, but the possibility that they exist has not been investigated in a systematic way. A useful first step would involve presenting males with a choice either of differently-sized females or differently-sized dummy females (subsection 3.5.4).

3.9.2 COPULATION AND INSEMINATION

A copulating male penetrates the female and releases an ejaculate that comprises sperm and products from the accessory glands (Leopold, 1976). The anatomy and function of the male genitalia in *Nasonia* were studied by Sanger and King (1971). A comparison of the distance from the external opening of the vagina to the opening of the spermathecal duct with the length of the external, protruded male genitalia suggests that sperm are deposited at the duct's opening (King, 1961). Sperm are found in the spermatheca within one minute of the termination of genital contact. Transport of sperm from the spermatheca's duct opening to its interior may occur by peristalsis, or it may be guided by a chemical gradient. Because there is a movement in the opposite direction once sperm are recruited for fertilisation (King, 1962, King and Ratcliffe, 1969), the directional stimuli must be reversible. Hogge and King (1975) studied the ultrastructure of individual spermatozoa in Chalcidoidea and measured sperm size. The sperm are of relatively large size (head 50 μm wide), tail 100 μm long), so they can pass through the spermathecal canal only one at a time. This implies an extremely efficient use of the supply, approaching the absolute minimum of one sperm to one egg. A mass of sperm swimming along the spermatheca can be observed under a light microscope at 600–1000 × magnification (Nadel and Luck, 1985); the spermatheca has to be removed from the female's abdomen and kept in insect saline solution (section 2.3.7).

Immediately following an insemination a female *Nasonia* seems incapable of using the sperm. This presents few, if any, problems to freshly emerged and newly inseminated females, because they usually have no genuinely mature eggs (eggs that can be instantly fertilised and laid) available in their ovaries. During the period required for full egg maturation, the fertilisation routines become operational. However, for more mature females, i.e. ones that have a supply of fully mature eggs at the moment of insemination (or reinsemination) the situation is different. Sperm use by female *Nasonia* was investigated in a simple experiment (van den Assem and Feuth de Bruijn, 1977) that is outlined in Figure 3.38. The sex of progeny produced after the second copulation indicates for how long females are unable to fertilise eggs. Our experiments revealed a considerable degree of variation among females; in some the fertilisation routines were not operational for as much as 24 hours. For the males we used different eye colour mutants (Saul *et al.*, 1965). This enabled us to show that a second male's sperm becomes mixed with that of the first male. The proportion of daughters sired by either the first or the second male remained constant following the second insemination; the proportions depended on how many sperm of the first male were still present, and how many sperm the second male could add. Order effects, such as 'last in, first out' (subsection 3.9.4), were not found.

Within pteromalid and eulophid species, there is no correlation between the male's or

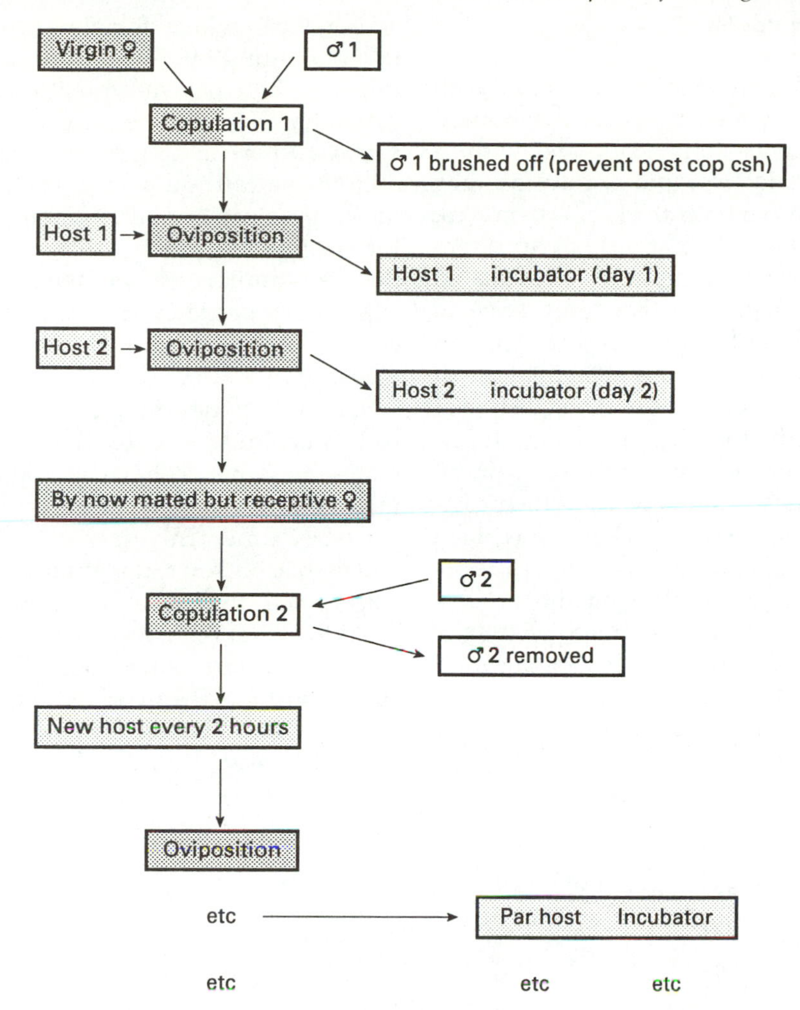

Figure 3.38 An experimental set-up for testing a female *Nasonia's* ability to fertilise eggs following a second insemination. Unshaded and shaded parts of boxes denote contribution by female and male, respectively.

female's body size and the duration of copulation. The durations of successive copulations of individual males were similar. Furthermore, the duration of copulation is not related to the degree of sperm depletion (subsection 3.9.3). Between species however, considerable differences are apparent (e.g. *Nasonia* 15–20 s, *Lariophagus* 40–80 s, *Melittobia* 5–10 s, all at 22°C, durations being temperature dependent). Obviously, a minimum period of time is required for delivery of an ejaculate; the lower time limit to copulations may indicate just that. It seems unlikely that copulations of longer duration reflect the transmission of a more substantial volume of sperm, or a slower flow; rather they suggest that only part of the time period is needed for sperm transmission. In investigating this, the experimenter would need to interrupt copulations at various points in time and count the numbers of daughters produced following such treatments.

3.9.3 SPERM DEPLETION

In general, males are able to serve more than one female. However, sperm are a limited commodity: males that copulate with a number of females in rapid succession may deplete their immediate stock of sperm. Are males able to allocate (ration) sperm donations in relation to the number of available females, i.e. donate less than they have to offer at any moment in time, and so stock up for future use? Anatomical examination of the male's reproductive system suggests that they can and do (Sanger and King, 1971). Sperm pass from the testes into two pairs of chambered vesicles; the vesicles of the proximate pair are smaller and thicker-walled than those of the distal pair and open into the ejaculatory duct. Between the chambers is a sphincter muscle. The proximate chambers become empty immediately after mating, and their combined volumes can be taken to correspond to the full potential quantity of sperm with which females can be inseminated. Judging by the minimum durations of intervals between inseminations, the proximate chambers can be refilled rapidly. Only in cases where mating is a genuinely once-in-a-lifetime affair is a male expected to expend his entire store at one go. Honeybee drones, for example, 'explode' while copulating and die; a similar phenomenon would be expected in predator species where males are certain to be consumed when copulating (subsection 3.9.7).

Sperm depletion may be a laboratory artefact, since under natural conditions most males may never have an opportunity to mate with many females within a short period of time. Only species with extremely female-biased sex ratios might merit investigation, but the males of these parasitoids seem to be more or less depletion-proof (see below). *Nasonia* males can be depleted of sperm within 15–25 copulations, depending on body size (Figure 3.39). During a 'rest'

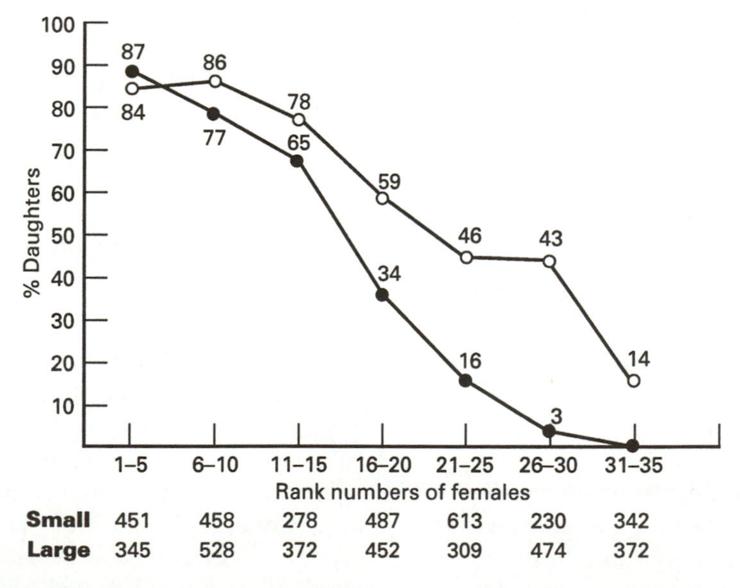

Figure 3.39 Percentage of fertilised eggs (daughters) in the progeny of 35 *Nasonia* females that had copulated with either a large male or a small one. Both types of male copulated 35 times in rapid succession. The data points are averages of five consecutive females. Indicated is the mean egg production of the females of the successive groups

period of 24 h, the sperm supply is replenished somewhat, but males never reach the initial level of inseminatory powers, and full depletion reappears rapidly (Figure 3.40). Males that copulate infrequently are depleted of sperm less rapidly than males that copulate frequently. A state of equilibrium may be reached where production equals donation, as demonstrated with males of *Pachycrepoideus vindemmiae* that copulated at 30 min intervals (Nadel and Luck, 1985).

Investigating total sperm production in individual males poses no special problems, but it is time-consuming. Individual males need to be paired with a succession of females to ensure that depletion occurs. This procedure needs to be repeated at intervals if one is interested in recovery processes. If a measure of the effective use of sperm by females is required, then all females have to be provided with a daily surplus of hosts, and all their progeny need to be accounted for. Only those females having progenies which show a transition from mixed sexes to all sons can be used for measuring the effective use of an inseminator's sperm. This is because females that cease to oviposit for one reason or another before the switch to all-male progeny has occurred may still harbour an (unknown) amount of sperm. Nadel and Luck (1985) offer an alternative approach: dissection of spermathecae in Ringer's solution following insemination, and measuring the band-width of the sperm layer (Figure 3.41). By themselves, estimates of the quantity of sperm used by females are of limited interest, but

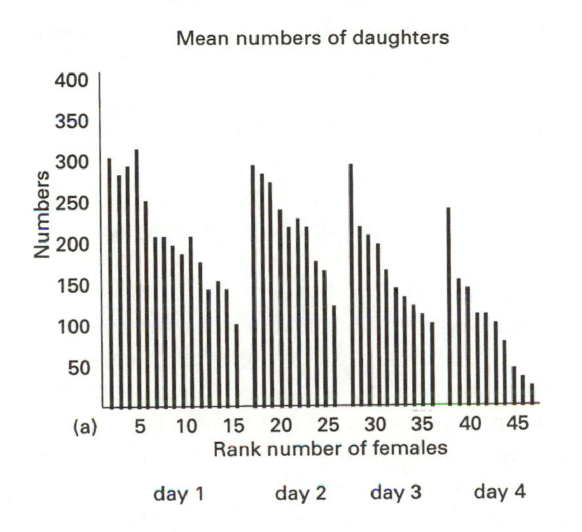

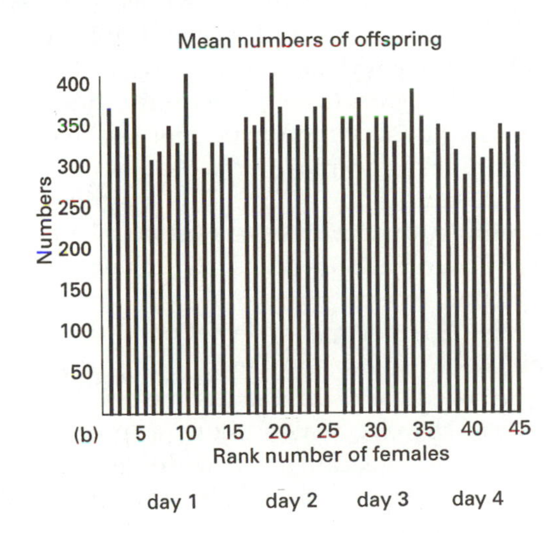

Figure 3.40 Nine *Nasonia* males copulated (always in rapid succession) with 15 females each on day 1, and with 10 per day on days 2, 3, and 4: (a) average numbers of daughters produced by the females, showing evidence of sperm depletion; (b) average clutch sizes. Rank numbers of females on X-axis.

    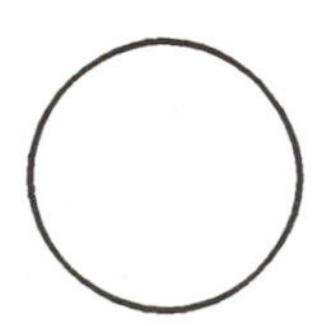

Figure 3.41 Band width of the sperm layer inside the spermatheca of *Pachycrepoideus vindemmiae* (Pteromalidae) after insemination by males at various stages of depletion. (Source: Nadel and Luck, 1985.)

are indispensable for an assessment of reproductive success (section 3.10).

It is unlikely that rapid sperm depletion occurs in species with extremely female-biased primary sex ratios. Any *Melittobia* species may serve as an example. Males are the minority sex (forming 5% or less of a population), and the operational sex ratio is even more skewed because males are mutually very aggressive and may kill potential competitors (subsection 3.6.5). Unlike those of *Nasonia* (or *Pachycrepoideus*), males of *Melittobia* deliver sperm effectively, over very long sequences of females, and all females that emerge from a host are inseminated. Such males are either very well stocked with sperm or have a high rate of replenishment or deliver strictly rationed and equal portions of sperm or spread inseminations over time so that sperm supply and demand are equal. Probably a combination of these factors is involved in *Melittobia*. With finite displays (subsection 3.7.1), the male, not the female, determines the timing of a copulation (van den Assem *et al.*, 1982a). This may help in keeping sufficient intervals between successive inseminations so as to prevent early depletion.

The degree of sperm depletion in the male appears to affect the tendency to court in some species, e.g. *Lariophagus* species, but not others. *Nasonia* males show a practically ineradicable readiness to court and mate. Barrass (1961) records a male performing 157 displays, including 154 copulations, in less than 5 hours. Is the seemingly unlimited capacity of *Nasonia* males to serve females an adaptation or is it a laboratory artefact?

3.9.4 SPERM COMPETITION, DISPLACEMENT AND PRECEDENCE

In females that are inseminated by more than one male successive sperm donations may become mixed, and so sperm competition among the males will result. We would therefore expect the evolution of counter-measures that either uphold the monopoly of sperm or at least minimise sperm competition. An overview of the diversity of **copulatory mate guarding** and other counter-measures is given in Thornhill and Alcock (1983).

In parasitoid wasps, post-copulatory displays probably serve to minimise sperm competition. The incidence of such behaviour varies among species; displays can be obligatory (e.g. *Nasonia* spp.), absent (e.g. all *Melittobia* spp.), or optional (e.g. *Lariophagus distinguendus*). Usually, post-copulatory displays are of short duration but extreme cases have been observed. For example, in *Encarsia asterobemisiae* display and copulation take 15 seconds, whereas a post-copulatory display takes about half an hour (Viggiani and Battaglia, 1983). The question of whether post-copulatory displays serve a biological function can be investigated by preventing males from performing the display by removing them. The reproductive output (number of eggs, number of daughters) of females that have or have not been subjected to post-copulatory courtship can be compared using inexperienced males as courters. Most likely, being subjected or not subjected to post-copulatory courtship influences the probability of mating anew (with another male). This point can be investigated by presenting females of both categories with a second male, always taking care that first and second males are genetically different and are distinguishable by some means (e.g. colour), so as to allow the respective shares in progeny to be distinguished in those cases where second males achieve a copulation.

Mate-guarding is a prominent feature of mating in dragonflies (Figure 3.42) (Miller, 1987, gives a review, and Conrad and Pritchard, 1990, describe a recent study). Although females may obtain sufficient sperm from a single copulation to fertilise many egg batches, they commonly mate with several males. There may be a selective advantage to this behaviour, the females trading copulations for: (a) access to oviposition sites which are in the territories set up by

Figure 3.42 A copulating pair of the dragonfly species *Sympetrum striolatum*. Prior to copulation, male dragonflies and damselflies transfer sperm to the secondary genitalia located in the anterior part of the abdomen; hence the female's genitalia are applied to that region of the male's body. Following insemination, males remain for some time 'in tandem' with the female, guarding her (see text). (Premaphotos Wildlife)

males; and (b) being guarded from disturbance by other males whilst ovipositing. Copulation in dragonflies can last for several minutes, even hours (up to 8 h in *Ischnura elegans*). Such prolonged copulations must reduce a male's opportunities for mating with other females and reduce the time available for a female to oviposit, suggesting that there must be some selective advantages in the behaviour to the *male*. Guarding could: (a) provide time to allow more sperm to be donated; (b) prevent other males from mating with the female; or (c) allow the male to compete more effectively with other males through sperm competition.

Sperm competition in Odonata has been investigated by a variety of authors, including Waage (1979), Miller (1982), Fincke (1984), Siva-Jothy (1984, 1987), McVey and Smittle (1984) and Michiels and Dhont (1988). The penis of some Odonata bears a variety of barbs and hooks on structures capable of entering the females's sperm storage organs (the penis of Odonata is not homologous to the penis of other insects, as it is housed, along with the rest of the secondary genitalia, in the second abdominal segment). A male uses these barbs and hooks to remove the sperm of rival males from a female's sperm storage organs, before inseminating her himself. Males of other Odonata may place their sperm in the most advantageous position within the female, so that it takes precedence when eggs are fertilised (i.e. 'last in, first out'). This may involve pushing the sperm of rivals deep into the females' sperm storage organs. The blunt penis with large inflatable sacs found in some *Libellula* species may be an adaptation for such re-positioning of rival sperm.

Sperm precedence and sperm displacement in Odonata have largely been studied (a) *directly*, by irradiating males (gamma rays and X-rays sterilise the sperm they contain) (McVey and Smittle, 1984; Michiels and Dhont, 1988); and (b) *indirectly*, by comparing sperm volumes in each sex at various stages of copulation, and by counting the number of sperm in females before, during, and after copulation (Siva-Jothy, 1984, 1987; Michiels and Dhont, 1988). Michiels and Dhont (1988) applied the **irradiated male technique** to *Sympetrum danae* (Libellulidae). Field-collected males were divided into two groups: those that were irradiated (for details of appropriate doses, see Michiels and Dhont's paper and the paper by McVey and Smittle) and those that were not exposed to radiation. To test the effectiveness of the irradiation procedure in sterilising sperm, virgin females were mated with the irradiated males. Two types of mating were carried out: single matings between normal (previously mated) females and either irradiated or normal males, and double matings between normal females and both irradiated and normal males (irradiated male first, normal male second, or normal male first and irradiated male second). With the double matings, females were mated for the second time before they could oviposit. When oviposition eventually took place, the eggs were collected and their embryonic development monitored (after three weeks, fertilised eggs can be distinguished from unfertilised ones). P_r, the proportion of eggs fathered by the irradiated male in double matings was computed as follows:

$$P_r = \frac{(n-x)}{(n-r)}$$

where n is the proportion of eggs developing after a single mating with a normal male, x is the proportion developing after a double mating, r is the proportion developing after a single mating involving an irradiated male and a virgin female.

P_2, the proportion of offspring sired by the last male to copulate in double matings, can be computed from P_r when the irradiated male is the second (last) mate, and from $1-P_r$ when it is the first (Michiels and Dhont's paper describes further aspects of these computations).

Michiels and Dhont also used an indirect method to study sperm displacement. Sperm loads of males and females were determined before, during, and after copulation. Copulating pairs were interrupted at various stages (5, 10, 15 and 20 or more minutes). The sperm load of males was determined by dissecting the penis. The basal segment was held under water and squeezed with blunt forceps. When sperm release stopped, a drawing was made and the cube of the radius (mean of width and length) of the sperm droplet was taken as the male sperm volume index. Female sperm storage organs were likewise drawn and a sperm volume index calculated (albeit in a different manner). The degree of sperm opacity was also recorded, and used as a measure of sperm density (Siva-Jothy, 1987, describes a method of actually counting sperm).

Michiels and Dhont plotted sperm volume index against the stage of copulation and oviposition, for both males and females (field-collected, in both cases). In males, significant changes in sperm volume were recorded between the pre-copulatory stage to the post-copulatory stage. After translocation of the sperm from the primary to the secondary genitalia, the sperm volume index tripled. It remained constant during the first five minutes of copulation, but then started to decrease when sperm release (into the female) began. At the end of copulation, the mean sperm load of unmated males was reached again. Most females examined prior to copulation had a low-opacity bursa copulatrix (part of the sperm storage apparatus), indicating the presence of sperm from a previous mating. In females examined after the first five minutes of copulation, the bursa copulatrix decreased in volume, indicating that the male had either removed or packed the sperm (*S. danae* actually removes the sperm). In females examined after the five minute stage the sperm volume index rose, peaking immediately after copulation. After one oviposition bout, the original sperm volume was reached again, showing that females use most (about three-quarters) of the ejaculate volume to fertilise one clutch of eggs. Observations on sperm opacity indicated that replacement of sperm was accomplished during the first five minutes of copulation.

High levels of sperm precedence, from 80% to 100%, have been recorded in several species of Odonata. Whereas in some Odonata copulation mainly comprises handling of previously deposited sperm, in *Sympetrum danae* sperm removal occurs over a relatively short period. In those species that pack, rather than remove sperm, mixing of the two lots of sperm eventually occurs if the female does not oviposit soon after the second copulation. McVey and Smittle (1984), using the irradiated male technique, showed that in *Erythemis simplicicollis* complete sperm mixing occurs within 24–48 hours of the second copulation.

3.9.5 MATING FREQUENCY AND LIFE SPAN

Frequent mating may shorten an individual's life span. For both female and male fruit-flies (*Drosophila melanogaster*) significant, negative correlations between mating frequency and life span have been recorded (Kummer, 1960; Partridge and Farquhar, 1981). In parasitoid wasps, a similar relationship may exist. Gülel (1988) reported a significant decrease in longevity of *Dibrachys boarmiae* males (Pteromalidae) that copulated with five females per day and eventually became sperm-depleted, compared with males that mated only once per day. However, a control treatment ensuring equal levels of general motor activity by wasps was not included in Gülel's (1988) experiments.

Males of several parasitoid species continue mating 'to the bitter end', and may literally die in the act. I have observed this with *Melittobia acasta*, and Domenichini (1967) has done the same with *Tetrastichomyia clisiocampae*. As such, these observations tell us little about mating frequency and life span. Experiments aimed at investigating the effects of mating procedures on life span are easy to devise but are very time-consuming to conduct: males of a standard size are presented with variable numbers of virgin females, every day until death. The now inseminated females are kept alive and are supplied daily with a surplus of hosts. The time spent by males in courting and mating and the frequency of copulations are determined by the experimenter, while the amounts of sperm delivered to the females are assessed from the number of ensuing daughters. Controls should involve males that court as much (on average) but are prevented from copulating.

3.9.6 MATING AND EGG PRODUCTION

For diploid insects (i.e. in which all eggs need to be fertilised to produce progeny) being inseminated is likely to have a significant effect upon the behaviour of the female, releasing host-finding behaviour. However, for haplo-diploid insects such as parasitoid wasps, being inseminated may not necessarily be expected to alter female behaviour significantly, although as the following examples show, it sometimes has and sometimes has not been found to do so:

No differences in oviposition behaviour were apparent between virgin and mated females of *Lariophagus* and *Nasonia*. Female *Melittobia acasta*, once mated, can oviposit as soon as a suitable host has been found, and oviposition reaches a maximum in one or two days. By contrast, unmated females, when presented with a host will sting it and host-feed but will not oviposit. It is only after a day or two that one or very few eggs are laid, despite there being eggs available in the ovaries. Such females will mate as soon as a son has emerged, and only then start ovipositing. Thus, being inseminated affects the oviposition behaviour of females. *Cotesia glomerata* virgin females lay consistently smaller clutches of eggs than mated females, both in the field and the laboratory. Virgin females that were inseminated mid-way during an experiment began laying large egg clutches (Tagawa, 1987). The adaptive significance of the virgin female/mated female clutch size difference is unclear. According to Tagawa (1987), the difference is not a reflection of female egg load, since he dissected females and counted eggs. Egg-staining methods, however, were not used to distinguish between fully mature and nearly mature eggs in this study (subsection 2.2.3).

3.9.7 SEXUAL CANNIBALISM

A problem for the males of some predatory insects is that they may be treated by females as prey and be killed before a copulation is secured. Two strategies may be adopted by males in order to maximise their inclusive fitness in such circumstances:

1. they can attempt to appease the female with either food, some other object or special signals. This strategy is exemplified by members of the Empididae (Diptera). Within this family, there is a gradation of behaviour: (a) prey is devoured independently of courtship, i.e. no appeasement (e.g. *Platypalpus, Hybos, Empis trigramma, E. punctata*); (b) the prey offered by the male is fed upon by the female during copulation (e.g. many *Empis* species); (c) the prey or object offered by the male is not fed upon by the female but acts as a releaser stimulus for copulation. In *Hilara*, inanimate objects may be enclosed in a silken web (spun by the male, from glands in his fore tarsi) or the web may be an empty 'balloon' (Kessell, 1955, 1959).
2. they can take the high risk of establishing a genital contact without appeasement. The latter strategy can be selected for if matings that result in the male being eaten lead to him leaving more progeny than in matings that are not fatal. This strategy is exemplified by praying mantids. The mating behaviour of mantids has long been known for the sexual cannibalism that can occur, for example read the colourful descriptions given by Howard (1886) of *Stagomantis carolina* and by Fabre (1907) of *Mantis religiosa*. Roeder (1935) describes the mating behaviour of *M. religiosa* in more objective terms. A male will approach a female from behind by making almost imperceptible steps in her direction, unlike normal locomotion. He jumps suddenly on to her back and clasps her with the raptorial fore legs. Once out of reach of her own raptorial legs, he positions his genitalia so they are close to the female's. The female needs to raise her genitalia before the male can copulate successfully. Mating proper lasts for 4 to 5 hours. Upon completing copulation, the male drops off the female, out of her reach, and may copulate again later on. Roeder witnessed several cases of sexual cannibalism: whilst on the female's back a male may lose his grip, in which case the female may strike, capture the male and consume him. Being eaten at this stage does not exclude establishing a genital contact and insemination, but the conventional wisdom that assumes decapitation to be a necessary prerequisite for ejaculation is incorrect, as intact males perform just as well as decapitated males.

Doubt has been cast on the function of sexual cannibalism in mantids. Some authors (Liske and Davis, 1984, 1987) suggest that sexual cannibalism is a laboratory artefact resulting from either confinement or feeding conditions. However, Birkhead *et al.* (1988) obtained evidence that the female may gain significantly in fitness by eating the male. They found that in *Hierodula membranacea* the probability of the male being eaten correlated strongly with the dietary regime of the female; females maintained on 0.1 g (dry weight) of cricket prey per day (a non-starvation diet) ate males in 12 out of 14 cases, whereas those maintained on 0.42 g of cricket prey per day ate males in only one out of five cases. The fecundity (as measured by ootheca weight, see Birkhead *et al.* (1988) for details) of females that had been maintained on a poor cricket prey diet but then consumed their mate was significantly higher than that of females that were similarly maintained but did not consume their mate. Also, it was observed that males act with extreme caution; they mount the female by leaping on to her back, invariably initiated from well outside the female's grasp.

3.10 MATING SUCCESS

3.10.1 MEASURING SUCCESS

Mating success is a relative measure that cannot be equated with total reproductive output. Strictly speaking, the mating success of individuals should be measured and *compared* in terms of their inclusive fitness, i.e. the number of reproducing children, grandchildren and so on. That is, a male that leaves more progeny, measured over several generations, than a rival that operates in the same mating pool can be considered to be the more fit genetically speaking; his mating success can be expressed as the differences between him and his rival. The same reasoning can be applied to females. No workers, so far as I am aware, have attempted to measure mating success in these terms. This comes as no surprise, in view of the amount of work that would need to be put into such a study. Therefore, it is wise to be modest in one's initial aims, concentrating on identification of those factors that contribute to eventual success. Competition between individuals lies at the root of mating success; so look for easily quantifiable aspects of factors that influence the outcome of this competition. I use my own studies on *Nasonia vitripennis* as an example. The most simple experimental set-up for investigating the mating success of males of this species involves the use of a small vial, two males and one inseminable female. Introduce the males either simultaneously or consecutively after a predetermined period of time, and record which male is more successful at securing an insemination. With this set-up, one can test for the possible effects of factors such as prior residence, previous experience of courtship and/or mating, ability to mount court and/or copulate ahead of the rival, body size, eye colour etc.

Clearly, the above-set up is rather artificial. An experiment of a less artificial nature that can be carried out with *Nasonia* and many other parasitoids involves the following procedures: puparia of parasitised hosts are cut approximately in half and the parasitoid pupae removed and identified to sex. The remains of the host should also be removed (if these remains are large, as they can be in the case of *Nasonia*, take care not to puncture them and thereby contaminate the puparium). Re-introduce the number of wasp pupae required for the experiment, e.g. one male and several females, into a now empty and dry half-puparium, and place the latter, cut end down, on a small piece of Plasticine. Wasps will eclose from their pupae, gnaw their way out of the half-puparium, and mate, usually on the puparium's exterior. The behaviour of different individuals can be monitored by using readily identifiable genotypes such as eye colour morphs (Saul *et al.*, 1965). One individual at a time is assigned for observation, and his behaviour recorded, e.g. the number and duration of his displays, the number of copulations he achieves. Under natural conditions, a female may remain unattended, and so leave the natal patch as a virgin. However, with the experimental set-up all females will, sooner or later, be mated. Therefore it will be necessary to define a certain boundary and remove, with an aspirator, all females that move beyond it. All females should be presented with hosts at the end of the observation period, and their progeny examined. The effects of different factors on mating success (see previous experiment) can then be quantified in terms both of the probability of obtaining copulations and of the number of progeny produced.

The above experiments can be extended by adding more male wasps and/or puparia, arranging the latter in different spatial patterns, e.g. in compact clusters or in rows, so as to make it either easier or more difficult for certain individuals to monopolise the collection of females that will emerge from the puparia in due course. It is necessary to use a large enough 'arena' for the more complicated experiments on competition, its size depending on the mobility of the competitors. *Nasonia* has the additional advantage that males are territorial under some conditions, remaining in the same place for long periods of time (subsection 3.10.2). The arena I used was a glass container 100 × 50 cm, 4 cm high sides and covered with a sheet of glass.

3.10.2 TERRITORIALITY IN MALE *NASONIA*

Nasonia males make good experimental animals for investigations of territoriality. They emerge in advance of the females. The first one to appear remains on the exterior of the puparium, continuously patrolling it. Intruders are warded off; other males that come close are confronted directly, the resident male holding his wings vertically erect and his mandibles agape (Figure 3.43). An assault may follow, involving one male mounting the other. Males may bite one another with the result that parts of the antennae or legs may be removed. If the rivals driven away by dominant males remain in the vicinity of the puparium, they become **satellite males** and they may act as **sneakers** (subsection 3.8.6) and steal a copulation. The benefit obtained by being territorial must be traded against the risk that a female may emerge from the host puparium whilst the resident male is occupied with driving a rival away, and the opportunity to mate with her may be lost.

If there is only one exit-hole in a host puparium, females will emerge from a host one by one. A male on the exterior has no means of influencing the timing of emergences. Females may appear so rapidly in succession that the dominant male fails to inseminate them all. Those females that are missed may be intercepted and inseminated by satellite males.

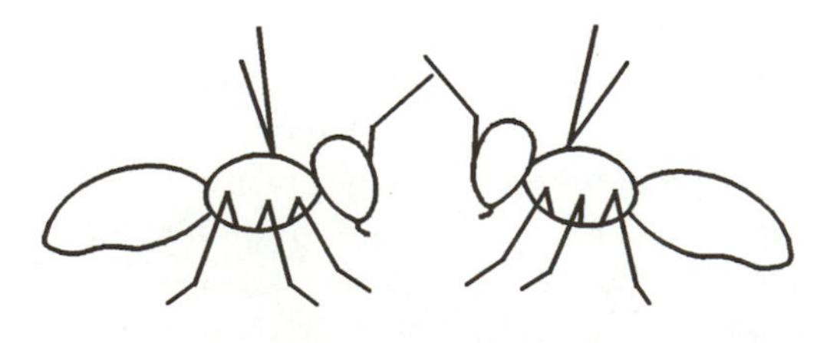

Figure 3.43 Threat posture of *Nasonia* males, with mandibles agape.

The territorial system of *Nasonia* breaks down when too many rival males are present on a host patch, a situation that may occur with carrion-feeding species of hosts such as *Calliphora* or *Sarcophaga* (Werren, 1983). Under these circumstances **grabbling** between males becomes the profitable strategy (van den Assem *et al.*, 1982b). Grabbling is the usual strategy of Pteromalidae whose males emerge, usually in large numbers, from large hosts, e.g. the males of *Pteromalus puparum* from the pupae of pierid butterflies. A switch from a territorial towards a grabbling system can be investigated in an experimental arena by varying the number of parasitised puparia. At a certain critical density of competing males, the 'pressure' on a dominant, territorial male prohibits a continued successful defence of his position at an emergence hole, and his monopoly breaks down. The consequences of such a switch for mating success will be extremely difficult to measure in practice.

3.10.3 MARKING BEHAVIOUR

Presumably, males remain in an area for as long as females can be expected to emerge, and will leave when this is no longer the case. In several species this is what happens, whereas in others it is not, one reason being that they lack the means to disperse, e.g. eulophids such as *Melittobia* or *Tachinobia*, whose males have very short wings or no wings at all. However, there seem to be factors other than flightlessness that keep a male on the spot, even where females are absent. Is it possible that remaining in one place is a profitable strategy? This question can be partly answered for *Nasonia vitripennis* (which is short-winged) using the following procedure:

Inexperienced males (one per vial, vials are stoppered with a plug of cotton wool) are either presented with a virgin female or are kept solitary. Those presented with a female are allowed to copulate, and the female is removed immediately afterwards. Following copulation males, whilst walking, repeatedly drag the tip of the abdomen over a vial's inner surface, depositing a very fine streak of a whitish substance. Once the male has performed a certain amount of this **marking behaviour** (which may be quantified by keeping the vial stoppered for various periods of time), he will not leave the vial if the cotton wool stopper is removed. By contrast, an unmated male that has been kept isolated for a similar period will leave as soon as the stopper has been removed (Figure 3.44).

It may be more convenient to have males make marks on a horizontal glass surface. A female and a male of *Nasonia* can be confined within a chamber comprising an inverted Petri dish placed over a glass plate. Following copulation the female is removed. The male will commence marking. After some time the Petri dish can be removed (males do not stop marking, although the rate of marking will decrease) and the male's walking movements then monitored. The male will restrict his movements to the area within which he had previously been confined (Figure 3.45).

Thus, in *Nasonia*, performance of a copulation triggers marking behaviour. So does the performance of a display, but with a markedly less profound effect. Even unmated males that are kept isolated for a long enough period (a couple of days) occasionally mark the wall of their vial. Once a site has been marked to some extent, a male will no longer move away. In *Nasonia* the continued presence of one or more females (unemerged females; females are quick to disperse once they have emerged) is also not a relevant factor (King *et al.*, 1969a). Marks do not relate to certain individuals (owners), nor does an individual need to perform marking behaviour to become a resident. An inexperienced male that is introduced into a well-marked vial will no longer walk out, whereas it will walk out of an unmarked vial as soon as the stopper is removed.

Virgin females appear to react to marks to some extent: their walking paths over a marked substrate are noticeably influenced. However, the reaction is short-lived; the

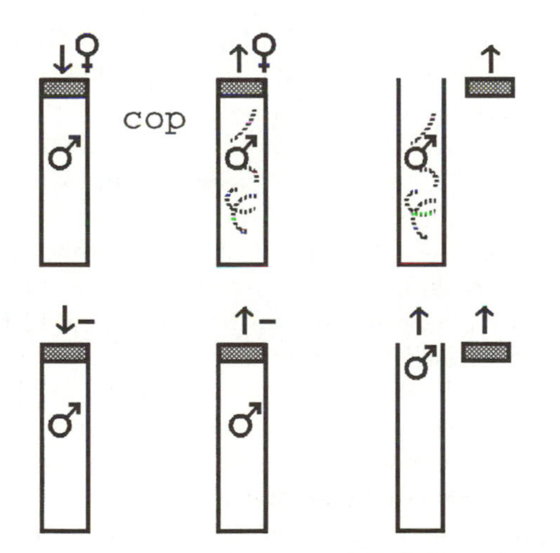

Figure 3.44 Experimental design for an investigation into the tendency of male *Nasonia* to remain in an area. In this example males are either presented with a virgin female (and copulate, **cop**), or remained isolated. After various periods of time vials are opened and the numbers of males staying and leaving are recorded.

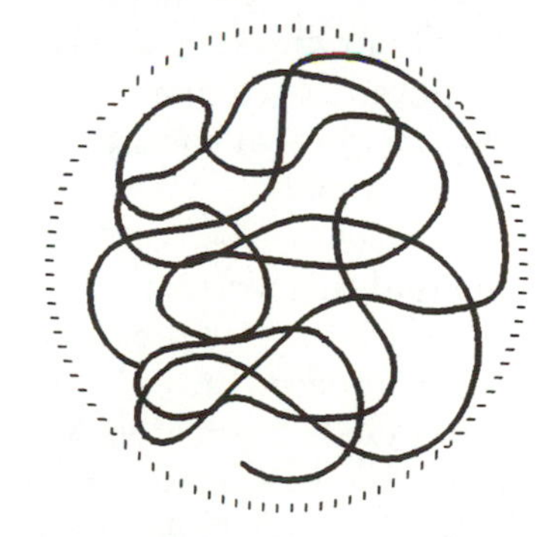

Figure 3.45 The tendency for a male *Nasonia* to remain in an area after he has performed mating behaviour. Males show restricted walking even when the Petri dish lid is removed.

females do not wait for a male's arrival. Virgin females that were exposed to a male's display until they became receptive (the male being brushed off before a genital contact can be established) do not react to male marks, whereas controls (virgins in which receptivity has not been induced) do (unpublished observations). Mated females do not react to marks in any noticeable way.

3.11 COMPARATIVE STUDIES OF MATING BEHAVIOUR

3.11.1 DISPLAYS AS A SOURCE OF TAXONOMIC CHARACTERS

The immense diversity of parasitoid wasps poses a major challenge to taxonomists. Sibling species abound, as do various degrees of incompatibility between field populations or laboratory strains, and in many cases it is difficult to establish the species status of populations. Eventually, DNA-sequencing techniques will have the final say in problems of relatedness, but because they are expensive to use and require special skills, investigating characteristics of courtship display behaviour may remain an acceptable alternative. Many courtship displays exhibit combinations of features that can be used for identification purposes (and for more ambitious taxonomic work, subsection 3.11.2), and neither sophisticated equipment nor an expensive laboratory set-up are required to compare them, provided that environmental conditions can be kept reasonably constant. Many parasitoid species, particularly wasps, will court and mate normally under laboratory conditions.

When looking for species-diagnostic characters, one can compare entire displays (van den Assem *et al.*, 1982a,b; van den Assem and Gijswijt, 1989). Records made with video equipment have proved to be very useful (subsection 3.2.2) as they allow rapid viewing, as often as is desirable, of the homologous patterns of different species. Species differences and similarities in courtship behaviour can be detected that might otherwise remain unnoticed. Besides the motor patterns of the appendages, it is the temporal organisation of displays (how they differ with respect to the order of appearance of components, and how they differ with respect to the number and lengths of intervals between components) that are particularly species-diagnostic. The study by van den Assem and Povel (1973) of courtship in *Muscidifurax* illustrates this point. Furthermore, this study also shows that what was originally believed to be a single, variable species is in reality a complex of sibling species.

For a long time *Muscidifurax* was considered to be a monotypic genus with *M. raptor*, a cosmopolitan parasitoid of the house-fly (*Musca domestica*), its only representative. However, Legner (1969) reported that several of the many strains in his field-derived laboratory cultures appeared to be reproductively isolated, and later on a taxonomist recognised them as four species (the fifth being *raptor sensu stricto*) on the basis of morphological characters. These characters are, however, difficult to observe and are very variable (Kogan and Legner, 1970). I obtained material of these species and presented males with females of the same strain and of different strains. Interspecific differences in display movements (patterns of antennal movements (Figure 3.46), the duration of the intervals between successive display cycles) were readily observed and so the species could be characterised behaviourally (van den Assem and Povel, 1973). The species also differed in the way in which males treated either conspecific females or conspecific-like dummies compared with heterospecific ones.

The case of *Nasonia vitripennis*, another pteromalid, is very similar, but it is included here because it illustrates the state of our knowledge of the group as a whole. *N. vitripennis* is the most intensively studied species of parasitoid wasp (Cousin, 1933; Whiting, 1967; Holmes, 1976). The genus was until recently considered to be monotypic, its

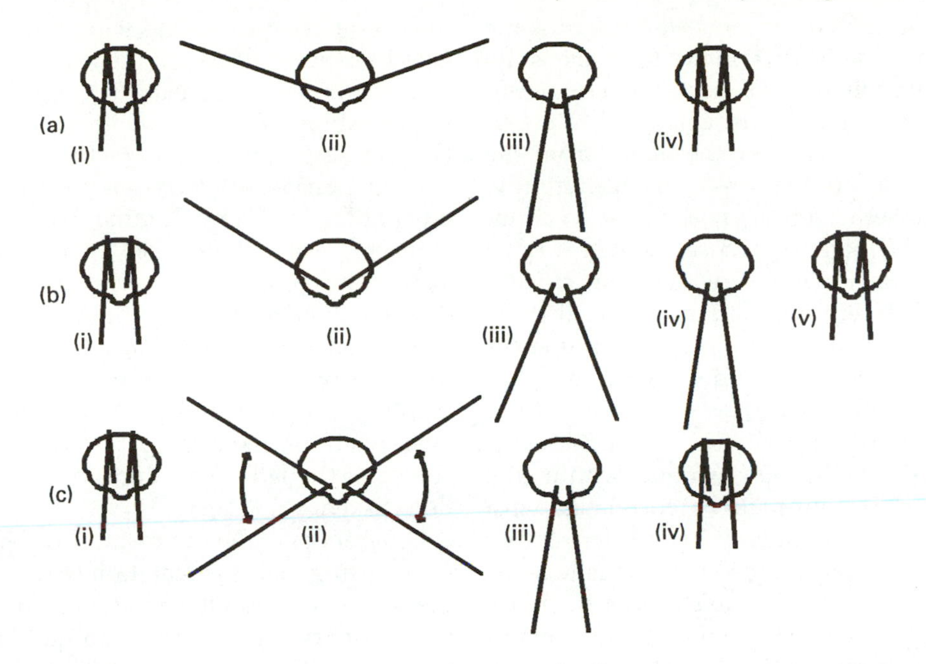

Figure 3.46 Species-characteristic display movements with the antennae in three sibling species of *Muscidifurax* (Pteromalidae): (a) *M. raptor*, (i) start position, (ii) spreading the antennae, (iii) extreme low position, (iv) end position; (b) *M. zaraptor*, (i) start position, (ii) spreading the antennae, (iii) intermediate position during the slow downward movements, (iv) end position; (c) *M. raptorellus*, (i) start position, (ii) waving episode, (iii) extreme low position, (iv) end position.

sole representative being a cosmopolitan parasitoid of the pupae of cyclorrhaphous flies in a variety of habitats including manure, decaying carcasses, and birds' nests (Werren, 1983). Through extensive collecting of fly puparia from birds' nests over a large area of the USA, three reproductively isolated strains were obtained (J.Werren, personal communication). Members of the different strains were sometimes collected from the same nest, or even from the same puparium. A morphological investigation led to the description of three species (Darling and Werren, 1990) but simple diagnostic differences (except for wing length in males) could not be provided. As with *Muscidifurax*, museum taxonomists were sceptical of the existence of more than one species, but again, studies of courtship behaviour revealed reliable diagnostic characters, namely differences both in the overall temporal organisation of behaviour and in the details of motor patterns (van den Assem and Werren, 1994).

Complexes of sibling species will, no doubt, be found in many taxa as intensive investigations of courtship behaviour are carried out. A parasitoid species may be suspected of being a complex for two major reasons:

1. it is consistently found that when males are taken from one population and presented with females from another, the females produce all-male progeny. In this case, cultures of isofemale lines are required in studies of mating behaviour, to establish whether the different populations represent biological species. From the first generation progeny of wild-caught (inseminated) specimens, select a

number of virgin females. Present some of the virgins with males from the same population and present the remaining virgins with males from a different population. See whether the males from the different populations court differently, and whether the females refuse to copulate; if females do refuse, then this can be taken as evidence of heterospecificity. The different populations should then be examined for other biological differences and also hitherto undiscovered morphological differences;

2. morphological variation (other than size variation) that is evident within the species is found to be correlated with some environmental factor such as the species of plant inhabited by the host, or the species of host itself. Present males reared from one kind of host with females reared from a different kind, observe the males' courtship behaviour and see whether the females refuse to mate.

3.11.2 MATING BEHAVIOUR AND PHYLOGENETICS

The aims of comparative studies of the mating behaviour of parasitoids and predators go beyond rendering assistance in problems of identification. Such studies, if they use the appropriate analytical techniques, can also enable us to establish how behaviour evolved, i.e. changed during phylogeny. Parasitoids and predators, because of their high biological diversity, provide excellent material for comparative work.

It is the author's experience that morphological characters that are considered to be 'ancestral' or 'primitive' correlate with behavioural traits that can be similarly labelled, even where the particular morphological characters do not seem to bear directly upon behavioural traits (e.g. venation of wings, number of antennal segments etc, with orientation of the male during courtship, timing of release of receptivity-inducing stimuli etc.). Goodpasture (1975) arrived at the same conclusion when considering phylogenetic relationships of *Monodontomerus* species (Torymidae), using character sets from mating behaviour, karyology and external morphology. The Comparative Method (subsection 1.2.3) can be used to test the significance of the above correlations between data sets. The texts by Brooks and McLennan (1991) and Harvey and Pagel (1991) provide information on how to proceed in these matters once the necessary observations have been made on the appropriate material. The choice of the latter will depend on the questions asked.

Differences among parasitoid wasp taxa in the courting male's orientation with respect to the female can be viewed as indicators of an evolutionary gradient (Hölldobler and Wilson, 1983). In some groups, the male adopts the same position in courtship as in copulation, i.e. to the rear (*Trichogramma evanescens*, Hase, 1925; *Brachymeria intermedia*, Leonard and Ringo, 1978; *Spalangia nigra*, Parker and Thompson, 1925; *Spalangia endius*, van den Assem, 1986). In the majority of parasitoid wasps, however, positions during courtship and copulation are distinctly different; males court away from the mating position, and mostly perform either on top of the female or near to her on the substratum (Gordh and DeBach, 1978; Bryan, 1980; Grissell and Goodpasture, 1981; Orr and Borden, 1983). An hypothesised phylogenetic sequence of positional changes is summarised in Figure 3.47. The hypothesis remains to be tested despite it appearing obvious which direction evolution has taken. For example, the tendency towards a more anterior courtship position is apparent in several families and subfamilies and may represent a response to a general selection pressure, perhaps more efficient communication at the front due to the presence of more diverse or more dense sense organs at the anterior ends

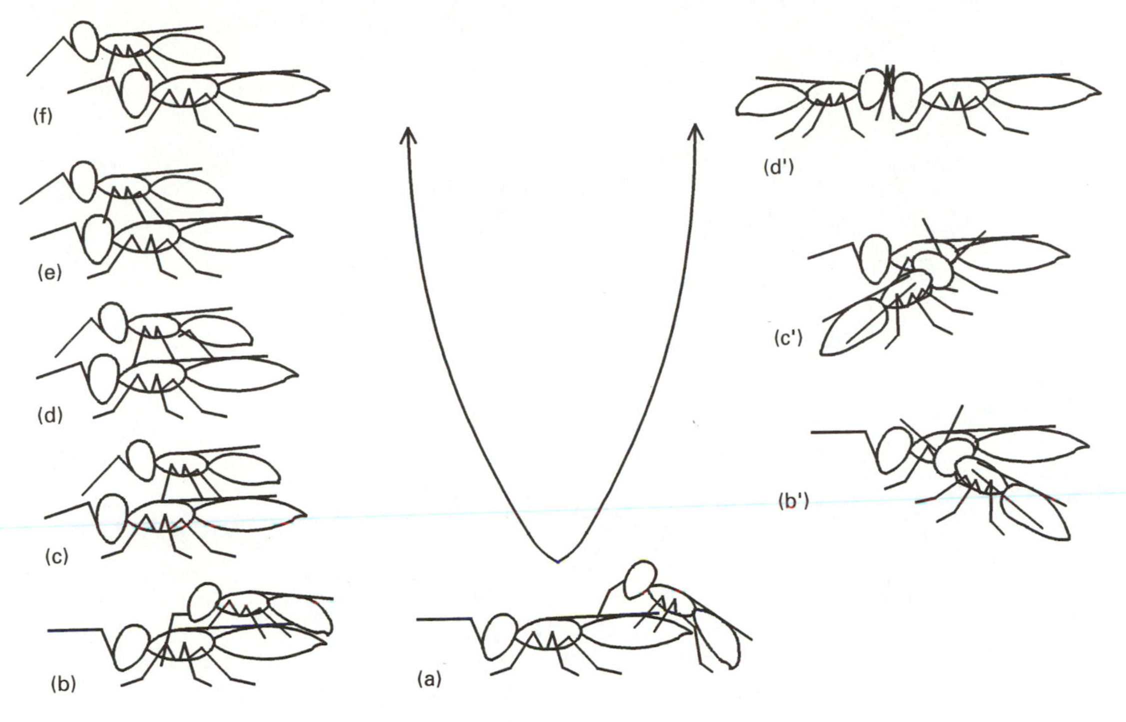

Figure 3.47 Suggested evolutionary trends of the male's courtship position in Chalcidoidea. The ancestral courtship position is similar to the mating position in: (a) *Trichogramma evanescens* (Trichogrammatidae), *Choetospila elegans* (Pteromalidae). In the left hand branch the male mounts on the female, and his position moves gradually to the front; (b) *Spalangia cameroni* (Pteromalidae); (c) *Asaphes vulgaris* (Pteromalidae), *Sympiesis sericeicornis* (Eulophidae), *Aceratoneuromyia granularis* (Eulophidae); (d) *Pachycrepoideus vindemmiae* (Pteromalidae), *Vrestovia fidenas* (Pteromalidae); (e) *Nasonia vitripennis* (Pteromalidae), *Hobbya stenonota* (Pteromalidae), *Eupelmus spongipartus* (Eupelmidae), *Tetrastichus sesamiae* (Eulophidae), *Systole albipennis* (Eurytomidae); (f) *Anagyrus pseudococci* (Encyrtidae). In the right hand branch, the male courts on the substrate, (b') *Achrysocharoides* species (Eulophidae) (Source: Bryan, 1980.); (c') *Pediobius* species (Eulophidae), *Tachinaephagus zelandicus* (Encyrtidae); (d') *Microterys ferrugineus* (Encyrtidae) (Source: Parker and Thompson, 1925). There is probably a third direction of development: males staying at the rear but having peculiarly elongated antennae that reach out to the front (Source: van den Assem, 1986). Examples were chosen arbitrarily. Categories can be extended with more species. Where no references are provided, the information is based on the author's own observations.

of both sexes. For a recent phylogenetic study of mating in parasitoid wasps, see Eggleton (1991).

3.12 ACKNOWLEDGEMENTS

My co-workers, to whom I owe much, are mentioned in the list of references. Three of them are acknowledged specifically: my *collega proximus* F.A. Putters, for constructive criticism over the years, M.J. Gijswijt, for expert taxonomic assistance, and F.X. van Berge Henegouwen for all kinds of technical assistance and unlimited patience with composing the computer-made line drawings. T.A. Hooft produced the photographs. Lastly, I should like to thank Mark Jervis for his constructive comments on the chapter.

POPULATIONS AND COMMUNITIES 4

W. Powell, M.P. Walton and M.A. Jervis

4.1 INTRODUCTION

In nature, any particular habitat contains animal and plant species which exist together in both time and space. Many of these species will interact with each other, for example when one species feeds on another or when two species compete for the same food or other resource. A group of species having a high degree of spatial and temporal concordance, and in which member species mutually interact to a greater or lesser extent, constitute a **community** (Askew and Shaw, 1986). The size and complexity of a community will depend upon how broadly that community is defined. For example, we could consider as a community the organisms which interact with each other within a particular area of woodland, the herbivore species which compete for a particular food plant or the complex of natural enemies associated with a particular prey or host species. The animal species of a community obtain their food directly or indirectly from plants which are the primary producers of the community. Herbivores feed directly on plants whilst predators and parasitoids are either primary carnivores, feeding on herbivores, or secondary or tertiary carnivores, feeding on other predators or parasitoids. The successive positions in this feeding hierarchy are termed **trophic levels**. Thus, green plants occupy the first trophic level, herbivores the second level, carnivores which eat herbivores the third level, secondary carnivores the fourth level, and so on, although a species may occupy more than one level. For example, some insect parasitoids are facultative hyperparasitoids, thus having the potential to occupy two trophic levels. Similarly, some carabid beetles eat both insect prey and plant seeds, and because they are polyphagous predators, the insect prey consumed may consist both of herbivores and of other carnivores.

When studying natural enemies it is important to determine the nature of their interactions with other members at all trophic levels. Techniques useful in discovering who eats whom are outlined in section 4.3. Phytophagy by natural enemies is discussed in Chapter 6.

Species within a community exist as **populations**. In its broadest sense the term population can be applied to any group of individuals of the same species occupying a particular space. This space may vary greatly in size, for example from a single tree to a wide geographical area, depending upon how the population is defined. It is important to define the spatial scale over which the population is to be studied at the start of an ecological investigation since the principal factors influencing the population dynamics of a species may vary depending upon the spatial scale used. For example, immigration and emigration may have a much greater influence on population persistence at small spatial scales than it does at larger ones (Dempster, 1989). The term population is sometimes incorrectly used to refer to the combined numbers of a range of related species occupying a discrete area, for example the 'carabid population' of a field. Great care must be taken in the interpretation of changes in the abundance of such a 'population' since it will comprise a mixture of species with differing ecologies, and different species will be affected in different ways by the same environmental factors. The concept of a metapopulation as a collection of sub-populations, each occupying a discrete habitat patch but with some level of genetic interchange, is receiving increasing attention in the study of species population dynamics (Gilpin and Hanski, 1991). This concept is discussed further in section 4.4.

In order to study a natural enemy population or community it is usual to select at random a representative group – a **sample** – of individuals on which to make the appropriate observations or measurements, so that valid generalisations concerning the population or community as a whole can be made. Often, but not always, the sampled individuals need to be removed from their natural habitat using an appropriate collection technique. Most sampling methods are destructive, involving the physical removal of organisms from the study area. However, it is sometimes possible to sample by counting organisms *in situ* (subsection 4.2.6). It is essential that before starting a sampling programme, sampling techniques are chosen that are appropriate for both the type of ecological problem being investigated and the particular natural enemy species being investigated. Section 4.2 is devoted to a description of the most commonly used sampling techniques and their limitations, with reference to examples from the literature on natural enemies. Table 4.1 lists some applications of these techniques.

In summary, this chapter is concerned with the sampling and monitoring of natural enemy populations and communities, and describes a number of techniques that can be used for measuring or estimating the abundance of natural enemies, determining the structure and composition of communities and examining the spatial distribution of natural enemies in relation to their host or prey populations. Most of the techniques discussed can be used to obtain qualitative data relating to the predator or parasitoid species present in a community or the prey/host species attacked by a particular natural enemy. Some of the techniques can also be used in obtaining quantitative estimates of natural enemy abundance or predation and parasitism, an aspect of natural enemy biology taken further in Chapter 5.

Estimates of animal numbers may be expressed in terms of either density per unit area or unit of habitat, and the unit of habitat can be an area of ground or a unit of vegetation such as a leaf or a whole plant. Estimates of this type are termed absolute estimates of abundance and must be distinguished from relative estimates of abundance which are not related to any defined habitat unit (Southwood, 1978). Relative estimates are expressed in terms of trapping units or catch per unit effort and are influenced by other factors (e.g. climatic conditions) besides the abundance of the insect being sampled. When considering absolute estimates of insect abundance, the term population density may be applied to numbers per unit area of habitat and the term population intensity applied to numbers per leaf or shoot or host (Southwood, 1978).

This chapter is concerned with practical techniques and we say little about the statistical analysis and interpretation of sampling data. Information on these topics can be found either elsewhere in this book or in the following publications: Southwood (1978); Cochran (1983); Eberhardt and Thomas (1991); McDonald *et al.* (1989); Perry (1989); Crawley (1993).

4.2 FIELD SAMPLING TECHNIQUES

The discussion of techniques is mainly confined to the sampling of insects from terrestrial habitats and from air. Southwood (1978) and Williams and Feltmate (1992) give a fuller account of sampling methods for use in aquatic habitats.

Table 4.1 Applications of different field sampling methods

Data required	*Sampling technique*	*Comments*
Absolute abundance	Pitfall traps	Do not provide data on absolute abundance
	Vacuum net	When used to sample a defined area or unit of habitat, calibration is necessary
	Sweep net	Estimates of absolute abundance difficult to obtain
	Knock-down	For chemical knock-down, unit of habitat (e.g. whole plant) needs to be enclosed; calibration necessary
	Visual Count	Labour-intensive; insects need to be conspicuous if census walk method used; efficiency varies with insect activity and observer
	Mark-release-recapture	Important to satisfy a number of assumptions; choose appropriate calculation methods
Relative abundance	Pitfall traps	Factors affecting locomotor activity need to be taken into account
	Vacuum net	Efficiency can change with height and density of vegetation
	Sweep net	A wide range of factors cause sampling variability; significant variability between operators
	Knock-down	Except for chemical knock-down, may not sample all species with the same efficiency
	Visual count	Labour-intensive; insects need to be conspicuous if census walk method used; efficiency varies with insect activity and observer
	Attraction	Difficult to define area of influence; insect responses may change with time
Dispersion pattern	Pitfall traps	Trap spacing important in minimising inter-trap interference; vegetation around individual traps can affect capture rates; some carabids aggregate in traps
	Vacuum net	Vegetation type can affect efficiency
	Sweep net	Disturbance can change dispersion pattern during sampling
	Knock-down	Chemical knock-down needs to be confined to defined sampling areas
	Visual count	Detectability needs to be constant over study area
	Examination of hosts for parasitoid immatures	Identification of immature stages may prove problematical
Phenology	Pitfall traps	Cannot detect immobile insects (e.g. during cold weather)
	Vacuum net	
	Sweep net	Changes in vertical distribution within the vegetation may result in non-detection
	Malaise trap	Provides useful information on flight periods
	Knock-down	
	Visual count	Changes in behaviour may affect ease of detection
	Examination of hosts for parasitoid immatures, and/or rearing of parasitoids	Rearing provides information on diapause characteristics

Table 4.1 *continued.*

Data required	*Sampling technique*	*Comments*
Phenology – *continued*		
	Attraction	Responses to visual or chemical stimuli may be restricted to certain periods or behavioural states
	Sticky and window traps	Only valid during active flight periods
Species composition	Pitfall traps	May not sample all species with the same efficiency; provide useful presence/absence data
	Vacuum net	Night samples need to be taken of nocturnal insects
	Sweep net	Only efficient for groups active in the vegetation canopy
	Malaise trap	
	Knock-down	Very active fliers may escape
	Visual count	Most efficient for very conspicuous groups
	Attraction	Different species may not respond to the same visual or chemical stimuli
	Sticky and window traps	Species without an active flying stage will not be recorded
Relative abundance of spp.	Pitfall traps, Vacuum Net, Sweep Net, Malaise trap, Knock-down, Attraction	Except for chemical knock-down, may not sample all species with the same efficiency
	Visual count	
Locomotor activity	Pitfall traps	Linear pitfall traps provide useful information on population movements, especially direction of movements; best to combine with a marking technique
	Visual count	Can provide useful information; best to combine with a marking technique
	Attraction	Attractant properties of trap can interfere with insect behaviour
	Mark-release-recapture	Provides useful information such as minimum distance travelled
	Sticky and window traps	Provide useful information on height of flight, as well as on direction

4.2.1 PITFALL TRAPPING

A pitfall trap is a simple interception device consisting of a smooth-sided container which is sunk into the ground so that its open top lies flush with the ground surface (Figure 4.1a). Invertebrates moving across the soil surface are caught when they fall into the container.

Pitfall traps are the most commonly used method of sampling ground-dwelling predators such as carabid and staphylinid beetles, spiders and predatory mites (Figure 4.1b). The containers that can be used as pitfall traps are many and varied, but round plastic pots and glass jars with a diameter of 6–10 cm are the most popular. It is important to remember, however, that both trap size and trap material are known to influence trap catches, sometimes very strongly (Luff, 1975; Adis, 1979; Scheller, 1984).

A liquid preservative is often placed in the trap both to kill and to preserve the catch,

(a)

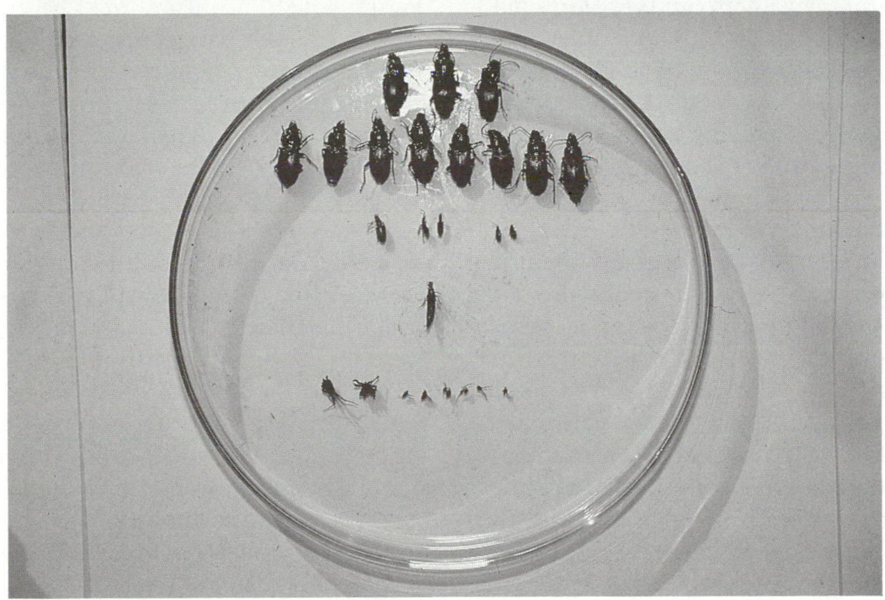

(b)

Figure 4.1 (a) A simple pitfall trap (with rain cover) for sampling predators moving across the soil surface, e.g. carabid and staphylinid beetles, spiders; (b) carabid beetles, staphylinid beetles and spiders caught in a single pitfall trap over a 7-day period in a cereal crop in southern England.

thereby reducing the risk of escape and preventing predation within the trap, particularly of small individuals by larger ones (conspecifics or heterospecifics). Amongst the preservatives that have been used are formaldehyde, alcohol, ethylene glycol, acetic acid and picric acid, but there is evidence that some preservatives have an attractant or repellent effect on some insects and such effects can differ between the two sexes of the same species (Luff, 1968; Skuhravý, 1970; Adis and Kramer, 1975; Adis, 1979; Scheller, 1984). It is sometimes necessary, for example when carrying out mark-release-recapture studies, to keep specimens alive within the trap. In such cases it is advisable to place some kind of material in the bottom of the trap to provide a refuge for smaller individuals. Compost, small stones, leaf litter, moss or even polystyrene granules may be used for this purpose.

At the end of a trapping period it is advisable to replace traps with fresh ones so that the catch can be taken *en masse* back to the laboratory. Plastic pots with snap-on lids are readily available commercially and are convenient when large numbers of traps need to be transported. To prevent the sides of the hole from collapsing during trap replacement, a rigid liner is useful, and liners can be readily made from sections of plastic drainpipe of an appropriate diameter. In open habitats it is also advisable to place covers over traps in order to prevent birds from preying on the catch and to reduce flooding problems during wet weather. Covers for simple, round traps can be made from inverted plastic plant pot trays, supported by wire. Soil-coring tools, including bulb planters, can be used to make the initial holes when setting pitfall traps but it is essential to ensure that the lip of the trap is flush with the soil surface. When traps are operated over a prolonged period of time, some maintenance work is often necessary. For example, in hot, dry weather some soils crack and shrink, creating gaps around the edge of traps, thus reducing their efficiency.

A number of variations on the conventional pitfall trap have been developed in attempts to improve their efficiency. Several workers have used linear traps made from lengths of plastic or metal guttering to increase catches and to obtain directional information on insect movements (subsection 4.2.11). Alternatively, two conventional traps may be placed a short distance apart and joined by a solid barrier which diverts walking insects into the traps at either end (Wallin, 1985; Jensen *et al.*, 1989). Mechanical devices have been used to allow automatic, time-based sorting of trap catches (Williams, 1958; Ayre and Trueman, 1974; Barndt, 1976), whilst Heap (1988) suspended UV-emitting, fluorescent light tubes above his traps to increase the catch rate. Several traps, which may be regarded as equivalents of pitfall traps, have been developed for sampling aquatic invertebrates (Southwood, 1978); that designed by James and Redner (1965) is particularly useful for collecting predatory water beetles.

A common practice in recent years has been the erection of physical barriers in the field, usually to enclose defined treatment areas within which pitfall trapping is carried out (Powell *et al.*, 1985; Desender and Maelfait, 1986; Holopäinen and Varis, 1986) (Figure 4.2a,b). Caution must be exercised in the use of such barriers in arable crop fields because some predators invade the fields from field boundaries during spring and early summer (Pollard, 1968; Coombes and Sotherton, 1986). The erection of full barriers too early in the year would exclude these species from the enclosed areas, resulting in erroneous data on predator communities. The use of physical barriers can reduce catches of carabid beetles by as much as 35–67% over a growing season (Edwards *et al.*, 1979; de Clercq and Pietraszko, 1982; Holopäinen and Varis, 1986). In contrast, some carabid beetle species emerge as adults from the soil within arable fields and if emerging populations are high many individuals will disperse away from overcrowded areas. Barriers may prevent this dispersal, resulting in artificially high predator densities within enclosed areas (Powell and Bardner, 1984). Further aspects of pitfall trap methodology are discussed in Luff (1975, 1987); Adis (1979); Scheller (1984).

Pitfall trap catches are a function of both predator abundance and predator activity and so changes in either will affect the capture rate. Locomotor activity in natural enemies and methods for its investigation are

Figure 4.2 (a) Polyethylene barriers surrounding experimental plots containing pitfall traps to restrict immigration by carabid beetles. (b) Close-up of polyethylene barriers.

discussed in subsection 4.2.11. Also, the number of individuals present in a trap at the end of a trapping period is determined by both the capture rate (the rate at which individuals fall into the trap) and by the escape rate (the rate at which individuals manage to escape from the trap). Both capture rate and escape rate depend on the predator species and they are also influenced by a variety of factors including trap size, trap material, trap spacing, presence of a preservative, vegetation density, soil type and surface texture, weather conditions and food availability. There is also evidence that some carabid beetles aggregate in pitfall traps, probably in response to aggregation or sex pheromones or to defensive secretions, and this can result in considerable inter-trap variability in catches (Luff, 1986). All these influences must be considered when interpreting pitfall trap data.

Pitfall traps do not provide absolute measures of predator density because the number of individuals caught depends partly on their locomotor activity. Therefore, the abundance of a species as measured by pitfall trap catches has been termed its **activity density** or **activity abundance** (Heydemann, 1953; Tretzel, 1955; Thiele, 1977). The activity density of a species provides a measure of the predator's role in an ecosystem (for example in catching prey), since this role depends on its mobility as well as on its frequency (Thiele, 1977). Therefore, pitfall traps can be used either to compare activity densities for the same species in different habitats and at different times of year or to assess the impact of agricultural practices on predation pressures experienced by prey species within crops. It must be remembered, however, that the activity densities of different species are not necessarily comparable (Bombosch, 1962). Although it is difficult to separate the influences of activity and abundance on trap catches, there is evidence that, when trapping is continued throughout the year, whole year catches are linearly related to density for individual species (Baars, 1979a).

It is also difficult to obtain from pitfall trap catches an accurate picture of the relative abundance of different species within a community, because different species are caught at different rates and also escape at different rates (Jarošík, 1992). This has been elegantly demonstrated using video recording techniques both in laboratory arenas (Halsall and Wratten, 1988) and in the field (N. Paling, personal communication). Jarosík (1992) used pitfall trap data to compare patterns of species abundance in communities of carabid beetles from different habitats, but concluded that pitfall trap data were inadequate for this purpose. Nevertheless, Desender and Maelfait (1986) have shown that intensive trapping within enclosed arenas can give reliable information on the relative abundances of coexisting carabid species. Pitfall trapping does provide useful presence/absence data, and species lists derived from extensive trapping programmes can be used, with the aid of modern ordination techniques, to classify different habitats based on their carabid communities, or to identify environmental factors which are influencing species distributions (Eyre and Luff, 1990; Eyre *et al.*, 1990).

Locomotor activity, and therefore capture rate, can vary between the two sexes of a species, so that reliable estimates of sex ratio within predator populations are also difficult to obtain from pitfall data. For example, most of the spiders which are active on the ground in cereal crops are males and so more males than females are caught in pitfall traps (Sunderland, 1987).

Information on the spatial dispersion patterns of predators living on the soil surface can be obtained using pitfall traps either spaced in a grid system (Ericson, 1978; Niemelä, 1990) or laid out as transects through heterogeneous habitats (Wallin, 1985). Traps within a regular grid system may interfere with each other, the central traps catching fewer individuals than the outer traps (Scheller, 1984). Trap spacing is therefore important, interference increasing as between-trap distances decrease. Consideration also needs to be given to aggregation that may occur in response to pheromones or defensive secretions; such aggregation may be independent of trap positions and can vary with between-trapping periods (Luff, 1986). Similarly, within heterogeneous habitats differences in vegetation type and density around traps will affect capture rates, hindering the investigation of dispersion patterns.

Despite their limitations, pitfall traps remain a very useful sampling tool for obtaining both qualitative and quantitative data on those predators which are active on the soil surface, providing that the objectives of the sampling programme are clearly defined and that the many factors which can influence the catch rate are considered during data interpretation. Advice on the interpretation of pitfall data is given by Luff (1975, 1987); Adis (1979); Ericson (1979); Baars (1979a).

Some small-sized natural enemies live on or in leaf-litter (e.g. some diapriid and eucoilid parasitoids). Pitfall trapping is unlikely to be useful in sampling such insects, and a leaf-litter sampling technique (subsection 4.2.7 and Southwood, 1978) might be attempted.

4.2.2 VACUUM NETTING

Although several different types of vacuum net have been developed, they all operate on the same principle. They employ a fine mesh net enclosed in a rigid sampling head which is attached to a flexible tube. The tube is connected to a fan which is driven by an electric or petrol-fuelled motor. The fan draws air through the flexible tube via the net in the sampling head, sucking small arthropods on to the net from the vegetation enclosed by the sampling head. The sampling tube and its head are quite narrow in some machines (Johnson *et al.*, 1957; Southwood and Pleasance, 1962) (Figure 4.3b) but much wider in others (Dietrick *et al.*, 1959; Dietrick, 1961; Thornhill, 1978) (Figure 4.3a). Machines with wide tubes require a more powerful motor in order to attain the required air speed of at least 90 km/h through the collecting head. Many of the machines currently in use in the UK are of the wide tube variety and are driven by two-stroke, lawn-mower, petrol-fuelled engines which are mounted on rucksack frames so that they can be carried on the back of the operator (Thornhill, 1978). Taubert and Hertl (1985) have designed a small vacuum net driven by a battery-powered, electric motor, presenting a lighter load for the operator.

Vacuum nets can be used in a number of different ways to collect arthropods – either natural enemies or hosts and prey – from vegetation. A commonly used method involves pressing the sampling head to the ground over the vegetation, providing this is not too tall or dense, and holding it in place for a defined period of time (e.g. 10 s), a process which may be repeated several times within a specified area to form a single sample. This method is appropriate for wide-nozzled machines, and by measuring the size of the sampling head the catch can be related to a finite area of vegetation, so giving an absolute measure of species densities.

An alternative method of using a wide-nozzled suction sampler is to hold the sampling head at an angle to the ground whilst pushing it through the vegetation over a defined distance. As it brushes through the vegetation the advancing sampling head dislodges arthropods which are then sucked into the net. This method is particularly useful when collecting insects from crops planted in discrete rows as it allows a specified length of row to be sampled.

Another approach involves enclosing an area of vegetation with a bottomless box or cylinder and using a narrow-nozzled sampler to remove insects from the enclosed vegetation, the soil surface and the interior walls of the box (Smith *et al.*, 1976; Henderson and Whitaker, 1977; Wright and Stewart, 1992). If the vegetation is tall, it can then be cut and removed before sampling a second time. Wright and Stewart (1992) adapted a commercial garden leaf-blower (Atco Blow-Vac) as a narrow-nozzled suction sampler and used it to extract insects from areas of grassland enclosed by an acetate sheet cylinder. Compared with a wide-nozzled suction sampler, the sampling head of which covered the same area of ground as the acetate cylinder, the narrow-nozzled apparatus

(a)

(b)

Figure 4.3 (a) Wide-nozzled (0.1 m^2), petrol engine-driven vacuum sampler; (b) entomologists vacuum-sampling rice insects in a paddy in Indonesia. ((b) reproduced by kind permission of Anja Steenkiste.)

extracted significantly more predatory beetles and spiders from the vegetation.

After each sample is taken, the collecting net is usually removed from the sampling head. The net can either be tied off (to secure the catch) and replaced with a fresh one or the catch can be transferred to a polythene bag so that the net can be re-used. Moreby (1991) described a simple modification to the net and sampling head which speeds up the transfer of samples to polythene bags. In order to reduce the risk of losses due to predation within the net or bag it is useful to kill the catch whilst still in the field. This can be done by placing the tied-off nets into bags already containing a chemical killing agent or by adding wads of cloth or paper, impregnated with killing agent, to the catch once it has been placed in the polythene bag. For some purposes however, for example the rearing of parasitoids from hosts (subsection 4.3.6), the catch may need to be kept alive.

On returning to the laboratory, catches may be sorted by hand but it is often useful to pass the sample through a series of sieves before hand-sorting, especially if organisms of a fixed size are being counted. Hand-sorting can be extremely time-consuming, particularly if very small insects are being sampled (Figure 4.4). Consequently, some workers have used Berlese funnels or flotation techniques to speed up the process (Dietrick *et al.*, 1959; Marston, 1980). Further details on the construction of vacuum nets and on their use may be found in Johnson *et al.* (1957); Dietrick *et al.* (1959); Dietrick (1961); Southwood and Pleasance (1962); Weekman and Ball (1963); Arnold *et al.* (1973); Thornhill (1978); Southwood (1978); Kogan and Herzog (1980).

Figure 4.4 Contents of a vacuum sample taken in a cereal field in July in southern England. The catch shown is the result of five 10 s samples, each taken from an area of 0.1 m^2. The catch is preserved in 70% ethanol and is divided into Hemiptera (mainly aphids), Coleoptera (including larvae), Diptera, Hymenoptera 'Parasitica', spiders and soil and plant debris.

The efficiency of vacuum nets varies considerably for different natural enemy groups and is also affected by the density and type of vegetation being sampled. Generally, small, winged insects such as Diptera and adult Hymenoptera 'Parasitica' are sampled with the greatest efficiency, and Poehling (1987), working in cereal crops, concluded that it is a suitable method for sampling these types of insect. Powell (unpublished data) investigated the efficiency of a wide-tubed vacuum net (area of sampling head: 0.1 m^2) for sampling adult parasitoids in flowering winter wheat. Known numbers of parasitoids were released into large field cages (9 m × 9 m × 6 m) and allowed to settle for several hours before sampling. Comparisons of expected and actual catches indicated a sampling efficiency of over 90%. Henderson and Whitaker (1977) investigated the efficiency of a narrow-tubed vacuum net used in conjunction with a bottomless box which enclosed a 0.5 m^2 area of grassland. Again, efficiency was highest for Diptera (79–98%) and Hymenoptera (60–83%) and was lowest for Acarina (12–40%). Sampling efficiency tended to decline with increasing grass height for most of the groups sampled. Vacuum nets are also regarded as an efficient means of sampling adult parasitoids in soybean crops (Marston, 1980).

The time of day when samples are taken can also influence vacuum net catches, because the locomotor activity of most natural enemies varies during the day. Vickerman and Sunderland (1975) used a vacuum net and sweep nets to compare the diurnal and nocturnal activity of predators in cereal crops. More adults and larvae of the staphylinid beetles *Tachyporus* spp. were caught during the hours of darkness than in the daytime, as were spiders, syrphid larvae and the earwig *Forficula auricularia*. Whitcomb (1980) used a vacuum net to sample spiders in soybean fields, but Sunderland (1987) advocates a combination of vacuum netting and ground-searching to estimate spider densities in crops, because some species are more accurately assessed by the latter technique than by the former whereas the reverse is true for other species. More robust predators such as adult beetles and those which can rapidly move out of the way of the advancing operator are less efficiently sampled by a vacuum net (Sunderland *et al.*, 1987b).

Because of the variability in sampling efficiency in relation to different natural enemy groups, vacuum netting is not useful for comparing the relative abundances of different taxa in communities. When using a vacuum net in a sampling programme it is important to calibrate the data collected by comparison with data obtained using an absolute sampling method whenever possible (Smith *et al.*, 1976; Whitcomb, 1980; Dewar *et al.*, 1982). Usually, once this is done the vacuum net can be used to obtain absolute estimates of predator population densities by sampling discrete units of vegetation. A series of samples taken in a regular grid pattern can then be used to investigate the spatial dispersion pattern of individual natural enemy species.

A further limitation on the use of vacuum nets ought to be mentioned. Moisture on the vegetation being sampled severely reduces their efficiency and so their use either during wet weather or following a heavy dew is not advised, since large numbers of insects tend to stick to the sides of the net and to the inner surfaces of the collecting head. Not only does this make their efficient removal from the net difficult, but delicate insects will be damaged, making their identification to sex or species either difficult or impossible.

A new type of vacuum sampler, manufactured by the Burkard Manufacturing Company Ltd, of Rickmansworth, UK, has recently come on to the market. *Vortis* is a lightweight (7.8 Kg), petrol-engined device that lacks a net of any kind. Air, instead of being drawn in through the sampling nozzle, enters via another inlet located near the opposite end of the apparatus. Insects are deposited directly into a detachable, transparent collecting vessel. As well as being lightweight and therefore very portable, the device has several advantages over vacuuum nets:

1. Suction pressure remains constant even after many hours of operation. Because of the way the insects are accumulated, a fall-off in suction pressure does not occur (the nets in other devices need to be repeatedly unclogged of accumulated insects and debris in order to avoid a reduction in suction pressure);
2. Because of the type of suction mechanism the sampler employs, the nozzle can be applied continuously to the ground during sampling, thus saving time (with vacuum nets, the nozzle has to be repeatedly lifted from the ground during sampling, to allow the entry of air);
3. The manufacturers claim that insects collected with the device suffer much less damage than with vacuuum nets. If this is the case, the device could prove very valuable in various ways, including the collection of hosts from which parasitoids need to be reared (subsections 4.2.9 and 4.3.6), the collection of live samples of predators and parasitoids, e.g. for mark-release-recapture studies (subsection 4.2.10), and the collection of well preserved material for identification and other taxonomic purposes.

Vacuum sampling of natural enemies from air has also been carried out. Perry and Bowden (1983) used Johnson-Taylor suction traps (see Southwood, 1978, for design) situated at different heights from the ground to study the phenology of the green lacewing, *Chrysoperla carnea*, in woodland and open-field habitats.

4.2.3 SWEEP NETTING

A sweep net comprises a fine-meshed, cone-shaped net mounted on a rigid, circular or D-shaped frame which is attached to a short handle. It is commonly used for collecting arthropods from vegetation (especially herbaceous vegetation) because it is inexpensive, highly portable and easy to use and also because it allows the rapid collection of a large number of insects. As the name suggests, sampling with the net involves sweeping it rapidly through the vegetation so that the rigid frame dislodges insects which are caught in the moving net. It may be swept backwards and forwards in a simple arc or it may be made to follow a more sinuous track, for example a figure of eight. However, it is important to define a standard sweeping technique before commencing a sampling programme because the method and pattern of sweeping can significantly affect capture rates (DeLong, 1932; Kogan and Pitre, 1980; Gauld and Bolton, 1988). For example, Gauld and Bolton (1988) point out that with Hymenoptera 'Parasitica' the particular sweeping technique used can account for as much as a twenty-fold difference in catch size and a corresponding difference in diversity of wasps caught. It is advisable to practice the sweeping technique in order to achieve a reasonable level of consistency.

Each sample will normally consist of a fixed number of sweeping movements of fixed speed and fixed duration over a predetermined path through the vegetation. After each sample has been taken, less active natural enemies may be selectively extracted from the net using an aspirator or, if the species being collected are active fliers, the sample can be transferred to a polythene bag and extracted later in the laboratory. As with vacuum net sampling, it is advisable to kill the catch immediately after capture to prevent loss from predation, unless, that is, the insects are specifically required to be kept alive.

Sweeping is a method particularly suitable for the collection of small hymenopteran parasitoids, especially chalcidoids, proctotrupoids, cynipoids and braconids (Noyes, 1989), and a sweep net has been designed specifically for this purpose (see Noyes, 1982, for details).

Although they are frequently used for sampling arthropods in crop fields, sweep nets are subject to considerable sampling variability because their efficiency is affected by a range of factors. These include the distribution, density, activity and life stage of the organism being sampled (Ellington *et al.*, 1984) as well as vegetation height and density, and climatic conditions. In addition, the proficiency of different operators often varies significantly. It is also very difficult to relate a sweep net sample to a finite unit or area or volume of vegetation, making absolute estimates of population densities almost impossible to obtain using this method alone. Nevertheless Tonkyn (1980) developed a formula which he used to express sweep net data as the number of insects caught per unit volume of vegetation sampled, thereby facilitating comparisons with other sampling methods.

Sedivy and Kocourek (1988), studying variation in the species composition of herbivore, predator and parasitoid communities in alfalfa crops, compared the sampling efficiency of a sweep net with that of a vacuum net. They concluded that larval lacewings (Chrysopidae) were caught equally well by both methods but that the sweep net was more efficient in capturing adult ladybirds (Coccinellidae) and adult hover-flies (Syrphidae) whereas the vacuum net was more efficient at capturing adult lacewings, nabid bugs and hymenopteran parasitoids. The same two methods have also been compared in relation to the sampling of spiders in soybean fields, where 34% fewer spiders were collected by the sweep net than by the vacuum net (LeSar and Unzicker, 1978). In cereal crops, Poehling (1987) found the sweep net to be more suitable than the vacuum net for sampling ladybird and hover-fly larvae. As with vacuum nets, different natural enemy species are caught with differing efficiencies by the sweep net, and the time of day when sampling is carried out affects sweep net catches in the same way as it affects vacuum net catches (Vickerman and Sunderland, 1975) (subsection 4.2.2). Wilson and Gutierrez (1980) assessed the efficiency of a sweep net in the sampling of predators in cotton, comparing this method with visual counting carried out on whole plants. The sweep net was only 12% efficient in estimating total predator numbers compared with the visual counts. When individual species were considered, the sweep net was the more efficient method only for detecting lacewing adults because these insects were easily disturbed and tended to fly away during visual counting. In addition, the vertical distribution of predators on the cotton plants affected the efficiency of the sweep net which was most efficient for catching those species with a distribution biased towards the top of the crop canopy. Fleischer *et al.* (1985) also concluded that the sweep net is an inefficient method for sampling predators in cotton.

In summary, sweep nets, while frequently used as a tool for sampling arthropods in field crops, are limited in their usefulness for investigations of natural enemy populations, and so should be used with caution particularly if quantitative information is being sought. Also, like vacuum nets, they perform poorly in wet conditions. Further descriptions of sweeping techniques and the factors which influence sweep net catches are given by DeLong (1932); Saugstad *et al.* (1967); Southwood (1978); Kogan and Pitre (1980).

4.2.4 MALAISE TRAPPING

Malaise traps are tent-like interception devices that are particularly useful for obtaining large quantities of insect material for faunal surveys, studies of the relative abundance of species, phenological studies, studies of diurnal activity patterns, and taxonomic investigations (Steyskal, 1981; Gressitt and Gressitt, 1962; Townes, 1962; Butler, 1965; Ticehurst and Reardon, 1977; Matthews and Matthews, 1971, 1983). They are especially recommended for the collection of groups such as Tachinidae, Pipunculidae, Dolichopodidae, Empididae, Syrphidae and certain parasitoid Hymenoptera (Benton, 1975; Owen, 1981; Owen *et al.*, 1981; Gilbert and Owen, 1990; Belshaw, 1993). Traps are nowadays constructed of fine mesh fabric netting, and incorporate a vertical panel (matt black in colour) that serves to direct insects upwards to the roof apex (a commercial supplier of this type of trap is Marris House Nets, of Bournemouth, UK) (Figure 4.5). Insects accumulate at the highest point within the roof and pass eventually into a collecting bottle or jar that contains preservative (usually 70–95% ethanol). For most kinds of insect, siting and orientation of traps is likely to be crucial; boundaries between different vegetation types, e.g. the edges of forest clearings, should be used, to exploit the fact that insect flight paths tend to be concentrated in such areas, while the collecting chamber end of the trap ought to point towards the sun's zenith to exploit the positively phototactic responses of the insects. Malaise traps are normally positioned on the ground, but traps with rigid frames can be hoisted into tree canopy habitats.

Unless diurnal activity patterns are being investigated, Malaise traps can be left for several days before emptying, although some workers report catches so large that daily

Figure 4.5 A Malaise trap of the type manufactured by Marris House Nets, UK.

collection is necessary (Gilbert and Owen, 1990). Malaise traps do not provide data on the absolute abundance of insects.

An interception device similar in operation to the Malaise trap was devised by Masner and Goulet (1981). It incorporates a vertical polyester net treated with a synthetic pyrethroid insecticide. Intercepted insects crawling on the net are killed by the insecticide and fall on to a plastic tray placed beneath. A clear polythene roof minimises rain damage and deflects positively phototactic insects back on to the net. In the field, this design provided a larger catch of small-sized Hymenoptera than did a Malaise trap (Masner and Goulet, 1981). Masner (in Noyes, 1989) later improved the efficiency of the trap by setting a yellow trough into the ground below the intercepting vertical net. Noyes (1989) used a Masner-Goulet trap without treating the vertical net with insecticide, and obtained poor catches of hymenopteran parasitoids compared with a Malaise trap, in tropical rain forest.

Malaise traps need to be located away from ant's nests, as the ants can severely reduce catches.

4.2.5 KNOCK-DOWN

Introduction

Knock-down involves the dislodgement of insects from their substratum (usually vegetation), causing them to fall on to either a tray, a funnel, a sheet or a series of such devices situated beneath.

Mechanical Knock-down

A common method employed in sampling insects from vegetation is to either shake plants or beat them with a stick, causing the insects to fall on to either a white cloth laid on the ground or a beating tray (a device resembling an inverted white umbrella, Figure 4.6) (Jervis, 1979, 1980a). The fallen insects are either collected by hand or with an aspirator.

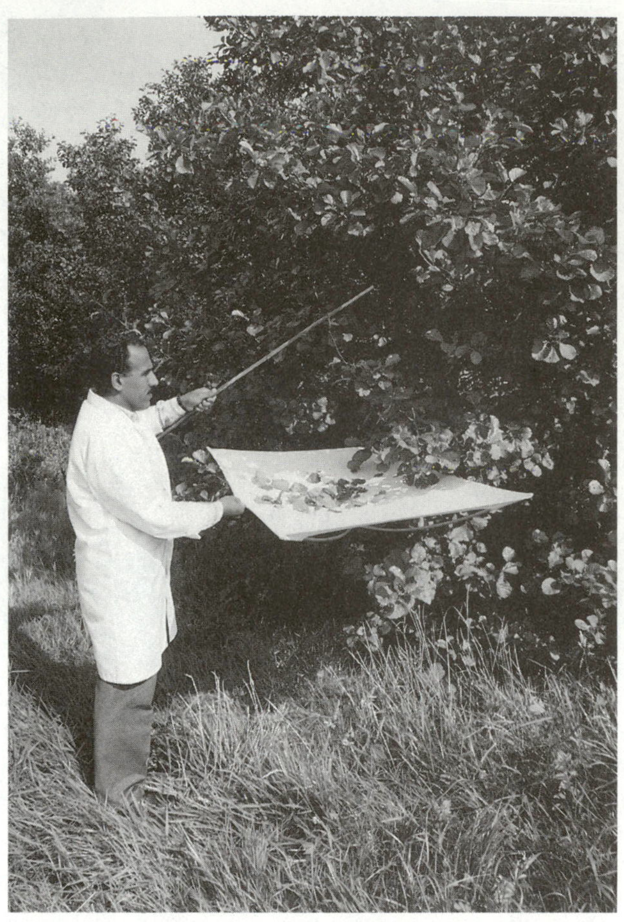

Figure 4.6 Use of a beating tray to collect insects from tree canopy.

This method is useful for estimating numbers of slow-moving arthropods which are easily dislodged from the plant, but it is advisable to calibrate the technique against other absolute sampling methods (Kogan and Pitre, 1980). Fleischer *et al.* (1985) found that the beating method gave estimates of predator density in cotton that were less than one-third the size of those obtained by removing bagged plants to the laboratory. Beating is more commonly used than sweep netting to sample arthropods in trees, but Radwan and Lövei (1982) found that beating ladybirds from apple trees collected only 30% of the beetles observed by visual searching. Beating has also been used to sample spiders from apple trees (Dondale *et al.*, 1979) and to assess numbers of adult anthocorid bugs, lacewings and ladybirds feeding on pysllids in pear trees (Hodgson and Mustafa, 1984).

Pale-coloured insects may prove difficult to locate on white fabric such as that in a standard beating tray. This problem may be overcome by using a dark-coloured fabric.

When beating is carried out in warm, sunny conditions, winged insects that fall on to the collecting sheet or tray surface tend to fly off from the latter very readily. It is therefore advisable in such circumstances to have a small team of workers who can remove insects as they land.

Belshaw (1993) used a beating technique, together with a box trap (Figure 4.7) to collect resting Tachinidae from ground vegetation.

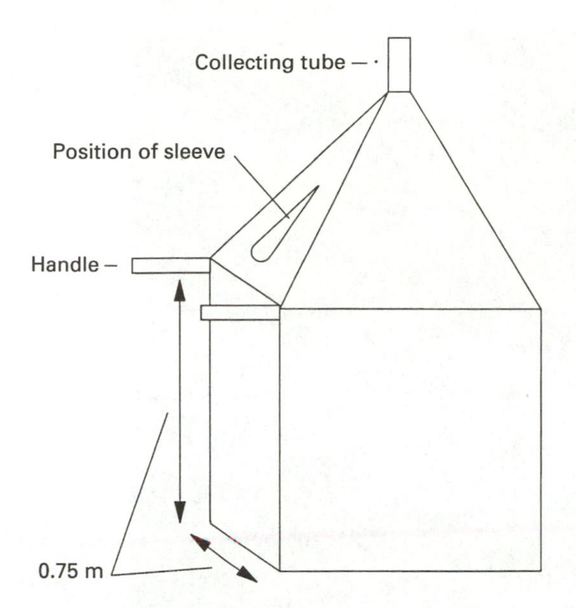

Figure 4.7 Box trap used by Belshaw (1993) to collect insects, disturbed by beating, from ground vegetation. Reproduced by permission of The Royal Entomological Society of London.

The box trap comprises a muslin-covered, open-bottomed metal frame. Using the handles, the device is placed quickly on the ground and the vegetation thus enclosed beaten with a stick. Winged insects (Tachinidae in the case of Belshaw's study) disturbed in this way are collected either in the removable tube situated at the apex of the trap or (via the sleeve, using an aspirator) from the insides of the cage. Because the box trap covers a known area of vegetation, Belshaw's technique can be used to obtain absolute measures of the density of insects. Because the insects are not collected from beneath the plant, it could be argued that this is not, strictly speaking, a knock-down technique.

Chemical Knock-down

Insecticide knock-down, popularly known as **fogging,** is increasingly being used in faunistic and other investigations of parasitoids and predators, particularly in tree canopy habitats (Neuenschwander and Michelakis, 1980; Noyes, 1984; Askew, 1985; Noyes, 1989). A pyrethroid insecticide (e.g. Resilin E, Resilin 50E; Noyes, 1989) fog is released from a device that is either held by the operator (Figure 4.8a) or is hoisted into the tree canopy using a system of ropes and pulleys. Insects can be collected by placing, beneath the canopy, either a sheet on the ground (this method may, however, lead to problems with ants plundering the catch) or several funnel-shaped trays slung from ropes (Figure 4.8b).

Fogging can also be used for extracting insects from field crops; for example, Kogan and Pitre (1980) used a large, plastic fumigation cage in the collection of arthropods from soybeans.

Knock-down, whether mechanical or chemical, although a very effective method compared with some other collecting methods, is unlikely to retrieve all individuals present in the unit of habitat being sampled, and it remains to be determined how useful the method is as a means of estimating absolute abundance. Knock-down can, however, provide useful data on relative abundance (Noyes, 1989).

4.2.6 VISUAL COUNTING

Although relative estimates of natural enemy densities are useful when investigating the effects of experimental treatments, it is often desirable to obtain absolute density assessments for natural enemy populations in the field (Chapters 1, 5 and 6). The most common way of doing this, especially for conspicuous life stages, is to count visually, individuals *in situ* on the vegetation, on either a defined area or period-of-search basis. For the counting of less conspicuous stages or those in concealed locations, plants or other parts of the habitat can be removed to the laboratory for either extraction or examination (Southwood, 1978). Visual counting is also a good method for examining spatial dispersion patterns of natural enemy species, provided that detectability of insects remains constant across the study area.

(a)

(b)

Figure 4.8 Chemical knock-down of insects from forest canopy, using the 'fogging' technique: (a) application of the fog; (b) devices used to collect insects as they drop from the canopy. (Reproduced by kind permission of Nigel Stork.)

Lapchin *et al.* (1987) compared three methods of estimating densities of hoverflies and ladybird beetles in winter wheat crops:

1. an observer walked at a steady pace through the crop and counted all adults and larvae seen within a period of 2 min (**census walk method**);

2. an observer made a detailed search of both plants and the soil surface in a defined area and immediately afterwards a second observer searched the same area. Predator density was then calculated using DeLury's method (Laurent and Lamarque, 1974);
3. plants were cut and removed to the laboratory for detailed examination.

The visual searches carried out in the field (1. and 2.) proved to be unsuitable for estimating the densities of hover-fly larvae because of their cryptic colouration, their relative immobility and their habit of resting in concealed positions on plants. The population density of hover-fly larvae estimated by detailed searches in the field was less than 1% of that estimated from plant removal and examination. However, the detailed searches using DeLury's method were adequate for assessing absolute numbers of ladybird adults and final-instar larvae, although they were not reliable for earlier instars. The census walks provided less accurate estimates of absolute density but they were satisfactory for determining seasonal trends in predator numbers, especially numbers of adult ladybirds.

Chambers and Adams (1986) also used detailed visual searches to estimate numbers of hover-fly and ladybird eggs and larvae in winter wheat. Plant shoots were searched *in situ* and additionally the shoots in 0.1 m^2 quadrats were cut and counted and the resulting stubble, together with the soil surface and any weeds present, was searched. However, detailed visual searching is labour-intensive and Poehling (1987) concluded that, although it was the most accurate method of estimating ladybird and hover-fly numbers in cereal crops, it was not appropriate when large numbers of experimental plots needed to be assessed in a short time.

Frazer and Gilbert (1976) used a census walk method to assess adult ladybird densities in alfalfa fields. This entailed walking along either side of each crop row and counting all beetles that were visible. However, because these counts were influenced by the weather, beetles being more active and therefore more easily seen when temperatures were high, they also searched 30 cm lengths of crop row more thoroughly. Even so, they found that these counts never exceeded 25% of the beetles actually present because the ladybirds spend most of their time in the stubble at the base of the crop. This finding highlights the influence of insect activity on the efficiency of visual searches and census walks in the field, active individuals being far easier to see than inactive ones. Both the degree of predator hunger and the time of day can affect the activity of ladybirds (Frazer and Gill, 1981; Davis and Kirkland, 1982). Time of day can also affect counts of hover-fly larvae; twice as many were found during detailed night-time searches compared with day-time searches of plants and the soil surface in an oat crop (Helenius, 1990). A further factor influencing visual field assessments is the observer's experience with the method. When assessing adult ladybirds in strawberry crops using census walks Frazer and Raworth (1985) noted significant differences between the numbers recorded by different observers. Similarly, variation among observers accounted for a large proportion of the error in predator density estimates from visual searches in cotton crops (Fleischer *et al.*, 1985). Because of the effect of hunger on activity, Frazer (1988) advocates the use of sampling procedures which are specifically designed to estimate only the numbers of active, hungry individuals where predictions of future predation rates are required.

Weseloh (1993) used a census walk method to determine the abundance of ants in different plots.

It is possible to obtain data on the densities of adult parasitoids using visual counting in the field, if the parasitoids are either large and conspicuous or slow-moving. Many parasitoids search the food plants of their hosts by walking, facilitating visual observations in the field. The numbers of adult *Diadegma* foraging over Brussels sprout plants in an experimental plot were recorded by a pair of observers using binoculars (Waage, 1983). The observers checked all sides of each plant but did not approach closer than 5 m, to avoid disturbing the foraging wasps. Hopper *et al.* (1991) counted adult *Microplitis croceipes* (Braconidae) along rows of cotton plants 60 m in length, catching each individual with a hand net in order to establish its sex.

Obviously, census walk methods are only appropriate for highly conspicuous insects and the methodology that has been developed for monitoring butterfly populations in Britain (Pollard, 1977) could be applied to certain natural enemies, such as ladybirds, hoverflies, dragonflies or large parasitoid species. Because weather affects flight activity, butterfly census walks are not done when the temperature is below 13°C, and between 13°C and 17°C they are only done in sunny conditions (Pollard, 1977). It is also important to define the distance limits, on either side of the observer, within which the counts are made during a census walk. During the butterfly counts use is made of natural features such as footpaths or forest rides, but for census walks through crops boundaries can be defined in terms of crop rows or by placing markers, such as bamboo canes, along the route. The butterfly census walk data are used to calculate indices of abundance for individual species at each census site. The mean count per census walk is calculated for each week and these are summed to give the index of abundance for the season. Population changes from year to year can then be assessed by comparing these indices.

In some cases parasitised insects can be readily distinguished from healthy individuals, either in the later stages of parasitism or immediately after death, and their numbers assessed by visual counting. An obvious example is the 'mummification' of parasitised aphids. Aphid 'mummies' (Figure 4.9) act as protective cells within which larval parasitoids complete their metamorphosis to the adult stage. Mummies are conspicuous within aphid colonies and are therefore readily counted *in situ*. Consequently, mummy counts are frequently used to estimate aphid parasitoid abundance (Lowe, 1968; van den Bosch, *et al.*, 1979; Carter and Dixon, 1981; Messing and Aliniazee, 1989; Feng and Nowierski, 1992). However, it is advisable to combine this sampling technique with other methods of assessing parasitoid populations, such as the rearing or dissection of parasitoids from samples of live aphids (subsections 4.2.9, 4.3.5, 4.3.6), since mummy counts alone can give misleading results for several reasons:

Figure 4.9 Aphid 'mummies': *Sitobion avenae* parasitised by *Aphidius rhopalosiphi* (Braconidae) on ear of wheat.

1. Some parasitised aphids leave their colonies, and may even leave the food

plant altogether, prior to mummification (Powell, 1980; Dean *et al.*, 1981; Brodeur and McNeil, 1992);

2. Many aphid parasitoids diapause in the aphid mummy and the proportion doing so increases as the season progresses (Singh and Sinha, 1980), leading to the accumulation of mummies within the study site so that the same individuals are counted in successive samples;
3. Heavy rain or strong wind may dislodge mummies from plant leaves.

The removal of whole plants or parts of plants for subsequent close examination is the most efficient way of counting life stages which are not highly mobile. Harris (1973) assessed several methods for estimating numbers of aphidophagous cecidomyiid fly larvae attacking aphid colonies on a variety of plants. Visual searches in the field proved difficult when aphid colonies were very dense or occurred in protected situations such as curled leaves or galls. An alternative method was to place samples of aphid-infested plant material into polythene bags and keep them in the laboratory for 2–3 days. Subsequently, older predator larvae left the aphid colonies and could be counted as they crawled on the inner surface of the bag. However, the most efficient method was to shake samples with 70% ethyl alcohol in polythene bags and then wash them into a plastic dish where insects could be brushed from the plant material with a soft brush. After removal of larger individuals the samples were washed through a series of filters to retrieve the eggs and small larvae. Field searches and incubation in polythene bags were respectively only 11% and 32% as efficient as the alcohol-washing method.

Visual searching of plants in the field or of cut plants in the laboratory has been used to assess predators on a variety of other field crops including soybeans (Kogan and Herzog, 1980), sorghum (Kirby and Ehler, 1977), brassicas (Smith, 1976; Horn, 1981) and cotton (Fleischer *et al.*, 1985). Fleischer *et al.* (1985) placed cylindrical cloth bags over whole cotton plants, tying the bottom securely around the base of the main stem. The bags were left collapsed on the ground for a week and then two people rapidly pulled the bags up over the entire plants, tying off the top before removal of the bagged plants to the laboratory for subsequent visual examination.

Similarly, units of aquatic vegetation can be collected and examined visually in the laboratory for freshwater predators. Over the years, various devices have been designed for extracting samples of vegetation from aquatic habitats (Hess, 1941; Gerking, 1957; McCauley, 1975; James and Nicholls, 1961). An extensive review of these and other techniques for sampling aquatic habitats can be found in Southwood (1978). The substratum from areas of stream beds, delineated by quadrats, can be removed with the aid of a Surber sampler (see Surber, 1936, for details) and taken to the laboratory for extraction of invertebrates, including predators (Hildrew and Townsend, 1976, 1982). Since the Surber sampler does not catch all the invertebrates disturbed from the stream bed, it cannot be used for absolute population estimates (Southwood, 1978).

Soil surface-dwelling, polyphagous predators are more difficult to assess visually in the field because many of them are active mainly at night and spend much of the daytime concealed in the litter on the soil surface, under stones or in the soil itself. Brust *et al.* (1986a,b) estimated ground predator numbers in corn by placing metal quadrats (13 × 75 cm) over plant rows and visually searching the surface litter and the soil to a depth of 0.5–1.0 cm.

Sunderland *et al.* (1986a) estimated linyphiid spider densities in cereal fields by searching the crop and the top 3 cm of soil within 0.1 m^2 quadrats, and collecting the spiders with an aspirator. Nyffeler and Benz (1988) used a different method to estimate numbers of micryphantid spiders in winter wheat fields and hay meadows; they counted the number of webs within randomly placed quadrats but conceded that their counts were probably underestimates because some spiders occupied cracks in the soil.

Sunderland *et al.* (1987b) attempted to achieve an accurate estimate of total predator density in cereal crops by combining a range of sampling techniques. Insects were first extracted using a vacuum net with a sampling head which covered 0.1 m^2. The sampled area was then isolated by means of a metal cylinder which was driven into the ground to a depth of 8–10 cm, and the plants within this area were cut and removed to the laboratory for close visual examination. The soil surface was searched visually and predators collected with an aspirator. Next, pitfall traps were set in the enclosed area which was further isolated by sealing the top of the metal cylinder with a fine mesh net. It was concluded that any single sampling method will underestimate predator density and that a more accurate estimate is obtained by combining different methods. However, in any system the advantages of intensive sampling methods must be balanced against the disadvantages of excessive labour requirements.

4.2.7 EXTRACTION OF INSECTS FROM PLANTS, LEAF-LITTER AND SOIL

Some natural enemies spend part of their life cycle concealed, along with their prey or hosts, within plants. The prey- or host-infested parts of the plants can be removed for subsequent dissection. In some cases, the plants infested by the hosts or prey, e.g. gall-causing insects and leaf-miners, are relatively easy to distinguish from uninfested ones, and these alone need to be collected. However, plants containing the early stages of gall-causers, and those containing borers can be difficult or impossible to recognise.

Predators or parasitoids concealed in soil or leaf-litter can be studied by taking a sample of the concealing medium and extracting the insects either by hand, by a Tullgren funnel apparatus, or by flotation (Desender *et al.*, 1981; Sotherton, 1984). Sotherton (1984) used a spade to dig out soil cores (20 cm × 20 cm), to a depth of 35 cm or to bed-rock, which were extracted by breaking them up into a large container of saturated salt solution. Organic matter, which floated, was removed using a fine-mesh sieve. Predators were then extracted by hand.

Adult insects emerging from soil (from pupae contained therein) can be collected using an emergence trap (Southwood, 1978, describes various designs). Such traps comprise a metal, plastic or opaque cloth box that covers a known area of the soil surface. Insects are collected in glass vials situated at either the sides or the top of the box. If emergence traps are to be used in studying phenologies, it should be borne in mind that the construction of the trap will influence the microclimate above the enclosed area of soil. All traps tend to reduce daily temperature fluctuations (the deeper the trap, the smaller the fluctuations), and the development rate of pupae may be affected (Southwood, 1978). Emergence traps can also be used to collect insects emerging from pupae contained in plant tissues (Williams and Feltmate, 1992).

4.2.8 SAMPLING BY ATTRACTION

Introduction

It is possible to exploit the attraction responses shown by natural enemies towards certain stimuli (section 1.5) in order to develop sampling techniques which can provide information on phenology and relative abundance. However, data collected by attraction for comparative purposes need to be treated with caution because different species will vary in their level of response to the same stimulus. Obviously, absolute estimates of insect abundance cannot be obtained using these methods.

Visual Attraction

Many winged insects are attracted to certain colours during flight, most commonly to yellow. Consequently, shallow, coloured trays or bowls filled with water in which the attracted insects drown are frequently used as traps to sample flying insects (Southwood, 1978) (Figure 4.10a,b). Principally designed to attract phytophagous insects, yellow-coloured water traps also catch some groups of natural enemies, with varying levels of efficiency. They have been used in cereal crops where adult hover-flies were the most abundant predators caught, although adult lacewings and adult parasitoids also occurred in the traps (Storck-Weyhermüller, 1988; Helenius, 1990). The relative abundance of adult hover-flies, anthocorid bugs and parasitoids has been assessed in brussels sprouts fields using yellow water traps (Smith, 1976). The attractiveness of yellow water traps to individual insects will depend to some extent on the latter's physiological condition; adult hover-flies for example, are probably caught more readily when they are newly emerged and hungry and when food sources are scarce (Schneider 1969). White traps and blue traps, in addition to yellow ones, have been used to catch adult hover-flies (Dixon, 1959; Sol, 1961, 1966), and Schneider (1969) suggested that colour preference may be influenced by the colour of the most abundant flowers in bloom at the time of sampling. Sol (1966) defined the colours used in his water trap studies on the basis of human interpretations of colour, whereas other workers such as Kirk (1984) defined colours in terms of ultra violet reflectance spectra. Kirk (1984) recorded two

(a)

(b)

Figure 4.10 Yellow water traps: (a) trap used by entomologists at Rothamsted to sample insects such as adult hover-flies and parasitoids in a potato crop; (b) traps arranged in an experimental cereal field (arranged perpendicularly to the field edge), used by entomologists at the University of Southampton to measure the within-field distribution and abundance of adult hover-flies.

species of predatory fly of the genus *Medetera* (Dolichopodidae) in water traps of seven different colours and found most in white and blue traps, with very few in yellow traps. Although the rank order of the catches for each colour was similar in both species, the proportions caught by each colour of trap were different. Obviously, great care must be taken when estimating the relative abundances of different species from coloured trap catches.

The attraction of hover-flies to coloured water traps is based on their visual attraction to flowers as pollen and nectar sources. Many adult parasitoids also feed on nectar and pollen (Chapter 6) and are therefore likely to respond to coloured traps when foraging for such resources, although it has been suggested by Helenius (1990) that some parasitoids such as those attacking aphids, may lack a behavioural response to yellow traps, because they feed on host blood or honeydew (but see Jervis *et al.*, 1993). Visual cues such as certain colours may very well be used in host habitat location by adult parasitoids. When a glass prism was used to split a beam of light entering a clear plastic box which contained a group of adult aphid parasitoids (*Aphidius rhopalosiphi*), the parasitoids all moved to the region of the box illuminated by the visible yellow-green band of the spectrum (Budenberg and Powell, unpublished data). Goff and Nault (1984) tested the pea aphid parasitoid, *Aphidius ervi*, for its response to transmitted light and recorded the strongest response to green (wavelength 514 nm).

When used in field crops, water traps are usually positioned level with or just above the top of the crop canopy, but they have also been used successfully when placed near the ground between the trees in orchards, forest plantations or natural forest habitats (Noyes, 1989). Traps of an unspecified colour were used to sample adult anthocorid bugs which attack psyllids in pear orchards (Hodgson and Mustafa, 1984). For some of the species caught there was a significant relationship between the numbers of anthocorids in the water traps and the numbers present on the trees, as estimated by beating. Yellow pan traps are particularly efficient in the sampling of hymenopteran parasitoids such as Ceraphronidae, Scelionidae, Platygasteridae, Diapriidae, Mymaridae, and Encyrtidae (Masner, 1976; Noyes, 1989).

A few drops of detergent are normally added to the water in coloured pan traps, to decrease the surface tension and so facilitate drowning of the insects. If samples are not collected daily, a saturated salt solution may be used as an insect preservative.

The attractiveness of certain colours to insects has also been employed in the sampling of predators and parasitoids with sticky traps (Wilkinson *et al.*, 1980; Weseloh, 1981, 1986; Neuenschwander, 1982; Moreno *et al.*, 1984; Trimble and Brach, 1985; Ricci, 1986; Samways, 1986). Trimble and Brach (1985) examined the effect of colour on sticky trap catches of *Pholetesor ornigis*, a braconid parasitoid of leaf-miners. Yellow (of various shades) and orange traps were significantly more attractive than red, white, blue or black traps, and spectral composition appeared to be more important than reflectance. In contrast, high reflectance regardless of colour seemed to be important in the attraction of tachinid parasitoids to sticky traps placed in forests (Weseloh, 1981). The aphelinid parasitoid *Aphytis melinus* was caught in greater numbers on green and yellow sticky cards than on white, blue, fluorescent yellow, black, red or clear cards (Moreno *et al.*, 1984), whilst eight times as many *Aphidius ervi* (Braconidae Aphidiinae) were attracted to green sticky cards than to gold, blue, white, red, black or orange cards (Goff and Nault, 1984). Ricci (1986) examined catches of ladybird beetles on yellow sticky traps which were being used to monitor pest populations in olive groves and in safflower and sunflower fields. Attached to some of these traps were infochemical (section 1.5) lures used to increase further their attractiveness to target pest species, but curiously in the oilseed crops more ladybirds were caught on unbaited traps than on those with lures.

Light traps have also been investigated as a means of monitoring the numbers of predators such as adult lacewings and ladybirds (Bowden, 1981; Perry and Bowden, 1983; Honěk and Kocourek, 1986). Over a ten year period annual light trap catches were positively correlated with temperature for some species but negatively so for others, whilst the abundance of some species was directly related to aphid abundance (Honěk and Kocourek, 1986). However, it should be noted that light trap efficiency varies with illumination so that changes in activity and abundance may be obscured if catches are not corrected for variations in the intensity of background illumination provided by moonlight (Bowden, 1981).

Nocturnally active Hymenoptera 'Parasitica' are attracted to **blacklight traps** (emitting UV light of wavelength 320–280 nm) and these may be useful for determining relative abundance and the seasonal distribution of insect parasitoids (Burbutis and Stewart, 1979). Gauld and Bolton (1988) note that light trapping is a valuable collecting method for hymenopteran parasitoids in tropical habitats, where a significant proportion of the (mostly ichneumonid) fauna is nocturnal.

Olfactory Attraction

The McPhail trap (McPhail, 1939; Steyskal, 1977) (Figure 4.11), a device commonly used to monitor the numbers of olive pests such as olive fly, can also be used to collect the adults of certain lacewings (Neuenschwander and Michelakis, 1980; Neuenschwander *et al.*, 1981). The trap is suitable only for lacewing species where adults are non-predatory (Neuenschwander *et al.*, 1981). The attractant used is either protein hydrolysate (to which borax is added as an insect preservative) or ammonium sulphate. Protein hydrolysate solutions, and perhaps even tryptophan solutions (van Emden and Hagen, 1976; Jervis *et al.*, 1992b; McEwen *et al.*, 1994) might in future also be employed as attractants, in studies of lacewings and other natural enemies, for use in conjunction with visually-based trapping devices, e.g. coloured sticky traps, to enhance trapping efficiency.

Figure 4.11 A McPhail trap in olive tree canopy.

There is obvious scope for the use of info-chemicals (section 1.5) to monitor natural enemy populations in the same way as they are currently used to monitor pests (Pickett, 1988). Delta traps baited with virgin females of aphid parasitoids (Braconidae, Aphidiinae) caught large numbers of conspecific males when placed in cereal crops (Decker, 1988; Decker *et al.*, 1993) (Figure 4.12a). The use of pheromone traps could provide information

Figure 4.12 (a) Delta-shaped pheromone trap baited with virgin female aphidiine braconid parasitoids. Captured male parasitoids, that have been attracted by the female sex pheromone, can be seen in the water tray; (b) Petri dish pheromone trap with aphid sex pheromone lure consisting of a small glass vial containing aphid sex pheromone. Female parasitoids are attracted by the pheromone and are caught in the dish which contains water with a small amount of detergent added.

on seasonal population trends and on the dispersal behaviour of parasitoid species.

There is also recent evidence of aphid parasitoids being attracted by the sex pheromones of their hosts. Adult female aphid parasitoids of the genus *Praon* were caught in large numbers in water traps that were combined with a source of synthetic aphid sex pheromones (Hardie *et al.*, 1991) (Figure 4.12b). Similarly, aphelinid parasitoids were collected on sticky traps baited with either synthetic sex pheromones or virgin females of the San Jose scale, *Quadraspidiotus perniciosus* (Rice and Jones, 1982; McClain *et al.*, 1990b). Sticky traps baited with 'Multilure', an aggregation pheromone for the bark beetle *Scolytus multistriatus*, attracted a range of hymenopterous parasitoids which attack both eggs and larvae of the beetle (Kennedy, 1979). Other possibilities for exploiting infochemicals include the use of aggregation pheromones for monitoring Coleoptera, particularly carabid beetles (Thiele, 1977; Pickett, 1988) and the use of plant-derived chemicals as attractants. Caryophyllene ($C_{15}H_{24}$), a volatile sesquiterpene given off by cotton plants, has been used in delta traps to monitor lacewing adults and the predacious beetle *Collops vittatus* (Malachiidae) (Flint *et al.*, 1979, 1981).

Using Hosts and Prey as 'Bait'

Hosts of parasitoids can be used as 'bait' to detect the presence and level of activity of adult parasitoids in the field. Although this method is considered under the heading of sampling by attraction, the natural enemies may not locate the hosts/prey by attraction *per se*.

Potted cereal seedlings infested with cereal aphids were used by Vorley (1986) to detect winter activity of aphid parasitoids in pasture and winter cereal crops. The pots were left in the field for 14 days after which they were retrieved and the surviving aphids reared in the laboratory until parasitised individuals mummified. Hyperparasitoids can be investigated in the same way by exposing secondary hosts that contain primary parasitoids.

4.2.9 SAMPLING THE IMMATURE STAGES OF PARASITOIDS

Methods for detecting the presence of the immature stages of parasitoids (eggs, larvae and pupae) occurring upon or within hosts, are usually employed when estimating the impact of parasitism on host populations (Chapter 5), but they are also used in studies of foraging behaviour (Chapter 1), parasitoid life cycles and phenologies, parasitoid fecundity (section 2.7), spatial distribution of parasitism (subsection 5.3.10), parasitoid-host trophic relationships (subsection 4.3.5), and host physiological defence mechanisms (subsection 2.10.2).

Parasitised and unparasitised hosts can sometimes be easily distinguished visually. Insects parasitised by ectoparasitoids can often be easily recognised by the presence of eggs and larvae on their exterior. Hosts parasitised by endoparasitoids are generally less easily distinguishable from unparasitised ones, especially during the early stages of parasitism. During the later stages of parasitism, hosts may undergo alterations in integumental colour; for example some green leafhoppers become orange or yellow in colour (Waloff and Jervis, 1987). Aphid mummies, which contain the late larval stages and pupae of parasitoids, are normally a distinctly different colour from live aphids (Figure 4.9).

If parasitised hosts cannot be distinguished at all from unparasitised ones by external examination, then host dissection can be used to detect the presence of parasitoid immature stages (section 2.6). However, host dissection is a time-consuming activity and, for species attacked by a complex of closely related parasitoids, identification of larvae, and particularly eggs, may prove either difficult or impossible. The immature stages of insect parasitoids have received little taxonomic study.

Some information is available concerning the larvae of the following groups: Hymenoptera (Finlayson and Hagen, 1977); ichneumonoids (Short, 1952, 1959, 1970, 1978); pimpline ichneumonids (Finlayson, 1967); ichneumonine ichneumonids (Gillespie and Finlayson, 1983); braconids (Čapek, 1970, 1973), eurytomids (Roskam, 1982; Henneicke *et al.*, 1992); aphidiine braconids (O'Donnell, 1982; O'Donnell and Mackauer, 1989; Finlayson, 1990); pipunculids (Benton, 1975: several genera; Albrecht, 1990: *Dorylomorpha*; Jervis, 1992: *Chalarus*); Rhinophoridae (Bedding, 1973). In a few groups, some species can be distinguished on the basis of structural differences in their eggs, e.g. *Eurytoma* spp. (Claridge and Askew, 1960) and Rhinophoridae (Bedding, 1973). If one takes a large taxonomic group such as the family Tachinidae, published descriptions of the immature stages are available for several species; however, there is a dearth of synthetic taxonomic studies that provide identification keys.

The colour and form of aphid mummies, the shape of the parasitoid exit-hole, and characteristics of the meconial pellets (waste products deposited by parasitoid larvae prior to pupation) can be used to identify the parasitoids involved, at least to generic level (Johnson, *et al.*, 1979; Powell, 1982). Certain parasitoids which attack hosts feeding in concealed locations within plant tissues often leave clues to their identity such as exuviae or meconial pellets within the hosts' feeding cells.

Host dissection is less reliable as a method for the detection of parasitoid immatures in cases where small parasitoid stages occur inside relatively large hosts, and accuracy can vary considerably between different people performing dissections (Wool *et al.*, 1978). Therefore, the most popular method for the detection of parasitoids in a sample of hosts is to maintain the latter alive in the laboratory until the adult parasitoids emerge (subsection 4.3.6). This approach avoids the problem of identifying the parasitoid immatures but, from the standpoint of obtaining estimates of either percentage parasitism or parasitoid adult density, the problem arises of a time delay being introduced between sampling and obtaining the population estimate – parasitised hosts are removed from the influence of other mortality factors, e.g. predation, fungal pathogens, multiparasitism, host physiological defence mechanisms (subsection 2.10.2), operating in the field after the sampling date. Therefore, the number of emerging adult parasitoids may not give an accurate estimate of the density of adults emerging in the field. The problem of estimating percentage parasitism is further discussed in Chapter 5.

The reliability of the rearing method as a means of obtaining information on the occurrence of parasitoid immatures in hosts depends upon the ease with which the hosts can be cultured. Any hosts which die during the rearing process before parasitoid emergence is expected should be dissected and examined for the presence of parasitoids.

Recently, electrophoretic and serological techniques have been developed for detecting parasitoid immature stages within their hosts (Powell and Walton, 1989; Stuart and Greenstone, 1995) (subsection 4.3.11).

4.2.10 MARK-RELEASE-RECAPTURE METHODS FOR ESTIMATING POPULATION DENSITY

Many of the sampling methods discussed above provide only relative estimates of predator and parasitoid population densities, and of those that provide absolute estimates some only work well for conspicuous or less active species. Inconspicuous or very active insects such as carabid beetles are more difficult to count and an alternative method of estimating population levels is to estimate densities using mark-release-recapture data. These data are obtained by live-trapping a sample of individuals from the population, marking the insects so that they can be distinguished from uncaptured individuals, releasing them back into the population, and then re-sampling the population. An absolute estimate of population density can be calculated from the proportion of re-captured, marked individuals in the second sample, provided that a number of assumptions are satisfied. The main ones are:

1. that marking neither hinders the movement of an individual nor makes it more susceptible to predation or any other mortality factor;
2. that marks are retained throughout the trapping period;
3. that marked and unmarked individuals have an equal chance of being captured;
4. that, following release after marking, marked individuals become completely and randomly mixed into the population before the next sample is taken.

The original method of estimating total population size from mark-release-recapture data was devised by Lincoln (1930) for the study of waterfowl populations. The standard **Lincoln Index** formula is:

$$N = \frac{a \cdot n}{r} \tag{4.1}$$

where N is the population estimate, a is the number of marked individuals released, n is the total number of individuals captured in the sample, and r is the number of marked individuals captured.

The Lincoln Index relies upon the sample of marked individuals which is released back into the population becoming diluted in a random way, so that the proportion of marked individuals in a subsequent sample is the same as the proportion of marked individuals originally released within the total population. However, the original Lincoln Index method unrealistically assumes that the population is closed, with no losses or gains either from emigration and immigration or from deaths and births, in the period between the taking of consecutive samples. Since the Lincoln Index was devised, a number of modifications and alternative methods of calculating the population density have been developed. The best known are those of Fisher and Ford (1947); Bailey (1951, 1952); Craig (1953); Seber (1965); Jolly (1965); Parr (1965); Manly and Parr (1968); Fletcher *et al.* (1981). These methods, together with the formulae used to estimate population sizes, are described and discussed in detail by Seber (1973); Southwood (1978); Begon (1979, 1983); Blower *et al.* (1981) and Pollock *et al.* (1990). Begon (1983) reviews the use of Jolly's method, based on 100 published studies, and discusses potential alternatives.

Among natural enemies, carabid beetles are the group most often subjected to mark-release-recapture studies, but mainly for the purpose of investigating activity and dispersal (subsection 4.2.11). However, there have also been several attempts at estimating carabid populations using mark-release-recapture data (Ericson, 1977; den Boer, 1979; Brunsting *et al.*, 1986, Nelemans *et al.*, 1989; Hamon *et al.*, 1990). In a ten year study of the carabid *Nebria brevicollis*, 11 521 beetles were individually marked using a branding technique (Nelemans *et al.*, 1989) (marking methods are discussed more fully in subsection 4.2.11). The beetle population was sampled continuously using pitfall traps, and two methods were used to calculate population sizes: Jolly's method, as modified by Seber (1973), and Craig's (1953) method. Of the two, Craig's method gave the better estimates, those obtained using Jolly's (1965) method being much too low. All the methods used to estimate population sizes from mark-release-recapture data assume constant catchability over the trapping period, but Nelemans *et al.* (1989) found significant between-year differences in recapture probabilities. Also, the frequency distribution of recaptures deviated significantly from that predicted by a Poisson distribution, more beetles than expected failing to be recaptured and more than expected being recaptured three or more times. Consequently, there is a danger of significant errors in estimates of carabid numbers based on mark-release-recapture data from pitfall trapping, Jolly's method in particular tending to underestimate populations (Nelemans *et al.*, 1989; den Boer, 1979). The size of errors arising from variability in the chances of recapture will depend to some extent on the species being studied. Since the behaviour of individuals affects catchability in pitfall traps, it is sometimes advisable to treat the sexes separately (Ericson, 1977). Furthermore, the handling procedure during marking and release will affect the activity of beetles following release. This can be counteracted by leaving an interval of a few days between successive trapping periods to allow marked beetles time to redistribute themselves within the population before being recaptured (Ericson, 1977).

Because carabid beetles are highly mobile, some marked individuals are likely to leave and re-enter the trapping area, thereby biasing population estimates (Ericson, 1977). To avoid this problem, enclosures have been used during some carabid mark-release-recapture studies (Brunsting *et al.*, 1986; Nelemans *et al.*, 1989) (subsection 4.2.1). In a similar way, a field cage was used to assess the accuracy of Jolly's method for estimating populations of the ladybird beetles *Coccinella californica* and *Coccinella trifasciata* in alfalfa and oat crops (Ives, 1981b). Beetles were captured in experimental plots using a visual searching method, these insects were marked at the site of capture with spots of enamel paint and they were then released immediately after marking. Estimates of population density were calculated using Jolly's method and, in the caged plots, these proved to be very accurate, although estimates of populations in open plots were likely to be less precise. When there was no limitation on flight, it was considered necessary that aphid densities in the study area should be high enough to provide adequate food for the ladybirds, thus preventing dispersal. A limitation noted in Ives' study was the amount of time spent marking captured beetles which restricted the numbers that could be caught in a sampling period, thereby reducing the accuracy of the mark-release-recapture method. To alleviate this problem, visual counts were done whilst walking through the plots, in addition to the counts made whilst marking. A similar method was used to estimate wolf spider population densities in an estuarine salt marsh (Greenstone, 1979a).

Some mark-release-recapture studies have been done on hover-fly populations, but on phytophagous and saprophagous rather than predatory species (Neilsen, 1969; Conn, 1976). Adult narcissus bulb flies, *Merodon equestris* were marked on the tibiae with cellulose paint applied with a fine blade of grass (Conn, 1976), a method that could also be used with aphidophagous syrphids. In this study population size was estimated using Jolly's method and the Fisher-Ford method. Despite recapture rates of 30–50%, both methods gave large errors in daily estimates of population size because samples were small, and the total amount of variation in estimates using Jolly's method was usually two to three times higher than that obtained by the Fisher-Ford estimates. Multiple regression analyses revealed that 35–40% of the variation in the Fisher-Ford estimates and 35–45% of the variation in the Jolly estimates was attributable to the effect of variation in temperature.

Adult parasitoids are much more difficult to mark for mark-release-recapture studies because many of them are small and difficult to handle. However, some success has recently been achieved by labelling braconids and mymarids with trace elements (Jackson *et al.*, 1988; Jackson and Debolt, 1990; Hopper and Woolson, 1991). The trace elements were added to artificial diets on which the hosts of the parasitoids were reared. Fleischer *et al.* (1986) studied patterns of uptake and elimination of rubidium by the mirid bug *Lygus lineolaris* by spraying mustard plants with varying rates of Rb. A concentration of 200 ppm RbCl added to the diet of its *Lygus* spp. hosts provided Rb-labelled *Leiophron uniformis* which could be distinguished from wasps collected from field populations for 6–8 days (Jackson and Debolt, 1990). Similarly, *Microplitis croceipes* adults labelled with rubidium or strontium via the diet of their hosts, *Helicoverpa* spp., at a concentration of 1000–2000 ppm, were distinguishable from field-collected, unlabelled wasps for up to 20 days (Hopper and Woolson, 1991). However, there is a danger that high levels of these elements could affect the biology of the labelled insects. Jackson *et al.* (1988) noted that labelling *Anaphes ovijentatus* with high doses of rubidium (1000 ppm RbCl), via the eggs of its *Lygus* spp. hosts, tended to reduce longevity and fecundity slightly. The radioisotope ^{32}P was used to label *Trichogramma dendrolimi*, a parasitoid of lepidopteran eggs, in order to evaluate the impact of mass releases against a tortricid moth (Feng *et al.*, 1988). In this case, no adverse effects of the radioisotope labelling on longevity, reproduction or sex ratio of the parasitoid were detected.

By applying unique marks to each individual insect in a mark-release-recapture study, additional information on longevity and survival rates can be obtained. Conrad and Herman (1990) used data from a mark-release-recapture study of dragonflies to calculate daily survival estimates using the Manly-Parr method (Manly, 1971; Southwood, 1978), which were then converted to daily expected lifespan estimates using the formula of Cook *et al.* (1967): Expected lifespan = $-1/\log_e$ (Survival).

Conn (1976) calculated the longevity of phytophagous hover-flies from mark-release-recapture data.

4.2.11 METHODS USED IN INVESTIGATING INSECT MOVEMENTS

Predators need to locate their prey in order to feed and parasitoids must find their hosts in order to lay eggs on/in them. Prey/host location generally requires spatial displacement on the part of the predator or parasitoid, and this may be either directed (e.g. movement towards the source of a stimulus) or random. Such locomotor activity is a major factor influencing the population dynamics of natural enemies because it affects the rate of encounters with prey and host patches and so influences the amount of predation or parasitism. It occurs as a result of behavioural responses to stimuli which may be internal, physiological stimuli, e.g. hunger (Mols, 1987), or external, environmental stimuli, e.g. infochemicals (section 1.5). In analysing insect behaviour, it is useful to distinguish between **trivial movements** which are restricted to the habitat normally occupied by the insects, and **migratory movements** that take the insect away from its original habitat (Southwood, 1962). The techniques described here can be applied to either type of movement, although so far as trivial movements are concerned, foraging behaviour at a low level of host and prey patchiness is not dealt with (Chapter 1 discusses this aspect of behaviour).

The study of insect movement in the field is often difficult; in general, it is easier to measure the consequences of movement than it is to examine the process itself. For example, it is easier to record that an individual has shifted from one location to another during the period between two sampling occasions than it is to record the path taken by that individual or the speed at which it travelled.

Some measure of the level of locomotor activity within a population can be obtained by intercepting moving individuals. For a given population, the higher the level of activity the more individuals will be intercepted within a given time. Pitfall traps for example (subsection 4.2.1), can be used to study the activity of ground-dwelling predators such as carabid beetles. In order to study the diurnal activity patterns displayed by carabid beetles and other surface-living predators, automatic, time-sorting pitfall traps have been developed and used in a variety of habitats (Williams, 1958; Ayre and Trueman, 1974; Barndt, 1976; Luff, 1978; Desender *et al.*, 1984; De Keer and Maelfait, 1987; Kegel, 1990; Alderweireldt and Desender, 1990). Individuals falling into the trap are channelled into fresh collecting tubes or compartments after predetermined time intervals, which may be as short as two hours (Ayre and Trueman, 1974; Kegel, 1990). Linear pitfall traps can be used to gain information on the direction of movement (Duelli *et al.*, 1990). If such traps are positioned along the boundary between two distinct habitats, they will detect major population movements across the boundary, and this could be related to changing conditions, such as levels of prey availability, within habitats. Linear traps placed at increasing distances from a habitat boundary will give some indication of rates of immigration into that habitat, e.g. rates of colonisation of arable fields from overwintering refuges (Pausch *et al.*, 1979).

Interception traps such as window traps and sticky traps can be used to monitor the activity and movements, including the direction of movement, of flying insects. A window trap consists of a sheet of transparent material, e.g. glass, plastic, supported in a vertical position, at an appropriate height above the ground, by means of a rigid frame (Figure 4.13a). Flying insects hit the window and fall into a water- filled tray fixed to its lower edge. Lengths of plastic guttering, fitted with end stops, make convenient trays. A drainage hole bored into the base of the tray and closed by means of a rubber or plastic bung allows it to be emptied at the end of each sampling period. Catches can then be sorted in the laboratory. Separate trays are normally fitted to each side of the window trap to provide information on flight direction. Directional information can be increased by using two traps positioned at right angles to one another. Data from window traps have been used to relate the flight activity of carabid beetles to wind direction and to the reproductive state of females (van Huizen, 1990), to monitor the immigration of carabids into newly-formed polders in the Netherlands, and thereby detect the establishment of new populations (Haeck, 1971), to monitor levels of predator activity in cereal fields (Storck-Weyhermüller, 1988) and to detect periods of aerial dispersal by spiders (De Keer and Maelfait, 1987).

Sticky traps work on a similar principle to window traps but consist of a sticky surface to which the intercepted insects adhere upon impact (Figure 4.13b). Any suitable surface, either flat or curved, can be coated with a weatherproof adhesive, several of which are commercially available. Sticky traps supported on poles at various heights were used to monitor the activity of predatory anthocorid bugs in pear orchards (Hodgson and Mustafa, 1984). Very few anthocorids were caught on traps placed within the orchard, but those placed at its edges successfully detected two peaks of flight activity during the year, and the peaks are believed to represent movement to and from hibernation sites. The airborne dispersal of the predatory mite *Phytoseiulus persimilis* and its prey *Tetranychus urticae* was investigated using sticky traps made from microscope slides coated with silicon grease (Charles and White, 1988). These were clipped in a vertical position on to pieces of wooden dowelling which were arranged in the form of a horizontal cross at the top of a supporting pole.

Sticky trap catches of parasitoids and predators can sometimes be increased by using colour as an attractant (Neuenschwander, 1982; Moreno *et al.*, 1984; Trimble and Brach, 1985; Samways, 1986; Antolin and Strong, 1987) (see 4.2.8), but it is questionable whether meaningful information on natural patterns of movement can be obtained using coloured sticky traps, because of the likelihood of the traps' attractant properties interfering with the natural behaviour of insects. The material comprising interception traps should be tested for its degree of UV reflectance, if such traps are to be used for investigations of insect movements.

The direction and height of flight of green lacewings in alfalfa fields was recorded using wire-mesh sticky traps attached to poles, as were data on both diel and seasonal patterns of flight activity (Duelli, 1980). Similarly, wire-mesh traps, measuring 1 m^2 and placed at a

(a)

(b)

Figure 4.13 Traps for interception of flying insects: (a) window trap; (b) sticky trap. The window trap consists of a sheet of transparent plastic; the insects are caught in sections of guttering containing water with a small amount of detergent.

range of heights, were used to monitor insect flight activity in cereal crops (Kokubu and Duelli, 1986). The main predators caught were adult ladybirds, hover-flies and lacewings, but there was evidence that different species preferred to fly at different heights, so that the choice of trap height can be very important. On a larger scale, sticky nets measuring 2.5 m high by 1.5 m wide were used to monitor the flight activity of two ladybird species around experimental plots of alfalfa and oats (Ives, 1981b). Captures of marked beetles in these nets helped to demonstrate that reductions in ladybird numbers recorded in certain plots were due to the emigration of beetles rather than to beetle mortality.

Nets are more efficient than solid interception traps for catching weak flyers because they interfere much less with wind flow. It was estimated that the wire-mesh sticky traps used to catch green lacewings reduced wind speeds by only 10% (Duelli, 1980). Therefore, Malaise traps constructed of netting, and the Masner-Goulet trap (subsection 4.2.4), might be useful for monitoring the flight activity and movement patterns of insect parasitoids.

If the vertical aerial distribution of natural enemies is being investigated, then it is important to bear in mind that the number of insects trapped at a particular trap elevation will be a function of both the spatial density of the insects at that elevation and the downwind component of their ground speed (Duelli, 1980). The latter depends upon the wind velocity (V_W) and the flight velocity (air speed) (V_L) of the insects. If the flight course coincides with the wind direction, the two velocities sum, but if the insects fly on a course at an angle (α) to the wind direction, the downwind component of the ground speed is $V_W + V_L \cdot \cos \alpha$ (see Duelli, 1980). If, as is likely, wind velocity increases with height from the ground (Figure 4.14a), the numbers of trapped insects (N_T) need to be corrected for each trap elevation, to obtain the relative densities per unit volume of air, as follows (see Figure 4.14b):

$$N_F = N_T / (V_W + V_L \cdot \cos \alpha) \qquad (4.2)$$

Interception traps can also be used to study the movement of aquatic predators which are carried along in flowing water (Elliott, 1970; Southwood, 1978). These devices generally take the form of tapering nets, with or without a collecting vessel attached, which are either positioned on the stream bed (Waters, 1962) or are designed to float (Elliott, 1967).

Valuable information on natural enemy activity in the field can be gained by monitoring the movements of marked individuals (Scott, 1973). The commonest marking technique used for this purpose is the application of spots of enamel paint (many other types of paint are unsuitable because of potentially toxic effects), which has been used to mark carabid beetle adults (Jones, 1979; Perfecto *et al.*, 1986) and larvae (Nelemans, 1988), lycosid spiders (Dondale *et al.*, 1970) and adult ladybirds (Ives, 1981b). However, paint marks may be lost, particularly by carabid species which burrow in the soil or squeeze themselves into cracks or under stones. It is advisable, therefore, to test the durability of any marking system in preliminary trials. More robust marking systems have been used with carabid beetles, including branding with a surgical cautery needle (Figure 4.15a,b) or a fine-pointed electric soldering iron (Ericson, 1977; Nelemans *et al.*, 1989), etching the elytral surface with a pin held in a high-speed drill (Best *et al.*, 1981; Wallin, 1986), cutting notches in the edges of the elytra and thorax with a small medical saw (Benest, 1989) and cutting the tips off the elytra (Wallin, 1986). It is important, especially when using a technique which involves physical mutilation, to ensure that marked individuals have the same survival rate as unmarked individuals and that their subsequent behaviour is unaffected.

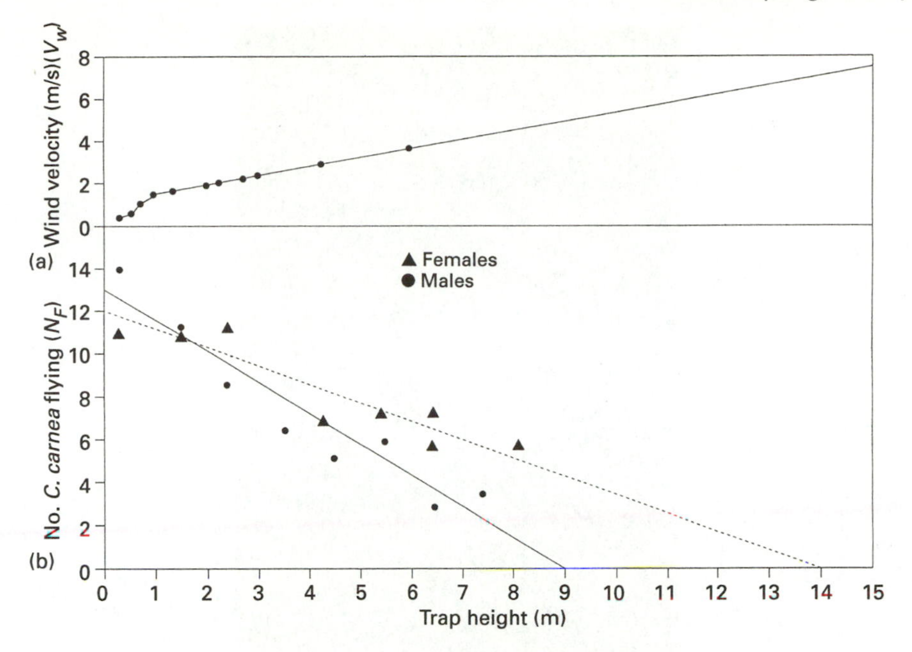

Figure 4.14 Correcting the results of sticky trapping (traps set at varying heights above the ground) for the effects of wind velocity, in the green lacewing, *Chrysoperla carnea*, in alfalfa fields. This figure shows: (a) wind velocity (V_W, see text) as a function of height in an alfalfa field. In a night with average wind speed, the 'boundary layer' (the layer within which the insect's air speed exceeds wind speed) is at 50 cm. Above 1 m the ground speed is mainly a function of the wind speed; (b) the relative numbers (N_F, see text) of insects flying at a particular elevation. The numbers can be estimated from the vertical distribution of trapped insects by correcting for wind speed, air speed of the lacewing, and the angle between wind direction and flight course (see text for details). The corrected vertical distribution of females in the oviposition phase and of males, are treated here as linear regressions. (Source: Duelli, 1980.) Reproduced by permission of Springer-Verlag GmbH & Co. Kg.

Information on minimum distances travelled and speed and direction of movement can be obtained from the recapture data of marked individuals (Evans *et al.*, 1973; Ericson, 1978; Greenstone, 1979a; Best *et al.*, 1981; Nelemans, 1988; Mader *et al.*, 1990; Lys and Nentwig, 1991). Evans *et al.* (1973) studied the movements of adult mutillid wasps in an open sandy habitat. The area was divided into a grid using wooden stakes and patrolled daily by an observer. Adult wasps were caught, marked with spots of coloured paint (model aircraft dope) using the head of an insect pin, and then released. Individuals could be recognised by the colour and positions of paint spots. The activity of wolf spiders in an estuarine salt marsh was investigated in a similar way; the spiders were caught using an aspirator and individually marked with enamel paint (Greenstone, 1979a). During an investigation of the population dynamics of the dragonfly *Calopteryx aequabilis*, Conrad and Herman (1990) studied movement patterns in relation to adult territoriality by marking insect wings with a drawing pen so that individuals could be recognised. The study was carried out along a 635 m section of a stream which was divided into 5 m sectors delineated with small numbered flags. Observers patrolling the stream noted the sex, age category and sector location of all individuals sighted. Population

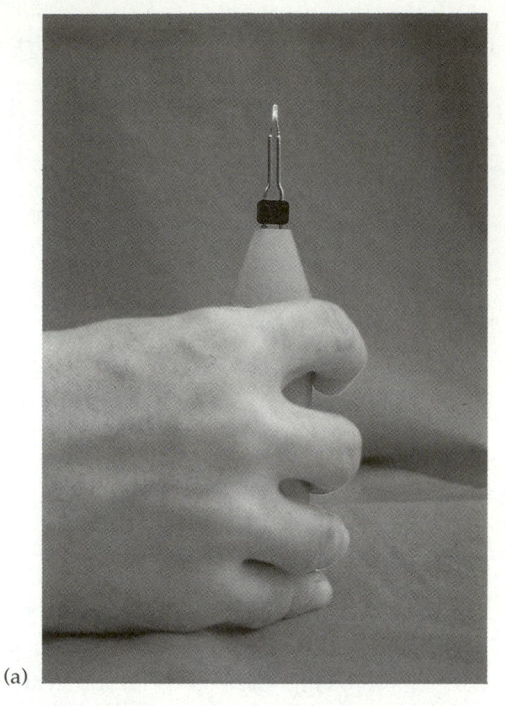
(a)

(b)

Figure 4.15 (a) Surgical cautery needle used for marking the elytra of carabid beetles; (b) carabid beetle (*Pterostichus melanarius*) marked using the needle.

movements were analysed using a modification of Scott's (1973) method, which allows separation of the velocity and distance components of movements.

Some attempts have been made to obtain information on the movement of individuals without interrupting that movement (Baars, 1979b; Mascanzoni and Wallin, 1986; Wallin and Ekbom, 1988). Baars (1979b) labelled adult carabids with the isotope Iridium 192 by mixing it with quick-drying enamel paint. Using a portable scintillation detector it was possible to track the movements of up to ten individuals over the same period of time in the field. However, the beetles frequently lost their paint marks and there was some evidence from laboratory tests that the radiation affected beetle mortality rates. A harmonic radar system has been used to track the movements of large carabids (Mascanzoni and Wallin, 1986; Wallin and Ekbom, 1988). A tiny electronic diode, glued to the beetle's dorsum, reflected microwave beams emitted by a portable detection unit.

Other marking techniques which have been used to monitor the movements of natural enemies include numbered plastic discs glued to the dorsum of adult or larval carabids (Nelemans, 1988; Lys and Nentwig, 1991), spots of typewriter correction fluid applied to carabids (Wallin, 1986; Hamon *et al.*, 1990), spots of acrylic paint applied to the pronotum of adult eucoilid parasitoids with a fine brush (Driessen and Hemerik, 1992), numbers written on the wings of damselflies with a drawing pen (Conrad and Herman, 1990), fluorescent powder applied to green lacewings (Duelli, 1980) and stable isotopes fed to coccinellid adults in drinking water (Iperti and Buscarlet, 1986). Driessen and Hemerik (1992) attempted to mark adult *Leptopilina clavipes*, a eucoilid parasitoid of *Drosophila* flies, with micronised fluorescent dust. Although the dust worked well with the drosophilids, it was unsuccessful as a means of marking the parasitoids because it was rapidly lost from the smooth surface of the exoskeleton and elicited prolonged preening behaviour in the wasps. Fluorescent dusts have also been used to detect visits by parasitoids to specific plants by applying the dusts to the leaves of the plant and then searching for traces of the dust on parasitoids under UV illumination (Ledieu, 1977). Topham and Beardsley (1975) labelled tachinid parasitoids of sugarcane weevils with a radioactive marker in an attempt to monitor the distance travelled by the adult flies between oviposition sites within crops and nectar food sources at field margins (section 6.5). Southwood (1978) provides further information on marking techniques and the use of codes which allows the recognition of large numbers of individuals.

4.3 DETERMINING TROPHIC RELATIONSHIPS

4.3.1 INTRODUCTION

In nature, the trophic interrelationships of insects within a community rarely, if ever, consist of simple food chains (section 4.1). They often comprise an extensive feeding web composed of several trophic levels. The comparative ease with which parasitoids of endophytic hosts can be reared from, and their remains (egg chorions, larval exuviae) located within structures such as leaf mines, stem bores and galls, together with the consequent certainty with which parasitoid-host relationships can be discerned, has meant that the most detailed studies of food webs involving insects have been on gall-formers, leaf-miners and stem-borers (Askew and Shaw, 1986; Redfern and Askew, 1992; Claridge and Dawah, 1993). The food webs associated with endophytic insects are also attractive to researchers because of their greater complexity. Askew (1984) has documented more than fifty species of cynipid gall wasps and their associated parasitoids on oak and rose galls in Britain whilst Redfern and Askew (1992), Claridge and Dawah (1993) and Memmott *et al.* (1994) give diagrammatic representations of a range of food webs based on several gall-formers, stem-borers and leaf-miners.

Described below are several methods currently used (see Table 4.7 for a summary of their particular uses and relative merits) for elucidating the trophic relationships between insects and their natural enemies from the standpoint of: (a) the host or prey, i.e. the species composition of its natural enemy complex and (b) the natural enemy, i.e. its host or prey range. Some of these methods may also be used when the trophic relationships are already known, either to confirm that a relationship exists in a particular locality and/or to record the amount of predation and parasitism. Their use in quantitative studies is largely dealt with in Chapter 5 (subsections 5.2.11–14).

False negatives are a problem common to all the methods; that is, due to undersampling, some parasitoid or predator species comprising an insect's natural enemy complex may be overlooked, as may certain insect species comprising a natural enemy's prey/host range. More serious, however, is the problem of **false positives** with several of the methods (tests of prey and host acceptability, serology, prey/host labelling, electrophoresis).

4.3.2 FIELD OBSERVATIONS ON FORAGING NATURAL ENEMIES

This method is the simplest and most unequivocal for gathering evidence on the trophic relationships within both predator-prey and parasitoid-host systems. It falls generally into two areas; first, **direct observation** with the naked eye and binoculars in the field (Holmes, 1984), and second, **remote observation** using video recording apparatus. However, there are a number of practical drawbacks to these techniques: they are often very labour intensive, data often accumulate very slowly and data may be of limited value if the trophic relationship itself is disturbed during observations. Furthermore, many natural enemies are nocturnal, live concealed in the soil or are very small, creating difficulties for observation. These drawbacks are especially important when studying minute, but often very fast moving, parasitoids. An added complication when attempting to record oviposition by endoparasitoids is that it is sometimes difficult to assess whether or not an ovipositor insertion has in fact resulted in an egg being laid (subsection 1.5.6).

Despite the drawbacks outlined above, direct field observation has been the technique used in a number of studies of predation and parasitism. Workers have either used prey that has deliberately been placed out in a habitat or they have used naturally occurring populations of prey.

Predation of the eggs of cotton bollworms, *Helicoverpa zea*, was investigated by attaching batches of eggs to the upper surface of cotton leaves, using fresh egg white and a small paint brush (Whitcomb and Bell, 1964). The eggs were observed directly for 12 hour periods during daylight and all acts of predation were recorded. Predators were identified whenever possible, and the type of feeding damage inflicted was noted. Ants tended to remove eggs completely, explaining the disappearance of eggs from cotton plants recorded in earlier experiments. Using a similar approach Buschman *et al*. (1977) allowed adults of the velvetbean caterpillar, *Anticarsia gemmatalis*, to lay eggs on soybean plants in laboratory cages and then placed the egg-laden plants out in soybean fields. The plants were continuously observed through the daylight hours by a team of observers working in 2 hour shifts.

Weseloh (1989b) identified potential ant predators of gypsy moth larvae by placing moth larvae on the forest floor in an area where a particular ant species was seen to be foraging and then noting the ants' behavioural reaction to the larvae.

Nyffeler *et al*. (1992) observed diurnal predation of the cotton fleahopper, *Pseudatomoscelis seriatus*, by walking through cotton crops (subsection 4.2.6). During 1 hour observation periods the number of predators without prey, the number of predators with fleahopper prey and the number of predators with alternative prey were counted. Predators carrying prey were collected and both the predator and its prey were later identified in the laboratory. Lavigne (1992) observed the predatory behaviour of Australian robberflies (Diptera, Asilidae) in pastures, using two main approaches: continuous observation of single flies over extended periods and transect walks to record as much behaviour as possible. Feeding flies were collected, identified and finally released once the prey had been removed for identification.

Holmes (1984) examined individual colonies of the cereal aphid *Sitobion avenae* on selected ears of wheat identified by a plastic label, and re-examined them at regular intervals to observe the searching behaviour of staphylinid beetles. The data not only confirmed that aphids were consumed but also provided quantitative information on the rate of prey consumption by predators.

Griffiths *et al.* (1985) studied nocturnal carabids in cereals, using arenas placed out in the field at night and illuminated with red light, to which the carabids do not respond. Red light illumination is used to minimise disturbance to the predator and prey but allow observation. It is, however, important to note that not all arthropods are insensitive to red light (Sunderland, 1988). Harvestmen (Arachnida: Opiliones) and springtails (Collembola) are known exceptions. Many such studies do not yield quantitative data since the prey has to be removed to be identified and the predator's normal behaviour disturbed (subsection 5.2.11 discusses quantitative aspects of this technique).

Time-lapse video recording techniques which reduce the arduous and time-consuming nature of direct observation are now available (Wratten, 1994). Recording is possible at very low light intensities or under infra-red light, thus reducing disturbance (Howling and Port, 1989, Howling, 1991). Laboratory studies of the foraging behaviour of some slug species have been carried out by several workers (Bailey, 1989; Howling and Port, 1989; Howling, 1991). Security and protection of video equipment is an important consideration in field studies.

Because an individual act of predation is often completed in a relatively short period of time and predators spend only part of their time foraging for food, the number of records of predatory acts occurring during an observation period can be very low (Sterling, 1989). Consequently, the likelihood of observing any acts of predation is often low for many predator species. It is sometimes possible to increase the rate of field data accumulation by artificially increasing prey densities. In an investigation of spider predation on the green rice leafhopper, *Nephotettix cincticeps*, Kiritani *et al.* (1972) placed, in paddies, rice plants that had been artificially infested with unnaturally high numbers of leafhopper eggs. Observers patrolled the rice plots after the eggs had begun to hatch and recorded ocurrences of spiders feeding on leafhoppers, noting the species of both the predator and the prey involved.

4.3.3 COLLECTION OF PREY AND PREY REMAINS FROM WITHIN AND AROUND THE NESTS, BURROWS AND WEBS OF NATURAL ENEMIES

With the workers of some predatory ant species it is possible to remove food items from the insects' jaws as the ants return to the nest (Stradling, 1987). Rosengren *et al.* (1979), in a detailed study of wood ants (*Formica rufa*-group) in Finland, used this method to investigate the diet of *Formica polyctena* (Table 4.2). Unfortunately, the manual method of food item collection disturbs the ants and some prey items may be missed. Skinner (1980) overcame these problems when studying the feeding habits of the wood-ant *Formica rufa* (Formicidae), by using a semi-automatic sampling device which collected the ant 'booty' as it was carried back to the nest. With this method, an ant nest is surrounded by a barrier which forces the ants to use the exit and entry ramps provided. The incoming ants are then treated in either of two ways:

1. they are periodically directed through a wooden box with an exit-hole sufficiently large to allow the ant, but not its booty, to pass through. Prey left in the box may then be retrieved and identified;
2. they are allowed to fall off the end of the entry ramp into a solution of 70% alcohol.

The latter treatment, because it kills the ants, can be used for short periods only.

Table 4.2 Pooled data from four nests showing number of prey items (*n*) expressed as a percentage of total items (%) carried to the nest of *Formica polyctena* during two periods in 1978 (Source: Rosengren *et al.*, 1979.) Reproduced by permission of the International Organisation for Biological and Integrated Control.

	30/6 – 3/7		27/7 – 31/7	
	n	%	*n*	%
Homoptera (mostly aphids)	135	20.4	15	2.4
Adult diptera (mostly Nematocera)	122	18.4	147	23.4
Sawfly larvae (Tenthredinidae)	73	11.0	50	7.9
Lepidoptera larvae	74	11.2	43	6.8
Unidentified carrion	53	8.0	78	12.4
Adult Coleoptera (mostly Cantharidae)	51	7.7	34	5.4
Ant workers	33	5.0	52	8.3
Ant reproductives	2	0.3	38	6.0
Lumbricidae (no. of pieces)	23	3.5	21	3.3
Homoptera Auchenorrhyncha	19	2.9	22	3.5
All other groups	76	11.5	129	20.5
Total number of items	661	99.9	629	99.9

The remains of the prey of larval ant-lions (Neuroptera: Myrmeleontidae) can be located either buried in the predator's pit or on the sand surface close to the pit, and can be identified, as shown by Matsura (1986). The diet of first instar larvae of *Myrmeleon bore* was far less varied than that of second and third instar larvae.

Food items may be either observed *in situ* on, or manually collected from, the webs of spiders. This method was used by Middleton (1984) in a study of the feeding ecology of web-spinning spiders in the canopy of Scots Pine (*Pinus sylvestris*) and by Sunderland *et al.* (1986b) when measuring predation rates of aphids in cereals by money spiders (Linyphiidae).

Prey and hosts of fossorial wasps such as Sphecidae and Pompilidae may be analysed by examining the prey remains within the nest. Larval provisioning by the wasp *Ectemnius cavifrons*, a predator of hover-flies, was investigated by Pickard (1975) who removed prey remains from the terminal cell of a number of burrows in a nest. Other fossorial Hymenoptera may be investigated using artificial nests: already hollow plant stems or pieces of dowelling or bamboo that have been hollowed out by drilling (Cooper, 1945; Danks, 1971; Fye, 1965; Parker and Bohart, 1966; Krombein, 1967; Southwood, 1978). The stem or dowelling is split and then bound together again, e.g. with string or elastic bands. The binding may subsequently be removed to allow examination of the nest contents. Details of one of the aforementioned methods, together with photographs of artificial nests and individual cells, are contained in Krombein (1967). Thiede (1981) describes a trap-nesting method involving the use of transparent acrylic tubes.

4.3.4 TESTS OF PREY AND HOST ACCEPTABILITY

Information on host or prey range in a natural enemy species can be obtained by presenting the predators or parasitoids, in the laboratory, with potential prey or hosts and observing whether the latter are accepted (Goldson *et al.*, 1992; Schaupp and Kulman, 1992, for studies on parasitoids).

Such laboratory assessments of prey and host acceptability should always be treated with caution because of the artificial conditions under which they are conducted. They can only indicate *potential* trophic relationships, and other factors need to be taken into consideration when extrapolating to the field situation. For example, it is necessary to establish that the natural enemy being investigated actually comes into contact with the potential prey or host species under natural conditions. The predator or parasitoid may only be active within the habitat for a limited period during the year and this may not coincide with the presence of vulnerable stages of the prey or host in that habitat. It is thus advisable to establish whether there is both spatial and temporal synchrony before carrying out laboratory acceptability tests, so as to avoid testing inappropriate predator/prey or parasitoid/ host combinations (Arnoldi *et al.*, 1991). The placing of insects in arenas such as Petri dishes and small cages also raises serious questions regarding the value of the method. Confining potential prey or hosts in such arenas alters prey and host dispersion patterns, and in particular it is likely to reduce the opportunities for prey to escape from the natural enemy, so increasing the risk of false positives being recorded. Some insect species may rarely, if ever, be attacked in the field by a particular natural enemy species because they are too active to allow capture by the latter.

Even when a parasitoid stabs the host with its ovipositor, the host may still be rejected for oviposition. It is therefore important that egg release is confirmed during host acceptability tests involving parasitoids. This can be done either by dissecting hosts shortly after exposure to parasitoids to locate eggs within the host body (section 2.6) or by rearing hosts until parasitism becomes detectable. Furthermore, just because an insect species is accepted for oviposition, this does not mean that that species is suitable for successful parasitoid development. Under laboratory conditions many parasitoids will, if given no alternative, oviposit in unsuitable hosts, but subsequent monitoring of the parasitoids' progeny will show them to be killed, e.g. by the host's physiological defences (subsection 2.10.2).

When the introduction of an exotic natural enemy species into a new geographical area is being contemplated, i.e. in a classical biological control programme, it is important to determine whether the natural enemy is potentially capable of parasitising or preying on members of the indigenous insect fauna. This can only safely be done through laboratory acceptability tests. The braconid parasitoid *Microctonus hyperodae* was recently collected from South America and screened for introduction into New Zealand as a biological control agent of the weevil *Listronotus bonariensis*, a pest of pasture (Goldson *et al.*, 1992). Whilst in quarantine, the parasitoid was exposed to as many indigenous weevil species as possible, giving priority to those of a similar size to the target host. In the tests, 25 to 30 weevils of each species were exposed to a single parasitoid for 48 hours in small cages, after which the weevils were held in a larger cage to await the emergence of parasitoids. In a second series of trials the parasitoids were given a choice of target hosts (*L. bonariensis*) and test weevils in the same cages. It is important to carry out such a choice test, since in non-choice laboratory trials natural enemies may attack insect species which they would normally ignore in field situations, thereby giving a false impression of the natural enemy's behaviour following field release.

4.3.5 EXAMINATION OF HOSTS FOR PARASITOID IMMATURES

The principle behind this technique is that by examining (dissecting in the case of endoparasitoids) field-collected hosts, parasitoid immatures or their remains located upon and within can be identified to species, thus providing information on the host range of parasitoids and the species composition of parasitoid complexes. Unfortunately, the requirement that the parasitoid taxa involved be identified can rarely be satisfied. Few genera and even fewer higher taxa have a published taxonomy, i.e. keys and descriptions to the species, developed for their immature stages (subsection 4.2.9), and where such taxonomies are available, they are rarely complete. Nevertheless, informal taxonomies can be developed in conjunction with the rearing method described in the next subsection (4.3.6), by associating immature stages or their remains with reared adult parasitoids.

4.3.6 REARING PARASITOIDS FROM HOSTS

One of the most obvious ways of establishing host-parasitoid trophic associations is to rear the parasitoids from field-collected hosts (Smith, 1974 and Gauld and Bolton, 1988, give general advice, and Jervis, 1978, and Starý, 1970, describe methods of rearing parasitoids of leafhoppers and aphids respectively). The ease with which this can be done varies with parasitoid life history strategy. It is usually far easier to rear idiobionts (subsection 1.5.7) from eggs, pupae or paralysed larvae, than it is to rear koinobionts (subsection 1.5.7), since the hosts of koinobionts usually require feeding. When supplying plant material to phytophagous hosts of koinobionts during rearing it is very important to ensure that the material does not contain individuals of other insect species from which parasitoids could also emerge, as this can lead to erroneous host–parasitoid records. The risk of associating parasitoids with the wrong hosts is particularly acute when parasitoids are reared from hosts in fruits, seed heads and galls, as these may contain more than one herbivore, inquiline or parasitoid host species. Such material should be dissected after the emergence of any parasitoids so that the host remains can be located and identified, and the parasitoid-host association correctly determined.

Whenever possible, hosts should be reared individually in containers so that emerging adult parasitoids can be associated with particular host individuals. This allows accurate data on any parasitoid preferences for particular host developmental stages or sexes to be recorded, and it also avoids potential problems arising from the failure of the entomologist to distinguish between closely related host species. The choice of suitable containers will depend upon the host involved but very often simple boxes, tubes or plastic bags will suffice (Smith, 1974; Jervis, 1978; Starý, 1970; Grissell and Schauff, 1990). The addition of materials such as vermiculite, Plaster of Paris or wads of absorbent paper to rearing containers is recommended to avoid problems, e.g. growth of moulds, arising from the accumulation of excessive moisture.

Some parasitoids, including several Braconidae (Shaw and Huddleston, 1991), cause changes in host behaviour that often result in the movement of parasitised individuals from their normal feeding sites prior to death, e.g. the aphid parasitoid, *Toxares deltiger* (Powell, 1980). A significant proportion of the mummies formed by several aphid parasitoid species occur some distance away from the aphid colony and even away from the food plant. They tend to be missed during collection, so leading to inaccuracies in measurements of parasitism and of the composition of the parasitoid complex (Powell, 1980). To avoid the possible complicating effects of parasitoid-induced changes in host behaviour, it is advisable to collect and rear a random sample of apparently healthy hosts in addition to obviously parasitised individuals. With aphids, it is also advisable to rear both live aphids and mummies because some hyperparasitoids oviposit into the mummy stage whereas others oviposit into the larval parasitoid prior to host mummification (Dean *et al.*, 1981).

Parasitoids may also be collected from the field after they have emerged from and killed the host. The larvae of some species leave the body of their host to pupate either upon the host remains or upon the surrounding vegetation. The host remains (for subsequent identification) and the parasitoid pupae (which may be in cocoons) may be collected and the parasitoids reared from the latter in individual containers. Where gregarious species are concerned, the pupae from a particular host individual should be kept together.

The importance of keeping detailed records of all parasitoid rearings cannot be stressed enough. At the time of collection from the field the identity and developmental stage of the host should be recorded, along with habitat and food plant data, location and date (Gauld and Bolton, 1988). Where possible, the date of host death and of parasitoid emergence should be recorded. When adult parasitoids are retained as mounted specimens the remains of the host and parasitoid pupal cases and cocoons should also be retained, to aid identification. Such remains are best kept in a gelatin capsule which can be impaled on the pin of the mounted parasitoid. If a previously unrecorded host-parasitoid association is recorded during rearing, it is very important to provide 'voucher' specimens for deposition in a museum collection.

In order to ensure as far as possible that all the species in the parasitoid complex of a host are reared, samples of parasitised hosts need to be taken from over a wide area, from both high and low density host populations, from a variety of host habitats, and over the full time span of the host life cycle or at least the time span of the host stage that one is interested in (e.g. the pupal stage if only pupal parasitoids are the subject of interest). A parasitoid species may be present in some local host populations and not in others; its absence may be due to climatic unsuitability or because of the parasitoid's inability to locate some populations. Ecologists refer to the **constancy** of a parasitoid species (Zwölfer, 1971) – the probability with which a species may be expected to occur in a host sample (constancy is expressed as the percentage or proportion of samples taken that include the species, e.g. Volkl and Starý, 1988). Host species with a wide distribution require more extensive sampling to reveal the total diversity of their parasitoid complex than host species with a restricted distribution (Hawkins, 1994). We can reasonably assume that the relationship between the extent of the host's distribution and parasitoid species richness is linear (B.A. Hawkins, personal communication).

Major temporal changes may occur in the species composition of a parasitoid complex. Askew and Shaw (1986) refer to the example of the lepidopteran *Xestia xanthographa*; samples of larvae of this host taken 10 weeks apart yielded very different species of parasitoid (Table 4.3). Another factor to consider is sample size; the probability of a parasitoid species being reared from a host will depend on the percentage parasitism inflicted and also on the size of the host sample taken. The

Table 4.3 Parasitism of larvae of *Xestia xanthographa* collected near Reading, UK, on two dates during spring in 1979. The figures take no account of larval-pupal parasitoids (Source: Askew and Shaw, 1986.) Reproduced by permission of Academic Press Ltd.

		Sampling dates	
		2 March	*12 May*
Numbers of unparasitised larvae		188	50
Numbers of larvae parasitised by			
Tachinidae:	*Periscepsia spathulata* (Fallén)	24	0
	Pales pavida (Meigen)	1	0
Braconidae:	*Glyptapanteles fulvipes* (Haliday)	8	0
	Cotesia hyphantriae Riley	1	0
	Meteorus gyrator (Thunberg)	1	0
	Aleiodes sp. A	46	0
	Aleiodes sp. B	10	0
Ichneumonidae:	*Hyposoter* sp.	1	0
	Ophion scutellaris Thomson	0	6
Total larvae sampled		280	56
Percentage parasitism		33	11

larger the sample of hosts from a locality the greater the probability that all the parasitoid species in the local complex will be reared. This relationship is probably asymptotic, and sample sizes of 1000 hosts or larger provide good estimates of parasitoid species richness that are not so strongly dependent on sample size (B.A. Hawkins, personal communication). Martinez *et al.* (1995) discuss the amount of sampling effort required to reveal the structural properties of parasitoid food webs.

As well as providing information on host range, parasitoid complex and community structure, rearing of parasitoids can also yield valuable data on parasitoid phenologies. For example, by noting the times of emergence of larval parasitoids from the hosts and the time of adult emergence. Waloff (1975) and Jervis (1980b) obtained valuable information on the timing and duration of adult flight period and on the incidence of diapause in several species of Dryinidae and Pipunculidae.

Rearing can also be a valuable source of material for taxonomic study. If a larval taxonomy of the parasitoids is to be developed (subsection 4.2.9), association of parasitoid adults with larvae can usually only be reliably achieved through rearing. By associating reared adults with their puparial remains, Benton (1975) and Jervis (1992) developed a larval taxonomy for a number of Pipunculidae. Some parasitoids are more easily identified through examination of their larval/puparial remains than they are through examination of the adult insects (Jervis, 1992).

4.3.7 FAECAL ANALYSIS

With those predators having faeces which contain identifiable prey remains, faecal analysis can be used to determine dietary range. This method has been applied to aquatic insects (Lawton, 1970; Thompson, 1978; Folsom and Collins, 1984). Folsom and Collins (1984) collected larvae of *Anax junius* and kept them until their faecal pellets were egested. The pellets were subsequently placed in a drop of glycerin-water solution on a microscope slide and teased apart. Prey remains were identified by reference to faecal pellets obtained through feeding *Anax* larvae on single species of prey, by examination of whole prey items and by examination of previously published illustrations of prey parts. The data obtained were used to compare the proportion of certain prey types in the diet with the proportions found in the aquatic environment.

4.3.8 GUT DISSECTION

If a predator is of the type that ingests the hard, indigestible, parts of its prey, simple dissection of the gut might easily disclose the prey's identity, as is indeed the case for a number of predators. The main advantages of this method for investigating trophic relationships are the simplicity of equipment required and the immediacy of results.

Dissection is usually performed using entomological micropins which are used to remove and transfer the digestive tract (the crop, proventriculus, mid gut and hind gut) to a microscope slide where it can be teased apart and the prey fragments identified. As with faecal analysis (subsection 4.3.7), the prey remains found in field-collected predators can be compared with those on reference slides prepared by feeding a predator with a single, known type of prey. Predators need to be killed as soon as possible after collection, otherwise the gut contents may be lost due to the prey defecating or regurgitating. *Sialis fuliginosa* loses a proportion of its gut contents by regurgitation when placed in preservative (Hildrew and Townsend, 1982). Not all invertebrates lose gut contents in this manner but this possibility must be borne in mind when carrying out gut dissection. Some aquatic predators collected in nets continue feeding in the net, so need to be anaesthetised immediately upon capture, e.g. *Chaoborus* larvae (Pastorok, 1981).

While gut dissection has been used to study the diet of a range of predators, the method has mostly been applied to aquatic predators (Fedorenko, 1975; Bay, 1974; Hildrew and Townsend, 1976, 1982; Pastorok, 1981) and predatory terrestrial beetles (Sunderland and

Vickerman, 1980; Hengeveld, 1980; Chiverton, 1984; Luff, 1987; Sunderland *et al.*, 1987b). Sunderland (1975) examined the crop, gizzard and hind gut of a variety of predatory beetles (Carabidae, Staphylinidae). Although many fragments found in the gut could be placed in general prey categories, identification of prey remains to species was often only possible if distinctive pieces of prey cuticle (e.g. aphid siphunculi) remained intact. Recognisable fragments generally found in the guts of carabid and staphylinid beetles include: chaetae and skin of earthworms; cephalothoraces of spiders; claws, heads and/or antennae from Collembola; aphid siphunculi and claws; sclerotised cuticle, mandibles and legs from beetles and the head and tarsal claws of some Diptera (Figure 4.16). Hengeveld (1980) and Chiverton (1984), in their studies of Carabidae, also had to group most dietary

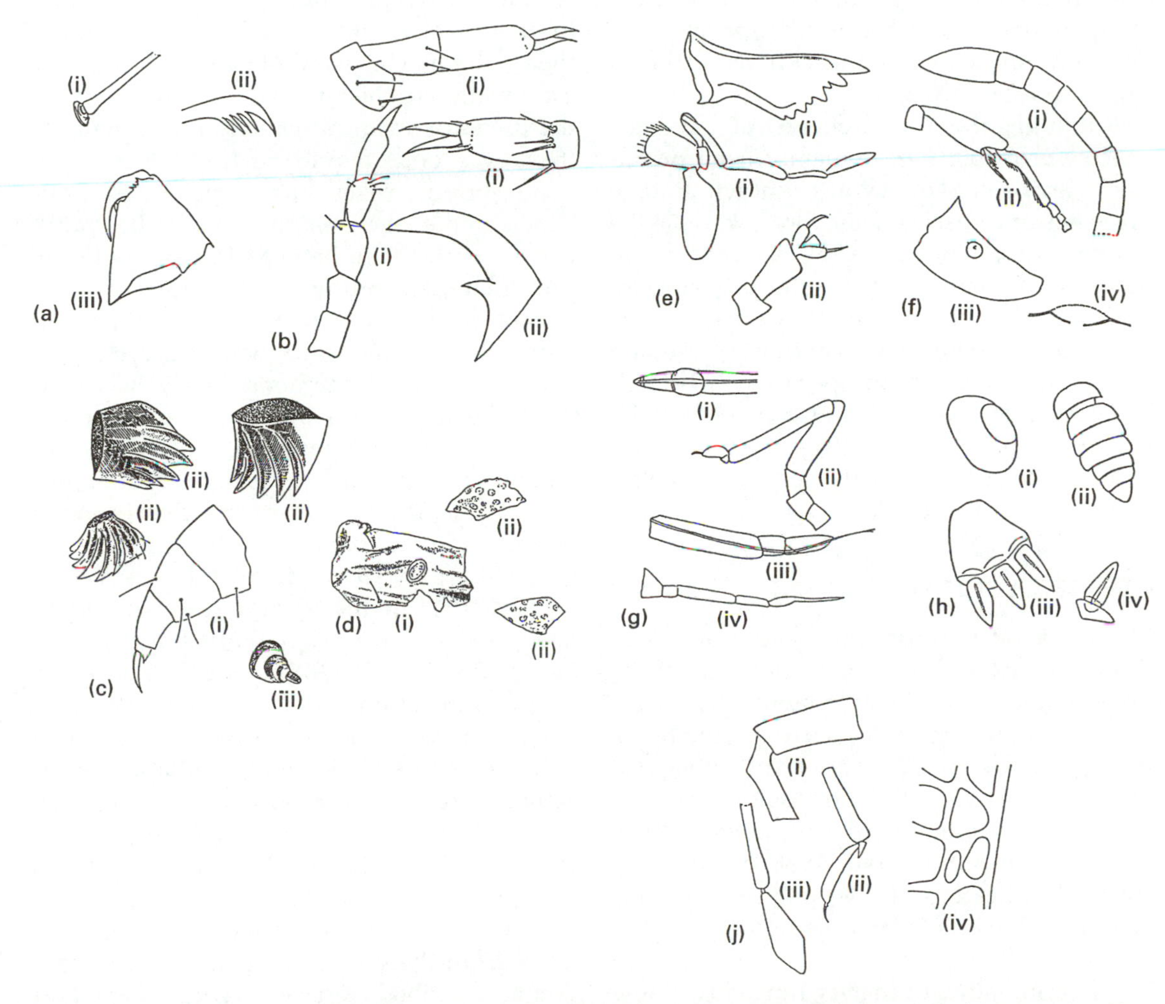

Figure 4.16 Examples of prey fragments found in dissections of carabid beetle guts: (a) lycosid spider, (i) bristle, (ii) claw, (iii) chelicera; (b) carabid or staphylinid beetle larva, (i) legs, (ii) mandible; (c) lepidopterous (?) larva, (i) leg, (ii) mandibles, (iii) unidentified part; (d) fragment of exoskeleton of (i) lepidopterous (?) larva, (ii) heteropteran bug; (e) components of ant's mouthparts (i) and tarsus (ii); (f) ant's (i) antenna, (ii) leg, (iii) exoskeleton (fragment), (iv)ocellus, (g) components of aphid's (i,iii) mouthparts, (ii) leg, (iv) antenna; (h) bibionid fly, (i) antennal segment, (ii) antenna, (iii, iv) parts of leg; (j) *Acalypta parvula* (Hemiptera: Tingidae), (i,ii) parts of leg, (iii) part of antenna, (iv) fragment of forewing. (Source: Hengeveld, 1980.) Reproduced by permission of E.J. Brill (Publishers) Ltd.

components, assigning only a few prey remains to species. One of the food categories of Hengeveld (1980) was 'liquid food', covering any amorphous gut content with too few solid remains present for positive identification by visual inspection.

An obvious difficulty with the gut dissection method is the digestion of prey prior to dissection. The abilities of the different investigators to recognise the prey remains will influence comparative studies, as will the ability to distinguish between fragments from the same or different individuals within a prey species or group.

Although the gut contents of predator species probably derive mainly from predation, carrion-feeding is also known to take place in some species. Therefore, a knowledge of carrion availability may also be required in determining aspects of the diet of a species. The acquisition by a predator of prey materials through **secondary predation** (i.e. feeding upon another predator species which itself contains prey items) can also produce misleading results. An awareness of these problems is also required with the faecal analysis method described earlier.

4.3.9 SEROLOGICAL METHODS

Many predators are either solely or partly fluid-feeders and so their diet cannot be investigated using the aforementioned faecal analysis and gut dissection methods. Ecologists have used a variety of serological methods to study predation (see reviews by Boreham and Ohiagu, 1978; Boreham, 1979; Frank, 1979; Sunderland, 1988; Sopp *et al.*, 1992). Applications of serological methods dealt with here include:

1. determination of dietary range in a single predator species;
2. determination of the composition of a single prey species' associated predator complex (this includes the detection of predation by suspected predator species).
3. detection and identification of parasitoid immatures within or upon hosts.

Serological methods are based on the common principle that antibodies raised in a mammal (often a rabbit), against antigens of an insect prey species, can be used to detect antigens from that species in the gut contents of the predator. In determining dietary range in a predator species (1. above), a first step would involve establishing which invertebrate species might form the natural diet of the predator. This would be done through an examination of the invertebrate community in the predator's natural habitat. Next, material (tissues, blood) would be taken from each invertebrate species and used to raise antibodies in the laboratory mammal. These antibodies would then be used to react with the prey materials present in the guts of field-caught predators, to test for the presence of molecules specific to particular invertebrate taxa. Serological methods have, however, rarely been used for this purpose, the main reason being that potential cross-reactivity among prey species can be high, and the procedures required to overcome this problem become more complicated the greater the number of prey species involved.

Serological methods are more likely to be used, at least in the qualitative sense used here (subsection 5.2.13 includes a discussion of the application of serological methods in studies of population dynamics), to establish which of an assemblage of predator species actually preys upon a particular crop pest (2. above). In this case a survey of the predator community within the prey's habitat is carried out, so that likely natural predators can be identified. Materials would then be taken from the prey species under investigation and antibodies raised against them, and these antibodies then used to react with prey remains in the guts of field-caught predators, to test whether those predators feed naturally on the pest species, to determine what proportion of predator individuals contain prey remains and where possible, to measure the quantity of such remains.

Applying serological methods to the detection and identification of parasitoids (3. above) is relatively straighforward. Antibodies are raised to the immature stage(s) of the parasitoid, and field-collected hosts are screened for the presence of antigens.

The major attractiveness of serological studies is their potential for extreme specificity due to biological recognition at the molecular level. Cross-reactions with antibodies produced to molecular configurations common to 'target' and 'non-target' species – an important drawback to some serological methods – may be overcome by absorption (Sunderland and Sutton, 1980; Symondson and Liddell, 1993a), a process whereby unwanted antibodies are precipitated out of the antiserum by incubation with extracts of the unwanted, cross-reacting, species or by passing through a column to which cross-reacting proteins have been bound. Alternatively, since absorption techniques reduce the titre of the antiserum, other methods such as electrophoresis (subsection 4.3.11) may be used to identify and separate out, from a 'target' prey or parasitoid species extract, any protein component specific to that species. This is then used to raise a more specific antiserum. Laboratory procedures for many immunochemical techniques are described by Rose and Friedman (1980); Wilson and Goulding (1986); Hillis and Moritz (1990).

Production of Antibodies

Introduction

The response 'triggered' in mammalian tissues by the invasion of a foreign organism or compound (Roitt *et al.*, 1989) may be either **non-specific**, involving inflammation and phagocytosis, or **specific**, involving the induction and interaction of a variety of cell types which include cells producing antibodies specific for the antigen. In a specific response, the antigen is processed by macrophages. Activated lymphocytes are then produced which proliferate, differentiate and produce antibody which can react with the configuration responsible for its production. Antibodies possess specific binding sites capable of binding to the invading antigen, thus causing its precipitation, neutralisation or death. Antibodies belong to five major classes of proteins (IgG, IgM, IgA, IgD and IgE immunoglobulins). Although all classes are present in the immune system, with the levels of each mediated by the strength, timing and type of 'antigenic challenge' posed, IgG is generally the major component.

Raising Antibodies

Antibodies are usually raised by repeated injections of an antigen into a rabbit, followed by the taking of a small quantity of blood from the immunised animal after a suitable period. The blood is allowed to clot (at 37°C), the clot is then removed from the container wall, the serum cleared by centrifugation and the proteases and complement inactivated by heat treatment. Inoculation using a single injection of antigen in this manner stimulates the primary humoral response with the production of IgM antibodies. A secondary response is induced by further injections, using the same antigen, at a later date. This secondary response is faster and stronger than the first and yields much higher levels of IgG antibody. When using weakly antigenic compounds, the levels of antibodies produced may be increased either by increasing the duration of the antigenic challenge or by increasing antigenicity of the compound by conjugation with larger proteins (see below).

Antibodies are readily raised to large proteins and polysaccharides. They can also be raised against haptens (small-molecule antigens which, by themselves, are incapable of eliciting an immune response), but before this can be done the haptens must first be coupled to, or conjugated with, larger, more strongly immunogenic proteins to form complexes. Even a small hapten, when complexed, may induce a mammal to produce a variety of antibodies to each available site, or antigenic determinant, with different affinities and specificities. Therefore, a large antigen such as a high molecular weight protein induces the production of a multitude of different antibodies. Hence the term **polyclonal antiserum** is used. The relative amounts of each antibody clone in the serum are determined by a variety of different factors, thus no two polyclonal antisera, even to the same antigen, are the same and so results are not entirely consistent. Polyclonal antisera therefore cannot be used to quantify predation upon one pest species within a habitat shared by related species and genera, because of problems of cross-reaction and variable antigenic reponse related to taxonomic distance between species. To overcome this problem, separation of a single molecular clone (i.e. **monoclonal antibody**) is required.

Monoclonal antibody techniques offer several advantages over those employing a polyclonal serum. One of these is the reproducibility of assays over time. In contrast to polyclonal sera, monoclonal cell lines can produce almost limitless supplies of monoclonal antibody and clones may be stored under liquid nitrogen and used to produce further supplies of the specific antibody at a later date (Liddell and Cryer, 1991). Another advantage is that monoclonal antibodies can confer extreme specificity on the assay to prey genus, species or even instar (Greenstone and Morgan, 1989; Hagler *et al.*, 1991; Symondson and Liddell, 1993c,d; Hagler *et al.*, 1993). The comparatively high production costs of monoclonal antibodies have meant that, in the past, polyclonal antibodies have preferentially been used. However, by using monoclonal antibodies, problems of cross-reactivity and antibody supply are largely overcome and greater specificity is often achieved. These advantages have been exploited by Greenstone and Morgan (1989) and Symondson and Liddell (1993c) in predation studies of *Helicoverpa zea* (Lepidoptera) and slugs respectively.

The standard methodology for production of monoclonal antibody is described by Blann (1984), and illustrated in Figure 4.17, while a practical and up to date account of monoclonal antibody techniques, including methods of antibody preparation, may be found in Liddell and Cryer (1991). The spleen from a mouse immunised with an antigen is used as a source of sensitised *B* lymphocytes. These cells, which are capable of producing the enzyme hypoxanthine guanine phosphoribosyl transferase (HGPRT$^+$), are fused with myeloma cells, cancerous cell lines capable of indefinite growth *in vitro* but lacking the enzyme (HGPRT$^-$). Fusion is usually effected with polyethylene glycol, and the mixture of fused and unfused cells seeded into vessels containing HAT culture medium (hypoxanthine, aminopterin and thymidine). Cells can normally synthesise DNA either by the main biosynthetic pathway (*de novo*) or by the 'salvage pathway' using pre-formed bases in the presence of HGPRT. Aminopterin, in the HAT medium, is a metabolic poison which completely inhibits the *de novo* pathway. Thus, only hybrid cells can survive in HAT medium: myeloma cells will die, because they lack genes for the production of the HGPRT and cannot use the salvage pathway; lymphocytes will die because they cannot be cultured *in vitro*; hybridomas will survive because they contain the enzyme genes (i.e. they are HGPRT$^+$), allowing DNA synthesis, and inherit the ability to grow indefinitely *in vitro* from their myeloma parent. Therefore, after 10–14 days hybridomas are the only surviving cells. Over the next 7–14 days individual cultures are tested for the production of the required specific antibody. Non-productive cultures are rejected at an early stage and aliquots from productive ones frozen down at each stage, to allow selection of other useful clones or in case of later contamination.

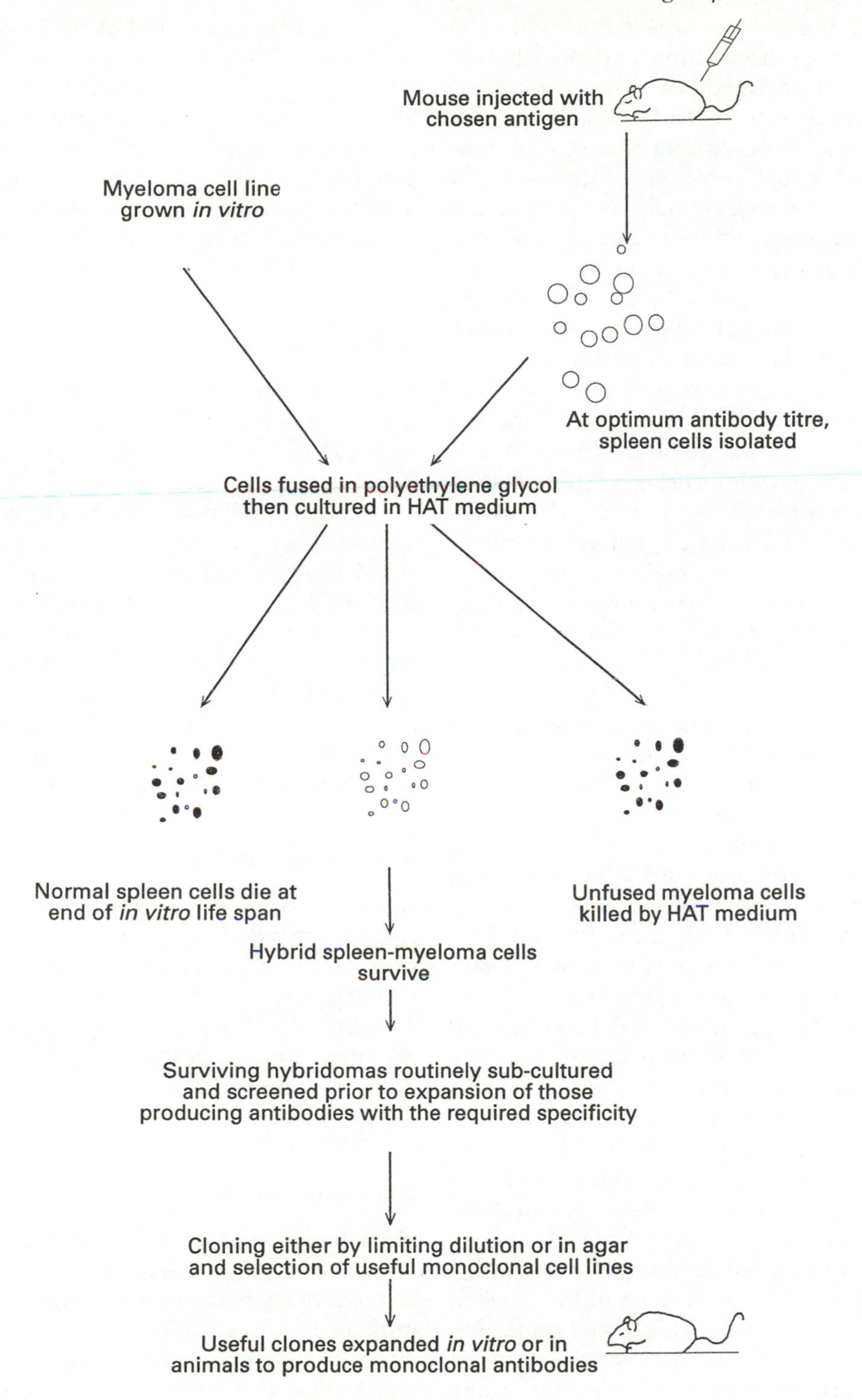

Figure 4.17 General principles of monoclonal antibody production.

Finally, single clones are obtained either by diluting down the cultures and distributing them so that each well of a microtitre plate contains only one hybridoma cell, or else diluting in a weak nutrient agar so that individual clones may be observed growing and then picked out individually using a Pasteur pipette. Larger quantities of monoclonal antibody are then produced either by large scale culture of cells *in vitro* or by growing the cells within the peritoneal cavity of a mouse and harvesting antibody-rich ascitic fluid.

One potential area for the further development of conventional antibody procedures lies in the use of recombinant phage antibodies and antibody engineering (Liddell and Symondson, 1995). These techniques allow the more rapid and less labour intensive production of monoclonal antibodies. Furthermore, they also provide the opportunity to 'engineer' (i.e. manipulate genetically) antibodies for specific purposes, such as improving prey species specificity.

Once antibodies have been raised to a specific prey antigen, immunoassays fall into four categories: precipitation, agglutination, complement fixation and labelled antibody tests. The latter exploit the signal amplification properties of a label to increase sensitivity. In predation and insect feeding studies, precipitation, agglutination and complement fixation have been used with polyclonal antisera whereas labelled antibody techniques have been employed with both polyclonal and monoclonal antibodies. The first two groups of tests have been used in a variety of natural enemy studies (Table 4.4) whilst complement fixation tests have been largely confined to the analysis of blood meals in tsetse-flies and mosquitoes. However, all three have to a large extent been replaced by labelled antibody assays which offer greater sensitivity. Much of the labelled antibody work to date has centred on testing polyclonal antibodies using enzyme-linked immunosorbent assays (ELISA), but more recently some predation studies have been carried out involving ELISA and monoclonal antibodies (Symondson and Liddell, 1995) which gives a highly sensitive and specific test. However, it must be recognised that monoclonal antibody production is still comparatively costly and time-consuming, so the need for specificity in any particular study must be balanced against the extra costs involved with using monoclonal antibodies.

Precipitation Methods

Precipitation tests rely on the fact that antigens may be reacted with specific antibodies to form complexes sufficiently large to be precipitated out of solution. Such precipitation may be performed either in liquids, or gels such as agarose or polyacrylamide, and then visualised using protein stains such as Coomassie Brilliant Blue. A variety of suitable methods is described in Crowle (1980), and examples of their use in studies of insect predators are given by Frank (1979). This technique is comparatively simple, inexpensive and gives a clear result, but in practice often lacks both sensitivity and specificity (Greenstone, 1979b).

Agglutination Methods

Agglutination occurs when antigens, and specific antibodies to those antigens, are present. Antibodies may be attached to particles, e.g. latex or erythrocytes both of which have been used in predation studies. Many of the tests used with erythrocytes have been described and discussed by Nichols and Nakamura (1980).

Complement Fixation

Most antibodies fix/bind **complement** when reacting with the corresponding antigen (complement is a general term for a group of serum proteins involved in the control of inflammation, the activation of phagocytes and lytic attack on cell membranes). The presence of complement is easily detected by mixing the

Table 4.4 Examples of the application of precipitation, agglutination and complement fixation, in studies of predacious and other insects

Method	*Test*	*Predator/prey system*	*References*
Precipitation	Ring Test (Inter-face test)	Ground-living beetles (predominantly Carabidae and Staphylinidae)	Dennison and Hodkinson (1983)
		Predators of *Aedes cantans*	Service (1973)
	Single Radial Immunodiffusion	Lycosid spider *Pardosa sternalis*	McIver (1981)
		Cereal aphid *Rhopalosiphum padi*	Pettersson (1972)
	Double diffusion (Ouchterlony technique)	Predators of *Lygus lineolaris*	Whalon and Parker (1978)
		Carabid predation of *Eurygaster integriceps*	Kuperstein (1979)
	Immunoelectrophoresis	Cereal aphid *Rhopalosiphum padi*	Pettersson (1972)
	Rocket Immunoelectrophoresis	Predation of *Dendroctonus frontalis*	Miller *et al.* (1979)
	Crossover Immunoelectrophoresis	Predation of	
		Dendroctonus frontalis	Miller *et al.* (1979)
		Eldana saccharina	Leslie and Boreham (1981)
		Inopus rubriceps	Doane *et al.* (1985)
Agglutination	Latex Agglutination	Predation of *Acyrthosiphon pisum* by *Coccinella septempunctata*	Ohiagu and Boreham (1978)
	Passive Haemagglutination Inhibition Assay	Lycosid spider *Pardosa ramulosa*	Greenstone (1977, 1983)
Complement Fixation	Complement fixation	Blood meals of tsetse-flies	Staak *et al.* (1981)

test sample with sensitised sheep red blood cells (e.g. sheep erythrocytes sensitised with antibodies from another organism such as a rabbit) to see if haemoglobin is liberated. The rabbit antibodies bonded to the sheep red blood cells cannot lyse in the absence of complement. However, the presence of an antigen which reacts with a complement-fixing antibody can be demonstrated by mixing them together in the presence of complement. If they react, complement is fixed and on addition of red blood cells lysis does not occur. If the specific antigen is absent, the complement is not fixed and the red blood cells will be lysed. Serial increase of concentration will produce a standard curve from which the concentration of the unknown antigen may be read.

Complement fixation is a sensitive and semi-quantitative technique for detecting small amounts of antigen.

Palmer (1980) and Staak *et al.* (1981) provide practical details of this test. However, the latter authors also point out that the specificity of the antiserum used cannot be increased by absorption because complement-fixing immune complexes, resulting from absorption, interfere with the test.

Labelled Antibody Immunoassays

Introduction

The complement fixation technique is more sensitive than either precipitation or agglutination assays. However, the increased sensitivity conferred by the signal amplification of antibody labelling has meant that all the methods described above have recently been superseded to a large degree by labelled antibody immunoassays, such as ELISA. It is, however, important to appreciate that the extreme sensitivity of tests such as ELISA does not necessarily result in increased specificity, indeed the reverse is sometimes the case. Increased specificity is largely conferred by the use of monoclonal rather than polyclonal antibodies.

In ELISA, prey antigen is detected when it binds with labelled antibodies, hence detection of the label is indicative of the presence of the antigen. Since the amount of label trapped is directly proportional to the amount of antigen in the sample, the test may be regarded as quantitative, a major advantage over the tests previously described. A variety of labels are available and they include radioisotopes, fluorescent compounds and enzymes which catalyse specific reactions to give a coloured product. Rose and Friedman (1980) describe the use of radio-immunoassays in medicine but to date this method has not been widely applied in entomology, due to expense and safety considerations.

Fluorescence Immunoassay

In this method, the immunoglobulin fraction of antiserum can be coupled with a fluorescent compound whilst the test antigen is held within a matrix such as cellulose acetate discs. Following incubation in the conjugated antiserum, the discs may be read under UV light. More commonly, however, the antigen is bound to a solid phase such as a polystyrene microtitration plate. This assay is both very sensitive and quantitative, although problems of contrast were encountered by Hance and Gregoire-Wibo (1983) in their studies of aphid predation by carabid beetles.

Enzyme-linked Immunosorbent Assay

The basis of ELISA is that some enzymes, commonly horseradish peroxidase or alkaline phosphatase, may be coupled with antibody in such a way as to produce conjugates which retain both immunological and enzymatic activity. Such conjugates are stable for a period of many months and can be reacted with prey antigen, which has been previously fixed to a solid phase, to form a complex. Antigen fixation may be either direct or through a non-conjugated antibody for the same antigen. The non-reacting components are then flushed away. The fixed complex is then incubated with a substrate under standardised conditions and the enzyme part catalyses the conversion of the substrate to form a coloured product, the intensity of which can be measured (Figure 4.18). ELISA has been widely modified for differing studies; many of the variations are reported and described by Voller *et al.* (1979) and references dealing with their entomological applications are cited in Sunderland (1988). The latter author discusses three variations of this method used in studies of predation. For purely qualitative studies, positive and negative controls are usually sufficient but for quantitative studies (subsection 5.2.13) an **antigen dilution series** should be included on each ELISA plate to take account of any interplate variation in results. The message amplification effect of the enzyme component makes ELISA a much more sensitive test than those tests based upon precipitation (Miller, 1981). Examples of its use for analysis of field-collected invertebrates are given by Sunderland (1988) who emphasises its suitability for routine testing of large numbers of individuals. The amount of antigen required per test is small enough to allow an individual predator meal to be tested against a wide range of antisera.

The most straightforward method of using ELISA – known as **Direct ELISA** – is to bind

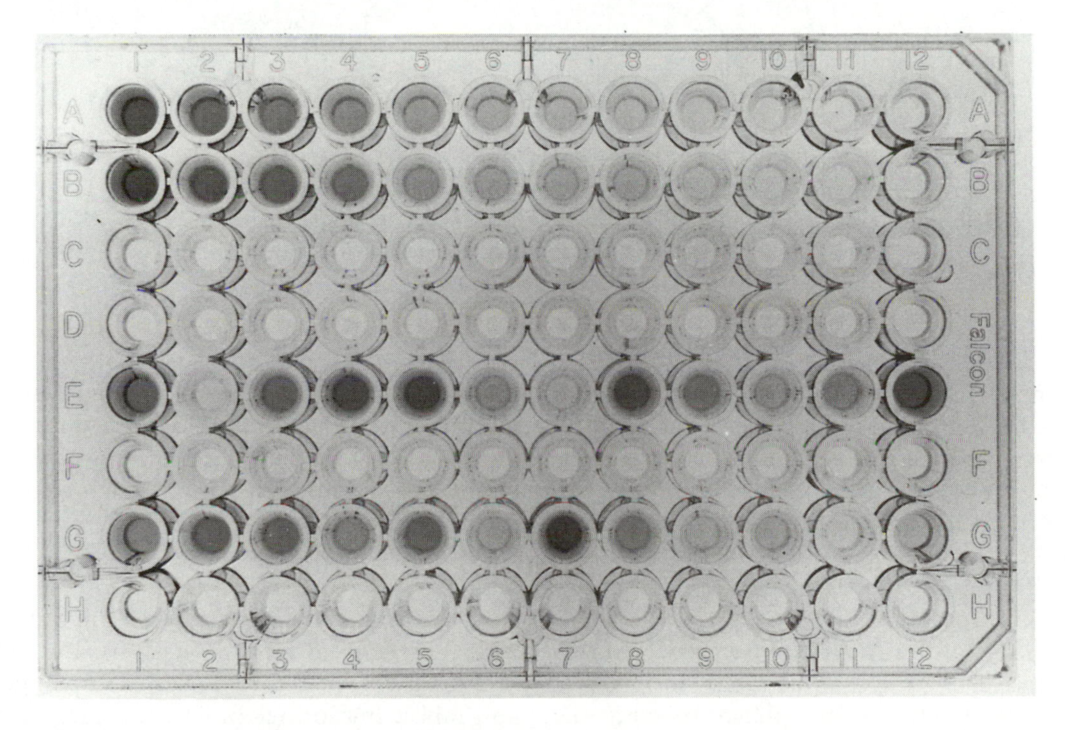

Figure 4.18. An ELISA plate used by Symondson and Liddell (1993b). Contained in rows A-D: a 1.5 x dilution series of antigen (slug haemolymph), with the highest concentrations towards on the left; E–G (replicated in rows F and H): diluted samples extracted from the crops of carabid beetles. Rows A, B, E and G were treated with an anti-haemolymph antiserum, and rows C, D, F and H with non-immune serum. Background readings, and heterologous reactions between non-immune serum and the antigen, can be subtracted from the readings obtained with the specific antiserum. Regression of $\log_e$ haemolymph concentration versus $\log_e$ absorbance readings produces an equation by which the concentration of haemolymph in the unknown samples can be calculated. Reproduced by kind permission of Bill Symondson.

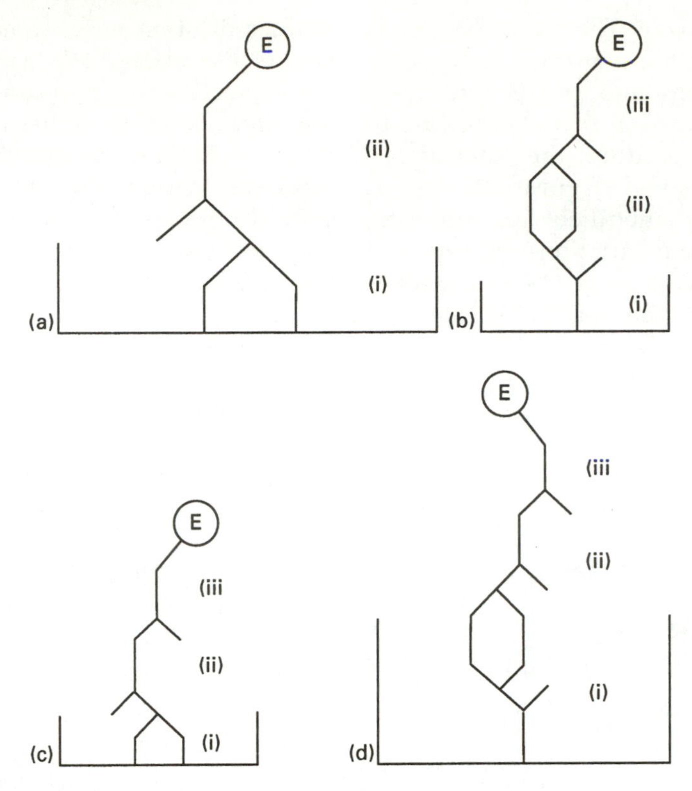

Figure 4.19 (a) Direct ELISA; (b) Double Antibody Sandwich ELISA; (c) Indirect ELISA; (d) Indirect Double Antibody Sandwich ELISA.

antigen (Figure 4.19a) directly to a solid phase ((i) in the Figure), such as a polystyrene microtitration plate, where it is reacted with enzyme-labelled antibody (ii). The intensity of colour produced following the addition of enzyme substrate is then measured, and is proportional to the amount of antigen present.

In the **Double Antibody Sandwich ELISA** (Figure 4.19b), non-conjugated antibody is attached to the solid phase (i) and the unknown antigen solution added. The captured antigen (ii) then reacts with enzyme-labelled antibody (iii) and, on addition of the enzyme substrate, the intensity of the coloured product formed is again proportional to the amount of unknown antigen. This method is more specific than Direct ELISA, since it relies on matching two epitopes (an epitope is a single antigenic determinant) rather than one. Specificity may be increased still further if the epitopes are different and the antigen has to match two different binding sites.

With the **Indirect ELISA** (Figure 4.19c), antibody (ii) is produced in a mammal such as a rabbit and is then reacted with antigen bound directly to a solid phase (i). The antibody is then reacted with, for example, goat anti-rabbit immunoglobulin conjugated with enzyme (iii) and the appropriate substrate added. This method is essentially similar to the direct method but utilises commercially-available anti-species conjugate, thus saving on specific antiserum. The coloured product formed results from enzyme activity which is proportional to the quantity of antigen present.

The qualitative aspects of all the above methods have been tested and compared by Crook and Payne (1980) who found the indirect method to be the most sensitive and the double antibody sandwich method to be the most specific.

A fourth method of ELISA sometimes used is the **Indirect Double Antibody Sandwich ELISA** (Figure 4.19d). As with the double antibody sandwich, non-conjugated antibody is first bound to the substrate (i) then antigen added. Thereafter, the captured antigen is reacted with antibody from another species (ii) which in turn reacts with a goat anti-rabbit immunoglobulin conjugated with enzyme (iii). The coating antibody (i) must not be the same species as the second antibody (ii), otherwise the conjugate (iii) will bind to (i) as well as (ii). For example, (i) may be a polyclonal antiserum raised in a rabbit, (ii) a mouse monoclonal antibody raised against the same antigen and (iii) a goat anti-mouse IgG conjugate. The intensity of the colour of the resulting product is again used to measure the amount of antigen present. Once again, specificity is increased by use of a two-site assay.

The major advantage offered by ELISA lies in the fact that, when calibrated, it can be used quantitatively. However, the major disadvantage for field entomologists is that there is always at least some colour development, even in the control, so that a quantitative colour reading must be taken, using a spectrometer or plate reader. Statistical procedures are then applied to these readings and used to set confidence limits for determination of positive or negative results (Fenlon and Sopp, 1991; Symondson and Liddell, 1993a,b). This drawback prompted Stuart and Greenstone (1990) to apply an **immunodot assay**, based on ELISA, to their studies of predator gut contents. Although based on ELISA, this technique obviates the need for expensive apparatus and it is claimed that tests can be unequivocally scored by eye as positive or negative. However, inherent in this approach is a risk of reduced sensitivity, when compared with statistical treatment of standard ELISA results, thus leading to 'false negatives'. Therefore, current research is aimed at optimising conditions to minimise this risk. Despite the slight danger of false negatives, these authors found the approach to be species- and instar-specific for *Helicoverpa zea* remains in predators, and the results obtained compared well with those gained using a standard ELISA for the same predator-prey system. If correct, this method would appear to offer some advantages over standard ELISA techniques, and may make immunoassay techniques more accessible to those workers wishing to study feeding relationships in arthropod predators in the future. The results obtained so far with parasitoids are also encouraging. Stuart and Greenstone (1995), used an immunodot assay with a monoclonal antibody for the detection of second instar *Microplitis croceipes* in *H. zea*.

ELISA, because it combines speed, specificity and sensitivity, has been widely adopted for arthropod predation studies. Examples of applications in entomology can be found in Fichter and Stephen (1979); Ragsdale *et al.* (1981); Crook and Sunderland (1984); Lövei *et al.* (1985); Service *et al.* (1986); Sunderland *et al.* (1987a); Greenstone and Morgan (1989); Symondson and Liddell (1995). ELISA has so far been applied to parasitoids in only two cases. Stuart and Burkholder (1991), using indirect ELISA, developed monoclonal antibodies specific to two parasitoids of stored products pests. The antibodies reacted with all life stages and both sexes of the parasitoids. The use of immunodot assay by Stuart and Greenstone (1995) has already been mentioned.

Sunderland *et al.* (1987a) compared ELISA with gut dissection in the detection of feeding by polyphagous predators on cereal aphids. They found that when the same individuals were tested using both methods agreement between the two tests was poor for those individuals which had eaten aphids. This suggests that the rate of voiding solids from the gut is not directly proportional to the rate of disappearance of prey antigens during digestion. Furthermore, the relative efficiencies varied according to the species of predator tested, and secondary predation was found to be a potential problem with both techniques.

Preparation of Insect Material for Serological Analysis

The method of preparing insect material for analysis depends on the size of the predator being studied. In the case of smaller predators, a whole-body homogenate is prepared, whereas with larger predators such as carabid beetles, the gut (or even just the fore gut) is required. The contents of the fore gut are likely to be most useful for analysis, since the fore gut is used mainly for storage and so less digestion (i.e. biochemical alteration) of prey materials takes place. The excised gut or gut region is opened under a suitable buffer solution (e.g. phosphate buffered saline), agitated and then centrifuged. The resulting supernatant is then used as a stock solution which may be stored frozen for subsequent use. Of course, when using serological methods, it is essential that laboratory starved predators are used to calibrate the assay.

For protocols involved in preparing insect material for studies of parasitism, see Stuart and Burkholder (1991) and Stuart and Greenstone (1995).

4.3.10 LABELLING OF PREY AND HOSTS

With this method potential prey are labelled with a chemical which remains detectable in the predator or parasitoid (Southwood, 1978, gives a general review of labelling methods). By screening different predators within the prey's habitat for the presence of the label, the species composition of the predator complex can be determined.

Various labels are available for studying predation and parasitism. These include radioactive isotopes, rare elements and dyes (usually fluorescent) which are introduced into the food chain where their progress is monitored. The label is injected directly into the prey or it is put into the prey's food source. Parasitoid eggs can be labelled by adding a marker to the food of female parasitoids. Appropriate field sampling can then reveal whether suspected predators have eaten labelled prey or whether suspected hosts have been oviposited in by a particular parasitoid species.

Radioactive elements have been used to monitor predation of *Helicoverpa armigera and H. punctigera* eggs by injecting adult females with ^{32}P so that radiolabelled eggs are produced (Room, 1977). The fate of these eggs was then monitored in the food chain. Both adults and larvae of the bean weevil *Sitona lineatus* were labelled with ^{32}P by allowing them to feed on broad bean plants which had their roots immersed in distilled water containing the radiolabel (Hamon *et al.*, 1990). The labelled weevils were then exposed to predation by carabid beetles within muslin field cages and the levels of radioactivity shown by the carabids subsequently measured using a scintillation counter.

Monitoring of radioactive labels may be done using a Geiger counter or scintillation counter to measure α and or β emissions (Hagstrum and Smittle, 1977, 1978) or by autoradiography (McCarty *et al*, 1980). In autoradiography the sample containing the potentially labelled individuals is brought into contact with X-ray sensitive film, and dark spots on the developed film indicate the position of labelled individuals. However, although simpler to perform than either serological or electrophoretic methods, hazards to both the environment and the operator posed by radiolabelling mean that the method is confined to laboratory studies, or is used in such a way that the labelled insects can be safely and reliably recovered at the end of the experiment.

Rare elements such as rubidium or strontium may be used as markers in a similar way to radioactive elements (Shepard and Waddill, 1976). The path of the rare element through the food chain is monitored with the aid of an atomic absorption spectrophotometer. Unfortunately, although such markers are generally retained for life, and self-marking is possible using labelled food sources, the equipment necessary for detection is expensive both to purchase and to run as well as requiring trained operators.

Fluorescent dyes offer another means of marking prey. Hawkes (1972) marked lepidopteran eggs with such dyes during studies of predation by the European earwig *Forficula auricularia*. An alcoholic suspension of dye was sprayed on to eggs which were then eaten by the earwigs. Thereafter, the earwig guts were dissected out and examined under UV light. Although such dyes are useful, being both simple and inexpensive to use, they are relatively short-lived and possible repellant effects due to either the dye or the carrier must be taken into account. The dye is usually voided from the predator gut within a few days. Hawkes (1972) found no evidence of dye retention in any of the internal structures of earwigs dissected.

A major potential problem with the use of labels is secondary predation and scavenging; the label may be recorded in several predatory insect species within the prey's habitat, but only some of these may be directly responsible for prey mortality.

4.3.11 ELECTROPHORETIC METHODS

Introduction

As with serology, both the dietary range of a predator species and the composition of a prey's predator complex can be determined using electrophoresis. Electrophoresis is defined as the migration of colloids (usually proteins) under the influence of an electric field. The rate of ion migration is proportional to the electric field strength, net charge on the ion, size and shape of the ion and the viscosity of the medium through which it passes (Sargent and George, 1975). Therefore, application of an electric field to a protein mixture in solution, e.g. a predator's gut contents, will result in differential migration towards one or other of two (positive + and negative –) electrodes. The direction of migration is, of course, dependent on charge and, since proteins can exist as zwitterions (ions whose net charge depends upon the pH of the medium) conventional electrophoresis is performed under buffered conditions (i.e. constant pH). **Zone electrophoresis** is a modification of this idea whereby the mixture of molecules to be separated is sited as a narrow zone or band at a suitable distance from the electrodes so that during electrophoresis proteins of different mobilities travel as discrete zones which then gradually resolve as electrophoresis proceeds.

Development of this basic concept has led to a wide range of electrophoretic techniques (vertical and horizontal slab, thin layer, two-dimensional, gradient pore and isoelectric focusing) utilising a variety of supporting media (cellulose acetate, alumina, agarose, starch and polyacrylamide). **Isoelectric focusing**, the second most important method after conventional electrophoresis, separates proteins which differ mainly in their charge. Proteins tested migrate in a pH gradient and separate at their respective isoelectric points (Sargent and George, 1975; Hames and Rickwood, 1981). Thus, given that prey species will differ in the proteins they contain, the pattern produced by electrophoresing the gut contents of a polyphagous predator will, with either of these methods, provide information on the composition of the predator's diet.

Here, electrophoretic methodology is briefly described, followed by examples of how the techniques outlined have been used in some natural enemy studies.

Electrophoretic Methodology

Many texts are available which deal with the general methods and techniques used in electrophoresis, e.g. Chrambach and Rodbard (1971); Gordon (1975); Sargent and George

(1975); Hames and Rickwood (1981); Richardson *et al.* (1986); Pasteur *et al.* (1988); Hillis and Moritz (1990). The reader is referred to these texts for detailed descriptions and experimental protocols.

Since coming into common usage in many laboratories, standard protein electrophoresis has been extensively modified to suit the requirements of individual studies. Many aspects of the methodology used are dependent upon the choice of medium. Various media are commonly used (e.g. polyacrylamide, starch, cellulose acetate and agarose), since no single medium is ideal under all circumstances. Therefore, the advantages and disadvantages of each should be recognised and taken into account when choosing a medium for any particular study. A discussion of the relative merits of each may be found in Richardson *et al.* (1986) (see also Menken and Raijmann, 1995). Methodologies for **polyacrylamide gel electrophoresis (PAGE)** will be described below since it has been used by a variety of workers on a range of different insect groups. The recipes and procedures are based upon those used by Loxdale *et al.* (1983) in their studies of cereal aphid population genetics.

Most enzyme groups studied using electrophoresis are involved in the fundamental functions of all living organisms. Therefore, failure to detect a particular enzyme is usually a technical problem related to sample preparation, storage or electrophoretic conditions, rather than absence of the enzyme in the organism. Also, since electrophoresis is primarily a comparative technique, it is important that all samples are standardised and treated in exactly the same way. Preparation of insect material for analysis is dealt with later in this section. Suffice to say at this point that the material to be analysed (gut contents, gut regions, whole insects) needs to be homogenised in a solution of buffer, to produce a solution containing all the constituent soluble proteins.

When using vertical electrophoresis, the sample density is increased by the addition of either glycerol or sucrose to the homogenising solution. This ensures that the sample will sink into, and remain at the bottom of, the wells. A few crystals of bromophenol blue may also be used as a tracking dye. A detergent such as sodium dodecyl sulphate (SDS) or Triton X-100 will solubilise membrane-bound proteins and eliminate aggregates, and should be added to both the homogenising solution and the gel buffer. Triton X-100, unlike ionic surfactants (e.g. SDS), does not cause denaturing of proteins which results in a concomitant loss of activity in the case of enzymes. Nicotinamide-adenine dinucleotide phosphate (NADP) and β-mercaptoethanol may also be added to the homogenising solution in some cases, to stabilise NADP-dependent dehydrogenases and to reduce oxidative changes respectively. Finally, it may sometimes be necessary to buffer the homogenising solution with Tris buffer (Tris (hydroxymethyl) aminomethane).

After homogenisation, the sample is centrifuged to remove any solid matter and the supernatant either electrophoresed immediately or stored deep-frozen. If stored, the sample should be thawed rapidly to 1°C immediately prior to electrophoresis. Following centrifugation, the supernatant is transferred to sample wells in the gel medium using a microsyringe, capillary tube or by absorption into small squares of filter paper.

A simple electrophoresis apparatus consists of a DC power supply connected to electrodes in buffer-filled tanks (Figure 4.20).

Electrical continuity is established and maintained through a buffer-impregnated gel. The samples to be separated are applied to the gel at the origin immediately before the current is applied. Power units supplying constant voltage whilst limiting current are used to protect gels from excessive power levels. Commercially produced electrophoresis tanks are widely available (Figure 4.21) and should be used according to the manufacturer's instructions.

All chemicals/enzymes used in electrophoretic work should be high purity grades, all solutions should be made up using high quality, freshly prepared deionised

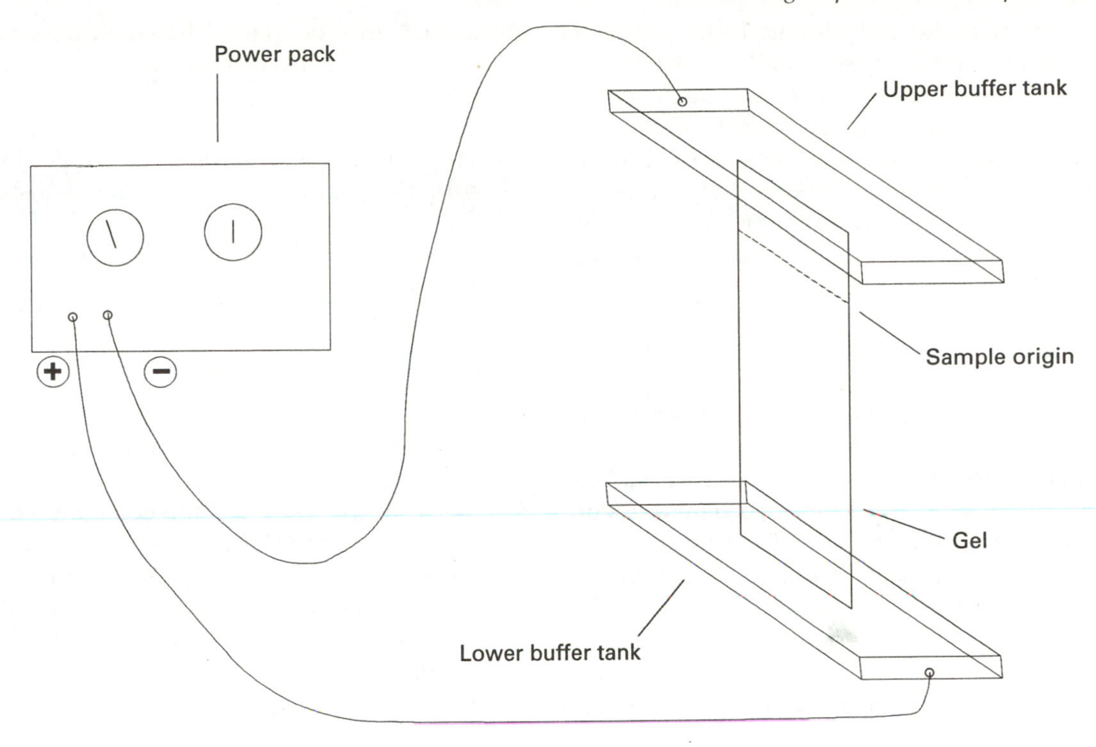

Figure 4.20 Diagram of a simple vertical electrophoresis system.

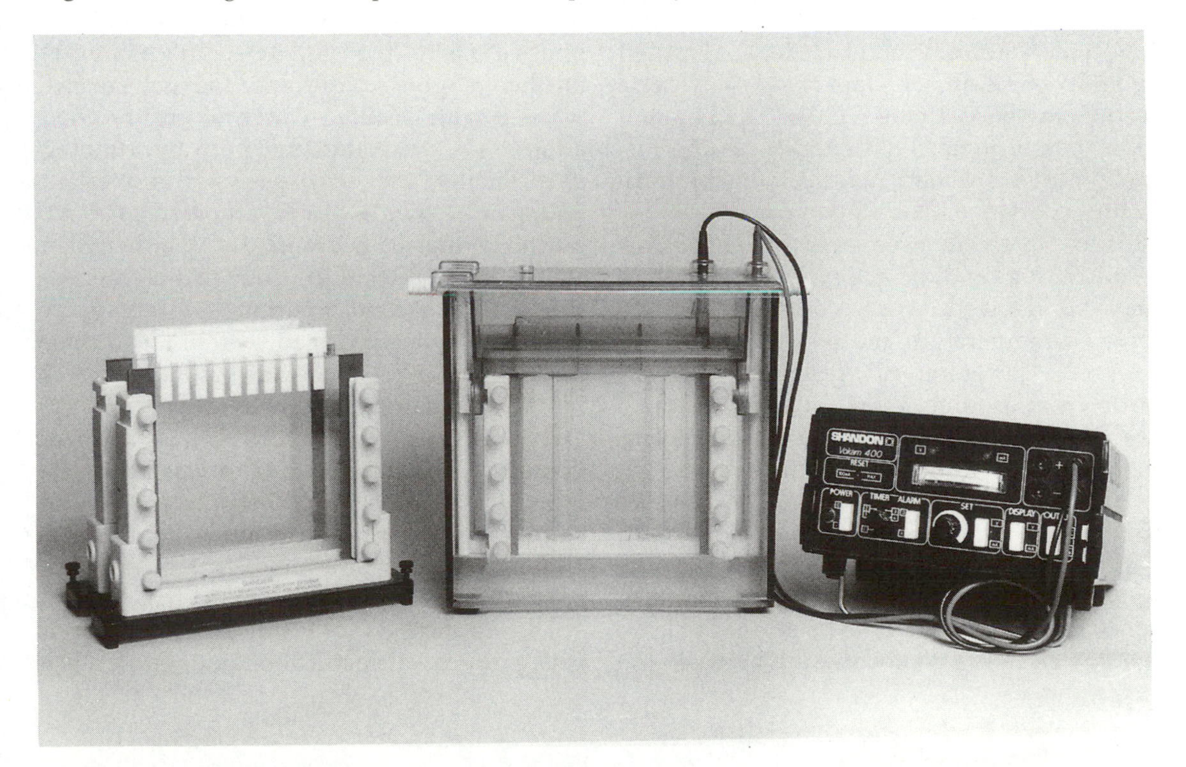

Figure 4.21 Cooled vertical slab gel electrophoresis apparatus available from Pharmacia LKB Ltd.

water, and the pH should be adjusted as necessary by adding either HCl or NaOH.

It is important to note that many of the compounds used in these procedures are known carcinogens, poisons and/or skin irritants. Therefore, all compounds must be treated as potentially dangerous and handled with *extreme* care. Furthermore, since high voltages are used in the running of gels, *extreme* care must be exercised, to avoid electrical hazards.

Buffers are used to resist electrolytically induced pH changes and to maintain electrical continuity in the electrophoresis circuit. Depending on the enzyme(s) being stained for, the buffer employed may be either continuous (e.g. Tris/HCl) or discontinuous (e.g. Tris/HCl – Tris/barbitone). Discontinuous systems employ a different buffer/pH in the gel and electrode solutions whereas in continuous systems they are the same. Quantities of buffer constituents are given in Table 4.5, and these should be made up to the appropriate volumes using high quality deionised water.

Polymerisation of acrylamide is catalysed by the addition of both N,N,N^1,N^1-tetra methylethylenediamine (TEMED) and ammonium persulphate: TEMED induces the formation of persulphate radicals, which in turn induce polymer cross-linking by bis-acrylamide. Since the free base of TEMED is required for catalytic action, polymerisation may be delayed at low or high pH, by lowered temperature and by the presence of oxygen. Therefore, all reactions are H^+-buffered at around neutral pH. Gel mixtures are degassed under reduced pressure and then left to set at room temperature.

Unlike starch and agarose, batches of several acrylamide gels may be cast in clean glass cassettes held within a 'Perspex' box with a sliding lid. For optimal results, it is essential for these plates to be perfectly clean, thereby ensuring good adhesion between the gel surface and the glass. The small-pore gel solution is cast first, and it is made as follows:

1. the appropriate quantities of buffer constituents are dissolved in protogel (Table 4.5);
2. the solution is transferred to a measuring cylinder and TEMED added;
3. the pH is adjusted to 7.4 using 10.2 M HCl and ammonium persulphate added;
4. The solution is then made up to the appropriate volume with deionised water and degassed for two minutes using a vacuum pump;
5. the solution is then mixed with cold Triton X-100 (1.6% aqueous solution) and quickly poured into the glass cassettes;
6. the gel solution is overlaid with a thin layer (several mm) of iso-butyl alcohol (to exclude oxygen) and left to polymerise on a horizontal surface for about 45 min.

Polymerisation has occurred when a sharp line appears between the butanol and gel phases. Once the gel sets, the iso-butyl alcohol is poured off and the gel surface washed with distilled water, to ensure good adhesion between the large pore and running gels. Finally, the small pore gel is overlaid with a large pore solution, prepared as above but adjusting the pH with 1.0 M rather than 10.2 M HCl (Table 4.5). Prior to polymerisation of this second solution, sample spacer combs are quickly placed therein to produce pockets in the gel. Once this second gel layer has set, the combs may be removed and the isobutyl alcohol again poured off, although at this stage it is unnecessary to wash the gels with deionised water. Gels may then be stored at 4°C for several days, if required, although the best resolution is gained if gels are used immediately.

The application of samples to prepared gels, prior to an electrical potential difference

Table 4.5 Preparation of gels and buffers (Source: Loxdale *et al.*, 1983.)

(a) Components of Tris/barbitone gels and buffer		
	Electrode buffer	
pH 7.45		
Tris	5.0 g	(8.3 mM)
Barbitone	27.6 g	(30 mM)
(made up to 5 litres)		
	Gel constituents (including buffer)	
pH 7.80	*small pore gel (per 1000 ml total)*	*large pore gel (per 100 ml total)*
Protogel *	200.00 ml	10.33 ml
Tris	8.57 g (70.6 mM)	0.86g (70.6 mM)
TEMED	370 μl	60 μl
10.2 M HCl	to pH 7.80	to pH 7.80
Ammonium persulphate	0.7 g	0.07 g
Triton X-100 (1.6% aqueous soln.)	130 ml	25 ml

(b) Components of Tris/citrate gels and buffer (Source: Loxdale *et al.*, 1983.)		
	Electrode buffer	
pH 6.8		
Tris	8.174 g	(13.5 mM)
Citric acid**	4.515 g	(4.3 mM)
NADP+	25 mg	(6 μM)
(made up to 5 litres)		
	Gel constituents (including buffer)	
pH 6.8	*small pore gel (per 1000 ml total)*	*large pore gel (per 100 ml total)*
Protogel	200.00 ml	10.33 ml
Tris	1.63 g (13.5 mM)	0.163 g (13.5 mM)
Citric acid	0.906 g (4.33 mM)	0.090 g (4.33 mM)
Magnesium chloride***	1.014 g (5 mM)	0.102 g (5 mM)
TEMED	370 μl	60 μl
1.0 M HCl	to pH 6.8	to pH 6.8
Ammonium persulphate	0.70 g	0.07 g
NADP+	11 mg (13 μM)	1 mg (13 μM)
Triton X-100 (1.6% aqueous soln.)	130 ml	25 ml

* 30% (w/v) acrylamide and 0.8% (w/v) bisacrylamide stock solution.
** monohydrate.
*** hexahydrate.

being applied, is referred to as **loading.** In the absence of a stacking gel the initial position of the samples is termed the 'origin' (position of proteins at time zero), whereas when a stacking gel is employed the origin is at the interface between the stacking and running gels. Homogenate samples are transferred to the gel using a fine microsyringe, although it is important not to overload the pockets, as excess material in one track may cause distortion in band running or create enzyme band shadows (possibly due to lateral diffusion), in adjacent tracks. Such shadows may be recorded as false positives. It is advisable to perform test runs with differing volumes and concentrations of sample to determine the best values for optimum sharpness and clarity in any particular case. Also, it may be necessary to run a standard (e.g. a parthenogenetic aphid, Loxdale *et al.*, 1983), of known mobility, on each gel so that a mobility ratio for each band can be calculated with respect to this standard. To prevent 'carry-over' (and the possibility of false positives) from one sample to another, the microsyringe should always be flushed out with reservoir buffer between each sample addition. When loaded, dense sample solution appears as a sharply defined layer at the bottom of each well.

Electrophoretic running conditions for a selection of enzyme systems are listed in Table 4.6. When assaying general proteins, gels are usually run at room temperature whilst gels separating isoenzymes are run under refrigerated conditions (5–10°C). In both cases, buffer may be drip-circulated between anodal and cathodal reservoirs. In the case of isoenzymes, buffer cooling is essential since temperatures above approximately 40°C denature enzymes and rising temperature during electrophoresis results in a concomitant fall in resistance and hence faster migration through the gel matrix. If a series of gels is run in parallel, the resultant decrease in resistance must be countered by a proportional rise in the current supplied at constant voltage.

The electrophoresis kit is connected to the power supply, the appropriate voltage set, and the apparatus switched on. For most purposes the voltage is kept constant (usually automatically by the power supply) and the current allowed to fall as electrophoresis proceeds. The electrophoretic run is allowed to continue for a constant time, determined by

Table 4.6 Electrophoretic running conditions (Source: Loxdale *et al.*, 1983.)

Enzyme	*Gel*	*Volts*	*mA/gel*	*Run time (hrs)*	*Stain time (hrs)*	*Buffer system*	*Homogenising solution*
EST	6% total acrylamide	150	40	2	0.5	electrode Tris/barb (pH 7.45) gel (pH 7.8) Tris/HCl	15% (w/v) sucrose soln. + bromophenol blue
GOT	"	"	"	"	1	"	"
ME	"	"	"	"	0.5	"	"
PEP	"	"	"	"	2	"	"
MDH	"	100	25	6	1	electrode Tris/ citrate (pH 6.8) gel Tris/ citrate (pH 6.8)	15% (w/v) sucrose in 0.5% (w/v) Triton X-100 + 50 mM Tris/HCl (pH 7.1)
FUM	"	"	30	4	2		

Table 4.7 Comparison of available methods for investigating qualitative aspects of trophic relationships in natural enemy complexes and communities. Disadvantages and advantages of each method are discussed in more detail in the text (section 4.2).

Method	*Advantages*	*Disadvantages*
Field Observation	Immediate and usually unequivocal results obtained with minimal equipment	Time-consuming and labour-intensive; small, hidden or fast-moving insects difficult to observe; possibility of disturbance to insects during observation; high cost of video equipment
Collection of Prey and Remains from within and around Burrows and Webs	Prey items conveniently located in or around nest/burrow/web	Often involves disturbing predators; applicable to few natural enemies
Tests of Prey/host Acceptability	Little field sampling involved	High risk of false positives
Examination of Hosts for Parasitoid Immatures	Results usually unequivocal	Identification of parasitoid immatures to species level often not possible due to lack of descriptions and keys; false negatives a problem
Rearing Parasitoids from Hosts	Results usually unequivocal	False negatives a problem, if sampling is not sufficiently extensive and intensive; hosts of koinobionts need to be maintained until parasitoid larval development is complete
Gut Dissection and Faecal Analysis	Simple and quick to perform with minimal equipment; samples can be stored either deep-frozen or in preservative for long periods prior to testing	Can only be used if predator or parasitoid ingests solid, identifiable parts of prey or; host identification of prey remains not always possible; possibility of misleading results due to scavenging and secondary predation
Serology	Includes accurate, very sensitive and reproducible techniques; with monoclonal antibodies, the cell lines can be retained for future use, giving reproducibility of assays over time; sample can be stored deep-frozen for long periods prior to testing	Time-consuming to calibrate, requires specialist equipment and expertise; risk of false positives through cross-reactions, scavenging and secondary predation
Prey Labelling	Simpler to perform than either electrophoretic serological tests; dyes easily detected; samples can be stored deep-frozen for long periods prior to testing	Expensive equipment required to detect some labels; radioactive labels a hazard to operator and environment; usually involves severe disturbance to the system under study
Electrophoresis	Can be accurate and sensitive Once standardised, large numbers of samples can be rapidly analysed	Expensive, time consuming to calibrate, and requires specialist equipment and expertise; risk of false positives through scavenging and secondary predation

the power supplied, apparatus used, gel, buffer and the mobility of protein bands under examination.

Following electrophoresis, gels are removed from the cassettes and placed in square plastic Petri dishes containing 50 ml of freshly prepared staining solution. Many of the staining reactions for enzyme visualisation are based on histochemical stains which have been extensively modified for use with a wide variety of gel media. Recipes for a wide range of stains may be found in Shaw and Prasad (1970); Harris and Hopkinson (1977); Loxdale *et al.* (1983); Richardson *et al.* (1986); Pasteur *et al.* (1988). However, it must be noted that in many laboratories these basic recipes have been extensively modified to suit individual requirements. Gels stained with reaction mixtures containing either 3-(4,5-dimethylthiazole)-2,5-diphenyltetrazoliumchloride (MTT) or N-methyl-phenozine-methylsulphate (PMS) should be stained in the dark since both compounds are light-sensitive. Other staining can be performed in daylight and at room temperature, although incubation at around 37°C will accelerate the process.

The appearance and intensity of bands on a gel may change during the staining process and so, at least until standardised conditions are established, the appearance, position and intensity of bands should be monitored. Also the isoenzymes responsible for different bands may be present at different levels and/or with different levels of activity in different samples. Therefore, it is important that detailed records of sample preparation, electrophoresis conditions and staining are kept for later use.

Following staining, gels are destained for five minutes in freshly prepared 7% acetic acid and then recorded. Samples are usually recorded in one or more of four ways: a scoring diagram, gel preservation, photography (Figure 4.22) and tracing. Light-sensitive gels should be scored immediately and discarded whereas all other types of gel can be scored and then stored refrigerated in sealed, evacuated polythene bags. In scoring gels, the number and mobility of bands produced is

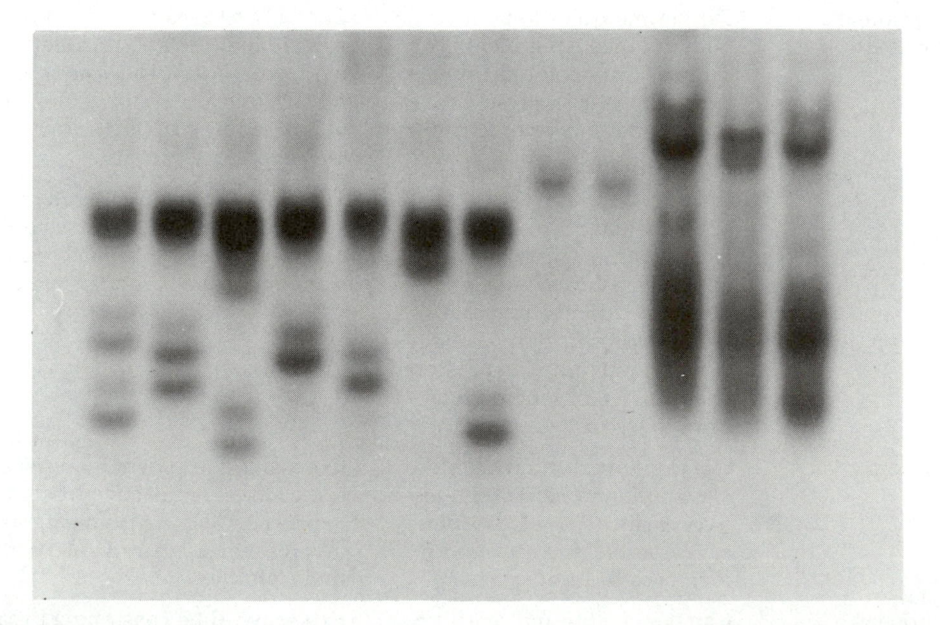

Figure 4.22 Electrophoretic gel, showing esterase banding patterns of three parasitoid species attacking the grain aphid, *Sitobion avenae*. Left to right: 1-7 *Aphidius picipes*; 8-9 *Ephedrus plagiator*; 10-12 *Praon volucre*. (Source: Powell and Walton, 1989.) Reproduced by permission of Oxford University Press.

recorded. This involves measuring the migration distance of individual bands (in mm) from the gel origin. A mobility ratio (MR) can then be calculated by comparison with a band of standard mobility.

Some Applications of Electrophoresis

Determination of Dietary Range

Murray and Solomon (1978) used PAGE gradient gels to determine dietary range in several mite and insect predators, including the bugs *Anthocoris nemoralis, Orius minutus* and *Blepharidopterus angulatus*. Giller (1982, 1984, 1986) used disc gel electrophoresis in examining the dietary range of the water boatman, *Notonecta glauca*. Giller concluded that provided 'diagnostic' bands were available for prey, the method was a rapid means of determining dietary range. Van der Geest and Overmeer (1985) evaluated the technique for predaceous mite-prey studies. They found that different prey spider mite species could be detected in predators, but that there were problems with the detection of eriophyid and tydeid mites. With the eriophyids, 30 to 40 mites have to be present in a predator's gut for detection to be successful, whereas with the tydeids the esterase patterns were not very distinct and the patterns sometimes overlapped with those of the predator. In a later study, Dicke and De Jong (1988) determined prey range and measured the preference of the phytoseiid mite *Typhlodromus pyri* for different mite prey. Electrophoresis could, in principle, also be used to determine dietary range in host-feeding and pollen-feeding adult parasitoids (Jervis *et al.*, 1992a).

Detection and Identification of Parasitoids in Hosts

Electrophoresis is proving very useful for the detection and identification of parasitoids in hosts. Wool *et al* (1978) used PAGE to detect the presence of *Aphidius matricariae* (Braconidae: Aphidiinae) in a glasshouse population of the aphid *Myzus persicae*. These authors obtained the most promising results when staining for the enzymes esterase (EST), malate dehydrogenase (MDH) and malic enzyme (ME) since all these enzymes gave diagnostic banding patterns for parasitised aphids. MDH was later used on starch gels to assess levels of parasitism in field populations of the rose aphid *Macrosiphum rosae* (Tomiuk and Wöhrmann, 1980). A more comprehensive study of parasitism of cereal aphids was carried out by Castañera *et al.* (1983) who examined the enzyme banding patterns produced by the parasitoid *Aphidius uzbekistanicus* in laboratory populations of the grain aphid *Sitobion avenae*. They tested 14 enzyme systems and agreed with Wool *et al.* (1978) that the most useful systems were EST, MDH and ME, but also obtained reasonable results with glutamate oxaloacetate transaminase (GOT). This work was followed up by Walton *et al.* (1990) who used EST to monitor parasitism in field populations of *S. avenae*.

EST, aconitase (ACON), leucine amino peptidase (LAP) and 6-phosphoglucose dehydrogenase (6PGDH) proved useful for detecting the endoparasitoid *Glypta fumiferanae* (Ichneumonidae) in larvae of the western spruce budworm *Choristoneura occidentalis* (Castrovillo and Stock, 1981). In this case, ACON and 6PGDH produced monomorphic bands, whereas EST and LAP were polymorphic. For the routine screening of host populations in order to assess percentage parasitism, enzyme systems giving single host bands are quicker and easier to interpret (Castrovillo and Stock, 1981).

Wool *et al.* (1984) looked at six enzyme systems for detecting the endoparasitoids *Encarsia lutea* and *Eretmocerus mundus* (Aphelinidae) in the whitefly *Bemisia tabaci*. EST again produced the best results, although weakly staining bands were also produced using MDH and aldolase (AO).

Höller and Braune (1988) used isoelectric focusing to assess percentage parasitism of

Sitobion avenae by *Aphidius uzbeckistanicus*. Both host and parasitoid showed specific bands for MDH, so allowing clear detection of parasitism.

In order to interpret electrophoretic estimates of percentage parasitism, it is necessary to know which developmental stages of the parasitoid can be reliably detected by the technique. When establishing which enzyme systems were best for detecting parasitism of *S. avenae*, Castañera *et al.* (1983) deliberately used aphids containing final-instar larvae, but they stressed the importance of determining the earliest detectable stage of the parasitoid for the purposes of field monitoring. In laboratory studies, Walton *et al.* (1990), using three parasitoid species, *A. rhopalosiphi*, *Praon volucre* and *Ephedrus plagiator*, failed to detect the egg stage and they concluded that estimates of percentage parasitism should be corrected to take this into account.

In most cases, comparison of results using electrophoresis with those from mummy-counting and host rearing have shown electrophoretic estimates to be very similar to estimates obtained from host rearing. This is to be expected, since both methods sample the same parasitoid developmental stages and both involve the removal of live aphids from the field and hence from the risk of further parasitism. However, Höller and Braune (1988) compared estimates of percentage parasitism by *A. uzbeckistanicus* obtained with isolectric focusing with measurements obtained from the rearing of field-collected hosts, and found poor agreement between the two methods. The possible reasons for this discrepancy are discussed in their paper.

As a sampling tool in the detection and measurement of parasitism in insect pest populations, electrophoresis has a number of obvious advantages over rearing and host dissection. Parasitoids can be detected at an early stage in their development and the species involved can usually be identified without the need to rear parasitoids through to the adult stage. This enables accurate measurements of parasitism to be obtained very rapidly, so enhancing their value in forecasting and decision-making in pest management. Large numbers of host individuals can be processed within a short time, and this facility is likely to increase in the future as improvements in the design and operation of commercial equipment take place. Walton (1986), in screening cereal aphid populations, was able to test around 100 aphids per day when staining for esterase activity alone. In their evaluation of electrophoresis, Castrovillo and Stock (1981) established that a single technician could process 150 budworm larvae per day.

In studies concerned with the detection and measurement of parasitism, it is necessary to identify those enzyme systems which produce distinctive banding patterns for host and parasitoid. In some cases diagnostic parasitoid bands may have similar mobilities to bands produced by the host, in which case the parasitoids will be difficult to detect. The existence of enzyme systems with clearly defined parasitoid bands which can easily be distinguished from those of the host allows the rapid detection of parasitoid immature stages within a sample of hosts. In many cases there is the added advantage of parasitoid species identification based on the banding patterns.

Taxonomic Characterisation and Separation

Electrophoresis is also being used in taxonomy to discriminate between parasitoid species. For example, Castañera *et al.* (1983) examined the adults of five aphidiid wasp species (*Aphidius rhopalosiphi*, *A. picipes*, *Ephedrus plagiator*, *Praon volucre* and *Toxares deltiger*) associated with the cereal aphid *S. avenae*, and established that they could be identified on the basis of their EST banding patterns. Distinctive MDH patterns were also obtained by Tomiuk and Wöhrmann (1980) for two aphidiid parasitoids of the rose aphid *Macrosiphum rosae*: *Aphidius rosae* and *Ephedrus* species.

Höller *et al.* (1991), using isoelectric focusing with EST, found good supportive evidence for the existence of a species closely related to *Aphidius rhopalosiphi*.

The use of electrophoresis in solving taxonomic problems relies upon the fact that even closely related insect species may have homologous proteins which, over the course of time have evolved different amino acid sequence structures. If these proteins have different net charges, they will separate under the influence of an electric field. Unfortunately, complete discrimination of all protein/enzyme bands does not necessarily occur, since polypeptides with different amino acid sequences may possess the same net charge and consequently display the same electrophoretic mobility (Berlocher, 1980, 1984). This means that enzyme bands separated electrophoretically effectively represent only phenotypes and not genotypes. Hence the term **electromorph** (a polypeptide characterised by its electrophoretic mobility) is often preferred to **allozyme** (defined as any of one or more variants of an enzyme coded by different alleles at the same gene locus), especially when the genetic basis of the bands seen on the gel is complex and not easily elucidated by formal crossing experiments.

Isoenzymes are alternative forms of an enzyme produced by different loci and may exist in a number of different allozyme states. At any given locus with segregating alleles producing monomeric enzymes, homozygotes display either slow or fast (SS, FF) bands whereas heterozygotes (SF) have both bands. Where the coded enzyme is polymeric, (i.e. di-, tri-, tetra-), very complex heterozygous banding patterns may be produced involving multiple bands (five in the case of a tetramer) (Harris and Hopkinson, 1977; Richardson *et al.*, 1986). Any, or all, of such banding patterns may be diagnostic in species recognition (Berlocher, 1980; Ayala, 1983). Additional apparent differences in electrophoretic mobility may also arise from artefacts in the gel, or from post-translational modifications of the enzyme (Ferguson, 1980). Post-translational modification may result from a variety of mechanisms, such as the diet-induced changes found in *Drosophila* alcohol dehydrogenase (Schwartz and Sofer, 1976), enzyme modification occurring during enzyme extraction (Ferguson, 1980) and mutations arising from transposable element insertion. Transposable elements (a defined length of DNA within a sequence where it had not previously been detected) may alter the gene sequence in a manner detectable by electrophoresis (Calos and Miller, 1980) and may influence the level of enzyme variation within natural populations (Burkhart *et al.*, 1984). Such post-translational modifications may be a potential source of the considerable enzyme polymorphism reported in the literature and could, as a consequence, affect the usefulness of banding pattern differences in species discrimination (Finnerty and Johnson, 1979). However, other authors such as Coyne (1979) and Tsuno *et al.* (1984) maintain that the frequency of modifications within natural populations is probably low.

Preparation of Insect Material for Analysis

To minimise modification of enzyme structure, organisms should be frozen as quickly as possible after death, and once frozen should not be allowed to thaw out until ready for processing. Although Castañera *et al.* (1983) found storage at –25°C had no discernible effect on enzymatic properties over a short period, samples ideally should be stored at, or below, –50°C either in an ultra-low temperature freezer or under liquid nitrogen.

Sampling programmes are usually designed to allow the maximum number of loci of prey, hosts and parasitoids to be scored at a later date, and so often require the collection of large numbers of individuals. When dealing with larger insects, it is advisable to dissect out and freeze-store specific tissue, rather than the whole organism, for analysis. With small insects, concentration on specific tissues is likely to prove impracticable if not impossible. The advantage of concentrating on a specific tissue is that it minimises the introduction of different allozymes, as different tissues vary qualitatively in their allozyme content. It is generally desirable to perform pilot studies to determine which tissues should be sampled in any given study; in gut content analysis, such studies might involve determining which parts of the prey the predator actually ingests.

For the purpose of detecting prey remains in the guts of large predators, the gut or gut region should be excised and macerated in buffer to produce an homogenate (preferably the fore gut should be used; Giller, 1984, investigated rates of digestion and found that prey material could be readily distinguished in the fore gut (or crop) of *Notonecta glauca* but not after it has reached the mid gut, where most digestion takes place). With small predators, removal or separation of the gut are impracticable, and therefore a whole-body homogenate is prepared. To minimise enzyme degradation within a sample (especially important when dealing with very small amounts of tissue or small insects), it is advisable to place the homogenising block in a tray of crushed ice, thus keeping the sample as cool as possible. Usually, insects are homogenised with the aid of a flame-sealed Pasteur pipette or, in the case of large, whole insects, a solid glass rod with a ground glass tip. However, if dealing with a large number of samples, a variety of multiple homogenising devices have been designed (Brookes and Loxdale, 1985). Because of the difficulty of homogenising very small insects such as aphids and whiteflies, the insects can, as an alternative to homogenisation, be individually squashed with a plastic rod in the middle of a narrow strip of filter paper and then placed in a pocket of the gel (Wool *et al.*, 1984). Sample preparation is discussed in detail in Richardson *et al.* (1986).

When analysing the gut contents of predators, an electrophoretic run should be carried out using starved predators, in order to take account of any bands of predator origin that may show up in the gels obtained from the field-caught individuals. If one is interested in determining the dietary range of a predator species, then the suspected prey species need to be individually characterised electrophoretically.

For the detection and identification of parasitoids in hosts, field-collected whole hosts can be homogenised in buffer, and a sample of the homogenate used for the analysis. As noted above, the electrophoretic banding patterns of host individuals known to be unparasitised need to be examined so that any bands of host origin appearing in gels obtained from field-collected hosts can be taken account of. A necessary prequisite for the detection and identification of parasitoids in hosts is, of course, the electrophoretic characterisation of each suspected parasitoid species.

4.4 GENETIC VARIABILITY IN FIELD POPULATIONS OF NATURAL ENEMIES

Individuals within a population usually vary genetically, and this variation is often expressed both in the insects' morphology and in a range of biological attributes such as behaviour (Roush, 1989). Genetic variation may thus have considerable influence on the killing efficiency of predators and parasitoids. Genetic variability appears to be the rule rather than the exception within animal populations (Ayala, 1983). For example, studies of proteins using electrophoretic techniques have shown that very high levels of allelic genetic variation exist in many populations (McDonald, 1976). Relative assessments of genetic variability within populations can be obtained using electrophoresis, and allozyme frequencies and levels of heterozygosity can be used to compare different populations (Unruh *et al.*, 1989).

Most populations are patchily distributed due to discontinuities in the distribution of food or other essential resources. When investigating the population dynamics of a predator or parasitoid, it is important to determine whether one is dealing with a local population or a metapopulation (subsection 5.3.10). In a local population, although its distribution may be discontinuous, individuals regularly move between patches and the population is largely redistributed every generation (Taylor, 1991). In other words, there is considerable interaction between individuals throughout the population. A metapopulation consists of a series of local populations between which movements of individuals are infrequent and random, so that the local populations have largely independent dynamics (Taylor, 1991). Hence, there is a high rate of genetic exchange between groups of individuals occupying separate habitat patches within a local population but a low rate of genetic exchange between local populations. Electrophoretic analysis of allozyme frequencies within samples of individuals collected from discrete habitat patches, or geographic areas, could provide indirect evidence of the levels of genetic exchange between populations. Identical or very similar allozyme frequency scores would suggest high rates of genetic exchange and *vice versa*.

However, some caution is necessary because differences in allozyme frequencies may not always be indicative of low rates of movement between two populations. To illustrate this, let us consider a hypothetical parasitoid attacking three different host species which occupy different local habitats within the same area (e.g. an aphid parasitoid attacking pest aphids on two different crops and a non-pest aphid on plants in neighbouring hedgerows). When examined electrophoretically, allozyme frequencies are consistently different between samples of parasitoids collected from different hosts, some allozymes only appearing in samples from one of the hosts. This might suggest that discrete parasitoid populations exist on each host species with very little movement of individuals between hosts. However, the same result would occur if only a proportion of the individuals within the population, those with certain genotypes, moved regularly between host species. Allozyme frequency differences would still be maintained despite high rates of movement between populations. Powell and Wright (1992) have suggested that this situation exists for some oligophagous aphid parasitoids, their populations consisting of a mixture of specialist and generalist individuals. The genotype of a generalist allows it to recognise and attack all the host species, whereas that of a specialist allows it to recognise and attack only part of the host range, possibly a single species. This hypothesis agrees with the concept of **population diversity centres** proposed for aphid parasitoids by Němec and Starý (1984, 1986, who advocate that the 'main', or original host of a parasitoid is the one on which it shows the greatest degree of genetic diversity. On this basis, oligophagous and evolutionarily 'young' parasitoid species will show the greatest genetic variability, whilst monophagous and evolutionarily 'ancient' species will show the least (Němec and Starý, 1983).

Although the existence of intraspecific genetic variation has long been recognised, considerable confusion still exists over terminology, with 'subspecies', 'variety', 'race', 'strain' and 'biotype' all being widely used but rarely defined clearly (Berlocher, 1984). Gonzalez *et al.* (1979) proposed that the term **biotype** alone should be used to designate genetic variants in parasitoids, but Claridge and den Hollander (1983) argue that the term has no biological basis and is therefore redundant. Because of this confusion in terminology, it is obviously important to describe and quantify any intraspecific variability with precision, and electrophoresis appears to be a very useful tool in this respect.

Genetic variability has considerable implications when natural enemies are being used as pest control agents. The example of the aphid parasitoid *Trioxys pallidus* has already been described in subsection 2.9.3. More recently in California, the same parasitoid species has been used to control the filbert aphid, *Myzocallis coryli*, but a new variant had to be introduced from Europe because the original introduced population would not attack *M. coryli* (Messing and Aliniazee, 1988). Biochemical techniques such as electrophoresis can be useful for distinguishing between morphologically identical genetic variants (Němec and Starý, 1985; Unruh *et al.*, 1986; Smith and Hubbes, 1986).

Powell and Walton (1989) review the use of electrophoresis in the study of intraspecific variability in hymenopteran parasitoids whilst Roush (1989, 1990) and Hopper *et al.* (1993) discuss genetic considerations in the use of entomophagous insects as biological control agents.

Recently, other techniques based on DNA analysis (e.g. restriction site analysis), have been employed in population studies. At present, the large number of individuals and loci required for the majority of intraspecific studies have made sequencing techniques prohibitively expensive. Nevertheless, recent developments in the precision and power of molecular techniques will almost guarantee them an ever increasing importance in future research (Loxdale *et al.*, 1985).

ACKNOWLEDGEMENTS

We wish to thank Bill Symondson for suggesting useful improvements to the section on trophic relationships, in particular our account of serological methods; Brad Hawkins for sharing his thoughts concerning parasitoid communities; Hugh Loxdale, Nigel Stork and Anja Steenkiste for providing photographic material; Jan Cawley and Vyv Williams for photographic assistance; and Annette Walker for taxonomic advice.

POPULATION DYNAMICS 5

N.A.C. Kidd and M.A. Jervis

5.1 INTRODUCTION

The reasons for studying the population dynamics of insect natural enemies are basically two-fold. First, predators and parasitoids are an important component of terrestrial communities (LaSalle and Gauld, 1994), so they are of central interest to the ecologist who attempts to unravel the complexity of factors driving the dynamics of species interactions. Second, the knowledge gained from a study of natural enemies may be of immense practical value in insect pest management. The manipulation of natural enemies for this purpose has a long pedigree (DeBach and Rosen, 1991), but it is only relatively recently that population dynamics theory has been brought to bear seriously on 'biological control' practice.

In this chapter, we aim to demonstrate how the role of natural enemies in insect population dynamics can be assessed, and how this information can be put to use by biological control workers. We begin by reviewing methods for demonstrating and quantifying predation and parasitism (section 5.2). We then examine the different techniques for determining the effects of natural enemies on insect population dynamics (section 5.3). Finally, we look at ways in which this and other information can be used in the selection of biological control agents (section 5.4). The reader should note that we make no attempt to provide a comprehensive review of insect population dynamics – that would require a book in itself. Thus, there are a number of topics which we either do not discuss or make only brief reference to (e.g. chaos theory, ratio-dependence). In most cases, the reasons for omission are simply that the topic has either received adequate coverage in other texts, or that it is not specifically related to natural enemy problems.

5.2 DEMONSTRATING AND QUANTIFYING PREDATION AND PARASITISM

5.2.1 INTRODUCTION

In this section we are concerned with a variety of techniques that can be applied to field, and in some cases laboratory, populations of natural enemies and their prey for either of the following purposes:

1. to demonstrate that natural enemies have significant impact upon host and prey populations (subsections 5.2.3 to 5.2.9);
2. to measure rates of predation and parasitism in the field and/or the laboratory and to provide indices of predation and parasitism (subsections 5.2.10 to 5.2.15).

The techniques used for 1. and 2. provide respectively: a preliminary assessment of the impact of parasitoids and predators on host and prey populations, and quantitative information which can be used to further our understanding of the role of natural enemies in insect population dynamics. In the latter case, further methodologies are required for the implementation of this objective; these are discussed in detail in section 5.3.

We begin by discussing the introduction of natural enemies in classical biological control.

Strictly speaking, this is not a technique *per se* for demonstrating that natural enemies have a significant impact on host/prey populations. However, we include it because: (a) it can provide dramatic results, and (b) it can be simulated in the laboratory.

5.2.2 NATURAL ENEMY INTRODUCTIONS IN CLASSICAL BIOLOGICAL CONTROL

Some of the best demonstrations of the effectiveness of natural enemies are provided by cases of so-called 'classical' biological control. Classical biological control may be defined as the importation of a specific natural enemy into a geographic region where an exotic insect species has previously become established without its adapted natural enemy complex and, in the absence of effective local natural enemies, has become a pest. For a classical biological control programme to be completely successful, the natural enemy has to reduce the pest populations to a level where the latter no longer inflict economic damage. In ecological terms, the natural enemy is required both to depress the pest population below a certain level and to prevent it from again reaching that level by promoting stability (Figures 5.1 and 5.2b) (for a dissenting view regarding the need for stability in natural enemy-pest population interactions, see Murdoch *et al.*, 1985 and section 5.4). Successful biological control resulting from parasitoid introduction can be simulated in the laboratory, as shown in Figure 5.2b (also Figure 5.3).

Classical biological control programmes can be viewed as ecological experiments on a grand scale, allowing us to compare under field conditions the population dynamics of insect species in the presence and absence of natural enemies. As pointed out by Waage (1992), there is no fundamental difference between the successes achieved using exotic natural enemies in classical biological control and the action of indigenous species. Classical biological control simply isolates a process that is taking place around us all the time. Thus, studies of classical biological control introductions can shed considerable light on

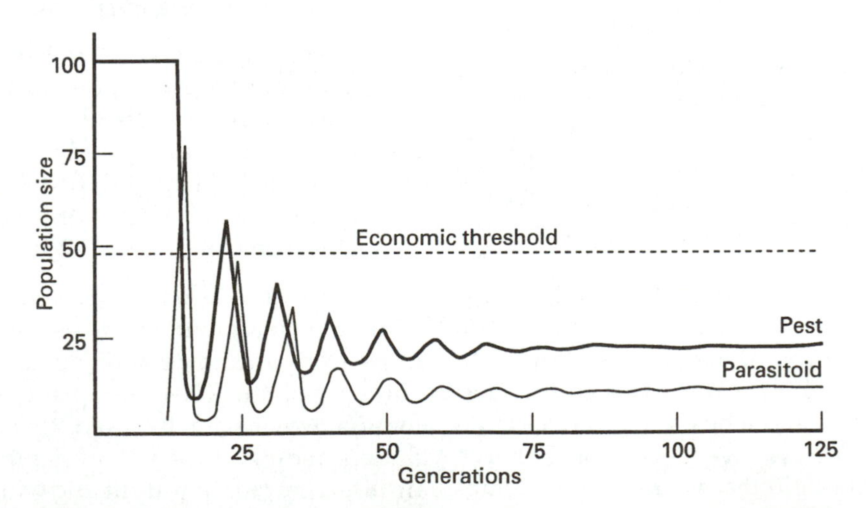

Figure 5.1. Hypothetical example of successful control of a pest population by an introduced parasitoid or predator. In this case, the pest population in the first generation is at an outbreak level (100), well above the economic threshold. The predator or parasitoid is introduced after 10 generations. As its numbers increase, the pest population declines in abundance. Initial oscillations in both populations decrease with time to a stable equilibrium at which the pest population is depressed to well below the economic threshold. (Source: Greathead and Waage, 1983.) Reproduced by permission of The World Bank.

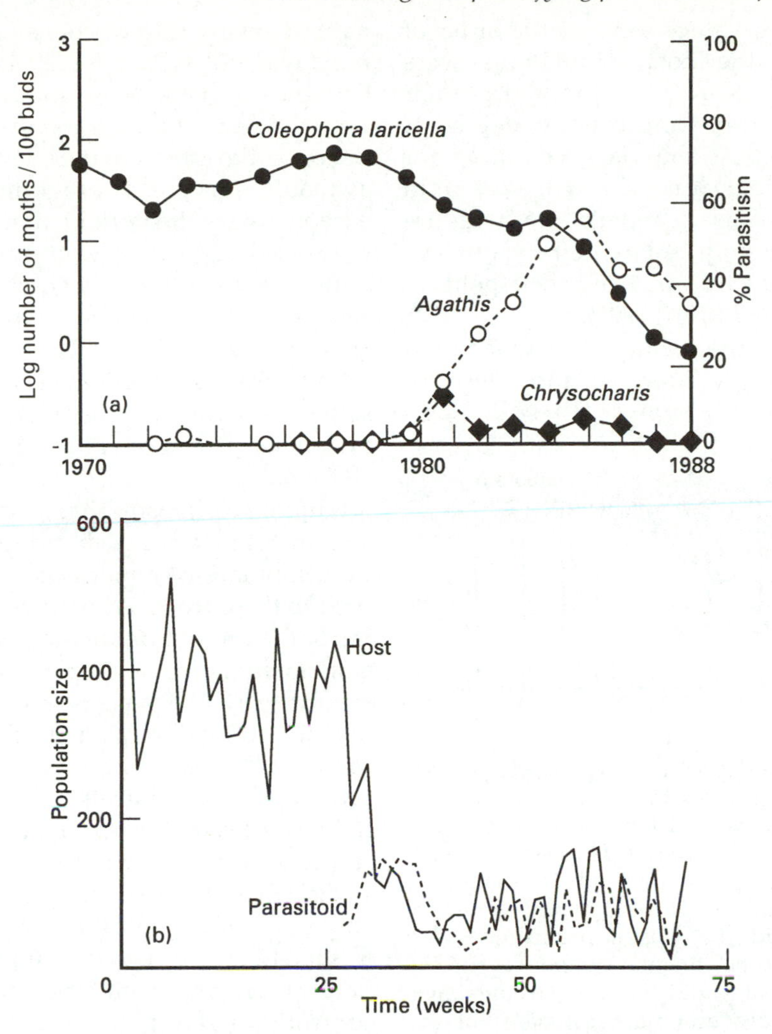

Figure 5.2. Examples, one from the field, the other from the laboratory, of the successful biological control of an insect pest following the introduction of parasitoids: (a) two parasitoid species, *Agathis pumila* (Braconidae) and *Chrysocharis laricella* (Eulophidae), were introduced into Oregon, USA, against the larch casebearer, *Coleophora laricella* (Lepidoptera). The Figure shows the combined data for 13 plots over 18 years; (Source: Ryan (1990) (b) the pteromalid *Anisopteromalus calandrae* was introduced into a laboratory culture of the bruchid beetle *Callosobruchus chinensis* 26 weeks after the beetle culture was started. (Source: May and Hassell, 1988.) (a) Reproduced by permission of The Entomological Society of America, (b) by permission of The Royal Society.

'natural control' (Solomon, 1949) by predators and parasitoids.

The degree to which a host population may be reduced in abundance by an introduced parasitoid was examined by Beddington *et al.* (1978), who used a simple measure, $q = N^*/K$, where N^* is the average abundance of the host in the presence of the parasitoid (i.e. post-introduction), and K is the average abundance of the host prior to introduction of the parasitoid. Beddington *et al.* calculated q-values for six different field parasitoid–host systems (cases of successful biological control) and four laboratory systems. Figure 5.3 shows the calculated q-values to be of the order of 0.01; that is, the host populations were depressed to about one hundredth their former abundance. Note that the degree of depression of pest abundance required for successful biological control will vary from case to case, because economic thresholds are determined not only by pest density, but also by pest impact, crop value and socio-political value (Waage and Mills, 1992).

In the past, the value of detailed and precise quantitative data on the effects of natural enemy introductions on pest populations was not fully appreciated (Waage and Greathead, 1988; May and Hassell, 1988), with the result that often only anecdotal evidence on the effects of introductions has been available for some programmes. Notable exceptions are the introduction of the tachinid parasitoid *Cyzenis albicans* to control the winter moth in Canada (Embree, 1966), the release of two hymenopteran parasitoids against the larch casebearer in the USA (Ryan, 1990) (Figure 5.2a) and the release of *Encarsia partenopea* against the whitefly *Siphoninus phillyreae* in the USA (Gould *et al.*, 1992a,b).

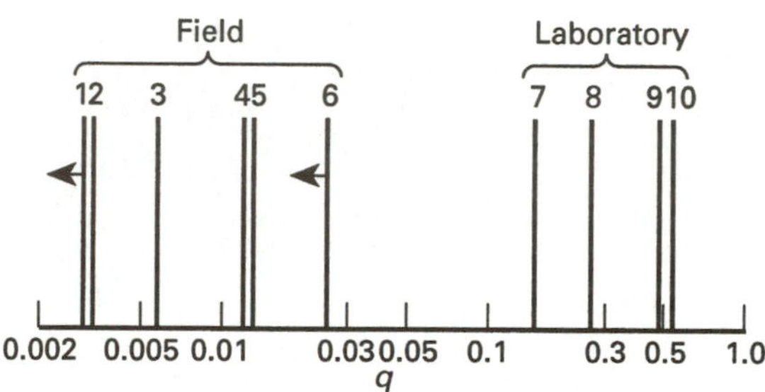

Figure 5.3. The degree to which a pest population may be depressed by an introduced parasitoid, in six field and four laboratory parasitoid — host systems. The degree of depression in each case is expressed as a q-value, q being defined as the average abundance of the host in the presence of the parasitoid (i.e. post-introduction = N^*) divided by the the average abundance in the absence of the parasitoid (i.e. pre-introduction = K). Arrows imply minimum estimates of the degree of depression. Note that doubts have recently been cast upon the role of *Cyzenis albicans* in the direct control of the winter moth in Canada (subsection 5.3.4). Host—parasitoid systems: (1) *Aonidiella aurantii* — *Aphytis melinus*; (2) *Chromaphis juglandicola* — *Trioxys pallidus*; (3) *Parlatoria oleae* — *Aphytis maculicornis*; (4) *Aonidiella aurantii* — *Aphytis melinus*; (5) *Operophtera brumata* — *Cyzenis albicans*; (6) *Pristiphora erichsonii* — *Olesicampe benefactor*; (7) *Musca domestica* — *Nasonia vitripennis*; (8) *Callosobruchus chinensis* — *Neocatolaccus mamezophagus*; (9) *Anagasta kuehniella* — *Venturia canescens*; (10) *Callosobruchus chinensis* — *Heterospilus prosopoidis*. (Source: Beddington *et al.*, 1978.) Reproduced by permission of Macmillan Magazines Ltd.

Another criticism that can be aimed at some programmes is that depression of the pest population cannot necessarily be attributed to the introduced natural enemy; that is, introduction and depression may be merely coincidental. For example, in programmes involving whitefly pests, reductions in pest density following parasitoid release were reported but the workers concerned failed to provide proper controls to demonstrate that the introduced natural enemies were indeed responsible for the depression (Gould *et al.*, 1992a). Even where experimental controls are employed in biological control programmes, it is likely that, due to the rapid spread of the natural enemy, comparisons of test and control plots are possible for a brief period only. This problem arose with the monitoring of releases of *Encarsia partenopea* against the whitefly *Siphoninus phillyreae* in California (Figure 5.4). Parasitoids were released in May, and by midsummer had appeared at all control sites) (4–11 km away from the nearest release sites (Gould *et al.*, 1992a)). A problem of this type could be difficult to overcome, since too wide a separation of control and test sites makes the validity of comparisons questionable

Nowadays there is an increasing awareness of the need for detailed and precise quantitative data on the effects of introductions, and classical biological control programmes are tending to be much more carefully documented through the routine collection of population data. However, this usually applies only to biological control programmes with good funding and well-trained staff. Another constraint upon the gathering of pre- and post-release population data is the great rapidity with which many pest problems arise, recent examples of insects very rapidly becoming serious pests being the mealybug *Rastrococcus invadens* in west Africa, and the psyllid *Heteropsylla cubana* in the Pacific region, Asia and elsewhere.

Traditionally, most quantitative studies measure population density over a number of seasons before and after release (section 5.3). Recently, however, detailed studies of within-season changes in population density (Gould *et al.*, 1992b) and also of changes in pest age structure immediately following parasitoid introduction (Figure 5.4) have been carried out.

At present, there is no standard protocol (at least not a sufficiently detailed one; see Neuenschwander and Gutierrez, 1989, and Waterhouse, 1991) for the quantification of the impact of natural enemies in classical biological control programmes. Such a protocol, if developed, would probably be restrictive, given the diversity in ecology that exists among insect pests (J.K. Waage, personal communication).

5.2.3 EXCLUSION OF NATURAL ENEMIES

Exclusion methods have been widely employed in assessing the impact of insect natural enemies on host and prey populations under field conditions. The principle behind their use is that prey populations in plots (any habitat unit, from part of a plant to a whole plant or a group of plants) from which natural enemies have been eradicated and subsequently excluded will, compared with populations in plots to which natural enemies are allowed access: (a) suffer lower predator-induced mortality or parasitism, and (b) if the experiment is continued for a long enough period, increase more rapidly and reach higher levels. The results of some exclusion experiments are shown in Figure 5.5.

Usually, the starting densities of prey are made equal in both the test and the control plots (for consistency's sake, we refer here to the exclusion plots as 'test' plots, and the non-exclusion plots as 'control' plots, since what is being tested is the effect of excluding natural enemies, not of including them; not all authors use the same nomenclature). Exclusion experiments may be conducted for periods of several days to several weeks. A long experimental period will be required if test-control differences in prey equilibrium densities (section 5.3) are to be compared.

Various exclusion devices have been employed; they include mesh cages placed over individual plants or groups of plants (Figure 5.6), mesh cages (sleeves) placed over branches or leaves, clip cages attached to leaves, greased plastic bands tied around tree branches and trunks, and vertical barriers (walls constructed of polythene, wood or hardboard) around plants; for details of construction, consult the references cited below. The precise type of device used will depend upon the natural enemies being investigated, and whether the aim is to exclude all natural enemies (so-called **total exclusion**) or to exclude particular species or groups of species (so-called **partial exclusion**). For example, a terylene mesh/gauze cage placed over a plant ought, if the mesh size is sufficiently small, to exclude all aerial and surface-dwelling insect natural enemies, from the largest to the smallest. By increasing the mesh size slightly, small hymenopteran parasitoids may be allowed in, while increasing the mesh size further will allow larger types of natural enemy to enter also, and so on. By first examining the ability of tiny *Anagrus* parasitoids (Mymaridae) to pass through terylene meshes of different mesh sizes, colleagues at Cardiff were able to decide on the appropriate size of mesh for excluding all

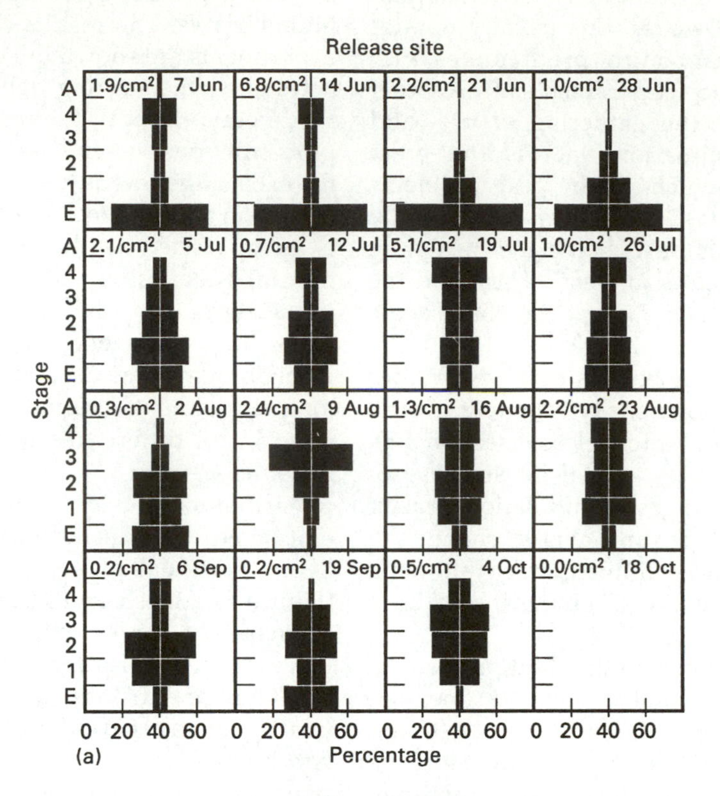
Release site
1.9/cm² 7 Jun
6.8/cm² 14 Jun
2.2/cm² 21 Jun
1.0/cm² 28 Jun
2.1/cm² 5 Jul
0.7/cm² 12 Jul
5.1/cm² 19 Jul
1.0/cm² 26 Jul
0.3/cm² 2 Aug
2.4/cm² 9 Aug
1.3/cm² 16 Aug
2.2/cm² 23 Aug
0.2/cm² 6 Sep
0.2/cm² 19 Sep
0.5/cm² 4 Oct
0.0/cm² 18 Oct
Stage
A
4
3
2
1
E
0 20 40 60
Percentage
(a)

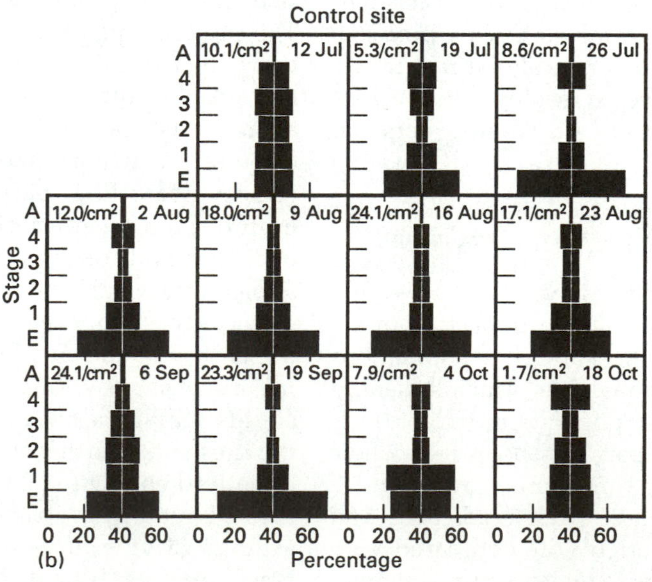
Control site
10.1/cm² 12 Jul
5.3/cm² 19 Jul
8.6/cm² 26 Jul
12.0/cm² 2 Aug
18.0/cm² 9 Aug
24.1/cm² 16 Aug
17.1/cm² 23 Aug
24.1/cm² 6 Sep
23.3/cm² 19 Sep
7.9/cm² 4 Oct
1.7/cm² 18 Oct
Stage
A
4
3
2
1
E
0 20 40 60
Percentage
(b)

Figure 5.4. An example of classical biological control, where changes in pest age structure were monitored. The parasitoid wasp *Encarsia partenopea* (Aphelinidae) was introduced into California for the control of the whitefly *Siphoninus phillyreae*. The whitefly, a pest of ornamental shrubs and fruit trees, was first recorded in the USA in 1988, and has since spread rapidly within California and into other states, Gould *et al.* used several study sites; these were randomly divided into release (i.e. test) and non-release (control) sites. In the former, parasitoids were released in large numbers over a period of several weeks, commencing 10th May. In all sites, densities of the immature stages (eggs, nymphs) and of the adults of the whitefly were monitored, while in the release sites parasitism by *E. partenopea* was monitored. Densities of the pest (shown in top left hand corner of each graph) remained at low levels at the release sites, whereas at the control sites they were increasing by the beginning of the summer. Shown here is information on age structure changes, for one release site (a) and one control site (b). After the parasitoid became abundant at a site (parasitoids eventually dispersed to, and became established in, control sites), the pest population contained a decreasing proportion of young stages, as a result of *E. partenopea* killing, through parasitism, fourth instar whiteflies, so reducing recruitment of eggs to the whitefly population. Observe that in (a) the decline in the proportion of immature stages was much more marked, and took place much earlier than in (b). The initial increase in density of immature whiteflies in (a) can be attributed to oviposition by whitefly females present at the time of parasitoid release. (Source: Gould *et al.*, 1992a.) Reproduced by permission of Blackwell Scientific Publications Ltd.

natural enemies of rice brown planthopper other than the egg parasitoids (Mymaridae and Trichogrammatidae) whose impact upon planthopper populations was being investigated (A. Steenkiste, J.C. Morgan and M.F. Claridge, personal communication). By having a cage with its sides raised slightly above the ground, predators such as carabid beetles and ants may be allowed access to insect prey such as aphids on cereals, whereas adult hover-flies and many types of parasitoid will be denied access. Conversely, a trench or a wall may prevent access to prey by ground-dwelling predators but allow access by aerial predators and parasitoids.

The exclusion devices can be placed around or over already existing populations of prey, in which case the density of the prey at the start of the experiment will need to be recorded. Preferably, prey-free individual plants, plant parts or plots of several plants (any prey already present are cleared by hand removal or by using low-persistence insecticides) can be loaded with set numbers of prey. The latter approach has the advantage that equivalent starting densities of prey/hosts in test and control plots can be more easily ensured, and also any parasitoid immature stages present within hosts can be eliminated from within the test plots. It may also be necessary to employ a systemic insecticide when clearing prey such as leafhoppers and planthoppers from a plot, in order that any prey eggs present within plant tissues are killed; of course, loading with prey cannot take place until one can be sure that the plant is free of the insecticide.

In some cases, simply comparing prey densities on caged plants with prey densities on uncaged plants may produce misleading results, because:

1. prey within the cages may be protected to some extent from the mortality or other deleterious effects of weather factors such as rainfall or wind;
2. In the two treatments microclimatic conditions may be very different. Cages, even ones constructed largely of nylon or terylene mesh, may alter the microenvironment (light intensity, humidity, wind speed, temperature) surrounding the plant (Hand and Keaster, 1967) and thus influence the impact of natural enemies, either: (a) directly by affecting the physiology, the behaviour and consequently the searching efficiency of the predators or parasitoids, or (b) indirectly by affecting

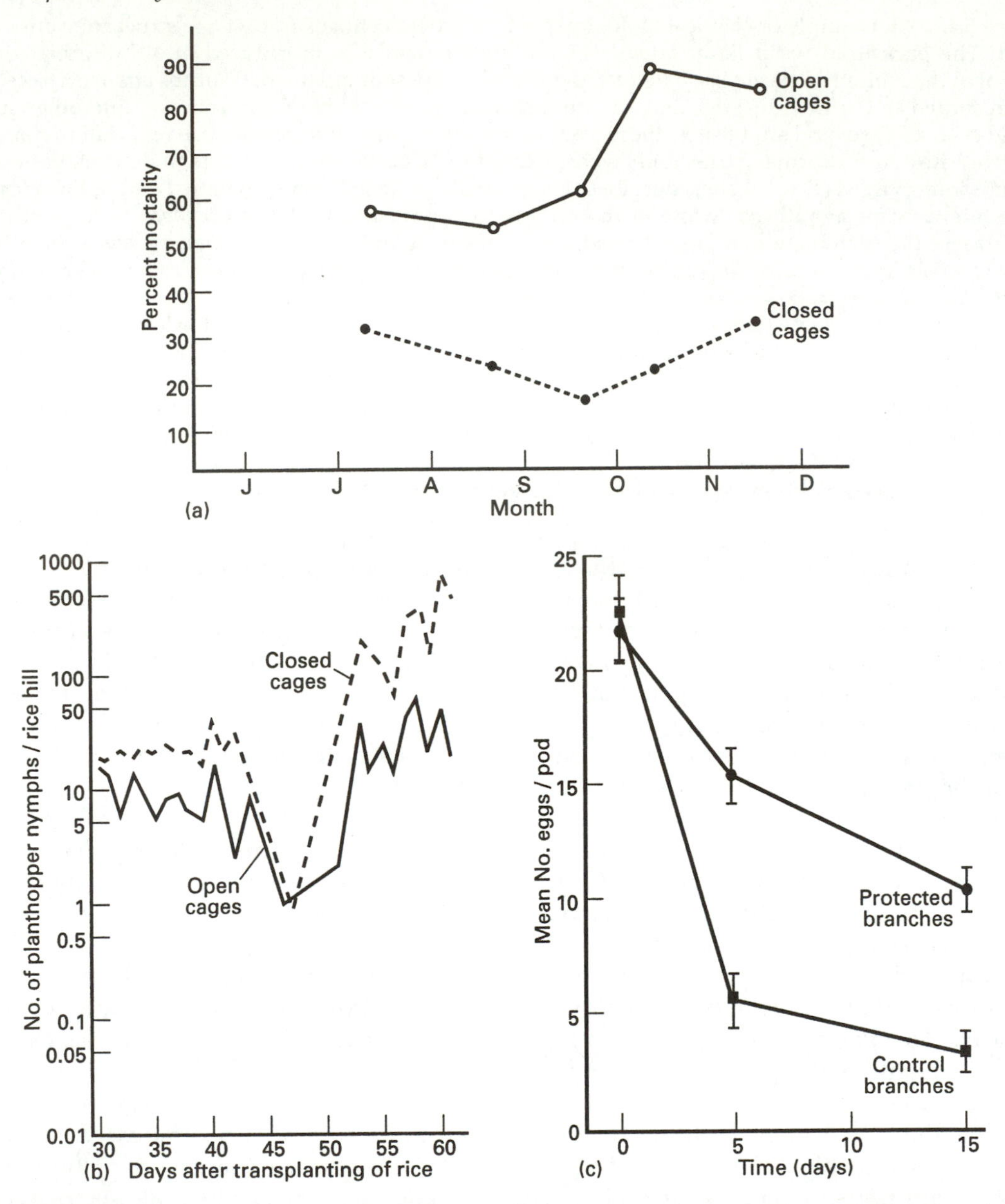

Figure 5.5. Effects of excluding predators from prey: (a) percentage mortality of California red scale (*Aonidiella aurantii*) on *Hedera helix*, in open clip cages that permitted the entry of parasitoids, and in closed clip cages that excluded them; (Source: DeBach and Huffaker, 1971.) (b) population changes in the rice brown planthopper (*Nilaparvata lugens*) in closed and opened field cages after initially removing all the arthropods and stocking with 25 first instar planthopper nymphs per rice plant. Exclusion was not perfect; large numbers of predators were recorded in the 'closed' towards the end of the experiment, so the experiment ought to be considered as an 'interference' experiment; (Source: Kenmore *et al*., 1985.) (c) mean number of bruchid beetle eggs per pod of *Acacia farnesiana* on control and protected branches at days 0, 5 and 15 of the experiment. Protection of branches was achieved by wrapping a 10 cm-wide band of tape around the base and applying a sticky resin to the tape. (Source: Traveset, 1990.) (a) Reproduced by permission of Plenum Publishing Corporation, (b) by permission of The Malaysia Plant Protection Society, (c) by permission of Blackwell Scientific Publications Ltd.

Figure 5.6. Exclusion cages in use in a rice paddy in Indonesia. (Reproduced by kind permission of Anja Steenkiste.)

the behaviour, e.g. spatial distribution, and physiology, e.g. rate of development, longevity, fecundity, of the prey. Changes in prey behaviour and physiology can be brought about by microclimate-induced alterations in plant physiology.

In order to determine whether microclimatic effects on prey are likely to confound the results of an exclusion experiment, the effects of caging on prey population parameters such as fecundity and survival should be investigated. Frazer *et al.* (1981b), for example, investigated whether the observed increase in densities of pea aphid (*Acyrthosiphon pisum*) in exclusion cages (to as much as five times the levels recorded in uncaged plots) was due to an effect of caging upon aphid fecundity. No significant differences in the fecundity of aphids were found between caged and uncaged insects. Furthermore, a simulation model (subsection 5.3.8) showed that for a change in fecundity alone to be responsible for the difference in densities of prey between test and control plots, fecundity would have to have been three times the maximum rate ever observed.

In order to separate the effects of microclimate and natural enemy exclusion upon prey populations, exclusion devices that are either: (a) as similar as possible in construction, or (b) very different in construction but which nevertheless provide similar microclimatic conditions in their interiors, may have to be employed in both test and control treatments, with the obvious proviso that predators need to be allowed adequate access to prey in the control treatment. For example, in assessing the impact of the egg parasitoids *Anagrus* spp. (Mymaridae) and *Oligosita* (Trichogrammatidae) upon planthopper populations, exclusion cages can be constructed that have a very small mesh size to prevent such tiny parasitoids from entering (see above), while almost identical cages with a slightly larger mesh size can be constructed to allow the parasitoids to enter but prevent entry of larger types of natural enemy (Fowler, 1988). In assessing the impact of parasitoids on insect herbivores on trees, gauze sleeve cages can be used on tree branches, the test cages being tied at both ends to exclude parasitoids, and the control cages being left open at both ends to allow parasitoids to enter (DeBach and Huffaker, 1971). However, insects such as hoverflies are deterred from ovipositing on branches in open-ended sleeves (Way and Banks, 1968). Way and Banks used rather dissimilar test and control cages in controlling for the effects of microclimate. The test cages had walls of terylene mesh, whereas the control cages had walls of wooden slats. Despite the major difference in construction, microclimate was similar in the two cage types.

One solution to the problem of achieving a closely similar cage design in the different treatments is not to bother providing natural enemies with access routes to the interior of the control cages, but to carry out an exclusion/inclusion experiment. Such an experiment would involve the use of identical cages in the two treatments and the caging of a known number of predators and parasitoids with prey/hosts in the 'control' treatment (Lingren *et al.* 1968). This type of experiment has the added advantage that the densities of natural enemies will be more precisely known. *Per capita* predation and parasitism rates can be calculated (Dennis and Wratten, 1991) and provided the densities used reflect those normally recorded in the field (this includes taking account of aggregative responses; Dennis and Wratten, 1991), useful estimates of *per population* rates of predation and parasitism can be obtained. A major disadvantage of exclusion/inclusion experiments is that in the control cages the dispersal of natural enemies is likely to be severely restricted or prevented. Long-distance approach behaviour of foraging predators and parasitoids to prey and hosts e.g. in response to kairomones (subsection 1.5.1), may also be interfered with;

3. If the prey are mobile, both immigration and emigration of prey/hosts may be different between the test and the control treatments (restricted or prevented altogether in the test treatment, normal in the control). In order to rule out the possibility that aphid population numbers in fully-caged cereal plots were augmented as a result of emigrant alatae re-infesting the plots, Chambers *et al.*(1983) removed all alate (winged) aphids that settled on the insides of some of the test cages whilst allowing the aphids to remain in another. Removal of alatae was found not to alter the pattern of population change in the cages. Therefore, re-infestation of shoots inside cages was unlikely to have been a cause of the cage/open plot differences in population numbers observed by Chambers *et al.* in their study (Figure 5.7a).

Exclusion methods have a number of other important potential limitations:

4. Even where the microclimate is the same in different treatments, it may be so different from ambient conditions that prey/host populations are severely affected, and any results obtained bear little relation to natural processes. The effects of caging upon microclimate can be assessed using instrumentation of various kinds (see Unwin and Corbet, 1991, for a review). If caging is found to affect microclimate significantly, then it may be possible to provide some means of ventilation, e.g. an electrical fan, to maintain ambient temperature and humidity. Effects on light intensity may be minimised by choosing the appropriate type of screening material;
5. Whilst it is often possible to establish whether a particular **guild** (a group of

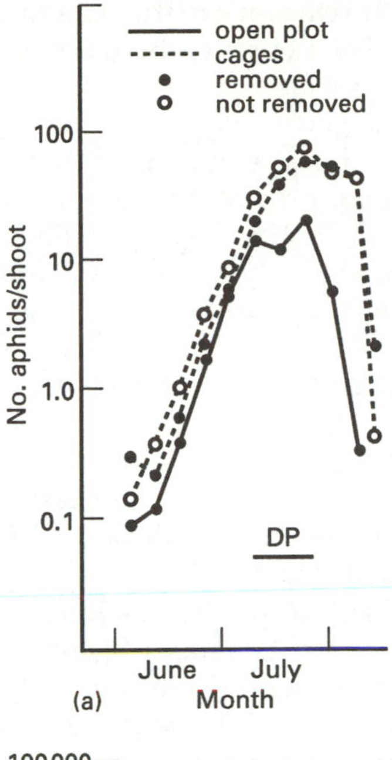

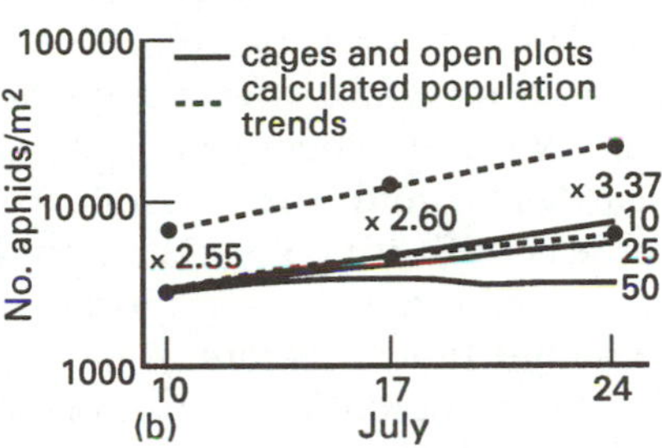

Figure 5.7. (a) Total aphid numbers in two cages where alate cereal aphids were removed or not removed from the cage roof and walls, compared with the adjacent open plot (DP denotes the period of divergence between treatments); (b) aphid populations in cages and open plots, and calculated population trends for different *per capita* predation rates; also shown is the difference, expressed as a multiple, between populations in cages and in open plots during the divergence period (DP). Note log scales. (Source: Chambers *et al.*, 1983.) Reproduced by permission of Blackwell Scientific Publications Ltd.

species attacking the same host or prey stage(s)) of natural enemies has a significant impact on prey populations, it may not be possible to determine which particular species of that guild is mainly responsible for the effect. Unless direct observations shed light on which species is responsible, information will therefore be needed on the relative abundance of different natural enemy species within a locality. Where the immature stages of parasitoids can be identified to genera or species, dissections of hosts (subsection 4.2.9) in the control plots may allow determination of the parasitoid species that usually contributes most to parasitism and whether an increase in host numbers in the test plots is due to the exclusion of that species. The problem of attribution of predatory or parasitic impact is a minor one where the natural enemy complex is known to comprise only one or two species in a locality;

6. If, in the test plots, prey numbers (e.g. of aphids) increase, they may do so to such an extent that predator species (e.g. coccinellids, hover-flies) other than the ones that are excluded (e.g. carabid beetles) are attracted preferentially into the test plots, i.e. through an aggregative response (subsection 1.14.7) by the predator or parasitoid. The impact of the excluded natural enemy species may thus be underestimated. This limitation also applies to the use of barriers and trenches, where the enclosed plants are exposed to invasion by a variety of aerial predators.
7. Whilst exclusion methods can reveal that natural enemies have a significant impact upon prey populations, other methods generally need to be applied before the predator–prey interaction can be properly quantified. The results need to be related to the density of predators present in the habitat, if realistic estimates of predation rates are to be obtained. Exclusion experiments provide minimal information, if any, on the *dynamics* of the predator–prey or parasitoid–host interaction, a limitation that applies also to several of the methods described below;
8. One hundred per cent exclusion of natural enemies is sometimes difficult to achieve,

with the result that zero predation or parasitism in test plots is not recorded (e.g. see Kenmore *et al.*, 1985). Either during or at the end of an exclusion experiment, it is important to check for the presence of natural enemies in the test plots (see caption to Figure 5.5b), and to count the numbers of any such insects that have succeeded in gaining entry to the latter. Exclusion methods employing devices that are far from 100% effective in excluding natural enemies are, strictly speaking, interference methods (see below).

Other serious problems that may be encountered by experimenters include: (a) plants outgrowing their cages – expanding cages can be devised to counter this problem (Nicholls and Bérubé, 1965), and (b) plants in test cages deteriorating very rapidly due to the abnormally high prey densities reached – little can be done to remedy this problem, which can severely limit the duration of the experiment.

Exclusion methods have been used to assess the impact of predators and parasitoids on populations of a wide variety of prey and host insects including: aphids (Way and Banks, 1968; Campbell, 1978; Edwards *et al.*, 1979; Aveling, 1981; Frazer *et al.*, 1981b; Chambers *et al.*, 1983; Carroll and Hoyt, 1984; de Clercq, 1985; Kring *et al.*, 1985; Hance, 1986; Dennis and Wratten, 1991); armoured scale insects (DeBach and Huffaker, 1971); soft scale insects (Smith and DeBach, 1942); mealybugs (Neuenschwander and Herren, 1988); planthoppers (Kenmore *et al.*, 1985; Rubia and Shepard, 1987; Fowler, 1988; Rubia *et al.*, 1990); pond-skaters (water-striders) (Spence, 1986); aquatic stoneflies and chironomids (Lancaster *et al.*, 1991); beetles (Sotherton, 1982; Sotherton *et al.*, 1985; Traveset, 1990); flies (Burn, 1982); moths (Sparks *et al.*, 1966; Lingren *et al.*, 1968; van den Bosch *et al.*, 1969; Irwin *et al.*, 1974; Rubia and Shepard, 1987; Steward *et al.*, 1988; Rubia *et al.*, 1990); butterflies (Ashby, 1974).

The results of exclusion experiments can be quite dramatic. For example, the numbers of brown planthopper nymphs on rice plants in test cages reached twelve times the level attained in control cages, even though exclusion of predators proved to be imperfect (Kenmore *et al.*, 1985; Figure 5.5b). In Campbell's (1978) study of the hop aphid (*Phorodon humuli*), aphid numbers reached around 1000/0.1 m^2 in test cages, whereas in uncaged control plots they declined virtually to zero. In exclusion/inclusion experiments carried out by Lingren *et al.* (1968), adult bollworm moths were introduced into test and control cages, and in the control cages different types of predator were subsequently introduced. The number of moth eggs in the test cages reached a level ten times higher than that recorded in the control cages containing the predators *Geocoris punctipes* (Lygaeidae) and *Chrysoperla* spp. (Chrysopidae).

Even where a marked difference in prey numbers is observed between test and control treatments, and the possible confounding effects of factors other than predation can be discounted, it is important to establish whether the predators in question really do have the potential to produce the test/control plot difference observed. This requirement was appreciated by Chambers *et al.* (1983). As well as testing for the effects of aphid emigration, parasitism, fungal disease and cage microclimate, they sought to determine whether the *per capita* daily predation rates of aphid-specific predators were sufficiently high to have accounted for the differences in aphid numbers they recorded between fully-caged and open plots (Figure 5.7a and above). Using information on: (a) aphid rate of increase in the absence of predators (i.e. data were obtained from aphids in the cages), (b) predator densities in the open (control) plots, and (c) *per capita* daily predation rates of predators (published values), Chambers *et al.* (1983) were able to calculate population trends for aphid populations exposed to predation (Figure 5.7b, see Chambers *et al.* for method of calculation). They established that the predation rate that would be required to bring about the observed cage/open plot difference lay within published values.

5.2.4 INSECTICIDAL INTERFERENCE

The phenomenon of pest resurgence brought about by the application of insecticides and inadvertent elimination of a pest's natural enemies reveals dramatically the significant impact the latter normally have (DeBach and Rosen, 1991; Shepard and Ooi, 1991). With this effect in mind, insecticides have been used as a method of assessing the effectiveness of natural enemies.

With the insecticidal interference method, the test plots are treated with an insecticide, so as to eliminate the natural enemies, and the control plots are untreated. The insecticide used is either a selective one, or a broad-spectrum one that is applied in such a way as to be selective, affecting only the natural enemies. Depending on the duration of the experiment, repeated applications of the insecticide may be required, to prevent immigrating natural enemies from exerting an impact upon prey in the test plots. Drift of insecticides on to control plots also needs to be carefully avoided. The results of some insecticidal interference experiments are shown in Figure 5.8.

Some limitations of the method are that:

1. In the test plots not only the natural enemies but also the prey may be affected by the insecticides, so confounding the results of the experiment. The numbers of prey may be inadvertently reduced due to the toxic effects of the insecticide (i.e. either the insecticide turns out not to be selective in action, or drift of a broad spectrum insecticide has occurred) or they may be inadvertently increased due to some stimulatory, sublethal, effect of the insecticide upon prey reproduction (e.g. prey fecundity may be increased). Insecticides can be tested in the laboratory for their possible sublethal effects upon prey reproduction (Mueke *et al.*, 1978; Kenmore *et al.*, 1985);
2. In the test plots, 100% elimination of natural enemies is often not achieved, and so the full potential of natural enemies to reduce prey numbers is underestimated;
3. Limited information is provided on the dynamics of the predator–prey interaction, even where the densities of natural enemies are known (see **Exclusion Methods**).

The main advantages of the method are that the possible confounding effects of microclimate can be ruled out, and very large experimental plots can be used.

As an alternative to blanket spraying of test plots, an insecticide trap method can be used. Ropes of plaited straw treated with insecticide, trenches dug in the soil and containing formalin solution or insecticide-soaked straw, or some other insecticide-impregnated barrier can severely reduce the numbers of natural enemies entering test plots. One treatment used by Wright *et al.* (1960) and Coaker (1965) in studying beetle predators of the cabbage root-fly *Delia radicum* , involved the placing of insecticide-soaked straw ropes along the perimeters of test plots. Whilst it was not 100% efficient, the latter treatment had a dramatic effect upon predator numbers, and also significantly affected prey numbers in test plots.

The insecticide interference method has been used to assess the impact of parasitoids and predators upon populations of aphids (Bartlett, 1968); armoured scale insects (DeBach, 1946; 1955; Huffaker *et al.*, 1962; Huffaker and Kennett, 1966); leafhoppers and planthoppers (Kenmore *et al.*, 1985; Ooi, 1986); flies (Wright *et al.*, 1960; Coaker, 1968);

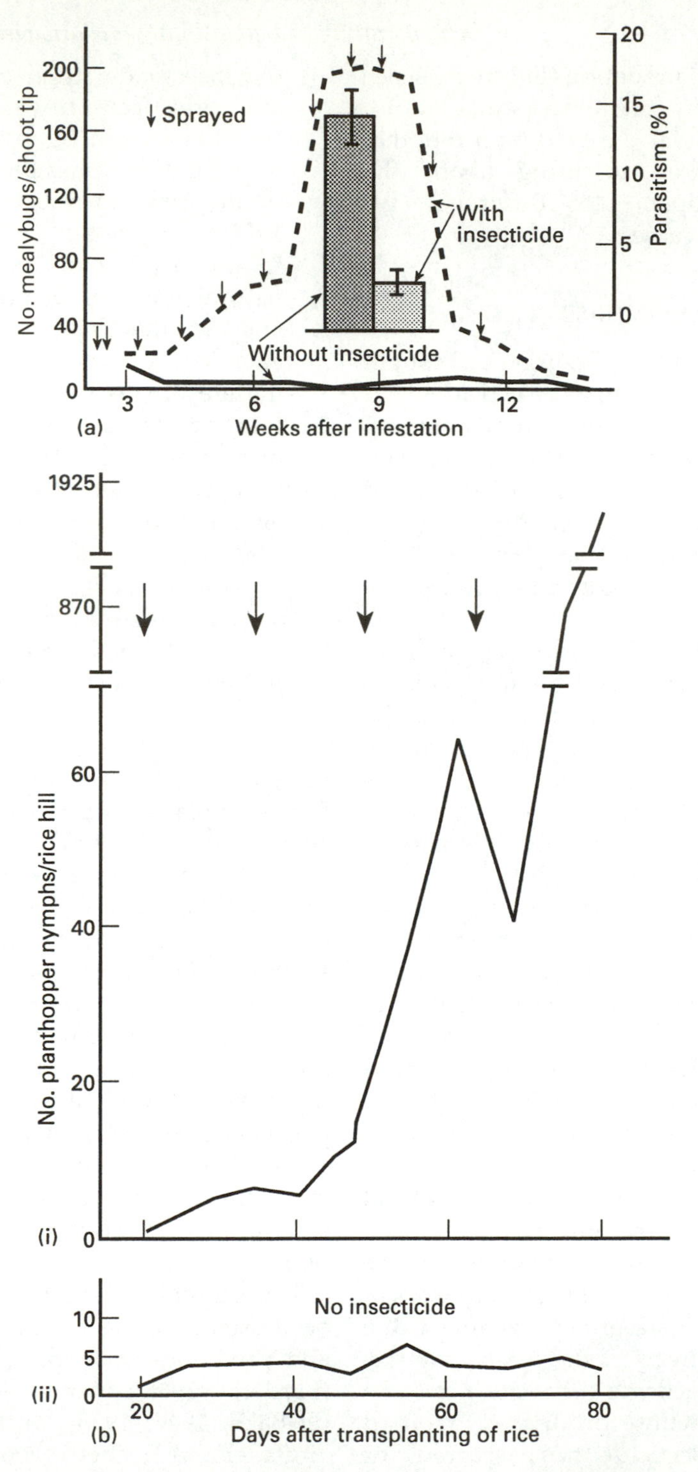

Figure 5.8. (a) Cassava mealybug (*Phenacoccus manihoti*) population development in insecticide-treated and untreated plots, together with mean levels of parasitism (histograms); (Source: Neuenschwander and Herren, 1988.) (b) population changes in rice brown planthopper (*Nilaparvata lugens*) in (i) plot treated with four sprays (arrowed) of decamethrin, and (ii) untreated plot. Planthoppers were sampled from 40 rice hills per plot using a vacuum net (subsection 4.2.2). (Source: Heinrichs *et al.*, 1982.) (a) Reproduced by permission of The Royal Society, (b) by permission of The Entomological Society of America.

moths (Ehler *et al.*, 1973; Eveleens *et al.*, 1973); and spider mites (Plaut, 1965; Readshaw, 1973).

5.2.5 PHYSICAL REMOVAL OF NATURAL ENEMIES

As its name suggests, physical removal involves just that; predators are removed either by hand or with a hand-operated device, each day from test plots. The method is a variation on exclusion, described above. Large, relatively slow-moving predators can simply be picked off plants by hand, while small, very active predators and parasitoids can be removed using an aspirator. This method has advantages in that microclimatic confounding effects can be ruled out (since cages are not used), and the contribution of particular natural enemy species to parasitism and predation can be relatively easily assessed. However, the method also has disadvantages in that:

1. Removal of natural enemies is very labour intensive; for the method to provide more than just a crude measure of natural enemy effectiveness, a 24 h/day watch needs to be kept on plants, and several observers need to be involved in removing insects;
2. Removal of natural enemies may involve disturbance to prey and thereby increase prey emigration;
3. Predators and parasitoids, before they are detected and removed, may have the opportunity to kill or parasitise hosts;
4. Like exclusion, the method provides limited information on the dynamics of the predator–prey interaction, even where densities of natural enemies are known (see **Exclusion Methods**).

Hand removal has been used to evaluate the effectiveness of aphid predators (Way and Banks, 1968 (Figure 5.9); Pollard, 1971). Luck *et al.* (1988) suggest it can be used as a calibration method for interference and exclusion methods.

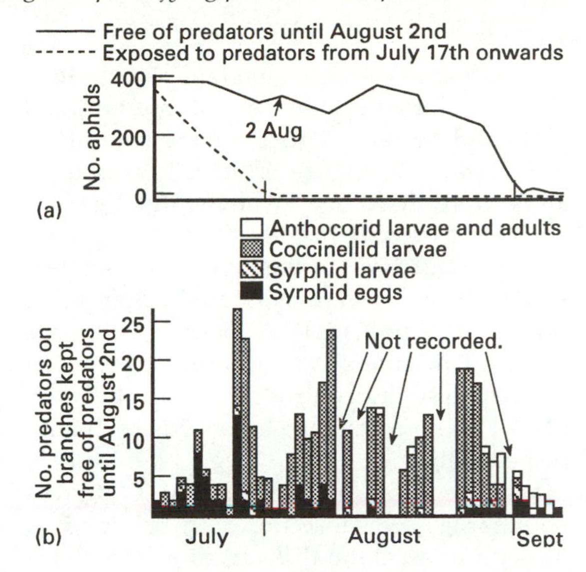

Figure 5.9. Effect of hand-removal of predators from colonies of *Aphis fabae*: (a) numbers of aphids on six branches of *Euonymus europaeus* kept free of predators until August 2nd compared with numbers of aphids on branches exposed to predators from July 17th onwards; (b) predators removed from branches that were kept free of predators until August 2nd (after that time predators were counted but not removed). In this experiment, crawling predators were excluded from the branches by a grease band. (Source: Way and Banks, 1968.) Reproduced by permission of The Association of Applied Biologists.

5.2.6 BIOLOGICAL 'CHECK' METHOD

This interference method exploits the fact that honeydew-feeding ant species, when foraging for honeydew sources and tending homopteran prey, interfere with non-ant predators and parasitoids, either causing them to disperse or killing them. In one set of plots, ants are allowed to forage over plants, whereas in the other set they are excluded. Natural enemies have access to both types of plot, but they are subject to interference by ants in the former. The method can be used with prey that do not produce honeydew, provided either natural or artificial honeydew is made available to the ants. This method has several of the disadvantages of other interference methods and exclusion methods.

5.2.7 EGRESS BOUNDARIES

Egress boundaries are simple devices which allow predators to move out of, but not into, plots and thereby reduce predator numbers (see caption to Figure 5.10). Egress boundaries were used by Wratten and Pearson (1982) in assessing the effectiveness of predators of sugar beet aphids. Predator numbers within the plots were monitored using pitfall traps (subsection 4.2.1). A 150-fold difference in aphid numbers was eventually recorded between test and control plots (40 aphids per plant and no more than 0.3 aphids per plant, respectively).

With egress boundaries (and ingress boundaries, see subsection 5.2.8 below), it is difficult to attribute differences in prey mortality between test and control plots to particular densities of predators, as the densities vary continuously over time. Therefore, it is difficult to calculate predation rates.

5.2.8 PREDATOR ENRICHMENT

With this method, the numbers of predators in the test plots are artificially boosted whereas the numbers in the control plots are not. Predator numbers in the test plots are enhanced by means of ingress boundaries, devices which allow predators to move into but not out of the plots. This method was employed by Wright *et al.* (1960) and Coaker (1965) in assessing the effectiveness of predatory beetles attacking cabbage root-fly (Figure 5.10), and by Wratten and Pearson (1982) in assessing the effectiveness of various predators of sugar beet aphids. Wratten and Pearson found that, using their ingress boundaries, total numbers of Opiliones captured (by pitfall-trapping) in test plots were 45% higher than in control plots, whereas the numbers of staphylinids, coccinellids and lycosids were increased by a maximum of 14%. However, the numbers of aphids eventually recorded in the test and control plots did not differ greatly.

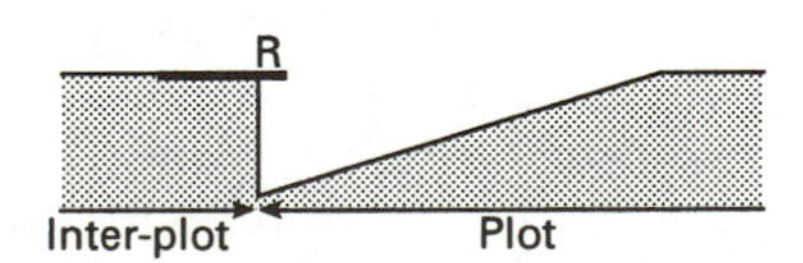

Figure 5.10. Cross-section through boundary that was used by Wright *et al.* (1960) to allow predators into but not out of plots (i.e. **ingress** boundary), in a study of predation of cabbage root-fly (*Delia radicum*). R denotes roofing felt. The device can easily be converted into an **egress** boundary, if the roofing felt is suspended from the plot margin and the sloping part of the trench is on the margin of the interplot. Reproduced by permission of The Association of Applied Biologists.

5.2.9 PREDATION AND PARASITISM OF PLACED-OUT PREY

With this method, known densities of prey are placed out in the field for a set period of time, and the numbers of dead or parasitised individuals recorded. The main conditions applying to the method are that prey ought to be placed out in as natural a fashion as possible, using natural densities, locations and spatial arrangements, so that they are neither more nor less susceptible to predation than usual. Also, an alteration in the overall density of prey in the field habitat (and therefore a perturbation to the system) ought to be avoided by having the artificially placed prey replace an equivalent number of prey simultaneously removed from the habitat. To enable the prey to be identified at the end of the experimental period, they may need to be either marked in some way or, if they are mobile, tethered. The marking or tethering technique ought not to either increase or decrease the acceptability of prey to predators. Burn (1982) placed out eggs (stained with Bengal Rose) of the carrot fly (*Psila rosae*) to measure predation by beetles, and Weseloh (1990) placed out larvae (tethered with long thread) of the gypsy moth (*Lymantria dispar*) to measure predation by a complex of predators. Burn (1982) determined beforehand whether staining of eggs affected the readiness of predators to eat treated eggs. Weseloh (1990), using a type of cage that allowed ants to enter but prevented moth larvae from escaping, compared the degree of predation recorded for tethered larvae with that recorded for untethered larvae. He found that tethered larvae are more susceptible to predation by ants than untethered ones, and so he used a correction factor to apply to the mortality rates he recorded for tethered larvae placed out in open sites.

Ôtake (1967, 1970) devised a method, involving the use of artificially infested plants containing eggs of known age, to measure field parasitism of planthopper eggs by *Anagrus* (Mymaridae) wasps. The plants were exposed in the field for a set time period, and were then returned to the laboratory and dissected to determine the numbers of parasitised and unparasitised eggs. This 'trap plant' method was used by Fowler *et al.* (1991) to investigate various aspects of egg parasitism of planthoppers and leafhoppers, including the spatial distribution of parasitism.

Provided the conditions set out above are satisfied, or some correction for bias in results can be applied, this method can provide useful data on the effectiveness of natural enemies. Weseloh (1990) concluded that the estimates of daily *per population* predation rates that he obtained by placing out tethered larvae, if suitably corrected for bias, were comparable with estimates obtained from other methods.

The main usefulness of the method, however, lies in providing comparative data, especially indices of predation and parasitism. For example, it can shed light on the relative effectiveness of different natural enemy species within a habitat, or on the effectiveness of a particular natural enemy species in different parts of a habitat (Fowler *et al.*, 1991; Speight and Lawton, 1976). Speight and Lawton (1976) used the method to examine the influence of weed cover on predation by carabid beetles within a habitat. Their study is also interesting in that artificial prey, *Drosophila* pupae killed by deep-freezing, were used.

The term 'prey enrichment' has been used to describe experiments involving the placing out of prey without the removal of existing prey.

5.2.10 LABELLING OF PREY

With this method, prey are labelled with a dye, a radioactive isotope or a rare element (subsection 4.3.10 describes labelling methods) and released into the field to expose them to natural predation. After a suitable period of time has elapsed, predators are collected from the field, screened for the label, and the amount of label present quantified. The *per capita* consumption rates of predators are calculated by measuring the label 'burdens' of the insects, and if the field density of predators is known, *per population* estimates of predation can also be estimated. For details, see McDaniel and Sterling (1979).

The technique has little to recommend it, in view of the following:

1. Radioactive labels can be hazardous to health;
2. It is difficult to ensure that all prey carry the same amount of label – there is normally considerable variation;
3. The same level of radioactivity can result from consumption of different numbers of prey;
4. The rate of excretion of the label from an individual predator appears to depend upon the the quantity of food subsequently eaten;
5. Labelling can affect the susceptibility of prey to predation. Earwigs (*Forficula auricularia*), for example, prefer to feed on undyed as opposed to dyed eggs of the cinnabar moth (*Tyria jacobaeae*) (Hawkes, 1972).
6. The label can easily and rapidly spread through the insect community by various routes, including excretion, honeydew

production, trophallaxis (i.e. by ants) moulting, scavenging on dead prey, and secondary predation.

7. The protocol can be very labour-intensive and, where rare elements and radioisotopes are used, specialised equipment is required;
8. Field populations of prey are disturbed.

Prey labelling has been used to quantify predation by natural enemies of aphids (Pendleton and Grundmann (1954), moths (Buschman *et al.*, 1977; McDaniel and Sterling, 1979; Gravena and Sterling, 1983) and isopods (Paris and Sikora, 1967).

5.2.11 FIELD OBSERVATIONS

With this method, predation is quantified by making field observations, either directly or using video recording techniques, of predators *in situ* (subsections 4.2.6 and 4.3.2 describe methods). Kiritani *et al.* (1972) estimated the number (n) of rice leafhoppers killed by spiders per rice hill per day as follows:

$$n = F \cdot C / P \tag{5.1}$$

where F is the number of predators seen feeding per rice hill during the observation period, C is the total amount of feeding activity in 24 hours expressed in terms of the specified period of observation, and P is the probability of observing predation (the average amount of time, in hours, taken to eat a prey individual, divided by 24 hours). A series of values of n were plotted against time and the area under the curve taken as the total number of prey killed. As noted by Southwood (1978), this method relies upon a high degree of accuracy in observing all instances of predation at a given moment and on values of C and the time taken to eat prey being fairly constant.

Edgar (1970) measured predation by wolf spiders (Lycosidae) in a similar manner to Kiritani *et al.* (1972), while Sunderland *et al.* (1986b) quantified predation of web-spinning money spiders (Liniphyiidae) differently, as follows:

$$n = p \cdot r \cdot k \tag{5.2}$$

where n is the number of aphids killed/m^2/day, p is the proportion of ground covered by webs, r is the rate of aphid falling/m^2, and k is the proportion of aphids entering webs that are killed or die (determined from field observations and laboratory experiments). Using this approach, it was shown that aphid populations could be reduced by spider predation by up to approx. 40%.

Waage's (1983) work on foraging by ichneumonid parasitoids (subsection 4.2.6) shows direct observation to have potential as a method for measuring rates of parasitism.

As noted in subsection 4.3.3 the prey 'booty' collected by ants can be taken from the insects upon their return to the nest. A mechanical or photoelectric counter, as suggested by Sunderland (1988), or video recording equipment can enable predation rates to be calculated. The particular prey population being exploited by ants can easily be located by following the insects' trails, and the prey's population density measured.

Video recording of predation and parasitism is likely to prove most fruitful if either the prey are relatively sedentary (e.g. predators of aphids) or the predators are sedentary (e.g. ant-lion larvae, tiger-beetle larvae).

5.2.12 GUT DISSECTION

Gut dissection (subsection 4.3.8), is one of the simplest techniques for measuring ingestion and predation rates. Also, being a 'post-mortem method' like serology and electrophoresis discussed below, it has the advantage over methods involving experimental manipulations of the predator–prey system that the results apply directly to an undisturbed, natural system (Sunderland, 1988).

The proportion of dissected predators containing remains of a particular prey in their guts can provide a crude index of *per population* predation rates. More meaningful measures can be obtained by counting the number of prey remains, corresponding to prey individuals, present within the guts of predators (e.g. number of prey head capsules), and recording also the through-put time of prey in the gut.

Sunderland and Vickerman (1980) used gut dissection in evaluating the relative effectiveness of different predators of cereal aphids, by multiplying the proportion of such insects that contained aphid remains during the aphid population increase phase by the mean density of predators (ground examination samples) at this time. The species with the highest indices were considered to be the most valuable in constraining the build-up of aphids in cereal fields.

5.2.13 SEROLOGY

Field Predation Rates

The serological methods discussed in subsection 4.3.9 in relation to the determination of dietary range in field-collected predators have mainly been employed to estimate field predation rates. Various models (Table 5.1) are available, based upon predator density, the proportion of predators testing positive for the particular prey, and usually also the detection period of the prey remains in the predator's gut. Earlier models were based on the assumption that each positive result represents a fixed number of prey consumed, but with the advent of enzyme-linked immunosorbent assay (ELISA), the predation rates can be estimated much more accurately.

Dempster (1960) used the precipitin test to calculate the predation rate of predators of the broom beetle, *Phytodecta olivacea*. Dempster's method assumes that each positive reaction represents one prey item consumed. Whilst this is probably a correct assumption when applied to his study (the prey in question were taken in small numbers), it could well be an incorrect assumption when applied to other predator–prey systems. If a second prey individual is taken by the predator within the detection period of the first prey individual, Dempster's method will underestimate the amount of predation that takes place.

Rothschild (1966) also used the precipitin test in investigating predation upon a planthopper, *Conomelus anceps*. He used the proportion of predators testing positive as a measure of the encounter rates between predator and prey, and assumed that at each encounter a set number of prey was consumed, this number being independently measured in an insectary. Rothschild's method takes no account of the time period over which a meal is detectable in the predator's gut after ingestion. Thus, when predator species are being compared, the method tends to introduce bias in favour of those species with long detection periods. Kuperstein (1974, 1979) modified Rothschild's model to take account of detection period, and Kuperstein's model has been used by Ashby (1974), Doane *et al.* (1985) and others. It can give an estimate below that of the Dempster model (Sopp *et al.*, 1992).

Nakamura and Nakamura (1977), also using the precipitin test, investigated spider predation of gall wasp (*Dryocosmus kuriphilus*) adults. They used a model that both took account of detection period and assumed predation to be a random process, with the proportion of spiders not testing positive forming the zero term of the Poisson distribution. This model, however, can lead to overestimates of predation rate if the predators feed on small numbers of large prey or if they have a long detection period (Sopp *et al.*, 1992).

The model of Sopp *et al.*(1992), developed for use with ELISA (but also applicable to electrophoresis), represents a significant departure from, and advance upon, previous models, in that p, the proportion of predators testing positive, is replaced by Q_o/f, where Q_o is the quantity of prey biomass recorded in predators and f the average proportion of the meal remaining. Q_o is calculated from ELISA optical densitometric readings (Figure 4.18, Fenlon and Sopp, 1991, describe the calibration method). The value of f is calculated from the digestion decay function (discussed below).

Sopp *et al.* (1992) compared the estimate of r (see Table 5.1) obtained using their model with: (a) the observed rate obtained in a laboratory experiment (a predator-aphid system), and (b) estimates obtained using previously published models. Compared with the latter, the new model was found to give a more accurate estimate.

The model of Sopp *et al.* (1992), though more accurate than previous ones, still involves several assumptions (several of which are common to other models) (Sopp *et al.*, 1992):

1. the detection periods measured in the laboratory are realistic estimates of those in the field;
2. the term $f.t_{DP}$ relates to the mean time since ingestion of prey materials for the population under study;
3. it is assumed that only a single meal is taken during the detection period and that feeding is random with respect to time. If the meal comprises several prey individuals or, in the case of those predators with long detection periods, several meals are taken in rapid succession, they may be regarded for practical purposes as a single meal;
4. it is assumed that the predator densities are accurately known;
5. the predator sample and the amount of prey biomass present are representative of the population as a whole. This is related to sample size and the sampling regime adopted;
6. the presence of prey remains is the result of predation and not of scavenging or secondary predation.

Serious errors may arise if: (a) these assumptions are not satisfied, (b) the weight of prey in the field is not accurately known, (c) there is a variable degree of partial consumption of prey, and (d) there is cross-reactivity between the prey species and non-target species (this problem can be overcome by the use of monoclonal antibodies, subsection 4.3.9).

It is important to know how the detection period (t_{DP}) can vary with: (a) meal size, (b) temperature, and (c) the presence of non-target prey items in the gut. Sopp and Sunderland (1989) demonstrated the effects of (a) and (b) on the detection period and antigen decay rate (the rate of disappearance

Table 5.1 Predation rate models employed with serological methods

Model	Source
$r = p.d/t_{DP}$	Dempster (1960, 1967)
$r = p.r_i.d$	Rothschild (1966)
$r = p.r_i.d/t_{DP}$	Kuperstein (1974, 1979)
$r = [\log_e(1-p)]d/t_{DP}$	Nakamura and Nakamura (1977)
$r = Q_o.d/f.t_{DP}$	Sopp *et al.* (1992)

r = *per population* field ingestion or predation rate (biomass or numbers of prey; to convert the former to the latter the mean weight of individual prey in the field needs to be known).
r_i = *per capita* ingestion or predation rate measured in an insectary.
p = proportion of field-collected predators found to contain prey remains.
d = predator population density.
t_{DP} = detection period of prey in predator's gut.
Q_o = quantity of prey recorded in gut (note that the immunodot assay technique (subsection 4.3.9) cannot be used to record this [Greenstone, 1995]).
f = proportion of food remaining in gut.

of detectable food) in the beetles *Bembidion lampros* (Carabidae) and *Tachyporus hypnorum* (Staphylinidae) and the spider *Erigone atra* (Liniphyiidae). Previously starved predators were fed freshly-killed aphids and were then kept at one of a range of temperatures for varying periods. The proportion of prey remaining in the gut at intervals after feeding was measured and plotted (Figure 5.11a). Curves were fitted to the (transformed) data (Figure 5.11b) and the detection period estimated (this is just one method of detection period measurement; Symondson and Liddell, 1993e, provide a review).

Sopp and Sunderland (1989) concluded the following from their study and other studies:

1. usually within a predator species, the detection period declines with increasing temperature; larger species tend to have longer detection periods, possibly because of the larger meal sizes, but within a species meal size appears to have little effect upon the detection period. Spiders have very long detection periods, even at high temperatures, perhaps because of their ability to store partially digested food in the gut diverticula;
2. In most predators the antigen decay rate follows a negative exponential form, i.e. the majority of the detectable antigens disappear within one-third of the detection period, but exponential declines have been recorded for the carabid *Pterostichus cupreus*.

Symondson and Liddell (1993e) point out that the detectability of invertebrate remains in the crops of predators such as carabid beetles is influenced not only by the antigen decay rate but also the residence time of a meal in the crop and gizzard (i.e. the fore gut). If the rate of through-put of prey material happens to be less than the antigen decay rate, then there will be a discrepancy between the true proportion of prey material present and the amount estimated from an ELISA. Quantification of this discrepancy would provide a means of estimating original meal size, when the time since feeding can be esti-

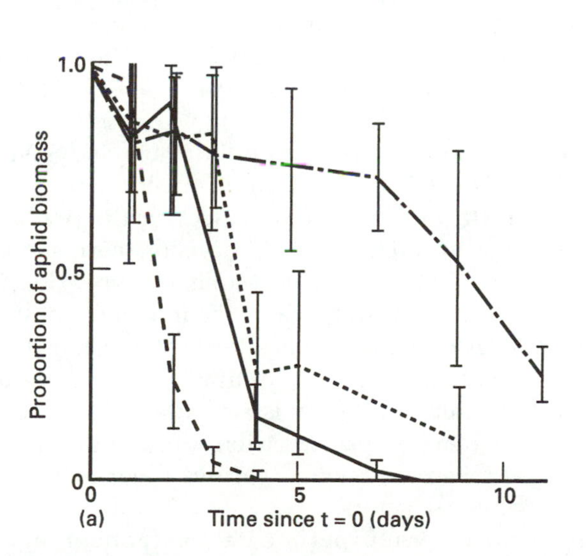

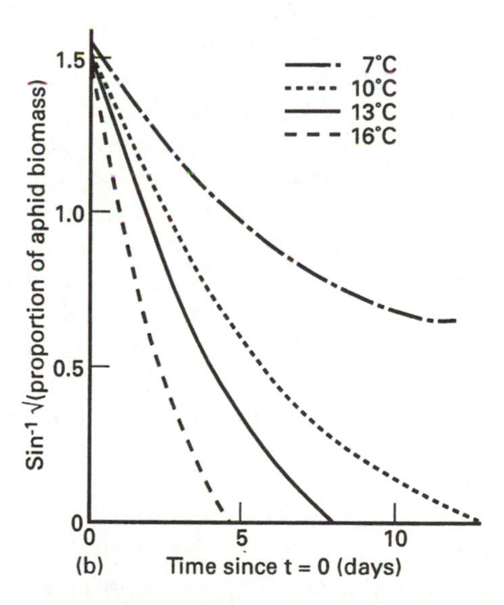

Figure 5.11. (a) The proportion of aphid biomass present in the gut immediately after feeding that is subsequently detected at various time intervals in the carabid beetle *Bembidion lampros*; (b) antigen decay rate curve. Symondson and Liddell (1993e) expressed the antigen decay rate differently, and also took the loss in weight of the predator's crop into account in estimating meal size (see their paper for details). (Source: Sopp and Sunderland, 1989.) Reproduced by permission of Kluwer Academic Publishers.

mated; such quantification requires crop weight loss and antigen decay rate to be measured as separate variables (for protocol, see Symondson and Liddell's paper).

Predation Indices

Sunderland *et al.* (1987a) compared different polyphagous predators in terms of their probable value as cereal aphid predators, by calculating, for each predator species, the following index:

$$P_g \cdot d / D_{max} \tag{5.3}$$

where P_g is the percentage of predators testing positive using ELISA, D_{max} is the maximum period over which prey antigens can be detected in any individual of a given species, and d is the mean predator density. Spiders tended to have the highest indices.

5.2.14 ELECTROPHORESIS

Electrophoresis, like serology, has been used in quantifying predation by fluid-feeding arthropod predators, albeit less commonly (for a recent review, see Solomon *et al.*, 1995). As with serology and gut dissection, the proportion of predators testing positive for prey contents can easily be determined (subsection 4.3.11), but to obtain meaningful quantitative information on predation, the quantity and detection period of prey materials ingested also need to be known. We have little further to say about electrophoresis, as it has been superseded by ELISA. The latter is not only a far more sensitive method for determining the quantity of prey proteins in the guts of predators, but also it requires less time to test gut contents.

With the aforementioned indirect and post-mortem methods (electrophoresis, serology, gut dissection, labelling of prey) converting ingestion rate to predation rate can generate serious errors. For example, if scavenging occurs, the true predation rate will be overestimated. Sunderland (1995), in discussing sources of potential error in estimating predation rates, points out that the latter (if predation is defined loosely) can also be *underestimated*, because predators may kill or wound prey without ingestion occurring (e.g. 'wasteful killing' by satiated predators [Johnson *et al.*, 1975]). For a discussion of the multiplicity of factors that can lead to inaccurate estimates of predation rates, see Sunderland's (1995) review.

5.2.15 EXPERIMENTAL COMPONENT ANALYSIS

The impact that natural enemies have on their prey depends upon the *density of predators* (or parasitoids) present in the habitat, and their *ability to locate and consume (or parasitise) their prey*. Population ecologists therefore distinguish between those factors that affect natural enemy abundance and those affecting natural enemy searching efficiency, and recognise two natural enemy responses: the **functional response** and the **numerical response** (Solomon, 1949). A functional response is the change in attack rate of *individual* natural enemies with changing prey density (section 1.10), and the numerical response is the change in the number of predators (i.e. the predator *population*) with changing prey density. Hassell *et al.* (1976) and Beddington *et al.* (1976b) developed this concept further, and distinguished between factors affecting the **prey death rate** and factors affecting the **predator rate of increase**. This division corresponds broadly to the previous one, but prey density is viewed as only one of the independent variables or **components** affecting prey consumption and predator numbers. For example, searching efficiency is also affected by predator density, prey distribution, the density of alternative prey and various other biotic and physical factors.

The term **experimental component analysis** refers to the combined theoretical and empirical approach, initiated by Holling (1959a,b, 1961, 1965, 1966), to analysing, by means of carefully controlled experiments, the components of predation. Experiments on the response of predators to prey density, predator density and prey distribution were discussed in Chapter 1 (sections 1.10, subsections 1.14.3 and 1.14.2 respectively). These experiments provide measurements of *per capita* rates of predation and parasitism. The use to which information from such experiments is put in assessing the role of natural enemies in insect population dynamics, is discussed in section 5.3.

5.3 THE ROLE OF NATURAL ENEMIES IN INSECT POPULATION DYNAMICS

5.3.1 INTRODUCTION

Having reviewed some of the methods by which insect mortality due to natural enemies can be quantified, we now turn our attention to the more difficult task of assessing its dynamic significance. Mortality factors acting on an insect population can cause three possible dynamic changes. They can:

1. affect the average population density;
2. induce fluctuations in numbers;
3. help **regulate** population numbers.

Of the three, it is undoubtedly the contribution that natural enemies make to population regulation which has most occupied the minds of ecologists over the years.

Factors which regulate population numbers can act either by:

1. returning populations towards a notional **equilibrium** number after some perturbation (i.e. **stabilising** population numbers);
2. restricting population numbers within certain limits, but allowing fluctuations in numbers (e.g. cycles) within those limits (Murdoch and Walde, 1989).

For a factor such as parasitism or predation to regulate, the strength of its action must be dependent on the density of the population affected. That is, it needs to be **density-dependent**, its *proportional* effect being greater at high population densities and smaller at low densities (Figure 5.12; cf. **density-independent** factors). Density-dependence operates through negative feedback on population numbers, which may involve changes in the rates of reproduction, dispersal and immigration as well as changes in mortality. If the proportion of hosts parasitised varies with changing host density, either temporally or spatially (subsection 5.3.10), this can profoundly affect the dynamics of the interaction. As we shall see (subsections 5.3.4, 5.3.7), density-dependent factors can also affect average population levels and can, under certain conditions, induce perturbations too (subsection 5.3.4).

We begin by addressing the problems associated with using percentage parasitism estimates to assess the impact of parasitoids on host populations (subsection 5.3.2). We then discuss perhaps the simplest (but least insightful) technique of assessing the impact of natural enemies, that of comparing their numbers with those of the prey or host populations (subsection 5.3.3). We then review the more conventional methods of life-table analysis (subsection 5.3.4) and show how simple population models can be derived from the information obtained. The limitations of the life-table approach are discussed, showing the need for supplementary field experiments (e.g. convergence and factorial

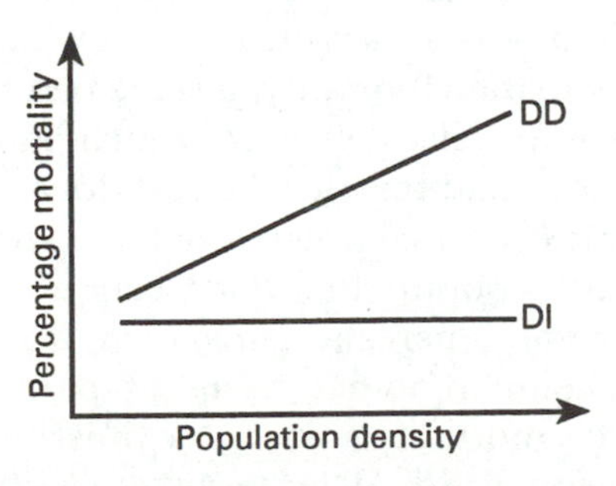

Figure 5.12 The negative feedback effect of a density-dependent mortality factor (DD) in which proportional mortality increases with population density (cf. density-independent factors (DI) in which proportional mortality is unrelated to population density).

experiments) (subsection 5.3.5). Next, we discuss how the important methodology of Experimental Component Analysis can be applied using both analytical and simulation models (subsections 5.3.6, 5.3.7, 5.3.8, 5.3.9), and go on to examine some of the more contentious issues which have developed out of this approach (subsection 5.3.10).

5.3.2 THE PROBLEM OF 'PERCENT PARASITISM'

A point which is perhaps worth stressing at this stage is that the importance of natural enemies in host or prey population dynamics may have little to do with the degree of mortality which they cause *per se*, a fact which is often misunderstood by researchers in pest management. Many publications, for example, have reported high 'percent parasitism' in insect pest populations, the clear implication being that this mortality is likely to contribute, in a major way, to reducing average population levels and/or to regulating populations. Unfortunately, such inferences may not be justified, for reasons which will become apparent.

'Percent parasitism' may also be a poor measure of the impact of parasitoids on host population dynamics for a number of other reasons. First, as van Driesche (1983) has pointed out, the number and timing of samples taken are usually inadequate for the task. To assess a parasitoid's contribution to host population mortality, it is the percentage attacked for the *generation* which must be determined and this may be best done within the context of a complete life table study of the host population (subsection 5.3.4). Furthermore, percent parasitism does not take account of other forms of parasitoid-induced mortality, such as host-feeding (Jervis and Kidd, 1986; Jervis *et al.*, 1992a), which may sometimes outweigh parasitism in their contribution to host mortality. The synchrony of parasitoid and host population in time may also be an important factor in determining how well sampling estimates generational levels of parasitism. Using a series of simple theoretical models van Driesche (1983) was able to establish that:

1. where susceptible hosts are all present before parasitoids begin ovipositing, and the parasitoid oviposition period does not overlap with the start of parasitoid emergence (Figure 5.13a), then the peak percent parasitism sampled can give a good estimate of generational percent parasitism;
2. where the situation in 1. prevails, but hosts begin to move to the next (unsusceptible) stage before all parasitoids have emerged, this will cause the peak percent parasitism to overestimate generational parasitism (Figure 5.13b);
3. where the situation in 1. prevails, but parasitoids begin to emerge before all parasitoid oviposition is complete, then peak percent parasitism will underestimate generational parasitism (Figure 5.13c);
4. if hosts enter the susceptible stage gradually and concurrently with parasitoid oviposition, and if host entry and exit do not overlap appreciably and parasitoid oviposition and emergence do not overlap appreciably (point X in Figure 5.13d), then a sample of per cent parasitism at this point can accurately estimate generational parasitism;
5. if hosts enter the susceptible stage gradually and concurrently with parasitoid oviposition, but if both hosts and parasitoids enter and leave the system at rates other than in 4., then samples of percent parasitism will bear little relation to generational percentage parasitism.

All of the above conclusions are based on the assumption that host mortality is caused solely by parasitism. Where this restriction does not apply (possibly most cases!), correspondence between samples and generational parasitism levels will be even harder to determine. Also, if the sampling method is in any way selective towards either parasitised or

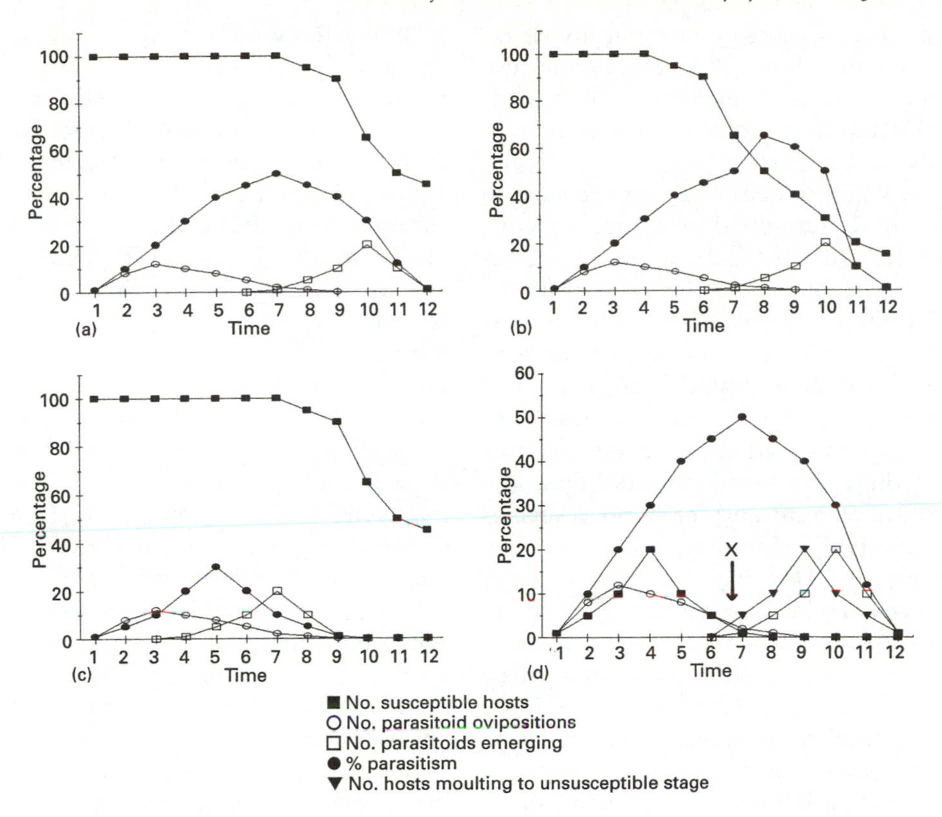

Figure 5.13 Synchrony of parasitoid and host populations may affect the accuracy of estimates of generational percentage parasitism (see text for explanation). (Adapted from: van Driesche, 1983.)

unparasitised hosts (subsection 4.2.9), this will introduce a further error into the estimate (van Driesche, 1983). van Driesche *et al.* (1991) suggested some ways of circumventing the above problems. One is to measure recruitment to both the host and the parasitoid (parasitised hosts) populations continuously, total recruitment to both populations being found by summing the recruitment values for all intervals. The ratio of total parasitoid recruitment to total host recruitment provides an unbiased estimate of total losses to parasitism. Another method uses death rate measurements from field samples. If individuals are collected at frequent intervals, reared under field temperatures, and the proportion dying from each cause recorded from one sample to the next, then the original percentage of the sample that was parasitised can be estimated. Gould *et al.* (1990) and Buonaccorsi and Elkinton (1990) provide equations for the calculations. The method requires that all hosts have entered the susceptible stage before the first sample and that no host recruitment occurs during the sampling period. Details and examples of these and other techniques can be found in van Driesche and Bellows (1988), Bellows *et al.*, (1989a) and van Driesche *et al.* (1991).

5.3.3 CORRELATION METHODS

In field populations a useful preliminary indication of the impact of natural enemies can often be obtained by statistically correlating their numbers against those of their prey or hosts. Significant positive or negative correlations may imply some causative association, which can then be tested by further investigation. Correlation alone, of course, should not be taken as proof of causation. A high **positive correlation** may indicate a degree of prey specificity on the part of the predator (Kuno and Dyck, 1985), which might be expected to show a rapid numerical response to variations in prey density (subsection 5.3.7 gives a definition). Heong *et al.* (1991), for example, found that the numbers of heteropteran bugs and spiders, which are major predators of Homoptera Auchenorrhyncha in rice, correlated positively with the numbers of Delphacidae and Cicadellidae. A positive correlation would also be accentuated by a low predator attack rate and/or a prey species with a relatively slow rate of population growth (Figure 5.14a).

Negative correlations, on the other hand, may indicate a slow or delayed numerical response by the predator to changing prey density. These responses are commonly shown by highly polyphagous predators which may 'switch' to feeding on a prey type only after it has increased in relative abundance in the environment (section 1.11). Negative correlations are also more likely to be associated with prey species which tend to show rapid changes in abundance, or with predators possessing a high attack rate (Figure 5.14b). For example, negative correlations between aphids and coccinellid beetles are frequently found on lime trees during the summer, and can be explained by the rapid rate of increase in the aphid population in the spring, coupled with the slow rate of response by the coccinellids (Dixon and Barlow, 1979). Later in the season predator numbers increase, forcing the already declining aphid population to crash (Figure 5.15a). Syrphid predators, on the other hand, can show a rapid numerical response to increasing cereal aphid populations, producing a positive within-season correlation (Chambers and Adams, 1986) (Figure 5.15b).

The tentative conclusions afforded by correlation techniques should only be drawn with extreme caution, and then only with a detailed appreciation of the biologies of the species involved. In particular, it must be remembered that correlations can be created just as easily by predator populations *following* changes in prey numbers, as by *bringing about* those changes. Also, absence of any cor-

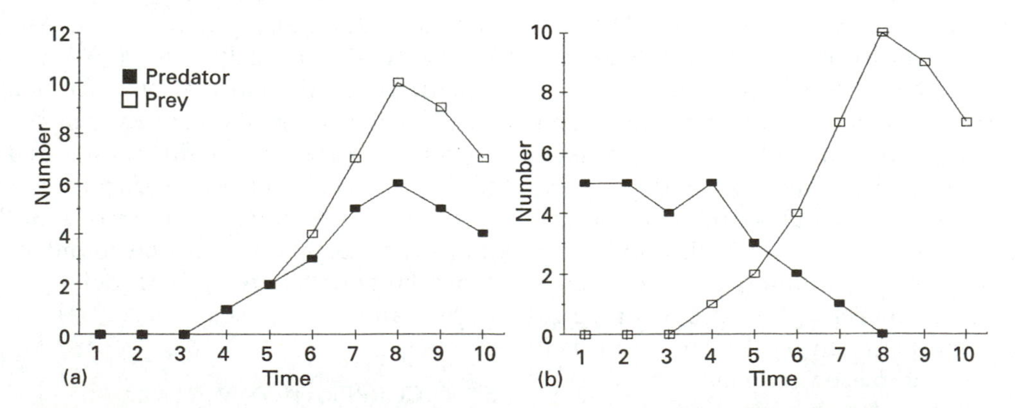

Figure 5.14 Relationships between predator and prey population numbers which produce either positive or negative correlations: (a) a positive correlation between predator and prey numbers produced by a slow rate of prey increase coupled with a relatively low predator attack rate, such that prey numbers are not reduced while predator numbers are still rising; (b) a negative correlation between predator and prey numbers caused by predators depressing prey numbers, which only increase after predator numbers have declined.

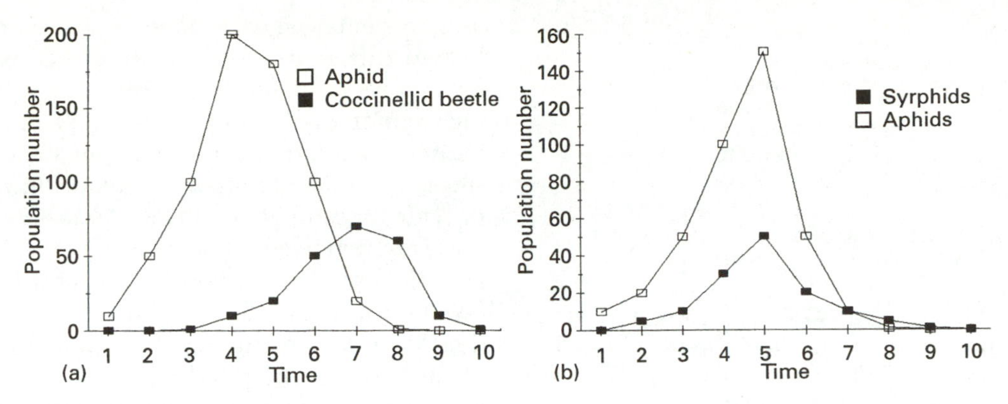

Figure 5.15 (a) A negative relationship between aphid and coccinellid beetle numbers on lime trees. Predator numbers increase slowly in response to aphid numbers and only reach their highest densities after aphid numbers have already declined; (Schematic representation based on information in Dixon and Barlow, 1979.) (b) syrphids show a rapid numerical response to increasing cereal aphid numbers, declining as aphid numbers decline. This produces a positive correlation between predator and prey numbers. (Schematic representation based on Chambers and Adams, 1986.)

relation should not be taken to imply that predators do not have any impact. We might expect a lack of any correlation in cases where predators have features intermediate to the aforementioned extremes.

5.3.4 LIFE-TABLE ANALYSIS

The concept of the life-table has already been introduced in section 2.11.2, in relation to the calculation of intrinsic rates of increase. Here, we are concerned with using life tables of a somewhat different nature to determine how specific mortality factors (e.g. a particular natural enemy species) affect prey or host population dynamics. For example, is the mortality density-dependent or density-independent? Is there evidence for delayed or over-compensating density-dependence? In short, does mortality from this source tend to regulate numbers at, or disturb numbers from, a certain level? To answer these and related questions, we need to take life tables apart and analyse the specific mortalities separately. Because some insect populations (e.g. aphids) tend to have generations which overlap in time, while others do not, two quite different approaches have been developed for each category, respectively the time-specific life table and the age-specific life-table.

Age-Specific Life-tables

The life-table approach pioneered by Pearl and Parker (1921), Pearl and Miner (1935) and Deevey (1947) was extended to insects with discrete generations by the single-factor analysis of Morris (1959) and the key-factor analysis of Varley and Gradwell (1960) (the latter sometimes incorrectly referred to as *k*-factor analysis). Of the two methods, the latter has been most widely used in population ecology (Podoler and Rogers, 1975) and will be the one concentrated upon here. For those readers interested in the Morris method, details are provided by Southwood (1978). Varley and Gradwell's method is given a very detailed treatment suitable for the beginner in Varley *et al.* (1973). As this book is now, alas, out of print, we feel it is worthwhile discussing the procedures in detail, especially since there have been some new developments in recent years.

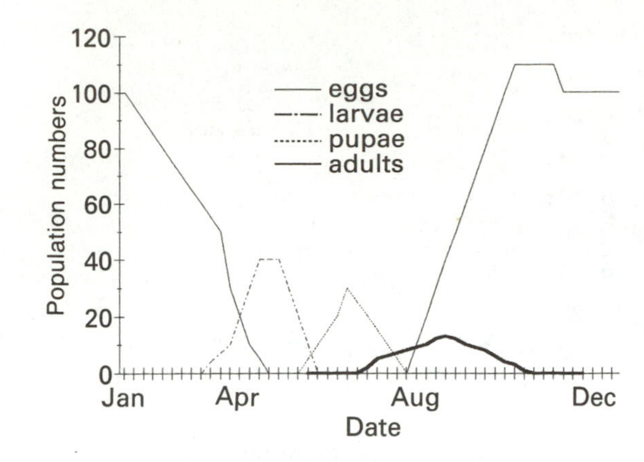

Figure 5.16 Schematic life cycle of a typical temperate-zone univoltine insect population.

The usefulness of the Varley and Gradwell approach depends on the availability of sequential life tables for a number of generations of a univoltine population. In temperate regions, for example, it is commonly the case that insect populations overwinter as eggs and develop through a number of discrete stages in the spring and summer (Figure 5.16). The adults then mature in the autumn to lay a new generation of overwintering eggs before dying. In this situation, generations remain completely separate. By obtaining population density estimates for the numbers entering each stage in the life cycle, it is then possible to construct a composite life-table, consisting of a sequence of independent life-tables for each generation (Table 5.1). The numbers entering each stage can be estimated in two different ways: (a) by direct assessment of recruitment (for example, by measuring fecundity or fertility, section 2.7), or (b) by indirect calculation from counts of stage densities. Several techniques are available which provide an estimate by the second route, and these are reviewed by Southwood (1978). The graphical method of Southwood and Jepson (1962), for example, involves plotting the density of a stage against time and dividing the area under the plot by the average duration of the stage (mean development time). This yields an unbiased estimate of the number entering the stage if there is either no mortality, or the mortality occurs only at the end of the stage. Any mortality during the stage will result in underestimation. Bellows *et al.* (1989b) provide an extension to this method which can be used for interacting host and parasitoid populations. A number of other methods are discussed by Manly (1990).

It should be noted from Table 5.2 that the actual density estimates of numbers entering each stage are retained in the life-table, rather than corrected to a common starting number (cf. Table 2.4). The reason for this will become clear. Where stage mortalities can be partitioned into a number of definable causes, these are quantified separately in the table. In this way it may be possible to build similar life-tables for particular natural enemies. Varley *et al.* (1973) provide a number of rules to follow in the construction of the table. These are:

Table 5.2 Composite life-tables for six generations of a hypothetical insect population with discrete generations. Each *k*-value is calculated as $k = \log_{10}$ before mortality–$\log_{10}$ after mortality. $K = k_1 + k_2 + k_3$ (Note: whilst such life tables have traditionally been presented in columns, putting them in rows (as is done here) makes spreadsheet regression calculations easier.)

Year	*Eggs*	k_1	*Larvae*	k_2	*Pupae*	k_3	*Adults*	*K*
1	1000	0.824	150	0.398	60	1.080	5	2.302
2	800	0.426	300	0.685	62	1.190	4	2.301
3	1200	0.681	250	0.455	80	0.824	12	1.960
4	700	0.942	80	0.204	50	0.699	10	1.845
5	500	0.553	140	0.301	70	0.766	12	1.620
6	1200	1.000	120	0.150	85	1.230	5	2.380

1. Where mortalities are reasonably well separated in time, they are treated as if they are entirely separated with no overlap;
2. Where events overlap significantly in time, they can be considered as if they are exactly contemporaneous;
3. All insects must be considered either as alive and healthy or, alternatively, as dead or certain to die from some cause. For example, parasitised larvae are scored as certain to die, with the parasite recorded as the cause of death;
4. No insect can be killed more than once. Where hosts are attacked by two parasitoid species, death of the host is credited to the first parasitoid. If the second parasitoid emerges as the victor, it is taken as the cause of death of the first parasitoid. The second attack is thus entered in the life-table of the first parasitoid but not in that of the host.

Although somewhat arbitrary, rules such as these are necessary to balance the budget. However, as we explain below, conclusions from the analysis may unfortunately be sensitive to the rules adopted.

By converting the data in Table 5.2 to logarithms ($\log_{10}$), we can calculate for each successive mortality, in any generation:

k = log number before mortality – log number after mortality

where k is a measure of the proportion dying from the action of the mortality factor. In practice these calculations are easily carried out using a spreadsheet programme (Table 5.2 caption), which can also be used for the regression analyses (see below). Within each generation, we can thus determine a sequence of k-values, k_1, k_2, k_3 ...k_n, corresponding to each successively-acting defined mortality up to the adult stage (Table 5.2). Strictly speaking, this should be up to the stage before reproduction begins, any pre-reproductive mortality being counted as separate k-factors. Mortality during the adult stage can be counted as one or more k-factors acting on the adults, or alternatively as a k-mortality acting on the next generation of eggs (Varley *et al.*, 1973). The final post-reproductive mortality to act on a generation, i.e. that which brings generation numbers to zero, contributes nothing to between-generation variation in numbers and is not included in the analysis. To do so would cause two problems. First, we are dealing here with the $\log_{10}$ of numbers, so how would we treat zero values? Second, the final reduction in adult numbers to zero, is by its nature density dependent. In a sense, the ultimate extreme of regulation is to return a population to an equilibrium of zero! We illustrate the point by including this spurious density-dependence in our analysis (Figure 5.19). The sum of all the k-values up to, but not including this last mortality, provides us with a measure of total generation mortality K, i.e.

$$k_1 + k_2 + k_3 \ldots\ldots k_{n-1} = K \qquad (5.8)$$

The advantages of using k-values instead of percentage mortalities lie in the ease of calculation and the fact that k-values can be added to give a measure for total generation mortality (K) (adding percentages would have no meaning).

Two basic questions can be answered from an analysis of the table at this stage:

1. Which factor or factors contribute most to variations in mortality from generation to generation, i.e. the so-called **key factors** causing population change?
2. Which factors contribute to regulation of population numbers?

Key Factors

The answer to the first question can often be obtained from a graphical representation of the data. Plotting the k-values against generation may be enough to reveal the key factor(s) causing population change (Figure 5.17). Here, varations in k_3 between generations most

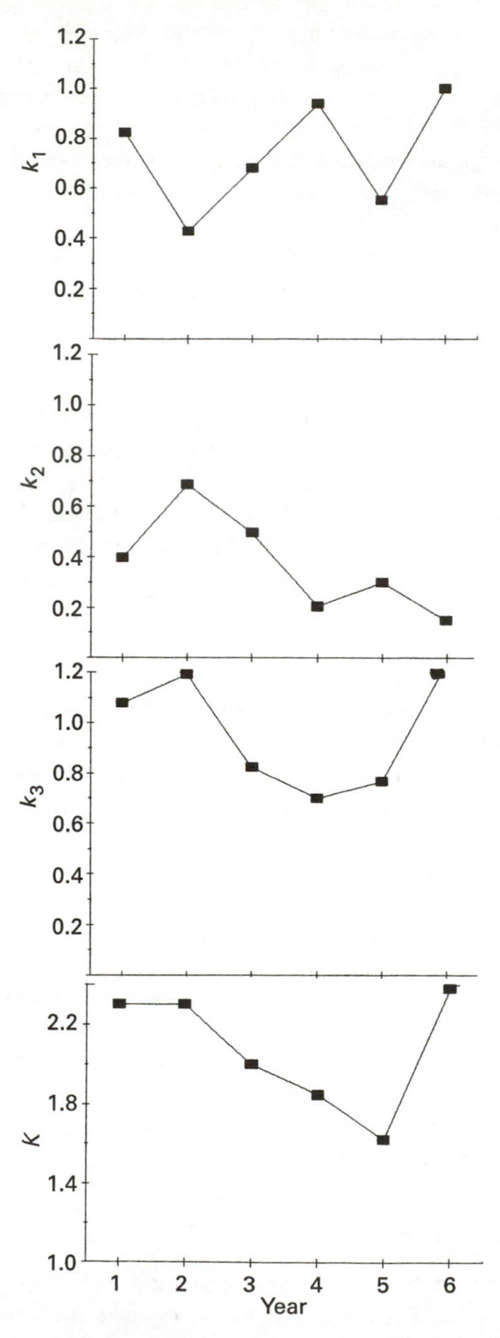

Figure 5.17 Key factor analysis of the mortalities acting on an hypothetical insect population (see Table 5.2 for data).

closely follow variations in overall mortality (K), indicating that k_3 is the key factor. Note that the key factor is not necessarily the factor causing greatest mortality (k_1 in this case).

Sometimes, a simple graphical inspection may not be enough to reveal the key factor, in which case the statistical method of Podoler and Rogers (1975) can be employed. This involves regressing each k-value against total generation mortality (K), the mortality with the greatest slope (b) being the key factor. In our example k_3 is confirmed as the only significant key factor (Figure 5.18). Where more than one factor is found to contribute, a hierarchy of significance can be constructed.

Strictly speaking, the Podoler and Rogers' procedure for identifying key factors is not statistically valid, in that it contravenes the basic rules of regression. These are that the axes should be independent of each other and the independent variable should be error-free. Clearly, K consists of the k-values against which it is being regressed, and is also subject to sampling error. Where the regression relationship of the putative key factor is not clear cut, a simpler expedient may be to use the

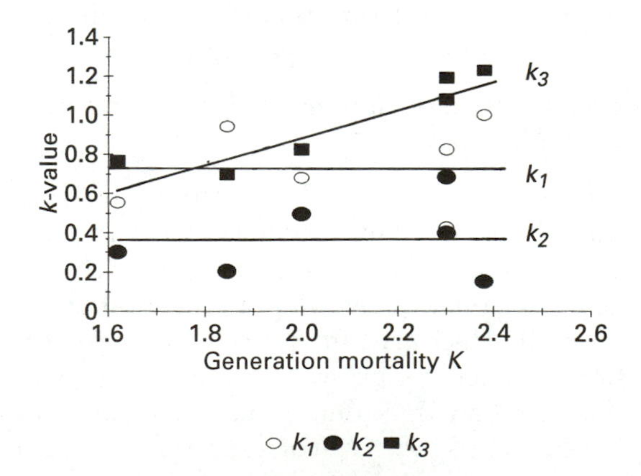

Figure 5.18 Podoler and Rogers' method for identifying key factors. The factor with the greatest slope (k_3 in this case) is the key factor causing population change. ($k_3 = 0.68K - 0.46$; $R^2 = 0.84$)

correlation coefficients, which are not subject to the same restrictions. In this case, the key factor would be the one with the highest correlation between k and K, (maximum $r = 1$). Manly (1977) devised an alternative method based on multiple regression analysis, whilst the problems of sampling error have also been considered by Kuno (1971). As we shall now see, a similar problem with regression is confronted in the detection of density-dependence from life-table data.

Detecting Density-dependence

Assessing which factors contribute to regulation of the population again involves plotting each k-value, this time against the log density on which it acts (i.e. *before* the mortality). In our example (Figure 5.19) the plot of k_1 against log density of eggs contains six data points, corresponding to each generation. Similarly, k_2 is plotted against $\log_{10}$ density of new larvae, again with six data points, and so on. Remembering that each k-value is a measure of *proportional* mortality, positive relationships for any of these plots would indicate that mortality is acting in a density-dependent fashion. A horizontal slope would indicate density-independence, while a negative slope would indicate inverse density-dependence. Regression analysis is generally employed to calculate the significance of the slopes. Here, the only significant density-dependence is found in k_2. However, the

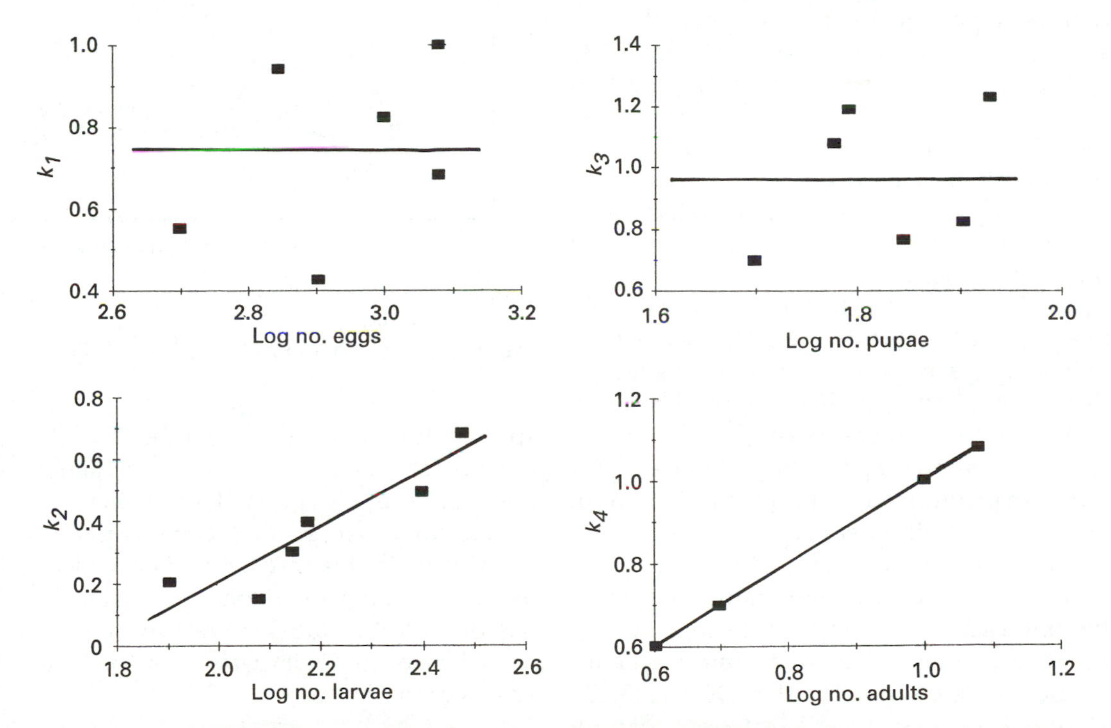

Figure 5.19 The identification of density-dependent factors from life table data. k-values for the different mortalities are plotted against the population densities on which they acted. In this case, only k_2 is significantly density-dependent. ($k_2 = 0.86L - 1.52$; $R^2 = 0.84$; $k_1 = 0.74$; $k_3 = 0.96$.) k_4 is the last mortality to act, bringing numbers down to 0 (or in this case 1, which was used to make the log calculations workable). This mortality is, by its nature, always density-dependent (see text), but is not included in the analysis, as it contributes nothing to population variation or regulation.

problem of statistical validity (mentioned above in relation to Podoler and Roger's method) again arises. As *k*-values are calculated in the first place from $\log_{10}$ densities, the two axes are not independent. Neither is the independent variable ($\log_{10}$ density), estimated from population samples, free from error. To overcome the problem, Varley and Gradwell (1968) suggest a 'two-way regression' test, which involves both the regression of $\log_{10} N_t$ (initial density) on $\log_{10} N_{t+1}$ (final density) and $\log_{10} N_{t+1}$ on $\log_{10} Nt$. If both regressions yield slopes significantly different from $b = 1$ and are on the same side of the line, then the density-dependence can be taken as real. This method may be unnecessarily stringent (Hassell *et al.*, 1987; Southwood *et al.*, 1989), requiring that density-dependence remains apparent when all sampling errors are assumed to lie firstly in the estimates of N_t, then in N_{t+1}. Bartlett (1949) provided an alternative regression method in which sampling errors are distributed between both axes.

If density-dependence is accepted, then the regression coefficients can be taken as a measure of the *strength* of the density-dependence. The closer *b* is to 1, the greater the stabilising effect of the mortality. A slope of $b = 1$ will compensate perfectly for any changes in density at this stage, while a slope of $b < 1$ will be unable to compensate completely for any changes (**undercompensation**). Slopes of $b > 1$ imply **overcompensation**, the significance of which will become clear later.

A further insight into the nature of density-dependence can also be obtained by again plotting each *k*-value against the $\log_{10}$ density on which it acts, but in a **time sequence** (Varley and Gradwell, 1965; Figure 5.20). Different factors trace a different pattern depending on their mode of action; density-independent factors show an irregular, zigzag pattern (Figure 5.20a), while direct density-dependent factors show a more discernible straight-line pattern of points clustered within a narrow band (Figure 5.20b). A spiral pattern (Figure 5.20c) indicates delayed density-dependence, in which the action of the *k*-mortality is not felt until one or two generations hence. Insect parasitoids frequently act in this way for reasons which will be explained in subsection 5.3.7. Manly (1988) has provided a statistical test for spiral patterns based on a comparison of the internal angles of the spiral.

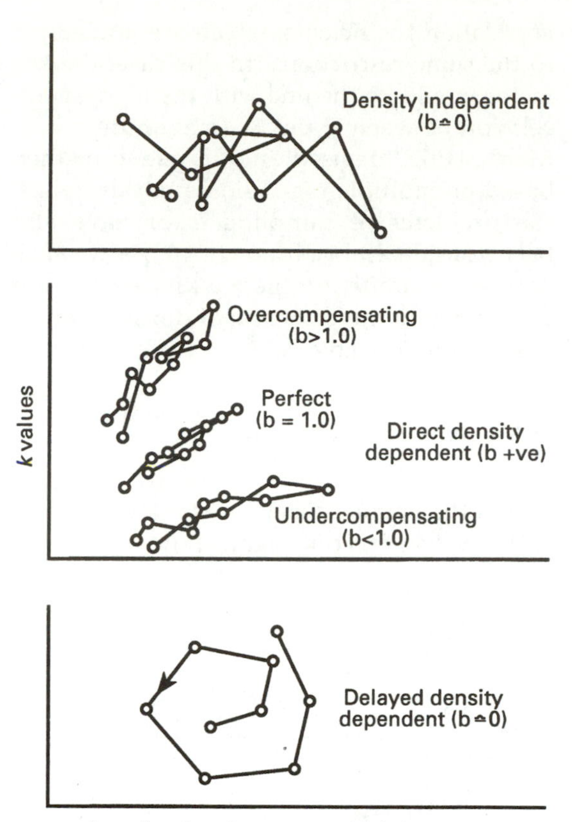

Figure 5.20 Time sequence plots showing how density relationships can be identified from the patterns produced. (Source: Southwood, 1978.)

A Simple Inductive Model

At this point the 'formal' methodology associated with key factor analysis has been fully described, but further insights into how *k*-mortalities affect population dynamics can be derived from a simple **inductive model** constructed using the information obtained above (inductive models are those based on particular case studies, which yield general insights into population dynamics; philosophically, induction is the process of arguing from the particular case to the general case (cf. deduction, deductive models, subsection 5.3.7).

We begin by linking the numbers in each life stage to the next, through the mortalities expressed by $k_1 \ldots k_{n-1}$ as follows:

$$k_1 = m \cdot E_t + c_1 \tag{5.9}$$

$$L_t = E_t - k_1 \tag{5.10}$$

$$k_2 = m \cdot L_t + c_2 \tag{5.11}$$

$$P_t = L_t - k_2 \tag{5.12}$$

$$k_3 = m \cdot P_t + c_3 \tag{5.13}$$

$$A_t = P_t - k_3 \tag{5.14}$$

where E_t, L_t, P_t and A_t are the $\log_{10}$ numbers of eggs, larvae, pupae and adults respectively at time t (c values = constants). Assuming a 50:50 sex ratio, we can find the $\log_{10}$ number of females (F) from:

$$10^{F_t} = 10^{A_t}/2 \tag{5.15}$$

or

$$F_t = A_t - 0.30103 \tag{5.16}$$

In our example, $k_1 = 0.74$, $k_2 = 0.86L_t - 1.52$ and $k_3 = 0.96$. The number of eggs laid by adults can be estimated from either: (a) cohort fecundity experiments (performed in the laboratory and/or in the field), (b) dissection of females and estimating potential fecundity (subsection 2.7.3), or (c) regression of eggs in year t+1 against the estimated number of females (A/2) in year t. Assuming the following relationship between female numbers and eggs deposited (Figure 5.21):

$$E_{t+1} = 0.86F_t + 2.1 \tag{5.17}$$

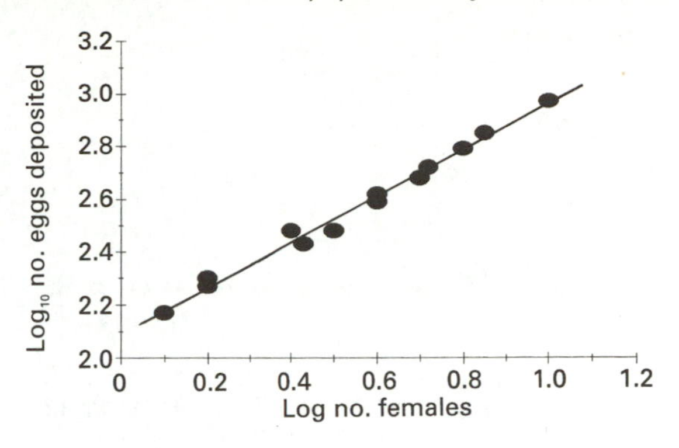

Figure 5.21 The relationship between female numbers and reproduction used in model 5.3.4. ($E = 0.86F + 2.1$; $R^2 = 0.99$)

we now have a series of equations which can be used sequentially to simulate dynamic changes from one generation to the next, over as many years as we require. The BASIC listing of a computer program which can be used to do this is provided in Appendix 5A, and can be executed on either a suitable programmable hand calculator or a personal computer. Note that the model as it stands is completely **deterministic** in that it takes no account of the potential variation in the relationships, i.e. particular values for variables on the right hand side of the equations produce only one possible value for the variable on the left hand side. **Stochastic** models, on the other hand, *do* take account of the variability in the relationships, by including mathematical terms to describe chance events which may affect one or more of the relationships in the model. In this case, particular values for variables on the right hand side of the equations may produce a number of possible values for the variables on the left hand side. The methodology of stochastic modelling is discussed further in subsection 5.3.8, and a good introductory treatment can also be found in Shannon (1975).

Simulations of the model with different starting densities of eggs show that numbers approach an equilibrium within 2–3 generations, i.e. are strongly regulated (Figure 5.22a).

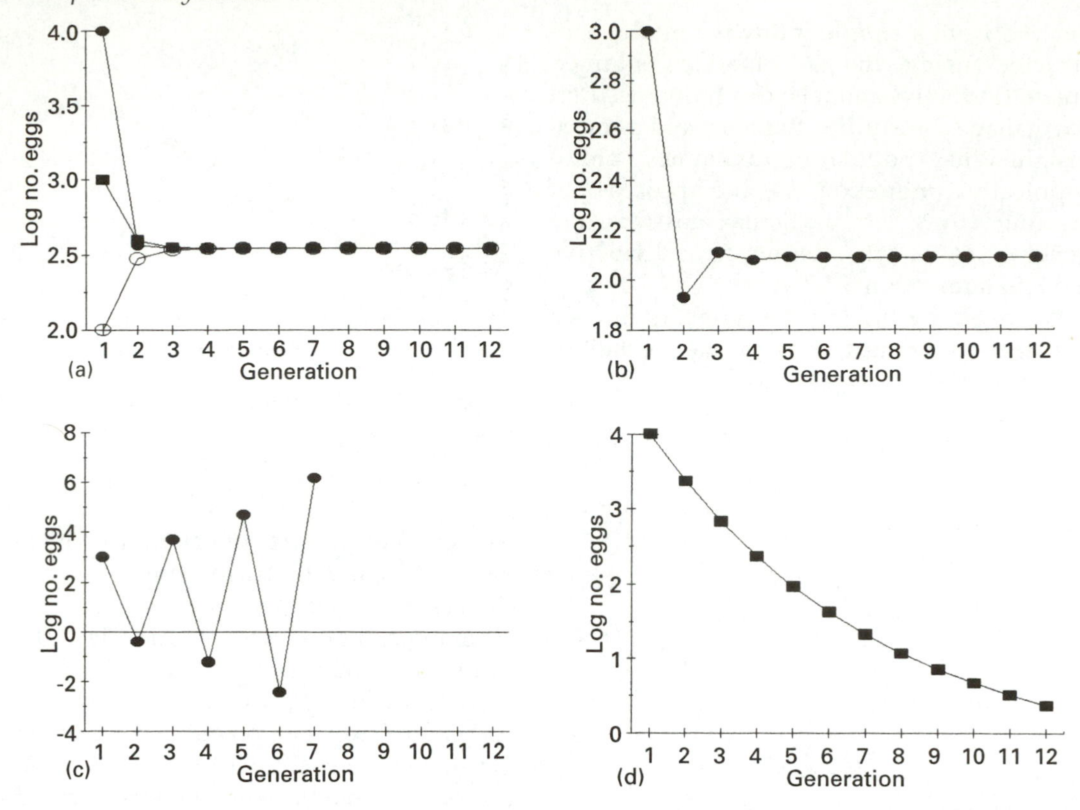

Figure 5.22 Predicted egg numbers over 12 generations: (a) with different starting densities of eggs: (b) with density dependence of larval mortality increased from b=0.86 to b=1.2; (c) with density dependence of larval mortality increased from b=0.86 to b=2.4, and (d) with density dependence of larval mortality removed (b=0).

Proof that regulation is provided by k_2 can be obtained by altering equation (5.12) such that k_2 becomes density independent ($k_2 = c_2$). Here, numbers either increase indefinitely or decrease to zero, depending on the other parameter values (Figure 5.22d), i.e. regulation is removed. Alternatively, increasing the strength of density-dependence by increasing the slope of the regression relationship between k_2 and L_t (e.g. b = 1.2 in equation (5.12), can produce oscillations of decreasing amplitude which eventually return to equilibrium (Figure 5.22b). Increasing the b-value even further (e.g. b = 2.4), however, can result in oscillations of increasing amplitude leading to the extinction of the population (Figure 5.22c). Thus, density-dependence is confirmed to be potentially either stabilising or destabilising in its effect, depending on its strength. It is also apparent that the weaker the density-dependence, the higher the equilibrium value becomes.

A Case Study: The Winter Moth

To appreciate the considerable number of studies on which key factor analysis has been performed, the reader is referred to Podoler and Rogers (1975), Dempster (1983), Price (1987) and Stiling (1987, 1988). There is no doubt, however, that it is Varley and Gradwell's own study (1968, 1970) of the winter moth (*Operophtera brumata*), together with the various follow-up studies in England

and Canada, which have made this perhaps the best understood and most widely-quoted example. It is worth reviewing briefly some of the features of this study, as it serves to illustrate some of the potential problems in using key factor analysis, which we shall discuss shortly.

The winter moth feeds on a wide range of mainly deciduous trees, occasionally defoliating oaks. The life cycle at Wytham Wood, near Oxford, UK, where Varley and Gradwell's study was carried out, is as follows: eggs are laid in early winter in the tree canopy and hatch in spring to coincide with bud-burst; the caterpillars feed on the foliage until fully grown, whereupon they descend to the forest floor on lines of silk and pupate in the soil; adults emerge in November and December to ascend the trees and mate, the females ovipositing in crevices on the bark. There is therefore one generation each year.

Data collected between 1950 and 1962 reveal that 'winter disappearance' (k_1), during the period between the egg stage and that of the fully grown larvae, is the key factor inducing population variation between years. Parasitism, disease, and predation (k_2–k_6) are relatively insignificant in this respect (Figure 5.23). The only significant regulating factor to be detected, however, was predation on pupae (k_5, Figure 5.24), subsequently shown to be caused mainly by shrews and ground beetles (Frank, 1967a,b; East, 1974; Kowalski, 1977). Parasitism showed no sign of being density-dependent, either at the larval stage (k_2) or at the pupal stage (k_6), leading the authors to suggest that the wide variations in densities from year to year, caused by the key factor winter disappearance, may be obscuring a possible delayed density-dependent relationship. The lack of any detectable regulating potential by the larval parasitoid

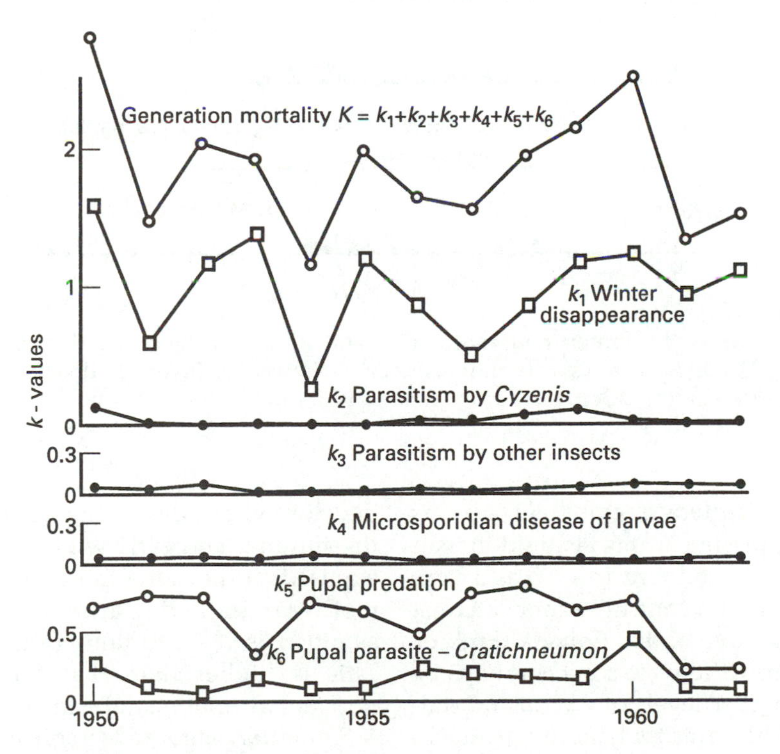

Figure 5.23 Key factor analysis of the mortalities acting on the winter moth. (Source: Varley *et al.*, 1973.) Reproduced by permission of Blackwell Scientific Publications Ltd.

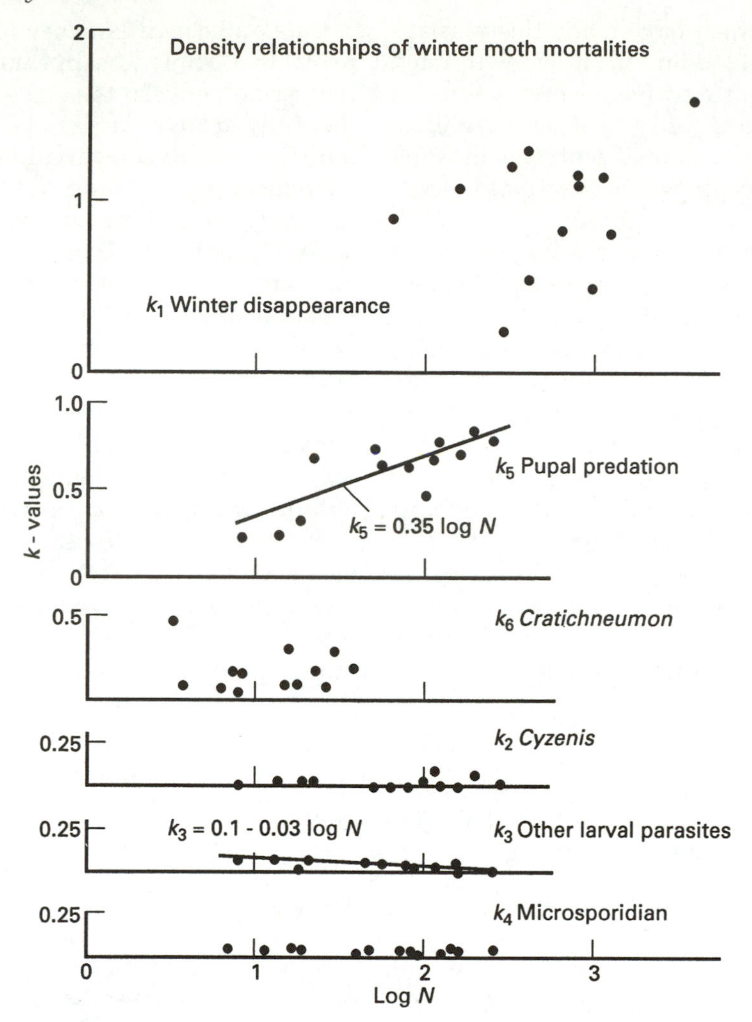

Figure 5.24 *k*-values of the winter moth mortalities plotted against the population densities on which they acted. k_1, k_2, k_4 *and* k_6 are density-independent; k_3 is weakly inversely density-dependent; k_5 is strongly density-dependent. (Source: Varley *et al.*, 1973.) Reproduced by permission of Blackwell Scientific Publications Ltd.

Cyzenis albicans (Diptera: Tachinidae) (k_2) was particularly surprising as this tachinid fly had previously been introduced in 1955 as a very effective biological control agent against winter moth in Nova Scotia, Canada (Embree 1966, 1971). This difference could perhaps be explained by higher levels of *Cyzenis* mortality in the UK. The parasitoid, although attacking the moth in the larval stage, continues to develop within the moth puparia throughout the summer and early winter and is therefore exposed to the same mortality factors as the moth pupae. Varley and Gradwell recorded as much as 98% mortality of *Cyzenis* puparia. This is higher than that for winter moth pupae, but understandable as *Cyzenis* spends 4–5 months longer in the soil, emerging in the spring.

A population model for the winter moth and its main parasitoids, *Cyzenis* (k_2) and the ichneumonid wasp *Cratichneumon culex* (k_6), was developed by Varley *et al.* (1973), using basically the same approach which we elaborated above, but with two important differences. First, the variations in k_1 could not be predicted, so the observed values were used instead. Second, parasitism (k_2 and k_6) were modelled using the 'area of discovery' concept (subsection 5.3.7) rather than the simple regression relationships shown in Figure 5.24. There was good agreement between the model output and estimated field densities of the winter moth and its two parasitoids (Figure 5.25), although it has to be pointed out that testing the accuracy of a population model against the same data from which it is constructed, is not considered to be good modelling practice (subsection 5.3.8). However, collection of independent field data for acceptable validation of such life table models is likely in many cases to prove impracticable, possibly involving years of extra work. This is one of a number of drawbacks associated with the Varley and Gradwell approach, which we shall now consider in detail.

Disadvantages of the Approach

The difficulty of obtaining additional field data for model validation highlights the single biggest problem of the whole approach, namely that of securing a long enough sequence of data to perform the analysis with a reasonable likelihood of detecting statistically significant relationships (Hassell *et al.*, 1987). For insect populations having one generation a year, we may be contemplating the commitment of 15–20 years of time and resources to a study, with no guarantees of success. The population processes affecting the main species may also change over the period of the study, with the result that key factors or density-dependent factors

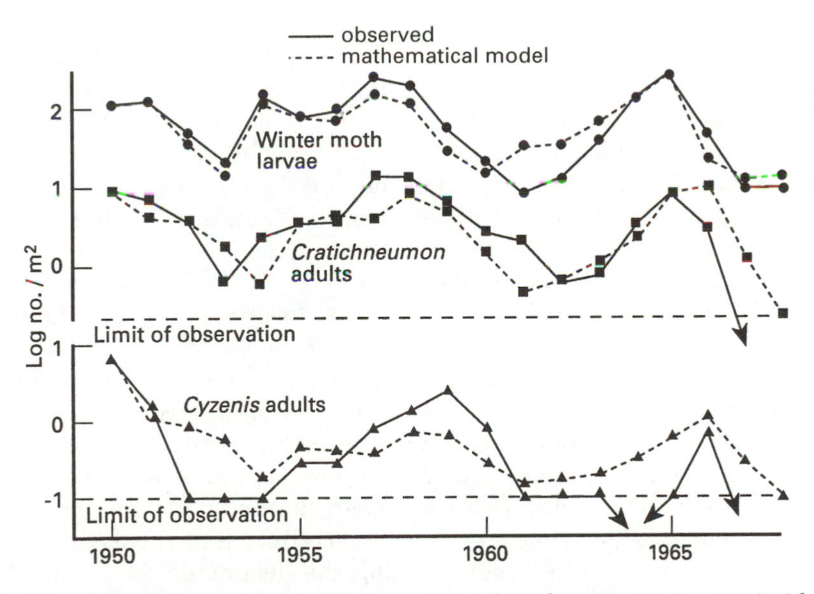

Figure 5.25 Observed changes in density of the winter moth and its two main parasitoids, and the densities predicted by the mathematical model. (Source: adapted from Varley *et al.*, 1973.) Reproduced by permission of Blackwell Scientific Publications Ltd.

may alter or become obscured. Moreover, the method depends heavily on knowing all of the important factors to include in the study at the outset. There is not much scope for incorporation of new components at a later stage. There are a number of other problems as outlined in A to F below:

A: Contemporaneous and sequential mortalities Difficulties can arise when more than one agent acts contemporaneously on a stage or when the precise sequence in which they act is unclear. Clearly, changes in the proportion killed by one agent will affect the number available to be attacked by other agents. Whether they are assumed to act concurrently or sequentially will have an important bearing on the results of the analysis. Buonaccorsi and Elkinton (1990) provide methods for estimating contemporaneous mortality factors using **marginal attack rates**. The marginal attack rate of a mortality factor is equivalent to the proportion of the population which would be killed by the factor acting alone, instead of in combination with other factors (Bellows *et al.*, 1992). The methodology can also be extended to give estimates of *k*-values (see Bellows *et al.*, 1992 for a review). Assumptions about the sequence in which the mortalities act can strongly affect conclusions drawn, a point made forcefully by Putman and Wratten (1984) who audaciously illustrated their argument with a re-analysis of Dempster's (1975) study of the cinnabar moth (*Tyria jacobaeae*). In the original study, which assumed starvation of larvae to precede predation, Dempster's analysis showed starvation to be the key factor. Putman and Wratten reversed the sequence of these mortalities and found that predation became the key factor instead. We may question the justification for Putman and Wratten's re-ordering of the sequence of mortalities in this study, but we cannot ignore the point of the demonstration!

B: Composite mortalities Some of the mortality categories in the life-table may contain or mask a number of others which could be important key or regulating factors. This is particularly likely to be the case with poorly understood, wide categories, such as 'winter disappearance' in the winter moth example. Varley *et al.* (1973) account for this variable mortality as being mainly due to asynchrony between egg hatch and tree bud burst. Late opening of buds deprives young larvae of leaves to feed on, leading to death or emigration. However, other unstudied processes may also have had a part to play, for example, variations in adult fecundity, egg mortality from a number of possible sources, predation of early instars, etc.

C: Proving causation As Price (1987) has pointed out, the methods of life-table analysis, based as they are on correlation, do not always provide an unambiguous picture of cause and effect relationships. We can distinguish between **Type A density-dependence**, which is *causally* related to changes in population density and **Type B density-dependence** which is only *statistically* related (Royama, 1977), and to prove the former we need ideally to obtain life-table data both in the presence and absence of the suspected agent. Biological control introductions offer one potentially productive source of information on the causative role of natural enemies, but few 'before and after' life-table studies have in fact been carried out (subsection 5.2.2). Ryan (1990), however, has described a good example in the larch casebearer, *Coleophora laricella* (Lepidoptera), in North America, against which two parasitoid species were introduced from Europe. Life-tables were carried out both before and after the introductions and from those it became clear that one of the parasitoids, *Agathis pumila* (Braconidae) was the key factor inducing the decline of the moth. The parasitoid also appeared to be acting in delayed density-dependent fashion, stabilising moth numbers at low densities. Another successful example, which we have already alluded to, is the study of Embree (1966, 1971) on the introduction of two winter moth parasitoids, *Cyzenis albicans* and *Agrypon flaveolatum* (Ichneumonidae) into Nova Scotia. An opportunity to carry out a similar study was more recently afforded by the introduction of the winter moth to western Canada (Roland, 1990; see also Roland, 1994). The same two parasitoid species were introduced to British Columbia from 1979 to 1981, host populations subsequently declining to one-tenth of their peak density. Parasitism, mostly from *C. albicans*, rose from zero before introduction to around 80% in 1984 and declined thereafter to 47% in 1989. Mortality of pupae, interestingly, rose during the same period to a level higher (>90%) than that caused by parasitism, suggesting a strong interaction between parasitism and subsequent mortality of unparasitised pupae. This effect was subsequently found to be present also in the Nova Scotia data. Roland suggested three possible explanations:

1. pupae parasitised by *Cyzenis* are present in the soil for twice as long as unparasitised pupae, so the greater availability of pupae after parasitoid introduction may be attracting higher numbers of pupal predators;
2. predation and parasitism do not act independently of each other, predation rising in the presence of parasitism;
3. pupal mortality factors in the soil have a minor effect at high population density and only exert a major effect after populations have declined.

To unravel the factors responsible Roland carried out experiments with placed-out moth pupae (subsection 5.2.8), pitfall traps (subsection 4.2.1) to measure predator activity and exclusion cages of different mesh size (subsection 5.2.3) to determine which predator sizes, if any, account for pupal mortality. Staphylinid beetles were found to be the most likely contenders, being also important predators of winter moth pupae in Britain. The results of Roland's experiments suggest that explanations 2. and 3. both have a part to play, predators showing a preference for unparasitised pupae (loading survival in favour of parasitoids and against the moth), predation becoming a major factor only after the parasitoid-induced decline (for an update on the winter moth analysis, see Roland, 1994).

D: Interaction effects Roland's analysis highlights the difficulties created when interaction effects occur between mortality factors. This is a problem which conventional life-table analysis is not equipped to deal with, assuming as it does that factors operate independently of each other. The only effective solution is to carry out factorial exclusion experiments (subsection 5.3.5) both in the presence and in the absence of the suspected interacting agents, under a range of relevant conditions, e.g. population density.

E: Compensatory effects An alternative method for evaluating the role of natural enemies from life table data, discussed by Price (1987), might be to develop survivorship curves for cohorts of insects and to subtract from these the effects of specific natural enemies. A comparison could then be made to assess the important contribution of natural enemies to mortality (see Figure 5.26). As Price himself points out, however, this type of analysis is likely to lead to very misleading conclusions, as it fails to recognise the possibility of compensation in the system. For example, removal of a high mortality due to natural enemies, may be compensated for by a relatively higher mortality from other factors, such as starvation or adverse weather conditions. An understanding of these potential compensatory mechanisms is crucial and again can only be gained adequately by experimentation.

F: Difficulties in detecting density-dependence It is possible for strongly regulated populations to show little variation from

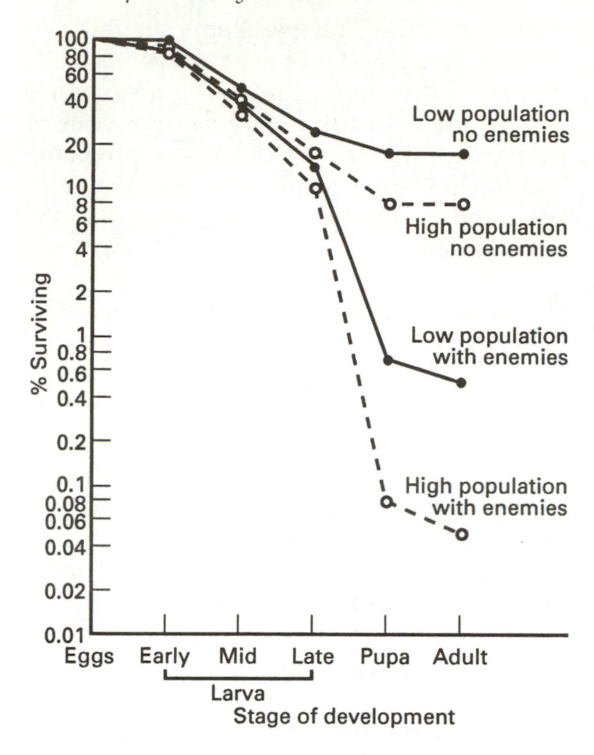

Figure 5.26 Survivorship curves for low and high populations of spruce budworm, *Choristoneura fumiferana*, together with those in which the effects of natural enemies have been removed. (Source: Price, 1987.) Reproduced by permission of Academic Press.

equilibrium and this may make statistical detection of the processes of regulation difficult using traditional life-table methods (Gould *et al.*, 1990). Equally, stochastic variation may also obscure underlying density-dependent processes. Dempster (1983), for example, analysed 24 sets of data on Lepidoptera and could find in only three cases evidence of density-dependent mortality from natural enemies. He concluded that most insect populations are unlikely to be regulated by predators or parasitoids. However, Hassell (1985), using a simple model, showed that density-dependence can be present but remain undetected because of natural stochastic variation obscuring the relationships (see, however, Mountford, 1988). Hanski (1990) provides a useful review of the various problems inherent in detecting density-dependence from life-table and time-series data, together with a number of the statistical methods which have been proposed (see also Pollard *et al.*, 1987).

Being aware of the possible pitfalls outlined above is crucial before embarking on any population study based on age-specific life-tables, but it is in the nature of such long-term studies that unforeseen problems are likely to arise and may be difficult to correct after starting. For further more detailed treatments of age-specific or stage-structured life-table analysis, the reader is referred to Manly (1990).

Time-specific Life-tables

Time-specific (or vertical) life-tables are more suitable for use with populations in which the generations overlap, due to a short development time of the immature stages relative to the reproductive period of the adults (Kidd, 1979). Such populations (humans and aphids being examples) tend, after a period of time, to achieve a stable age distribution (Lotka, 1922) in which the proportion of the population in each age group or stage remains constant. In this situation, all the ecological processes affecting the population are, at least in theory, operating concurrently. This means that the relative numbers in each age group at any instant in time provides an indication of the proportional mortality from one age group to the next. However, we cannot deduce from this what mortality factors are operating, or whether any regulation is occurring, so the value of a time-specific life-table is limited in this respect.

Estimating mortality from parasitism may be easy to do with discrete generations (Varley, *et al.*, 1973; van Driesche and Taub, 1983; but see subsection 5.3.2), but is more difficult when generations overlap. Van Driesche and Bellows (1988) provide an analytical method for doing this.

Hughes (1962, 1963, 1972) developed a technique based on the time-specific life-table approach, which could be used for analysing aphid populations with a stable age (i.e. instar) distribution. Using a graphical method to compare population profiles at successive physiological time intervals, Hughes was able to partition the mortalities acting on the different instars, for example, parasitism, fungal disease and 'emigration'. As Hughes (1972) pointed out, however, there is no easy way of estimating errors in the construction of these life table diagrams. In fact, the whole technique is critically dependent on the assumption of a stable age distribution. Although Hughes provided a simple statistical (χ^2) method to test the validity of the assumption, Carter *et al.* (1978), have shown it to be insensitive to significant changes in the age distribution. Applying a more stringent test to Hughes' own field data for the cabbage aphid, *Brevicoryne brassicae* (L.), upon which his technique was developed (Hughes, 1962, 1963), Carter *et al.* (1978) found that these populations never achieved a stable instar distribution. Although Hughes' method has been widely used and was recommended for the International Biological Programme's study of the aphid *Myzus persicae* (Mackauer and Way, 1976), it should now only be used with extreme caution. Readers interested in the detailed methodology should consult Hughes (1972) and Carter *et al.* (1978).

While Hughes' method is now considered to be of limited applicability, his work did lead directly to the development of the earliest simulation models for analysing insect populations with relatively complex population processes. For field populations with overlapping or partially overlapping generations, the use of such models is now the only sensible way forward. These techniques are discussed in detail below (subsection 5.3.8).

Variable Life-tables

The term 'variable life-table' (or 'time-varying life-table') has been used to describe a particular class of computer-based, age-structured population model, in which the birth and survival rates experienced by each age-class change in a realistic way (Gilbert *et al.*, 1976). The population life-table is in fact computer-generated from reproduction and survival relationships obtained in the field or laboratory, and as such becomes the output of the exercise rather than forming the basis of the analysis. The technique has therefore more in common with the methodology of simulation modelling than with that of life-table analysis, and will be discussed further in subsection 5.3.8.

5.3.5 MANIPULATION EXPERIMENTS

Convergence Experiments

The problems of detecting density-dependence from life-table data have already been discussed (subsection 5.3.4). One way of testing directly whether density-dependent mechanisms are operating is to carry out a 'convergence experiment' (Nicholson, 1957) in which densities of comparable subpopulations are manipulated to achieve artificially high or low levels and are then monitored through time. Convergence to a common density is then taken as evidence for density-dependent regulation. What constitutes an 'artificially' high or low population density in the context of this type of experiment will vary according to the species under study, and can only be adequately assessed from some historical knowledge of past densities. Practical difficulties in manipulating densities of some species may also limit the usefulness of this technique. Amongst successful studies, Brunsting and Heessen (1984) manipulated densities of the carabid predator *Pterostichus oblongopunctatus* within enclosures in the field and found evidence for convergence within two years. Criticisms can be levelled at this technique in that enclosures may prevent emigration or immigration of beetles, leading to spurious mortality from 'artificial' sources. In this particular study, however, care was taken to note that beetle motility was naturally low and remained low even at the enhanced densities, there being no evidence for density-induced emigration. Gould *et al.* (1990) manipulated densities of gypsy moth by artificially loading eight forest areas with different densities of egg masses to achieve a wide range of infestation levels. This method revealed previously undetected density-dependent mortality in the larval stage, primarily due to two parasitoid species. Orr *et al.* (1990) also provide a good example of a convergence experiment carried out in the laboratory using the freshwater predatory bug, *Notonecta*.

Factorial Experiments

Factorial experiments are used to determine whether factors potentially capable of limiting population numbers combine in a simple additive way, or show more complex patterns of interaction (Hilborn and Stearns, 1982; Arthur and Farrow, 1987; Mitchell *et al.*, 1992). There are three criteria for successfully carrying out factorial experiments:

1. at least two factors need to be manipulated to at least two levels each;
2. a sufficiently long time series of data must be available to assess equilibrium population levels around which numbers fluctuate;
3. there must be replication (Mitchell *et al.*, 1992).

Mitchell *et al.* (1992) examined the interaction between resource levels (food and food/water ratio) and three population levels (zero, low, high) of the parasitoid *Leptopilina heterotoma* on laboratory populations of *Drosophila melanogaster*. This provided 12 different experimental combinations of the three potentially interacting factors. Both food and wasps showed significant effects on equilibrium levels, but without any significant interaction. With this type of experiment involving census data collected over time, the problem of serial autocorrelation is encountered (Arthur and Farrow, 1987), which makes the use of analysis of variance inappropriate. This can be circumvented using GLIM (see Mitchell *et al.*, 1992, and Crawley, 1993).

5.3.6 EXPERIMENTAL COMPONENT ANALYSIS

As explained in subsection 5.2.15, this approach is based on the belief that the complexities of ecological interactions, such as those involving predation and parasitism, can be quantified in terms of a relatively small number of dynamic processes (Southwood, 1978). Each process, reduced to its component parts, can be investigated experimentally and described by a series of equations. The equations describing all the component processes can then be incorporated into a system or population model, the accuracy of which can then be assessed by comparing its behaviour with real observations.

The components of predation can be investigated experimentally using the important distinction between functional and numerical responses (subsection 5.2.15). To assess the significance of these responses, particularly to predator–prey and parasitoid–host population equilibrium levels and stability, two different modelling approaches can be adopted; one based on simple analytical models, the other involving the construction of more elaborate simulation models.

5.3.7 PUTTING IT TOGETHER: ANALYTICAL MODELS

Incorporating the Components of Predation

To assess the impact of parasitism or predation on an insect population, the information on functional and numerical responses needs to be incorporated into population models. Analytical models are usually based on systems of relatively simple equations which can be 'solved', usually by rearrangement, to provide straightforward answers. Some population models, however, have systems of equations which are too complex for simple solution and the only way of obtaining useful insights is to perform simulations with the model under differing conditions, for example by changing parameter values. Of course, there is no reason why models capable of analytical solution cannot also be used for simulation. Analytical solutions tend to be more tractable when a simple **deductive** modelling approach is adopted (deductive models are those based on very general, often intuitive, concepts, which can be useful in providing insights which might apply to particular case studies; philosophically, deduction is the process of arguing from the general case to the particular case (cf. induction, inductive models, subsection 5.3.4).

Whilst populations with overlapping generations and stable age distributions can be modelled analytically in continuous time using differential equations, this method is less suitable for the bulk of insect populations, at least in temperate regions, which have **discrete** generations, i.e. separated in time. A more appropriate modelling format is provided by difference equations which model population change in discrete time steps, i.e. $N_{t+1} = f(N_t)$. The discrete time model which has been most widely used in insect population ecology is the host–parasitoid model of Nicholson and Bailey (1935), hereafter referred to as the **Nicholson– Bailey model**. Originally developed to explore the dynamic implications of parasitoid searching behaviour, the model has been extensively elaborated in recent years to examine other features of parasitoid (and by implication, predator) biology (Hassell, 1978; Waage and Hassell, 1982; Hassell and Waage, 1984; Godfray and Hassell, 1988; May and Hassell, 1988; Hassell and Godfray, 1992). No doubt part of the appeal of this model lies in the simplicity with which it purports to capture the essence of the parasitoid–host interaction. Using a time step of one generation, the model takes the following form:

$$N_{t+1} = F \cdot \exp(-aP_t) \qquad (5.18a)$$

$$P_{t+1} = N_t[1 - \exp(-aP_t)] \qquad (5.18b)$$

where N_t and P_t are the numbers of hosts and parasitoids respectively at time t, F is the host net rate of increase in the absence of parasitism and a is the parasitoid's **area of discovery**, which is essentially the proportion of the habitat which is searched in the parasitoid's lifetime. A number of assumptions about the parasitoid and its host are implicit in these equations:

1. generations of both populations are completely discrete and fully synchronised; the time-step (t) is therefore one generation;
2. one encountered host leads to one new parasitoid in the next generation;
3. the parasitoid is never egg-limited;
4. the area of discovery (= searching efficiency) is constant;
5. each parasitoid searches the habitat at random.

This latter assumption is catered for in the model by using the Poisson distribution to distribute attacks at random between hosts. The zero term of the distribution (e^{-x}, where x is the mean number of attacks per host) defines the proportion of the host population escaping attack, in this case e^{-aP}, or $\exp(-aP)$. The proportion attacked is therefore $1-\exp(-aP)$. A more detailed description of the derivation of these equations is not provided here as it has already been covered in a number of texts (e.g. Varley *et al.*, 1973; Hassell, 1978).

Following Hassell (1978), the equilibrium populations N^* and P^* can now be found by setting $N_{t+1} = N_t = N^*$ and $P_{t+1} = P_t = P^*$ giving the analytical solution:

$$N^* = F \qquad (5.19a)$$

$$P^* = \frac{\log_e}{a} \qquad (5.19b)$$

The equilibrium levels of both populations thus change with respect to both F and a.

However, the Nicholson–Bailey model *is inherently unstable*, a fact that can be confirmed either by simulation (Figure 5.27a) or by **stability analysis** (Hassell and May, 1973; Hassell, 1978) (stability analysis is a technique which has been widely used in the analysis of deductive models, but the mathematics are beyond the scope of this book; we refer readers to the appendices in Hassell and May, 1973 and Hassell, 1978). When perturbed from equilibrium, the model produces oscillations of increasing amplitude, which in the real world would result in the extinction of one or both populations (Figure 5.27a). Stability in the model could easily be produced, however, by the incorporation of a density-dependence component into *F*, to simulate, for example, competition between hosts for food resources. Progressively increasing the degree of density-dependence produces in the first instance stable limit cycles (Figure 5.27b) followed by damping oscillations (Figure 5.27c). Whilst density-dependence in *F* would be a reasonable component to incorporate, it does not advance our understanding of how features of parasitoid biology such as searching behaviour influence dynamics. To achieve that, more detailed descriptions of, for example, functional and numerical responses need to be incorporated into the model.

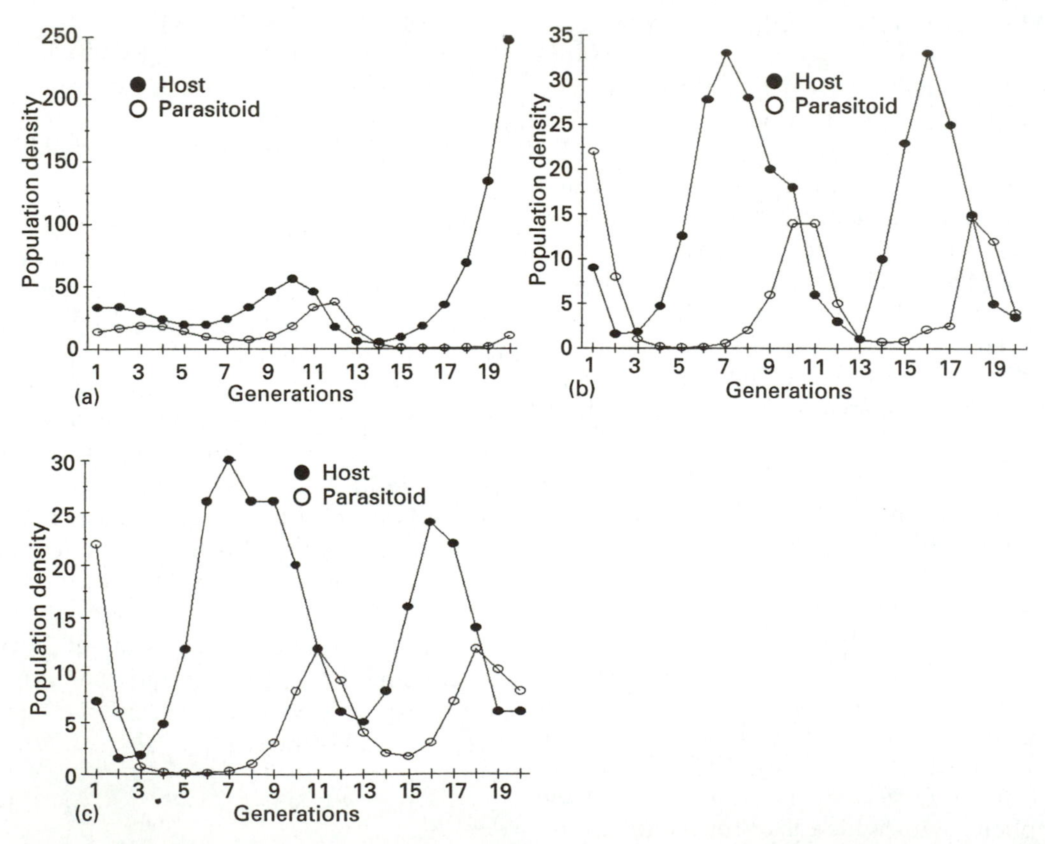

Figure 5.27 (a) Typical numerical changes predicted by the Nicholson–Bailey model. The incorporation of (increasing) density-dependence into the model results (b) in cyclical oscillations within an upper and lower boundary (limit cycles), followed by (c) damping oscillations which approach an equilibrium.

To facilitate the incorporation of the components of predation into the Nicholson–Bailey model, it is useful to begin with a more generalised form of the model (Hassell, 1978):

$$N_{t+1} = FN_t\, f(N_t, P_t) \qquad (5.20a)$$

$$P_{t+1} = N_t[1 - f(N_t, P_t)] \qquad (5.20b)$$

in which survival of hosts is a function of both host and parasitoid numbers. This survival function can now be explored in relation to the following components:

A: Functional responses The assumption of a constant searching efficiency described by one component, *a* or *a*′, is of course an oversimplification, given what we know of the way in which parasitoid and predator attack rates change with prey density (section 1.10). The way in which handling time affects this relationship have already been discussed (section 1.10) and can be described by **Holling's 'disc' equation**:

$$\frac{N_e}{P_t} = \frac{a'TN_t}{1 + a'T_hN_t} \qquad (5.21)$$

where N_e is the number of hosts encountered. This Type 2 functional response (section 1.10) can now be incorporated into the Nicholson–Bailey model using the so-called **random parasite' equation** first described by Rogers (1972):

$$N_a = N_t\left[1 - \exp\left\{-\frac{a'TP_t}{1 + a'T_hN_t}\right\}\right] \qquad (5.22)$$

Predator versions of equation 5.22 were also developed by Royama (1971) and Rogers (1972) to take account of gradual prey depletion during each interval *t*. Prey eaten by predators do not remain exposed to further encounters, in contrast to hosts which may be re-encountered and thus incur additional T_h costs to the parasitoid. Reproducing the Royama (1971) equation:

$$N_a = N_t\left[1 - \exp\left\{-a'P_t\left(T - T_h(N_a/P_t)\right)\right\}\right] \qquad (5.23)$$

Note that here N_a is present on both sides of the equation. The simplest way of finding N_a for given values of the other parameters is by iteration (i.e. by repeatedly subsituting different values of N_a, until both sides of the equation balance. Detailed mathematical derivations for both equations (5.22) and (5.23) are given by Hassell (1978). As the Type 2 functional response can be seen to be inversely density-dependent when percentage parasitism is plotted against prey density (Figure 1.15b), it is perhaps not surprising that when incorporated into the Nicholson–Bailey model as:

$$f(N_t, P_t) = \exp\left[-\frac{-a'TP_t}{1 + a'T_hN_t}\right] \qquad (5.24)$$

for parasitoids, its effect is to further destabilise the model. In general, the greater the ratio T_h/T the greater the destabilising effect, while the original Nicholson–Bailey model is re-established when $T_h = 0$ (Hassell and May, 1973). Although T_h thus determines both the degree of destabilisation and the maximum attack rate (= the asymptote), both could equally well be influenced by egg-limitation (Hassell and Waage, 1984; section 1.10).

To explain the sigmoid Type 3 functional response, Hassell (1978) suggested a model which assumes that only *a*′ varies with prey density, such that $a' = bN_t\,(1 + cN_t)$, with *b* and *c* constants. This gives a sigmoid analogue to the disc equation:

$$N_e/P_t = \frac{bN_t^2T}{1 + cN_t + bT_hN_t^2} \qquad (5.25)$$

where N_e is the number of prey encountered. For parasitoids, where hosts remain to be re-encountered:

$$N_a = N_t\left[1 - \exp\left(-\frac{bTN_tP_t}{1 + cN_t + bT_hN_t^2}\right)\right] \qquad (5.26)$$

which can easily be incorporated into the Nicholson–Bailey model. Hassell *et al.* (1977) also provide an alternative equation for predators, where prey are gradually depleted with time:

$$N_a = N_t\left[1 - \exp\left\{-\frac{bP_t}{c}\left(T - \frac{T_h N_a}{P_t} - \frac{N_a}{bN_t P_t (N_t - N_a)}\right)\right\}\right] \quad (5.27)$$

N_a can again be found by iteration (cf. equation 5.23). Both equations (5.26) and (5.27) produce similar sigmoid relationships, but (5.27) is easier to use. Intuitively, such sigmoid responses might be expected to have a stabilising influence on population interactions, where the equilibrium falls within the density-dependent part of the response (Figure 1.15c). This has been demonstrated by Murdoch and Oaten (1975) using continuous time differential equations (see below) with no time delays. However, the time delay of one generation inherent in the Nicholson–Bailey model is sufficient to prevent any sigmoid functional response, of the form of equation above, from stabilising an interaction (Hassell and Comins, 1978).

The above conclusion is, of course, restricted to predators or parasitoids which are prey- or host-specific. For generalist predators, which are more loosely coupled to their prey, the situation is somewhat different (Hassell, 1986). Here, predators may display switching behaviour (section 1.11). Hence, neither the predator numerical response nor its population density is likely to be dependent on the abundance of any one prey type. The interaction between a generalist predator and a *single* prey species was modelled by Hassell and Comins (1978) using equation (5.26) above for a parasitoid population at equilibrium:

$$P_{t+1} = P_t = P^* \quad (5.28)$$

The model was found to have two equilibria (Figure 5.28), the lower (S) being locally stable, while the upper (R) was unstable, such that prey exceeding R escaped parasitoid control and increased indefinitely. The

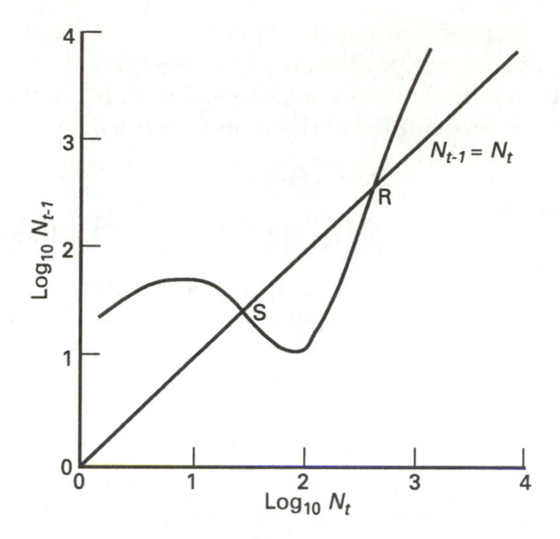

Figure 5.28 A population growth curve for equation 5.26. The intersections with the 45° line are the lower potentially stable equilibrium (S) and the upper unstable equilibrium (R). (Source: Hassell and Comins, 1978.) Published by permission of Elsevier Science.

important point in this model is that the total response of the parasitoid shows a sigmoid relationship with prey density. This can be achieved by a sigmoid functional response and a constant parasitoid density, as in the model, or, alternatively, by a rising numerical response to prey density, coupled with either a Type 2 or a Type 3 functional response (Figure 5.29).

At first sight the incorporation of functional responses into simple models seems to be fairly straightforward, but there are a number of complications which the reader needs to be aware of:

1. for a particular predator, the functional response is likely to vary with age or size of both predator and prey, and the model may have to be modified to incorporate the effects of age structure (see **Other Analytical Modelling Approaches**, below);
2. simple laboratory experiments to assess functional responses over a short time

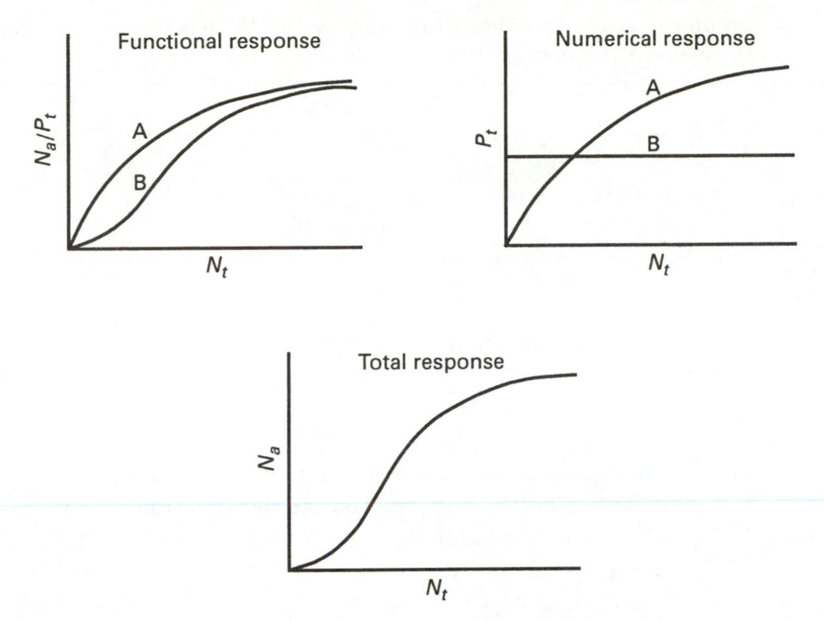

Figure 5.29 Alternative ways of achieving a sigmoid total functional response between prey eaten (N_a) by P_t predators and prey density N_t. The total response may be achieved by combining either response A or B with numerical response A or from functional response B with no numerical response. (Source: Hassell, 1978.) Reproduced by permission of Princeton University Press.

period (e.g. 24 hours) should be used in *generation-based* models with caution, as they may give a misleading impression of the predator's *lifetime* functional response. This in the context of the Nicholson–Bailey model is the relevant component if we are interested in understanding the effects of functional responses on population dynamics (Waage and Hassell, 1982; Kidd and Jervis, 1989). The problem may be avoided, however, when we consider the functional response in relation to predator aggregation in patchy environments (see below).

B: Aggregative responses Although the Nicholson–Bailey model assumes random search by parasitoids (i.e. each host has the same probability of being parasitised), in reality natural enemies tend to show an aggregative response (defined in subsection 1.14.2). Examples have been widely reported and reviewed by a number of authors (Hassell *et al.*, 1976; Hassell, 1978; Krebs and Davies, 1978; Lessells, 1985; Walde and Murdoch, 1988). This behaviour has already been discussed in relation to foraging behaviour (section 1.14), but here we are concerned with its implications for population dynamics.

Hassell and May (1973) modelled the effects of parasitoid aggregation in a simple way by first distributing hosts and parasitoids between n patches, then considering each patch as a sub-model of the Nicholson–Bailey model (Table 5.3). Thus, in each patch i, there is a proportion α_i of hosts and β_i of parasitoids.

As can be seen in Table 5.3, a greater proportion of parasitoids is placed in the high density patches of prey than in the low density ones. The parasitism function now becomes:

$$f(N_t, P_t) = \sum_{i=1}^{n}\left[\alpha_i \exp(-a\beta_i P_t)\right] \qquad (5.29)$$

Table 5.3 The proportional distribution of hosts (α) and parasitoids (β) between n patches to incorporate the aggregative response into the Nicholson–Bailey model

	Patch				
	1	*2*	*3*	*4*	*5*
α	0.5	0.2	0.1	0.1	0.1
β	0.8	0.1	0.05	0.03	0.02

which redistributes hosts and parasitoids as in the above scheme at the beginning of each generation. In the case where parasitoids are distributed evenly over all patches we regain the property of random oviposition as in the original model. The stability analysis of Hassell and May (1973) shows that the model may now become stable with a sufficiently uneven prey distribution and enough parasitoid aggregation in high density host patches. To allow easier analysis of the properties of the model, Hassell and May (1973) used a single high host-density patch (α) and distributed the rest of the host population evenly amongst the other patches [$(1 - \alpha)/(n - 1)$]. The parasitoid distribution was defined by a single 'aggregation index' μ where:

$$\beta_i = c\alpha_i^{\mu} \tag{5.30}$$

with c a normalisation constant. The degree of parasitoid aggregation is now governed by μ, $\mu = 0$ corresponding to random search and $\mu = \infty$ to the situation where all parasitoids are in the high host density patch. Stability is now affected by precise values of μ, F, α and $n - 1$ (Figure 5.30).

A more general model has also been developed by May (1978), to capture the essential features of parasitoid aggregation without the detail. This model uses the negative binomial

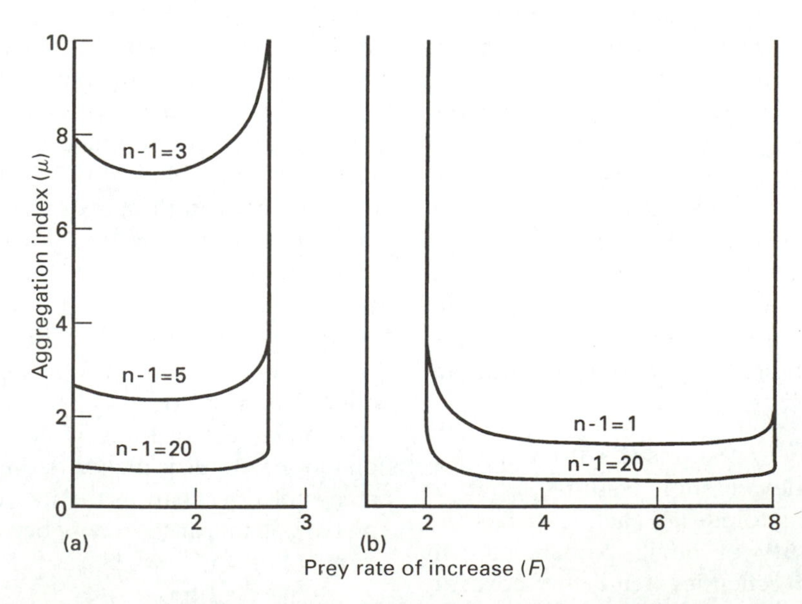

Figure 5.30 Stability boundaries between the aggregation index μ and the prey rate of increase F, for different values of n - 1: (a) $\alpha = 0.3$; (b) $\alpha = 0.7$. (Source: Hassell and May, 1973.) Reproduced by permission of Blackwell Scientific Publications Ltd.

to distribute parasitoid encounters between hosts. Thus:

$$N_{t+1} = FN_t\left[1 + (aP_t/k)\right]^{-k} \qquad (5.31a)$$

$$P_{t+1} = N_t\left[1 - \left(1 + (aP_t/k)\right)^{-k}\right] \qquad (5.31b)$$

Here the parameter k (the exponent of the negative binomial) describes parasitoid aggregation, strongest when $k \rightarrow 0$ and weakest when $k \rightarrow \infty$ (random). May's model, and variants thereof, has been used by a number of authors to include an aggregative response component into population models (Beddington *et al.*, 1975, 1976a, 1978; Hassell, 1980b). Hassell (1978) provides a good account of the development and application of this model.

The stabilising potential of aggregation by predators and parasitoids must therefore temper our previous conclusions regarding the significance of functional responses, as measured in single-patch experiments (section 1.10). We can envisage the Type 2 and Type 3 response curves as essentially a 'within patch' phenomenon, with searching between patches defined by the aggregative response (Hassell, 1980) (see also section **What is Searching Efficiency?**, below).

C: Mutual interference The incorporation of mutual interference (subsection 1.14.3 gives a detailed discussion) into the Nicholson–Bailey model was first carried out by Hassell and Varley (1969), using the inverse relationship between parasitoid searching efficiency and the density of searching parasitoids shown in equation (1.6). Removing the logarithms, this becomes:

$$a = QP_t^{-m} \qquad (5.32)$$

which when substituted into the Nicholson-Bailey model gives the equations:

$$N_{t+1} = FN_t \exp\left(-QP_t^{1-m}\right) \qquad (5.33a)$$

$$P_{t+1} = N_t\left[1 - \exp\left(-QP_t^{1-m}\right)\right] \qquad (5.33b)$$

This modification has the effect of producing a stable equilibrium given suitable values of m and F (Figure 5.31). The higher the mutual interference constant m, and the lower F, the

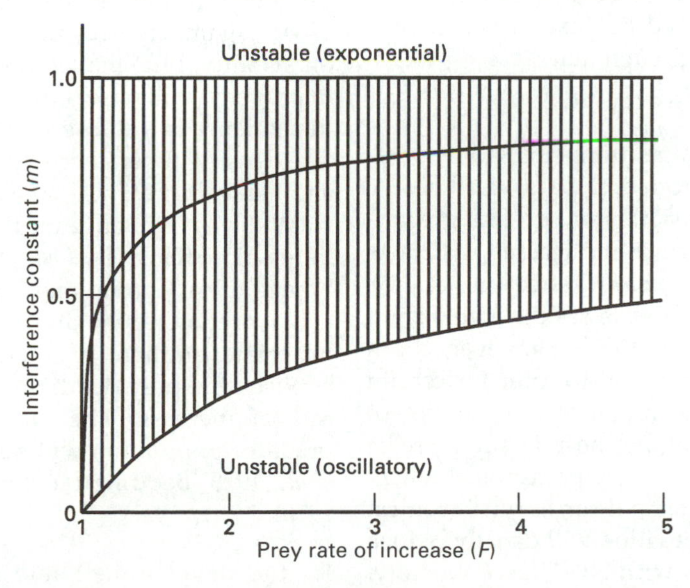

Figure 5.31 Stability boundaries for equation (5.33) in terms of the interference constant m and the prey rate of increase F. The hatched area denotes the conditions for stability, approached through exponential damping above the central curve and oscillatory damping below the curve. (Source: Hassell and May, 1973.) Reproduced by permission of Blackwell Scientific Publications Ltd.

more likely stability becomes. Q has no effect on stability, but does affect the equilibrium level.

More elaborate, but behaviourally more meaningful, mathematical descriptions of mutual interference have also been developed (Rogers and Hassell, 1974; Beddington, 1975) and explored in deductive models. While the precise stability conditions in these models may differ from the earlier version, the essential conclusion, that mutual interference can be a powerful stabilising force, remains intact.

D: Numerical responses In strict terms, the aggregative response is also a form of numerical response, but here we use the term to refer specifically to changes in predator numbers from one generation to the next. In the Nicholson–Bailey model this is simply achieved by making the proportion of hosts killed by parasitoids in each generation $[1 - exp(- aP_t)]$ equivalent to the number of parasitoids in the next generation. Each host killed therefore produces one live parasitoid. Gregarious larval development can, however, be easily catered for by incorporating an additional component c into equation 5.18b, such that:

$$P_{t+1} = cN_t[1 - \exp(-aP_t)] \qquad (5.34)$$

Here, c is the average number of adult parasitoids to emerge from each parasitised host ($c = 1$ for solitary parasitoids and $c > 1$ for gregarious parasitoids (Waage and Hassell, 1982). As incorporated above, c has no effect on stability, but raising its value depresses the equilibrium (Waage and Hassell, 1982). However, c can also be further elaborated to cater for certain factors which may influence the number of parasitoids emerging per host. For conspecific superparasitism in solitary parasitoids, where only one larval parasitoid can survive (contest competition), the situation will usually reduce to equation (5.18b) with $c = 1$ (exceptionally, superparasitism results in the death of both rivals in a host, in which case, $c < 1$). If there is mortality from either multiparasitism, hyperparasitism or encapsulation, c will be < 1. Where clutch size affects larval survival adversely in gregarious species (scramble competition), this effect can be incorporated by replacing c in equation (5.34) with $c(1-\delta F)$, for values = >0, where F is clutch size and δ a constant defining the strength of density-dependence. This expression assumes a negative linear relationship between clutch size and progeny survival (although a negative exponential relationship may be more realistic), and will have a regulating effect both on the parasitoid population and the parasitoid–host interaction. To obtain realistic values for parameter c would require the construction of detailed life-tables for the parasitoid (Hassell, 1969; Escalante and Rabinovich, 1979).

An additional parameter s can also be incorporated into equation (5.34) to take account of variation in the sex ratio of the parasitoid progeny. Thus:

$$P_{t+1} = scN_t[1 - \exp(-aP_t)] \qquad (5.35)$$

where s is the proportion of parasitoid progeny that are female (Hassell and Waage, 1984). Again, changes in s will have no effect on stability, but smaller values (i.e. male-bias in progeny) will raise the equilibrium. Density-dependence in s (subsection 1.9.2) (to incorporate density-dependence, s has to be altered to another form, i.e. $s = f$ [parasitoid density]), has a stabilising influence on the parasitoid–host interaction (Hassell *et al.*, 1983; Hassell and Waage, 1984; Comins and Wellings, 1985).

For predators, the above models are inappropriate, as there is no simple relationship between the prey death rate and the predator rate of increase. The rate of increase of a predator population will depend on (Lawton *et al.*, 1975; Beddington *et al.*, 1976b; Hassell, 1978):

1. the development rate of the immature stages;
2. the survival rate of each instar;
3. the fecundity of the adults.

The factors affecting each of these components are considered in detail in sections 2.9, 2.10 and 2.7 respectively. To build a general model of the predator rate of increase we would need to incorporate the effects of prey consumption on development and survival of the different instars, and on adult fecundity. This task would be beyond the scope of analytical modelling (Hassell, 1978), being more suited to simulation (subsection 5.3.8). Beddington *et al.* (1976a), however, took a simpler approach whilst retaining some of the features of predator reproduction. Adult fecundity F was related to the number of prey eaten during the predator's life by the equation:

$$F = c\left[\left(N_a/P_t\right) - \beta\right] \qquad (5.36)$$

where N_a is the number of prey attacked, c is the efficiency with which consumed prey are converted to new predators and β is the threshold prey consumption needed for reproduction to start (see also equation (2.3) and related discussion). The model therefore takes account, in a simple way, of the predator's need to use some prey for growth and maintenance (2.8.3). Incorporating this equation into the Nicholson–Bailey model yields the equations:

$$N_{t+1} = N_t \exp\left[\rho\left(1 - N_t/K\right) - aP_t\right] \qquad (5.37a)$$

$$P_{t+1} = c\left[\left\{N_t\left[1 - \exp(-aP_t)\right]\right\} - \beta P_t\right] \qquad (5.37b)$$

with the rate of increase of the prey population, in the absence of predation, defined by $\rho(1-N_t/K)$, which includes a density-dependent feedback component. Of course, handling time, predator aggregation and mutual interference are not included. Nevertheless, the model can be used to show the stability differences between predator–prey models where $\beta > 0$ and those of parasitoid–host models where $\beta = 0$. The important effect of increasing β is to reduce the range of stable parameter space in the model. Furthermore, where $\beta = 0$, the model is **globally stable**, i.e. it returns to equilibrium irrespective of the degree of perturbation. With $\beta > 0$, only **local stability** is apparent, i.e. equilibrium is re-attained only when perturbation is within certain limits. Thus, as predators need to eat more prey before reproducing, the chances of a stable interaction diminish.

E: Other components The number of relevant components of predator–prey and parasitoid–host interactions which could be examined in simple analytical (Nicholson–Bailey-type) models is potentially very large. We have attempted to summarise the approach with reference to some of the more widely discussed examples. Others which have been examined include: (a) differential susceptibility of hosts to parasitism (i.e. variability in host escape responses or physiological defences, temporal asynchrony between parasitoid and host) (Kidd and Mayer, 1983; Hassell and Anderson, 1984; Godfray *et al.*, 1994); (b) parasitoid host-feeding (Kidd and Jervis, 1989); (c) competing parasitoids (Hassell and Varley, 1969; May and Hassell, 1981; Taylor, 1988b); (d) hyperparasitism (Beddington and Hammond, 1977); (e) multiple prey systems (Comins and Hassell, 1976); (f) host–generalist–specialist interactions (Hassell, 1986; Hassell and May, 1986); (g) combinations of, parasitoids and pathogens (May and Hassell, 1988; Hochberg *et al.*, 1990).

What is Searching Efficiency?

Having reviewed the essential behavioural components involved in searching by natural enemies (functional and aggregative responses, mutual interference), we are perhaps in a better position to answer the question of what is meant by the term 'searching efficiency'. This term has most often been used synonymously with 'attack rate', i.e. a more efficient predator kills more prey per unit time than a less efficient one. However, particularly in the context of population models, the detailed usage of the term has varied considerably. In the original Nicholson–Bailey model the area of discovery a defines a lifetime searching efficiency, whereas the 'attack' coefficient a' in equation (5.21), defines an 'instantaneous' searching efficiency in terms of numbers of prey attacked per unit time, T. In a patchy environment, however, searching efficiency is sensitive to two factors: (a) the patch-specific searching ability of the predators, and (b) the extent to which the distribution of the predators is non-random (Hassell, 1982a). To take account of this, Hassell (1978, 1982b) proposed a model for overall searching efficiency of predators (or parasitoids) where prey are gradually depleted:

$$a' = \frac{1}{n}\sum_{i=1}^{n}\left[\frac{1}{P_t T_i}\log_e\left(\frac{N_i}{N_i - N_{ai}}\right)\right] \qquad (5.38)$$

where n is the number of patches and N_a the number of prey attacked, and N_i, N_{ai} and T_i are the number of hosts available, number of hosts parasitised, and the time spent searching, respectively on the ith patch. This equation (which has already been discussed in Chapter 1, equation (1.8)) represents an important step forward, as it views the functional response as essentially a within-patch phenomenon, i.e. occurring on a small spatial and temporal scale. Laboratory experiments to measure functional responses have, of course, been carried out on exactly this scale (section 1.10), so the within-patch interpretation affords a more realistic correspondence between experiment and modelling. It also circumvents the need for an average *lifetime* functional response measure to use in single-patch models, such as the original Nicholson–Bailey model (see above). Potentially, the patch model defined by equation (5.38) could be further expanded to include variation in the functional response with prey size, predator age and a range of other components, but the complexity involved would make this procedure more appropriate for the simulation approach discussed below (subsection 5.3.8).

An experimental measure of overall searching efficiency of a predator can be obtained by:

1. estimating the average amount of time spent per predator in each patch (this can be calculated from the average number of predators found in each patch during a predetermined time period or over the course of the experiment (e.g. 24 h);
2. recording the average number of prey killed in each patch (providing a range of prey densities between patches).

These estimated parameter values can then be substituted in equation (5.38) to find a'. A suitable patch scale and experimental arena will need to be chosen, and this will depend on the prey species involved. Wei (1986), for example, used five clusters of rice plants, each set into a glass tube of water and interconnected by slender wooden strips to facilitate searching by the mirid bug *Cyrtorhinus lividipennis* for its prey, the brown planthopper *Nilaparvata lugens*. Ideally, a lifetime measure of searching efficiency could be obtained by repeating the procedure for each day of the predator's immature and adult life, and then either totalling or averaging the values of a' obtained. In practice, this may be very difficult due to the time involved and the large number of prey needed during the experiment.

Hassell and Moran (1976) proposed for parasitoids a measure of 'overall performance', A, that takes account of larval survival.

$$A = \frac{1}{P_t}\log_e\left[N_t/(N - P_{t+1})\right] \qquad (5.39)$$

where P_t and P_{t+1} are the densities of searching parasitoids in successive generations. Clearly, a major constraint on the effectiveness of a parasitoid is likely to be mortality during the immature stages (subsection 2.10.2). 'Overall performance' may thus provide a more useful measure of the relative usefulness of different parasitoid species in biological control (Hassell, 1982b) (subsection 5.4).

Other Analytical Modelling Approaches

So far, we have concentrated our attention on ways of expanding the Nicholson–Bailey model to include more realistic components of predation. Although space does not permit a detailed discussion, it should be mentioned that other analytical modelling frameworks are available and have been used successfully to gain insights into the dynamics of predator–prey interactions. We now briefly discuss two of these to provide the reader with a lead into the literature.

Models structured in terms of differential equations, to encompass continuous time changes, have a long pedigree, beginning with the predator–prey interactions of Lotka and Volterra (Lotka, 1925; Volterra, 1931) (Berryman, 1992 gives a recent review). Host–parasitoid versions are also available (Ives, 1992). Recently, it has been shown that, where parasitoids aggregate in patches of high host density, such models are usually unstable (Murdoch and Stewart-Oaten, 1989), in contrast to their discrete time counterparts. This serves to highlight the fact that there are important differences between the two model types, and also emphasises the point that the behaviour of analytical models can be highly sensitive to minor variations in their construction (Ives, 1992; Berryman, 1992).

Continuous-time models incorporating age structure have been developed by Nisbet and Gurney (1983), Gurney and Nisbet (1985) and extended to cover host–parasitoid systems (Murdoch *et al.*, 1987; Godfray and Hassell, 1989; Gordon *et al.*, 1991). Murdoch *et al.* (1987) have shown that the incorporation of an age class which is invulnerable to parasitism into such a model with overlapping generations can, under certain circumstances, promote stability (subsection 5.3.2).

Discrete-time models incorporating age structure have also been developed (Bellows and Hassell, 1988; Godfray and Hassell, 1987, 1989), using an elaboration of equation 5.31. Godfray and Hassell (1987, 1989) used this model form to demonstrate that parasitism can act to separate the generations of a host population, when otherwise they would tend to overlap. Whether host generations were separated in time depended on the relative lengths of host and parasitoid life cycles.

An alternative modelling format which also allows for the incorporation of age structure in populations is provided by matrix algebra (Leslie, 1945, 1948). Matrices can also be used to incorporate quite complex age-specific variations in fecundity, survival, development and longevity, thus encompassing populations which show either discrete or overlapping generations. This flexibility also makes the matrix approach extremely suitable for simulation modelling (see below). The matrix methodology is also easy to use and has the advantage over computer-based models of having an easily communicated, mathematical notation. Williamson (1972) provides a good introduction to the use of the technique in population dynamics.

5.3.8 PUTTING IT TOGETHER: SIMULATION MODELS

The main value of analytical models has been to provide insights into the general possibilities of population dynamics and how they might alter with changing conditions. Simulation models, on the other hand, attempt to mimic the detailed dynamics of particular systems and involve a somewhat different methodology. Simulation models of population dynamics can be constructed at different levels of complexity, from a relatively simple expansion of the analytical approach described in the previous section (Godfray and Hassell, 1989), to extremely elaborate systems of interlocking equations involving large numbers of components. However, they all share a common methodology for testing their accuracy (**validation**) and for assessing their behaviour (**experimentation and sensitivity analysis**) which will be discussed in detail. Whilst they can be formulated in conventional mathematical notation, simulation models are often constructed in practice as computer programs, which facilitate the complex calculations involved and also present the output in a readily accessible format. The models can be constructed in continuous or discrete time, again using systems of differential or difference equations respectively. In recent years, with the wide availability of powerful and inexpensive digital computers, the discrete time format has tended to be favoured by modellers interested in simulating the most complex insect population systems, involving age or stage-specific fecundities and mortalities (Stone and Gutierrez, 1986 and Crowley *et al.*, 1987 give continuous time examples). The aim of such inductive models is to encapsulate the detail of the particular system in question, with the emphasis on realism and accuracy. If successful, the model can often be used in decision-making, for example, in integrated pest management programmes.

To illustrate the way in which age structure can be incorporated into a relatively simple simulation model, the model of Kidd (1984), which can be used to simulate populations with either discrete or overlapping generations, is discussed. The model considers an hypothetical population reproducing asexually and viviparously. Each individual is immature for the first three days, becomes adult at the beginning of the fourth day, reproduces on the fifth and sixth days, then dies (Figure 5.32).

The population is divided, therefore, into six one-day age groups, each adult producing, say, two offspring per day. To simulate population change from day to day, the computer:

1. dimensions an array with six elements;
2. places the initial number in each element of the array (initial age structure);
3. calculates the total reproduction [REPMAX = (2* number of adults in age group 4) + (2* number of adults in age group 5) + (2* number of adults in age group 6)];
4. ages the population by one day (this is done by moving the number in each age group N into age group N + 1);
5. puts the number reproduced (REPMAX) into the one-day-old age group.

The model operates with a time-step of one day and stages 3.–5. can be repeated over as many days as required to simulate population growth. This very simple model can then be further elaborated to include additional components, such as mortality acting on each age group (e.g. either a constant proportional mortality, a uniformly-distributed random mortality or a density-dependent function) (Figure 5.33). By changing the length of immature life relative to reproductive life the

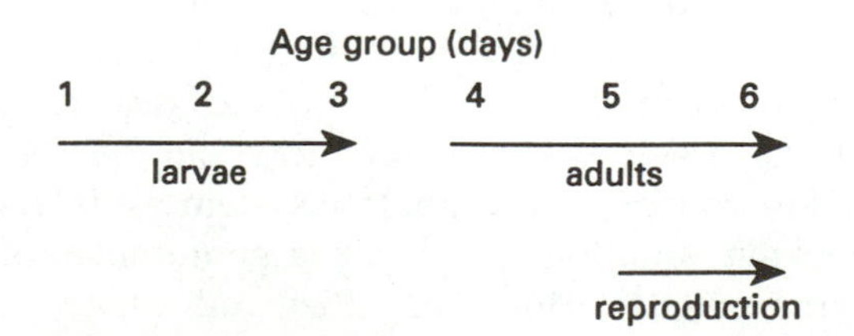

Figure 5.32 Life history characteristics of an hypothetical population with discrete generations. (Source: Kidd, 1979.) Reproduced from *Journal of Biological Education* by permission of The Institute of Biology.

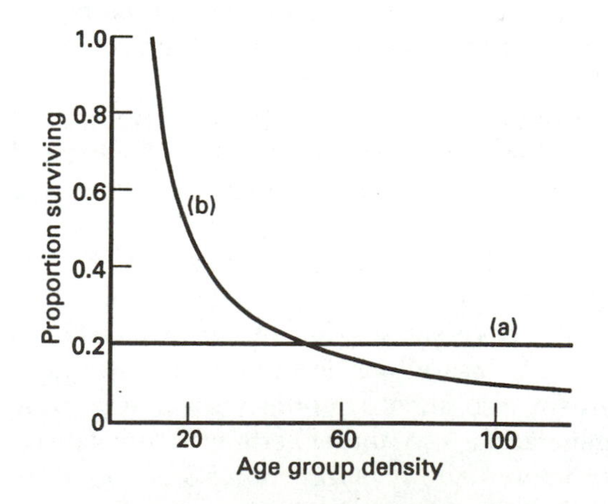

Figure 5.33 The effects of: (a) a density-independent factor: (b) a density-dependent factor used in the simulation model. (Source: Kidd, 1979.) Reproduced from *Journal of Biological Education* by permission of The Institute of Biology.

model can be used to explore the behaviour of populations with either discrete, partially overlapping, or fully overlapping generations (Kidd, 1979, 1984 for details and for a BASIC program).

This very simple deterministic simulation model provides the basic format for a number of modelling approaches, including the 'variable life-table' models of Hughes and Gilbert (1968) and Gilbert *et al.* (1976) (see above). Variable life-table models have generally been used to determine how the relationships governing birth and death processes affect population dynamics in the field, and rely on intensive laboratory and field observations and experiments during the model development and validation stages (Gilbert *et al.*, 1976; Gutierrez *et al.*, 1990; see Getz and Gutierrez, 1982 for an historical review and examples). The sequence of steps involves:

1. estimating intrinsic relationships, such as development rates, growth rates;
2. estimating extrinsic biotic relationships, including density-dependent effects, effects of natural enemies etc.;
3. estimating abiotic effects such as weather factors.

Ideally, the accuracy of the model needs to be tested at each stage, before the modeller can confidently proceed with the incorporation of more complex components. In this way, the model increases progressively in realism and complexity, without sacrificing accuracy.

To illustrate the process, including the important techniques of validation and sensitivity analysis, the population study of Kidd (1990a,b) on the pine aphid, *Cinara pinea* is used as an example. This species infests the shoots of pine trees (*Pinus sylvestris*) and can be cultured in the laboratory as well as studied in the field. The first task was to construct a relatively simple model to simulate the changing pattern of aphid numbers on small trees in the laboratory. This model incorporated a number of relationships obtained from observation and experiment, including: (a) an increase in the production of winged migratory adults with increasing population density; (b) a decrease in growth rates (adult size) and development rates with crowding and poor nutrition; (c) a decline in fecundity with smaller adult size. The effect of variable temperature on development was included by accumulating day-degrees above a thermal development threshold of 0°C (subsection 2.9.3).

Model Validation

The output from this prototype model (Figure 5.34) was compared with the population changes on four small saplings using a 'least-squares' goodness-of-fit test. Testing the accuracy of model output against real data is the process of **model validation** and should involve some statistical procedures (Naylor, 1971), although many population modellers have in the past relied on subjective assessment of similarity (e.g. Dempster and Lakhani, 1979). One frequently used statistical method is to compare model output with means and their confidence limits for replicated population data (Holt *et al.*, 1987). If the model predictions fall within the confidence range, then it can be assumed that the model provides an acceptable description of the data. Frequently, however, the replicates of population data show divergent behaviour, and for the model to be useful it needs to take this variability into account. This was the case with the pine aphid; the four populations on the small trees behaved differently and simply to average the data for each sampling occasion would have, at best, lost valuable information and, at worst, been statistically meaningless. The 'least-squares' method used (Kidd 1990a for details) took into account the variability both within and between the trees and, in fact, the model was able to explain 83% of the variation in the data. The model thus seemed to provide an acceptable description of aphid population behaviour in the laboratory, so is it adequately validated at this stage? The answer is *no* for another

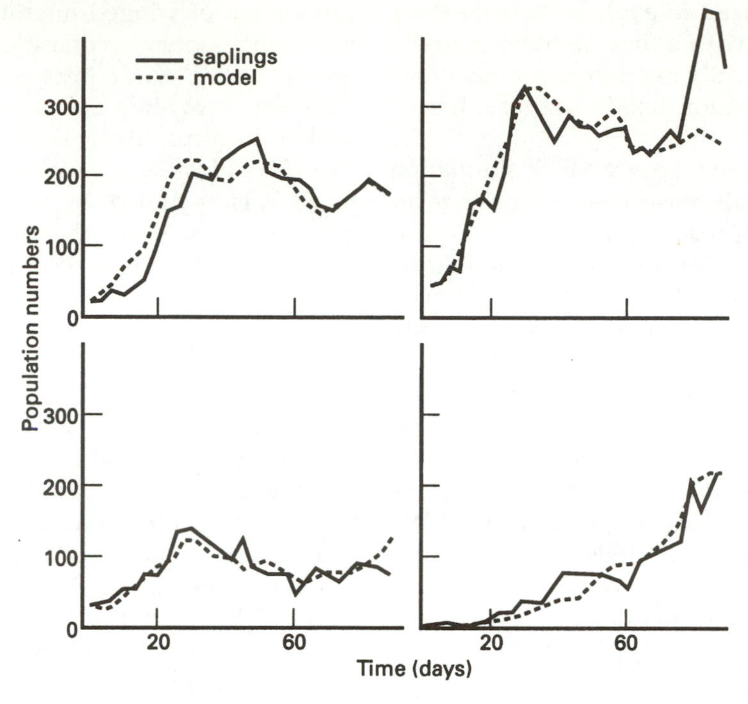

Figure 5.34 Population dynamics of pine aphids on four laboratory saplings and population numbers predicted by the simulation model. (Source: Kidd, 1990a.) Reproduced by permission of the Society for Population Biology.

important reason: for acceptable validation, models need to be compared with population data which have been independently collected and not used to provide data for the construction of the model. The four sapling populations, in this case, had yielded data which had been used in the model. For acceptable validation an independent set of populations on four trees was used and here the model explained 78% of the varability within and between trees (Figure 5.35).

Experimentation and Sensitivity Analysis

At this stage it was possible to use the model to assess the relative contribution of each component to the aphid's population dynamics in the laboratory. This involved manipulating or removing particular components and observing the behaviour of the system. It was also possible to reveal those components to which model behaviour was particularly sensitive and which might repay closer investigation. For example, changing the reproductive capabilities of the adults not surprisingly affected the rate of population increase, but this effect was extremely sensitive to the nature of density-dependent nymphal mortality. Population growth rates and the periodicity of fluctuations were also found to be sensitive to changing development rates, mediated through changes in plant quality.

Having achieved a sufficient degree of accuracy in simulating laboratory populations, it was then possible to incorporate the complexities associated with the field environment. In the first instance, this meant: (a) revising some components of the aphid/ plant interaction to make them more appropriate for field trees; (b) including the more extreme variations in

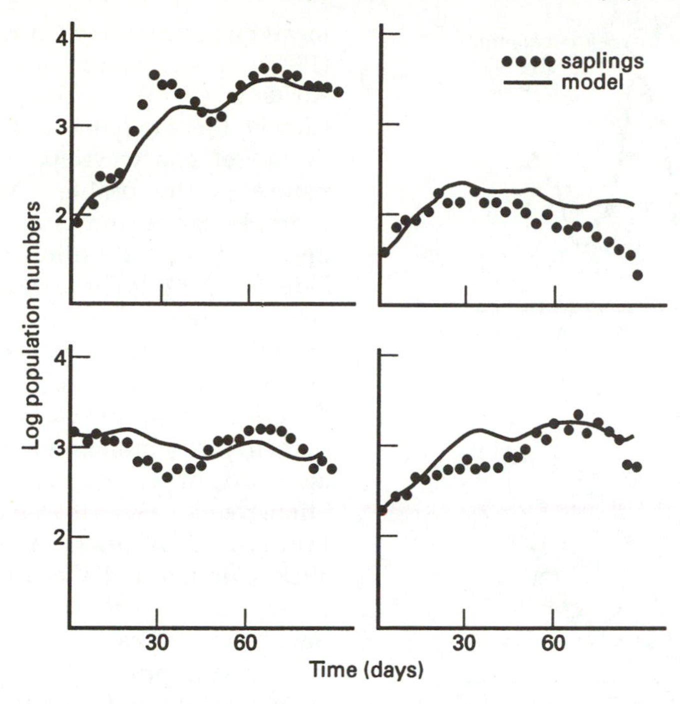

Figure 5.35 Population dynamics of pine aphids on four independent laboratory saplings and population numbers predicted by the simulation model. (Source: Kidd, 1990a.) Reproduced by permission of the Society for Population Biology.

temperature associated with the field; (c) incorporating weather effects. At this stage mortalities due to natural enemies were excluded. Output from the revised model was then compared with populations on three field saplings covered by cages designed to exclude predators (Kidd, 1990b). At this stage the model was able to account for 52% of the numerical variation within and between trees (Figure 5.36); this is acceptable given the greater innate variability of field data. The model also predicted a pattern of numbers which was very close indeed to that of aphid populations on mature field pine trees, at least in the early season (Figure. 5.37). Where predictions diverged from reality later in the season, this could probably be taken to reflect the impact of natural enemies which only become apparent after June.

While the model as it stands is purely deterministic in its construction, a stochastic element could have been included in any of the components by defining one or more parameters, not as constants, but in terms of their mean values and standard deviations. A random number generator could then have been used to produce a normally-distributed random number each time a parameter was used in the model. In this way, biologically meaningful variation could be reproduced. A number of standard computer programs are available for generating random numbers for a range of possible distributions. Sometimes it is desirable in a simulation model to reproduce an entire or partial distribution of data, rather than a single random value. This technique, known as **Monte Carlo simulation**, can easily be carried out using the random number generator in an iterative fashion.

We have dwelt at some length on this example in order to show the general procedures involved in a simulation study. Similar

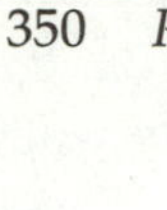

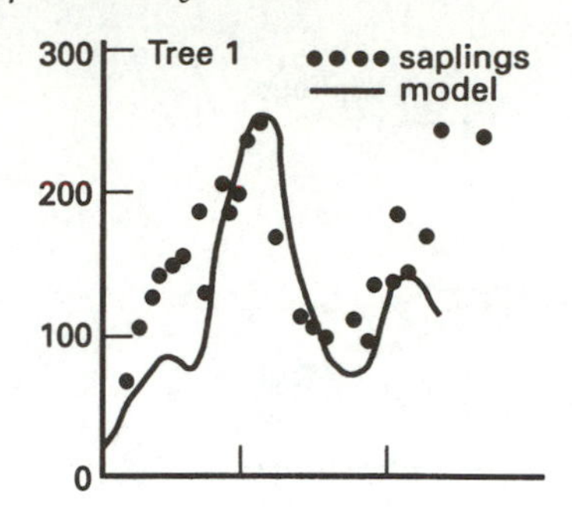

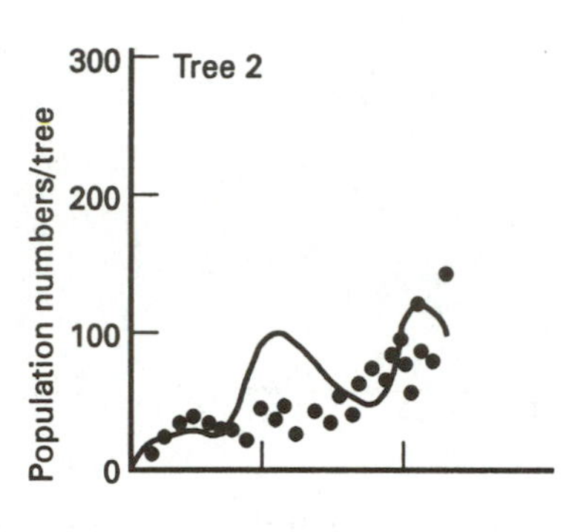

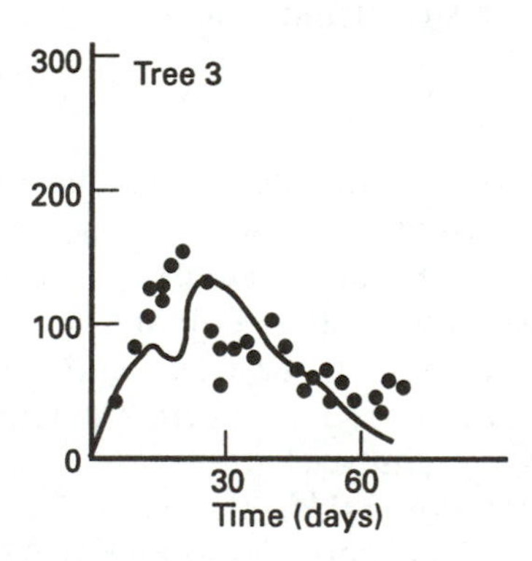

Figure 5.36 Pine aphid population dynamics on three field saplings from which predators were excluded and population numbers predicted by the simulation model. (Source: Kidd, 1990b.) Reproduced by permission of The Society for Population Biology.

examples concerning aphid populations are provided by Hughes and Gilbert (1968), Gilbert *et al.* (1976), Barlow and Dixon (1980), and Carter *et al.* (1982) and for other insect groups by Gutierrez *et al.* (1988a,b) (cassava mealybug and cassava green mite), Holt *et al.* (1987) (brown planthopper), and Sasaba and Kiritani (1975) (green rice leafhopper). Clearly, there may be considerable variations in model construction, depending on the nature of the problem to be solved. For example, Stone and Gutierrez (1986) developed a model to simulate the interaction between pink bollworm and its host plant cotton, which also included a detailed plant growth submodel (see also Gutierrez *et al.*, 1988a,b; Gutierrez *et al.*, 1994). Incorporating the effects of variable temperatures on development into simulation models has also been achieved in a number of different ways Stinner *et al.*, 1975; Frazer and Gilbert, 1976; Pruess, 1983; Wagner *et al.*, 1984, 1985; Nealis, 1988; Comins and Fletcher, 1988; Weseloh, 1989a; Kramer *et al.* 1991). With models based on a physiological time scale, for example, it is clearly impractical to use a time-step as short as one day-degree (one day may be the equivalent of about 20 day-degrees). In this situation, a physiological time-step corresponding to a convenient developmental period may be used. For *Masonaphis maxima*, Gilbert *et al.* (1976) used the quarter-instar period or 'quip' of 13.5 day-degrees Farenheit.

Incorporating the Components of Predation

Having constructed a basic simulation model, how can we best incorporate the components of predation or parasitism? This can be done in each of three ways:

1. by applying simple field-derived mortality estimates for predation (or parasitism) (subsections 5.3.2, Figure 5.21) in the prey (host) model; this tells us nothing of the dynamic interaction between predator and prey, however;
2. by constructing submodels for predators and parasitoids, involving age-structure if necessary, and including components

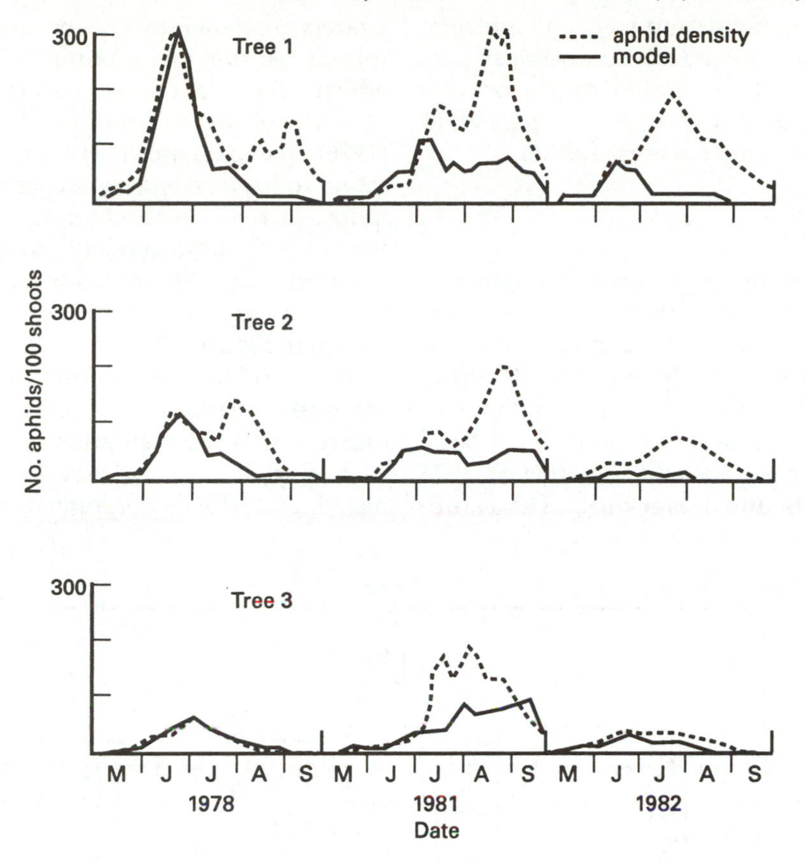

Figure 5.37 Pine aphid population densities on three mature field trees during three years, together with densities predicted by the model. (Source: Kidd, 1990b.) Reproduced by permission of The Society for Population Biology.

of searching behaviour (e.g. functional and aggregative responses and numerical responses as already described in subsections 2.7.3, 2.9.2, 5.3.6 and 5.3.7);

3. by constructing natural enemy submodels, but in this case using empirically-derived estimates of predation and its effects.

These methods are now discussed in turn:

Method 1

Vorley and Wratten (1985) used a variable life-table model of the cereal aphid, *Sitobion avenae* to predict population growth in the absence of natural enemies. The difference between predicted and actual numbers in the field was attributed to 'total mortality', i.e. parasitism plus 'residual mortality'. By discounting the effects of parasitism (estimated from dissections), residual mortality could then be calculated. By running the model with only residual mortality acting, the effects of parasitism on the dynamics of the aphid population could finally be estimated. Clearly, this technique, in common with others in this category, is only capable of assessing the 'killing power' of a mortality source during the period for which data have been collected. It has no reliable predictive power (i.e. it is an interpolative rather than an extrapolative method) and yields no information on the dynamic interaction between parasitoid and host. A similar approach was adopted by Carter *et al.* (1982) for parasitoid and fungal mortality of cereal aphids.

Method 2

(a) **Age-specific:** Age-specific submodels have frequently been developed for parasitoids and predators for use in both theoretical models (Kidd and Jervis, 1989; Godfray and Hassell, 1989 for parasitoids) and in field or crop-based simulation studies (Yano, 1989a,b for parasitoids and Gilbert *et al.*, 1976 for parasitoids and hover-flies). These submodels are generally constructed in the same format as the main population model on which they act. For example, in the *Masonaphis maxima* model of Gilbert *et al.* (1976), the parasitoid *Aphidius rubifolii* was found to have an egg development time of six quips. In the theoretical model of Kidd and Jervis (1989), time and age groups were measured in days, the life history features of the host and parasitoid having the structure shown in Figure 5.38.

(b) **Searching behaviour:** As functional response relationships have usually been derived from short duration experiments (e.g. 24 h; section 1.10), they are likely to be more meaningful when incorporated into simulation

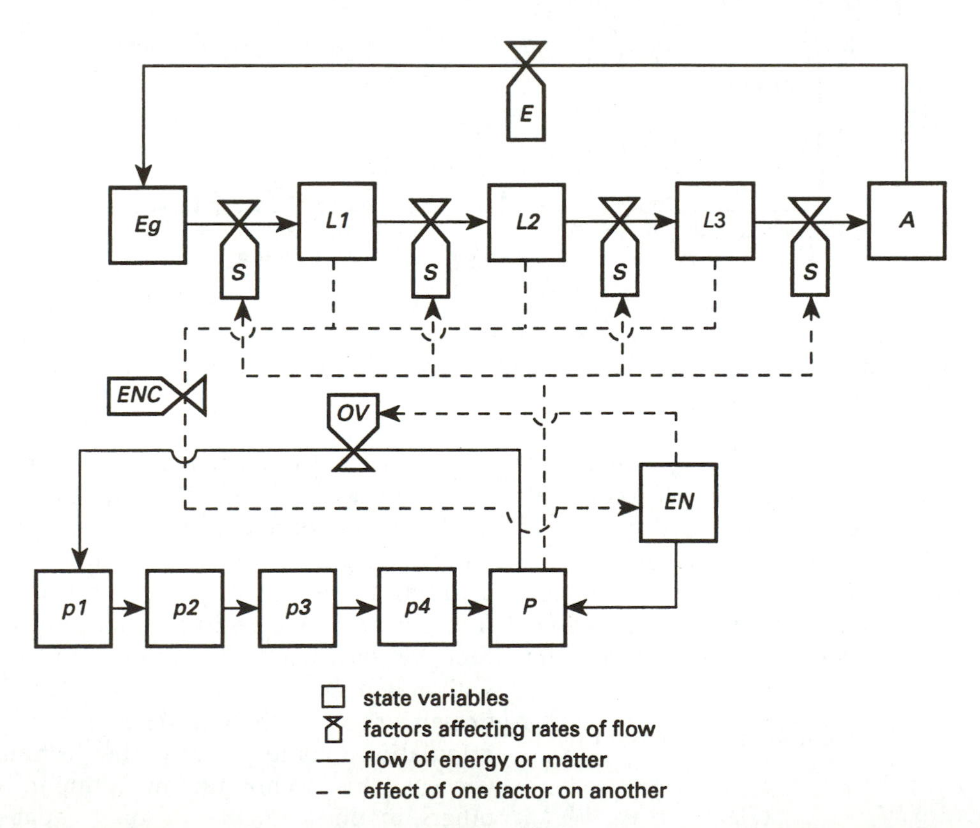

Figure 5.38 Simplified relational diagram of the parasitoid–host system: *Eg*, host egg age group; *L1–L3*, host larval age group; *A*, host adults; *S*, survival rate; *E*, host reproductive rate; *ENC*, encounter rate; *p1–p4*, immature parasitoid age groups; *P*, adult parasitoids; *OV*, rate of parasitoid oviposition; *EN*, individual parasitoid energy stocks. (Source: Kidd and Jervis, 1989.) Reproduced by permission of The Society for Population Biology.

models with a short time step of, for example, one day, rather than the one-generation time step of the Nicholson–Bailey model (subsection 5.3.7). In the model of Kidd and Jervis (1989), parasitoids searched for hosts sequentially, either at random between patches, or in selected high host density patches. Type 2 functional responses were incorporated for both feeding and oviposition in a biologically realistic way, by allowing each parasitoid a maximum available search time per day (10 h), the 'efficiency of search' being constrained by feeding handling time, oviposition handling time and egg limitation. A full BASIC program listing for this model is provided by Kidd and Jervis (1989), to which we refer readers interested in developing the individual-based queueing techniques described.

Few attempts have been made to measure the searching efficiency of natural enemies in the field (Young, 1980; Hopper and King, 1986; Jones and Hassell, 1988), probably due in large part to the obvious technical difficulties of confining particular densities of predators/parasitoids and prey/hosts within localised patches. Where the Holling disc equation (equation (5.21)) has been used to model a parasitoid–host interaction, parameter values have sometimes been estimated from field data. This can be done by iteration, i.e. by using a range of alternative parameter values in repeated simulations to find which fit the data best (Ravlin and Haynes, 1987). The difficulty here is that values of a' and T_h which produce accurate simulations may bear no resemblance to laboratory estimates of these parameters, casting some doubt on the realism of at least some components in the model. An alternative approach has been used by a number of workers (Godfray and Waage, 1991; Barlow and Goldson, 1993) to capture the essence of parasitoid search in a way that can potentially be used to describe their dynamic interactions with host populations in the field. This is done using the simple Nicholsonian component [$1-\exp(-aP^m)$] to describe the proportion of hosts attacked during a defined time period, where P is the number of parasitoids, a is searching efficiency and m defines the strength of density-dependence in parasitoid attack (cf. equation (5.33)). Barlow and Goldson (1993) used this term to model the interaction between the weevil pest of legumes, *Sitona discoideus* and an introduced braconid parasitoid, *Microctonus aethiopoides*. Thus:

$$q = 1 - \exp\left(-aP^m\right) \qquad (5.40)$$

where q is the estimated proportion of hosts parasitised. Leaving aside life cycle complications, parameter values for a and m were simply estimated from the relationship between parasitoid densities (estimated from percentage parasitism) with percentage parasitism in the following generation (Figure 5.39). To do this, equation (5.40) was first linearised by rearrangement using the following steps:

$$1 - q = \exp\left(-aP^m\right) \qquad (5.41a)$$

$$\log_e(1-q) = -aP^m \qquad (5.41b)$$

$$-\log_e(1-q) = aP^m \qquad (5.41c)$$

$$\log_{10}\left[-\log_e(1-q)\right] = \log_{10} a + m\log_{10} P \qquad (5.41d)$$

Thus, $\log_{10}[-\log_e(1-q)]$ can be regressed against $\log_{10}P$ with slope m and intercept $\log_{10}a$ to find values of m and a. The model derived by Barlow and Goldson (1993) gave acceptable predictions of parasitoid and host numbers over a ten year period. It seems likely that more extensive use will be made of such **models of intermediate complexity** (Godfray and Waage, 1991; see also subsection 5.4.3) in the future.

(c) **Numerical responses**: As indicated in subsection 5.3.7, the rate of increase of a predator population depends on a number of components, namely, the development rate of the immature stages, the survival rate of each instar, and the fecundity of the adults. To build a comprehensive submodel of the

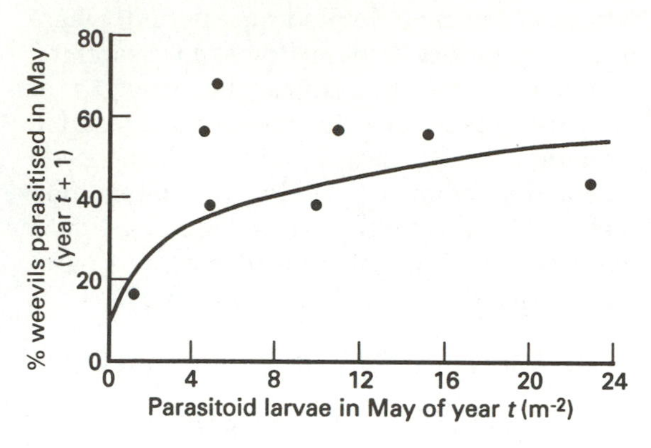

Figure 5.39 The relationship between the density of parasitised legume weevils in year t and the percentage parasitised in year $t + 1$. (Source: Barlow and Goldson, 1993.) Reproduced by permission of Blackwell Scientific Publications Ltd.

predator rate of increase we would need therefore to incorporate the various factors which might affect prey consumption (e.g. prey density, handling time, aggregative response, mutual interference), together with the effects of prey consumption on development and survival of the different instars, and on adult fecundity. The influence of prey consumption on predator growth and development, survival and fecundity has already been discussed at length, together with experimental methodologies and descriptive equations (subsections 2.9.2, 2.10.2, 2.7.3). Although the modelling of these processes is technically feasible with computer simulation, to obtain sufficiently detailed information on all of the components involved would be a potentially time-consuming operation. To our knowledge the task has not yet been fully carried out for any single predator species in the formal detail prescribed by Beddington *et al.* (1976b). However, Crowley *et al.* (1987) describe a detailed submodel applicable to damselfly predators, but without an equally detailed model for the *Daphnia* prey. Nevertheless, this clearly remains an important challenge for those interested in modelling predator–prey processes, as a computer submodel, based on general theory and incorporating all of the components of predation in this way, would have immense practical as well as theoretical value.

Method 3

For simulation models specifically designed to mimic field population dynamics, the components of predation and parasitism have usually been incorporated into submodels in a more pragmatic way relevant to the particular system and problem in question. As a wide variety of approaches has been adopted, with no single unified methodology, we can only present a number of examples to illustrate the possibilities. For parasitism acting in their *Masonaphis maxima* model, for example, Gilbert *et al.* (1976) first established the duration (in quips) of the four developmental stages of *Aphidius rubifolii* (egg, larva, pupa and adult). Adult females were assumed to search at random, ovipositing a constant (average) number of eggs per quip, in susceptible third and fourth instar and young adult aphids. No account was taken of how parasitoid oviposition might vary with aphid density, so it is perhaps not surprising that the model, at least in its early versions, provided poor predictions of aphid parasitism.

For syrphid predation, on the other hand, Gilbert *et al.* (1976) assigned predator larvae to four developmental periods, each of 10 quips duration and consuming 1, 2, 6 and 15 average-sized aphids per quip. This provided a consumption rate of 240 aphids for each syrphid to complete development, a figure which broadly agreed with observations. In the model, syrphid eggs were simply laid in proportion to aphid densities, thus generating the number of larvae attacking the aphids. Aphids were assumed to be attacked at random and were subtracted from the model during each time-step in proportion to the voracity and number of each syrphid age group present. No account was taken of predator starvation or survival, but, even so, Gilbert *et al.* found this simple model gave reasonably accurate predictions of predation in the field.

Some variations on this 'maximum consumption' method have been carried out on other systems. Wratten (1973) and Glen (1975) respectively examined the effectiveness of the coccinellid, *Adalia bipunctata* and the mirid bug, *Blepharidopterus angulatus* as predators of the lime aphid, *Eucallipterus tiliae* using the methodology developed by Dixon (1958, 1970). The sequence of tasks involved in this approach are:

1. to monitor the age distribution and population densities of prey and predators at regular intervals (of, say, one week) over a period of months or years;
2. to measure the efficiency of capture (% encounters resulting in capture) of each predator instar in relation to the different prey instars in the laboratory;
3. to estimate the amount of time spent searching per day (for *A. bipunctata* this meant time exceeding a minimum temperature threshold for activity, corresponding approximately to the 16 hours of daylight);
4. to estimate the percentage of searching time spent on areas with prey (44% for *A. bipunctata*);
5. to calculate the area traversed/day (= distance travelled in 1 h at mean temperature *16*0.44*R [R being the width of perception of the predator]);
6. to find the % of time spent on areas already searched;
7. to calculate the 'area covered' from 5. and 6., taking into account time wasted by recrossing areas already searched;
8. to estimate the time spent feeding as a proportion of total time available.

Components 3.–5. are all determined empirically from observations in the laboratory. It can be seen that some of the above procedures follow Nicholson and Bailey's (1935) methodology for calculating the 'area of discovery' (subsection 5.3.7) and effectively take into account handling time and aggregation of predation, but not mutual interference. Negative effects on the predators are again simply estimated from experiments to determine the minimum density of aphids required for predator survival at different stages (see reference to Gilbert *et al.*, 1976 above).

For *A. bipunctata* (Wratten, 1973), the number of aphids of a particular instar consumed by each coccinellid instar per week (N_a) could be found from the equation:

$$N_a = 0.01A * D * 0.01E * C_o * n * C_f \qquad (5.42)$$

where A is the area covered/larva/week, D is the density of aphids, E is the capture efficiency, C_o is a correction factor for aphid distribution, n is density of larvae, C_f is the correction factor for the proportion of available time spent feeding up to satiation. This calculation was made for every combination of coccinellid and aphid instar in the populations on each sampling date to arrive at an overall number of aphids in each instar removed per week. Assuming that these aphids would have remained in the population in the absence of predation, Wratten (1973) was able to estimate the potential aphid population size and structure in the absence of predation from information on the development times and reproductive patterns of the aphids, obtained by confining them in leaf cages. The difference between the observed aphid population and the predicted population in the absence of predation, thus demonstrated the effect of coccinellid mortality. A later study incorporating these components of coccinellid predation, together with similar ones for the mirid predator (Glen, 1975), into a simulation model of the lime aphid (Barlow and Dixon, 1980), showed that both species can have a destabilising effect on aphid numbers.

Clearly, where prediction of the prey death rate over a limited period (e.g., one season) is the sole aim of the model, fairly crude representations of predation may suffice. However, where a longer time-frame is being simulated, especially one involving tightly coupled monophagous predator–prey interactions, a more accurate submodel incorporating a predator numerical response will be needed. A number of approaches which could be used to achieve this end are described in detail for aphid–coccinellid interactions by Frazer *et al.* (1981a), Gutierrez *et al.* (1981), Mack and Smilowitz (1982) and Frazer and Raworth (1985). In practice, none of these studies attempted to simulate the interaction beyond one field season. To take just one example in detail, Frazer *et al.* (1981a) derived empirical relationships for coccinellid reproduction and survival by confining adult beetles with aphids in screen-walled cages in alfalfa fields. Ives (in Frazer *et al.*, 1981a) had found in the laboratory that female coccinellids required 1.3 mg live weight of aphids/quip for maintenance, additional prey being converted to eggs at a rate of 0.7 eggs/mg of aphid. A direct relationship could then be established between the predation rate and the reproductive rate of adult females. Overall survival from egg to adult was estimated by comparing the expected numbers of eggs laid in the cages with the number of beetles recovered at emergence, and was subsequently found to show a sigmoid relationship with aphid density (using total aphid density during the first larval instar of the beetle, although a running average could also have been used).

5.3.9 COMBINING TESTABILITY AND GENERALITY

We have emphasised above the distinction between analytical and simulation models and also the difference between the deductive and inductive approaches. The latter, essentially philosophical, distinction is equivalent to that proposed by May (1974), who referred instead to **strategic** and **tactical models.** Strategic models attempt to describe and abstract the general features of population dynamics while ignoring the detail. Tactical models, on the other hand, are developed to explain the complex dynamics of particular systems, being particularly useful in population management programmes. Tactical models, however, may not readily lead to general conclusions about population dynamics. Although they can readily be tested against real system behaviour (subsections 5.3.4 and 5.3.8), the testing of strategic models is more problematical, as more than one model can often be invoked as a plausible explanation of a particular phenomenon. Murdoch *et al.* (1992) suggest a way of making models both testable and general, by first building tactically-orientated models, then progressively stripping out the detail, whilst testing the new models at each stage to determine the loss in predictive capacity. Murdoch *et al.* (1992) provide an example of how the methodology might be used by referring to the work of Murdoch and McCauley (1985) and McCauley and Murdoch (1987 and related papers) on *Daphnia*–algae interactions. A similar approach has also been advocated by Berryman (1990) and Kidd (1990c).

5.3.10 SPATIAL HETEROGENEITY AND PREDATOR–PREY MODELS

One of the most important conceptual advances in recent years has been the realisation of the importance of spatial heterogeneity in the dynamics of predator–prey and parasitoid–host interactions. It is now well established from deductive modelling that direct density-dependent relationships from patch to patch, resulting from the aggregative responses of natural enemies, can be a powerful stabilising influence (Hassell, 1978, 1980b). Whilst optimality theory (section 1.12) predicts that such patterns of **direct spatial density-dependence** ought to be common (Comins and Hassell 1979; Lessells, 1985), surveys of published studies suggest that they are in fact in the minority. Examples of the opposite, i.e. **inverse spatial density-**

dependent relationships, where predation or parasitism are concentrated in the lowest density patches, are just as common, as are examples with no relationship at all, i.e. **density-independent** relationships (Morrison and Strong, 1980, 1981; Lessells, 1985; Stiling, 1987; Walde and Murdoch, 1988). An inverse spatial density-dependent relationship may be found if hosts or prey in high density patches are less likely to be located than those in low density patches, perhaps due to greater host or prey concealment. Price (1988) showed this for parasitism by *Pteromalus* of the stem-galling sawfly *Euura lasiolepis* on willow (*Salix lasiolepis*). Even without this concealment effect, it is still theoretically possible for parasitism among patches to be density-independent or inversely density-dependent (Lessells, 1985). A mechanistic explanation (subsection 1.2.1) for this lies in the balance between two counteracting processes (Hassell *et al.*, 1985): (a) the spatial allocation of searching time by parasitoids in relation to host density per patch, and (b) the degree to which exploitation is constrained by a relatively low maximum attack rate per parasitoid within a patch. Inverse density-dependent parasitism can theoretically result from insufficient aggregation of searching time by female parasitoids in high density patches to compensate for any within-patch constraints on host exploitation imposed per parasitoid by time-limitation, egg limitation or imperfect information on patch quality. Density-independent relationships, on the other hand, can result if processes (a) and (b) are in balance.

These observations would appear to undermine the importance of spatial density-dependence as a regulating factor in natural predator–prey and parasitoid–host systems, but Hassell (1984) was able to demonstrate, using the approach encapsulated in equations (5.30) and (5.31) above, that patterns of inverse spatial density-dependence can be just as stabilising as patterns of direct spatial density-dependence; whether direct or inverse relationships have the greater effect depends upon the characteristics of the host's spatial distribution (Hassell, 1985; also Chesson and Murdoch, 1986). In fact, it is now apparent that even where parasitism between patches is density-independent, the spatial distribution of parasitism, if sufficiently variable, can also promote stability (Chesson and Murdoch, 1986; Hassell and May, 1988). The biological interpretation seems to be that as long as some patches of hosts or prey are protected in **refuges** from natural enemy attack, whether in high density or low density patches, stability remains possible. This may be so, but we should exercise caution in making instant biological inferences from the behaviour of such models – other interpretations may be equally plausible or preferable (McNair, 1986; Murdoch and Reeve, 1987).

The conclusion derived from deductive models that different spatial patterns of density-dependence have a powerful stabilising potential, has opened up a whole new area of study and debate in population ecology in recent years. Consequently new methods for studying spatial patterns of density-dependence in the laboratory and field have appeared. These are now reviewed in turn before we address some of the more contentious issues surrounding the subject.

Detecting Spatial Density-dependence

Laboratory Methods

The general methodology for studying spatial variation in parasitism or predation in the laboratory is described in subsection 1.14.2. Hassell *et al.* (1985) provide a good example of how two species of parasitoid attacking bruchid beetles (*Callosobruchus chinensis*) in such experiments can show inverse spatial density-dependence. The beetle itself is a pest of legumes and commonly breeds in stored dried pulses. Black-eyed beans (*Vigna unguiculata*) were used in these experiments, which were carried out in clear perspex arenas

(460 mm × 460 mm × 100 mm). Twenty–five patches of equal size were marked out on white paper sheets in an hexagonal grid, with 75 mm spacing between the centres. Different densities of beans containing 13-day-old hosts were allocated at random to patches and the number of beans in each patch made up to 32 with the addition of the required number of uninfested beans. Twenty–five parasitoids of one species were introduced into each arena and, after 24 h, infested beans transferred to vials to await parasitoid emergence. The variable pattern of density-dependence found might be explained by both: (a) the allocation of parasitoid searching time in patches of different host density, and (b) the maximum attack rate per parasitoid constraining the extent of exploitation within patches. Parasitoids showed no tendency to aggregate in some patches over others, while their maximum attack rates per patch were limited by handling time constraints (see above) (Hassell *et al.*, 1989b give further details; see both papers for methods of culturing the beetle and its parasitoids).

Field Methods

Traditional techniques for analysing field population data for density-dependence seldom take account of spatial variation in patterns of mortality. The Varley and Gradwell method (subsection 5.3.4), for example, explores variations in *k*-mortalities over several generations, but takes no account of spatial variation amongst subunits of the population within a generation (Hassell,1987; Hassell *et al.*, 1987). Stiling (1988) surveyed 63 life table studies on insects of which about 50% failed to detect any density-dependence acting on the populations. Hassell *et al.* (1989a), however, argued that density-dependence may still have been present but undetected due to: (a) the inadequate length of time over which some studies were conducted, and (b) the inability of the analyses to take account of spatial variations in patterns of mortality amongst subpopulations.

If traditional methods are inappropriate, how do we go about detecting spatial density-dependence in the field? The first problem is to decide on a suitable spatial scale (e.g. leaf, twig, tree) for the collection of data. The most appropriate scale is that at which natural enemies recognise and respond to variations in host density (Heads and Lawton, 1983; Waage, 1983). As this may be difficult to determine initially, samples are best taken in an hierarchical manner, so that analyses can be carried out at a number of different scales afterwards (Ruberson *et al.* 1991). Within each level of patchiness, patch density can then be related by regression analysis to percentage mortality or *k*-value, although the statistical validity of regression in this context is questionable (subsection 5.3.4). Hails and Crawley (1992) have proposed an alternative logistic regression analysis based on generalised linear interactive modelling (GLIM) (Crawley, 1993). Using the cynipid wasp, *Andricus quercuscalicis*, which forms galls on Turkey oak, as a test system, the method was able to detect spatial density-dependence in 15% of cases, with 66% of those being inversely density-dependent. Hails and Crawley manipulated patch densities by controlling the oviposition of adults on the buds of the trees. A similar study was carried out by Cappuccino (1992), who manipulated densities of another gall-making insect, the tephritid fly *Eurosta solidaginis*. Spatial variation in predation of the fly by a beetle, *Mordellistena* (*Mordellidae*), was noted at three scales, together with parasitism of the beetle. Interestingly, spatial variation in beetle parasitism depended, not on beetle patch density, but on the density of the fly.

Recently, Pacala *et al.* (1990) and Hassell *et al.* (1991) have shown, again using the simple deductive models of host–parasitoid systems, that the contribution of spatial heterogeneity in parasitism to stability can be assessed using a simple rule. This states that the coefficient of variation squared (CV^2 = variance/mean^2) of the density of searching parasitoids close to each host must exceed approximately unity for the heterogeneity in parasitism to stabilise the interaction, i.e. $CV^2>1$. Moreover, CV^2 can be partitioned into the component of heterogeneity that is independent of host density (C_i) and the component that is dependent on host density (C_D)., such that $CV^2 = C_iC_D - 1$. To estimate CV^2 directly the local density of searching parasitoids needs to be known. In most field systems, however, this is impracticable and consequently little information on this parameter is available (Waage, 1983). However, a considerable body of information is already available on percentage parasitism as a function of local host density and a procedure is provided by Hassell and Pacala (1990) to estimate the relevant parameters required from these data. The reader is warned, however, that the calculations require some mathematical facility, so it is probably best to await further developments of the approach, especially as it is only applicable to restricted types of host–parasitoid interaction, as Hassell and Pacala are careful to point out. Readers interested in applying the technique should consult Pacala *et al.* (1990), Pacala and Hassell (1991), and Hassell *et al.* (1991).

While Hassell and Pacala (1990) provide a detailed account of how to derive CV^2 from field data using 65 examples, it needs to be pointed out that most studies on spatial distribution of parasitoids have only been conducted over a very short time span (e.g. one generation) (reviews by Lessells, 1985; Stiling, 1987; Walde and Murdoch, 1988). Observed spatial distribution patterns may not therefore be typical of the interaction (Redfern *et al.*, 1992) and data collected over a number of generations or years is more likely to give a representative picture. Redfern *et al.* (1992) studied patterns of spatial density-dependence amongst parasitoids of two tephritid fly species over a period of seven years. CV^2 values were calculated for total parasitism and each parasitoid species separately. The CV^2 values (together with their statistical significance) were found to be highly variable from year to year, making it difficult to draw conclusions. What drives fluctuations in CV^2 from generation to generation is not yet known, but will need to be investigated, if techniques such as the CV^2 rule are to have wider applicability. A further problem with the method is that it only applies to interactions where parasitism is of overriding importance and other regulating effects of the host are negligible. How the latter might affect the stability conditions of the interaction are not known and may be a major limitation of the technique.

Spatial Density-Dependence vs. Temporal Density-Dependence in Regulation

As spatial (i.e. within-generation) density-dependence or heterogeneity in the pattern of natural enemy attack appears to be a potentially powerful stabilising mechanism, we need to examine its relationship with temporal (i.e between-generation) density-dependence, upon which conclusions about regulation have traditionally been based. We have already seen that conventional methods for the analysis of life-table data are unsuitable for the detection of regulation resulting from spatial density-dependence (Hassell, 1985), so that many previous conclusions regarding the failure of natural enemies to regulate in particular systems may eventually need to be revised. Hassell (1985) and May (1986) have postulated that much of the regulation in natural populations is likely to arise from within-generation variation in parasitism and predation, with only weak dependence on between-generation variations.

Not all authors have agreed with this view, however. Dempster and Pollard (1986) have argued that regulation must ultimately depend on temporal density-dependence and should therefore be detectable, at least in principle, by conventional analyses. They doubt that spatial density-dependence leads necessarily to temporal density-dependence. This raises interesting questions about the relationship between spatial and temporal density-dependence and the way the former operates. For regulation to occur some temporal feedback is required. Using the discrete generation model of DeJong (1979), Hassell (1987) was able to show that, in the absence of stochastic variation, spatial density-dependence acting within generations translates directly into temporal density-dependence acting between generations. With variability added to the parameters governing spatial distribution and survival, however, temporal density-dependence becomes obscured and less likely to be detected by the conventional method of plotting mortality against population density from generation to generation. Murdoch and Reeve (1987) also point out that we should not necessarily expect to detect spatial density-dependence from life-table data on prey or host populations, as spatial density-dependence acts on the *natural enemy* population, *not* the host or prey. The authors demonstrate their point by a closer analysis of the terms of the Nicholson–Bailey model which indicate that stability results from a decline in parasitoid efficiency as parasitoid density increases. Unless life-tables also contain data for the relevant natural enemies, then this information would be overlooked.

Hassell *et al.* (1987) analysed 16 generations of life-table data for the viburnum whitefly (*Aleurotrachelus jelinekii*) and found no evidence for temporal density-dependence. Spatial density-dependence of egg mortality between leaves was apparent in eight generations. Again using a variant of the DeJong model to simulate the whitefly population, Hassell *et al.* (1987) claimed to show that this spatial density-dependence could regulate the population in the absence of temporal density-dependence. Stewart-Oaten and Murdoch (1990), however, disputed this conclusion, arguing that the DeJong model has implicit temporal density-dependence which Hassell *et al.* (1987) had overlooked. With this temporal density-dependence removed Stewart-Oaten and Murdoch were able to show that the spatial density-dependence in the resulting model can indeed lead to stability, but that destabilisation is also a strong feature at low population levels and when whitefly clumping increased. The Stewart-Oaten and Murdoch model thus demonstrates the important point that stability in population models is often sensitive to both changes in parameter values and to any subtle variations incorporated (Murdoch and Stewart-Oaten, 1989; Reeve *et al.*, 1989).

We have tried to provide a brief overview of the subject of spatial heterogeneity and aggregation in predator–prey systems, hopefully conveying some sense of the speed with which new developments in the subject are taking place, and the excitement that they generate. Much of the progress is directly attributable to the use of deductive models, but as we have seen, the conclusions which can be derived from these models are to some extent dependent both on how the models are constructed and on the parameter values used. Paradoxically, while progress using this approach can be rapid, with new ideas and hypotheses being generated, often as many questions are raised as are answered. Thus, much of the subject remains speculative at his stage, and firm conclusions regarding the precise role of spatial heterogeneity in population dynamics remain elusive.

Metapopulation Dynamics

Any consideration of spatial heterogeneity in population dynamics inevitably impinges on the concept of the metapopulation, which was introduced and defined in section 4.1. To recap briefly, **local populations** can be defined as units within which local population processes (e.g. reproduction, predation etc.) occur, and within which movements of most individuals are confined. **Regional populations** (or **metapopulations**) are collections of local populations linked by dispersal. Between-patch variations in parasitism and predation, as discussed above, are deemed to influence dynamics at the local population level, but regional metapopulation effects may also be important, although the distinction between such 'within-population' and 'between-population' processes may be somewhat artificial in many systems (Taylor, 1990).

Dispersal between local populations has frequently been proposed to account for the persistence of regional populations despite unstable fluctuations or extinctions at the local level (DeAngelis and Waterhouse, 1987 and Taylor, 1990 give reviews). For predator–prey systems, two different approaches have been used to model this situation (see Taylor, 1990 for details):

1. those in which extinctions and recolonisations of local 'cells' (= populations) occur frequently – the so-called **cell occupancy models**;
2. those in which within-cell dynamics are described explicitly by standard predator–prey models.

These models consistently show that persistence can be enhanced by dispersal between local populations, provided that: (a) populations fluctuate asynchronously between cells, (b) predator rates of colonisation are not too rapid relative to those of the prey, and (c) some local density-dependence is present.

These conclusions are consistent with the results of a number of laboratory studies exploring the effects of spatial structure and dispersal on persistence of predator–prey systems (Pimentel *et al.*, 1963; Huffaker, 1958). Pimentel *et al.* (1963), for example, examined the interaction between a parasitic wasp and its fly host in artificial environments consisting of small boxes connected by tubes. The interaction persisted longer with more boxes and with reduced parasitoid dispersal. While agreement with theory may be encouraging, the small scale on which these experiments, by necessity, have to be carried out is unlikely to reflect processes at the regional metapopulation level. The results may be equally well explained by local population, between-patch, spatial dynamics. Similarly, Murdoch *et al.* (1985) invoked metapopulation processes to explain persistence of a number of field predator–prey and parasitoid–host systems, despite apparent local extinctions. However, as Taylor (1990) points out, extinction in these examples was either not proven or occurred at a scale more consistent with local population processes.

Evidence for the importance of metapopulation processes in field predator–prey systems is thus lacking, chiefly because few studies have been conducted specifically to obtain such evidence (see, however, Walde, 1994). This in turn may be due to the scale at which experiments and observations need to be conducted and the technical problems of measuring dispersal. One type of possible compromise system for facilitating study may be provided by glasshouse populations (e.g. *Trialeurodes-Encarsia*), interconnected by chance dispersal of pest insects and their biocontrol agents. Reeve (1990) discusses specific experimental protocols which could be used in this situation. A recent comprehensive review of metapopulation dynamics is provided by Gilpin and Hanski (1991). Hanski (1994) describes a 'practical model' of metapopulation dynamics.

5.4 SELECTION CRITERIA IN BIOLOGICAL CONTROL

5.4.1 INTRODUCTION

In this section, we deal mainly with the criteria used in the selection of natural enemy species for introduction in **classical** biological control programmes (defined in subsection 5.2.2). Some of these criteria will, however, apply to agent selection in other types of biological control programme. We realise that the majority of readers may never actively practise biological control, but we hope that what follows will provide a framework for research such that those readers can nevertheless make an indirect contribution to the subject.

Several steps are involved in a classical biological control programme (Beirne, 1980; DeBach and Rosen, 1991; Waage and Mills, 1992). The first is the **evaluation** of the pest problem in the region targeted by the programme. Information should ideally be gathered on: (a) the distribution and abundance of the pest; (b) crop yield loss and economic thresholds of pest numbers; (c) the taxonomic identity and probable region of origin of the pest; (d) the identity and ecology of indigenous natural enemies that may have become associated with the pest in its exotic range; (e) the climate in the target region.

The next step is **exploration** in the pest's region of origin. This should ideally involve: (a) determination, through quantitative surveys, of the composition of the natural enemy complex associated with the pest in its natural habitat(s) (see section 4.3 for methods); (b) quantification of the impact of these natural enemies on populations of the pest (see sections 5.2 and 5.3 for methods); (c) determination of their degree of specificity, i.e. their host/prey ranges (see section 4.3 for methods). Exploration may also involve determining which natural enemies are associated with taxonomically related pests (the search for 'new associations', see below). The process of selection – the selection of agents for subsequent importation and establishment in the target region – usually begins during the exploration phase. Of the range of species available, one or more are chosen on the basis of their potential to control pest populations. Usually, only a small fraction of the natural enemy complex of a pest is used in a typical classical biological control programme. The constraints upon using a larger fraction include: (a) the limited resources (finances, time) available to practitioners; (b) the unsuitability of certain species due to their polyphagous habits or potential to interfere with other species (facultative hyperparasitoids (section 4.1) may raise pest populations by attacking an effective primary parasitoid to a greater extent than they lower them by attacking the pest itself); (c) the difficulty sometimes encountered, in multiple parasitoid introductions, of achieving establishment of a species, following establishment of another (Ehler and Hall, 1982; Waage, 1990; Waage and Mills, 1992). If 'new associations' are to be employed, then one or more members of the natural enemy complex of a species taxonomically related to the pest, but not of the complex associated with the pest itself, will be selected in a programme.

The next step is the holding of imported agents in **quarantine**, to eliminate any associated hyperparasitoids, parasitoids of predators or insect and plant pathogens that have inadvertently been introduced into cultures along with the agents. Further selection of agents may occur during this phase (see **Ease of Handling and Culturing**), which is followed by the **release** of agents cleared through quarantine. Release is followed by **monitoring** of changes brought about in the pest populations in the target region (see sections 5.2 and 5.3 for discussion of monitoring methods).

A detailed discussion of the steps involved in biological control programmes aimed at aphid pests is provided by Hughes (1989).

5.4.2 WHAT WORKED (OR DID NOT WORK) PREVIOUSLY

The history of biological control shows that it has been largely an 'art', aided by the knowledge of what worked successfully in the past either against the same pest species or a taxonomically related species (van Lenteren, 1980; Waage and Hassell, 1982). Biological control is becoming much more of a science, and data on past introductions can still prove useful to practitioners in helping them to make generalisations informed by ecological theory (Greathead, 1986; Waage and Greathead, 1988). A database (called BIOCAT) of introductions recently developed at the CAB International Institute of Biological Control is the most comprehensive database on biological control developed to date. BIOCAT has been used to test ecological hypotheses such as that on 'new associations'. Unfortunately, like other databases, it is composed of records whose reliability is in some cases questionable; as Waage and Mills (1992) point out: 'the record of classical biological control is troubled by erroneous identifications of pests and natural enemies, occasional errors in dates and places and the arbitrary (and sometimes entirely incorrect) interpretation of success'. Owing to this and other shortcomings, database analysis is not considered to be a sound basis for actively promoting universal selection criteria (Waage, 1990).

5.4.3 NATURAL ENEMY ECOLOGY AND BEHAVIOUR

Introduction

During the exploration phase of a biological control programme, a decision will need to be made as to whether the natural enemies are to be collected from pest species or from other, taxonomically closely related, species. The theory of **new associations** (Pimentel, 1963; Hokkanen and Pimentel, 1984) states that natural enemy-host interactions will tend to evolve towards a state of reduced natural enemy effectiveness, and that natural enemies not naturally associated with the pest (i.e. species presumed to be less coevolved with the target pest) either because they do not come from the native area of the pest or because they come from a related pest species, may prove more successful in biological control. Hokkanen and Pimentel (1984) analysed 286 successful introductions of biological control agents (insects and pathogens) against insect pests and weeds, using data from 95 programmes, to estimate the anticipated frequency of new and old associations. They concluded that new associations were 75% more successful than old associations. However, the validity of this conclusion has been called into question by more refined analyses (Waage and Greathead 1988; Waage, 1990). The latter show that new associations can be as effective as old ones once the natural enemy is established, but that establishment of a natural enemy on a new host is very difficult. Therefore, in practice, studies on the target pest in its region of origin remain the most promising approach to finding an effective biological control agent rapidly, but the potential usefulness of new associations should also be considered (Waage and Mills, 1992).

If only a fraction of the natural enemy complex of a pest can be used in classical biological control, it is essential that the most important species (i.e. important from the standpoint of their potential impact on pest populations) are used from among the candidates available (Waage and Mills, 1992). Decisions on the relative merits of natural enemies can be based on **reductionist criteria** and **holistic criteria** (Waage, 1990). As we shall see, many of these criteria are based mainly or entirely on theory. Guttierrez *et al.* (1994) question this reliance on theory, arguing that it has contributed little either to increasing the rate of success of introductions or to an understanding of reasons for failures.

Reductionist Criteria

Introduction

The reductionist approach involves selecting agents on the basis of particular biological attributes e.g. searching efficiency, aggregative response. Reductionist criteria are mostly derived from the parameters of the analytical parasitoid–host or predator–prey population models discussed in 5.3, in particular those parameters which are important to the lowering of host or prey population equilibria and/or which promote population stability (stability of the natural enemy-pest interaction reduces the risk that the pest population will be driven to extinction by the control agents, which themselves would otherwise become extinct). Despite their frequent use in theoretical studies, reductionist criteria have rarely been used in practice, one reason being the difficulty of estimating parameter values (Godfray and Waage, 1991). As Waage and Mills (1992) point out, laboratory measures of parameters are unlikely to reflect the values of parameters in the field, where the environment is more complex (although, as shown in Chapter 1 of this book, with a little thought experiments can in some cases be designed which take environmental complexity more into account). Field measures of parameters can be very difficult to obtain (subsection 5.3.7), due to an inability to control the various biotic and physical environmental factors.

A further criticism that can be aimed at the reductionist approach concerns the validity of separating a parasitoid's or predator's biology into components (and of assuming that there are natural enemies that have 'ideal combinations' of such components, see below). The whole organism, not its component parts, is what forms the basis for prediction of success in biological control (Waage and Mills, 1992). However, models have been developed (see subsections 5.3.7, 5.3.8) in which biological attributes are assembled in a more realistic fashion (Hassell, 1980a; Murdoch *et al.*, 1987; Gutierrez *et al.*, 1988a, 1994; Godfray and Waage, 1991). Even so, a major difficulty with some models is that a very large number of parameters have to be estimated in order for the model to be operated. A notable exception to this is the Godfray and Waage (1991) model of the mango mealybug–parasitoid system, incorporated in which were important but easily measurable parameters such as the stage of the host attacked by the different parasitoids (*Gyranusoidea tebygi* and *Anagyrus* sp.), age-specific development rates for host and parasitoids, age-specific survivorship of hosts in the field, adult longevities and daily oviposition rates. Some parameters such as searching efficiency were more difficult to measure, and so a range of realistic values was tested in each case using sensitivity analysis. Godfray and Waage (1991) categorise their model as one of intermediate complexity, in that it is more complex than the analytical models of theoretical ecology e.g. the Nicholson–Bailey model (subsection 5.3.7) but less complex than many detail-rich simulation models, e.g. that of Gilbert *et al.*, (1976) (subsection 5.3.8).

The Godfray and Waage (1991) model was **prospective** (i.e. constructed prior to introduction) as opposed to **retrospective** (i.e. constructed following introduction). Prospective models can be a very useful tool in decision-making (Godfray and Hassell, 1991), but an important constraint upon their development and use in biological control programmes is the need for practitioners to achieve control of the pest as rapidly as possible (Waage, 1990).

Thus, selection of agents based on comparisons of species' attributes will probably never be a top priority; indeed so far it has been very rare for candidate species available in culture for introduction and cleared through quarantine, not to be introduced (Waage, 1990). Nevertheless, pre-introduction studies on candidate species, aided by modelling, still have a significant contribution to make to classical biological control programmes; at the very least they could bring about a re-ordering of the sequence of introduction of species destined for release, i.e. help to move the more effective agents to the front of the queue. Bearing in mind that programmes usually end before all candidate agents have been released, prioritising agents on the basis of their likely efficacy would ensure that the 'best' species are released (Waage, 1990).

Before we go on to discuss which attributes of candidates are considered to be particularly desirable, it is important to remind the reader that most biological control models are based on equilibrium population dynamics (Gutierrez *et al.*, 1994, describe a recent exception), and that in using such models, one seeks to determine to what extent a low, stable pest equilibrium can be achieved. The assumption that low, stable host equlibrium populations result in good pest control was questioned by Murdoch *et al.* (1985), who argue that instead of considering biological control in terms of local population dynamics, theoreticians and practitioners should view it terms of metapopulation dynamics (subsection 5.3.10). Murdoch *et al.* argue that local instability (and thus a high risk of extinction of local populations) is not necessarily bad news for biological control practitioners; persistence may be possible in the metapopulation.

Another point concerning biological control models based on equilibrium population dynamics is that they will be more appropriate for some pests than for others (Godfray and Waage, 1991). In some agroecosystems, e.g. arable crops and glasshouse systems, cultural practices and/or seasonal factors will prevent the population ever coming close to equilibrium. Thus, some of the selection criteria discussed below have little or no relevance to certain crop systems.

Biological control models have mainly been of the deterministic analytical type. Murdoch *et al.* (1985) suggest the use of **stochastic boundedness models** as a preferable alternative. With the latter, emphasis is placed on estimating the probabilities that either host or parasitoid (or prey and predators) may become extinct, or that the pest exceeds an economic threshold, rather than on stability analysis (Chesson, 1978b; 1982).

Listed below are several attributes (some of which are common model parameters) of natural enemies considered to be among the most desirable for biological control, based on theoretical modelling, practical considerations and past experience. Implicit to the reductionist approach is the notion that any combination of life history parameters is possible (Waage, 1990). Ideal combinations may not, however, be available. It is becoming increasingly apparent, through studies on natural enemies, that there are trade-offs between one attribute and another, for example, that between adult reproductive capacity and larval competitive ability in parasitoids (see **Holistic Criteria**). It would therefore be better to concentrate on 'suites' of often counterbalancing attributes. Waage argues that for reductionist criteria to be useful, they need to be derived at a higher level where the traits are integrated in the patterns or strategies that we observe in nature, and he cites one approach as being the computation of intrinsic rates of natural increase (r_m) (section 2.11 for methods).

Murdoch (1990) argues that a low mean pest density relative to the economic threshold and a low degree of temporal variability can be traded off to some extent: higher temporal variability is acceptable if the mean is very low. Thus, a parasitoid may have a weakly stabilising effect upon the parasitoid–host population interaction, but as long as it reduces the pest population equilibrium to a sufficiently low level, there is a reduced likelihood of the pest becoming a problem.

High Searching Efficiency and Aggregative Response

The density responsiveness of candidate biological control agents has been compared through short-term (24 h) functional response experiments (section 1.10 for design). For example, if one is comparing the potential effectiveness of two parasitoid or predator species, the species with the higher maximum attack rate (which is set by handling time and/or egg limitation) may be selected, as it will, all other things being equal, depress the pest equilibrium to a greater extent. Sigmoid functional responses, because they result in density dependent parasitism or predation, are potentially stabilising. However, responses have to be very pronounced, and the destabilising time delays in the population interaction small, for the stabilisation to be marked (Hassell and Comins, 1978; subsection 5.3.7).

As pointed out by Waage and Greathead (1988), the natural enemy functional response offers a good conceptual framework for understanding the action of agents in inundative releases.

Aggregation by predators and parasitoids in a spatially heterogeneous environment (subsection 1.14.2) is recognised as being one of the major stabilising factors that lead to the persistence of parasitoid–host and predator–prey populations in models of parasitoid–host interactions (subsection 5.3.10) (but see Murdoch and Stewart-Oaten, 1989). This applies to both aggregation in patches of high or low host or prey density and aggregation in patches independent of host or prey density. Heterogeneity in parasitism, whether the result of aggregation or other factors, can have a stabilising effect at even the lowest population levels, unlike host resource limitation or mutual interference (Hassell and Waage, 1984; May and Hassell, 1988).

Aggregation has been proposed as a selection criterion for parasitoids (Murdoch *et al.*, 1985 and Waage, 1990), but it is questionable whether aggregation, or any stabilising property of an agent for that matter, ought to be promoted as an independent selection criterion. It remains to be demonstrated that any failure in biological control has been the result of a lack of the persistence of an established agent, rather than from the failure of that agent to become established in the first place (Waage, 1990).

Aggregation has generally been shown in models to raise equilibrium population levels (but see Murdoch and Stewart-Oaten, 1989), so it could be argued that aggregation is an undesirable feature in biological control agents, given that it is possible for the latter to persist despite local extinction (see above).

Van Lenteren (1986) discourages detailed study of the 'density responsiveness' (which encompasses factors such as functional responses, mutual interference and aggregation) in the selection of natural enemies for use in the control of greenhouse pests, and recommends that tests should be carried out in the greenhouse, to determine whether the species is able to:

1. discover pest patches throughout the greenhouse before the economic threshold of pest numbers has been exceeded;
2. reduce pest numbers sufficiently after having found a pest patch;
3. keep the overall pest density below the economic threshold during the growing season.

High Degree of Seasonal Synchrony With the Host

Populations of hosts and parasitoids with discrete generations frequently show imperfect phenological synchrony, with the result that some host individuals experience a reduced or even zero risk of parasitism, i.e. there is a partial refuge effect. Compared with perfect synchrony, imperfect synchrony will result in a raising of the host equilibrium level. Models developed by Münster-Swendsen and Nachman (1978) and Godfray *et al.* (1994) have shown that it will also stabilise the parasitoid–host population interaction.

The degree of 'phenological matching' between parasitoid and pest can be investigated in the laboratory, by subjecting the insects to a range of temperatures and/or photoperiods (Goldson and McNeill, 1992). Parasitoid populations taken from different localities from within the host's native range may show different diapause characteristics and therefore will show different degrees of synchrony with pest populations that originate from only one locality.

Seasonal synchrony with the host is not an important selection criterion in biological control programmes involving inoculative release and inundative release, since synchrony can be achieved by the grower through the release of parasitoids when most hosts are in the susceptible stage for parasitism (van Lenteren, 1986).

Predators, being mostly polyphagous, are unlikely to show a high degree of seasonal synchrony with their prey.

Short Development Time Relative to That of the Host

Murdoch *et al.* (1987) developed a stage-structured parasitoid–host model in which either the adults or juveniles of the pest can be made invulnerable to attack by the parasitoid. The model also incorporates a developmental delay in both the host and the parasitoid (Figure 5.40). The stability of this model depends on the length of the parasitoid time lag, T_2, relative to the duration of the invulnerable stage, T_A. The parasitoid's time lag is destabilising; the longer the developmental period of the parasitoid is relative to that of the host, the more difficult it is to obtain stability (i.e. high T_2 / T_A). A longer development time also leads to exponential increases in the pest equilibrium. Therefore, Murdoch (1990) considers a short parasitoid development time to be a desirable attribute of a parasitoid species for biological control.

For methods of measuring development times of parasitoids, see section 2.9.2

A High Intrinsic Rate of Natural Increase (r_m)

Candidate species can be preferentially selected over others if they have a higher r_m (see section 2.11 for method of calculating r_m for a parasitoid) under temperature conditions similar to those in the area of introduction. As pointed out by Huffaker *et al.* (1977), it is a common error to conclude that a natural enemy having a lower r_m than that of its host or prey would be a poor biological control agent. The parasitoid (or predator) need only possess an r_m high enough to offset that part of the host's r_m that is not negated by parasitism (and host-feeding and host-mutilation).

Rapid Numerical Response

The calculation of r_m values yields no insights into the dynamic interaction between natural enemy and the pest, whereas the numerical response does (subsection 5.3.7). Whilst some biological control workers have assumed that a rapid numerical response is desirable in a natural enemy, this conclusion needs some qualification. While it is likely that a slow numerical response to a pest with a high population growth rate will lead to delayed density-dependence and limit cycles (subsection 5.3.7), a rapid numerical response to slow pest population growth may result in overcompensating fluctuations and decreased stability. Therefore the potential value of a numerical response needs to be assessed in relation to the population characteristics of the target pest.

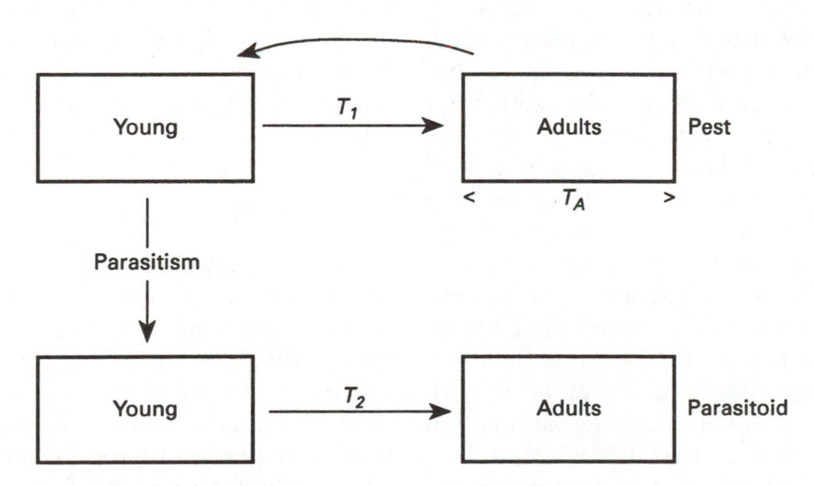

Figure. 5.40 Diagram of a parasitoid–pest model in which both species have an immature stage lasting for T_1 and T_2 days respectively. The adult pest is invulnerable to attack by the parasitoid and lives for an average of T_A days (Source: Murdoch *et al.*, 1990.) Reproduced by permission of Intercept Ltd.

Mode of Reproduction

Modes of reproduction in parasitoids are discussed in section 2.5.

Stouthamer (1993) considered the merits of arrhenotoky and thelytoky in parasitoid wasps from the standpoint wasps from the stand point of classical biological control; some of his conclusions were that:

1 arrhenotokous forms (species or 'strains') will be able to adapt more rapidly to changed circumstances. If environmental conditions in the area of introduction are different from those in the form's native range, arrhenotokous wasps may have the advantage;
2 assuming that a thelytokous form and an arrhenotokous form produce the same number of progeny, the thelytokous form will: (a) have a higher rate of population increase and (b) depress pest populations to a lower level;
3 arrhenotokous forms must mate to produce female offspring; therefore, in situations where the wasp population density is very low, males and females may have problems encountering one another. Thelytokous forms are therefore better colonisers.

The production of diploid males by Ichneumonidae was discussed in subsection 2.5.2. The occurrence of such males in a population will potentially reduce the population growth rate because a proportion of the eggs that are fertilised become males that will either die during development or become male adults. Females inseminated by the latter produce only haploid sons or triploid (sterile) daughters. The extent to which population growth rate is influenced depends on the number of alleles at the sex locus and the mating system (Stouthamer *et al.*, 1992). Stouthamer *et al.*, using computer simulation models, also showed that diploid male production can reduce rates of establishment. Diploid male production therefore presents problems for biological control introductions and for mass-rearing of agents. Stouthamer *et al.* suggest that these problems can be alleviated by rearing wasps either as one large population (this will reduce the rate at which sex alleles are lost) or as a large number of smaller subpopulations (this will result in many alleles being lost from each subpopulation but at least two sex alleles will be retained; all the different alleles will probably be present in the amalgam of the subpopulations). For further practical advice on culturing practices, see Stouthamer *et al.*'s paper.

As far as the problem of establishment is concerned, Stouthamer *et al.* recommend the release of many wasps per colonisation site, so as to avoid the introduction of only a subpopulation of the sex alleles. Releases should also be made in a grid pattern, so as to reduce the frequency of sib-mating; because relatively few females are released in an introduction, the number of foundresses per host patch will be low and the frequency of sib-mating will be high (see subsection 1.9.1). However, as alleles are expected to spread evenly with time. Stouthamer *et al.* also recommend increasing allelic diversity by the importation and release of additional alleles (this can be done by introducing wasps taken from different geographic areas, although crossing incompatibility may prove to be a severe constraint upon allele addition), in which case fewer individuals need to be released.

Destructive Host-feeding Behaviour

In the literature on biological control, the occurrence of non-concurrent destructive host-feeding behaviour (the use of different host individuals for either destructive host-feeding or oviposition, see section 1.7) has been given as a reason for thinking that a particular parasitoid species is a potentially effective biological control agent. This is not an unreasonable assumption to make, in view of the fact that several parasitoid species have been shown to kill far more hosts by host-feeding than by parasitism. Flanders (1953) was the first to question the value of destructively host-feeding parasitoids in biological control, reasoning that the parasitoids' high feeding requirements at low host densities (section 1.7) will result in their low persistence when hosts are scarce, producing population oscillations of high amplitude. Later, Kidd and Jervis (1989) tackled the problem, using analytical models (modifications of the Nicholson–Bailey-model) and computer simulation models. The analytical models indicated that host-feeding behaviour is unlikely to contribute to stability, and that the host equilibrium will be raised and the parasitoid equilibrium lowered, compared with the (stabilised) Nicholson–Bailey model. An increase in the amount of energy extracted per host encounter lowered the host equilibrium number and raised that of the parasitoid, whereas increases in the parasitoids' maintenance, search or egg production costs produced the opposite effect. The simulation models showed that destructive host-feeding parasitoids have a destabilising effect on the parasitoid-host interaction, not through feeding *per se*, but through the effects of egg-limitation, egg resorption and variable longevity. Together, the latter three processes constitute an important destabilising mechanism that can override other supposedly very powerful stabilising processes such as survival of hosts in refugia. The mechanism operates as follows: when hosts are scarce, female parasitoids are able to survive using energy obtained from egg resorption (subsection 2.3.4), but this is possible only at the expense of oviposition. Host encounter rates thus remain relatively high, even at very low host densities. The females continue to feed on hosts, thereby forcing the host population to even lower levels. Eventually, when egg resorption and host-feeding attacks can no longer sustain the parasitoids, the parasitoids themselves die out, leaving no offspring to continue the interaction (Jervis and Kidd, 1991). The deleterious effects of this mechanism could, perhaps, be weakened if the parasitoids have an ample supply of non-host foods (section 6.1) such as honeydew (not necessarily that of the host) or nectar, during periods of host scarcity.

Methods of supplying field populations of parasitoids with non-host foods are discussed by Jervis *et al*. (1993).

High Rate of Dispersal

Techniques for studying dispersal by natural enemies are discussed in subsection 4.2.11.

If a natural enemy has a high ability to disperse (either as an adult or as an immature stage within the host), then it can be expected to spread rapidly from the initial release point. Thus, fewer resources (time, money) may need to be invested in large numbers of point releases over a region to ensure that the natural enemy becomes established over a wide area.

In biological control programmes that involve inundative releases of parasitoids, the parasitoids are used as a 'biopesticide', so it is important that the insects do *not* disperse rapidly away from the crop. Parasitoids can be encouraged to remain within the crop either by 'pre-treating' females with host kairomones so as to stimulate search following release (Gross *et al*., 1975), by applying kairomones directly to the crop to act as arrestants (defined in section 1.5) (Waage and Hassell, 1982) or by applying non-host foods (section 6.1) to the crop.

In greenhouses, the temperature threshold for flight may need to be considered in relation to the temperatures in the pest's environment (van Lenteren, 1986).

High Degree of Host Specificity

Practical approaches to studying host specificity in parasitoids and predators are discussed in subsections 1.5.7 and 2.10.2, and section 4.3.

One explanation for the poor performance, overall, of predators compared with parasitoids in classical biological control in perennial crop systems is their tendency to be more polyphagous. The rationale for this is that, because of polyphagy, the pest cannot be maintained at low equilibrium populations: the natural enemy will concentrate on the more abundant alternative prey or host species (see switching behaviour, section 1.11). However, as Murdoch *et al.* (1985) point out, a polyphagous natural enemy can survive in the absence of the pest should there be local extinction of the latter, and it can be ready to attack the pest when it re-invades. Therefore, polyphagy may not be as undesirable an attribute in classical biological control as it is commonly assumed to be.

A lack of host specificity is not a problem in programmes aimed at pests of protected crops i.e. in glasshouses, as the environment is a simple one, usually containing unrelated pest species at each of which different parasitoid species are targeted (van Lenteren, 1986).

High Degree of Climatic Adaptation

The optimum range of temperatures or humidities for development, reproduction and survival of a candidate biological control agent (subsections 2.5.2, 2.9.2, 2.7.3 and 2.10.2) may be different from that of the pest, and the parasitoid may either fail to establish or prove ineffective owing to the direct or indirect effects of climate in the region of introduction. The conventional wisdom is that a parasitoid species should be selected for which climatic conditions in the region of introduction are optimal (DeBach and Rosen, 1991) (subsection 2.9.3). This view is supported by the database analysis of Stiling (1990) which showed that the climatological origin of parasitoids has a large influence on establishment rate. However, the climatic adaptation criterion should not be rigidly applied; *Epidinocarsis lopezi*, which successfully controlled cassava mealybug in West Africa, originated from Paraguay, where the climate is very different (Gutierrez *et al.*, 1994).

Ease of Handling and Culturing

Greathead (1986) concluded from an analysis of the BIOCAT database, that the most important factors in choice of natural enemy in classical biological control programmes have, perhaps, been ease of handling and availability of a technique for culturing the insects. The recent case of biological control of the mango mealy bug, *Rastrococcus invadens*, is an illustration of how ease of rearing can influence selection. Two encyrtid parasitoids, *Gyranusoidea tebygi* and *Anagyrus* sp., were being considered for introduction into West Africa. Despite the latter species being the dominant parasitoid in rearings from field-collected mealybugs in India, the former species was selected as the first candidate for introduction. owing to the ease with which it could be cultured (Waage and Mills, 1992 for a discussion). A reason given by Waage (1990) for the more extensive use of Ichneumonidae compared with Tachinidae in programmes aimed at controlling exotic Lepidoptera is the greater difficulty encountered in culturing the latter parasitoids. It is noteworthy that the ranking of culturable agents for introduction usually follows the sequence in which they are established in culture (Waage, 1990).

Practical approaches to rearing parasitoids and predators are discussed by Waage *et al.* (1985).

Holistic Criteria

Introduction

The holistic approach to the selection of agents considers less the properties of the agent and more the possible interactions between candidate species and between agents and mortalities acting on the pest in its area of introduction. One important consideration in this approach is that the relationships between natural enemies in biological control releases need to be viewed as dynamic, not static. Examples of this approach are:

Collecting of Parasitoids From Non-outbreak Areas in the Native Range of the Pest

Selection of agents can begin during the exploration phase of a programme by confining exploration to low density populations of the host. The species composition of parasitoid complexes varies with the density of the host population (subsection 4.3.6). Parasitoids collected from host population outbreak areas may not necessarily be those best suited to maintain the pest at low densities in its exotic range (Pschorn-Walcher, 1977; Fuester *et al.*, 1983; Waage, 1990; Waage and Mills, 1992). To increase the likelihood of obtaining the better species, Waage (1990) and Waage and Mills (1992) recommend the use of experimental host cohorts placed out in the field (subsections 4.2.8 and 5.2.9).

Selection of Agents That Follow, Rather than Precede, Major Density Dependent Mortalities in the Pest Life Cycle

Additional density-dependent mortalities acting later in the pest's life cycle can influence the effect of mortality caused by a natural enemy that attacks the pest earlier on (May *et al.*, 1981; May and Hassell, 1988). Indeed, if the density-dependence is over-compensating, too high a level of mortality caused by an early-acting parasitoid can lead to an increase in the host population above the parasitoid-free level!

A density-dependent mortality (whether due to intraspecific competition [van Hamburg and Hassell, 1984] or the action of natural enemies [Hill, 1988]) acting upon a host can be described by the following model (Hassell, 1975):

$$S = N(1+aN)^{-b} \quad (5.43)$$

where S is the number of survivors, N is the initial prey density, a is a constant broadly indicating the densities at which survival begins to fall rapidly, and b is a constant that governs the strength of the density-dependence (b = 1 is perfect compensation, b < 1 is under-compensation, and b > 1 is over-compensation, subsection 5.3.4). Figure 5.41 shows, for the stem-borer *Chilo partellus*

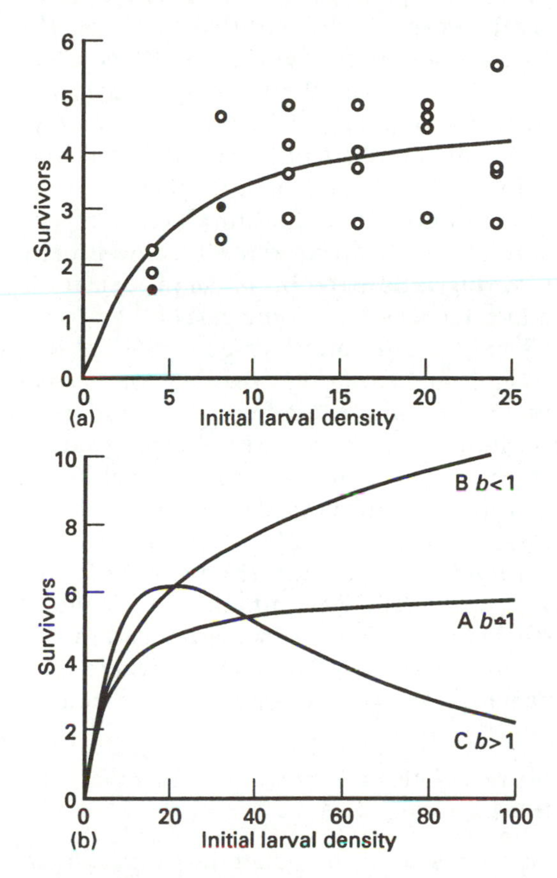

Figure. 5.41 (a) The density-dependent relationship between the number of surviving larvae on plants and the initial densities of first instar stem-borer, *Chilo partellus* (Lepidoptera: Pyralidae) larvae in a glasshouse experiment. The data are described by equation (5.43). Solid circles indicate duplicate points; (b) three hypothetical examples of the density-dependent relationship in equation (5.43) to show the effect of varying the parameter b which governs the strength of the density-dependence. Curve A uses the values estimated from the *C. partellus* data in (a) (b = 1.089); curve B shows how the number of survivors continues to increase with initial larval density when b < 1; curve C shows over-compensation when b>1 (Source: van Hamburg and Hassell, 1984). Reproduced by permission of Blackwell Scientific Publications Ltd.

(Lepidoptera), a hypothetical example where S is plotted against N, for three density-dependent functions with different values of b (N in this case refers to larval densities). When $b \simeq 1$ (curve A), the density-dependence tends to compensate for any early-acting parasitism, as long as the initial larval density is not reduced to lie on the steeply rising part of the curve. When $b < 1$ (curve B), however, there is only partial compensation and egg parasitism will always reduce the numbers of larvae ultimately surviving. When $b > 1$ (curve C) there is overcompensation, and the introduction of egg mortality will lead to *more* larvae eventually surviving unless the initial larval density is reduced to lie on the rising part of the curve.

Van Hamburg and Hassell (1984), in discussing augmentative releases of *Trichogramma* against stem-boring Lepidoptera, concluded that the success of a programme will be largely influenced by the level of egg parasitism, the level of the subsequent larval losses, and the degree to which the latter are density-dependent. Furthermore, these factors will vary between different agricultural systems and different pest species, and should be evaluated for each situation where augmentative releases are being contemplated.

Selection of Agents That are Constrained in Their Area of Origin

This is based on the notion that a parasitoid (or predator), in its area of origin, could be a more important factor in the dynamics of the host were it not for the fact that it is 'held back' by mortality (e.g. hyperparasitism) or competition (e.g. heterospecific superparasitism, section 2.10.2). For a discussion of this concept, see Myers *et al*. (1988).

Reconstructing Natural Enemy Communities on Exotic Pests

This is based on the notion that the most effective control of a pest is achieved through the use of the most complete community of agents (see section 4.3 for methods of determining the composition of natural enemy complexes). That is, if in the region of exploration the natural enemy community associated with a pest comprises n species of parasitoid and predator, all n species ought to be introduced, since they are presumed to have complementary characteristics.

There is support for the view taken by some biological control practitioners (Huffaker *et al*., 1971; DeBach and Rosen, 1991) that control of pests is enhanced by introductions of several parasitoid species (but see May and Hassell, 1981; Kakehashi *et al*., 1984). Hawkins (1993), using a global data set compiled from the literature on parasitoids, showed that by comparing (in native parasitoid communities) percentage parasitism of each host species with the size of the associated parasitoid complex, a highly significant, albeit 'noisy' positive correlation is obtained.

The presence of an invulnerable age-class in a pest is a factor that promotes stability in the parasitoid–host interaction (Murdoch *et al*. 1987; subsection 5.3.6). If one accepts Murdoch's (1990) argument that stability may be bought at the cost of a raised equilibrium level, then an invulnerable age class in the pest should be made vulnerable, by the use of two or more species that attack different stages of the pest.

For a recent study where reconstruction of a parasitoid complex is contemplated and is approached experimentally, see Patil *et al*. (1994)

Selection of Agents on the Basis of Their Complementary Interactions With Other Agents

May and Hassell (1981), using simple analytical Nicholson–Bailey models, studied two parasitoid species–one host species interactions involving parasitoids of the same guild, (subsection 5.2.3). They showed that the addition of a second parasitoid species (Q) to an existing interaction with an already established parasitoid (P) can have the following effects:

1. the most desired outcome of Q becoming established and coexisting with P, which leads to a further depression of the host equilibrium;
2. the satisfactory outcome of Q replacing P but then persisting in a stable interaction with the host population at a lower level than with P alone (an apparent example is shown in Figure 5.42);
3 the undesirable outcome of Q failing to become established (this frequently occurs in classical biological control, and is most likely to occur if Q has a lower searching efficiency than P);
4. the highly undesirable outcome of the interaction becoming unstable with periodic host outbreaks, either with Q and P co-existing or with Q having replaced P.

The analysis also showed that that Q and P are most likely to co-exist when one is the superior larval competitor (i.e. always wins in cases of heterospecific superparasitism) and the other is the superior adult competitor (i.e. has a higher searching efficiency). This phenomenon is known as **counterbalanced competition** (Zwölfer, 1971). Protocols for determining which of a pair of species is the superior larval competitor are discussed in subsection 2.10.2, while the measurement of searching efficiency is discussed in subsection 5.3.7. Investigators might also consider the possibility that parasitoid species may interfere with one another behaviourally, through patch marking and patch defence (subsection 1.5.3 and section 1.13 respectively).

It may be the case that a pest, in its exotic range, is attacked by one or more generalist natural enemies prior to the introduction of biological control agents. The latter may need to be selected on the basis of their potential for interaction with the former (May and Hassell, 1988). May and Hassell (1981, 1988),

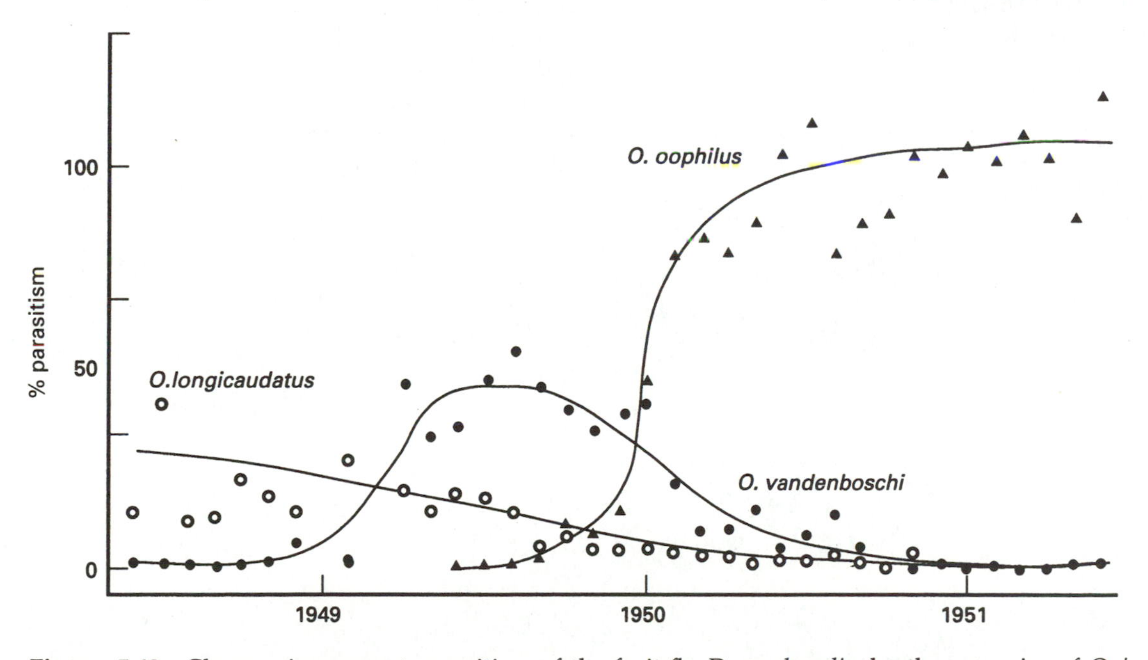

Figure. 5.42 Changes in per cent parasitism of the fruit-fly *Dacus dorsalis* by three species of *Opius* (Braconidae). Each successively introduced parasitoid species appears to cause a higher level of parasitism. (Source: Varley *et al.*, 1973, who used data from Bess *et al.* (1961).) Reproduced by permission of Blackwell Scientific Publications Ltd.

using a Nicholson–Bailey type model, have shown that an introduced parasitoid can 'invade' and co-exist more easily if it acts before the generalist in the host's life cycle. This is due to the larger pool of hosts available. However, if the host's net rate of increase is too low, or if the generalist's overall efficiency is too high relative to that of the parasitoid, the parasitoid will be unable to invade and a persistent three-species interaction is impossible. Specialist egg parasitoids may therefore be easier to establish than larval or pupal ones, particularly if the pest has a relatively low net rate of increase and already suffers signifcant mortality from generalist natural enemies. For other predictions of the model, see May and Hassell's (1981, 1988) papers.

ACKNOWLEDGEMENTS

We are very grateful to the following for providing useful advice and information: Mike Claridge, Peter McEwen, John Morgan, Anja Steenkiste, Keith Sunderland, Bill Symondson and Jeff Waage.

APPENDIX 5A

A BASIC program for running the simple inductive model developed from the life table data in Table 5.2.

```
10   PRINT "EGGS=";
20   INPUT E
30   PRINT "EGGS=";E
40   K=0.74                  ! find k1
50   L=E–K                   ! find number of larvae
60   K=0.86*L–1.52           ! find k2
70   P=L–K                   ! find number of pupae
80   K=0.96                  ! find k3
90   A=P–K                   ! find number of adults
100  F=A–0.30103             ! find number of females
110  E=0.86*F+2.1            ! calculate reproduction
120  GOTO 30
130  END
```

PHYTOPHAGY 6

M.A. Jervis and N.A.C. Kidd

6.1 INTRODUCTION

This chapter considers ways in which phytophagy (including mycophagy) by parasitoids and predators may be studied. Many insect natural enemy species, at some stage during their life cycle, exploit plant materials in addition to insect materials, feeding:

1. **directly** upon plants, consuming floral and extrafloral nectar, pollen (Figure 6.1), seeds (either whole seeds or specific tissues), and, less commonly, materials such as plant sap (including the juices of fruits), epidermis, trichomes and fungal spores, and/or
2. **indirectly**, consuming honeydew produced by Homoptera, e.g. aphids (Figure 6.2), mealybugs, scale insects and whiteflies, that feed on the plants.

It is often the case with predators that only the adults are phytophagous, and then only facultatively so, although in aphidophagous hover-flies, predatory ants and some lacewings the adults feed exclusively on plant materials. In *Coccinella septempunctata* (Coccinellidae) the larvae, as well as the predatory adults, have been observed to consume nectar. Amongst parasitoids, generally only the adults consume plant materials; exceptions include certain Eurytomidae which as larvae are 'entomophytophagous', developing initially as parasitoids and completing development by feeding upon plant tissues (Henneicke *et al.*, 1992).

Most investigations of predator or parasitoid foraging behaviour and population

Figure 6.1 A male of *Syrphus ribesii* (Syrphidae) feeding on the pollen of *Hypericum perforatum*. (Premaphotos Wildlife.)

dynamics have been concerned with a natural enemy's interaction with prey/hosts, and have ignored or overlooked its interaction with the non-prey/non-host food source. The foraging behaviour of predators and parasitoids in relation to honeydew has been

Figure 6.2 A male of *Episyrphus balteatus* feeding on aphid (*Pterocomma pilosum*) honeydew on a leaf (*Salix sp.*) surface. (Premaphotos Wildlife)

investigated (e.g. Ayal, 1987; Wilbert, 1977; Vinson *et al.*, 1978; Bouchard and Cloutier, 1984; Budenberg, 1990; Shimron *et al.*, 1992), but generally only the role of honeydew as a kairomone in host or prey location (subsection 1.5.3), not its role as a food, has been considered (but see Dreisig, 1988; Veena and Ganeshaiah, 1991). By the same token, information on the range of prey or host types attacked by natural enemies is far more comprehensive and detailed compared with information on the types of plant material that many consume. This difference in emphasis is to be expected, since workers have either been ignorant of the role of plant materials in the biology and ecology of natural enemies, or they have tended to regard the consumption of plant materials as somewhat peripheral to what is generally considered to be the most important aspect of natural enemy biology, namely entomophagy. Nevertheless, it represents a serious omission, since the consumption of plant materials is an important factor in the growth, development, survival and reproduction of natural enemies (subsections 2.7.3, 2.8.3, 2.9.2, 2.10.2; reviews by Hagen, 1986; Jervis and Kidd, 1986; van Lenteren, *et al.*, 1987; Evans, 1993). Furthermore, the effectiveness of predators and parasitoids as population control agents may depend greatly upon the availability and quality of non-prey and non-host foods (see reviews by Powell, 1986; Jervis *et al.*, 1993).

In this chapter we suggest how the source and identity of plant materials comprising the diet of natural enemies might be determined, and discuss some ways in which this information might be used.

6.2 WHAT PLANT MATERIALS DO NATURAL ENEMIES FEED UPON, AND FROM WHAT SOURCES?

6.2.1 INTRODUCTION

The range of plant materials exploited as food under field conditions is poorly known for most predators and almost all parasitoids. An obvious potential source of these foods to consider is flowers. There are numerous published records of predators and parasitoids visiting flowers, many of the earliest records being listed in works such as: Knuth (1906, 1908, 1909), Müller (1883) and Willis and Burkill (1895), which deal with various flower-visiting insects; Drabble and Drabble (1927) with various Diptera; Allen (1929) with Tachinidae in particular; Drabble and Drabble (1917) and Hamm (1934) with Syrphidae, including a few aphidophagous species. More recent records are in Parmenter (1956, 1961), Herting (1960), Karczewski (1967), Judd (1970), van der Goot and Grabandt (1970), Sawoniewicz (1973), Barendregt (1975), Maier and Waldbauer (1979), Toft (1983), Primack (1983), Haslett (1989a), de Buck (1990), Maingay *et al.* (1991), Cowgill *et al.* (1993), and Jervis *et al.* (1993).

There are several points to consider when referring to much of the existing literature on flower-visiting by natural enemies. First, some authors do not indicate what materials (whether nectar, pollen or both) the insects were recorded feeding upon, if they were feeding at all (natural enemies may visit flowers either accidentally or solely for purposes other than feeding, i.e. sheltering, meeting mates and locating prey and hosts). Müller (1883) is one notable exception. Second, some authors omit to mention the sex of the insects involved, thus greatly reducing the scientific value of the information obtained. Third, the identifications of the insects sometimes cannot be relied upon. The latter point applies particularly to the older literature on parasitoids; early publications probably contain a high proportion of misidentifications, as parasitoid taxonomy has advanced greatly over the years. Fourth, the taxonomic nomenclature of Hymenoptera 'Parasitica' also tends to be out of date, making it difficult to determine at a glance which parasitoid species the records actually refer to, even if the records themselves are reliable. Fifth, when considering a particular natural enemy, one should not attach too much significance to those plant species for which no record of visits has been obtained. If the insect species being investigated has a preference for the flowers of certain plant species (section 6.3), then the probability of the investigator recording that natural enemy on them will be significantly greater than on others. Because of this, a superficial field survey (i.e. one involving the taking of a small sample) could give the misleading impression that the less preferred plant species are not exploited at all. This is an important point to consider when dealing with almost all the published records of visits by insects to flowers.

Several of the above points also apply to natural enemies feeding at extrafloral nectaries (for records, see Nishida, 1958; Putman, 1963; Keeler, 1978; Bugg *et al.*, 1979; Beckmann and Stucky, 1981).

6.2.2 DIRECT OBSERVATIONS ON INSECTS

Generally, insects, including even very small parasitoid species (e.g. most Chalcidoidea, Cynipoidea and Proctotrupoidea) may be easily observed visiting flowers. It ought to be quite clear to the observer whether or not the insects are feeding, at least in the case of exposed floral nectaries (e.g. such as occur in *Hedera helix*, *Euphorbia* species and Umbelliferae, Figure 6.3) and flowers lacking nectaries (e.g. many grasses). If the insects can be seen (either with the naked eye or with magnifying optical equipment) to apply their mouthparts to the nectar or pollen source, then it is quite reasonable to infer from this that feeding is taking place (the same inference can be drawn from observations of parasitoids, ants and other natural enemies visiting extrafloral nectaries. However, caution needs to be applied to observations of some types of insect that apply their mouthparts to patches of honeydew on plant surfaces. Parasitoids for which honeydew is a kairomone in host location are known to test

Figure 6.3 An adult of an unidentified ichneumonid wasp, feeding on the nectar of the umbellifer *Angelica sylvestris*.

the substratum in this way, without ingesting the honeydew (Budenberg, 1990). Parasitoids and also predators such as sphecid and vespid wasps, anthocorid bugs, ants, and larval and adult chrysopids and coccinellids visit the flowers of Umbelliferae and feed upon nectar. Adults of *Chrysoperla carnea* can easily be observed applying their mouthparts either to the nectaries or to the anthers.

In plants with concealed nectaries, it is less easy to determine whether feeding is taking place and/or what materials are being fed upon, although in some cases it may be reasonably inferred from the insect's behaviour that nectar-feeding is taking place. For example, in the creeping buttercup (*Ranunculus repens*) and its close relatives, the nectaries are situated near the bases of the petals, and are each concealed by a flap or scale (Percival, 1965). The adults of a variety of small parasitoid wasps may be observed with their heads either at the flap edge or beneath the flap, suggesting nectar-feeding.

The activities of small Hymenoptera visiting some other flower types may be interpreted as nectar-seeking behaviour. For example, in the flowers of *Convolvulus repens* and *Calystegia sepium*, wasps and ants crawl down the narrow passages at the base of the corolla that lead to the nectaries (see also Haber *et al.*, 1981 on ants).

With many plants which have flowers with concealed nectaries, particularly species with narrow corollas (e.g. members of the daisy family [Compositae]), it is sometimes possible to ascertain directly what materials the insects are feeding on (this is generally not the case with Hymenoptera 'Parasitica'; Jervis *et al.*, 1993). Even in the absence of close observations, it may be reasonable either to deduce from the structure of the insects' mouthparts or to infer from the insects' behaviour that particular plant materials are being sought. For example, flies such as nemestrinids, phasiine tachinids and some conopids have long, slender probosces and are unlikely to be seeking any materials other than nectar from the flowers they visit. The same applies to the few hymenopteran parasitoids that have elongated mouthparts (e.g. species of *Lapton, Agathophiona, Rynchophion, Gonolochus, Agathilla, Certonotus* (Ichneumonidae) and *Agathis, Cardiochiles* and *Chelonus* (Braconidae)), and those that show other mouthpart adaptations for liquid-feeding (Perilampidae; Darling, 1988).

Examination of plants in the field for natural enemies carrying out nectar- and pollen-feeding may prove very difficult and time-consuming. Flowers may therefore be presented to the insects in the laboratory, and observations on behaviour carried out. This has been done for parasitoids (Györfi, 1945; Leius, 1960; Syme, 1975; Shahjahan, 1974).

However, the results of such tests need to be viewed with caution, because under field conditions the insects may not visit the same plant species as those with which they are presented in the laboratory. Even greater caution needs to be applied to the results of laboratory tests that involve presenting nectar extracted from different flowers to insects, as has been done for ants (see Feinsinger and Swarm, 1978; Haber *et al.*, 1981).

6.2.3 INDIRECT METHODS

Where direct evidence of floral nectar- and pollen-feeding is unavailable, other evidence needs to be sought. The body surface, including the mouthparts, of flower-visitors may be examined for the presence of pollen grains irrespective of whether the insects are collected at flowers (Holloway, 1976; Stelleman and Meeuse, 1976; Gilbert, 1981) or in other circumstances. The plant species source of such grains can be identified either:

1. by using identification works (e.g. Sawyer, 1981, 1988; Faegri and Iversen, 1989; Moore *et al.*, 1991; also Erdtman, 1969; Reitsma, 1966), and/or
2. by comparing the grains with those collected by the investigator from plants in the parasitoid's field habitat (flowering plant species, in some cases even closely related ones, differ with respect to pollen surface sculpturing).

However, the presence of pollen grains on the body surface, even on the mouthparts, does not necessarily prove that the insects have been consuming pollen itself, since the insects may become contaminated with grains whilst seeking nectar. Bear in mind also that when collecting insects for the purpose of examining their body surface for pollen grains, it is essential that insects are individually isolated, so as to avoid cross-contamination with (sticky type) grains.

The body surface of insects may also be examined for the presence of fungal spores. However, if spores are present, it does not necessarily mean that the insects feed on fungi. The spores may accidentally become attached to the insects as the latter brush against fruiting bodies.

Gut dissections may reveal the presence of pollen grains. This technique has been used with hover-flies (van der Goot and Grabant, 1970; Holloway, 1976; Haslett and Entwistle, 1980; Leereveld, 1982; Haslett, 1989a), hymenopteran parasitoids (Györfi, 1945; Leius, 1963; Hocking, 1967), coccinellid beetles (Hemptinne and Desprets, 1986) and green lacewings (Sheldon and MacLeod, 1971). The technique has been applied to both dried, pinned specimens, specimens preserved in ethanol, and deep-frozen insects (Holloway, 1976; Leereveld, 1982; Haslett, 1989a). Because the exines (outer coverings) of pollen grains are refractory structures, i.e. resistant to either decay or chemical treatment, they retain much of their original structure (hover-flies, at least, do not grind pollen [Gilbert, 1981; Haslett, 1983]), and the original plant source can thus often be identified, as indicated above.

Hunt *et al.* (1991) devised a pollen exine detection method in which the abdomens (or gasters in the case of wasps) of dried, preserved insects are cleaned, crushed, heated in a mixture of acetic anhydride and concentrated sulphuric acid, and the mixture centrifuged (for further details, see Hunt *et al.*, 1991; Lewis *et al.*, 1983). Note that with this method, the precise location of pollen grains within the abdomen/gaster cannot be determined.

Electrophoresis of the gut contents (subsection 4.3.11) also has potential as a method for detecting the presence of pollen in the diets of arthropods (van der Geest and Overmeer, 1985). However, given the ease with which dissection and visual examination may be carried out, it is unlikely that electrophoresis will ever be used as a general technique for pollen detection in natural enemies, and particularly for pollen identification.

The detection of pollen exines in the gut of a predator or parasitoid does not necessarily mean that the insect has been feeding directly at pollen sources (i.e. anthers). Pollen may fall or be blown from anthers and become trapped in nectar, honeydew or dew (Todd and Vansell, 1942; Townes, 1958; Hassan, 1967; Sheldon and MacLeod, 1971). For some species, the consumption of nectar, honeydew or dew may be the sole means of obtaining pollen for oögenesis. Also, with predators such as carabid beetles, pollen grains may enter the gut via the prey (Dawson, 1965).

Pollen grains do not have to be trapped in nectar, honeydew and dew for insects to consume them. Pollen grains that land upon the surfaces of leaves are deliberately taken by some insects, e.g. adults of the non-aphidophagous hover-fly genus *Xylota* (Gilbert, 1986, 1991).

The presence of honeydews in the guts of ants has been established chemically by the Anthrone Test (Skinner, 1980). This colorimetric test involves heating the gaster of the ant (the gaster is best crushed beforehand) with anthrone (9-oxyanthracene) in sulphuric acid. It has not, so far as we know, been used to detect the presence of nectars and honeydews in the guts of other insect natural enemies. The test, because it involves heating the reagent/test material mixture, is inconvenient to perform outside of the laboratory, and so the Cold Anthrone Test, developed by Van Handel (1972), is recommended for field use. The latter test, which is carried out at ambient temperatures, has been used to test for fructose moieties in the guts of haematophagous flies by various workers (Van Handel, 1972; Yuval and Schlein, 1986; Walsh and Garms, 1980; Young *et al.*, 1980; Killick-Kendrick and Killick-Kendrick, 1987), but can be applied to other insects. The rationale behind anthrone tests is that if fructose moieties are detected, the insect must have been feeding on materials (nectars, honeydews) rich in such substances. The Cold Anthrone test has the advantage that it is simple, convenient and rapid, so allowing many insects to be tested. It is the only qualitative sugar test which can be used easily on insects under field conditions.

The Cold Anthrone Test does not, however, allow the investigator to distinguish either between nectar and honeydew, or between floral nectar and extrafloral nectar. Chromatographic techniques have therefore been used to identify more precisely the sugars present in insect guts. Again applied to haematophagous Diptera, thin layer chromatography was used by Lewis and Domoney (1966) and the more sensitive techniques of high performance liquid chromatography (HPLC) and gas chromatography (GC) used by Moore *et al.* (1987) and MacVicker *et al.* (1990). With both HPLC and GC, the insects are cryopreserved whole and aqueous extracts analysed. MacVicker *et al.* concluded that the insects they studied had been consuming honeydew, because the trisaccharide melezitose was detected in their samples. Melezitose and another trisaccharide erlose (fructomaltose) are produced in the guts of phloem-feeding Homoptera as the result of gut enzyme action upon the sucrose component of the plant phloem. MacVicker *et al.* suggest that melezitose and erlose may be useful markers for honeydew. However, it needs to be borne in mind that melezitose occurs occasionally in nectars (Harborne, 1988).

Both the sugar spectra and the spectra of nitrogenous compounds present within nectar and honeydew tend to vary with source (Percival, 1961; Maurizio, 1975; Baker and Baker, 1983). Therefore, it ought to be possible in some cases (e.g. where a very narrow range of food sources is being exploited) to compare (using HPLC and GC) the chemical composition of the gut contents of parasitoids with that of potential food sources, and identify which nectars and honeydews are being fed upon.

A simpler indirect method of determining whether natural enemies consume nectar or honeydew is to mark potential food sources with dyes or other markers such as rare elements, and examine the guts of the insects for the presence of the marker (G. E. Heimpel and J. A. Rosenheim, personal communication).

However, there are problems associated with this approach. First, it may be necessary to apply the marker to the nectar of a large number of flower species and to a large number of honeydew sites, either at once or in turn, to establish whether feeding occurs and at which sources. Second, one has to be sure that the marker, particularly a dye, is non-repellent and non-toxic.

The source of whole diatoms, algae and fungal spores found in insect guts can be identified by comparison with plants in the habitat. Identifying the origin of plant sap or tissues, or even establishing in the first place that the natural enemy consumes these materials, is likely to prove difficult, unless the insect is directly observed feeding. Other methods therefore have to be used. Chant (1959) and Takafuji and Chant (1976) demonstrated that the gut of phytoseiid mites became coloured when the mites were confined on leaves that had absorbed acid fuschin stain. Porres *et al.* (1975) found that *Euseius hibisci* acquired radioactivity when placed on avocado leaves that had been labelled with $H_3{}^{32}PO_4$, but not when placed on labelled citrus leaves. Fleischer *et al.* (1986) recorded uptake of rubidium by the mirid bug *Lygus lineolaris* from leaves sprayed with liquid containing that element.

Seed predation by ants can be established by examining the 'booty' being carried in ant columns (subsection 4.3.3). The columns can also be followed to determine which plant species are being visited, although ants may take seeds from the ground instead of from the plants themselves.

6.3 DO NATURAL ENEMIES SHOW SPECIFICITY IN THE PLANT FOOD SOURCES THEY EXPLOIT?

Due to the present paucity of field records, only in a relatively small number of cases can we be confident that a predator or parasitoid species is discriminating in the food sources it visits under field conditions. Nevertheless, we can reasonably expect some degree of behavioural specificity (i.e. specificity other than that attributable to a lack of spatio-temporal synchrony between insects and food sources) to be shown generally among natural enemies.

So far as floral food sources are concerned, the range of plant species exploited will depend partly upon floral morphology. In hover-flies for example, species with short probosces are unable to exploit the nectar of flowers with long corollas (Gilbert, 1981; see also Allen, 1929 on Tachinidae). Similarly, with a few exceptions (subsection 6.2.2), most ichneumonid wasps, being relatively large insects and also mostly lacking elongated mouthparts, tend to be excluded from exploiting the nectar of plants having flowers/florets with narrow, tubular corollas (e.g. Compositae) and plants with flowers that have wide corollas but have their nectar concealed at the end of narrow passages, e.g. Convolvulaceae (subsection 6.2.2).

It would be interesting to relate the comparative attractiveness of different flower species for natural enemies to flower anatomy and insect mouthpart morphology. The attractiveness of different flowers has been investigated experimentally for certain hymenopteran parasitoids (Leius, 1960; Syme, 1975; Shahjahan, 1974), but it is difficult to relate, using the data obtained, the degree of attraction of wasps to the floral anatomy of the plants investigated. The range of flower species visited by hover-flies is clearly limited partly by flower anatomy (Gilbert, 1981) and partly by flower colour (defined according to reflectance spectrum; Haslett, 1989b). Gilbert (1991) considers that colour and/or odour probably form the proximate basis of flower selection by hover-flies. Gilbert (1991) also considers flower odour to play a probable role.

Selectivity by hover-flies (Gilbert, 1981; Haslett, 1989a; Cowgill *et al.*, 1993) and also by bee-flies (Toft, 1983) varies among species. Some species exploit the flowers of only a few flower species, in some cases only one. Toft, for example, found several bee-flies to be restricted to one flower species. At the other extreme are species such as the aphidophagous hover-fly *Episyrphus balteatus*, which exploits a much larger number. In the literature, the terms 'specialist' and 'generalist' are used to distinguish between the two types (Toft, 1983; Haslett, 1989a).

Generalist flower-visitors exploiting the flowers of a range of concurrently blooming plant species are very likely to behave as butterflies and some other insects do, visiting some flower types more frequently than would be expected on the basis of their respective abundances, thereby displaying a preference (defined in subsection 1.5.7). A preference is shown, for example, by *E. balteatus*. Preferences are unlikely to be fixed, however, altering:

1. as individual insects respond to the changing profitability of any of the plant species within its food plant range, for example due to: (a) exploitation by competitors for the resource (as shown by Toft, 1984b for bombyliid flies; section 6.4); (b) phenological changes in flower abundance and dispersion through the season; (c) changes in nectar secretion rates through the day, and/or;
2. as the nutritional requirements of the insects themselves change, either: (a) through their lifetimes, e.g. females may require pollen mainly during those stages of ovarian development when yolk deposition occurs (Haslett, 1989b on Syrphidae); or (b) through the day. Adult syrphids display flower **constancy** in the manner of bees (Gilbert and Owen, 1990; also discussion in Haslett, 1989a). That is, individuals specialise temporarily (e.g. for the duration of a foraging bout, or over several successive bouts) on one flower species. A testable hypothesis would be that this species provides, at that time, the greatest reward rate.

In flower-visiting nectarivores, food plant range and preferences are likely to be based upon the following floral characteristics that could affect the insect's foraging energetics: flower abundance and dispersion, and nectar volume, concentration and degree of accessibility. Since nectar is a source not only of energy (sugars) but also of nitrogenous materials (amino acids occur in a wide range of nectars; Baker and Baker, 1983), a preference could also be based upon the net rate of acquisition of these substances as well as of energy-providing ones (but see Willmer, 1980). The preference could have a more complex basis, such as the relationship between nectar sugar concentration and amino acid concentration. Clearly, any investigation that attempts to relate flower preference to the above-mentioned floral characteristics could prove very difficult, and if combined nectar- and pollen-feeding by flower visitors (e.g. hover-flies) is also considered, the problems are compounded.

A standard method of assessing (i.e. detecting and measuring) flower preferences in insects is to record the relative frequency with which an insect species visits the flowers of each of a range of available plant species – measured in terms of the number of sightings of individual foragers during a census walk (subsection 4.2.6) – and relating this to the abundances of the different flower types (Toft, 1983, 1984b; Cowgill *et al*., 1993). For example, in the study by Cowgill *et al*. (1993) of *Episyrphus balteatus*, a 50 m × 1 m sampling area containing a range of flowering plants was traversed at a constant speed on each observation day, and all sightings of hover-flies recorded. The behaviour of the insects at the time of first sighting was noted. Data gathered in this way can be analysed to determine whether or not the different flower types are visited in proportion to their abundances (i.e. detection of preference; subsection 1.5.7) and to determine to what degree each species is preferred. The results of Cowgill *et al*.'s analysis are expressed graphically in Figure 6.4, in which the different flower species are ranked. Clear preferences are indicated. It was also found that some

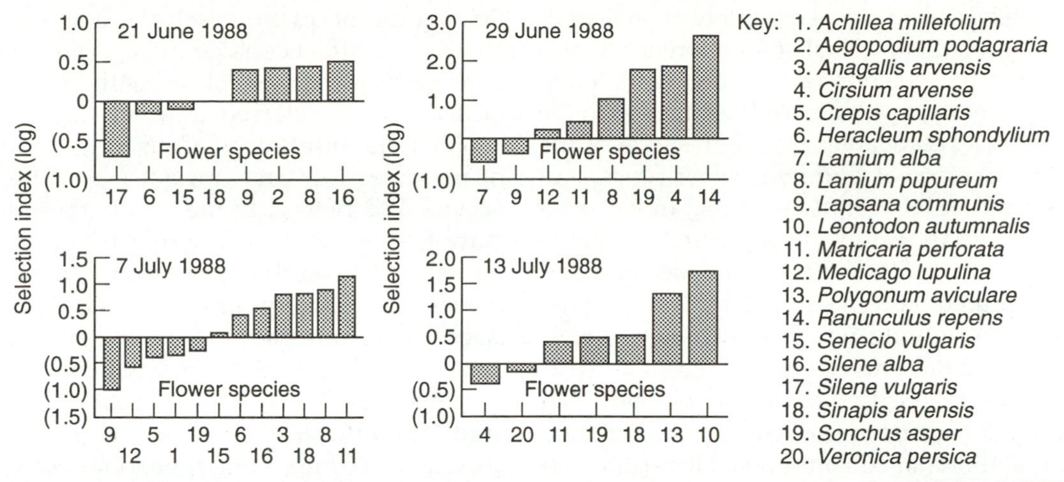

Figure 6.4 Changes in flower preferences of *Episyrphus balteatus* through the season. Murdoch's (1969) index *c* (section 1.11) was used to measure the degree of preference. This index produces asymmetrical scales in which selectivity is indicated by values from one to infinity, and non-selectivity by values from zero to one. Logarithms of the index were used to produce symmetrical scales, where values greater than zero indicate that the flowers were visited more often than would be expected from their relative abundance in the habitat. (Source: Cowgill *et al.*, 1993.) Reproduced by permission of The Association of Applied Biologists.

preferences changed through the season (see also Toft, 1983, on bombyliid flies).

In such studies, a number of problems need to be addressed. First, can the flowers of different plant species, particularly those of different structural types, be considered as equivalent foraging units? Observations of the insect's foraging behaviour upon particular flower types, made prior to the preference study, could be useful here; such an approach is better than making assumptions as to what the insect perceives as a patch (defined in subsection 1.4.3). Second, can the flowers of different plant species be considered as equivalent resource units? Published studies on flower preferences of hover-flies have considered preferences in terms of floral abundance and visitation rates, instead of more realistically in terms of the availability and quality of food materials and the consumption rates of the insects. Plant species may differ both in the rate at which they produce pollen or nectar, and in the biochemical content of their pollen and nectar. Also, as pointed out by Haslett (1989a), as far as the forager is concerned there may be differences in handling times between plant species. Handling time (defined in section 1.10) will depend upon the rate of nectar secretion, but it will also in part depend upon nectar viscosity (Harder, 1986). To be fair to previous workers, a number of interacting variables may be involved in flower preference, which make an objective assessment of the factors determining flower selectivity extremely difficult, especially if the insects utilise both pollen and nectar (Haslett, 1989a).

To overcome the drawbacks associated with methods involving behavioural observation, Haslett (1989a) examined the gut contents of hover-flies in his study of flower selectivity, counting the number of pollen grains found and identifying their source. Haslett measured pollen availability in terms of a 'floral area' index, which was calculated partly on the basis of flower diameter (Haslett's paper gives further details). Since this index is unlikely to provide even a remotely realistic measure of pollen availability, the results of Haslett's preference analyses are of questionable value. Toft (1983) however, argues that corolla diameter can be used as a reasonable relative measure of the nectar content of different flower types – an assumption that requires testing. In her preference analyses, Toft expressed resource availability in terms of flower number, unweighted and weighted by corolla diameter. Weighting was found to alter preference estimates only slightly.

Studies have yet to be carried out that combine realistic measurements of floral food availability and consumption. Hover-flies are perhaps not ideal subjects for such research, since they consume both pollen and nectar, while parasitoids, which rarely feed directly upon pollen (Jervis *et al.*, 1993), are less easy to observe in the field. Nevertheless, there is a pressing need for information on the flower preferences of natural enemies, due to an increasing awareness among biological control workers of the potential importance of non-prey and non-host foods in the population dynamics of natural enemies (Kidd and Jervis, 1989; Powell, 1986). It is widely appreciated that effective pest reduction may be achieved by the exploitation of naturally occurring enemies of indigenous or long-established insect pests. This approach to biological control involves efforts at conserving predators and parasitoids and enhancing their activity by manipulating their environment (Powell, 1986). One form of habitat management involves altering the availability (abundance and spatial distribution) of adult foods such as pollen and nectar. It has been shown in a number of natural enemy-insect pest systems that levels of parasitism or predation may be raised by encouraging flowering weeds in and around the crop (Powell, 1986; Jervis *et al.*, 1993 for parasitoid examples, and van Emden, 1965 for a hover-fly example). Clearly, quantitative assessments of flower selectivity are a necessary prerequisite for such work. However, basing the choice of plant species solely on the results of preference assessments, even where most of the pitfalls associated with preference measurements referred to above have been avoided, is insufficient. It is important to relate rates of consumption of different nectars and pollens to the effects these food materials have on components of insect fitness, particularly those components of fitness – fecundity and longevity – that largely determine searching efficiency (subsections 2.7.3, 2.8.3). We recommend that to quantify the effects of different foods on searching efficiency, workers ought, at the very least, to carry out functional response experiments (section 1.10), preferably lifetime ones (subsection 5.3.7), with each of a range of flower species. Ideally, measurements of overall searching efficiency (subsections 1.14.3, 5.3.7) ought to be carried out, with each flower species. Other considerations that need to be given to the choice of flowering plant species in habitat manipulation programmes are discussed in Jervis *et al.* (1993) and Cowgill *et al.* (1993).

Flower and other preferences are also considered further below (sections 6.4 and 6.5).

6.4 INTERPRETING PATTERNS OF RESOURCE UTILISATION

Coexisting hover-fly and bee-fly species divide up the resources available to them in the following (not necessarily exclusive) ways:

1. by being active at different times of the day (Gilbert, 1985a; Toft, 1984a);
2. by exploiting a different range of flower species (Gilbert, 1980, 1981; Haslett, 1989a; Toft, 1983);
3. by exploiting nectar and pollen in different proportions (Gilbert, 1981, 1985b).

Are species differences in resource exploitation the result of competition for limited resources? Gilbert (1985a) investigated in detail the diurnal activity patterns of several hover-fly species in the field. He carried out census walks (subsection 4.2.6) during which he noted what types of behaviour individual flies were performing, and used the data obtained to construct activity budgets for each species. With each observation of a fly, he measured ambient light intensity, temperature and humidity. From his observations and measurements, Gilbert concluded that species differences in diurnal activity pattern are partly due to thermal constraints (flies need to maintain a 'thermal balance'; Willmer (1983) discusses this concept, and Unwin and Corbet (1991) give details of techniques involved in the measurement of microclimate), and partly due to the need to synchronise their visits with the pollen and nectar production times of flowers. He did not invoke interspecific competition as a possible cause of the observed patterns of activity. Toft (1984a) constructed diurnal activity budgets for two coexisting species of bombyliids (*Lordotus* spp.) that each feed almost exclusively upon the flowers of *Chrysothamnus nauseosus* (Compositae). Toft found that one species, *Lordotus pulchrissimus*, engaged in aggressive interactions and visited flowers mainly in the morning, while the other, *L. miscellus*, performed these activities, each interspersed with brief periods of resting, over a much longer period of the day. Toft did not attempt to relate patterns of activity to changes in microclimate, and simply argued, on the basis that food resources can be limiting, that the interspecific differences she recorded corroborate the conclusion drawn in her other study on bee-fly communities (discussed below), namely that competition is the cause of the differences.

So far as species differences in food plant selectivity are concerned, these may be the result of interspecific competition for limited food resources. Several investigations aimed at testing whether competition is responsible for patterns of resource partitioning among taxonomic groups of flower-visitors have been carried out, but to date few have dealt with a group of insect natural enemies. Toft (1983, 1984a,b) in her study of bee-flies (which feed both on pollen and on nectar) adopted an observational approach to the problem. She tested for significant differences in resource use among different bee-fly species, by assessing preferences and measuring niche breadths. Species diversity was also measured at the two study sites (note that the communities at the different sites were not identical in species composition). See Toft (1983) for details of the methodology employed.

Toft (1983) found that the bee-flies divided the resources in a non-random fashion and in dissimilar ways, and concluded that patterns of resource utilisation were the result of interspecific competition largely for the following reasons:

1. niche breadths were smaller at one site when more bee-fly species were present (suggesting competition for limited resources such as food);
2. *S* (*N*) plots of the relative abundance of species (*s*, being the number of species, *n*, the number of individuals) altered from a log-normal to a log-series type at one site, as food resources became less available there (deviations from the log-normal distribution are interpreted as indicating that some single interaction, specifically competition among species, determines their abundances (May, 1975)).

Toft (1984b) also examined shifts in the food plant range of individual bee-fly species. She tested the hypothesis that the bee-fly *Geron* sp. used Compositae primarily because it was displaced from preferred flower species by the presence of *Pthiria* spp. and *Oligodranes* spp., and that if the latter two genera were 'removed' from those flower species, *Geron* sp. would use the Compositae less. An opportunity to test this hypothesis arose in a year (1982) when: (a) the abundance of the preferred flower species greatly increased, causing densities of *Pthiria* and *Oligodranes* relative to the resources to decrease; (b) a highly preferred flower species (*Dalea polyadenia*), not recorded late in the season in the previous study (carried out in 1981), was abundant late in the season. Toft observed from this 'natural experiment' that, as predicted, *Geron* sp. used the Compositae significantly less, and used preferred flower species significantly more, in 1982 than in 1981. *Geron* sp., *Pthiria* and *Oligodranes* all used *Dalea polyadenia*, the 'new' flower species. These shifts in resource use are presumed to have occurred because the preferred resources became less limiting for the bee-flies in 1982. The shifts also suggest that bee-flies of the division Homeopthalmae (the group to which all three genera belong) rank resources, from the most to the least preferred. *Geron* sp. does not use higher ranking plants on which densities of *Pthiria* and *Oligodranes* are high. When the latter two genera are 'removed' from from higher ranking resources, *Geron* sp. shifts back on to them. The competitive effect appears to be asymmetrical, with *Geron* sp. responding to the presence of the other two genera, but not *vice versa* on plants ranked in linear order by Homeopthalmae. Similar linear hierarchies have been reported for other taxonomic groups of flower-visitor, e.g. bumblebees (Morse, 1977).

If interspecific competition affects the choice of flower resources in bee-flies, how does it operate – through interference or exploitation? Toft found no convincing evidence that bee-flies of different species react to one another aggressively, suggesting that interference competition does not play a significant role. On the other hand, *Pthiria* and *Oligodranes* tend to have longer mouthparts than *Geron*, and may reduce nectar levels in flower corollas to such a level that they cannot be exploited by *Geron*, so suggesting competition through exploitation.

Gilbert and Owen (1990) used another observational approach in investigating resource partitioning among hover-flies; they looked for evidence of interspecific competition as indicated by population fluctuations and morphological relationships between species. They tested the prediction from community theory that competition will be stronger between species belonging to the same guild (i.e. species having similar ecological requirements) and that morphologically similar species will compete more strongly. If this hypothesis is correct, strong interspecific competition ought to be evident as reciprocal fluctuations in density among the species. Gilbert (1985b) had already established that morphological similarity (measured in terms of distance apart in multivariate space) and ecological similarity (foraging niche overlap) are correlated. Species with similar size and shape feed on similar types of flower and take similar floral food types. See Gilbert and Owen (1990) for a description of the methodology employed.

Gilbert and Owen found no convincing evidence that members of the same adult feeding guild compete. Reciprocal fluctuations in population density do not occur, and thus species appear to be 'tracking' resources independently of one another.

An alternative to the above observational approaches to studying competition for food resources among flies would be an experimental method involving manipulation of the insect populations. Inouye (1978) and Bowers (1985) have shown that this approach can be successfully applied to bumblebees. Inouye examined resource use by two species of bumblebee, *Bombus appositus* and *B . flavifrons*. Inouye found that when the two species occurred together at one study site, they exploited the same two plant species, but had significant preferences, *B. appositus* for *Delphinium barbeyi*, and *B. flavifrons* for *Aconitum columbianum*. The preferences were reflected in different visitation rates by the bees in relation to each flower species. When all the workers of *B. appositus* observed by Inouye upon *Delphinium* were removed by him, the other bee species, *B. flavifrons*, altered its preference and visited *Delphinium* more frequently than when manipulation was not performed. When the reciprocal manipulation was carried out, that is, workers of *B. flavifrons* were removed from *Aconitum*, *B. appositus* altered its preference, visiting *Aconitum* more frequently. It is interesting to note that these changes in preference took place over a period of a few days.

Inouye observed that at another site, where only *Delphinium* was present, and *B. appositus* was absent, *B. flavifrons* visited *Delphinium* much more frequently than in the situation where the two flower species were present.

Inouye's experimental and observational findings suggest each bee species experienced **competitive release** in the absence of the other.

Bowers (1985) examined resource use by two species of bumblebee, *Bombus flavifrons* and *B. rufocinctus*, in situations where the species occurred together and in situations where one of the species was absent through removal. Bowers found that at sites where only *B. rufocinctus* was allowed to be present, this species used the same range of flowers as *B. flavifrons* and exhibited a similar flower preference. However, at sites where the two species were sympatric, *B. rufocinctus* was relegated to foraging on secondary, less preferred flower species. The results indicate that the observed pattern of resource partitioning in situations of sympatry is the result of competition, and that there is a high degree of asymmetry in competitive effects, with one species the dominant competitor. Again, the changes in preference occurred over a short time period.

It would be interesting to see whether a removal technique similar to that practised by Inouye (1978) and Bowers (1985) can be used successfully with insects such as hover-flies. Gilbert and Owen (1990), however, are of the opinion that the conducting of manipulation experiments on hover-flies is an unrealistic goal, in view of the mobility of adult flies. One possible way around the latter problem, at least in the case of woodland species (F. Gilbert, personal communication), is to manipulate species densities by continually adding laboratory-reared hover-flies to, rather than by removing flies from, a site.

6.5 INSECT PREFERENCES AND FORAGING ENERGETICS

The relationship of insect predators and parasitoids to non-prey and non-host foods has hardly been considered from the point of view of foraging models (for discussions of models, see Pyke (1983), Stephens and Krebs (1986) and Pleasants (1989)). We are concerned here with the characteristics of: (a) nectar sources (extrafloral nectaries, as well as flowers) and (b) their insect visitors, that may need to be quantified in experimental or observational tests of foraging models, and how one may go about doing this.

So far as nectar-feeding is concerned, the literature on butterflies (e.g. Boggs, 1987; May, 1988) and bees (Heinrich, 1975, 1983; Hodges, 1985a,b; Pleasants, 1989; Waddington, 1987), provides useful information on approaches and techniques to adopt when planning investigations into the foraging behaviour, including flower preferences, of natural enemies. Such investigations, if carried out on nectarivores, may involve quantifying or simply recording some or all of the following **characteristics of the nectar sources:**

A: the mean energy content of individual flowers/florets of different flower types. Energy content is determined primarily by volume and sugar concentration. The latter two variables need to be measured during the periods of the day when the insects feed most frequently. For information on methodologies relating to the sampling of nectar, measurement of nectar volume and concentration, and the chemical analysis of nectar constituents, see Corbet (1978), Bolten *et al.* (1979), Corbet *et al.* (1979), McKenna and Thomson (1988), May (1988) and Dafni (1992). The concentration of nectars that are mixtures of glucose, fructose and sucrose can be expressed as the equivalent amount of sucrose. Given the concentration and volume of the nectar, milligrammes of sugar per flower or floret can be calculated and this quantity converted to energy per flower or floret, assuming 16.8 Joules/mg of sucrose (Dafni, 1992). Note that care must be exercised when converting from refractometer readings of nectar concentration to milligrammes of sucrose equivalents (Bolten *et al.*, (1979).

In studies of nectarivory, nectar has either been sampled from flowers/florets that have been protected against visits or it has been taken from unprotected flowers/florets. With the former method, problems of interpretation can arise because the nectar can:

1. accumulate to abnormally high levels;
2. decline in quantity due to nectar resorption by the plant (Burquez and Corbet, 1991), or
3. remain at around the original level (Pyke, 1991 showed that in some plant species removal of nectar increases net production of nectar).

Thus, we recommend that data are obtained from unprotected flowers only. Of course, diel variation in nectar availability and concentration also ought to be measured (for further discussion of the practical aspects of nectar sampling and analysis, see Dafni, 1992);

B: the relative accessibility of nectar in different flower types, e.g. distance from corolla opening to nectar. Insects, when extracting nectar from flowers/florets with long, tubular corollas, may incur a larger handling cost, in terms of both time and energy, than when extracting nectar from flowers with either short corollas or completely exposed nectaries;

C: the dispersion pattern of flowers/florets in each plant species. The degree of clumping of these plant parts will have an important bearing upon the insect's foraging behaviour, determining the amount of time spent travelling between the parts and the number of parts visited per unit of time;

D: the abundances (density per unit area) of the different flower types.

The following **characteristics of the insect's foraging behaviour,** in relation to certain flower types, may need quantifying:

A: the number of inflorescences, flowers or florets visited per foraging bout, and the number of foraging bouts performed per observation period, allowing the flower visitation rate (number of inflorescences, flowers or florets visited per unit time) to be calculated. A 'bout' may be difficult to define. May (1988) took it to be the time period between a butterfly's entrance and exit either from a flower patch (it is not clear whether he meant an inflorescence or a group of inflorescences) or from his field of view;

B: the mean time spent travelling (either walking or flying) between inflorescences, flowers or florets of a particular plant species during a foraging bout. It is important to observe closely the foraging behaviour of individual insects, as observations will reveal, for example, whether the insects fly between flowers but walk between the florets comprising an individual flower, and whether one or a series of flowers is visited during a foraging bout. The foraging movements of a non-aphidophagous hover-fly (*Eristalis tenax*) upon flower capitula of *Aster novae-angliae* (Compositae) were elucidated by Gilbert (1983). The flies systematically probe the nectar-producing disc florets which form a narrow ring, leaving the capitulum when they have circled it once. They probe a variable number of florets on each capitulum, behaviour which is thought to result from decisions, made by the insect after each floret is visited, about whether to probe another floret on the capitulum or move to another capitulum (Pyke, 1984; Hodges, 1985a, 1985b). The 'rule' nectarivores use in making these decisions may be a threshold departure rule, i.e. if the reward (nectar) received from the current floret is less than a certain threshold amount, then the insect should leave the capitulum and locate a new one;

C: the mean time spent dealing with each flower/floret of a flower type (probing it, perching on it, climbing into it, feeding, climbing out of it), i.e. the handling time;

D: the gross energy intake during foraging (E_{in}) (calculated either on a per bout or per observation period basis) of the insect for a given flower type. This can be calculated as follows:

$$E_{in} = R_v \times E_x \qquad (6.1)$$

where R_v is number of flowers visited per bout, and E_x is the mean quantity of energy (Joules) extracted per flower/floret as measured in unprotected flowers (see above) at the same time of day. The gross energy intake per second can be calculated by dividing the right side of the equation by the mean length, in seconds, of the foraging bouts.

In insects that are able to consume relatively large amounts of nectar during a foraging bout, the volume of nectar actually removed from each flower or floret can be measured and the mean quantity of energy extracted per flower or floret calculated by multiplying this amount by the nectar concentration, with suitable corrections (Bolten *et al.*, 1979). The amount of nectar removed can be measured perhaps most easily either by comparing the nectar volume in flowers/florets visited once by an individual insect with the volume in non-visited flowers or florets or any weight gain in the insect immediately after it has finished feeding at the flower/floret. Note that the quantity of nectar extracted may alter with any changes in nectar concentration that may occur, and may also alter as successive flowers/florets are visited, due to satiation or gut limitation. The mean quantity of energy extracted per flower may also vary with factors such as flower or floret density, the amount of nectar removed per flower decreasing with increasing flower or floret (and therefore nectar) availability (Heinrich, 1983) (this behaviour is predicted by optimal foraging theory). Average and maximum crop volumes can also be measured quite easily in insects such as hover-flies (Gilbert, 1983).

Insects such as small parasitoids will remove very small amounts of nectar from a flower/floret, so estimating the volume of nectar extracted by comparing visited and non-visited flowers is likely to be either impracticable or impossible. Therefore, measuring either crop volume or weight gain in insects that have recently fed may be the only alternatives. Again, measurements need to be taken taken at several stages during a foraging bout.

The energy expenditure (E_{cost}) of insects (calculated on a per bout or per observation period basis) in relation to a given flower type can, following May (1988), be calculated as follows:

$$E_{cost} = (T_{fl} \cdot E_{fl}) + (T_p \cdot E_p) \qquad (6.2)$$

where T_{fl} is the period of time spent in flight per foraging bout, T_p is the period of time spent handling each flower or floret of flower type per foraging bout, E_{fl} is the metabolic cost per second of travelling between nectar sources, and E_p is the metabolic cost per second of handling those sources. The energy expenditure per second can be calculated by dividing the right side of the equation by the mean length, in seconds, of the foraging bouts.

E_{fl} and E_p can be estimated in the laboratory using methods for measuring metabolic rates of insects when flying and stationary respectively, preferably at the range of temperatures experienced by the insects in the field (although in worker bumblebees the energy costs of flight are a constant whatever the speed of flight or ambient temperature

(Ellington *et al.*, 1990). To measure E_{fl}, insects can be allowed to fly, either tethered or free (e.g. see Dudley and Ellington (1990)), within a respirometer (for details of respirometers, see Southwood (1978), and their rate of oxygen consumption measured. The respirometer used by Gilbert (1983) to measure the oxygen consumption rate of loose-tethered *Eristalis tenax* was a paramagnetic oxygen analyser. Alternatively, a 'flight mill' can be used and the amount of flight fuel consumed calculated. Gilbert used such a device to measure the rate of energy consumption in flying *E. tenax*. With flight mills, insects are tight-tethered to a balanced, lightweight arm, and are made to fly continuously until they become exhausted. Gilbert (1983) simply recorded the amount of weight loss during flight, and, from the figure obtained, estimated the amount of carbohydrate (presumed to be the flight fuel) utilised and thus the amount of energy expended over the period of flight. Hocking (1953), by contrast, provided insects that had become exhausted on the flight mill with a known quantity of carbohydrate (glucose). The duration of the next 'flight' by the insect could then be measured and the rate of fuel consumption calculated. Tight-tethering in flight mills, because the insect does not have to support its own weight, may cause flight duration to be overestimated and thereby cause energy consumption to be underestimated. Even loose-tethering, as employed by Gilbert (1983) in respirometry, is likely to influence average flight performance, so affecting energy consumption.

Measuring E_p could prove to be a difficult task. Heinrich (1975) suggested that the resting metabolic rates of nectarivorous insects approximate to the metabolic costs of feeding and also the costs of walking, so the practical problems associated with estimating the metabolic costs of feeding and walking may be conveniently avoided. However, this assumption needs to be tested using a range of insect taxa. The metabolic rates of resting insects can easily be measured using respirometers (diurnal insects can be kept stationary in the dark).

If the nature of the fuel burned by the insects under study is known (i.e. carbohydrate, lipid, amino acid [e.g. proline]; the fuel used may vary with the type of activity and even during flight), the energy required to sustain the observed metabolic rate at particular activities can be calculated.

By quantifying E_{cost}, it is then possible to calculate the net amount of energy (E_{net}) that can be obtained from different flower types visited over a certain segment of the day as the difference between gross energy intake and energy expenditure:

$$E_{net} = E_{in} - E_{cost} \quad (6.3)$$

May (1988) explained observed preferences in terms of E_{net}. He used E_{net} as the measure of 'profitability' of flower types. E_{net} is in fact the net energy value of a flower type and *not* its profitability, which is $E_{net}/(T_{fl} + T_p)$ (see Stephens and Krebs, 1986). May asked whether the E_{net} of a flower type was largely a function of: (a) mean nectar volume; (b) mean nectar concentration, accessibility of the nectar (e.g. corolla length); (c) flower density; (d) flower dispersion; (e) the density of florets per inflorescence, (bear in mind that (a) and (b) may vary with the time of day). May used multivariate statistical analyses to establish the main ways in which flower species differed with respect to these variables. His analyses showed that nectar volume explained nearly all of the variation in energy content among different nectar sources, and so concluded that nectar volume was the best single predictor of E_{net}. By the same token, nectar concentration and flower density were poor predictors. Multivariate analyses could also be carried out to determine what factors mainly determine the profitability of a flower type; handling time and flight time should be among the variables considered.

As noted earlier, relating flower selectivity to foraging profitability in pollen- and pollen/nectar-feeding insects will prove more difficult than with nectarivores. One of the few possible advantages of studying pollen-feeders is that the amount of pollen consumed may be easier to quantify compared with the amount of nectar.

Pleasants (1989) should be consulted for a discussion of how one might test whether :(a) insects that probe a variable number of florets per capitulum use a threshold departure rule to decide whether to continue foraging on a capitulum or to leave it and locate another one (e.g. hover-flies, see above), and (b) whether the threshold used is optimal i.e. maximises the energy intake rate.

In considering the exploitation of non-prey/non-host food sources by natural enemies, it is also important to bear in mind that as well as having to decide which types of food source (e.g. flower species) to visit and what materials (e.g. nectar, pollen) to consume, foragers need to decide how to allocate their time and energy to foraging for non-prey or non-host food. Among parasitoids and aphidophagous hover-flies, females of many species forage for hosts or prey and food in distinctly different parts of a habitat, and as a consequence must be incurring a significant cost in both time and energy compared with insects that do not need to forage far afield for non-host and non-prey foods (some parasitoids take honeydew directly from the host at or around the time of oviposition). The problem of choosing between ovipositing or feeding in patches of differing profitability (in terms of net fitness gain) has been examined for parasitoids, using a state variable modelling approach (Jervis and Kidd, 1995). This method is based upon the technique of stochastic dynamic programming (Mangel and Clark, 1988) (subsection 1.2.2).

The models for parasitoids predict, not surprisingly, that feeding patches should preferentially be visited when energy reserves are low, and oviposition patches preferentially visited when reserves are high. The parasitoids are more likely to choose to feed: (a) the greater the cost associated with searching for food relative to host patches, (b) the lower the probability of finding food; (c) the lower the number of developing oöcytes or mature eggs the insect carries ((c) applies only to synovigenic parasitoids). However, as the end of life is approached, oviposition becomes increasingly favoured, especially when energy reserves are low, the insect 'cutting its losses' to deposit whatever remaining eggs it can.

These predictions have yet to be tested, a task which will involve monitoring the movements of individual parasitoids between previously defined feeding and oviposition patches. Such a test is likely to prove less of a problem in the laboratory than in the field. One possible experimental design which could be adopted in the laboratory involves providing individual parasitoids with an enclosed, elongated choice chamber containing a host patch at one end and a food patch (diluted honey or similar diet) at the other. The time spent within patches at either end of the chamber, and frequency of movements between the two, are monitored using video equipment. Of course, only non-host-feeding female parasitoids can be used. Host patches have to be continually replenished to avoid host-limitation, and a host patch also provided at the food end of the chamber, to test whether the parasitoid is food-searching or still host-searching. Once the parasitoid has commenced feeding, this patch is removed to force the parasitoid to return to the other end of the chamber.

Similar experimentation in the field is likely to be hampered by problems associated with tracking the movements of parasitoids, although some parasitoids such as large Pompilidae could be marked with paint and their movements tracked followed using binoculars.

The ability of natural enemies to move between prey- or host-containing areas and food-containing areas in the field needs to be examined before any empirical tests of these models are carried out. Marking insects with fluorescent dusts, radioactive elements or rare elements (e.g. caesium, dysprosium, rubidium), releasing them and tracking their movements is an obvious way of determining how far insects are able to travel, and where they travel to (subsection 4.2.11). For example, marked parasitoids can be released at various distances from a crop field margin, the distance chosen varying between, but not within, fields, and their visits to field margin flowers recorded.

A simpler and cheaper alternative to unnatural marker substances is pollen. The body surface of insects found within host-containing areas e.g. within a crop, can be examined for the presence of pollen grains from flower species that the insects are known to visit outside such areas. Recently, Hickman and Wratten (unpublished data), using pollen of *Phacelia tanacetifolia* as a marker, demonstrated that aphidophagous hover-flies dispersed into cereal fields up to 100 m from strips of *P. tanacetifolia* that had been sown around margin of each field. A similar effect has been recorded by Lövei *et al.* (1992) (Figure 6.5). The advantage of using *Phacelia* in such studies is that the pollen grains are both very large and very distinctive in appearance.

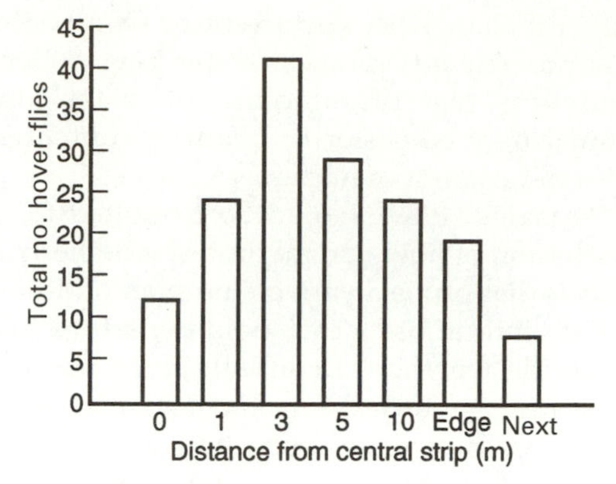

Figure 6.5 Total numbers of aphidophagous hover-flies (mostly *Melanostoma fasciatum*) caught in yellow pan traps at increasing distances from a central strip of *Phacelia tanacetifolia* within a wheat field in New Zealand, summer 1991–1992. Two parallel rows of traps were set: (a) within the flower strip; (b) in both directions, in rows 1, 3, 5 and 10 m away; (c) at the edge of the field (30–50 m from flower strip); (d) along the edges of adjacent fields (approximately 100 m away from flower strip). Catches in the central strip itself were low, probably because the syrphids preferred the flowers to the yellow trap colour. Lovei *et al.* (1992) concluded that the flower strip attracted hover-flies into the wheat crop and thereby increased their densities to above the average for the neighbouring fields. (Source: Lövei *et al.*, 1992.) Reproduced by permission of The New Zealand Plant Protection Society

Such marking of individuals can be used to determine whether the insects commute back and forth between food-containing and host-containing areas. Vespidae and ants certainly commute, as do some Sphecidae and at least one species of Tiphiidae (the latter travelling many kilometres between food and hosts located at markedly different altitudes; see Clausen *et al.*, 1933), but it is not known whether commuting is practised by either Hymenoptera 'Parasitica' or dipteran parasitoids (see Topham and Beardsley, 1975, for details of an unsuccessful attempt, using a radioactive labelling technique, at determining whether tachinid flies commute). Providing insects with a unique mark greatly facilitates investigations of commuting behaviour, as it provides information on the foraging movements of individuals.

With some natural enemies such as large hover-flies direct observations could possibly reveal whether commuting occurs. Wellington and Fitzpatrick (1981) succeeded in making detailed observations on the foraging and other movements of males of the non-aphidophagous species *Eristalis tenax*. They found that males, for part of any day, make frequent movements between the individual home territory and sites outside. In the latter areas, the males perform various 'off duty' activities including feeding at flowers.

It is important in habitat manipulation programmes (section 6.3) to know how far predators and parasitoids can move between prey/host-containing areas and food-containing ones. Van Emden (1988) raises one problem associated with the experimental demonstration of the effects of providing field-margin floral foods (Figure 6.6) upon the effectiveness of aphid predators. Due to the high mobility of most aphid predators (e.g. hover-flies), predator-induced mortality may be raised across the whole crop and thereby make a gradient in predation towards the edges of a field difficult to detect. Van Emden argues that this problem has led to many unwarranted negative conclusions in the literature. However, in habitat manipulation experiments, one should cater for the absence of edge effects by employing an appropriate experimental design involving control fields which lack flowers altogether.

Compared with hover-flies, most parasitoids are likely to be less mobile; edge effects are therefore more likely to be detected in habitat manipulation experiments involving parasitoids. If an obvious edge effect is detected, consideration will need to be given to the size of fields and/or to the possibility of growing flowers within, as opposed to just around, crops.

As noted above, not all natural enemies need to forage far afield for non-prey/non-host foods. For example, females of the aphelinid wasps *Coccophagus* sp. and *Encarsia formosa* consume honeydew obtained directly from their scale insect and whitefly hosts, respectively (Cendaña, 1937; Yamamura and Yano, 1988), while female aphidiine braconids probably feed mainly on honeydew situated close to their aphid hosts. Some staphylinid beetles (*Tachyporus* species) consume fungal materials, mainly spores (conidia), that are readily encountered on the wheat leaves they forage upon for aphid prey (Dennis *et al.*, 1991). It would be interesting to investigate how the frequency with which these natural enemies visit sources of non-prey/non-host materials alters with changes in host or prey abundance.

Figure 6.6 A flowering strip of coriander (*Coriandrum sativum*: Umbelliferae) sown along the margins of a cereal field by investigators at Southampton University, to determine whether the provision of such floral food sources resulted in increased oviposition, within the crop, by aphidophagous syrphids.

6.6 NATURAL ENEMIES AS POLLINATORS

When visiting flowers to feed, aphidophagous hover-flies become contaminated with pollen over their body surfaces (Stelleman and Meeuse, 1976), although the amount of pollen carried by a fly in this way is probably small compared with that carried by non-predatory and hairier relatives such as *Eristalis tenax* (Holloway 1976). Due to the flower constancy of individuals (section 6.3), even generalist hover-fly species may be effective pollinators (McGuire and Armbruster, 1991).

The role of most parasitoids in pollination is less clear. Such insects, particularly those visiting flowers with narrow tubular corollas, are likely to come into accidental contact with the anthers (if not actually feed on the pollen, (see Jervis *et al.*, 1993) and so pick up pollen grains. Parasitoids might therefore be important potential pollinators. Note that to be important as *cross*-pollinators, individual wasps have to visit more than one individual plant during a foraging bout and also show some degree of flower constancy. These are aspects of parasitoid behaviour about which we know very little (Fowler, 1989 describes observations on a sphecid parasitoid). The little we do know relates to wasp–orchid associations that involve pseudocopulation, the male parasitoids (certain Tiphiidae, Scoliidae and pimpline Ichneumonidae) visiting the orchid inflorescence solely for the purpose of mating, the orchids providing no food (e.g. see van der Pijl and Dodson, 1966; Stoutamire 1974; Kullenberg and Bergstrom, 1975). It is interesting to note that *The Flora of the British Isles* (Clapham *et al.*, 1989) does not mention parasitoids specifically as flower-visitors, but does mention other insects. The role of parasitoids in pollination is an area that clearly has been badly neglected, perhaps due to:

1 the very small size and relative inconspicuousness of many species;
2 the tendency of some of the larger parasitoids not to linger at inflorescences in the manner of some bees, butterflies, beetles and hover-flies (Hassan, 1967; Jervis *et al.*, 1993); and
3 the often immense difficulties associated with their identification.

This aspect of parasitoid ecology deserves much closer attention than it has received so far.

It is possible to demonstrate that an insect is capable of cross-pollinating, by depositing dyed or otherwise labelled pollen grains on inflorescences in anthesis, and seeing whether the grains appear on the stigmas of other inflorescences. Stelleman and Meeuse (1976) used this approach when attempting to establish whether the aphidophagous hover-flies *Melanostoma* and *Platycheirus* carry out pollen transfer between spikes of *Plantago lanceolata*. The flies do transfer pollen between spikes. As to the effectiveness of pollen transfer by a particular predator or parasitoid species, this can be determined by testing the ability a flower to produce seed after a single visit (Spears, 1983).

6.7 ACKNOWLEDGEMENTS

We are most grateful to Francis Gilbert and Rob Paxton for reading through a draft of this chapter and providing constructive comments. The chapter has also benefited from the advice of Steve Wratten, Janice Hickman, Flick Rothery and Robert Belshaw.

REFERENCES

Abdelrahman, I. (1974) Growth, development and innate capacity for increase in *Aphytis chrysomphali* Mercet and *A. melinus* DeBach, parasites of California red scale, *Aonidiella aurantii* (Mask.), in relation to temperature, *Australian Journal of Zoology*, **22**, 213–30.

Ables, J.R., Shepard, M. and Holman, J.R. (1976) Development of the parasitoids *Spalangia endius* and *Muscidifurax raptor* in relation to constant and variable temperature: simulation and validation. *Environmental Entomology*, **5**, 329–32.

van Achterberg, C. (1977) The function of swarming in *Blacus* species (Hymenoptera, Braconidae). *Entomologische Berichten*, **37**, 151–2.

Adis, J. (1979) Problems of interpreting arthropod sampling with pitfall traps. *Zoologischer Anzeiger*, **202**, 177–84.

Adis, J. and Kramer, E. (1975) Formaldehyd– Losung attrahiert *Carabus problematicus* (Coleoptera: Carabidae). *Entomologia Germanica*, **2**, 121–5.

Agarwala, B.K. and Dixon, A.F.G. (1992) Laboratory study of cannibalism and interspecific predation in ladybirds. *Ecological Entomology*, **17**, 303–9.

Alcock, J. (1979) The behavioural consequences of size variation among males of the territorial wasp *Hemipepsis ustulata* (Hym., Pompilidae). *Behaviour*, **71**, 322–35.

Albrecht, A. (1990) Revision, phylogeny and classification of the genus *Dorylomorpha* (Diptera, Pipunculidae). *Acta Zoologica Fennica*, **188**, 1–240.

Alderweireldt, M. and Desender, K. (1990) Variation of carabid diel activity patterns in pastures and cultivated fields, in *The Role of Ground Beetles in Ecological and Environmental Studies* (ed. N.E. Stork), Intercept, Andover, pp. 335–8.

Alexander, R.D. (1964) The evolution of mating behaviour in arthropods, in *Insect Reproduction* (ed. K.C. Higham). Royal Entomological Society. London, pp. 78–94.

Ali, A.H.M.. (1979) Biological Investigations on the Entomophagous Parasites of Insect Eggs Associated with *Juncus* species, University of Wales, Ph. D. Thesis.

Allen, H.W. (1929) An annotated list of the Tachinidae of Mississippi. *Annals of the Entomological Society of America*, **22**, 676–90.

Allsopp, P.G. (1981) Development, longevity and fecundity of the false wireworms *Pterohelaeus darlingensis* and *P. alternatus* (Coleoptera: Tenebrionidae). 1. Effect of constant temperature. *Australian Journal of Zoology*, **29**, 605–19.

van Alphen, J.J.M. (1980) Aspects of the foraging behaviour of *Tetrastichus asparagi Crawford* and *Tetrastichus* spec. (Eulophidae), gregarious egg parasitoids of the asparagus beetles *Crioceris asparagi* L. and *C. duodecimpunctata* L. (Chrysomelidae). *Netherlands Journal of Zoology*, **30**, 307–25.

van Alphen, J.J.M. and Drijver, R.A.B. (1982) Host selection by *Asobara tabida* Nees (Braconidae; Alysiinae) a larval parasitoid of fruit inhabiting *Drosophila* species. I. Host stage selection with *Drosophila melanogaster* as host. *Netherlands Journal of Zoology*, **32**, 215–31.

van Alphen, J.J.M and Galis, F. (1983) Patch time allocation and parasitization efficiency of *Asobara tabida* Nees, a larval parasitoid of *Drosophila*. *Journal of Animal Ecology*, **52**, 937–52.

van Alphen, J.J.M and Janssen, A.R.M. (1982) Host selection by *Asobara tabida* Nees (Braconidae; Alysiinae) a larval parasitoid of fruit inhabiting *Drosophila* species. II. Host species selection. *Journal of Animal Ecology*, **32**, 194–214.

van Alphen, J.J.M and Nell, H.W. (1982) Superparasitism and host discrimination by *Asobara tabida* Nees (Braconidae; Alysiinae) a larval parasitoid of Drosophilidae. *Journal of Animal Ecology*, **32**, 232–60.

van Alphen, J.J.M and Vet, L.E.M. (1986) An evolutionary approach to host finding and selection, in *Insect Parasitoids* (eds J.K. Waage and D.J. Greathead), Academic press, London, pp. 23–61.

van Alphen, J.J.M. and Visser, M.E. (1990) Superparasitism as an adaptive strategy for insect parasitoids. *Annual Review of Entomology*, **35**, 59–79.

van Alphen, J.J.M., van Dijken, M.J. and Waage, J.K. (1987) A functional approach to superparasitism: host discrimination needs not to be learnt. *Netherlands Journal of Zoology*, **37**, 167–79.

van Alphen, J.J.M., Visser, M.E. and Nell, H.W. (1992) Adaptive superparasitism and patch time allocation in solitary parasitoids: searching in groups versus sequential patch visits. *Functional Ecology*, **6**, 528–35.

Anholt, B.R. (1990) An experimental separation of interference and exploitative competition in a larval dragonfly. *Ecology*, **71**, 1483–93.

Antolin, M.F. and Strong, D.R. (1987) Long-distance dispersal by a parasitoid (*Anagrus delicatus*, Mymaridae) and its host. *Oecologia*, **73**, 288–92.

Anunciada, L. and Voegele, J. (1982) The importance of nutrition in the biotic potential of *Trichogramma maidis* Pintureau and Voegele and *T. nagarkattii* Voegele et Pintureau and oösorption in the females. *Les Trichogrammes, Les Colloques de l'INRA*, **9**, 79–84 (in French).

Apple, J.W. (1952) Corn borer development and control on canning corn in relation to temperature accumulation. *Journal of Economic Entomology*, **45**, 877–9.

Arditi, R. (1983) A unified model of the functional response of predators and parasitoids, *Journal of Animal Ecology*, **52**, 293–303.

Arnold, A.J., Needham, P.H. and Stevenson, J.H. (1973) A self-powered portable insect suction sampler and its use to assess the effects of azinphos methyl and endosulfan in blossom beetle populations on oilseed rape. *Annals of Applied Biology*, 75, 229–33.

Arnoldi, D., Stewart, R.K. and Boivin, G. (1991) Field survey and laboratory evaluation of the predator complex of *Lygus lineolaris* and *Lygocoris communis* (Hemiptera: Miridae) in apple orchards. *Journal of Economic Entomology*, 84, 830–36.

Arthur, A.P. and Wylie, H.G. (1959) Effects of host size on sex ratio, development time and size of *Pimpla turionellae* (L.) (Hymenoptera: Ichneumonidae). *Entomophaga*, **4**, 297–301.

Arthur, W. and Farrow, M. (1987) On detecting interactions between species in population dynamics. *Biological Journal of the Linnean Society*, **32**, 271–9.

Ashby, J.W. (1974) A study of arthropod predation of *Pieris rapae* L. using serological and exclusion techniques. *Journal of Applied Ecology*, **11**, 419–25.

Askari, A. and Alishah, A. (1979) Courtship behavior and evidence for a sex pheromone in *Diaeretiella rapae*, the cabbage aphid primary parasitoid. *Annals of the Entomological Society of America*, **72**, 749–50.

Askew, R.R. (1968) A survey of leaf-miners and their parasites on laburnum. *Transactions of the Royal Entomological Society of London*, **120**, 1–37.

Askew, R.R. (1971) *Parasitic Insects*, Heinemann, London.

Askew, R.R. (1984) The biology of gall wasps, in *Biology of Gall Insects* (ed. T.N. Ananthakrishnan), Edward Arnold, London, pp. 223–71.

Askew, R.R. (1985) A London fog. *Chalcid Forum*, **4**, 17.

Askew, R.R. and Shaw, M.R. (1986) Parasitoid communities: their size, structure and development, in *Insect Parasitoids* (eds J. Waage and D.J. Greathead), Academic Press, London, pp. 225–64,

van den Assem, J. (1971) Some experiments on sex ratio and sex regulation in the pteromalid *Lariophagus distinguendus*. *Netherlands Journal of Zoology*, **21**, 373–402.

van den Assem, J. (1975) The temporal pattern of courtship behaviour in some parasitc Hymenoptera, with special reference to *Melittobia acasta*. *Journal of Entomology*, **50**, 137–46.

van den Assem, J. (1976) Queue here for mating: waarnemingen over het gedrag van ongepaarde *Melittobia* vrouwtjes ten opzichte van een mannelijke soortgenoot. *Entomologische Berichten*, **36**, 74–8.

van den Assem, J. (1986) Mating behaviour in parasitic wasps, in *Insect parasitoids*, (eds J.K. Waage and D.J. Greathead) 13th Symposium of Royal Entomological Society of London, Academic Press, pp. 137–167.

van den Assem, J. and Feuth de Bruijn, E. (1977) Second matings and their effect on the sex ratio of the offspring in *Nasonia vitripennis* (Hym. Pteromalidae). *Entomologia Experimentalis et Applicata*, **21**, 23–8.

van den Assem, J. and Gijswijt, M.J. (1989) The taxonomic position of the Pachyneurini (Chalc. Pteromalidae) as judged by characteristics of courtship behaviour. *Tijdschrift voor Entomologie*, **132**, 149–54.

van den Assem, J. and Jachmann, F. (1982), The coevolution of receptivity signalling and body size in the Chalcidoidea. *Behaviour*, **80**, 96–105.

van den Assem, J. and Povel, G.D.E. (1973) Courtship behaviour of some *Muscidifurax* species (Hym., Pteromalidae): a possible example of a recently evolved ethological isolating mechanism. *Netherlands Journal of Zoology*, **23**, 465–87.

van den Assem, J. and Putters, F.A. (1980), Patterns of sound produced by courting chalcidoid males and its biological significance. *Entomologia Experimentalis et Applicata*, **27**, 293–302.

van den Assem, J. and Werren, J.H. (1994) A comparison of the courtship and mating behaviour of three species of Nasonia (Hym., Pteromalidae). *Journal of Insect Behaviour*, **7**, 53–66.

van den Assem, J., in den Bosch, H.A.J. and Prooy, E. (1982a), *Melittobia* courtship behaviour, a comparative study of the evolution of a display. *Netherlands Journal of Zoology*, **32**, 427–71.

van den Assem, J., Gijswijt, M.J. and Nübel, B.K. (1980a) Observations on courtship and mating strategies in a few species of parasitic wasps (Chalcidoidea). *Netherlands Journal of Zoology*, **30**, 208–27.

van den Assem, J., Gijswijt, M.J. and Nübel, B.K. (1982b), Characteristics of courtship and mating behaviour used as classificatory criteria in Eulophidae-Tetrastichinae. *Tijdschrift voor Entomologie*, **125**, 205–20.

van den Assem, J., van Iersal, J.J.A. and los den Hartogh, R.L. (1989) Is being large more important for female than for male parasitic wasps? *Behaviour*, **108**, 160–95.

van den Assem, J., Jachmann, F. and Simbolotti, P. (1980b) Courtship behaviour of *Nasonia vitripennis* (Hym. Pteromalidae): some qualitative evidence for the role of pheromones. *Behaviour*, **75**, 301–7.

van den Assem, J., Putters, F.A. and Prins, T.C. (1984) Host quality effects on sex ratio of the parasitic wasp *Anisopteromalus calandrae* (Chalcidoidea, Pteromalidae). *Netherlands Journal of Zoology*, **34**, 33–62.

van den Assem, J., Putters, F.A. and van der Voort-Vinkestijn, M.J. (1984) Effects of exposure to an extremely low temperature on recovery of courtship behaviour after waning in the parasitic wasp *Nasonia vitripennis*. *Journal of Comparative Physiology*, **155**, 233–7.

Austin, A.D. (1983) Morphology and mechanics of the ovipositor system of *Ceratobaeus* Ashmead (Hymenoptera: Scelionidae) and related genera. *International Journal of Insect Morphology and Embryology* , **12**, 139–55.

Austin, A.D. and Browning, T.O. (1981) A mechanism for movement of eggs along insect ovipositors. *International Journal of Insect Morphology and Embryology*, **10**, 93–108.

Aveling, C. (1981) The role of *Anthocoris* species (Hemiptera: Anthocoridae) in the integrated control of the damson–hop aphid. *Annals of Applied Biology*, **97**, 143–53.

Avilla, J. and Albajes, R. (1984) The influence of female age and host size on the sex ratio of the parasitoid *Opius concolor*. *Entomologia Experimentalis et Applicata*, **35**, 43–7.

Avilla, J. and Copland, M.J.W. (1987) Effects of host age on the development of the facultative autoparasitoid *Encarsia tricolor* (Hymenoptera: Aphelinidae). *Annals of Applied Biology*, **110**, 381–9.

Ayal, Y. (1987) The foraging strategy of *Diaeretiella rapae*. I. The concept of the elementary unit of foraging. *Journal of Animal Ecology*, **56**, 1057–68.

Ayala, F.J. (1983) Enzymes as taxonomic char-acters, in *Protein Polymorphism: Adaptive Significance and Taxonomic Significance* (eds G.S. Oxford and E. Rollinson), Systematics Association Special Volume No. 24, Academic Press, London.

Ayre, G.L. and Trueman, D.K. (1974) A battery operated time-sort pitfall trap. *The Manitoba Entomologist*, **8**, 37–40.

Baars, M.A. (1979a) Catches in pitfall traps in relation to mean densities of carabid beetles. *Oecologia*, **41**, 25–46.

Baars, M.A. (1979b) Patterns of movement of radioactive carabid beetles. *Oecologia*, **44**, 125–40.

Bai, B. and Mackauer, M. (1990) Oviposition and host-feeding patterns in *Aphelinus asychis* (Hymenoptera: Aphelinidae) at different aphid densities. *Ecological Entomology*, **15**, 9–16.

Bai, B. and Mackauer, M. (1992) Influence of superparasitism on development rate and adult size in a solitary parasitoid *Aphidius ervi*. *Functional Ecology*, **6**, 302–7.

Bai, B. and Smith, S.M. (1993) Effect of host availability on reproduction and survival of the parasitoid wasp *Trichogramma minutum*. *Ecological Entomology*, **18**, 279–86.

Bailey, N.T.J. (1951) On estimating the size of mobile populations from recapture data. *Biometrika*, **38**, 293–306.

Bailey, N.T.J. (1952) Improvements in the interpretation of recapture data. *Journal of Animal Ecology*, **21**, 120–27.

Bailey, P.C.E. (1986) The feeding behaviour of a sit-and-wait predator, *Ranatra dispar* (Heteroptera: Nepidae): optimal foraging and feeding dynamics. *Oecologia*, **68**, 291–7.

Bailey, S.E.R. (1989) Foraging behaviour of terrestrial gastropods: Integrating field and laboratory studies. *Journal of Molluscan Studies*, **55**, 263–72.

Baker, H.G. and Baker, I. (1983) A brief historical review of the chemistry of floral nectar, in *The Biology of Nectaries* (eds B. Bentley and T. Elias), Columbia University Press, New York, pp. 126–52.

Baker, R.L. (1981) Behavioural interactions and use of feeding areas by nymphs of *Coenagrion resolutum* (Coenagrionidae: Odonata). *Oecologia*, **49**, 353–8.

Baker, R.L. (1989) Condition and size of damselflies: a field study of food limitation. *Ecology*, **81**, 11–119.

Bakker, K., van Alphen, J.J.M., van Batenberg, F.H.D., van der Hoeven, N., Nell, N.W., van Strien-van Liempt, W.T.F.H. and Turlings, T.C. (1985) The function of host discrimination and superparasitism in parasitoids. *Oecologia*, **67**, 572–6.

Bakker, K. Peulet, Ph. and Visser, M.E. (1990) The ability to distinguish between hosts containing different numbers of parasitoid eggs by the solitary parasitoid *Leptopilina heterotoma*. (Hym., Cynip.). *Journal of Animal Ecology*, **40**, 514–20.

Bakker, K., Eysackers, H.J.P., van Lenteren, J.C. and Meelis, E. (1972) Some models describing the distribution of eggs of the parasite *Pseudeucoila bochei* (Hym., Cynip.) over its hosts, larvae of *Drosophila melanogaster*. *Oecologia*, **10**, 29–57.

Banks, M.J. and Thompson, D.J. (1987a) Lifetime reproductive success of females of the damselfly *Coenagrion puella*. *Journal of Animal Ecology*, **56**, 815–32.

Banks, M.J. and Thompson, D.J. (1987b) Regulation of damselfly populations: the effects of larval density on larval survival, development rate and size in the field. *Freshwater Biology*, **17**, 357–65.

Barendregt, A. (1975) Boemvoorkeur bij Zweefvliegen (Dipt., Syrphidae). *Entomologische Berichten*, **35**, 96–100.

Barfield, C.S., Bottrell, D.G. and Smith, J.W. Jr. (1977a) Influence of temperature on oviposition and adult longevity of *Bracon mellitor* reared on boll weevils. *Environmental Entomology*, **6**, 133–7.

Barfield, C.S., Sharpe, P.J.H. and Bottrell, D.G. (1977b) A temperature driven development model for the parasite *Bracon mellitor* (Hymenoptera: Braconidae). *Canadian Entomologist*, **109**, 1503–14.

Barlow, C.A. (1961) On the biology and reproductive capacity of *Syrphus corollae* Fab. (Syrphidae) in the laboratory. *Entomologia Experimentalis et Applicata*, **4**, 91–100.

Barlow, N. and Dixon, A.F.G. (1980) *Simulation of Lime Aphid Population Dynamics*, Pudoc, Wageningen.

Barlow, N.D. and Goldson, S.L. (1993) A modelling analysis of the successful biological control of *Sitona discoideus* (Coleoptera: Curculionidae) by *Microctonus aethiopoides* (Hymenoptera: Braconidae) in New Zealand. *Journal of Applied Ecology*, **30**, 165–8.

Barndt, D. (1976) Untersuchungen der diurnalen und siasonalen Activitat von Kafern mit einer neu entwickelten Elektro-Bodenfalle. *Verhandlungen des Botanischen Vereins der Provinz Brandenberg*, **112**, 103–22.

Barrass, R. (1960a) The courtship behaviour of *Mormoniella vitripennis*. *Behaviour*, **15**, 185–209.

Barrass, R. (1960b) The effect of age on the performance of an innate behaviour pattern in *Mormoniella vitripennis*. *Behaviour*, **15**, 210–18.

Barrass, R. (1976) Rearing jewel wasps *Mormoniella vitripennis* (Walker) and their use in teaching biology. *Journal of Biological Education*, **10**, 119–26.

Bartlett, B.R. (1964) Patterns in the host-feeding habit of adult Hymenoptera. *Annals of the Entomological Society of America*, **57**, 344–50.

Bartlett, B.R. (1968) Outbreaks of two-spotted spider mites and cotton aphids following pesticide treatment. I. Pest stimulation vs. natural enemy destruction as the cause of outbreaks. *Journal of Economic Entomology*, **61**, 297–303.

Bartlett, M.S. (1949) Fitting a straight line when both variables are subject to error. *Biometrics*, **5**, 207–12.

Bay, E.C. (1974) Predator–prey relationships among aquatic insects. *Annual Review of Entomology*, **19**, 441–53.

Beckage, N.E. and Riddiford, L.M. (1978) Developmental interactions between the tobacco hornworm *Manduca sexta* and its braconid parasite *Apanteles congregatus*. *Entomologia Experimentalis et Applicata*, **23**, 139–51.

Beckage, N.E. and Riddiford, L.M. (1983) Growth and development of the endoparasitic wasp *Apanteles congregatus*: dependence on host nutritional status and parasite load. *Physiological Entomology*, **8**, 231–41.

Beckmann Jr, R.L. and Stucky, J.M. (1981) Extrafloral nectaries and plant guarding in *Ipomoea pandurata* (L.) G.F.W. Mey (Convolvulaceae). *American Journal of Botany*, **68** (1), 72–9.

Bedding, R.A. (1973) The immature stages of Rhinophorinae (Diptera: Calliphoridae) that parasitise British woodlice. *Transactions of the Royal Entomological Society of London*, **125**, 27–44.

Beddington, J.R. (1975) Mutual interference between parasites or predators and its effect on searching efficiency. *Journal of Animal Ecology*, **44**, 331–40.

Beddington, J.R. and Hammond, P.S. (1977) On the dynamics of host–parasite–hyperparasite interactions. *Journal of Animal Ecology*, **46**, 811–21.

Beddington, J.R., Free, C.A. and Lawton, J.H. (1975) Dynamic complexity in predator–prey models framed in difference equations. *Nature*, **225**, 58–60.

Beddington, J.R., Free, C.A. and Lawton, J.H. (1976a) Concepts of stability and resilience in predator–prey models. *Journal of Animal Ecology*, **45**, 791–816.

Beddington, J.R., Free, C.A. and Lawton, J.H. (1978) Modelling biological control: on the characteristics of successful natural enemies. *Nature*, **273**, 513–19.

Beddington, J.R., Hassell, M.P. and Lawton, J.H. (1976b) The components of arthropod predation. II. The predator rate of increase. *Journal of Animal Ecology*, **45**, 165–85.

Begon, M. (1979) *Investigating Animal Abundance – Capture–recapture for Biologists*, Edward Arnold, London.

Begon, M. (1983) Abuses of mathematical techniques in ecology: applications of Jolly's capture–recapture method. *Oikos*, **40**, 155–8.

Begon, M., Harper, J.L. and Townsend, C.R. (1990) *Ecology: Individuals, Populations and Communities*, Blackwell, Oxford.

Beirne, B.P. (1980) Biological control: benefits and opportunities, in *Perspectives in World Agriculture*, Commonwealth Agricultural Bureaux, Slough, pp. 307–21.

Bell, G. and Koufopanou, V. (1986) The cost of reproduction. *Oxford Surveys in Evolutionary Biology*, **3**, 83–131.

Bellows Jr, T.S. (1985) Effects of host and parasitoid age on search behaviour and oviposition rates in *Lariophagus distinguendus* Forster (Hymenoptera: Pteromalidae). *Researches on Population Ecology*, **27**, 65–76.

Bellows Jr, T.S. and Hassell, M.P. (1988) The dynamics of age-structured host–parasitoid interactions. *Journal of Animal Ecology*, **57**, 259–68.

Bellows Jr, T.S., van Driesche, R.G. and Elkinton, J. (1989a) Life tables and parasitism: estimating parameters in joint host–parasitoid systems, in *Estimation and Analysis of Insect Populations* (eds L. McDonald, B. Manly, J. Lockwood and J. Logan). Springer-Verlag, Berlin, pp. 70–80.

Bellows Jr, T.S., van Driesche, R.G. and Elkinton, J.S. (1989b) Extensions to Southwood and Jepson's graphical method of estimating numbers entering a stage for calculating mortality due to parasitism. *Researches on Population Ecology*, **31**, 169–84.

Bellows Jr, T.S., van Driesche, R.G. and Elkinton, J.S. (1992) Life table construction and analysis in the evaluation of natural enemies. *Annual Review of Entomology*, **37**, 587–614.

Belshaw, R. (1993) Malaise traps and Tachinidae (Diptera): a study of sampling efficiency. *The Entomologist*, **112** (1), 49–54.

Ben-Dov, Y. (1972) Life history of *Tetrastichus ceroplastae* (Girault) (Hymenoptera: Eulophidae), a parasite of the Florida wax scale, *Ceroplastes floridensis* Comstock (Homoptera: Coccidae), in Israel. *Journal of the Entomological Society of South Africa*, **35**, 17–34.

Benest, G. (1989) The sampling of a carabid community. I. The behaviour of a carabid when facing the trap. *Revue d'Ecologie et de Biologie du Sol*, **26**, 205–11.

Benson, J.F. (1973) Intraspecific competition in the population dynamics of *Bracon hebetor* Say (Hymenoptera: Braconidae). *Journal of Animal Ecology*, **42**, 105–24.

Benson, M. (1989) The Biology and Specificity of the Host–parasitoid Relationship, With Reference to Aphidophagous Syrphid Larvae and their Associated Parasitoids, University of Nottingham, M.Phil. Thesis.

Benton, F. (1975), Larval Taxonomy and Bionomics of some British Pipunculidae, Imperial College, University of London, Ph.D. Thesis.

Berberet, R.C. (1982) Effects of host age on embryogenesis and encapsulation of the parasite *Bathyplectes curculionis* in the alfalfa weevil. *Journal of Invertebrate Pathology*, **40**, 359–66.

Berberet, R. C. (1986) Relationship of temperature to embryogenesis and encapsulation of eggs of *Bathyplectes curculionis* (Hymenoptera: Ichneumonidae) in larvae of *Hypera postica* (Coleoptera: Curculionidae). *Annals of the Entomological Society of America*, **79**, 985–8.

Bergelson, J. (1985) A mechanistic interpretation of prey selection by *Anax junius larvae* (Odonata: Aeschnidae). *Ecology*, **66**, 1699–705.

van Berjeijk, K.E., Bigler, F., Kaashoek, N.K. and Pak, G.A. (1989) Changes in host acceptance and host suitability as an effect of rearing *Trichogramma maidis* on a factitious host. *Entomologia Experimentalis et Applicata,* **52**, 229–38.

Berlocher, S. H. (1984) Insect molecular systematics *Annual Review of Entomology*, **29**, 403–33.

Berlocher, S.H. (1980) An electrophoretic key for distinguishing species of the genus *Rhagoletis* (Diptera: Tephritidae) as larvae, pupae or adults. *Annals of the Entomological Society of America*, **73**, 131–7.

Berry, I.L., Foerster, K.W. and Ilken, E.H. (1976) Prediction model for development time of stable flies. *Transactions of the American Society of Agricultural Engineers*, **19**, 123–7.

Berryman, A.A. (1990) Modelling Douglas-fir tussock moth population dynamics: the case for simple theoretical models, in *Population Dynamics of Forest Insects* (eds A.D. Watt, S.R. Leather, M.D. Hunter and N.A.C. Kidd), Intercept, Andover, pp. 369–80.

Berryman, A.A. (1992) The origins and evolution of predator–prey theory. *Ecology*, **73**, 1530–35.

Bess, H.A. (1936) The biology of *Leschenaultia exul* Townsend, a tachinid parasite of *Malacosoma distria* Hubner. *Annals of the Entomological Society of America*, **29**, 593–613.

Bess, H.A., van der Bosch, R. and Haramoto, F.H. (1961) Fruit-fly parasites and their activities in Hawaii. *Proceedings of the Hawaii Entomological Society*, **17**, 367–78.

Best, R.L., Beegle, C.C., Owens, J.C. and Oritz, M. (1981) Population density, dispersion, and dispersal estimates for *Scarites substriatus, Pterostichus chalcites*, and *Harpalus pennsylvanicus* (Carabidae) in an Iowa cornfield. *Environmental Entomology*, **10**, 847–56.

Beukeboom, L.W. & Werren, J.H. (1992) Transmission and expression of the parasitic paternal sex ratio (PSR) chromosome, *Heredity* **70**: 437–43.

Bigler, F., Meyer, A. and Bosshart, S. (1987) Quality assessment in *Trichogramma maidis* Pinteureau et Voegele reared from eggs of the factitious hosts *Ephestia kuehniella* Zell. and *Sitotroga cerealella* (Olivier). *Journal of Applied Entomology*, **104**, 340–53.

Birkhead, T.R., Lee, K.E. and Young, P. (1988) Sexual cannibalism in the praying mantis *Hierodula membranacea*. *Behaviour*, **106**, 112–18.

Blackburn, T.M. (1991a) A comparative examination of lifespan and fecundity in parasitoid hymenoptera. *Journal of Animal Ecology*, **60**, 151–64.

Blackburn, T.M. (1991b) Evidence for a 'fast-slow' continuum of life-history traits among parasitoid Hymenoptera. *Functional Biology*, **5**, 65–74.

Blackman, R.L. (1967) The effect of different aphid foods on *Adalia bipunctata* L. and *Coccinella 7-punctata* L.. *Annals of Applied Biology*, **59**, 207–19.

Blann, A.D. (1984) Cell fusion and monoclonal antibodies. *The Biologist*, **31**, 288–91.

Blower, J.G., Cook, L.M. and Bishop, J.A. (1981) *Estimating the Size of Animal Populations*, George Allen and Unwin, London.

Blumberg, D. (1991) Seasonal variations in the encapsulation of eggs of the encyrtid parasitoid *Metaphycus stanleyi* by the pyriform scale, *Protopulvinaria pyriformis*. *Entomologia Experimentalis et Applicata*, **58**, 231–7.

Blumberg, D. and DeBach P. (1979) Development of *Habrolepis rouxi* Compere (Hymenoptera: Encyrtidae) in two armoured scale hosts (Homoptera: Diaspididae) and parasite egg encapsulation by California red scale. *Ecological Entomology*, **4**, 299–306.

Blumberg, D. and Luck, R.F. (1990) Differences in the rates of superparasitism between two strains of *Comperiella bifasciata* (Howard) (Hymenoptera: Encyrtidae) parasitizing California red scale (Homoptera: Diaspididae): an adaptation to circumvent encapsulation? *Annals of the Entomological Society of America*, **83**, 591–7.

den Boer, P.J. (1979) The individual behaviour and population dynamics of some carabid beetles of forests, in *On the Evolution of Behaviour in Carabid Beetles* (eds P.J. den Boer, H.U. Thiele and F. Weber), Miscellaneous. Papers 18, University of Wageningen, H. Veenman and Zanen, Wageningen, pp. 151–66.

Boggs, C.L. (1977) Ovarian dynamics in heliconiine butterflies: programmed senescence versus eternal youth. *Science*, **197**, 487–90.

Boggs, C.L. (1987) Ecology of nectar and pollen feeding in Lepidoptera, in *Nutritional Ecology of Insects, Mites, Spiders and Related Invertebrates* (eds F. Slansky Jr and J.G. Rodriguez), Wiley-Interscience, New York.

Bolten, A.B., Feisinger, P., Baker, H.G. and Baker, I. (1979) On the calculation of sugar concentration in flower nectar. *Oecologia*, **41**, 301–4.

Bombosch, S. (1962) Untersuchungen über die Auswertbarkeit von Fallenfängen *Zeitschrift für Angewandte Zoologie* , **49**, 149–60.

Boreham, P.F.L. (1979) Recent developments in serological methods for predator–prey studies. *Entomological Society of America, Miscellaneous Publication*, **11**, 17–23.

Boreham, P.F.L. and Ohiagu, C.E. (1978) The use of serology in evaluating invertebrate prey–

predator relationships: a review. *Bulletin of Entomological Research*, **68**, 171–94.

van den Bosch, R., Leigh, T.F., Gonzalez, D. and Stinner, R.E. (1969) Cage studies on predators of the bollworm in cotton. *Journal of Economic Entomology*, **62**, 1486–9.

van den Bosch, R., Hom, R, Matteson, P., Frazer, B.D., Messenger, P.S. and Davis, C.S. (1979) Biological control of the walnut aphid in California: impact of the parasite *Trioxys pallidus*. *Hilgardia*, **47**, 1–13.

Bouchard, Y. and Cloutier, C. (1984) Honeydew as a source of host-searching kairomones for the aphid parasitoid *Aphidius nigripes* (Hymenoptera: Aphididae). *Canadian Journal of Zoology*, **62**, 1513–20.

Bouletreau, M. (1986) The genetic and coevolutionary interactions between parasitoids and their hosts, in *Insect Parasitoids* (eds J. Waage and D. Greathead), Academic Press, London, pp. 169–200.

Bowden, J. (1981) The relationship between light- and suction-trap catches of *Chrysoperla carnea* (Stephens) (Neuroptera: Chrysopidae) and the adjustment of light-trap catches to allow for variation in moonlight. *Bulletin of Entomological Research*, **71**, 621–9.

Bowers, M.A. (1985) Experimental analyses of competition between two species of bumble bees (Hymenoptera: Apidae). *Oecologia*, **67**, 224–30.

Brakefield, P. (1985) Polymorphic Müllerian mimicry and interactions with thermal melanism in ladybirds and a soldier beetle: a hypothesis, *Biological Journal of the Linnean Society*, **26**, 243–67.

Braman, S.K. and Yeargan, K.V. (1988) Comparison of developmental and reproductive rates of *Nabis americoferus*, *N. roseipennis*, and *N. rufusculus* (Hemiptera: Nabidae). *Journal of the Entomological Society of America*, **81**, 923–30.

Breeuwer, J.A.J. and Werren, J.H. (1990) Microrganisms associated with chromosome destruction and reproductive isolation between two insect species. *Nature*, **346**, 558–60.

Brodeur, J. and McNeil, J.N. (1992) Host behaviour modification by the endoparasitoid *Aphidius nigripes*: a strategy to reduce hyperparasitism. *Ecological Entomology*, **17**, 97–104.

Brookes, C.P. and Loxdale, H.D. (1985) A device for simultaneously homogenising numbers of individual small insects for electrophoresis. *Bulletin of Entomological Research*, **75**, 377–8.

Brooks, D.R. and McLennan, D.A. (1991) *Phylogeny, Ecology and Behavior. A Research Program in Comparative Biology*, University of Chicago Press, London.

Browning, H.W. and Oatman, E.R. (1981) Effects of different constant temperatures on adult longevity, development time, and progeny production of *Hyposoter exiguae* (Hymenoptera: Ichneumonidae). *Annals of the Entomological Society of America*, **74**, 79–82.

Brunsting, A.M.H. and Heessen, H.J.L. (1984) Density regulation in the carabid beetle *Pterostichus oblongopunctatus*. *Journal of Animal Ecology*, **53**, 751–60.

Brunsting, A.M.H., Siepel, H. and van Schaick Zillesen, P.G. (1986) The role of larvae in the population ecology of Carabidae, in *Carabid Beetles. Their Adaptations and Dynamics* (eds P.J. den Boer, M.L. Luff, D. Mossakowski and F. Weber), Gustav Fischer, Stuttgart, pp. 399–411.

Brust, G.E., Stinner, B.R. and McCartney, D.A. (1986a) Predation by soil-inhabiting arthropods in intercropped and monoculture agroecosystems. *Agriculture, Ecosystems and Environment*, **18**, 145–54.

Brust, G.E., Stinner, B.R. and McCartney, D.A. (1986b) Predator activity and predation in corn agroecosystems. *Environmental Entomology*, **15**, 1017–21.

Bryan, G. (1980) Courtship behaviour, size differences between the sexes and oviposition in some *Achrysocharoides* species (Hym., Eulophidae), *Netherlands Journal of Zoology*, **30**, 611–21.

de Buck, N. (1990) Bloembezoek en bestuivingsecologie van zweefvliegen in het bijzonder voor Belgie. *Studiedocumenten van het Koninklijk Belgisch Instituut voor Natuurwetenschappen*, **60**, Koninklijk Belgisch Instituut voor Natuurwetenschappen, Brussels, pp. 134.

Budenberg, W.J. (1990) Honeydew as a contact kairomone for aphid parasitoids. *Entomologia Experimentalis et Applicata*, **55**, 139–48.

Bugg, R.L., Ellis, R.T. and Carlson, R.W. (1989) Ichneumonidae (Hymenoptera) using extrafloral nectar of faba bean (*Vicia faba* L., Fabaceae) in Massachusetts. *Biological Agriculture and Horticulture*, **6**, 107–14.

Buonaccorsi, J.P. and Elkinton, J.S. (1990) Estimation of contemporaneous mortality factors. *Researches on Population Ecology*, **32**, 1–21.

Burbutis, P.P. and Stewart, J.A. (1979) Blacklight trap collecting of parasitic Hymenoptera. *Entomology News*, **90**, 17–22.

Burk, T. (1981) Signaling and sex in acalyptrate flies. *Florida Entomologist*, **64**, 30–43.

Burk, T. (1982) Evolutionary significance of predation on sexually signalling males. *Florida Entomologist*, **65**, 90–104.

Burkhart, B.D., Montgomery, E., Langley, C.H. and Voelker, R.A. (1984) Characterization of allozyme null and low activity alleles from two natural populations of *Drosophila melanogaster*. *Genetics*, **107**, 295–306.

Burn, A.J. (1982) The role of predator searching efficiency in carrot fly egg loss. *Annals of Applied Biology*, **101**, 154–15.

Burquez, A. and Corbet, S.A. (1991) Do flowers reabsorb nectar? *Functional Ecology*, **5**, 369–79.

Bursell, E. (1964) Environmental aspects: temperature, in *The Physiology of the Insecta* (ed. M. Rockstein), Academic Press, New York, pp. 283–321.

Burstone, M.S. (1957) Polyvinyl pyrrolidone. *American Journal of Clinical Pathology*, **28**, 429–30.

Buschman, L.L., Whitcomb, W.H., Hemenway, R.C., Mays, D.L., Nguyen Roo, Leppla, N.C. and Smittle. B.J. (1977) Predators of velvetbean caterpillar eggs in Florida soybeans. *Environmental Entomology*, **6**, 403–7.

van Buskirk, J. (1987) Density-dependent population dynamics in larvae of the dragonfly *Pachydiplax longipennis*: a field experiment. *Oecologia*, **72**, 221–5.

Butler, G.D. (1965) A modified Malaise insect trap. *Pan-Pacific Entomology*, **41**, 51–3.

Butts, R.A. and McEwen, F.L. (1981) Seasonal populations of the diamondback moth in relation to day-degree accumulation. *Canadian Entomologist*, **113**, 127–31.

Cade, W. (1975) Acoustically orienting parasitoids: fly phonotaxis to cricket song. *Science*, **190**, 1312–13.

Calos, M.P. and Miller, J.H. (1980) Transposable elements. *Cell*, **20**, 579–95.

Campbell, A., Frazer, B.D., Gilbert, N., Gutierrez, A.P. and Mackauer, M. (1974) Temperature requirements of some aphids and their parasites. *Journal of Applied Ecology*, **11**, 431–8.

Campbell, A. and Mackauer, M. (1975) Thermal constants for development of the pea aphid (Homoptera: Aphididae) and some of its parasites. *Canadian Entomologist*, **107**, 419–23.

Campbell, B.C. and Duffey, S.S. (1979) Tomatine and parasitic wasps: potential incompatibility of plant antibiosis with biological control. *Science, New York*, **205**, 700–02.

Campbell, C.A.M. (1978) Regulation of the Damson–hop aphid (*Phorodon humuli* [Schrank]) on hops (*Humulus lupulus* L.) by predators. *Journal of Horticultural Science*, **53**, 235–42.

Canard, M., Semeria, Y. and New, T.R. (1984) *Biology of the Chrysopidae*, W. Junk, The Hague.

Čapek, M. (1970) A new classification of the Braconidae (Hymenoptera) based on the cephalic structures of the final instar larva and biological evidence. *Canadian Entomologist*, **102**, 846–75.

Čapek,M. (1973) Key to the final instar larvae of the Braconidae (Hymenoptera). *Acta Instituti Forstalis Zvolenensis*, **1973**, 259–68.

Cappuccino, N. (1992) Adjacent trophic-level effects on spatial density dependence in a herbivore–predator–parasitoid system. *Ecological Entomology*, **17**, 105–8.

Carroll, D.P and Hoyt, S.C. (1984) Natural enemies and their effects on apple aphid, *Aphis pomi* DeGeer (Homoptera: Aphididae), colonies on young apple trees in central Washington, *Environmental Entomology*, **13**, 469–81.

Carter, M.C. and Dixon, A.F.G. (1984) Honeydew, an arrestment stimulus for coccinellids. *Ecological Entomology*, **9**, 383–7.

Carter, N. and Dixon, A.F.G. (1981) The 'natural enemy ravine' in cereal aphid population dynamics: a consequence of predator activity or aphid biology? *Journal of Animal Ecology*, **50**, 605–11.

Carter, N., Aikman, D.P. and Dixon, A.F.G. (1978) An appraisal of Hughes' time-specific life table analysis for determining aphid reproductive and mortality rates. *Journal of Animal Ecology*, **47**, 677–89.

Carter, N., Dixon, A.F.G. and Rabbinge, R. (1982) *Cereal Aphid Populations: Biology, Simulation and Prediction*, Pudoc, Wageningen.

Carton, Y. (1978) Olfactory responses of *Cothonaspis* sp. (parasitic Hymenoptera, Cynipidae) to the food habit of its host (*Drosophila melanogaster*). *Drosophila Information Service*, **53**, 183–4.

Casas, J. (1990) Foraging behaviour of a leafminer parasitoid in the field. *Ecological Entomology*, **14**, 257–65.

Casas, J., Gurney, W.S.C., Nisbet, R. and Roux, O. (1993) A probabilistic model for the functional response of a parasitoid at the behavioural time-scale, *Journal of Animal Ecology*, **62**, 194–204.

Castañera, P., Loxdale, H. D. and Nowak, K. (1983) Electrophoretic study of enzymes from cereal aphid populations. II. Use of electrophoresis for identifying aphidiid parasitoids (Hymenoptera) of *Sitobion avenae* (F.) (Hemiptera: Aphididae). *Bulletin of Entomological Research*, **73**, 659–65,

Castrovillo, P. J. and Stock, M.W. (1981) Electrophoretic techniques for detection of *Glypta fumiferanae*, an endoparasitoid of western spruce budworm. *Entomologia Experimentalis et Applicata*, **30**, 176–80.

Cave, R.D. and Gaylor, M.J. (1988) Influence of temperature and humidity on development and survival of *Telenomus reynoldsi* (Hymenoptera: Scelionidae) parasitising *Geocoris punctipes* (Heteroptera: Lygaeidae) eggs. *Annals of the Entomological Society of America*, **81**, 278–85.

Cave, R.D. and Gaylor, M.J. (1989) Longevity, fertility, and population growth statistics of *Telenomus reynoldsi* (Hymenoptera: Scelionidae). *Proceedings of the Entomological Society of Washington*, **91**, 588–93.

Cendaña, S.M. (1937) Studies on the biology of *Coccophagus* (Hymenoptera), a genus parasitic on non-diaspidiine Coccidae. *University of California Publications in Entomology*, **6**, 337–400.

Chambers, R.J. and Adams, T.H.L. (1986) Quantification of the impact of hoverflies (Diptera: Syrphidae) on cereal aphids in winter wheat: an analysis of field populations. *Journal of Applied Ecology*, **23**, 895–904.

Chambers, R.J., Sunderland, K.D, Wyatt, I.J. and Vickerman, G.P. (1983) The effects of predator exclusion and caging on cereal aphids in winter wheat. *Journal of Applied Ecology*, **20**, 209–24.

Chan, M.S. and Godfray, H.C.J. (1993) Host feeding strategies of parasitic wasps. *Evolutionary Ecology*, **7**, 593–604.

Chant, D.A. (1959) Phytoseiid mites (Acarina: Phytoseiidae). i. Bionomics of seven species in south-eastern England, II. A taxonomic review of the family Phytoseiidae, with descriptions of thirty eight new species. *Canadian Entomologist*, **91**(12).

Chantarasa ard, S. (1984) Preliminary study on the overwintering of *Anagrus incarnatus* Haliday (Hymenoptera: Mymaridae), an egg parasitoid of the rice planthoppers. *Esakia*, **22**, 159–62.

Chapman, R.F. (1982) *The Insects: Structure and Function*, Hodder & Stoughton, London.

Charles, J.G. and White, V. (1988) Airborne dispersal of *Phytoseiulus persimils* (Acarina: Phytoseiidae) from a raspberry garden in New Zealand. *Experimental and Applied Acarology*, **5**, 47–54.

Charnov, E.L. (1976) Optimal foraging: attack strategy of a mantid. *American Naturalist*, **110**, 141–51.

Charnov, E.L. (1979) The genetical evolution of patterns of sexuality: Darwinian fitness. *American Naturalist*, **113**, 465–80.

Charnov, E.L. (1982) *The Theory of Sex Allocation*, Princeton University Press, New Jersey.

Charnov, E.L. and Skinner, S.W. (1985) Complementary approaches to the understanding of parasitoid oviposition decisions. *Environmental Entomology*, **14**, 383–91.

Charnov, E.L. and Stephens, D.W. (1988) On the evolution of host selection in solitary parasitoids. *American Naturalist*, **132**: 707–22.

Charnov, E.L., Los-den Hartogh, R.L., Jones, W.T. and van den Assem, J. (1981) Sex ratio evolution in a variable environment. *Nature*, **289**, 27–33.

Chelliah, J. and Jones, D. (1990) Biochemical and immunological studies of proteins from polydnavirus *Chelonus* sp. near *curvimaculatus*. *Journal of General Virology*, **71**, 2353–9.

Chesson, J. (1978a) Measuring preference in selective predation. *Ecology*, **59**, 211–15.

Chesson, P.L. (1978b) Predator-prey theory and variability. *Annual Review of Ecology and Systematics*, **9**, 323–47.

Chesson, P.L. (1982) The stabilizing effect of random environments. *Journal of Mathematical Biology*, **15**, 1–36.

Chesson, J. (1983) The estimation and analysis of preference and its relationship to foraging models. *Ecology*, **63**, 1297–304.

Chesson, P.L. (1984) Variable predators and switching behaviour. *Theoretical Population Biology*, **26**, 1–26.

Chesson, P.L. and Murdoch, W.W. (1986) Aggregation of risk: relationships among host–parasitoid models. *American Naturalist*, **127**, 696–715.

Chiverton, P.A. (1984) Pitfall-trap catches of the carabid beetle *Pterostichus melanarius*, in relation to gut contents and prey densities, in treated and untreated spring barley. *Entomologia Experimentalis et Applicata*, **36**, 23–30.

Chow, F.J. and Mackauer, M. (1985) Multiple parasitism of the pea aphid: stage of development of parasite determines survival of *Aphidius smithi* and *Praon pequodorum* (Hymenoptera: Aphidiidae). *Canadian Entomologist*, **117**, 133–34.

Chow, A. and Mackauer, M. (1991) Patterns of host selection by four species of aphidiid (Hymenoptera) parasitoids: influence of host switching. *Ecological Entomology*, **16**, 403–10.

Chow, F.J. and Mackauer, M. (1984) Inter- and intraspecific larval competition in *Aphidius smithi* and *Praon pequodorum* (Hymenoptera: Aphidiidae). *Canadian Entomologist*, **116**, 1097–107.

Chow, F.J. and Mackauer, M. (1986) Host discrimination and larval competition in the aphid parasite *Ephedrus californicus*. *Entomologia Experimentalis et Applicata*, **41**, 243–54.

Chrambach, A. and Rodbard, D. (1971) Polyacrylamide gel electrophoresis. *Science*, **172**, 440–51.

Clapham, A.R., Tutin, T.G. and Moore, D.M. (1989) *Flora of the British Isles*, Cambridge University Press, Cambridge.

Claridge, M.F. and Askew, R.R. (1960) Sibling species in the *Eurytoma rosae* group (Hym., Eurytomidae). *Entomophaga*, **5**, 141–53.

Claridge, M.F. and Dawah, H.A. (1993) Assemblages of herbivorous chalcid wasps and their parasitoids associated with grasses – problems of species and specificity, in *Plant Galls: Organisms, Interactions, Populations:* (ed. M.A.J. Williams) Systematics Association Special Volume, **49,** Oxford Clarendon Press, Oxford, pp. 313–329.

Claridge, M.F. and den Hollander, J. (1983) The biotype concept and its application to insect pests of agriculture. *Crop Protection*, **2**, 85–95.

Clarke, B.C. (1962) Balanced polymorphism and the diversity of sympatric species, in *Taxonomy and Geography* (ed. D. Nichols), Systematics Association Publication No. 4, Systematics Association, Oxford, pp. 47–70.

Clausen, C.P. (1940) *Entomophagous Insects*, McGraw-Hill Co., New York.

Clausen, C.P., Jaynes, H.A. and Gardner, T.R. (1933) Further investigations of the parasites of *Popillia japonica* in the Far East. *United States Department of Agriculture Technical Bulletin*, **366**, 1–58.

de Clercq, R. (1985) Study of the soil fauna in winter wheat fields and experiments on the influence of this fauna on the aphid populations. *Bulletin IOBC/WPRS*, **7**, 133–5.

de Clercq, R. and Pietraszko, R. (1982) Epigeal arthropods in relation to predation of cereal aphids, in *Aphid Antagonists* (ed. R. Cavalloro), Proceedings of the EC Experts' Meeting, Portici, A.A. Balkema, Rotterdam, pp. 88–92.

Cloutier, C. and Mackauer, M. (1979) The effect of parasitism by *Aphidius smithi* (Hymenoptera: Aphidiidae) on the food budget of the pea aphid, *Acyrthosiphon pisum*. *Canadian Journal of Zoology*, **57**, 1605–11.

Cloutier, C. and Mackauer, M. (1980) The effect of superparasitism by *Aphidius smithi* (Hymenoptera: Aphidiidae) on the food budget of the pea aphid, *Acyrthosiphon pisum* (Homoptera: Aphidiidae). *Canadian Journal of Zoology*, **58**, 241–4.

Coaker, T.H. (1965) Further experiments on the effect of beetle predators on the numbers of the cabbage root fly, *Erioischia brassicae* (Bouche), attacking crops. *Annals of Applied Biology*, **56**, 7–20.

Cochran, W.G. (1983) *Planning and Analysis of Observational Studies*, John Wiley and Sons, New York.

Cock, M.J.W. (1978) The assessment of preference. *Journal of Animal Ecology*, **47**, 805–16.

Coe, R.L. (1966) Diptera: Pipunculidae. *Handbooks for the Identification of British Insects*, **10**(2c), Royal Entomological Society of London, London.

Cohen, A.C. (1984) Food consumption, food utilization, and metabolic rates of *Geocoris punctipes* (Het.: Lygaeidae) fed *Heliothis virescens* (Lep.: Noctuidae) eggs. *Entomophaga*, **29**, 361–7.

Cohen, A.C. (1989) Ingestion efficiency and protein consumption by a heteropteran predator. *Annals of the Entomological Society of America*, **82**, 495–9.

Cole, L.R. (1967) A study of the life-cycles and hosts of some Ichneumonidae attacking pupae of the green oak-leaf roller moth, *Tortrix viridana* (L.) (Lepidoptera: Tortricidae) in England. *Transactions of the Royal Entomological Society of London*, **119**, 267–81.

Cole, L.R. (1981) A visible sign of a fertilization act during oviposition by an ichneumonid wasp, *Itoplectis maculator*. *Animal Behaviour*, **29**, 299–300.

Collins, M.D. and Dixon, A.F.G. (1986) The effect of egg depletion on the foraging behaviour of an aphid parasitoid. *Journal of Applied Entomology*, **102**, 342–52.

Collins, M.D., Ward, S.A and Dixon, A.F.G. (1981) Handling time and the functional response of *Aphelinus thomsoni*, a predator and parasite of the aphid *Drepanosiphum platanoidis*. *Journal of Animal Ecology*, **50**: 479–87.

Comins, H.N. and Fletcher, B.S. (1988) Simulation of fruit fly population dynamics, with particular reference to the olive fly, *Dacus oleae*. *Ecological Modelling*, **40**, 213–31.

Comins, H.N. and Hassell, M.P. (1976) Predation in multi-prey communities. *Journal of Theoretical Biology*, **62**, 93–114.

Comins, H.N. and Hassell, M.P. (1979) The dynamics of optimally foraging predators and parasitoids. *Journal of Animal Ecology*, **48**, 335–51.

Comins, H.N. and Wellings, P.W. (1985) Density-related parasitoid sex ratio: influence on host–parasitoid dynamics. *Journal of Animal Ecology*, **54**, 583–94.

Commonwealth Scientific and Industrial Research Organisation (1991) *The Insects of Australia*, Vol. I, Melbourne University Press, Carlton.

Conn, D.L.T. (1976) Estimates of population size and longevity of adult narcissus bulb fly *Merodon equestris* Fab. (Diptera: Syrphidae). *Journal of Applied Ecology*, **13**, 429–34.

Conrad, K.F. and Herman, T.B. (1990) Seasonal dynamics, movements and the effects of experimentally increased female densities on a population of imaginal *Calopteryx aequabilis* (Odonata: Calopterygidae). *Ecological Entomology*, **15**, 119–29.

Conrad, K.F. and Pritchard, G. (1990) Preoviposition mate-guarding and mating behaviour of *Argia vivida* (Odonata: Coenagrionidae). *Ecological Entomology*, **15**, 363–70.

Cook, D. and Stoltz, D.B. (1983) Comparative serology of viruses isolated from ichneumonid parasitoids. *Virology*, **130**, 215–20.

Cook, L.M., Bower, P.P. and Croze, H.J. (1967) The accuracy of a population estimation from multiple recapture data. *Journal of Animal Ecology*, **36**, 57–60.

Cook, R.M. and Cockrell, B.J. (1978) Predator ingestion rate and its bearing on feeding time and the theory of optimal diets. *Journal of Animal Ecology*, **47**, 529–48.

Cook, R.M. and Hubbard, S.F. (1977) Adaptive searching strategies in insect parasitoids. *Journal of Animal Ecology*, **46**, 115–25.

Coombes, D.S. and Sotherton, N.W. (1986) The dispersal and distribution of polyphagous predatory Coleoptera in cereals. *Annals of Applied Biology*, **108**, 461–74.

Cooper, B.A. (1945) Hymenopterist's Handbook. *The Amateur Entomologist*, **7**, 1–160.

Copland, M.J.W. (1976) Female reproductive system of the Aphelinidae (Hymenoptera: Chalcidoidea). *International Journal of Insect Morphology and Embryology* , **5**, 151–66.

Copland, M.J.W. and King, P.E. (1971) The structure and possible function of the reproductive system in some Eulophidae and Tetracampidae. *Entomologist*, **104**, 4–28.

Copland, M.J.W. and King, P.E. (1972a) The structure of the female reproductive system in the Pteromalidae (Chalcidoidea: Hymenoptera). *Entomologist*, **105**, 77–96.

Copland, M.J.W. and King, P.E. (1972b) The structure of the female reproductive system in the Eurytomidae (Chalcidoidea: Hymenoptera). *Journal of Zoology*, **166**, 185–212.

Copland, M.J.W. and King, P.E. (1972c) The structure of the female reproductive system in the Torymidae (Hymenoptera: Chalcidoidea). *Transactions of the Royal Entomological Society of London*, **124**, 191–212.

Copland, M.J.W. and King, P.E. (1972d) The structure of the female reproductive system in the Chalcididae (Hym.). *Entomologist's Monthly Magazine*, **107**, 230–39.

Copland, M.J.W., King, P.E. and Hill, D.S. (1973) The structure of the female reproductive system in the Agaonidae (Chalcidoidea, Hymenoptera). *Journal of Entomology* (A), **48**, 25–35.

Corbet, S.A. (1978) Bee visits and the nectar of *Echium vulgare* L. and *Sinapis alba* L., *Ecological Entomology*, **3**, 25–37.

Corbet, S.A., Unwin, D.M. and Prys-Jones, O.E. (1979) Humidity, nectar and insect visitors to flowers, with special reference to *Crataegus, Tilia* and *Echium*. *Ecological Entomology*, **4**, 9–22.

Cornell, H. and Pimentel, D. (1978) Switching in the parasitoid *Nasonia vitripennis* and its effect on host competition. *Ecology*, **59**, 297–308.

Cornelius, M. and Barlow, C.A. (1980) Effect of aphid consumption by larvae on development and reproductive efficiency of a flower-fly, *Syrphus corollae* (Diptera: Syrphidae). *Canadian Entomologist*, **112**, 989–92.

Corrigan, J.E. and Lashomb, J.H. (1990) Host influences on the bionomics of *Edovum puttleri* (Hymenoptera: Eulophidae): effects on size and reproduction. *Environmental Entomology*, **19**, 1496–502.

Cousin, G. (1933) Étude biologique d'un Chalcidien, *Mormoniella vitripennis*. *Bulletin Biologique de France et Belgique*, **67**, 371–400.

Cowgill, S.E., Wratten, S.D. and Sotherton, N.W. (1993) The selective use of floral resources by the hoverfly *Episyrphus balteatus* (Diptera: Syrphidae) on farmland. *Annals of Applied Biology*, **122**, 223–31.

Cox, D.R. (1972) Regression models and life tables. *Biometrics*, **38**, 67–77.

Cox, D.R. and Oakes, D. (1984) *Analysis of Survival Data*, Chapman & Hall, London.

Coyne, J.R. (1979) Structural vs. post-translational component of genic variation. *Genetics*, **92**, 679–82.

Craig, C.C. (1953) On the utilisation of marked specimens in estimating populations of flying insects. *Biometrika*, **40**, 170–76

Crawley, M.J. (1993) *GLIM for Ecologists*, Blackwell, Oxford.

Croft, P. and Copland, M. (1993) Size and fecundity in *Dacnusa sibirica* Telenga, *Bulletin OILB/SROP*, **16**(8), 53–6.

Crook, N.E. and Payne, C.C. (1980) Comparison of three methods of ELISA for baculoviruses. *Journal of General Virology*, **46**, 29–37.

Crook, N.E. and Sunderland, K.D. (1984) Detection of aphid remains in predatory insects and spiders by ELISA. *Annals of Applied Biology*, **105**, 413–22.

Crowle, A.J. (1980) Precipitin and microprecipitin reactions in fluid medium and in gels, in *Manual of Clinical Immunology* (eds N.R. Rose and H. Friedman), American Society for Microbiology, Washington DC, pp. 3–14.

Crowley, P.H. and Martin, E.K. (1989) Functional responses and interference within and between year classes of a dragonfly population, *Journal of the North American Benthological Society*, **8**, 211–21.

Crowley, P.H., Nisbet, R.M., Gurney, W.S.C. and Lawton, J.H. (1987) Population regulation in animals with complex life histories: formulation and analysis of a damselfly model. *Advances in Ecological Research*, **17**, 1–59.

Dafni, A. (1992) *Pollination Ecology: A Practical Approach*, IRL Press, Oxford.

Dahms, E.C. (1984a) Revision of the genus *Melittobia* (Chalc. Eulophidae) with descriptions of seven new species. *Memoirs of the Queensland Museum*, **21**, 271–336.

Dahms, E.C. (1984b) A review of the biology of species in the genus *Melittobia* (Hym., Eulophidae) with interpretations and additions using observations on *Melittobia australica*. *Memoirs of the Queensland Museum*, **21**, 337–60.

Danks, H.V. (1971) Biology of some stem-nesting Aculeate Hymenoptera. *Transactions of the Royal Entomological Society of London*, **122**, 323–99.

Darling, D.C. (1988) Comparative morphology of the labrum of Hymenoptera: the digitate labrum of Perilampidae and Eucharitidae (Chalcidoidea). *Canadian Journal of Zoology*, **66**, 2811–35.

Darling, D.C. and Werren, J.H. (1990) Biosystematics of two new species of *Nasonia* (Hym. Pteromalidae) reared from birds' nests in North America. *Annals of the Entomological Society of America*, **83**, 352–70.

Davey, K.G. (1965) *Reproduction in the Insects*, Oliver & Boyd, London.

Davidson, J. (1944) On the relationship between temperature and the rate of development of insects at constant temperatures. *Journal of Animal Ecology*, **13**, 26–38.

Davies, D.H., Burghardt, R.L. and Vinson, S.B. (1986) Oögenesis of *Cardiochiles nigriceps* Viereck (Hymenoptera: Braconidae): histochemistry and development of the chorion with special reference to the fibrous layer. *International Journal of Insect Morphology and Embryology* , **15**, 363–74.

Davies, I. (1974) The effect of age and diet on the ultrastructure of Hymenopteran flight muscle. *Experimental Gerontology*, **9**, 215–19.

Davis, J.R. and Kirkland, R.L. (1982) Physiological and environmental factors related to the dispersal flight of the convergent lady beetle *Hippodamia convergens. Journal of the Kansas Entomological Society*, **55**, 187–96.

Dawson, N. (1965) A comparative study of the ecology of eight species of fenland Carabidae (Coleoptera). *Journal of Animal Ecology*, **34**, 299–314.

Dean, G.J., Jones, M.G. and Powell, W. (1981) The relative abundance of the hymenopterous parasites attacking *Metopolophium dirhodum* (Walker) and *Sitobion avenae* (F.) (Hemiptera: Aphididae) on cereals during 1973–79 in southern England. *Bulletin of Entomological Research*, **71**, 307–15.

DeAngelis, D.L. and Waterhouse, J.C. (1987) Equilibrium and nonequilibrium concepts in ecological models. *Ecological Monographs*, 57, 1–21.

DeBach, P. (1943) The importance of host-feeding by adult parasites in the reduction of host populations. *Journal of Economic Entomology*, **36**, 647–58.

DeBach, P. (1946) An insecticidal check method for measuring the efficacy of entomophagous insects. *Journal of Economic Entomology*, **39**, 695–7.

DeBach, P. (1955) Validity of insecticidal check method as measure of the effectiveness of natural enemies of diaspine scale insects. *Journal of Economic Entomology*, **48**, 584–8.

DeBach, P. and Huffaker, C.B. (1971) Experimental techniques for evaluation of the effectiveness of natural enemies, in *Biological Control* (ed. C.B. Huffaker), Plenum, New York, pp. 113–40.

DeBach, P and Rosen, D. (1991) *Biological Control by Natural Enemies*, 2nd edn, Cambridge University Press, Cambridge.

Decker, U.M. (1988) Evidence for semiochemicals affecting the reproductive behaviour of the aphid parasitoids *Aphidius rhopalosiphi* De Stefani-Perez and *Praon volucre* Haliday (Hymenoptera: Aphidiidae) – A contribution towards integrated pest management in cereals, University of Hohenheim, Ph.D. Thesis.

Decker, U.M., Powell, W. and Clark, S.J. (1993) Sex pheromones in the cereal aphid parasitoids *Praon volucre* and *Aphidius rhopalosiphi*. *Entomologia Experimentalis et Applicata*, **69**, 33–39.

Deevey, E.S. (1947) Life tables for natural populations of animals. *Quarterly Review of Biology*, **22**, 283–314.

DeJong, G. (1979) The influence of the distribution of juveniles over patches of food on the dynamics of a population. *Netherlands Journal of Zoology*, **29**, 33–51.

De Keer, R. and Maelfait, J.P. (1987) Life-history of *Oedothorax fuscus* (Blackwall, 1834) (Araneae, Linyphiidae) in a heavily grazed pasture. *Revue d'Ecologie et de Biologie du Sol*, **24**, 171–85.

DeLong, D.M. (1932) Some problems encountered in the estimation of insect populations by the sweeping method. *Annals of the Entomological Society of America*, **25**, 13–17.

Dempster, J.P. (1960) A quantitative study of the predators on the eggs and larvae of the broom beetle, *Phytodecta olivacea* Forster, using the precipitin test. *Journal of Animal Ecology*, **29**, 149–67.

Dempster, J.P. (1967) The control of *Pieris rapae* with DDT. 1. The natural mortality of the young stages of *Pieris*. *Journal of Applied Ecology*, **4**, 485–500.

Dempster, J.P. (1975) *Animal Population Ecology*, Academic Press, London.

Dempster, J.P. (1983) The natural control of populations of butterflies and moths. *Biological Reviews*, **58**, 461–81.

Dempster, J.P. (1989) Insect introductions: natural dispersal and population persistence in insects. *The Entomologist*, **108**, 5–13.

Dempster, J.P and Lakhani, K.H. (1979) A population model for cinnabar moth and its food plant ragwort. *Journal of Animal Ecology*, **48**, 143–65.

Dempster, J.P. and Pollard, E. (1986) Spatial heterogeneity, stochasticity and the detection of density dependence in animal populations. *Oikos*, **46**, 413–16.

Dennis, P. and Wratten, S.D. (1991) Field manipulation of individual staphylinid species in cereals and their impact on aphid populations. *Ecological Entomology*, **16**, 17–24.

Dennis, P., Wratten, S.D. and Sotherton, N.W. (1991) Mycophagy as a factor limiting predation of aphids (Hemiptera: Aphididae) by staphylinid beetles (Coleoptera: Staphylinidae) in cereals. *Bulletin of Entomological Research*, **81**, 25–31.

Dennison, D.F. and Hodkinson, I.D. (1983) Structure of the predatory beetle community in a woodland soil ecosystem. I. Prey selection. *Pedobiologia*, **25**, 109–15.

Desender, K. and Maelfait, J.P. (1986) Pitfall trapping within enclosures: a method for estimating the relationship between the abundances of coexisting carabid species (Coleoptera: Carabidae). *Holarctic Ecology*, **9**, 245–50.

Desender, K., Maelfait, J.P., D'Hulster, M. and Vanhercke, L. (1981) Ecological and faunal studies on Coleoptera in agricultural land I. Seasonal occurrence of Carabidae in the grassy edge of a pasture. *Pedobiologia*, **22**, 379–84.

Desender, K., Mertens, J., D'Hulster, M. and Berbiers, P. (1984) Diel activity patterns of Carabidae (Coleoptera), Staphylinidae (Coleoptera) and Collembola in a heavily grazed pasture. *Revue d'Ecologie et de Biologie du Sol*, **21**, 347–61.

Dewar, A.M., Dean G.J. and Cannon, Royal (1982) Assessment of methods for estimating the numbers of aphids (Hemiptera: Aphididae) in cereals. *Bulletin of Entomological Research*, **72**, 675–85.

Dewsbury, D.A. (1982) Ejaculate cost and male choice. *American Naturalist*, **119**, 601–10.

Dicke, M. and De Jong, M. (1988) Prey preference of the phytoseiid mite *Typhlodromus pyri*. 2. Electrophoretic diet analysis. *Experimental and Applied Acarology*, **4**, 15–25.

Dicke, M. and Sabelis, M.W. (1988) Infochemical terminology: based on a cost benefit analysis rather than origin of compounds? *Functional Ecology*, **2**, 131–9.

Dietrick, E.J. (1961) An improved back pack motor fan for suction sampling of insect populations. *Journal of Economic Entomology*, **54**, 394–5.

Dietrick, E.J., Schlinger, E.I. and van den Bosch, R. (1959) A new method for sampling arthropods using a suction collecting machine and modified Berlese funnel separator. *Journal of Economic Entomology*, **52**, 1085–91.

van Dijken, M.J. (1991) A cytological method to determine primary sex ratio in the solitary parasitoid *Epidinocarsis lopezi*. *Entomologia Experimentalis et Applicata*, **60**, 301–4.

van Dijken, M.J. and van Alphen, J.J.M. (1991) Mutual interference and superparasitism in the solitary parasitoid *Epidinocarsis lopezi*. *Mededelingen van de Landbouwwetenschappen Rijksuniversiteit Gent*, **56**, 1003–10.

van Dijken, M.J., van Alphen, J.J.M. and van Stratum, P. (1989) Sex allocation in *Epidinocarsis lopezi*: local mate competition. *Entomologia Experimentalis et Applicata*, **52**, 249–55.

Dijkerman, H.J. (1990) Suitability of eight *Yponomeuta* species as hosts of *Diadegma armillata*. *Entomologia Experimentalis et Applicata*, **54**, 173–80.

Dijkerman, H.J. and Koenders, J.T.H. (1988) Competition between *Trieces tricarinatus* and

Triclistus yponomeutae in multiparasitized hosts, *Entomologia Experimentalis et Applicata*, **47**, 289–95.

Dijkstra, L.J. (1986) Optimal selection and exploitation of hosts in the parasitic wasp *Colpoclypeus florus* (Hym., Eulophidae). *Netherlands Journal of Zoology*, **36**, 177–301.

Dixon, A.F.G. (1958) The escape responses shown by certain aphids to the presence of the coccinellid *Adalia decempunctata* (L.). *Transactions of the Royal Entomological Society of London*, **110**, 319–34.

Dixon, A.F.G. (1959) An experimental study of the searching behaviour of the predatory coccinellid beetle *Adalia decempunctata* (L.). *Journal of Animal Ecology*, **28**, 259–81.

Dixon, A.F.G. (1970) Factors limiting the effectiveness of the coccinellid beetle, *Adalia bipunctata* (L.), as a predator of the sycamore aphid, *Drepanosiphum platanoides* (Schr.). *Journal of Animal Ecology*, **39**, 739–51.

Dixon, A.F.G. and Barlow, N.D. (1979) Population regulation in the lime aphid. *Zoological Journal of the Linnean Society*, **67**, 225–37.

Dixon, T.J. (1959) Studies on oviposition behaviour of Syrphidae. *Transactions of the Royal Entomological Society of London*, **111**, 57–80.

Doane, J.F., Scotti, P.D., Sutherland, O.R.W. and Pottinger, R.P. (1985) Serological identification of wireworm and staphylinid predators of the Australian soldier fly (*Inopus rubriceps*) and wireworm feeding on plant and animal food, *Entomologia Experimentalis et Applicata*, **38**,65–72.

Domenichini, G. (1953) Studio sulla morfologia dell'addome degli Hymenoptera Chalcidoidea. *Bolletino di Zoologie Agraria e Bachicoltura*, **19**, 1–117.

Domenichini, G. (1967) Contributo alla conoscenza biologica e tassinomica dei Tetrastichinae paleartici (Hym., Eulophidae) con particolare riguardo ai materiali dell'Istituto di Entomologia dell'Università di Torino. *Bollettino di Zoologia Agraria e Bachicoltura*, **2** (8), 75–110.

Donaldson, J.S. and Walter, G.H. (1988) Effects of egg availability and egg maturity on the ovipositional activity of the parasitic wasp, *Coccophagus atratus*. *Physiological Entomology*, **13**, 407–17.

Dondale, C.D., Parent, B. and Pitre, D. (1979) A 6-year study of spiders (Araneae) in a Quebec apple orchard. *Canadian Entomologist*, **111**, 377–80.

Dondale, C.D., Redner, J.H., Farrell, E., Semple, R. B. and Turnbull, A.L. (1970) Wandering of hunting spiders in a meadow. *Bulletin du Musée Nationale d'Histoire Naturelle (Canada)*, **41**, 61–4.

Donisthorpe, H. (1936) The dancing habits of some Braconidae. *Entomologist's Record*, **78**, 84.

Dowell, R. (1978) Ovary structure and reproductive biologies of larval parasitoids of the alfalfa weevil (Coleoptera: Curculionidae). *Canadian Entomologist*, **110**, 507–12.

Drabble, E. and Drabble, H. (1917) The syrphid visitors to certain flowers, *New Phytologist*, **16**, 105–9.

Drabble, E. and Drabble, H. (1927) Some flowers and their dipteran visitors. *New Phytologist*, **26**, 115–23.

Dransfield, R.D. (1979) Aspects of host–parasitoid interactions of two aphid parasitoids, *Aphidius urticae* (Haliday) and *Aphidius uzbeckistanicus* (Luzhetski (Hymenoptera, Aphidiidae). *Ecological Entomology*, **4**, 307–16.

Dreisig, H. (1981) The rate of predation and its temperature dependence in a tiger beetle, *Cicindela hybrida*. *Oikos*, **36**, 196–202.

Dreisig, H. (1988) Foraging rate of ants collecting honeydew or extrafloral nectar, and some possible constraints. *Ecological Entomology*, **13**, 143–54.

van Driesche, R.G. (1983) Meaning of 'per cent parasitism' in studies of insect parasitoids. *Environmental Entomology*, **12**, 1611–22.

van Driesche, R.G. (1988) Field levels of encapsulation and superparasitism for *Cotesia glomerata* (L.) (Hymenoptera: Braconidae) in *Pieris rapae* (L.) (Lepidoptera: Pieridae). *Journal of the Kansas Entomological Society*, **61**, 328–31.

van Driesche, R.G. and Bellows Jr, T.S. (1988) Use of host and parasitoid recruitment in quantifying losses from parasitism in insect populations. *Ecological Entomology*, **13**, 215–22.

van Driesche, R.G. and Taub, G. (1983) Impact of parasitoids on *Phyllonorycter* leafminers infesting apple in Massachusetts, USA. *Protection Ecology*, **5**, 303–17.

van Driesche, R.G., Bellotti, A., Herrera, C.J. and Castillo, J.A. (1986) Encapsulation rates of two encyrtid parasitoids by two *Phenacoccus* spp. of cassava mealybugs in Colombia. *Entomologia Experimentalis et Applicata*, **42**, 79–82.

van Driesche, R.G., Bellows Jr, T.S., Elkinton, J.S., Gould, J.R. and Ferro, D.N. (1991) The meaning of percentage parasitism revisited: solutions to the problem of accurately estimating total losses from parasitism in a host generation. *Environmental Entomology*, **20**, 1–7.

Driessen, G. and Hemerik, L. (1992) The time and egg budget of *Leptopilina clavipes*, a parasitoid of larval *Drosophila*. *Ecological Entomology*, **17**, 17–27.

Driessen, G.J., Bernstein, C., van Alphen, J.J.M. and Kacelnik, A., (1994) A count-down mechanism for host search in the parasitoid *Venturia canescens*. *Journal of Animal Ecology*, **64**, 117–25.

Drost, Y.C. and Carde, R.T. (1992) Influence of host deprivation on egg load and oviposition behaviour of *Brachymeria intermedia*, a parasitoid of gypsy moth. *Physiological Entomology*, **17**, 230–34.

Dudgeon, D. (1990) Feeding by the aquatic heteropteran, *Diplonychus rusticum* (Belostomatidae): an effect of prey density on meal size. *Hydrobiologia*, **190**, 93–6.

Dudley, R. and Ellington, C.P. (1990) Mechanics of forward flight in bumblebees. I. Kinematics and morphology. *Journal of Experimental Biology*, **148**, 19–52.

Duelli, P. (1980) Adaptive dispersal and appetitive flight in the green lacewing, *Chrysopa carnea*. *Ecological Entomology*, **5**, 213–20.

Duelli, P., Studer, M., Marchand, I. and Jakob, S. (1990) Population movements of arthropods between natural and cultivated areas. *Biological Conservation*, **54**, 193–207.

East, R. (1974) Predation on the soil dwelling stages of the winter moth at Wytham Wood, Berkshire. *Journal of Animal Ecology*, **43**, 611–26.

Eberhardt, L.L. and Thomas, J.M. (1991) Designing environmental field studies. *Ecological Monographs*, **61**(1), 53–73.

Edgar, W.D. (1970) Prey and feeding behaviour of adult females of the wolf spider *Pardosa amentata* (Clerk.). *Netherlands Journal of Zoology*, **20**, 487–91.

Edwards, R.L. (1954) The effect of diet on egg maturation and resorption in *Mormoniella vitripennis* (Hymenoptera, Pteromalidae), *Quarterly Journal of Microscopical Science*, **95**, 459–68.

Edwards, C.A., Sunderland, K.D. and George, K.S. (1979) Studies on polyphagous predators of cereal aphids. *Journal of Applied Ecology*, **16**, 811–23.

Eggleton, P. (1991) Patterns in male mating strategies of the Rhyssini: a holophyletic group of parasitoid wasps (Hymenoptera: Ichneumonidae). *Animal Behaviour*, **41**, 829–38.

Ehler, L.E. and Hall, R.W. (1982) Evidence for competitive exclusion of introduced natural enemies in biological control, *Environmental Entomology*, **11**, 1–4.

Ehler, L.E., Eveleens, K.G. and van den Bosch, R. (1973) An evaluation of some natural enemies of cabbage looper on cotton in California, *Environmental Entomology*, **2**, 1009–15.

Ekbom, B.S. (1982) Diurnal activity patterns of the greenhouse whitefly, *Trialeurodes vaporariorum* (Homoptera: Aleyrodidae) and its parasitoid *Encarsia formosa* (Hymenoptera: Aphelinidae). *Protection Ecology*, **4**, 141–50.

Eller, F.J., Tumlinson, J.H. and Lewis, W.J. (1990) Intraspecific competition in *Microplitis croceipes* (Hymenoptera: Braconidae), a parasitoid of *Heliothis* species (Lepidoptera: Noctuidae). *Annals of the Entomological Society of America*, **83**, 504–8.

Ellington, C.P., Machin, K.E. and Casey, T.M. (1990) Oxygen consumption of bumblebees in forward flight. *Nature*, **347**, 472–3.

Ellington, J., Kiser, K., Ferguson, G. and Cardenas, M. (1984) A comparison of sweepnet, absolute, and Insectavac sampling methods in cotton ecosystems. *Journal of Economic Entomology*, **77**, 599–605.

Elliott, J.M. (1967) Invertebrate drift in a Dartmoor stream. *Archives of Hydrobiology*, **63,** 202–37.

Elliott, J.M. (1970) Methods of sampling invertebrate drift in running water. *Annals of Limnology*, **6**, 133–59.

Elton, R.A. and Greenwood, J.J.D. (1970) Exploring apostatic selection. *Heredity*, **25**, 629–33.

Elton, R.A. and Greenwood, J.J.D. (1987) Frequency-dependent selection by predators: comparison of parameter estimates. *Oikos*, **48**, 268–72.

Elzen, G.W., Williams, H.J. and Vinson, S.B. (1986) Wind tunnel flight responses by hymenopterous parasitoid *Campoletis sonorensis* to cotton cultivars and lines. *Entomologia Experimentalis et Applicata*, **43**, 285–9.

Elzen, G.W., Williams, H.J., Vinson, S.B. and Powell, J.E. (1987) Comparative flight behaviour of parasitoids *Campoletis sonorensis* and *Microplitis croceipes*. *Entomologia Experimentalis et Applicata*, **45**, 175–80.

Embree, D.G. (1966) The role of introduced parasites in the control of the winter moth in Nova Scotia. *Canadian Entomologist*, **98**, 1159–68.

Embree, D.G. (1971) The biological control of the winter moth in eastern Canada by introduced parasites, in *Biological Control* (ed. C.B. Huffaker), Plenum, New York, pp. 217–26.

van Emden, H.F. (1965) The role of uncultivated land in the biology of crop pests and beneficial insects. *Scientific Horticulture*, **17**, 121–36.

van Emden, H.F. (1988) The potential for managing indigenous natural enemies of aphids on field crops, *Philosophical Transactions of the Royal Society of London, B*, **318**, 183–201.

van Emden, H.F. and Hagen, K.S. (1976) Olfactory reactions of the green lacewing, *Chrysopa carnea* to tryptophan and certain breakdown products. *Environmental Entomology*, **5**, 469–473.

Emlen, J.M. (1966) The role of time and energy in food preference. *American Naturalist*, **100**, 611–17.

Engelmann, F. (1970) *The Physiology of Insect Reproduction*, Pergamon Press, Oxford.

Erdtmann, G. (1969) *Handbook of Palynology*, Munksgaard, Copenhagen.

Ericson, D, (1977) Estimating population parameters of *Pterostichus cupreus* and *P. melanarius* (Carabidae) in arable fields by means of capture–recapture. *Oikos*, **29**, 407–17.

Ericson, D. (1978) Distribution, activity and density of some Carabidae (Coleoptera) in winter wheat fields. *Pedobiologia*, **18**, 202–17.

Ericson, D. (1979) The interpretation of pitfall catches of *Pterostichus cupreus* and *P. melanarius* (Coleoptera, Carabidae) in cereal fields. *Pedobiologia*, **19**, 320–28.

Ernsting, G. and Huyer, F.A. (1984) A laboratory study on temperature relations of egg production and development in two related species of carabid beetle. *Oecologia*, **62**, 361–7.

Ernsting, G. and Isaaks, J.A. (1988) Reproduction, metabolic rate and survival in a carabid beetle. *Netherlands Journal of Zoology*, **38**, 46–60.

Escalante, G. and Rabinovich, J.E. (1979) Population dynamics of *Telenomus fariai* (Hymenoptera; Scelionidae), a parasite of Chagas' disease vectors, IX. Larval competition and population size regulation under laboratory conditions. *Researches on Population Biology*, **20**, 235–46.

Ettifouri, M. and Ferran, A. (1993) Influence of larval rearing diet on the intensive searching behaviour of *Harmonia axyridis* (Col.: Coccinellidae) larvae. *Entomophaga*, **38**, 51–9.

Eubank, W.P., Atmar, J.W. and Ellington, J.J. (1973) The significance and thermodynamics of fluctuating versus static thermal environments on *Heliothis zea* egg development rates. *Environmental Entomology*, **2**, 491–6.

Evans, D.A., Miller, B.R. and Bartlett, C.B. (1973) Host searching range of *Dasymutilla nigripes* (Fabricius) as investigated by tagging (Hymenoptera: Mutillidae). *Journal of the Kansas Entomological Society*, **46,** 343–6.

Evans, E. (1993) Indirect interactions among phytophagous insects: aphids, honeydew and natural enemies, in *Individuals, Populations and Patterns in Ecology* (eds A.D. Watt, S. Leather, N.J. Mills and K.F.A. Walters) *Intercept*, Andover, pp. 287–298.

Evans, E.W. and Dixon, A.F.G. (1986) Cues for oviposition by ladybird beetles (Coccinellidae); response to aphids. *Journal of Animal Ecology*, **55**, 1027–34.

Evans, H.F. (1976) Mutual interference between predatory arthropods. *Ecological Entomology*, **1**, 283–6.

Eveleens, K.G., van den Bosch, R. and Ehler, L.E. (1973) Secondary outbreak induction of beet armyworm by experimental insecticide applications in cotton in California. *Environmental Entomology*, **2**, 497–503.

Ewing, A.W. (1983) Functional aspects of *Drosophila* courtship. *Biological Reviews*, **58**, 275–92.

Eyre, M.D. and Luff, M.L. (1990) A preliminary classification of European grassland habitats using carabid beetles, in *The Role of Ground Beetles in Ecological and Environmental Studies* (ed. N.E. Stork), Intercept, Andover, pp. 227–36.

Eyre, M.D., Luff, M.L. and Rushton, S.P. (1990) The ground beetle (Coleoptera, Carabidae) fauna of intensively managed agricultural grasslands in northern England and southern Scotland. *Pedobiologia*, **34**, 11–18.

Fabre, J.H. *Souvenirs Entomologiques. Études sur l'Instinct et les Moeurs des Insectes, 5ieme Series*, Delagrave, Paris.

Faegri, K. and Iversen, J. (1989) *A Textbook of Pollen Analysis*, 4th edn (revised by K. Faegri, P.E. Kaland and K. Kryzwinski), John Wiley and Sons, Chichester.

Farrar, R.R., Barbour, J.D. and Kennedy, G.G. (1989) Quantifying food consumption and growth in insects. *Annals of the Entomological Society of America*, **82**, 593–98.

Fedorenko, A.Y. (1975) Instar and species-specific diets in two species of *Chaoborus*. *Limnology and Oceanography*, **20**(2), 238–42.

Feinsinger, P. and Swarm, L.A. (1978) How common are ant-repellant nectars? *Biotropica*, **10**, 238–9.

Feng, J.G., Zhang, Y, Tao, X. and Chen, X.L. (1988) Use of radioisotope ^{32}P to evaluate the parasitization of *Adoxophyes orana* (Lep.: Tortricidae) by mass released *Trichogramma dendrolimi* (Hym.: Trichogrammatidae) in an apple orchard. *Chinese Journal of Biological Control*, **4**, 152–4.

Feng, M.G. and Nowierski, R.M. (1992) Spatial patterns and sampling plans for cereal aphids (Hom.: Aphididae) killed by entomophthoralean fungi and hymenopterous parasitoids in spring wheat. *Entomophaga*, **37**, 265–75.

Fenlon, J.S. and Sopp, P.I. (1991) Some statistical considerations in the determination of thresholds in ELISA. *Annals of Applied Biology*, **119**, 177–89.

Ferguson, A. (1980) *Biochemical Systematics and Evolution*, Blackie, London.

Ferran, A., Buscarlet, A. and Larroque, M.M. (1981) The use of $HT^{18}O$ for measuring the food consumption in aged larvae of the aphidophagous ladybeetle, *Semiadalia 11notata* (Col: Coccinellidae), *Entomophaga*, **26**, 71–7 (in French).

Fichter, B.L. and Stephen, W.P. (1979) Selection and use of host-specific antigens. *Entomological Society of America, Miscellaneous Publication*, **11**, 25–33.

Finch, S. and Coaker, T.H. (1969) Comparison of the nutritive values of carbohydrates and related compounds to *Erioischia brassicae*. *Entomologia Experimentalis et Applicata*, **12**, 441–53.

Fincke, O.M. (1984) Sperm competition in the damselfly *Enallagma hageni* Walsh (Odon., Coenagrionidae): Benefits of multiple mating to males and females. *Behavioural Ecology and Sociobiology*, **14**, 235–40.

Finlayson, T. (1967) A classification of the subfamily Pimplinae (Hymenoptera: Ichneumonidae) based on final-instar larval characteristics. *Canadian Entomologist*, **99**, 1–8.

Finlayson, T. (1990) The systematics and taxonomy of final-instar larvae of the family Aphidiidae (Hymenoptera). *Memoirs of the Entomological Society of Canada*, **152**, 1–74.

Finlayson, T. and Hagen, K.S. (1977) Final-instar larvae of parasitic Hymenoptera. *Pest Management Papers*, **10**, Simon Fraser University, 1–111.

Finnerty, V. and Johnson, G. (1979) Post-translational modification as a potential explanation of high levels of enzyme polymorphism: xanthine dehydrogenase and aldehyde oxidase in *Drosophila melanogaster*. *Genetics*, **91**, 695–722.

Fisher, R.A. (1930) *The Genetical Theory of Natural Selection*, Oxford University Press, Oxford.

Fisher, R.A. and Ford, E.B. (1947) The spread of a gene in natural conditions in a colony of the moth *Panaxia dominula* L.. *Heredity, London*, **1**, 143–74.

Fisher, R.C. (1961) A study in insect multiparasitism. II. The mechanism and control of competition for the host. *Journal of Experimental Biology*, **38**, 605–28.

Fisher, R.C. (1971) Aspects of the physiology of endoparasitic Hymenoptera. *Biological Reviews*, **46**, 243–78.

Flanders, S.E. (1934) The secretion of the colleterial glands in the parasitic chalcids. *Journal of Economic Entomology*, **27**, 861–2.

Flanders, S.E. (1942) Oösorption and ovulation in relation to oviposition in the parasitic Hymenoptera. *Annals of the Entomological Society of America*, **35**, 251–66.

Flanders, S.E. (1950) Regulation of ovulation and egg disposal in the parasitic Hymenoptera. *Canadian Entomologist*, **82**: 134–40.

Flanders, S.E. (1953) Predatism by the adult hymenopterous parasite and its role in biological control. *Journal of Economic Entomology*, **46**: 541–4.

Flanders, S.E. (1962) The parasitic Hymenoptera: specialists in population regulation, *Canadian Entomologist*, **94**, 1133–47.

Fleischer, S.J., Gaylor, M.J. and Edelson, J.V. (1985) Estimating absolute density from relative sampling of *Lygus lineolaris* (Heteroptera: Miridae) and selected predators in early to mid-season cotton. *Environmental Entomology*, **14**, 709–17.

Fleischer, S.J., Gaylor, M.J., Hue, N.V. and Graham, L.C. (1986) Uptake and elimination of rubidium, a physiological marker, in adult *Lygus lineolaris* (Hemiptera: Miridae). *Annals of the Entomological Society of America*, **79**, 19–25.

Fleming, J.-A.G.W. (1992) Polydnaviruses: mutualists and pathogens. *Annual Review of Entomology*, 37, 401–25.

Fletcher, B.S. and Kapatos, E.T. (1983) An evaluation of different temperature-development rate models for predicting the phenology of the olive fly *Dacus oleae*, in *Fruit Flies of Economic Importance* (ed. R. Cavalloro), CEC/IOBC Symposium, Athens, November, 1982, Balkema, Rotterdam, pp. 321–30.

Fletcher, B.S., Kapatos, E. and Southwood, T.R.E. (1981) A modification of the Lincoln Index for estimating the population densities of mobile insects. *Ecological Entomology*, **6**, 397–400.

Flint, H.M., Merkle, J.R. and Sledge, M. (1981) Attraction of male *Collops vittatus* in the field by caryophyllene alcohol. *Environmental Entomology* **10**, 301–4.

Flint, H.M., Salter, S.S. and Walters, S. (1979) Caryophylene: an attractant for the green lacewing. *Environmental Entomology*, **8**, 1123–5.

Folsom, T.C., and Collins, N.C. (1984) The diet and foraging behavior of the larval dragonfly *Anax junius* (Aeshnidae), with an assessment of the role of refuges and prey activity. *Oikos*, **42**, 105–13.

Force, D.C. and Messenger, P.S. (1964) Fecundity, reproductive rates, and innate capacity for increase in three parasites of *Therioaphis maculata* (Buckton). *Ecology*, **45**, 706–15.

Formanowicz Jr., D.R. (1982) Foraging tactics of larvae of *Dytiscus verticalis* (Coleoptera, Dytiscidae): the assessment of prey density. *Journal of Animal Ecology*, **51**, 757–67.

Forsythe, T.G. (1987) *Common Ground Beetles, Naturalists' Handbooks No. 8*, Richmond, Slough.

Fowler, H.G. (1989) Optimization of nectar foraging in a solitary wasp (Hymenoptera: Sphecidae: *Larra* sp.). *Naturalia*, **14**, 13–17.

Fowler, H.G. (1987) Field behaviour of *Euphasiopteryx depleta* (Diptera: Tachinidae): Phonotactically orienting parasitoids of mole crickets (Orthoptera: Gryllotalpidae: *Scapteriscus*). *Journal of the New York Entomological Society*, **95**, 474–80.

Fowler, H.G. and Kochalka, J.N. (1985) New records of *Euphasiopteryx depleta* (Diptera: Tachinidae) from Paraguay: attraction to broadcast calls of *Scapteriscus acletus* (Orthoptera: Gryllotalpidae). *Florida Entomologist*, **68**, 225–6.

Fowler, S.V. (1988) Field studies on the impact of natural enemies on brown planthopper populations on rice in Sri Lanka. *Proceedings of the 6th Auchenorrhyncha Meeting, Turin, Italy, 7–11 September, 1987*, 567–74.

Fowler, S.V., Claridge, M.F., Morgan, J.C., Peries, I.D.R. and Nugaliyadde, L. (1991) Egg mortality of the brown planthopper, *Nilaparvata lugens* (Homoptera: Delphacidae) and green leafhoppers, *Nephotettix* spp. (Homoptera: Cicadellidae), on rice in Sri Lanka. *Bulletin of Entomological Research*, **81**, 161–7.

Fox, L.R. and Murdoch, W.W. (1978) Effects of feeding history on short term and long term functional responses in *Notonecta hoffmani*. *Journal of Animal Ecology*, **47**, 495–960.

Fox, P.M., Pass, B.C. and Thurston, R. (1967) Laboratory studies on the rearing of *Aphidius smithi* (Hymenoptera: Braconidae) and its parasitism of *Acyrthosiphon pisum* (Homoptera: Aphididae). *Annals of the Entomological Society of America*, **60**, 1083–7.

Frank, J.H. (1967a) The insect predators of the pupal stage of the winter moth, *Operophtera brumata* (L.) (Lepidoptera: Hydriomenidae). *Journal of Animal Ecology*, **36**, 375–89.

Frank, J.H. (1967b) The effect of pupal predators on a population of winter moth, *Operophtera brumata* (L.) (Lepidoptera: Hydriomenidae). *Journal of Animal Ecology*, **36**, 611–21.

Frank, J.H. (1979) The use of the precipitin technique in predator–prey studies to 1975. *Entomological Society of America, Miscellaneous Publication*, **11**, 1–15.

Frank, S.A. (1984) The behavior and morphology of the fig wasps *Pegoscapus assuetus* and *P. jimenezi*: descriptions and suggested behavioral characters for phylogenetic studies. *Psyche*, **91**, 289–308.

Frank, S.A. (1985) Are mating and mate competition by the fig wasp *Pegoscapus assuetus* (Agaonidae) random within a fig? *Biotropica*, **17**, 170–72.

Frazer, B.D. (1988) Coccinellidae, in *World Crop Pests. Aphids, their Biology, Natural Enemies and Control, Vol. 2B* (eds A.K. Minks and P. Harrewijn), Elsevier, Amsterdam, pp. 231–48.

Frazer, B.D. and Gilbert, N. (1976) Coccinellids and aphids: a quantitative study of the impact of adult ladybirds (Coleoptera: Coccinellidae) preying on field populations of pea aphids (Homoptera: Aphididae). *Journal of the Entomological Society of British Columbia*, **73**, 33–56.

Frazer, B.D. and Gill, B. (1981) Hunger, movement and predation of *Coccinella californica* on pea aphids in the laboratory and in the field. *Canadian Entomologist*, 1025–33.

Frazer, B.D. and McGregor, R.R. (1992) Temperature-dependent survival and hatching rate of eggs of seven species of Coccinellidae. *Canadian Entomologist*, **124**, 305–12.

Frazer, B.D. and Raworth, D.A. (1985) Sampling for adult coccinellids and their numerical response to strawberry aphids (Coleoptera: Coccinellidae: Homoptera: Aphididae). *Canadian Entomologist*, **117**, 153–161.

Frazer, B.D., Gilbert, N., Ives, P.M. and Raworth, D.A. (1981a) Predator reproduction and the overall predator–prey relationship. *Canadian Entomologist*, **113**, 1015–24.

Frazer, B.D., Gilbert, N., Nealis, V. and Raworth, D.A. (1981b) Control of aphid density by a complex of predators, *Canadian Entomologist*, **113**, 1035–41.

Free, C.A., Beddington, J.R. and Lawton, J.H. (1977) On the inadequacy of simple models of mutual interference for parasitism and predation. *Journal of Animal Ecology*, **46**, 543–4.

Fuester, R.W., Drea, J.J., Gruber, F., Hoyer, H. and Mercardier, G. (1983) Larval parasites and other natural enemies of *Lymantria dispar* (Lepidoptera, Lymantriidae) in Burgenland, Austria, and Wurzburg, Germany. *Environmental Entomology*, **12**, 724–37.

Fye, R.E. (1965) Methods for placing wasp trapnests in elevated locations. *Journal of Economic Entomology*, **58**, 803–4.

Galis, F. and van Alphen, J.J.M. (1981) Patch time allocation and search intensity of *Asobara tabida* Nees (Hym.: Braconidae). *Journal of Animal Ecology*, **31**, 701–12.

Gaston, K.J. (1988) The intrinsic rates of increase of insects of different sizes. *Ecological Entomology*, **14**, 399–409.

Gauld, I. and Bolton, B. (1988) *The Hymenoptera*, Oxford University Press, Oxford.

Gauld, I.D. and Huddleston, T. (1976) The nocturnal Ichneumonidae of the British Isles, including a key to genera. *Entomologist's Gazette*, **27**, 35–49.

van der Geest, L.P.S. and Overmeer, W.P.J. (1985) Experiences with polyacrylamide gradient gel electrophoresis for the detection of gut contents of phytoseiid mites. *Mededelingen van de Faculteit voor Landbouwwetenschappen Rijksuniversiteit Gent*, **50**, 469–71.

Gehan, E.A. and Siddiqui, M.M. (1973) Simple regression methods for survival time studies. *Journal of American Statistical Association*, **68**, 848–56.

Gerking, S.D. (1957) A method of sampling the littoral macrofauna and its application. *Ecology*, **38**, 219–26.

Getz, W.M. and Gutierrez, A.P. (1982) A perspective on systems analysis in crop production and insect pest management. *Annual Review of Entomology*, **27**, 447–66.

Geusen-Pfister, H. (1987) Studies on the biology and reproductive capacity of *Episyrphus balteatus* Deg. (Dipt., Syrphidae) under greenhouse conditions. *Journal of Applied Entomology*, **104**, 261–70.

Gilbert, F.S. (1980) Flower visiting by hoverflies, *Journal of Biological Education*, **14**, 70–4.

Gilbert, F.S. (1981) Foraging ecology of hoverfies: morphology of the mouthparts in relation to feeding on nectar and pollen in some common urban species. *Ecological Entomology*, **6**, 245–62.

Gilbert, F.S. (1983) The foraging ecology of hoverflies (Diptera: Syrphidae): circular movements on composite flowers. *Behavioural Ecology and Sociobiology*, **13**, 253–7.

Gilbert, F.S. (1985a) Diurnal activity patterns in hoverflies (Diptera, Syrphidae). *Ecological Entomology*, **10**, 385–92.

Gilbert, F.S. (1985b) Ecomorphological relationships in hoverflies (Diptera, Syrphidae), *Proceedings of the Royal Society of London*, B, **224**, 91–105.

Gilbert, F.S. (1986) *Hoverflies, Naturalists' Handbooks No. 5*, Cambridge University Press, Cambridge.

Gilbert, F.S. (1990) Size, phylogeny and life-history in the evolution of feeding specialisation in insect predators, in *Insect Life Cycles: Genetics, Evolution and Co-ordination* (ed. F.S. Gilbert), Springer-Verlag, London, pp. 101–24.

Gilbert, F.S. (1991) Feeding in adult hoverflies. *Hoverfly Newsletter*, **13**, 5–11.

Gilbert, F.S. and Owen, J. (1990) Size, shape, competition, and community structure in hoverflies (Diptera: Syrphidae). *Journal of Animal Ecology*, **59**, 21–39.

Gilbert, N. (1984) Control of fecundity in *Pieris rapae*. II. Differential effects of temperature. *Journal of Animal Ecology*, **53**, 589–97.

Gilbert, N., Gutierrez, A.P., Frazer, B.D. and Jones, R.E. (1976) *Ecological Relationships*, W.H. Freeman and Co., Reading.

Giller, P.S. (1980) The control of handling time and its effects on the foraging strategy of a heteropteran predator, *Notonecta*. *Journal of Animal Ecology*, **49**, 699–712.

Giller, P.S. (1982) The natural diet of waterbugs (Hemiptera: Heteroptera):electrophoresis as a potential method of analysis. *Ecological Entomology*, **7**, 233–7.

Giller, P.S. (1984) Predator gut state and prey detectability using electrophoretic analysis of gut contents. *Ecological Entomology*, **9**, 157–62.

Giller, P.S. (1986) The natural diet of the Notonectidae: field trials using electrophoresis. *Ecological Entomology*, **11**, 163–72.

Gillespie, D.R. and Finlayson, T. (1983) Classification of final-instar larvae of the Ichneumoninae (Hymenoptera: Ichneumonidae). *Memoirs of the Entomological Society of Canada*, **124**, 81pp.

Gilpin, M. and Hanski, I. (1991) *Metapopulation Dynamics – Empirical and Theoretical Investigations*, Academic Press, London.

Glas, P.C.G. and Vet, L.E.M. (1983) Host habitat location by *Diachasma alloeum* Muesebeck (Hym.; Braconidae), a parasitoid of *Rhagoletis pomonella* Welsh (Dipt. Tephritidae). *Journal of Animal Ecology*, **33**, 41–54.

Glen, D.M. (1973) The food requirements of *Blepharidopterus angulatus* (Heteroptera: Miridae) as a predator of the lime aphid, *Eucallipterus tiliae*. *Entomologia Experimentalis et Applicata*, **16**, 255–67.

Glen, D.M. (1975) Searching behaviour and prey-density requirements of *Blepharidopterus angulatus* (Fall.) (Heteroptera: Miridae) as a predator of the lime aphid, *Eucallipterus tiliae* (L.), and leafhopper, *Alnetoidea alneti* (Dahlbom). *Journal of Animal Ecology*, **44**, 116–35.

Godfray, H.C.J. (1987a) The evolution of clutch size in invertebrates. *Oxford Surveys in Evolutionary Biology*, **4** ,117–54.

Godfray, H.C.J. (1987b) The evolution of clutch size in parasitic wasps. *American Naturalist*, **129**, 221–33.

Godfray, H.C.J. and Hassell, M.P. (1987) Natural enemies may be a cause of discrete generations in tropical systems. *Nature*, **327**, 144–7.

Godfray, H.C.J. and Hassell, M.P. (1988) The population biology of insect parasitoids. *Science Progress, Oxford*, **72**, 531–48.

Godfray, H.C.J. and Hassell, M.P. (1989) Discrete and continuous insect populations in tropical environments. *Journal of Animal Ecology*, **58**, 153–74.

Godfray, H.C.J. and Hassell, M.P. (1991) Encapsulation and host–parasitoid population biology, in *Parasite–Host Associations* (eds C.A. Toft, A. Aeschlimann and L. Bolis), Oxford University Press, Oxford, pp. 131–47.

Godfray, H.C.J. and Ives, A.R. (1988) Stochasticity in invertebrate clutch-size models. *Theoretical Population Biology*, **33**, 79–101.

Godfray, H.C. and Pacala, S.W. (1992) Aggregation and the population dynamics of parasitoids and predators, *American Naturalist*, **140**, 30–40.

Godfray, H.C.J. and Waage, J.K. (1991) Predictive modelling in biological control: the Mango Mealy bug (*Rastrococcus invadens*) and its parasitoids. *Journal of Applied Ecology*, **28**, 434–53.

Godfray, H.C.J., Hassell, M.P. and Holt, R.D. (1994) The population dynamic consequences of phenological asynchrony between parasitoids and their hosts. *Journal of Animal Ecology*, **63**, 1–10.

Goff, A.M. and Nault, L.R. (1984) Response of the pea aphid parasite *Aphidius ervi* Haliday (Hymenoptera: Aphidiidae) to transmitted light. *Environmental Entomology*, **13**, 595–8.

Goldson, S.L. and McNeill, M.R. (1992) Variation in the critical photoperiod for diapause induction in *Microctonus hyperodae*, a parasitoid of Argentine stem weevil. *Proceedings of the 45th New Zealand Plant Protection Conference, 1992*, pp. 205–9.

Goldson, S.L., McNeill, M.R., Phillips, C.B. and Proffitt, J.R. (1992) Host specificity testing and suitability of the parasitoid *Microctonus hyperodae* (Hym: Braconidae, Euphorinae) as a biological control agent of *Listronotus bonariensis* (Col.: Curculionidae) in New Zealand. *Entomophaga*, **37**, 483–98.

Gonzalez, D., Gordh, G., Thompson, S.N. and Adler, J. (1979) Biotype discrimination and its importance to biological control, in *Genetics in Relation to Insect Management* (eds M.A. Hoy and J.J. McKelvey), The Rockefeller Foundation, New York, pp. 129–39.

Gonzalez, J.M., Matthews, R.W. and Matthews, J.R. (1985) A sex pheromone in males of *Melittobia australica* and *M. femorata* (Hym., Eulophidae). *Florida Entomologist*, **68**, 279–86.

Goodenough, J.L., Hartsack, A.W. and King, E.G. (1983) Developmental models for *Trichogramma pretiosum* (Hymenoptera: Trichogrammatidae) reared on four hosts. *Journal of Economic Entomology*, **76**, 1095–1102.

Goodpasture, C. (1975) Comparative courtship and karyology in *Monodontomerus* (Hym., Torymidae). *Annals of the Entomological Society of America*, **68**, 391–7.

van der Goot, V.S. and Grabandt, R.A.J. (1970) Some species of the genera *Melanostoma, Platycheirus and Pyrophaena* (Diptera, Syrphidae) and their relation to flowers *Entomologische Berichten*, **30**, 135–43.

Gordh, G. and DeBach, P. (1976) Male inseminative potential in *Aphytis lingnanensis* (Hym., Aphelinidae). *Canadian Entomologist*, **108**, 583–9.

Gordh, G. and DeBach, P. (1978) Courtship behaviour in the *Aphytis lingnanensis* group, its potential usefulness in taxonomy, and a review of sexual behaviour in the parasitic Hymenoptera (Chalc., Aphelinidae). *Hilgardia*, **46**, 37–75.

Gordon, A.H. (1975) Electrophoresis of proteins in polyacrylamide and starch gels, in *Laboratory Techniques in Biochemistry and Molecular Biology* (eds T.S. Work and E. Work), North-Holland, Amsterdam, London.

Gordon, D.M., Nisbet, R.M., de Roos, A., Gurney, W.S.L. and Stewart, R.K. (1991) Discrete generations in host–parasitoid models with contrasting life cycles. *Journal of Animal Ecology*, **60**, 295–308.

Gould, J.R., Bellows Jr, T.S. and Paine, T.D. (1992a) Population dynamics of *Siphoninus phillyreae* in California in the presence and absence of a parasitoid, *Encarsia partenopea*. *Ecological Entomology*, **17**, 127–34.

Gould, J.R., Bellows, T.S. and Paine, T.D. (1992b) Evaluation of biological control of *Siphoninus phillyreae* (Haliday) by the parasitoid *Encarsia partenopea* (Walker) using life-table analysis. *Biological Control*, **2**, 257–265.

Gould, J.R., Elkinton, J.S. and Wallner, W.E. (1990) Density dependent suppression of experimentally created gypsy moth, *Lymantria dispar* (Lepidoptera: Lymantriidae), populations by natural enemies. *Journal of Animal Ecology*, **59**, 213–33.

Graham, M.W.R. de V. (1989) A remarkable secondary sexual character in the legs of male

Nesolynx glossinae (Waterston). *Entomologist's Monthly Magazine,* **125**, 231–32.

Graham, M.W.R. de V. (1993) Swarming in Chalcidoidea (Hym.) with the description of a new species of *Torymus* (Hym., Torymidae) involved. *Entomologist's Monthly Magazine,* **129**, 15–22.

Grasswitz, T.R. and Paine, T.D. (1993) Effect of experience on in-flight orientation to host-associated cues in the generalist parasitoid *Lysiphlebus testaceipes*. *Entomologia Experimentalis et Applicata*, **68**, 219–29.

Gravena, S. and Sterling, W.L. (1983) Natural predation on the cotton leafworm (Lepidoptera: Noctuidae). *Journal of Economic Entomology*, **76**, 779–84.

Greany, P.D. and Oatman, E.R. (1972) Analysis of host discrimination in the parasite *Orgilus lepidus* (Hymenoptera: Braconidae). *Annals of the Entomological Society of America*, **65**, 377–83.

Greany, P.D., Hawke, S.D., Carlysle, T.C. and Anthony, D.W. (1977) Sense organs in the ovipositor of *Biosteres (Opius) longicaudatus*, a parasite of the Caribbean fruit fly *Anastrepha suspensa*. *Annals of the Entomological Society of America*, **70**, 319–21.

Greathead, D.J. (1986) Parasitoids in classical biological control, in *Insect Parasitoids* (eds J. Waage and D. Greathead), Academic Press, London, pp. 289–318.

Greathead, D.J. and Waage, J.K. (1983) Opportunities for biological control of agricultural pests in developing countries. *World Bank Technical Paper 11*, 44 pp.

Greenblatt, J.A., Barbosa, P. and Montgomery, M.E. (1982) Host's diet effects on nitrogen utilization efficiency for two parasitoid species: *Brachymeria intermedia* and *Coccygomimus turionellae*. *Physiological Entomology*, **7**, 263–7.

Greenfield, M.D. (1981) Moth sex pheromones: an evolutionary perspective. *Florida Entomologist*, **64**, 4–17.

Greenfield, M.D. and Karandinos, M.G. (1976) Fecundity and longevity of *Synanthedon pictipes* under constant and fluctuating temperatures. *Environmental Entomology*, **5**, 883–7.

Greenstone, M.H. (1977) A passive haemagglutination inhibition assay for the identification of stomach contents of invertebrate predators. *Journal of Applied Ecology*, **14**, 457–64.

Greenstone, M.H. (1979a) A line transect density index for wolf spiders (*Pardosa* spp.) and a note on the applicability of catch per unit effort methods to entomological studies. *Ecological Entomology*, **4**, 23–9.

Greenstone, M.H. (1979b) Passive haemagglutination inhibition: a powerful new tool for field studies of entomophagous predators. *Entomological Society of America, Miscellaneous Publication*, **11**, 69–78.

Greenstone, M.H. (1983) Site-specificity and site tenacity in a wolf-spider: a serological dietary analysis. *Oecologia*, **56**, 79–83.

Greenstone, M.H. (1995) Serological analysis of predation: past, present and future, in *The Ecology of Insect Pests: Biochemical Approaches* (eds W.O.C. Symondson and J.E. Liddell), Chapman and Hall, London. (in press).

Greenstone, M.H. and Morgan, C.E. (1989) Predation on *Heliothis zea* (Lepidoptera, Noctuidae) – an instar-specific ELISA assay for stomach analysis. *Annals of the Entomological Society of America*, **82**, 45–9.

Greenwood, J.J.D. and Elton, R.A. (1979) Analysing experiments on frequency dependent selection by predators. *Journal of Animal Ecology*, **48**, 721–37.

Gresens, S.E., Cothran, M.L. and Thorp, J.H. (1982) The influence of temperature on the functional response of the dragonfly *Celithemis fasciata* (Odonata: Libellulidae). *Oecologia*, **53**, 281–4.

Gressitt, J.L. and Gressitt, M.K. (1962) An improved Malaise trap. *Pacific Insects*, **4**, 87–90.

Gribbin, S.D. and Thompson, D.J. (1990) Asymmetric intraspecific competition among larvae of the damselfly *Ischnura elegans* (Zygoptera: Coenagrionidae). *Ecological Entomology*, **15**, 37–42.

Griffith, D.M. and Poulson, T.L. (1993) Mechanisms and consequences of intraspecific competition in a carabid cave beetle. *Ecology*, **74**, 1373–83.

Griffiths, D. (1980) The feeding biology of ant-lion larvae: growth and survival in *Morter obscurus*. *Oikos*, **34**, 364–70.

Griffiths, D. (1982) Tests of alternative models of prey consumption by predators, using ant-lion larvae. *Journal of Animal Ecology*, **52**, 363–73.

Griffiths, D. (1992) Interference competition in ant-lion (*Macroleon quinquemaculatus*) larvae. *Ecological Entomology*, **17**, 219–26.

Griffiths, E., Wratten, S.D. and Vickerman, G. P. (1985) Foraging behaviour by the carabid *Agonum dorsale* in the field. *Ecological Entomology*, **10**, 181–9.

Grissell, E.E. and Goodpasture, C.E. (1981) A review of nearctic Podagrionini, with description of sexual behavior of Podagrion mantis (Hym., Torymidae). *Annals of the Entomological Society of America*, **74**, 226–41.

Grissell, E.E. and Schauff, M.E. (1990) *A Handbook of the Families of Nearctic Chalcidoidea (Hymenoptera)*, The Entomological Society of Washington, Washington.

Grosch, D.S. (1945) The relation of cell size and organ size to mortality in *Habrobracon*. *Growth*, **9**, 1–17.

Grosch, D.S. (1950) Starvation studies with the parasitic wasp *Habrobracon*. *Biological Bulletin, Marine Biological Laboratory, Wood's Hole*, **99**, 65–73.

Gross, H.R., Lewis, W.J., Jones, R.L. and Nordlund, D.A. (1975) Kairomones and their use for management of entomophagous insects. III. Stimulation of *Trichogramma achaeae, T. pretiosum* and *Microplitis croceipes* with host seeking stimuli at the time of release to improve efficiency. *Journal of Chemical Ecology*, **1**, 431–8.

Gross, P. (1993) Insect behavioural and morphological defences against parasitoids. *Annual Review of Entomology*, **38**, 251–73.

Guilford, T. and Dawkins, M.S. (1987) Search images not proven: a reappraisal of recent evidence. *Animal Behaviour*, **35**, 1838–45.

Gülel, A. (1988) Effects of mating on the longevity of males and sex ratio of *Dibrachys boarmiae* (Hym., Pteromalidae). *Doga Tu Zooloji*, **12**, 225–30 (In Turkish, with English summary).

Gurney, W.S.C. and Nisbet, R.M. (1985) Fluctuation periodicity, generation separation, and the expression of larval competition. *Theoretical Population Biology*, **28**, 886–923.

Gutierrez, A.P. (1970) Studies on host selection and host specificity of the aphid hyperparasite *Charips victrix* (Hymenoptera: Cynipidae). 6. Description of sensory structures and a synopsis of host selection and host specificity. *Annals of the Entomological Society of America*, **63**, 1705–9.

Gutierrez, A.P., Baumgaertner, J.U. and Hagen, K.S. (1981) A conceptual model for growth, development and reproduction in the ladybird beetle, *Hippodamia convergens* (Coleoptera: Coccinellidae). *Canadian Entomologist*, **113**, 21–33.

Gutierrez, A.P., Hagen, K.S. and Ellis, C.K. (1990) Evaluating the impact of natural enemies: a multitrophic perspective, in *Critical Issues in Biological Control* (ed. M. Mackauer L.E. Ehler and J. Roland), Intercept, Andover, pp. 81–107.

Gutierrez, A.P., Neuenschwander, P. and van Alphen, J.J.M. (1994) Factors affecting biological control of cassava mealybug by exotic parasitoids: a ratio-dependent supply–demand driven model, *Journal of Applied Ecology*, **30**, 706–21.

Gutierrez, A.P. Neuenschwander, P., Schulthess, F. Herren, H.R., Baumgartner, J.U., Wermelinger, B., Lohr, B. and Ellis, C.K. (1988a) Analysis of biological control of cassava pests in Africa. II. Cassava mealybug *Phenacoccus manihoti*. *Journal of Applied Ecology*, **25**, 921–40.

Gutierrez, A.P., Yaninek, J.S., Wermelinger, B. *et al.* (1988b) Analysis of biological control of cassava pests in Africa. III. Cassava green mite *Mononychellus tanajoa*. *Journal of Applied Ecology*, **25**, 941–51.

Guzo, D. and Stoltz, D.B. (1985) Obligatory multiparasitism in the tussock moth *Orgyia leucostigma*. *Parasitology*, **90**, 1–10.

Györfi, J. (1945) Beobachtungen über die ernahrung der schlupfwespenimagos. *Erdészeti Kisérletek*, **45**: 100–112.

Haber, W.A., Frankie, G.W., Baker, H.G. *et al.* (1981) Ants like flower nectar. *Biotropica*, **13**, 211–14.

Haccou and Meelis (1992) *Statistical Analysis of Behavioural Data, an Approach Based on Time-structured Models*, Oxford University Press, Oxford.

Haccou, P., de Vlas, S.J., van Alphen, J.J.M. and Visser, M.E. (1991) Information processing by foragers: effects of intra-patch experience on the leaving tendency of *Leptopilina heterotoma*. *Journal of Animal Ecology*, **60**, 93–106.

Haeck, J. (1971), The immigration and settlement of carabids in the new Ijsselmeer–polders. *Miscellaneous Papers Landbonwwetenschappen Hogeschule Wageningen*, **8**, 33–52

Hag Ahmed, S. E. M. K. (1989) Biological control of glasshouse *Myzus persicae* (Sulzer) using *Aphidius matricariae* Haliday. Wye College, University of London, PhD Thesis,.

Hagen, K. (1964a) Developmental stages of parasites, in *Biological Control of Insect Pests and Weeds* (ed. P. DeBach), Chapman & Hall, London, pp. 168–246.

Hagen, K. (1964b) Nutrition of entomophagous insects and their hosts, in *Biological Control of Insect Pests and Weeds* (ed. P. DeBach), Chapman & Hall, London, pp. 356–80.

Hagen, K.S. (1986) Ecosystem analysis: plant cultivars (HPR), entomophagous species and food supplements, in *Interactions of Plant Resistance and Parasitoids and Predators of Insects* (eds D.J. Boethel and R.D.Eikenbary), Ellis Horwood, Chichester/ John Wiley & Sons, New York, pp. 151–97.

Hagler, J.R., Cohen, A.C., Enriquez, F.J. and Bradley-Dunlop, D. (1991) An egg-specific monoclonal antibody to *Lygus hesperus*. *Biological Control*, **1**, 75–80.

Hagstrum, D.W. and Smittle, B.J. (1977) Host-finding ability of *Bracon hebetor* and its influence upon adult parasite survival and fecundity. *Environmental Entomology*, **6**, 437–9.

Hagstrum, D.W. and Smittle, B.J. (1978) Host utilization by *Bracon hebetor*. *Environmental Entomology*, **7**, 596–600.

Hågvar, E.B. (1972) The effect of intra- and interspecific larval competition for food (*Myzus persicae*) on the development at 20° of *Syrphus ribesii* and *Syrphus corollae* (Diptera: Syrphidae). *Entomophaga*, **17**, 71–7.

Hågvar, E.B. (1973) Food consumption in larvae of *Syrphus ribesii* (L.) and *Syrphus corollae* (Fabr.) (Dipt., Syrphidae). *Norsk Entomologisk Tiddskrift*, **20**, 315–21.

Hågvar, E.B. and Hofsvang, T. (1991) Effect of honeydew on the searching behaviour of the aphid parasitoid *Ephedrus cerasicola* (Hymenoptera, Aphidiidae). *Redia*, **74**, 259–64.

Hails, R.S. and Crawley, M.J. (1992) Spatial density dependence in populations of a cynipid gall-former *Andricus quercuscalicis*. *Journal of Animal Ecology*, **61**, 567–83.

Hall, J.C., Siegel, R.W., Tompkins, L. and Kyriacou, C.P. (1980), Neurogenetics of courtship in *Drosophila*. *Stadler Symposia, University of Missouri*, **12**, 43–82.

Halsall, N.B. and Wratten, S.D. (1988) The efficiency of pitfall trapping for polyphagous Carabidae. *Ecological Entomology*, **13**, 293–99.

Hämäläinen, M., Markkula, M. and Raij, T. (1975) Fecundity and larval voracity of four ladybeetle species (Col., Coccinellidae). *Annales Entomologici Fennici*, **41**, 124–7.

van Hamburg, H. and Hassell, M.P. (1984) Density dependence and the augmentative release of egg parasitoids against graminaceous stalkborers. *Ecological Entomology*, **9**, 101–8.

Hames, B.D. and Rickwood, D. (1981) *Gel Electrophoresis of Proteins: a Practical Approach*, I.R.L. Press Ltd., Oxford.

Hamilton, W.D. (1967) Extraordinary sex ratios. *Science*, **156**, 477–88.

Hamm, A.H. (1934) Syrphidae (Dipt.) associated with flowers. *Journal of the Society for British Entomology*, **1**, 8–9.

Hamon, N., Bardner, R., Allen-Williams, L. and Lee, J.B. (1990) Carabid populations in field beans and their effect on the population dynamics of *Sitona lineatus* (L.). *Annals of Applied Biology*, **117**, 51–62.

Hance, T. (1986) Experiments on the population control of *Aphis fabae* by different densities of Carabidae (Coleoptera: Carabidae). *Annals of the Royal Society of Zoology, Belgium*, **116**, 15–24 (in French).

Hance, T. and Grégoire-Wibo, C. (1983) Étude du régime alimentaire des Carabidae par voie sérologique, in *New Trends in Soil Biology* (eds P. Lebrun, H. André, A. DeMedts, C. Grégoire-Wibo and G. Wauthy), Imprimerie Dieu-Brichart, Ottignies-Louvain, pp. 620–22.

Hand, L.F. and Keaster, A.J. (1967) The environment of an insect field cage, *Journal of Economic Entomology*, **60**, 910–15.

Hanski, I. (1990) Density dependence, regulation and variability in animal populations. *Philosophical Transactions of the Royal Society of London, B*, **330**, 141–50.

Hanski, I. (1994) A practical model of metapopulation dynamics. *Journal of Animal Ecology*, **63**, 151–62.

Harborne, J.B. (1988) *Introduction to Ecological Biochemistry*, Academic Press, London.

Harder, L.D. (1986) Effects of nectar concentration and flower depth on flower handling efficiency of bumblebees. *Oecologia*, **69**, 309–15.

Hardie, J., Nottingham, S.F., Powell, W. and Wadhams, L.J. (1991) Synthetic aphid sex pheromone lures female parasitoids. *Entomologia Experimentalis et Applicata*, **61**, 97–9.

Hardy, I.C.W. and Blackburn, T.M. (1991) Brood guarding in a bethylid wasp. *Ecological Entomology*, **16**, 55–62.

Hardy, I.C.W. and Godfray, H.C.J. (1990) Estimating the frequency of constrained sex allocation in field populations of Hymenoptera, *Behaviour*, **114**, 137–47.

Hardy, I.C.W., van Alphen, J.J.M. and van Dijken, M.J. (1992) First record of *Leptopilina longipes* in the Netherlands, and its hosts. *Entomologische Berichten*, **52**, 128–130.

Hardy, I.C.W., Griffiths, N.T. and Godfray H.C.J. (1992) Clutch size in a parasitoid wasp: a manipulation experiment. *Journal of Animal Ecology*, **61**, 121–9.

Hariri, G.E. (1966) Laboratory studies on the reproduction of *Adalia bipunctata* (Coleoptera: Coccinellidae). *Entomologia Experimentalis et Applicata*, **9**, 200–204.

Harris, H. and Hopkinson, D.A. (1977) *Handbook of Enzyme Electrophoresis in Human Genetics*, Amsterdam, North-Holland.

Harris, K.M. (1973) Aphidophagous Cecidomyiidae (Diptera): taxonomy, biology and assessments of field populations. *Bulletin of Entomological Research*, **63**, 305–25.

Harvey, I.F. and Walsh, K.J. (1993) Fluctuating asymmetry and lifetime mating success are correlated in males of the damselfly *Coenagrion puella* (Odonata: Coenagrionidae). *Ecological Entomology*, **18**, 198–202.

Harvey, J.A., Harvey, I.F. and Thompson, D.J. (1993) The effect of superparasitism on development of the solitary parasitoid wasp, *Venturia canescens* (Hymenoptera: Ichneumonidae). *Ecological Entomology*, **18**, 203–8.

Harvey, J.A., Harvey, I.F. and Thompson, D.J. (1994) Flexible larval feeding allows use of a range of host sizes by a parasitoid wasp. *Ecology*, **75**, 1420–1428

Harvey, P. (1985) Intrademic group selection and the sex ratio, in *Behavioural Ecology. Ecological consequences of Adaptive Behaviour* (eds R.M. Sibly and R.H. Smith), Blackwell Scientific Publications, Oxford, pp. 59–73.

Harvey, P.H. and Pagel, M.D. (1991) *The Comparative Method in Evolutionary Biology*, Oxford University Press, Oxford.

Hase A. (1925) Beiträge zur Lebensgeschichte der Schlupfwespe *Trichogramma evanescens* Westwood. *Arbuten aus der biologischen Reichsanstalt für Land und Forstwirtschaft* **14**, 171–224.

Haslett, J.R. (1983) A photographic account of pollen digestion by adult hoverflies. *Physiological Entomology*, **8**, 167–71.

Haslett, J.R. (1989a) Interpreting patterns of resource utilization: randomness and selectivity in pollen feeding by adult hoverflies. *Oecologia*, **78**, 433–42.

Haslett, J.R. (1989b) Adult feeding by holometabolous insects: pollen and nectar as complementary nutrient sources for *Rhingia campestris* (Diptera: Syrphidae). *Oecologia*, **81**, 361–3.

Haslett, J.R. and Entwistle, P.F. (1980) Further notes on *Eriozona syrphoides* (Fall.) (Dipt., Syrphidae) in Hafren Forest, mid-Wales. *Entomologist's Monthly Magazine*, **116**, 36.

Hassan, E. (1967) Untersuchungen über die bedeutung der kraut- und strauchschicht als nahrungsquelle für imagines entomophager Hymenopteren. *Zeitschrift für Angewandte Entomologie*, **60**: 238–65.

Hassell, M.P. (1968) The behavioural response of a tachinid fly (*Cyzenis albicans* [Fall.]) to its host, the winter moth (*Operophtera brumata* [L.]). *Journal of Animal Ecology*, **37**, 627–39.

Hassell, M.P. (1969) A population model for the interaction between *Cyzenis albicans* (Fall.) (Tachinidae) and *Operophtera brumata* (L.) (Geometridae) at Wytham, Berkshire. *Journal of Animal Ecology*, **38**, 567–76.

Hassell, M.P. (1971a) Mutual interference between searching insect parasites. *Journal of Animal Ecology*, **40**, 473–86.

Hassell, M.P. (1971b) Parasite behaviour as a factor contributing to the stability of insect host–parasite interactions, in *Dynamics of Populations* (eds P.J. den Boer and G.R. Gradwell), Proceedings of the Advanced Study Institute on 'Dynamics of Numbers in Populations', Oosterbeck, 1970, Centre for Agricultural Publishing and Documentation, Wageningen, pp. 366–79.

Hassell, M.P. (1975) Density dependence in single-species populations. *Journal of Animal Ecology*, **44**, 283–95.

Hassell , M.P. (1978) The dynamics of arthropod predator–prey systems. *Monographs in Population Biology*, **13**, Princeton University press, Princeton.

Hassell, M.P. (1980a) Foraging strategies, population models and biological control: a case study. *Journal of Animal Ecology*, **49**, 603–28.

Hassell, M.P. (1980b) Some consequences of habitat heterogeneity for population dynamics. *Oikos*, **35**, 150–60.

Hassell , M.P. (1982a) Patterns of parasitism by insect parasites in patchy environments. *Ecological Entomology*, **7**, 365–77.

Hassell, M.P. (1982b) What is searching efficiency? *Annals of Applied Biology*, **101**, 170–75.

Hassell, M.P. (1984) Parasitism in patchy environments: inverse density dependence can be stabilising. *IMA Journal of Mathematics Applied in Medicine and Biology*, **1**, 123–33.

Hassell, M.P. (1985) Insect natural enemies as regulating factors. *Journal of Animal Ecology*, **54**, 323–34.

Hassell, M.P. (1986) Parasitoids and host population regulation, in *Insect Parasitoids* (eds J. Waage and D. Greathead), Academic Press London, pp. 201–24.

Hassell, M.P. (1987) Detecting regulation in patchily distributed animal populations. *Journal of Animal Ecology*, **56**, 705–13.

Hassell, M.P. and Anderson, R.M. (1984) Host susceptibility as a component in host–parasitoid systems. *Journal of Animal Ecology*, **53**, 611–21.

Hassell, M.P. and Comins, H.N. (1978) Sigmoid functional responses and population stability. *Theoretical Population Biology*, **14**, 62–7.

Hassell, M.P. and Godfray, H.C.J. (1992) The population biology of insect parasitoids, in *Natural Enemies* (ed. M.J. Crawley), Blackwell, Oxford, pp. 265–92.

Hassell, M.P. and May, R.M. (1973) Stability in insect host–parasite models. *Journal of Animal Ecology*, **42**, 693–736.

Hassell, M.P. and May, R.M. (1986) Generalist and specialist natural enemies in insect predator–prey interactions. *Journal of Animal Ecology*, **55**, 923–40.

Hassell, M.P. and May, R.M. (1988) Spatial heterogeneity and the dynamics of parasitoid–host systems. *Annales Zoologici Fennici*, **25**, 55–61.

Hassell, M.P. and Moran, V.C. (1976) Equilibrium levels and biological control. *Journal of the Entomological Society of South Africa*, **39**, 357–66.

Hassell, M.P. and Pacala, S.W. (1990) Heterogeneity and the dynamics of host–parasitoid interactions. *Philosophical Transactions of the Royal Society of London, B*, **330**, 203–20.

Hassell, M.P. and Varley, G.C. (1969) New inductive population model for insect parasites and its bearing on biological control. *Nature*, **223**, 1113–37.

Hassell, M.P. and Waage, J.K. (1984) Host–parasitoid population interactions. *Annual Review of Entomology*, **29**, 89–114.

Hassell, M.P., Latto, J. and May, R.M. (1989a) Seeing the wood for the trees: detecting density dependence from existing life table studies. *Journal of Animal Ecology*, **58**, 883–92.

Hassell, M.P., Lawton, J.H. and Beddington, J.R. (1976) The components of arthropod predation. I. The prey death rate. *Journal of Animal Ecology*, **45**, 135–64.

Hassell, M.P., Lawton, J.H. and Beddington, J.R. (1977) Sigmoid functional responses by invertebrate predators and parasitoids. *Journal of Animal Ecology*, **46**, 249–62.

Hassell, M.P., Lessells, C.M. and McGavin, G.C. (1985) Inverse density dependent parasitism in a patchy environment: a laboratory system. *Ecological Entomology*, **10**, 393–402.

Hassell, M.P., Southwood, T.R.E. and Reader, P.M. (1987) The dynamics of the viburnum whitefly (*Aleurotrachelus jelinekii* (Fraunf.)): a case study on population regulation. *Journal of Animal Ecology*, **56**, 283–300.

Hassell, M.P., Taylor, V.A. and Reader, P.M. (1989b) The dynamics of laboratory populations of *Callosobruchus chinensis* and *C. maculatus* (Coleoptera: Bruchidae) in patchy environments. *Researches on Population Ecology*, **31**, 35–51.

Hassell, M.P., Waage, J.K. and May, R.M. (1983) Variable parasitoid sex ratios and their effect on host–parasitoid dynamics. *Journal of Animal Ecology*, **52**, 889–904.

Hassell, M.P., May, R.M., Pacala, S.W. and Chesson, P.L. (1991) The persistence of host–parasitoid associations in patchy environments. I. A general criterion. *American Naturalist*, **138**, 568–83.

Hattingh, V. and Samways, M.J. (1990) Absence of interference during feeding by the predatory ladybirds *Chilocorus* spp. (Coleoptera: Coccinellidae). *Ecological Entomology*, **15**, 385–90.

Hawke, S.D., Farley, R.D. and Greany, P.D. (1973) The fine structure of sense organs in the ovipositor of the parasitic wasp, *Orgilus lepidus* Muesebeck. *Tissue and Cell*, **5**, 171–84.

Hawkes, R.B. (1972) A fluorescent dye technique for marking insect eggs in predation studies. *Journal of Economic Entomology*, **65**, 1477–8.

Hawkins, B.A. (1993) Refuges, host population dynamics and the genesis of parasitoid diversity, in *Hymenoptera and Biodiversity* (eds J. LaSalle and I.D. Gauld), CAB International, Wallingford, pp. 235–56.

Hawkins, B.A. (1994) *Pattern and Process in Host-Parasitoid Interactions*, Cambridge University Press, Cambridge.

Heads, P.A. (1986) The costs of reduced feeding due to predator avoidance: potential effects on growth and fitness in *Ischnura elegans* larvae (Odonata: Zygptera). *Ecological Entomology*, **11**, 369–77.

Heads, P.A. and Lawton, J.H. (1983) Studies on the natural enemy complex of the holly leaf-miner: the effects of scale on the detection of aggregative responses and the implications for biological control. *Oikos*, **40**, 267–76.

Heap, M.A. (1988) The pit-light, a new trap for soil-dwelling insects. *Journal of the Australian Entomological Society*, **27**, 239–40.

Heaversedge, R.C. (1967) Variation in the size of insect parasites of puparia of *Glossina* spp. *Bulletin of Entomological Research*, **58**, 153–8.

Heidari, M. (1989) Biological control of glasshouse mealybugs using coccinellid predators. Ph.D. Thesis, Wye College, University of London.

Heidari, M. and Copland, M.J.W. (1992) Host finding by *Cryptolaemus montrouzieri* (Col., Coccinellidae) a predator of mealybugs (Hom., Pseudococcidae). *Entomophaga*, **37**, 621–5.

Heidari, M. and Copland, M.J.W. (1993) Honeydew: a food resource or arrestant for the mealybug predator *Cryptolaemus montrouzieri*? *Entomophaga*, **38**, 63–8.

Heinrich, B. (1975) Energetics of pollination. *Annual Review of Ecology and Systematics*, **6**, 139–70.

Heinrich, B. (1985) Insect foraging energetics, in *Handbook of Experimental Pollination Biology* (eds

C.E. Jones and R.J. Little), Van Nostrand Reinhold, New York, pp. 187–214.

Heinrichs, E.A., Aquino, G.B., Chelliah, S., Valencia, S.L. and Reisseg, W.H. (1982) Resurgence of *Nilaparvata lugens* (Stål) populations as influenced by method and timing of insecticide applications in lowland rice. *Environmental Entomology*, **11**, 78–84.

Heinz, K.M. and Parrella, M.P. (1990) Holarctic distribution of the leafminer parasitoid *Diglyphus begini* (Hymenoptera: Eulophidae) and notes on its life history attacking *Liriomyza trifolii* (Diptera: Agromyzidae) in chrysanthemum. *Annals of the Entomological Society of America*, **83**, 916–24.

Helenius, J. (1990) Incidence of specialist natural enemies of *Rhopalosiphum padi* (L.) (Hom., Aphididae) on oats in monocrops and mixed intercrops with faba beans. *Journal of Applied Entomology*, **109**, 136–43.

Hemerik, L., Driessen, G.J. and Haccou, P. (1993) Effects of intra-patch experiences on patch time, search time and searching efficiency of the parasitoid *Leptopilina clavipes*. *Journal of Animal Ecology*, **62**, 33–44.

Hemptinne, J.L. and Desprets, A. (1986) Pollen as a spring food for *Adalia bipunctata*, in *Ecology of Aphidophaga* (ed. I. Hodek), Academia, Prague and Dr. W. Junk, Dordrecht, pp. 29–35.

Henderson, I.F. and Whitaker, T.M. (1977) The efficiency of an insect suction sampler in grassland. *Ecological Entomology*, **2**, 57–60.

Hengeveld, R. (1980) Polyphagy, oligophagy and food specialization in ground beetles (Coleoptera, Carabidae). *Netherlands Journal of Zoology*, **30**, 564–84.

Henneicke, K., Dawah, H.A. and Jervis, M.A. (1992) The taxonomy and biology of final instar larvae of some Eurytomidae (Hymenoptera: Chalcidoidea) associated with grasses in Britain. *Journal of Natural History*, **26**, 1047–87.

Heong, K.L., Aquino, G.B. and Barrion, A.T. (1991) Arthropod community structures of rice ecosystems in the Philippines. *Bulletin of Entomological Research*, **81**, 407–16.

Herard, F., Keller, M.A. and Lewis, W.J. (1988) Rearing *Microplitis demolitor* Wilkinson (Hymenoptera: Braconidae) in the laboratory for use in studies of semiochemical mediated searching behaviour. *Journal of Entomological Science*, **23**, 105–11.

Herrebout, W. (1969) Habitat selection in *Eucarcelia rutilla* Vil. (Diptera: Tachinidae). II. Experiments with females of known age. *Zeitschrift für Angewante Entomologie*, **63**, 336–49.

Herrera, C.J., van Driesche, R.G. and Bellotti, A.C. (1989) Temperature-dependent growth rates for the cassava mealybug, *Phenacoccus herreni*, and two of its encyrtid parasitoids, *Epidinocarsis diversicornis* and *Acerophagus coccois* in Colombia. *Entomologia Experimentalis et Applicata*, **50**, 21–7.

Herting, B. (1960) Biologie der wespaläarktischen Raupenfliegen (Diptera:Tachnidae). *Monographien zur Angewandte Entomologie*, **16**, 1–202.

Hertlein, M.B. and Thorarinsson, K. (1987) Variable patch times and the functional response of *Leptopilina boulardi* (Hymenoptera: Eucoilidae). *Environmental Entomology*, **16**, 593–8.

Hess, A.D. (1941) New limnological sampling equipment. *Limnological Society of America, Special Publication*, **6**, 1–15.

Heydemann, B. (1953) Agrarökologische Problematik. University of Kiel, Ph.D. Thesis.

Hilborn, R. and Stearns, S.C. (1982) On inference in ecology and evolutionary biology: the problem of multiple causes. *Acta Biotheoretica*, **31**, 145–64.

Hildrew, A.G. and Townsend, C.R. (1976) The distribution of two predators and their prey in an iron rich stream. *Journal of Animal Ecology*, **45**, 41–57.

Hildrew, A.G. and Townsend, C.R. (1982) Predators and prey in a patchy environment: a freshwater study. *Journal of Animal Ecology*, **51**, 797–815.

Hill, M.G. (1988) Analysis of the biological control of *Mythimna separata* (Lepidoptera: Noctuidae) by *Apanteles ruficrus* (Braconidae: Hymenoptera) in New Zealand. *Journal of Applied Ecology*, **25**, 197–208.

Hillis, D.M. and Moritz, C. (eds) (1990) *Molecular Systematics*, Sinauer., Sunderland.

Hochberg, M.E., Hassell, M.P. and May, R.M. (1990) The dynamics of host–parasitoid– pathogen interactions. *American Naturalist*, **135**, 74–94.

Hocking, B. (1953) The intrinsic range and speed of flight in insects. *Transactions of the Royal Entomological Society of London*, **104**, 223–345.

Hocking, H. (1967) The influence of food on longevity and oviposition in *Rhyssa persuasoria* (L.) (Hymenoptera: Ichneumonidae), *Journal of the Australian Entomological Society*, **6**, 83–8.

Hodek, I. (1973) *Biology of Coccinellidae*, W. Junk, The Hague/Czechoslovakian Academy of Sciences, Prague.

Hodges, C.M. (1985a) Bumblebee foraging: the threshold departure rule. *Ecology*, **66**, 179–187.

Hodges, C.M. (1985b) Bumblebee foraging: energetic consequences of using a threshold departure rule. *Ecology*, **66**, 188–97.

Hodgson, C.J. and Mustafa, T.M. (1984) Aspects of chemical and biological control of *Psylla pyricola*

Forster in England. *Bulletin IOBC/WPRS Working Group Integrated Control of Pear Psyllids*, **7**(5), 330–53.

van der Hoeven, N. and Hemerik, L. (1990) Superparasitism as an ESS: to reject or not to reject, that is the question. *Journal of Theoretical Biology*, **146**, 467–82.

Hofsvang, T. and Hågvar, E.B. (1975a) Duration of development and longevity in *Aphidius ervi* and *Aphidius platensis* (Hymenoptera, Aphidiidae), two parasites of *Myzus persicae* (Homoptera, Aphididae). *Entomophaga*, **20**, 11–22.

Hofsvang, T. and Hågvar, E.B. (1975b) Developmental rate, longevity, fecundity, and oviposition period of *Ephedrus cerasicola* Stary (Hym.: Aphidiidae) parasitizing *Myzus persicae* Sulz. (Hom.: Aphididae) on paprika. *Norwegian Journal of Entomology*, **22**, 15–22.

Hogge, M.A.F. and King, P.E. (1975) The ultrastructure of spermatogenesis in *Nasonia vitripennis* (Hym., Pteromalidae). *Journal of Submicroscopical Cytology*, **7**, 81–96.

Hohmann, C.L., Luck, R.F., Oatman, E.R. and Platner, G.R. (1989) Effects of different biological factors on longevity and fecundity of *Trichogramma platneri* Nagarkatti (Hymenoptera: Trichogrammatidae). *Anais da Sociedade Entomologica de Brasil*, **18**, 61–70.

Hokkanen, H. and Pimentel, D. (1984) New approach for selecting biological control agents. *Canadian Entomologist*, **116**, 1109–21.

Holdaway, F.T. and Smith, H.F. (1932) A relation between size of host puparia and sex ratio of *Alysia maducator* Panzer. Australian *Journal of Experimental Biology and Medical Science*, **10**, 247–59.

Hölldobler, B. and Bartz, S.H. (1985) Sociobiology of reproduction in ants. *Fortschritte Zoologie*, **31**, 237–57.

Hölldobler, B. and Wilson, E.O. (1983) The evolution of communal nestweaving in ants. *American Scientist*, **71**, 489–99.

Höller, C. (1991) Evidence for the existence of a species closely related to the cereal aphid parasitoid *Aphidius rhopalosiphi* De Stefani-Perez based on host ranges, morphological characters, isolectric focusing banding patterns, cross-breeding experiments and sex pheromone specificities (Hymenpotera, Braconidae, Aphidiinae). *Systematic Entomology*, **16**, 15–28.

Höller, C. and Braune, H.J. (1988) The use of isoelectric focusing to assess percentage hymenopterous parasitism in aphid populations. *Entomologia Experimentalis et Applicata*, **47**, 105–14.

Höller, C., Christiansen-Weniger, P., Micha, S.G., Siri, N. and Borgemeister, C. (1991) Hyperparasitoid–aphid and hyperparasitoid-primary parasitoid relationships. *Redia*, **74**, 153–61.

Holling, C.S. (1959a) The components of predation as revealed by a study of small mammal predation of the European sawfly. *Canadian Entomologist*, **91**, 293–320.

Holling, C.S. (1959b) Some characteristics of simple types of predation and parasitism. *Canadian Entomologist*, **91**, 385–98.

Holling, C.S. (1961) Principles of insect predation. *Annual Review of Entomology*, **6**, 163–82.

Holling, C.S. (1965) The functional response of predators to prey density and its role in mimicry and population regulation. *Memoirs of the Entomological Society of Canada*, **45**, 3–60.

Holling, C.S. (1966) The functional response of invertebrate predators to prey density. *Memoirs of the Entomological Society of Canada*, **48**, 1–86.

Holloway, B.A. (1976) Pollen-feeding in hover-flies (Diptera: Syrphidae). *New Zealand Journal of Zoology*, **3**, 339–50.

Holm, S.N. and Farkas, J. (1986) Control of the chalcid wasp *Pteromalus apum* Retzius, a parasite of *Megachile rotundata* (Fabr.) (Hymenoptera, Pteromalidae and Apidae). *Entomologiske Meddelelser*, **53**, 59–64.

Holmes, H.B. (1976) *Mormoniella Publications Supplement*. Mimeographed edn, circulated by the author.

Holmes, P.R. (1984) A field study of the predators of the grain aphid, *Sitobion avenae* (F.) (Hemiptera: Aphididae), in winter wheat. *Bulletin of Entomological Research*, **74**, 623–31.

Holopäinen, J.K. and Varis, A.L. (1986) Effects of a mechanical barrier and formalin preservative on pitfall catches of carabid beetles (Coleoptera, Carabidae) in arable fields. *Journal of Applied Entomology*, **102**, 440–45.

Holt, J., Cook, A.G., Perfect, T.J. and Norton, G.A. (1987) Simulation analysis of brown planthopper (*Nilaparvata lugens*) population dynamics on rice in the Phillipines. *Journal of Applied Ecology*, **24**, 87–103.

Honěk, A. and Kocourek, F. (1986) The flight of aphid predators to a light trap: possible interpretations, in *Ecology of Aphidophaga* (ed. I. Hodek), Academia, Prague, pp. 333–37.

Hooker, M.E., Barrows, E.M. and Ahmed, S.W. (1987) Adult longevity as affected by size, sex, and maintenance in isolation or groups in the parasite *Pediobius foveolatus* (Hymenoptera: Eulophidae). *Annals of the Entomological Society of America*, **80**, 655–9.

Hopper, K.R. and King, E.G. (1986) Linear functional response of *Microplitis croceipes* (Hymenoptera: Braconidae) to variation in *Heliothis* spp. (Lepidoptera: Noctuidae) density in the field. *Environmental Entomology*, **15**, 476–80.

Hopper, K.R. and Woolson, E.A. (1991) Labeling a parasitic wasp, *Microplitis croceipes* (Hymenoptera: Braconidae), with trace elements for mark–recapture studies. *Annals of the Entomological Society of America*, **84**, 255–62.

Hopper, K.R., Powell, J.E. and King, E.G. (1991) Spatial density dependence in parasitism of *Heliothis virescens* (Lepidoptera: Noctuidae) by *Microplitis croceipes* (Hymenoptera: Braconidae) in the field. *Environmental Entomology*, **20**, 292–302.

Hopper, K.R., Roush, R.T. and Powell, W. (1993) Management of genetics of biological control introductions. *Annual Review of Entomology*, **38**, 27–51.

Horn, D.J. (1981) Effect of weedy backgrounds on colonization of collards by green peach aphid, *Myzus persicae*, and its major predators. *Environmental Entomology*, **10**, 285–9.

Horne, P.A. and Horne, J.A. (1991) The effects of temperature and host density on the development and survival of *Copidosoma koehleri*. *Entomologia Experimentalis et Applicata*, **59**, 289–92.

Houck, M.A. and Strauss, R.E. (1985) The comparative study of functional responses: experimental design and statistical interpretation. *Canadian Entomologist*, **117**, 617–29.

Howard, L.O. (1886) The excessive voracity of the female mantis. *Science*, **8**, 326.

Howe, R.W. (1967) Temperature effects on embryonic development in insects. *Annual Review of Entomology*, **12**, 15–42.

Howling, G.G. (1991) Slug foraging behaviour: attraction to food items from a distance. *Annals of Applied Biology*, **119**, 147–53.

Howling, G.G. and Port, G.R. (1989) Time-lapse video assessment of molluscicide baits, in *Slugs and Snails in World Agriculture*. BCPC Monograph No. 41, pp. 161–6.

Hubbard, S.F., Marris, G., Reynolds A. and Rowe, G.W. (1987) Adaptive patterns in the avoidance of superparasitism by solitary parasitic wasps. *Journal of Animal Ecology*, **56**, 387–401.

Huffaker, C.B. (1958) Experimental studies on predation: dispersion factors and predator–prey oscillations. *Hilgardia*, **27**, 343–83.

Huffaker, C.B. and Kennett, C.E. (1966) Biological control of *Parlatoria oleae* (Colvee) through the compensatory action of two introduced parasites. *Hilgardia*, **37**, 283–335.

Huffaker, C.B., Kennett, C.E. and Finney, G.L. (1962) Biological control of olive scale, *Parlatoria oleae* (Colvee) in California by imported *Aphytis maculicornis* (Masi) (Hymenoptera: Aphelinidae). *Hilgardia*, **32**, 541–636.

Huffaker, C.B., Luck, R.F. and Messenger, P.S. (1977) The ecological basis of biological control. *Proceedings of the 15th International Congress of Entomology, Washington, 1976*, 560–86.

Huffaker, C.B., Messenger, P.S. and DeBach, P. (1971) The natural enemy component in natural control and the theory of biological control, in *Biological Control* (ed. C.B. Huffaker), Plenum, New York, pp. 16–67.

Hughes, R.D. (1962) A method for estimating the effects of mortality on aphid populations. *Journal of Animal Ecology*, **31**, 389–96.

Hughes, R.D. (1963) Population dynamics of the cabbage aphid, *Brevicoryne brassicae* (L.). *Journal of Animal Ecology*, **32**, 393–424.

Hughes, R.D. (1972) Population dynamics, in *Aphid Technology* (ed. H.F. van Emden) Academic Press, London, pp. 275–93.

Hughes, R.D. (1989) Biological control in the open field, in *World Crop Pests, Vol. 2C: Aphids, Their Biology, Natural Enemies and Control* (eds A.K. Minks and P. Harrewijn), Elsevier, Amsterdam, pp. 167–98.

Hughes, R.D. and Gilbert, N. (1968) A model of an aphid population – a general statement. *Journal of Animal Ecology*, **37**, 553–63.

Hughes, R.D. and Sands, P. (1979) Modelling bushfly populations. *Journal of Applied Ecology*, **16**, 117–39.

Hughes, R.D., Woolcock, L.T. and Hughes, M.A. (1992) Laboratory evaluation of parasitic Hymenoptera used in attempts to biologically control aphid pests of crops in Australia. *Entomologia Experimentalis et Applicata*, **63**, 177–85.

van Huizen, T.H.P. (1990) 'Gone with the Wind': flight activity of carabid beetles in relation to wind direction and to the reproductive state of females in flight, in *The Role of Ground Beetles in Ecological and Environmental Studies* (ed. N.E. Stork), Intercept, Andover, pp. 289–93.

Hunt, J.H., Brown, P.A., Sago, K.M. and Kerker, J.A. (1991) Vespid wasps eat pollen. *Journal of the Kansas Entomological Society*, **64**, 127–30.

Hurst, G.D.D., Majerus, M.E.N. and Walker, L.E. (1993) The importance of cytoplasmic male killing elements in natural poulations of the two spot ladybird, *Adalia bipunctata* (Linnaeus) (Coleoptera: Coccinellidae). *Biological Journal of The Linnean Society*, **49**, 195–202.

Ibrahim, M.M. (1955) Studies on *Coccinella undecimpunctata aegyptiaca* Reiche 2. Biology and life-history. *Bulletin de la Societé Entomologique d'Egypte*, **39**, 395–423.

Ichikawa, T. (1976) Mutual communication by substrate vibration in the mating behaviour of planthoppers (Homoptera: Delphacidae). *Applied Entomology and Zoology*, **11**, 8–23.

Inouye, D.W. (1978) Resource partitioning in bumblebees: experimental studies of foraging behaviour. *Ecology*, **59**, 672–8.

Iperti, G. (1966) Some components of efficiency in aphidophagous coccinellids, in *Ecology of Aphidophagous Insects* (ed. I. Hodek), Academia, Prague and W. Junk, The Hague, p. 253.

Iperti, G. and Buscarlet, L.A. (1986) Seasonal migration of the ladybird *Semiadalia undecimnotata*, in *Ecology of Aphidophaga* (ed. I. Hodek), Academia, Prague and Dr. W. Junk, Dordrecht, pp. 199–204,

Irwin, M.E., Gill, R.W. and Gonzales, D. (1974) Field-cage studies of native egg predators of the Pink Bollworm in southern California cotton. *Journal of Economic Entomology*, **67**, 193–6.

Ives, A.R. (1992) Continuous-time models of host–parasitoid interactions. *American Naturalist*, **140**, 1–29.

Ives, P.M. (1981a) Feeding and egg production of two species of coccinellids in the laboratory. *Canadian Entomologist*, **113**, 999–1005.

Ives, P.M. (1981b) Estimation of coccinellid numbers and movement in the field. *Canadian Entomology*, **113**, 981–97.

Iwasa, Y, Higashi, M. and Matsuda, H. (1984) Theory of oviposition strategy of parasitoids. I. Effect of mortality and limited egg number. *Theoretical Population Biology*, **26**, 205–27.

Iwata, K. (1959) The comparative anatomy of the ovary in Hymenoptera. Part 3. Braconidae (including Aphidiidae) with descriptions of ovarian eggs. *Kontyû*, **27**, 231–8.

Iwata, K. (1960) The comparative anatomy of the ovary in Hymenoptera. Part 5. Ichneumonidae. *Acta Hymenopterologica*, **1**, 115–69.

Iwata, K. (1962) The comparative anatomy of the ovary in Hymenoptera. Part 6. Chalcidoidea with descriptions of ovarian eggs. *Acta Hymenopterologica*, **1**, 383–91.

Jachmann, F. and van den Assem (1993) The interaction of external and internal factors in the courtship of parasitic wasps (Hym., Pteromalidae). *Behaviour*, **125**, 1–19.

Jachmann, F. and van den Assem (1995) A causal ethological analysis of the courtship behaviour of an insect (the parasitic wasp *Nasonia vitripennis*, Chalc., Pteromalidae). *Behavior*, **127** (in press).

Jackson, C.G. (1986) Effects of cold storage of adult *Anaphes ovijentatus* on survival, longevity, and oviposition. *Southwestern Entomologist*, **11**, 149–53.

Jackson, C.G. and Debolt, J.W. (1990) Labeling of *Leiophron uniformis*, a parasitoid of *Lygus* spp., with rubidium. *Southwest Entomologist*, **15**, 239–44.

Jackson, C.G., Cohen, A.C. and Verdugo, C.L. (1988) Labeling *Anaphes ovijentatus* (Hymenoptera: Mymaridae), an egg parasite of *Lygus* spp. (Hemiptera: Miridae), with rubidium. *Annals of the Entomological Society of America*, **81**, 919–22.

Jackson, D.J. (1928) The biology of *Dinocampus* (*Perilitus*) *rutilus* Nees, a braconid parasite of *Sitona lineata* L. Part 1. *Proceedings of the Zoological Society of London*, **1928**, 597–630.

James, H.G. and Nicholls, C.F. (1961) A sampling cage for aquatic insects. *Canadian Entomologist*, **93**, 1053–5.

James, H.G. and Redner, R.L. (1965) An aquatic trap for sampling mosquito predators. *Mosquito News*, **25**, 35–7.

Janssen, A. (1989) Optimal host selection by *Drosophila* parasitoids in the field. *Functional Ecology*, **3**, 469–79.

Janssen, A., van Alphen, J., Sabelis, M. and Bakker, K. (1991) Microhabitat selection behaviour of *Leptopilina heterotoma* changes when odour of competitor is present. *Redia*, **74**, 302–10.

Jarošik, V. (1992) Pitfall trapping and species-abundance relationships: a value for carabid beetles (Coleoptera, Carabidae). *Acta Entomologica Bohemoslovaca*, **89**, 1–12.

Jensen, T.S., Dyring, L., Kristensen, B., Nilesen, B.O. and Rasmussen, E.R. (1989) Spring dispersal and summer habitat distribution of *Agonum dorsale* (Coleoptera: Carabidae). *Pedobiologia*, **33**, 155–65.

Jervis, M.A. (1978) Homopteran Bugs, in *A Dipterist's Handbook* (eds A. Stubbs and P.J. Chandler), Amateur Entomologist's Society, Hanworth, pp. 173–6.

Jervis, M.A. (1979) Courtship, mating and 'swarming' in *Aphelopus melaleucus* (Hym., Dryinidae). *Entomologist's Gazette*, **30**, 191–3.

Jervis, M.A. (1980a) Life history studies of *Aphelopus* species (Hymenoptera: Dryinidae) and *Chalarus* species (Diptera: Pipunculidae), primary parasites of typhlocybine leafhoppers (Homoptera: Cicadellidae). *Journal of Natural History*, **14**, 769–80.

Jervis, M.A. (1980b) Ecological studies on the parasite complex associated with typhlocybine leafhoppers (Homoptera: Cicadellidae), *Ecological Entomology*, **5**, 123–36.

Jervis, M.A. (1992) A taxonomic revision of the pipunculid fly genus *Chalarus* Walker, with particular reference to the European fauna. *Zoological Journal of the Linnean Society*, **105**, 243–352.

Jervis, M.A and Kidd, N.A.C. (1986) Host feeding strategies in hymenopteran parasitoids. *Biological Reviews*, **61**, 395–434.

Jervis, M.A. and Kidd, N.A.C. (1991) The dynamic significance of host-feeding by insect parasitoids – what modellers ought to consider. *Oikos*, **62**, 97–9.

Jervis, M.A. and Kidd, N.A.C. (1995) The role of non-host foods in parasitoid foraging behaviour – a state variable approach (submitted).

Jervis, M.A., Kidd, N.A.C. and Almey, H.A. (1994) Post-reproductive life in the parasitoid *Bracon hebetor* (say) (Hym., Braconidae). *Journal of Applied Entomology*, **117**, 72–7.

Jervis, M.A., Kidd, N.A.C. and Walton, M. (1992a) A review of methods for determining dietary range in adult parasitoids. *Entomophaga*, **37**, 565–74.

Jervis, M.A., Kidd, N.A.C., McEwen, P., Campos, M. and Lozano, C. (1992b) Biological control strategies in olive pest management, in *Research Collaboration in European IPM Systems*, BCPC Monograph No. **52**, pp. 31–9.

Jervis, M.A., Kidd, N.A.C., M.G.Fitton, T., Huddleston, T. and Dawah, H.A. (1993) Flower-visiting by hymenopteran parasitoids. *Journal of Natural History*, **27**, 67–105.

Johnson, C.G., Southwood, T.R.E. and Entwistle, H.M. (1957) A new method of extracting arthropods and molluscs from grassland and herbage with a suction apparatus. *Bulletin of Entomological Research*, **48**, 211–18.

Johnson, D.M., Akre, B.C. and Crowley, P.H. (1975) Modeling arthropod predation: wasteful killing by damselfly naiads. *Ecology*, **36**, 1081–93.

Johnson, D.M., Bohanan, R.E., Watson, C.N. and Martin, T.H. (1984) Coexistence of *Enallagma divagans* and *Enallagma traviatum* (Zygoptera: Coenagrionidae) in Bays Mountain Lake: an in situ enclosure experiment. *Advances in Odonatology*, **2**, 57–70.

Johnson, J.W., Eikenbary, R.D. and Holbert, D. (1979) Parasites of the greenbug and other graminaceous aphids: identity based on larval meconia and features of the empty aphid mummy. *Annals of the Entomological Society of America*, **72**, 759–66.

Jolly, G.M. (1965) Explicit estimates from capture–recapture data with both death and immigration– stochastic model. *Biometrika*, **52**, 225–47.

Jones, M.G. (1979) The abundance and reproductive activity of common Carabidae in a winter wheat crop. *Ecological Entomology*, **4**, 31–43.

Jones, R.L. and Lewis, W.J. (1971) Physiology of the host–parasite relationship between *Heliothis zea* and *Microplitis croceipes*. *Journal of Insect Physiology*, **17**, 921–7.

Jones, T.H. and Hassell, M.P. (1988) Patterns of parasitism by *Trybliographa rapae*, a cynipid parasitoid of the cabbage root fly, under laboratory and field conditions. *Ecological Entomology*, **13**, 65–92.

Jones, W.T. (1982) Sex ratio and host size in a parasitic wasp. *Behavioural Ecology and Sociobiology*, **10**, 207–10.

de Jong, P.W. and van Alphen, J.J.M. (1989) Host size selection and sex allocation in *Leptomastix dactylopii*, a parasitoid of *Planococcus citri*. *Entomologia Experimentalis et Applicata*, **50**, 161–9.

Judd, W.W. (1970) Insects associated with flowering wild carrot, *Daucus carota* L., in southern Ontario. *Proceedings of the Entomological Society of Ontario*, **100**, 176–181.

Juliano, S.A. (1985) The effects of body size on mating and reproduction in *Brachinus lateralis* (Coleoptera: Carabidae). *Ecological Entomology*, **10**, 271–80.

Juliano, S.A. and Williams, F.M. (1987) A comparison of methods for estimating the functional response parameters of the random parasite equation. *Journal of Animal Ecology*, **56**, 641–53.

Kaas, J.P., Elzen, G.W., Ramaswamy, S.B. (1990) Learning in *Microplitis croceipes* Cresson (Hym., Braconidae). *Journal of Applied Entomology*, **109**, 268–273.

Kajita, H. and van Lenteren, J.C. (1982) The parasite–host relationship between *Encarsia formosa* (Hymenoptera: Aphelinidae) and *Trialeurodes vaporariorum* (Homoptera: Aleyrodidae), XIII. Effect of low temperatures on egg maturation of *Encarsia formosa*. *Zeitschrift für Angewandte Entomologie*, **93**, 430–9.

Kakehashi, N., Suzuki, Y. and Iwasa, Y. (1984) Niche overlap of parasitoids in host–parasitoid systems: its consequences to single versus multiple introduction controversy in biological control. *Journal of Applied Ecology*, **21**, 115–31.

Karczewski, J. (1967) The observations on flower-visiting species of Tachinidae and Calliphoridae (Diptera). *Fragmenta Faunistica*, **13**, 407–84.

Karowe, D.N. and Martin, M.M. (1989) The effects of quantity and quality of diet nitrogen on the growth, efficiency of food utilisation, nitrogen budget, and metabolic rate of fifth-instar *Spodoptera eridania* larvae (Lepidoptera: Noctuidae). *Journal of Insect Physiology*, **35**, 699–708.

Kazmer, D.J. and Luck, R.F. (1995) Field tests of the size-fitness hypothesis in the egg parasitoid *Trichogramma pretiosum*. *Ecology*, **76** (in press).

Keeler, K.H. (1978) Insects feeding at extrafloral nectaries of *Ipomoea carnea* (Convolvulaceae). *Entomological News*, **89**, 163–8.

Kegel, B. (1990) Diurnal activity of carabid beetles living on arable land, in *The Role of Ground Beetles in Ecological and Environmental Studies* (ed. N.E. Stork), Intercept, Andover, pp. 65–76.

Kenmore, P.E., Carino, F.O., Perez, C.A., Dyck, V.A. and Gutierrez, A.P. (1985) Population regulation of the rice brown planthopper (*Nilaparvata lugens* Stål) within rice fields in the Philippines, *Journal of Plant Protection in the Tropics*, **1**, 19–37.

Kennedy, B.H. (1979) The effect of multilure on parasites of the European elm bark beetle, *Scolytus multistriatus*. *ESA Bulletin*, **25**, 116–18.

Kennedy, J.S. (1978) The concepts of olfactory 'arrestment' and attraction'. *Physiological Entomology*, **3**, 91–8.

Kerkut, G.A. and Gilbert, L.I. (1985) *Comprehensive Insect Physiology, Biochemistry and Pharmacology. Vol 1. Embryogenesis and Reproduction*, Pergamon, Oxford.

Kessel, E.L. (1955) The mating activities of balloon flies. *Systematic Zoology*, **4**, 97–104.

Kessel, E.L. (1959) Introducing *Hilara wheeleri* Melander as a balloon maker, and notes on other North American balloon flies. *Wasmann Journal of Biology*, **17**, 221–30.

Kfir, R. (1981) Fertility of the polyembryonic parasite *Copidosoma koehleri*, effect of humidities on life length and relative abundance as compared with that of *Apanteles subandinus* in potato tuber moth. *Annals of Applied Biology*, **99**, 225–30.

Kfir, R. and van Hamburg, H. (1988) Interspecific competition between *Telenomus ullyetti* (Hymenoptera: Scelionidae) and *Trichogramma lutea* (Hymenoptera: Trichogrammatidae) parasitizing eggs of the cotton bollworm *Heliothis armigera* in the laboratory. *Environmental Entomology* **17**: 664–70.

Kfir, R. and Luck, R.F. (1979) Effects of constant and variable temperature extremes on sex ratio and progeny production by *Aphytis melinus* and *A. lingnanensis* (Hymenoptera: Aphelinidae). *Ecological Entomology*, **4**, 335–44.

Kidd, N.A.C. (1979) Simulation of population processes with a programmable pocket calculator. *Journal of Biological Education*, **13**, 284–90.

Kidd, N.A.C. (1982) Predator avoidance as a result of aggregation in the grey pine aphid, *Schizolachnus pineti*. *Journal of Animal Ecology*, **51**, 397–412.

Kidd, N.A.C. (1984) A BASIC program for use in teaching population dynamics. *Journal of Biological Education*, **18**, 227–8.

Kidd, N.A.C. (1990a) The population dynamics of the large pine aphid, *Cinara pinea* (Mordv.). I. Simulation of laboratory populations. *Researches on Population Ecology*, **32**, 189–208.

Kidd, N.A.C. (1990b) The population dynamics of the large pine aphid, *Cinara pinea* (Mordv.). II. Simulation of field populations. *Researches on Population Ecology*, **32**, 209–26.

Kidd, N.A.C. (1990c) A synoptic model to explain long-term population changes in the large pine aphid, in *Population Dynamics of Forest Insects* (eds A.D. Watt, S.R. Leather, M.D. Hunter and N.A.C. Kidd), Intercept, Andover, pp. 317–27.

Kidd, N.A.C. and Jervis, M.A. (1989) The effects of host-feeding behaviour on the dynamics of parasitoid–host interactions, and the implications for biological control. *Researches on Population Ecology*, **31**, 235–74.

Kidd, N.A.C. and Jervis, M.A. (1991) Host-feeding and oviposition strategies of parasitoids in relation to host stage. *Researches on Population Ecology*, **33**, 13–28.

Kidd, N.A.C. and Mayer, A.D. (1983) The effect of escape responses on the stability of insect host–parasite models. *Journal of Theoretical Biology*, **104**, 275–87.

Killick-Kendrick, R. and Killick-Kendrick, M. (1987) Honeydew of aphids as a source of sugar for *Phlebotomus ariasi*. *Medical and Veterinary Entomology*, **1**, 297–302.

Kiman, Z.B. and Yeargan, K.V. (1985) Development and reproduction of the predator

Orius insidiosus (Hemiptera: Anthocoridae) reared on diets of selected plant material and arthropod prey. *Annals of the Entomological Society of America*, **78**, 464–7.

King, B.H. (1989) Host-size dependent sex ratios among parasitoid wasps: does host growth matter. *Oecologia*, **78**, 420–426.

King, B.H. (1993) Sex ratio manipulation by parasitoid wasps, in *Evolution and Diversity of Sex Ratio in Insects and Mites*, (eds, D.L. Wrensch and M.A. Ebbert) Chapman and Hall, New York, pp. 418–441.

King, P.E. (1961) The passage of sperm to the spermatheca during mating in *Nasonia vitripennis*. *Entomologist's Monthly Magazine*, **97**, 136.

King, P.E. (1962) The structure and action of the spermatheca in *Nasonia vitripennis*. *Proceedings of the Royal Entomological Society of London, A*, **37**, 73–5.

King, P.E. (1963) The rate of egg resorption in *Nasonia vitripennis* (Walker) (Hymenoptera: Pteromalidae) deprived of hosts. *Proceedings of the Royal Entomological Society of London, A* **38**, 98–100.

King, P.E. and Copland, M.J.W. (1969) The structure of the female reproductive system in the Mymaridae (Chalcidoidea: Hymenoptera). *Journal of Natural History*, **3**, 349–65.

King, P.E. and Fordy, M.R. (1970) The external morphology of the 'pore' structures on the tip of the ovipositor in Hymenoptera. *Entomologist's Monthly Magazine*, **106**, 65–6.

King, P.E. and Ratcliffe, N.A. (1969) The structure and possible mode of functioning of the female reproductive system in *Nasonia vitripennis* (Hymenoptera: Pteromalidae). *Journal of Zoology*, **157**, 319–44.

King, P.E. and Richards, J.G. (1968) Oösorption in *Nasonia vitripennis* (Hymenoptera: Pteromalidae). *Journal of Zoology*, **154**, 495–516.

King, P.E. and Richards, J.G. (1969) Oögenesis in *Nasonia vitripennis* (Walker) (Hymenoptera: Pteromalidae). *Proceedings of the Royal Entomological Society, London, A*, **44**, 143–57.

King, P.E., Askew, R.R. and Sanger, C. (1969a) The detection of parasitised hosts by males of *Nasonia vitripennis* (Hym., Pteromalidae) and some possible implications. *Proceedings of The Royal Entomological Society of London*, A, **44**, 85–90.

King, P.E., Ratcliffe, N.A. and Copland, M.J.W. (1969b) The structure of the egg membranes in *Apanteles glomeratus* (L.) (Hymenoptera: Braconidae). *Proceedings of the Royal Entomological Society, London, A*, **44**, 137–42.

King, P.E., Ratcliffe, N.A. and Fordy, M.R. (1971) Oögenesis in a Braconid, *Apanteles glomeratus* (L.) possessing an hydropic type of egg. *Zeitschrift für Zellforschung und Mikroskopische Anatomie*, **119**, 43–57.

Kirby, R.D. and Ehler, L.E. (1977) Survival of *Hippodamia convergens* in grain sorghum. *Environmental Entomology*, **6**, 777–80.

Kiritani, K., Kawahara, S., Sasaba, T. and Nakasuji, F. (1972) Quantitative evaluation of predation by spiders on the green rice leafhopper, *Nephotettix cincticeps* Uhler, by a sight-count method. *Researches on Population Ecology*, **13**, 187–200.

Kirk, W.D.J. (1984) Ecologically selective coloured traps. *Ecological Entomology*, **9**, 35–41.

Kirkpatrick, R.L. and Wilbur, D.A. (1965) The development and habits of the granary weevil, *Sitophilus granarius* within the kernel of wheat. *Journal of Economic Entomology* **58**, 979–85.

Kitano, H. (1969) Experimental studies on the parasitism of *Apanteles glomeratus* L. with special reference to its encapsulation-inhibiting capacity. *Bulletin Tokyo Gakugei University*, **21**, 95–136.

Kitano, H. (1975) Studies on the courtship behavior of *Apanteles glomeratus*, 2. Role of the male wings during courtship and the release of mounting and copulatory behavior in the males. *Kontyû* **43**, 513–21. (In Japanese with English summary).

Kitano, H. (1986) The role of *Apanteles glomeratus* venom in the defensive response of its host, *Pieris rapae crucivora*. *Journal of Insect Physiology*, **32**, 369–75.

Kitching, R.C. (1977) Time resources and population dynamics in insects. *Australian Journal of Ecology*, **2**, 31–42.

Klomp, H. and Teerink, B.J. (1962) Host selection and number of eggs per oviposition in the egg parasite *Trichogramma embryophagum* Htg.. *Nature*, **195**, 1020–21.

Klomp, H., Teerink, B.J. and Ma, W.C. (1980) Discrimination between parasitized and unparasitized hosts in the egg parasite *Trichogramma embryophagum* (Hym: Trichogrammatidae): a matter of learning and forgetting. *Journal of Animal Ecology*, **30**, 254–77.

Knuth, P. (1906) *Handbook of Flower Pollination*, Vol I, Clarendon Press, Oxford.

Knuth, P. (1908) *Handbook of Flower Pollination*, Vol II, Clarendon Press, Oxford.

Knuth, P. (1909) *Handbook of Flower Pollination*, Vol III, Clarendon Press, Oxford.

Kogan, M. and Herzog, D.C. (eds) (1980) *Sampling Methods in Soybean Entomology*, Springer-Verlag, New York.

Kogan, M. and Legner, E.F. (1970) A biosystematic revision of the genus *Muscidifurax* (Hymn. Pteromalidae) with descriptions of four new species. *Canadian Entomologist*, **102**, 1268–90.

Kogan, M. and Parra, J.R.P. (1981) Techniques and applications of measurements of consumption and utilization of food by phytophagous insects, in *Current Topics in Insect Endocrinology and Nutrition* (eds, G. Bhaskaran, S. Friedman and J.G. Rodriguez), Plenum Press, New York, pp. 337–52.

Kogan, M. and Pitre, H.N. (1980) General sampling methods for above-ground populations of soybean arthropods, in *Sampling Methods in Soybean Entomology* (eds M. Kogan and D.C. Herzog), Springer-Verlag, New York, pp. 30–60.

Kokuba, H. and Duelli, P. (1980) Aerial population movement and vertical distribution of aphidophagous insects in cornfields (Chrysopidae, Coccinellidae and Syrphidae), in *Ecology of Aphidophaga* (ed. I. Hodek), Academia, Prague and W. Junk. Dordrecht pp. 279–84.

Kouame, K.L. and Mackauer, M. (1991) Influence of aphid size, age and behaviour on host choice by the parasitoid wasp *Ephedrus californicus*: a test of host-size models. *Oecologia*, **88**, 197–203.

Kowalski, R. (1977) Further elaboration of the winter moth population models. *Journal of Animal Ecology*, **46**, 471–82.

Kramer, D.A., Stinner, R.E. and Hain, F.P. (1991) Time versus rate in parameter estimation of nonlinear temperature-dependent development models. *Environmental Entomology*, **20**, 484–8.

Krebs, J.R. and Davies, N.B. (eds) (1978) *Behavioural Ecology*, Blackwell, Oxford.

Kring, T.J., Gilstrap, F.E. and Michels, G.J. (1985) Role of indigenous coccinellids in regulating greenbugs (Homoptera: Aphididae) on texas grain sorghum. *Journal of Economic Entomology*, **78**, 269–73.

Krishnamoorthy, A. (1984) Influence of adult diet on the fecundity and survival of the predator, *Chrysopa scelestes* (Neur.: Chrysopidae). *Entomophaga*, **29**, 445–50.

Krishnamoorthy, A. (1989) Effect of cold storage on the emergence and survival of the adult exotic parasitoid, *Leptomastix dactylopii* How. (Hym., Encyrtidae). *Entomon*, **14**, 313–18.

Krombein, K.V. (1967) *Trap-nesting Wasps and Bees: Life-histories, Nests and Associates*, Smithsonian Press, Washington, DC.

Kruse, K.C. (1983) Optimal foraging by predaceous diving beetle larvae on toad tadpoles. *Oecologia*, **58**, 383–8.

Kullenberg, B. and Bergström, G. (1976) Hymenoptera Aculeata males as pollinators of *Ophrys* orchids. *Zoologica Scripta*, **5**, 13–23.

Kummer, H. (1960) Experimentelle Untersuchungen zur Wirkung von Fortpflanzungsfaktoren auf die Lebensdauer von *Drosophila melanogaster* Weibchen. *Zeitschrift für Vergleichende Physiologie*, **43**, 642–79.

Kuno, E. (1971) Sampling and analysis of insect populations. *Annual Review of Entomology*, **36**, 285–304.

Kuno, E. and Dyck, V.A. (1985) Dynamics of Philippine and Japanese populations of the brown planthopper: comparison of basic characteristics, in *Proceedings of the ROC-Japan Seminar on the Ecology and the Control of the Brown Planthopper*, Republic of China, National Science Council, pp. 1–9.

Kuperstein, M.L. (1974) Utilisation of the precipitin test for the quantitative estimation of the influence of *Pterostichus crenuliger* (Coleoptera: Carabidae) upon the population dynamics of *Eurygaster integriceps* (Hemiptera: Scutelleridae). *Zoologicheski Zhurnal*, **53**, 557–62.

Kuperstein, M.L. (1979) Estimating carabid effectiveness in reducing the sunn pest, *Eurygaster integriceps* Puton (Heteroptera: Scutelleridae) in the USSR. Serology in insect predator–prey studies. *Entomological Society of America, Miscellaneous Publication*, **11**, 80–4.

Lack, D. (1947) The significance of clutch size. *Ibis*, **89**, 309–52.

Lancaster, J., Hildrew, A.G. and Townsend, C.R. (1991) Invertebrate predation on patchy and mobile prey in streams. *Journal of Animal Ecology*, **60**, 625–41.

Langenbach, G.E.J. and van Alphen, J.J.M. (1985) Searching behaviour of *Epidinocarsis lopezi* (Hymenoptera; Encyrtidae) on cassava: effect of leaf topography and a kairomone produced by its host, the cassava mealybug (*Phenacoccus manihoti*). *Mededelingen van de Faculteit voor Landbouwwetenschappen, Rijksuniversiteit Gent*, **51**, 1057–65.

Lapchin, L., Ferran, A., Iperti, G. *et al.* (1987) Coccinellids (Coleoptera: Coccinellidae) and syrphids (Diptera: Syrphidae) as predators of aphids in cereal crops: a comparison of sampling methods. *Canadian Entomologist*, **119**, 815–22.

LaSalle, J. and Gauld, I.D. (1994) Hymenoptera: their diversity, and their impact on the diversity of other organisms, in *Hymenoptera and Biodiversity* (eds J. LaSalle and I.D. Gauld), C.A.B. International, Wallingford, pp. 1–26.

Laughlin, R. (1965) Capacity for increase: a useful population statistic. *Journal of Animal Ecology*, **34**, 77–91.

Laurent, M. and Lamarque, P. (1974) Utilisation de la methode des captures successives (De Lury) pour L'evaluation des peuplements piscicoles. *Annals of Hydrobiology*, **5**, 121–32.

Lavigne, R.J. (1992) Ethology of *Neoaratus abludo* Daniels (Diptera: Asilidae) in South Australia, with notes on *N. pelago* (Walker) and *N. rufiventris* (Macquart). *Proceedings of the Entomology Society of Washington*, **94**, 253–62.

Lawrence, E.S. and Allen, J.A. (1983) On the term 'search image'. *Oikos*, **40**, 313–14.

Lawrence, P.O. (1981) Host vibration – a cue to host location by the parasite *Biosteres longicaudatus*. *Oecologia*, **48**, 249–51.

Lawton, J.H. (1970) Feeding and food energy assimilation in larvae of the damselfly *Pyrrhosoma nymphula* (Sulz.) (Odonata: Zygoptera). *Journal of Animal Ecology*, **39**, 669–89.

Lawton, J.H., Beddington, J.R. and Bonser, R. (1974) Switching in invertebrate predators, in *Ecological Stability* (eds M.B. Usher and M.H. Williamson), Chapman & Hall, London, pp. 141–58.

Lawton, J.H., Hassell, M.P. and Beddington, J.R. (1975) Prey death rates and rates of increase of arthropod predator populations. *Nature*, **255**, 60–62.

Lawton, J.H., Thompson, B.A. and Thompson, D.J. (1980) The effects of prey density on survival and growth of damsel fly larvae. *Ecological Entomology*, **5**, 39–51.

Leather, S.R. (1988) Size, reproductive potential and fecundity in insects: things aren't as simple as they seem. *Oikos*, **51**, 386–9.

Ledieu, M.S. (1977) Ecological aspects of parasite use under glass, in *Pest Management in Protected Culture Crops* (eds F.F. Smith and R.E. Webb), Proceedings of the 15th International Congress of Entomology, Washington, 1976, USDA, pp. 75–80.

Lee, J.D. and Lee, T.D. (1982) *Statistics and Numerical Methods in BASIC for Biologists*, Van Nostrand Reinhold, New York.

Lee, K.Y., Barr, R.O., Gage, S.H. and Kharkar, A.N. (1976) Formulation of a mathematical model for insect pest ecosystems — the cereal leaf beetle problem. *Journal of Theoretical Biology*, **59**, 33–76.

Leereveld, H. (1982) Anthecological relations between reputedly anemophilous flowers and syrphid flies. III. Worldwide survey of crop and intestine contents of certain anthophilous syrphid flies. *Tijdschrift voor Entomologie*, **125**, 25–35.

Legner, E.F. (1969) Reproductive isolation and size variations in the *Muscidifurax raptor* complex. *Annals of the Entomological Society of America*, **62**, 382–5.

Legner, E.F. (1987) Transfer of thelytoky to arrhenotokous *Muscidifurax raptor* Girault. *Canadian Entomologist*, **119**, 265–271.

Leius, K. (1960) Attractiveness of different foods and flowers to the adults of some hymenopterous parasites. *Canadian Entomologist*, **92**, 369–76.

Leius, K. (1962) Effects of the body fluids of various host larvae on fecundity of females of *Scambus buolianae* (Htg.) (Hymenoptera: Ichneumonidae). *Canadian Entomologist*, **94**, 1078–82.

Leius, K. (1963) Effects of pollens on fecundity and longevity of adult *Scambus buolianae* (Htg.) (Hymenoptera: Ichneumonidae). *Canadian Entomologist*, **95**, 202–7.

van Lenteren, J.C. (1976) The development of host discrimination and the prevention of superparasitism in the parasite *Pseudeucoila bochei* Weld (Hym.: Cynipidae). *Netherlands Journal of Zoology*, **26**, 1–83.

van Lenteren, J.C. (1980) Evaluation of control capabilities of natural enemies: does art have to become science? *Netherlands Journal of Zoology*, **30**, 369–81.

van Lenteren, J.C. (1986) Parasitoids in the greenhouse: successes with seasonal inoculative release systems. In *Insect Parasitoids* (eds J. Waage and D. Greathead, Academic Press, London, pp. 341–374.

van Lenteren, J.C. (1991) Encounters with parasitised hosts: to leave or not to leave the patch. *Netherlands Journal of Zoology*, **41**, 144–57.

van Lenteren, J.C. and Bakker, K. (1978) Behavioural aspects of the functional responses of a parasite (*Pseudeucoila bochei* Weld) to its host (*Drosophila melanogaster*). *Netherlands Journal of Zoology*, **28**, 213–33.

van Lenteren, J.C., Bakker, K. and van Alphen, J.J.M. (1978) How to analyse host discrimination. *Ecological Entomology*, **3**, 71–5.

van Lenteren, J.C., van Vianen, A., Gast, H.F. and Kortenhoff, A. (1987) The parasite–host relationship between *Encarsia formosa* Gahan (Hymenoptera: Aphelinidae) and *Trialeurodes vaporariorum* (Westwood) (Homoptera: Aleyrodidae), XVI. Food effects on oögenesis, life span and fecundity of *Encarsia formosa* and other hymenopterous parasites. *Zeitschrift für Angewandte Entomologie*, **103**, 69–84.

Leonard, S.H. and Ringo, J.M. (1978) Analysis of male courtship patterns and mating behavior of *Brachymeria intermedia*. *Annals of the Entomological Society of America*, **71**, 817–26.

Leopold, R.A. (1976) The role of male accessory glands in insect reproduction. *Annual Review of Entomology*, **21**, 199–221.

LeSar, C.D. and Unzicker, J.D. (1978) Soybean spiders: species composition, population densities, and vertical distribution. *Illinois Natural History Survey Biological Notes*, **107**, 1–14.

Leslie, G.W. and Boreham, P.F.L. (1981) Identification of arthropod predators of *Eldana saccharina* Walker (Lepidoptera: Pyralidae) by cross-over electrophoresis. *Journal of The Entomological Society of South Africa*, **44**, 381–88.

Leslie, P.H. (1945) On the use of matrices in certain population mathematics. *Biometrika*, **33**, 183–212.

Leslie, P.H. (1948) Some further notes on the use of matrices in population mathematics. *Biometrika*, **35**, 213–45.

Lessells, C.M. (1985) Parasitoid foraging: should parasitism be density dependent? *Journal of Animal Ecology*, **54**, 27–41.

Lewis, D.J. and Domoney, C.R. (1966) Sugar meals in Phlebotominae and Simuliidae (Diptera), *Proceedings of the Royal Entomological Society*, A, **41**, 175–9.

Lewis, W.H., Vinay, P. and Zenger, V.E. (1983) *Airborne and Allergenic Pollen of North America*, Johns Hopkins Press, Baltimore.

Liddell, E.J. and Cryer, A. (1991) *A Practical Guide to Monoclonal Antibodies*, Wiley and Sons Ltd., Chichester.

Liddell, J.E. and Symondson, W.O.C. (1995) The potential of combinatorial gene libraries in pest-predator relationship studies, in *The Ecology of Agricultural Pests: Biochemical Approaches* (eds W.O.C. Symondson and J.E. Liddell), Chapman and Hall, London. (in press).

Lincoln, F.C. (1930) Calculating waterfowl abundance on the basis of banding returns. *USDA Circular*, **118**, 1–4.

Lingren, P.D., Ridgway, R.L. and Jones, S.L. (1968) Consumption by several common arthropod predators of eggs and larvae of two *Heliothis* species that attack cotton. *Annals of the Entomological Society of America*, **61**, 613–18.

Liske, E. and Davis, W.J. (1984) Sexual behaviour in the Chinese praying mantis. *Animal Behaviour*, **32**, 916–18.

Liske, E. and Davis, W.J. (1987) Courtship and mating behaviour of the Chinese praying mantis *Tenodera aridifolia chinensis*. *Animal Behaviour*, **35**, 1524–37.

Liu, S.-S. (1985a) Development, adult size and fecundity of *Aphidius sonchi* reared in two instars of its aphid host, *Hyperomyzus lactucae*. *Entomologia Experimentalis et Applicata*, **37**,41–8.

Liu, S.-S. (1985b) Aspects of the numerical and functional responses of the aphid parasite, *Aphidius sonchi*, in the laboratory *Entomologia Experimentalis et Applicata*, **37**, 247–56.

Livdahl, T.P. and Stiven, A.E. (1983) Statistical difficulties in the analysis of predator functional response data. *Canadian Entomologist*, **115**, 1365–70.

Livdahl, T.P. and Sugihara, G. (1984) Non-linear interactions of populations and the importance of estimating *per capita* rates of change. *Journal of Animal Ecology*, **53**, 573–80.

Lloyd, J.E. (1971) Bioluminiscent communication in insects. *Annual Review of Entomology*, **16**, 97–122.

Lloyd, J.E. (1975) Aggressive mimicry in *Photuris:* signal repertoires by femmes fatales. *Science*, **187**, 452–3.

Logan, J.A., Wollkind, D.J., Hoyte, S.C. and Tanigoshi L.K. (1976) An analytical model for description of temperature dependent rate phenomena in arthropods. *Environmental Entomology*, **5**, 1133–40.

Lohr, B., Varela, A.M. and Santos, B. (1989) Life-table studies on *Epidinocarsis lopezi* (DeSantis) (Hym., Encyrtidae), a parasitoid of the cassava mealybug, *Phenacoccus manihoti* Mat.-Ferr. (Hom., Pseudococcidae). *Journal of Applied Entomology*, **107**, 425–34.

de Loof, A (1987) The impact of the discovery of vertebrate-type steroids and peptide hormone-like substances in insects. *Entomologia Experimentalis et Applicata*, **45**, 105–13.

Lopez, R., Ferro, D.N. and van Driesche, R.G. (1994) Two tachinid species discriminate between parasitised and non-parasitised hosts *Entomologia Experimentalis et. Applicata* , **74**, 37–45.

Lorenz, K. (1939) Vergleichende Verhaltensforschung. *Zoologische Anzeiger Supplementband* **12**, 69–102.

Lotka, A.J. (1922) The stability of the normal age distribution. *Proceedings of the National Academy of Sciences of the USA*, **8**, 339–45.

Lotka, A.J. (1925) *Elements of Physical Biology*, Williams and Wilkins, Baltimore, USA.

Lövei, G.L., McDougall, D., Bramley, G., Hodgson, D.J. and Wratten, S.D. (1992) Floral resources for natural enemies: the effect of *Phacelia tanacetifolia* (Hydrophylaceae) on within-field distribution of hoverflies (Diptera: Syrphidae). *Proceedings of the 45th New Zealand Plant Protection Conference, 1992*, pp. 60–61.

Lövei, G.L., Monostori, E. and Andó, I. (1985) Digestion rate in relation to starvation in the larva of a carabid predator, *Poecilus cupreus*. *Entomologia Experimentalis et Applicata*, **37**, 123–7.

Lowe, A.D. (1968) The incidence of parasitism and disease in some populations of the cabbage aphid (*Brevicoryne brassicae* L.) in New Zealand. *New Zealand Journal of Agricultural Research*, **11**, 821–8.

Loxdale, H.D., Brookes, C.P. and DeBarro, P.J. (1995) Novel molecular markers in agricultural entomology: the quest for DNA methods to replace isozymes, in *The Ecology of Agricultural Pests: Biochemical Approaches* (eds W.O.C. Symondson and J.E. Liddell), Chapman and Hall, London. (in press).

Loxdale, H.D., Castañera, P. and Brookes, C.P. (1983) Electrophoretic study of enzymes from cereal aphid populations. I. Electrophoretic techniques and staining systems for characterising isoenzymes from six species of cereal aphid (Hemiptera: Aphididae). *Bulletin of Entomological Research*, **73**, 645–57.

Luck, R.F. (1990) Evaluation of natural enemies for biological control: a behavioural approach. *Trends in Ecology and Evolution*, **5**, 196–2.

Luck, R.F., Podoler, H. and Kfir, R. (1982) Host selection and egg allocation behaviour by *Aphytis melinus* and *A. lingnanensis*: a comparison of two facultatively gregarious parasitoids. *Ecological Entomology*, **7**, 397–408.

Luck, R.F., Shepard, B.M. and Kenmore, P.E. (1988) Experimental methods for evaluating arthropod natural enemies. *Annual Review of Entomology*, **33**, 367–91.

Luck, R.F., Stouthamer, R. and Nunney, L.P. (1993) Sex determination and sex ratio patterns in parasitic Hymenoptera, in *Evolution and Diversity of Sex ratio in Insects and Mites* (eds D.L. Wrensch and M.A. Ebbert), Chapman & Hall, New York, pp. 442–76.

Luff, M.L. (1968) Some effects of formalin on the numbers of Coleoptera caught in pitfall traps. *Entomologist's Monthly Magazine*, **104**, 115–16.

Luff, M.L. (1975) Some features influencing the efficiency of pitfall traps. *Oecologia*, **19**, 345–57.

Luff, M.L. (1978) Diel activity patterns of some field Carabidae. *Ecological Entomology*, **3**, 53–62.

Luff, M.L. (1986) Aggregation of some Carabidae in pitfall traps, in *Carabid Beetles – Their Adaptations and Dynamics* (eds P.J. den Boer, M.L. Luff, D. Mossakowski and F. Weber), Fischer, Stuttgart, pp. 385–97.

Luff, M.L. (1987) Biology of polyphagous ground beetles in agriculture. *Agricultural Zoology Reviews*, **2**, 237–78.

Lum, P.T.M. and Flaherty, B.R. (1973) Influence of continuous light on oocyte maturation in *Bracon hebetor*. *Annals of the Entomological Society of America*, **66**, 355–7.

Lys, J.A. and Nentwig, W. (1991) Surface activity of carabid beetles inhabiting cereal fields. Seasonal phenology and the influence of farming operations on five abundant species. *Pedobiologia*, **35**, 129–38.

MacArthur, R.H. and Pianka, E.R. (1966) On optimal use of a patchy environment. *American Naturalist*, 100, 603–609.

Mack, T.P. and Smilowitz, Z. (1982) CMACSIM, a temperature-dependent predator–prey model simulating the impact of *Coleomegilla maculata* (DeGeer) on green peach aphids on potato plants. *Environmental Entomology*, **11**, 1193–201.

Mackauer, M. (1983) Quantitative assessment of *Aphidius smithi* (Hymenoptera: Aphidiidae): fecundity, intrinsic rate of increase, and functional response. *Canadian Entomologist*, **115**, 399–415.

Mackauer, M. (1986) Growth and developmental interactions in some aphids and their hymenopteran parasites. *Journal of Insect Physiology*, **32**, 275–80.

Mackauer, M. (1990) Host discrimination and larval competition in solitary endoparasitoids, in *Critical Issues in Biological Control* (eds M. Mackauer, L.E. Ehler and J. Roland), Intercept, Andover, pp. 14–62.

Mackauer, M. and Henkelman, D.H. (1975) Effect of light–dark cycles on adult emergence in the aphid parasite *Aphidius smithi*. *Canadian Journal of Zoology*, **53**, 1201–06.

Mackauer, M. and Sequeira, R. (1993) Patterns of development in insect parasites, in *Parasites and Pathogens of Insects, Vol. 1* (eds N.E. Beckage, S.N. Thompson and B.A. Frederici), Academic Press, London, pp. 1–23.

Mackauer, M. and Way, M.J. (1976) *Myzus persicae* (Sulz.) an aphid of world importance, in *I.B.P. vol.*

9, *Studies in Biological Control*. (ed. V.L. Delucchi), Cambridge University Press, London, pp. 51–120.

MacVicker, J.A.K., Moore, J.S. and Molyneux, D.H. (1990) Honeydew sugars in wild-caught Italian phlebotomine sandflies (Diptera: Psychodidae) as detected by high performance liquid chromatography. *Bulletin of Entomological Research*, **80**, 339–44.

Mader, H.J., Schell, C. and Kornacher, P. (1990) Linear barriers to arthropod movements in the landscape. *Biological Conservation*, **54**, 209–22.

Maier, C.T. and Waldbauer, G.P. (1979) Diurnal activity patterns of flower flies (Diptera, Syrphidae) in an Illinois sand area. *Annals of the Entomological Society of America*, **72**, 237–45.

Maingay, H.D., Bugg, R.L., Carlson, R.W. and Davidson, N.A. (1991) Predatory and parasitic wasps (Hymenoptera) feeding at flowers of sweet fennel (*Foeniculum vulgare* Miller var. *dulce* Battandier & Trabut, Apiaceae) and Spearmint (*Mentha spicata* L., Lamiaceae) in Massachusetts. *Biological Agriculture and Horticulture*, **7**, 363–83.

Mangel, M. (1989a) Evolution of host selection in parasitoids: does the state of the parasitoid matter? *American Naturalist*, **133**, 157–72.

Mangel, M. (1989b) An evolutionary explanation of the motivation to oviposit. *Journal of Evolutionary Biology*, **2**, 157–72.

Mangel, M., and Clark, C.W. (1988) *Dynamic Modeling in Behavioural Ecology*, Princeton University Press, Princeton, New Jersey.

Mani, M. and Nagarkatti, S. (1983) Relationship between size of *Eucelatoria bryani* Sabrosky females and their longevity and fecundity. *Entomon*, **8**, 83–6.

Manly, B.F.J. (1971) Estimates of a marking effect with capture–recapture sampling. *Journal of Applied Ecology*, **8**, 181–89.

Manly, B.F.J. (1972) Tables for the analysis of selective predation experiments. *Researches on Population Ecology*, **14**, 74–81.

Manly, B.F.J. (1973) A linear model for frequency-dependent selection by predators. *Researches on Population Ecology*, **14**, 137–50.

Manly, B.F.J. (1974) A model for certain types of selection experiments. *Biometrics*, **30**, 281–94.

Manly, B.F.J. (1977) The determination of key factors from life table data. *Oecologia*, **31**, 111–17.

Manly, B.F.J. (1988) A review of methods for key factor analysis, in *Estimation and Analysis of Insect Populations* (eds L. McDonald, B. Manly, J. Lockwood and J. Logan), Springer-Verlag, Berlin, pp. 169–90.

Manly, B.F.J. (1990) *Stage-structured Populations*, Chapman & Hall, London.

Manly, B.F.J. and Parr, M.J. (1968) A new method of estimating population size, survivorship, and birth rate from capture–recapture data. *Transactions of the Society for British Entomology*, **18**, 81–9.

Manly, B.F.J., Miller, P. and Cook, L.M. (1972) Analysis of a selective predation experiment. *American Naturalist*, **106**, 719–36.

Marks, R.J. (1977) Laboratory studies of plant searching behaviour by *Coccinella septempunctata* L. larvae. *Bulletin of Entomological Research*, **67**, 235–41.

van Marle, J. and Piek, T. (1986) Morphology of the venom apparatus, in *Venoms of the Hymenoptera. Biochemical, Pharmacological and Behavioural Aspects*. (ed. T.O. Piek), Academic Press, London, pp. 17–44.

Marston, N.L. (1980) Sampling parasitoids of soybean insect pests, in *Sampling Methods in Soybean Entomology* (eds M. Kogan and D.C. Herzog), Springer-Verlag, New York, pp. 481–504.

Martin, P. and Bateson, P.P.G. (1993) *Measuring Behaviour. An Introductory Guide*, 2nd edn Cambridge University Press, Cambridge.

Martinez, N.D., Hawkins, B.A., Dawah, H.A. and Feifarek, B. (1995) Accurate estimation of food web structure with moderate observation effort (submitted).

Mascanzoni, D. and Wallin, H. (1986) The harmonic radar: a new method of tracing insects in the field. *Ecological Entomology*, **11**, 387–90.

Masner, L. (1976) Yellow pan traps (Moreicke traps, Assiettes jaunes). *Proctos*, **2**(2), 2.

Masner, L. and Goulet, H. (1981) A new model of flight-interception trap for some hymenopterous insects. *Entomology News*, **92**, 199–202.

le Masurier, A.D. (1991) Effect of host size on clutch size in *Cotesia glomerata*. *Journal of Animal Ecology*, **60**, 107–18.

Matsura, T. (1986) The feeding ecology of the pit-making ant-lion larva, *Myrmeleon bore*: feeding rate and species composition of prey in a habitat, *Ecological Research*, **1**, 15–24.

Matsura, T. and Morooka, K. (1983) Influences of prey density on fecundity in a mantis, *Paratenodera angustipennis* (S.). *Oecologia*, **56**, 306–12.

Matthews, R.W. (1975) Courtship in parasitic wasps, in *Evolutionary Strategies of Parasitic Insects and Mites* (ed. P.W. Price) Plenum Press, New York, pp. 66–86.

Matthews, R.W. and Matthews, J.R. (1971) The Malaise trap: its utility and potential for sampling insect populations. *The Michigan Entomologist*, **4**, 117–22.

Matthews, R.W. and Matthews, J.R. (1983) Malaise traps: the Townes model catches more insects. *Contributions of the American Entomological Institute*, **20**, 428–32.

Maund, C.M. and Hsiao, T.H. (1991) Differential encapsulation of two *Bathyplectes* parasitoids among alfalfa weevil strains, *Hypera postica* (Gyllenhal). *Canadian Entomologist*, **123**, 197–203.

Maurizio, A. (1975) How bees make honey, in *Honey: A Comprehensive Survey* (ed. E. Crane), Heinemann, London, pp. 77–105.

May, P. (1988) Determinants of foraging profitability in two nectarivorous butterflies. *Ecological Entomology*, **13**, 171–84.

May, R.M. (1974) *Stability and Complexity in Model Ecosystems*, Princeton University Press, Princeton.

May, R.M. (1975) Patterns of species diversity, in *Ecology and Evolution of Communities* (eds M.L. Cody and J.M. Diamond), Belknap Press, Cambridge/London, pp. 81–120.

May, R.M. (1976) Estimating *r*: a pedagogical note. *American Naturalist*, **110**, 496–9.

May, R.M. (1978) Host–parasitoid systems in patchy environments: a phenomenological model. *Journal of Animal Ecology*, **47**, 833–43.

May, R.M. (1986) When two and two do not make four: nonlinear phenomena in ecology. *Proceedings of the Royal Society of London, B*, **228**, 241–66.

May. R.M. and Hassell, M.P. (1981) The dynamics of multiparasitoid–host interactions. *American Naturalist*, **117**, 234–61.

May, R.M. and Hassell, M.P. (1988) Population dynamics and biological control. *Philosophical Transactions of the Royal Society of London, B*, **318**, 129–69.

May, R.M., Hassell, M.P., Anderson, R.M. and Tonkyn, D.W. (1981) Density dependence in host–parasitoid models. *Journal of Animal Ecology*, **50**, 855–865.

Maynard Smith, J. (1972) *On Evolution*, Edinburgh University Press, Edinburgh.

Maynard Smith, J. (1974) The theory of games and the evolution of animal conflicts. *Journal of Theoretical Biology*, **47**, 209–21.

Maynard Smith, J. (1978) *The Evolution of Sex*, Cambridge University Press, Cambridge.

Maynard Smith, J. (1982) *Evolution and the Theory of Games*, Cambridge University Press, Cambridge.

McArdle, B.H. (1977) An investigation of a *Notonecta glauca-Daphnia magna* predator-prey system. University of York PhD. Thesis.

McBrien, H. and Mackauer, M. (1990) Heterospecific larval competition and host discrimination in two species of aphid parasitoids: *Aphidius ervi* and *Aphidius smithi*. *Entomologia Experimentalis et Applicata*, **56**, 145–53.

McCarty, M.T., Shepard, M. and Turnipseed, S.G. (1980) Identification of predacious arthropods in soybeans by using autoradiography. *Environmental Entomology*, **9**, 199–203.

McCauley, E. and Murdoch, W.W. (1987) Cyclic and stable populations: plankton as paradigm. *American Naturalist*, **129**, 97–121.

McCauley, V.J.E. (1975) Two new quantitative samplers for aquatic phytomacrofauna. *Hydrobiologia*, **47**, 81–9.

McClain, D.C., Rock, G.C. and Stinner, R.E. (1990a) Thermal requirements for development and simulation of the seasonal phenology of *Encarsia perniciosi* (Hymenoptera: Aphelinidae), a parasitoid of the San Jose Scale (Homoptera: Diaspididae) in North Carolina orchards. *Environmental Entomology*, **19**, 1396–1402.

McClain, D.C., Rock, G.C. and Woolley, J.B. (1990b) Influence of trap color and San Jose scale (Homoptera: Diaspididae) pheromone on sticky trap catches of 10 aphelinid parasitoids (Hymenoptera). *Environmental Entomology*, **19**, 926–31.

McDaniel, S.G. and Sterling, W.L. (1979) Predator determination and efficiency on *Heliothis virescens* eggs in cotton using ^{32}P. *Environmental Entomology*, **8**, 1083–87.

McDonald, I.C. (1976) Ecological genetics and the sampling of insect populations for laboratory colonization. *Environmental Entomology*, **5**, 815–20.

McDonald, L., Manly, B., Lockwood, J. and Logan, J. (1988) *Estimation and Analysis of Insect Populations*, Springer-Verlag, Berlin.

McEwen, P., Jervis, M.A. and Kidd, N.A.C. (1993) Influence of artificial honeydew on larval development and survival in *Chrysoperla carnea* (Neur.: Chrysopidae). *Entomophaga*, **38**, 241–4.

McEwen, P.K., Jervis, M.A. and Kidd, N.A.C. (1994) Use of a sprayed L-tryptophan solution to concentrate numbers of the green lacewing *Chrysoperla carnea* in olive tree canopy. *Entomologia Experimentalis et Applicata*, **70**, 97–99.

McGuire, A.D. and Armbruster, W.S. (1991) An experimental test for reproductive interactions between 2 sequentially blooming *Saxifraga* species (Saxifragaceae). *American Journal of Botany*, **78**, 214–19.

McIver, I.D. (1981) An examination of the utility of the precipitin test for evaluation of arthropod predator-prey relationships. *Canadian Entomologist*, **113**, 213–22.

McKenna, M.A. and Thomson, J.D. (1988) A technique for sampling and measuring small amounts of floral nectar. *Ecology*, 1306–7.

McNair, J.N. (1986) The effects of refuges on predator–prey interactions: a reconsideration. *Theoretical Population Biology*, **29**, 38–63.

McPeek, M.A. and Crowley, P.H. (1987) The effects of density and relative size on the aggressive behaviour, movement, and feeding of damselfly larvae (Odonata: Coenagrionidae). *Animal Behaviour*, **35**, 1051–61.

McPhail, M. (1937) Relation of time of day, temperature and evaporation to attractiveness of fermenting sugar solution to Mexican fruitfly. *Journal of Economic Entomology*, **30**, 793–9.

McPhail, M. (1939) Protein lures for fruit files. *Journal of Entomology*, **32**, 758–61.

McVey, M.E. and Smittle, B.J. (1984) Sperm precedence in the dragonfly *Erythemis simplicicollis*. *Journal of Insect Physiology*, **30**, 619–28.

Meelis, E. (1982) Egg distribution of insect parasitoids: a survey of models. *Acta Biotheoretica*, **31**, 109–26.

van den Meiracker, R.A.F., Hammond, W.N.O. and van Alphen, J.J.M. (1990) The role of kairomone in prey finding by *Diomus* sp. and *Exochomus* sp., two coccinellid predators of the cassava mealybug, *Phenacoccus manihoti*. *Entomologia Experimentalis et Applicata*, **56**, 209–17.

Memmott, J., Godfray, H.C.J. and Gauld, I.D. (1994) The structure of a tropical host-parasitoid community. *Journal of Animal Ecology*, **63**, 521–40.

Mendel, M.J., Shaw, P.B. and Owens, J.C. (1987) Life history characteristics of *Anastatus semiflavidus* (Hymenoptera: Eupelmidae), an egg parasitoid of the range caterpillar, *Hemileuca oliviae* (Lepidoptera: Saturniidae) over a range of temperatures. *Environmental Entomology*, **16**, 1035–41.

Menken, S.B. and Raijmann, L.E.L. (1995) Biochemical systematics: principles and perspectives for pest management, in *The Ecology of Agricultural Pests: Biochemical Approaches* (eds W.O.C. Symondson and J.E. Liddell), Chapman and Hall, London. (in press).

Messenger, P.S. (1964a) The influence of rhythmically fluctuating temperature on the development and reproduction of the spotted alfalfa aphid *Therioaphis maculata*. *Journal of Economic Entomology*, **57**, 71–6.

Messenger, P.S. (1964b) Use of life tables in a bioclimatic study of an experimental aphid–braconid wasp host–parasite system. *Ecology*, **45**, 119–31.

Messenger, P.S. (1970) Bioclimatic inputs to biological control and pest management programs. In *Concepts of Pest Management* (eds R.L. Rabb and F.E. Guthrie), North Carolina State University, Raleigh, pp. 84–99.

Messenger, P.S. and van den Bosch, R. (1971) The adaptability of introduced biological control agents. In *Biological Control* (ed. C.F. Huffaker), Plenum, New York, pp. 68–92.

Messing, R.H. and Aliniazee, M.T. (1988) Hybridization and host suitability of two biotypes of *Trioxys pallidus* (Hymenoptera: Aphidiidae). *Annals of the Entomological Society of America*, **81**, 6–9.

Messing, R.H. and Aliniazee, M.T. (1989) Introduction and establishment of *Trioxys pallidus* (Hym.: Aphidiidae) in Oregon, USA for control of filbert aphid *Myzocallis coryli* (Hom.: Aphididae). *Entomophaga*, **34**, 153–63.

Michels Jr, G.J. and Behle, R.W. (1991) Effects of two prey species on the development of *Hippodamia sinuata* (Coleoptera: Coccinellidae) larvae at constant temperatures. *Journal of Economic Entomology*, **84**, 1480–84.

Michelsen, A. (1983) Biophysical basis of sound communication, in *Bioacoustics: A Comparative Approach* (ed. B. Lewis), Academic Press, London, pp. 3–38.

Michelsen, A., Fink, F., Gogala, M. and Traue, D. (1982) Plants as transmission channels for insect vibrational songs. *Behavioural Ecology and Sociobiology*, **11**, 269–81.

Michiels, N.K. and Dhondt, A.A. (1988) Direct and indirect estimates of sperm precedence and displacement in the dragonfly *Sympetrum danae* (Odonata : Libellulidae). *Behavioural Ecology and Sociobiology*, **23**, 257–63.

Middleton, R.J. (1984) The distribution and feeding ecology of web-spinning spiders living in the canopy of Scots Pine (*Pinus sylvestris* L.). University College, Cardiff, Ph.D. Thesis.

Miles, L.R. and King, E.G. (1975) Development of the tachinid parasite, *Lixophaga diatraeae*, on various developmental stages of the sugar cane borer in the laboratory. *Environmental Entomology*, **4**, 811–14.

Miller, M.C. (1981) Evaluation of enzyme-linked immunosorbent assay of narrow- and broad-spectrum anti-adult Southern pine beetle serum. *Annals of the Entomological Society of America*, **74**, 279–82.

Miller, M.C., Chappell, W.A., Gamble, W. and Bridges, J.R. (1979) Evaluation of immunodiffusion and immunoelectrophoretic tests using a broad-spectrum anti-adult Southern Pine Beetle serum. *Annals of the Entomological Society of America*, **72**, 99–104.

Miller, P.L. (1982) Genital structure, sperm competition and reproductive behaviour in some African libellulid dragonflies. *Advances in Odonatology*, **1**, 175–92.

Miller, P.L. (1987) *Dragonflies, Naturalists' Handbooks No 7*, Cambridge University Press, Cambridge.

Mills, N.J. (1981) Some aspects of the rate of increase of a coccinellid, *Ecological Entomology*, **6**, 293–9.

Mills, N.J. (1982a) Satiation and the functional response: a test of a new model. *Ecological Entomology*, **7**, 305–15.

Mills, N.J. (1982b) Voracity, cannibalism and coccinellid predation. *Annals of Applied Biology*, **101**, 144–8.

Minkenberg, O. (1989) Temperature effects on the life history of the eulophid wasp *Diglyphus isaea*, an ectoparasitoid of leafminers (*Liriomyza* spp.), on tomatoes. *Annals of Applied Biology*, **115**, 381–97.

Mitchell, P., Arthur, W. and Farrow, M. (1992) An investigation of population limitation using factorial experiments. *Journal of Animal Ecology*, **61**, 591–8.

Miura, K. (1990) Life-history parameters of *Gonatocerus cincticipitis* Sahad (Hym., Mymaridae), an egg parasitoid of the green rice leafhopper, *Nephotettix cincticeps* Uhler (Hom., Cicadellidae). *Journal of Applied Entomology*, **110**, 353–7.

Mols, P.J.M. (1987) Hunger in relation to searching behaviour, predation and egg production of the carabid beetle *Pterostichus coerulescens* L.: results of simulation. *Acta Phytopathologica et Entomologica Hungarica.*, **22**, 187–205.

Moore, J.S., Kelly, T.B., Killick-Kendrick, R., Killick-Kendrick, M. and Molyneux, D.H. (1987) Honeydew sugars in wild-caught *Phlebotomus ariasi* detected by high performance liquid chromatography (HPLC) and gas chromatography (GC). *Medical and Veterinary Entomology*, **1**, 427–34.

Moore, P.D., Webb, J.A. and Colinson, M.E. (1991) *Pollen Analysis*, Blackwell, Oxford.

Morales, J. and Hower, A.A. (1981) Thermal requirements for development of the parasite *Microctonus aethiopoides*. *Environmental Entomology*, **10**, 279–84.

Moratorio, M.S. (1987) Effect of host species on the parasitoids *Anagrus mutans* and *Anagrus silwoodensis* Walker (Hymenoptera: Mymaridae). *Environmental Entomology*, **16**, 825–7.

Moreby, S. (1991) A simple time-saving improvement to the motorized insect suction sampler. *The Entomologist*, **110**, 2–4.

Moreno, D.S., Gregory, W.A. and Tanigoshi, L.K. (1984) Flight response of *Aphytis melinus* (Hymenoptera: Aphelinidae) and *Scirtothrips citri* (Thysanoptera: Thripidae) to trap color, size and shape. *Environmental Entomology*, **13**, 935–40.

Morris, R.F. (1959) Single factor analysis in population dynamics. *Ecology*, **40**, 580–88.

Morris, R.F. and Fulton, W.C. (1970) Models for the development and survival of *Hyphantria cunea* in relation to temperature and humidity. *Memoires de la Societé Entomologique du Canada*, **70**, 1–60.

Morrison, G. and Strong, D.R. Jr. (1980) Spatial variations in host density and the intensity of parasitism: some empirical examples. *Environmental Entomology*, **9**, 149–52.

Morrison, G. and Strong Jr, D.R. (1981) Spatial variation in egg density and the intensity of parasitism in a neotropical chrysomelid (*Cephaloleia consanguinea*). *Ecological Entomology*, **6**, 55–61.

Morse, D.H. (1977) Resource partitioning in bumblebees: the role of behavioural factors. *Science*, **197**, 678–79.

Mountford, M.D. (1988) Population regulation, density dependence, and heterogeneity. *Journal of Animal Ecology*, **57**, 845–58.

Mueke, J.M., Manglitz, G.R. and Kerr, W.R. (1978) Pea aphid: interaction of insecticides and alfalfa varieties. *Journal of Economic Entomology*, **71**, 61–65.

Mukerji, M.K. and LeRoux, E.J. (1969) The effect of predator age on the functional response of *Podisus maculiventris* to the prey size of *Galleria mellonella*. *Canadian Entomologist*, **101**, 314–27.

Müller, H. (1883) *Fertilisation of Flowers* (translated by W. D'Arcy Thompson). Macmillan & Co., London.

Münster-Swendsen, M. and Nachman, G. (1978) Asynchrony in insect host–parasite interaction and its effect on stability, studied by a simulation model. *Journal of Animal Ecology*, **47**, 159–71.

Murakami, Y. and Tsubaki, Y. (1984) Searching efficiency of the lady beetle *Coccinella septempunctata* larvae in uniform and patchy environments. *Journal of Ethology*, **2**, 1–6.

Murdoch, W.W. (1969) Switching in general predators: experiments on predator specificity and stability of prey populations. *Ecological Monographs*, **39**, 335–54.

Murdoch. W.W. (1990) The relevance of pest–enemy models to biological control, in *Critical Issues in Biological Control* (eds M. Mackauer, L.E. Ehler and J. Roland), Intercept, Andover, pp. 1–24.

Murdoch, W.W. and McCauley, E. (1985) Three distinct types of dynamic behaviour shown by a single planktonic system. *Nature*, **316**, 628–30.

Murdoch, W.W. and Oaten, A.A. (1975) Predation and population stability. *Advances in Ecological Research*, **9**, 1–131.

Murdoch, W.W. and Reeve, J.D. (1987) Aggregation of parasitoids and the detection of density dependence in field populations. *Oikos*, **50**, 137–41.

Murdoch, W.W. and Sih, A. (1978) Age-dependent interference in a predatory insect. *Journal of Animal Ecology*, **47**, 581–92.

Murdoch, W.W. and Stewart-Oaten, A. (1989) Aggregation by parasitoids and predators: effects on equilibrium and stability. *American Naturalist*, **134**, 288–310.

Murdoch, W.W. and Walde, S.J. (1989) Analysis of insect population dynamics, in *Towards a More Exact Ecology* (eds P.J. Grubb and J.P. Whittaker) Blackwell, Oxford, pp., 113–40.

Murdoch, W.W., Chesson, J. and Chesson, P.L. (1985) Biological control in theory and practice. *American Naturalist*, **125**, 344–66.

Murdoch, W.W., McCauley, E., Nisbet, R.M., Gurney, W.S.C. and de Roos, A.M. (1992) Individual-based models: combining testability and generality, in *Individual-based Models and Approaches in Ecology* (eds D.L. DeAngeles and L.J. Gross), Chapman & Hall, New York, pp. 18–35.

Murdoch, W.W., Nisbet, R.M., Blythe, S.P., Gurney, W.S.C. and Reeve, J.D. (1987) An invulnerable age-class and stability in delay-differential parasitoid–host models. *American Naturalist*, **129**, 263–82.

Murray, R.A. and Solomon, M.G. (1978) A rapid technique for analysing diets of invertebrate predators by electrophoresis. *Annals of Applied Biology*, **90**, 7–10.

Myers, J.H., Higgins, C. & Kovacs, E. (1989) How many insect species are necessary for the biological control of insects? *Environmental Entomology* **18**: 541–47.

Nadel, H. (1987) Male swarms discovered in Chalcidoidea (Hym., Encyrtidae, Pteromalidae). *Pan-Pacific Entomology*, **63,** 242–6.

Nadel, H. and van Alphen, J.J.M. (1987) The role of host and host plant odours in the attraction of a parasitoid, *Epidinocarsis lopezi*, to the habitat of its host, the cassava mealybug, *Phenacoccus manihoti*. *Entomologia Experimentalis et Applicata*, **45**, 181–6.

Nadel, H. and Luck, R.F. (1985) Span of female emergence and male sperm depletion in the female biased quasi gregarious parasitoid *Pachycrepoideus vindemiae* (Hym., Pteromalidae). *Annals of the Entomological Society of America*, **78**, 410–4.

Nagarkatti, S. (1970) The production of a thelytokous hybrid in an interspecific cross between two species of *Trichogramma*. *Current Science*, **39**, 76–8.

Nakamura, M. and Nakamura, K. (1977) Population dynamics of the chestnut gall wasp, *Dryocosmus kuriphilus* Yasumatsu (Hymenoptera: Cynipidae). *Oecologia*, **27**, 97–116.

Nakamuta, K. (1982) Switchover in searching behaviour of *Coccinella septempunctata* L. (Coleoptera: Coccinellidae) caused by prey consumption. *Applied Entomology and Zoology*, **17**, 501–6.

Naseer, M. and Abdurahman, U.C. (1990) Reproductive biology and predatory behaviour of the anthocorid bugs (Anthocoridae: Hemiptera) associated with the coconut caterpillar *Opisina arenosella* (Walker). *Entomon*, **15**, 149–58.

Naylor, T.H. (1971) *Computer Simulation Experiments with Models of Economic Systems*, John Wiley, New York.

Nealis, V.G. (1988) Weather and the ecology of *Apanteles fumiferanae* Vier. (Hymenoptera: Braconidae). *Memoirs of the Entomological Society of Canada*, **146**, 57–70.

Nealis, V.G. and Fraser, S. (1988) Rate of development, reproduction, and mass-rearing of *Apanteles fumiferanae* Vier. (Hymenoptera: Braconidae) under controlled conditions. *Canadian Entomologist*, **120**, 197–204.

Nealis, V.G., Jones, R.E. and Wellington, W.G. (1984) Temperature and development in host–parasite relationships. *Oecologia*, **61**, 224–229.

Nechols, J.R. and Tauber, M.J. (1977) Age specific interaction between the greenhouse whitefly and *Encarsia formosa*: influence of host on the parasite's oviposition and development. *Environmental Entomology*, **6**, 143–49.

Neilsen, T. (1969) Population studies on *Helophilus hybridus* Loew and *Sericomyia silentis* (Harris) (Dipt., Syrphidae) on Jaeren, S.W. Norway. *Norsk Entomologisk Tidsskrift*, **16**, 33–8.

Nelemans, M.N.E. (1988) Surface activity and growth of larvae of *Nebria brevicollis* (F.) (Coleoptera, Carabidae). *Netherlands Journal of Zoology*, **38**, 74–95.

Nelemans, M.N.E., den Boer, P.J. and Spee, A. (1989) Recruitment and summer diapause in the dynamics of a population of *Nebria brevicollis* (Coleoptera: Carabidae). *Oikos*, **56**, 157–69.

Němec, V. and Starý, P. (1983) Electromorph differentiation in *Aphidius ervi* Hal. biotype on *Microlophium carnosum* (Bkt.) related to parasitization on *Acyrrthosiphon pisum* (Harr.) (Hym., Aphidiidae). *Zeitschrift für Angewandte Entomologie*, **95**, 524–30.

Němec, V. and Starý, P. (1984) Population diversity of *Diaeretiella rapae* (McInt.) (Hym., Aphidiidae), an aphid parasitoid in agroecosystems. *Zeitschrift für Angewandte Entomologie*, **97**, 223–33.

Němec, V. and Starý, P. (1985) Genetic diversity and host alternation in aphid parasitoids (Hymenoptera: Aphidiidae). *Entomologia Generalis*, **10**, 253–58.

Němec, V. and Starý, P. (1986) Population diversity centers of aphid parasitoids (Hym: Aphidiidae): a new strategy in integrated pest management, in *Ecology of Aphidophaga* (ed. I. Hodek), Academia, Prague and W. Junk, Dordrecht, pp. 485–8.

Neuenschwander, P. (1982) Beneficial insects caught by yellow traps used in mass-trapping of the olive fly, *Dacus oleae*. *Entomologia Experimentalis et Applicata*, **32**, 286–96.

Neuenschwander, P. and Gutierrez, A.P. (1989) Evaluating the impact of biological control measures, in *Biological Control: A Sustainable Solution to Crop Pest Problems in Africa* (eds J.S. Yaninek and H.R. Herren), Proceedings of The Inaugral Conference and Workshop of the IITA Biological Control Program Center for Africa, 5–9 December 1988, Cotonou IITA, Ibadan, pp. 147–54.

Neuenschwander, P. and Herren, H. (1988) Biological control of the cassava mealybug, *Phenacoccus manihoti*, by the exotic parasitoid *Epidinocarsis lopezi* in Africa, *Philosophical Transactions of the Royal Society of London, B*, **318**, 319–33.

Neuenschwander, P. and Michelakis, S. (1980) The seasonal and spatial distribution of adult and larval chrysopids on olive-trees in Crete. *Acta Oecologia/Oecologia Applicata*, **1**, 93–102.

Neuenschwander, P., Canard, M. and Michelakis, S. (1981) The attractivity of protein hydrolysate baited McPhail traps to different chrysopid and hemerobiid species (Neuroptera) in a Cretan olive orchard. *Annals of the Entomological Society of France*, **17**, 213–20.

New, T.R. (1969) The biology of some species of *Alaptus* (Mymaridae) parasitizing eggs of Psocoptera. *Transactions of the Society for British Entomology*, **18**, 181–93.

Nichols, W.S. and Nakamura, R.M. (1980) Agglutination and agglutination inhibition assays, in *Manual of Clinical Immunology* (eds N.R. Rose and H. Friedman), American Society for Microbiology, Washington D.C., pp. 15–22.

Nicholls, C.F. and Bérubé, J.A.C. (1965) An expandable cage for feeding tests of coccinellid predators of aphids. *Journal of Economic Entomology*, **58**, 1169–70.

Nicholson, A.J. (1957) Comments on paper of T.B. Reynoldson. *Cold Spring Harbor Symposium on Quantitative Biology*, **22**, (ed. K. Brehme Warren), Cold Spring Harbor, New York, pp. 313–27.

Nicholson, A.J. and Bailey, V.A. (1935) The balance of animal populations. *Proceeding of the Zoological Society of London,*, **1935**, 551–598

Niemelä, J. (1990) Spatial distribution of carabid beetles in the southern Finnish taiga: the question of scale, in *The Role of Ground Beetles in Ecological and Environmental Studies* (ed. N.E. Stork), Intercept, Andover, pp. 143–55.

Nisbet, R.M. and Gurney, W.S.C. (1983) The systematic formulation of population models for insects with dynamically varying instar duration. *Theoretical Population Biology*, **23**, 114–35.

Nishida, T. (1958) Extrafloral glandular secretions, a food source for certain insects. *Proceedings of the Hawaiian Entomological Society*, **26**: 379–386.

Noldus, L.P.J.J., Lewis, W.J., Tumlinson, J.H. and van Lenteren, J.C. (1988) Olfactometer and wind tunnel experiments on the role of sex pheromones of noctuid moths in the foraging behaviour of *Trichogramma* spp., in *Trichogramma and Other Egg Parasites* (eds J. Voegele, J.K. Waage and J.C. van Lenteren), *Colloques d'INRA, Paris*, **43**, 223–38.

Nordlund, D.A., Strand, M.R., Lewis, W.J. and Vinson, S.B. (1987) Role of kairomones from host accessory gland secretion in host recognition by *Telenomus remus* and *Trichogramma pretiosum*, with partial characterisation. *Entomologia Experimentalis et Applicata*, **44**, 37–43.

Noyes, J.S. (1982) Collecting and preserving chalcid wasps (Hymenoptera: Chalcidoidea). *Journal of Natural History*, **16**, 315–34.

Noyes, J.S. (1984) In a fog. *Chalcid Forum*, **3**, 4–5.

Noyes, J.S. (1989) A study of five methods of sampling Hymenoptera (Insecta) in a tropical forest, with special reference to the Parasitica. *Journal of Natural History*, **23**, 285–98.

Nunney, L. and Luck, R.F. (1988) Factors influencing the optimum sex ratio in structured populations. *Journal of Theoretical Biology*, **33**, 1–30.

Nur, U., Werren, J.H., Eickbush, D.G., Burke, W.D. and Eickbush, T.H. (1988) A 'selfish' B chromosome that enhances its transmission by eliminating the paternal genome. *Science*, **240**, 512–14.

Nyffeler, M. and Benz, G. (1988) Prey and predatory importance of micryphantid spiders in winter wheat fields and hay meadows. *Journal of Applied Entomology*, **105**, 190–97.

Nyffeler, M., Sterling, W.L. and Dean, D.A. (1992) Impact of the striped lynx spider (Araneae: Oxyopidae) and other natural enemies on the cotton fleahopper (Hemiptera: Miridae) in Texas cotton. *Environmental Entomology*, **21**, 1178–88.

Obara, M. and Kitano, H. (1974) Studies on the courtship behavior of *Apanteles glomeratus* L. I. Experimental studies on releaser of wing vibrating behavior in the male. *Kontyû*, **42**, 208–4.

O'Donnell, D.J. (1982) Taxonomy of the immature stages of parasitic Hymenoptera associated with aphids. Imperial College, University of London, Ph.D. Thesis.

O'Donnell, D.J. and Mackauer, M. (1989) A morphological and taxonomic study of first instar larvae of Aphidiinae (Hymenoptera: Braconidae). *Systematic Entomology*, **14**, 197–219.

Ogloblin, A.A. (1924) The role of extra-embryonic blastoderm of *Dinocampus terminatus* Nees during larval development. *Memoires de la Société Royale des Sciences de Bohème*, **3**, 1–27 (in French).

Ohiagu, C.E. and Boreham, P.F.L. (1978) A simple field test for evaluating insect predator-prey relationships. *Entomologia Experimentalis et Applicata*, **23**, 40–47.

O'Neill, K.M. and Skinner, S.W. (1990) Ovarian egg size and number in relation to female size in five species of parasitoid wasps. *Journal of Zoology*, **220**, 115–22.

Ooi, P.A.C. (1986) Insecticides disrupt natural biological control of *Nilaparvata lugens* in Sekinchan, Malaysia, in *Biological Control in the Tropics* (eds M.Y. Hussein and A.G. Ibrahim), Universiti Pertanian Malaysia, Serdang, pp. 109–20.

Opp, S.B. and Luck, R.F. (1986) Effects of host size on selected fitness components of *Aphytis melinus* and *A. lingnanensis* (Hymenoptera: Aphelinidae). *Annals of the Entomological Society of America*, **79**, 700–704.

Orr, B.K., Murdoch, W.W. and Bence, J.R. (1990) Population regulation, convergence and cannibalism in *Notonecta* (Hemiptera). *Ecology*, **71**, 68–82.

Orr, D.B. and Borden, J.H. (1983) Courtship and mating behavior of *Megastigmus pinus* (Hym., Torymidae). *Journal of the Entomological Society of British Columbia* **80**, 20–24.

Orzack, S.H. (1993) Sex ratio evolution in parasitic wasps, in *Evolution and Diversity of Sex Ratio in Insects and Mites* (eds D.L. Wrensch and M.A. Ebbert), Chapman & Hall, New York, pp. 477–511.

Osborne, L.S. (1982) Temperature-dependent development of greenhouse whitefly and its parasite *Encarsia formosa*. *Environmental Entomology*, **11**, 483–85.

Ôtake, A. (1967) Studies on the egg parasites of the smaller brown planthopper, *Laodelphax striatellus* (Fallén) (Hemiptera: Delphacidae), I. A device for assessing the parasitic activity, and the results obtained in 1966. *Bulletin of the Shikoku Agricultural Experiment Station*, **17**, 91–103.

Ôtake, A. (1970) Estimation of parasitism by *Anagrus* nr. *flaveolus* Waterhouse (Hymenoptera, Mymaridae). *Entomophaga*, **15**, 83–92.

Ottenheim, M., Holloway, G.J. and de Jong, P.W. (1992) Sex ratio in ladybirds (Coccinellidae). *Ecological Entomology*, **17**, 366–8.

Owen, J. (1981) Trophic variety and abundance of hoverflies (Diptera, Syrphidae) in an English suburban garden. *Holarctic Ecology*, **4**, 221–8.

Owen, J., Townes, H. and Townes, M. (1981) Species diversity of Ichneumonidae and Serphidae (Hymenoptera) in an English suburban garden. *Biological Journal of the Linnean Society*, **16**, 315–36.

Pacala, S.W. and Hassell, M.P. (1991) The persistence of host–parasitoid associations in patchy environments. II. Evaluation of field data. *American Naturalist*, **138**, 584–605.

Pacala, S.W., Hassell, M.P. and May, R.M. (1990) Host–parasitoid associations in patchy environments. *Nature*, **344**, 150–53.

Pallewatta, P.K.T.N.S. (1986) Factors Affecting Progeny and Sex Allocation by the Egg

Parasitoid, *Trichogramma evanescens* Westwood. Imperial College, University of London, Ph.D. Thesis.

Palmer, A.R and Strobeck, C. (1986) Fluctuating asymmetry: measurement, analysis, patterns. *Annual Review of Ecology and Systematics*, **17**, 391–421.

Palmer, D.F. (1980) Complement fixation test, in *Manual of Clinical Immunology* (eds N.R. Rose and H. Friedman), American Society for Microbiology, Washington, pp. 35–47.

Pampel, W. (1914) Die weiblischen Geschlechtosorgane der Ichneumoniden. *Zeitschrift für Wissenschaftliche Zoologie*, **108**, 290–357.

Papaj, D.R. and Vet, L.E.M. (1990) Odor learning and foraging success in the parasitoid, *Leptopilina heterotoma*. *Journal of Chemical Ecology*, **16**, 3137–50.

Paradise, C.J. and Stamp, N.E. (1990) Variable quantities of toxic diet cause different degrees of compensatory and inhibitory responses by juvenile praying mantids. *Entomologia Experimentalis et Applicata*, **55**, 213–22.

Paradise, C.J. and Stamp, N.E. (1991) Abundant prey can alleviate previous adverse effects on growth of juvenile praying mantids (Orthoptera: Mantidae). *Annals of the Entomological Society of America*, **84**, 396–406.

Paris, O.H. and Sikora, A. (1967) Radiotracer analysis of the trophic dynamics of natural isopod populations, in *Secondary Productivity of Terrestrial Ecosystems* (ed. K. Petrusewicz), Panstwowe Wydawnictwo Naukoe, Warsaw, pp. 741–71.

Parker, F.D. and Bohart, R.M. (1966) Host–parasite associations in some twig-nesting Hymenoptera from western North America. *Pan-Pacific Entomology*, **42**, 91–8.

Parker, G.A. (1970a) Sperm competition and its evolutionary consequences in the insects. *Biological Reviews*, **45**, 525–67.

Parker, G.A. (1970b) The reproductive behaviour and the nature of sexual selection in *Scatophaga stercoraria* L. I. Diurnal and seasonal changes in population density around the site of mating and oviposition. *Journal of Animal Ecology*, **39**, 185–204.

Parker, G.A. (1970c) The reproductive behaviour and the nature of sexual selection in *Scatophaga stercoraria* L. II. The fertilization rate and the spatial and temporal relationships of each sex around the site of mating and oviposition. *Journal of Animal Ecology*, **39**, 205–28.

Parker, G.A. (1970d) The reproductive behaviour and the nature of sexual selection in *Scatophaga stercoraria* L. V. The female's behaviour at the oviposition site. *Behaviour*, **38**, 140–68.

Parker, G.A. (1971) The reproductive behaviour and the nature of sexual selection in *Scatophaga stercoraria* L. VI. The adaptive significance of emigration from the oviposition site during the phase of genital contact. *Evolution*, **40**, 215–33.

Parker, G.A. (1978) Searching for mates, in *Behavioural Ecology, an Evolutionary Approach* (eds J.R. Krebs and N.B. Davies), Blackwell, Oxford, pp. 214–44.

Parker, G.A. (1984) Evolutionarily stable strategies, in *Behavioural Ecology: An Evolutionary Approach* (eds J.R. Krebs and N.B. Davies), Blackwell, Oxford, pp. 30–61.

Parker, G.A. and Courtney, S.P. (1984) Models of clutch size in insect oviposition. *Theoretical Population Biology*, **26**, 27–48.

Parker, G.A., Baker, R.R. and Smith, V.G.F. (1972) The origin and evolution of gamete dimorphism and the male-female phenomenon. *Journal of Theoretical Biology*, **36**, 529–53.

Parker, H.L. and Thompson, W.R. (1925) Contribution a la biologie des Chalcidiens entomophages. *Annals de la Société Entomologique de France*, **97**, 425–65.

Parmenter, L. (1956) Flies and their selection of the flowers they visit. *Entomologist's Record and Journal of Variation*, **68**, 242–3.

Parmenter, L. (1961) Flies visiting the flowers of wood spurge, *Euphorbia amygdaliodes* L. (Euphorbiaceae). *Entomologist's Record and Journal of Variation*, **73**, 48–9.

Parr, M.J. (1965) A population study of a colony of imaginal *Ischnura elegans* (van der Linden) (Odonata: Coenagriidae) at Dale, Pembrokeshire. *Field Studies*, **2**, 237–82.

Partridge, L. and Farquhar, M. (1981) Sexual activity reduces lifespan of male fruitflies. *Nature*, **294**, 580–82.

Pasteur, N., Pasteur, G., Bonhomme, F., Catalan, J. and Britton-Davidian, J. (1988) *Practical Isozyme Genetics*, Ellis Horwood Ltd., Chichester.

Pastorok, R.A. (1981) Prey vulnerability and size selection by *Chaoborus larvae*. *Ecology*, **62**, 1311–24.

Patel, K.J. and Schuster, D.J. (1983) Influence of temperature on the rate of development of *Diglyphus intermedius* (Hymenoptera: Eulophidae) Girault, a parasite of *Lyriomyza* spp. (Diptera: Agromyzidae). *Environmental Entomology*, **12**, 885–7.

Patil, N.G., Baker, P.S., Groot, W. and Waage, J.K. (1994) Competition between *Psyllaephagus yaseeni and Tamarixia leucaenae*, two parasitoids of the leucaena psyllid (*Heteropsylla cubana*). *International Journal of Pest Management*, **40**, 211–15.

Pausch, R.D., Roberts, S.J., Barney, R.J. and Armbrust, E.J. (1979) Linear pitfall traps, a modification of an established trapping method. *The Great Lakes Entomologist*, **12**, 149–51.

Pearl, R. and Miner, J.R. (1935) Experimental studies on the duration of life, XIV. The comparative mortality of certain lower organisms. *Quarterly Review of Biology*, **10**, 60–79.

Pearl, R. and Parker, S.L. (1921) Experimental studies on the duration of life. I. Introductory discussion of the duration of life in *Drosophila*. *American Naturalist*, **55**, 481–509.

Pendleton, R.C. and Grundmann, A.W. (1954) Use of ^{32}P in tracing some insect–plant relationships of the thistle, *Cirsium undulatum*. *Ecology*, **35**, 187–91.

Percival, M.S. (1961) Types of nectar in angiosperms. *New Phytologist*, **60**, 235–81.

Percival, M.S. (1965) *Floral Biology*, Pergamon Press, Oxford.

Perera, H.A.S. (1990) Effect of host plant on mealybugs and their parasitoids, Wye College University of London Ph.D. Thesis.

Perfecto, I., Horwith, B., van der Meer, J., Schultz, B., McGuinness, H. and Dos Santos, A. (1986) Effects of plant diversity and density on the emigration rate of two ground beetles, *Harpalus pennsylvanicus* and *Evarthrus sodalis* (Coleoptera: Carabidae), in a system of tomatoes and beans. *Environmental Entomology*, **15**, 1028–31.

Perry, J.N. (1989) Review: population variation in entomology 1935–1950. *The Entomologist*, **108**, 184–98.

Perry, J.N. and Bowden, J. (1983) A comparative analysis of *Chrysoperla carnea* catches in light- and suction-traps. *Ecological Entomology*, **8**, 383–94.

Pettersson, J. (1970) An aphid sex attractant. I. Biological Studies. *Entomologia Scandinavica*, **1**, 63–73.

Pettersson, J. (1972) Technical data of a serological method for quantitative predator efficiency studies on *Rhopalosiphum padi* (L.). *Swedish Journal of Agricultural Research*, **2**, 65–69.

Pickard, R.S. (1975) Relative abundance of syrphid species in a nest of the wasp *Ectemnius cavifrons* compared with that in the surrounding habitat. *Entomophaga*, **20**(2), 143–51.

Pickett, J.A. (1988) Integrating use of beneficial organisms with chemical crop protection. *Philosophical Transactions of the Royal Society, London, B*, **318**, 203–11.

Pickup, J. and Thompson, D.J. (1990) The effects of temperature and prey density on the development rates and growth of damselfly larvae (Odonata: Zygoptera). *Ecological Entomology*, **15**, 187–200.

Piek, T. (1986) *Venoms of the Hymenoptera. Biochemical, Pharmacological and Behavioural Aspects*, Academic Press, London.

Pierce, C.L., Crowley, P.H. and Johnson, D.M. (1985) Behaviour and ecological interactions of larval Odonata. *Ecology*, **66**, 1504–12.

van der Pijl, L. and Dodson, C.H. (1966) *Orchid Flowers: Their Pollination and Evolution*, University of Miami Press, Coral Gables.

Pintureau, B. and Babault, M. (1981) Enzymatic characterisation of *Trichogramma evanescens* and *T. maidis* (Hym.: Trichogrammatidae); study of hybrids. *Entomophaga*, **26**, 11–22 (in French).

Pimentel, D. (1963) Introducing parasites and predators to control native pests. *Canadian Entomologist*, **95**, 785–92.

Pimentel, D., Nagel, W.P. and Madden, J.L. (1963) Space-time structure of the environment and the survival of parasite–host systems. *American Naturalist*, **97**, 141–66.

Plaut, H.N. (1965) On the phenology and control value of *Stethorus punctillum* Weise as a predator of *Tetranychus cinnabarinus* Boid. in Israel. *Entomophaga*, **10**, 133–7.

Pleasants, J.M. (1989) Optimal foraging by nectarivores: a test of the marginal value theorem. *American Naturalist*, **134**, 51–71.

Podoler, H. and Rogers, D. (1975) A new method for the identification of key factors from life-table data. *Journal of Animal Ecology*, **44**, 85–114.

Poehling, H.M. (1987) Effect of reduced dose rates of pirimicarb and fenvalerate on aphids and beneficial arthropods in winter wheat. *Bulletin IOBC/WPRS Working Group Integrated Control of Cereal Pests*, **10**(1), 184–93.

Pollard, E. (1968) Hedges. III. The effect of removal of the bottom flora of a hawthorn hedgerow on the Carabidae of the hedge bottom. *Journal of Applied Ecology*, **5**, 125–39.

Pollard, E. (1971) Hedges, IV. Habitat diversity and crop pests: a study of *Brevicoryne brassicae* and its syrphid predators. *Journal of Applied Ecology*, **8**, 751–80.

Pollard, E. (1977) A method for assessing changes in the abundance of butterflies. *Biological Conservation*, **12**, 115–34.

Pollard, E., Lakhani, K.H. and Rothery, P. (1987) The detection of density-dependence from a series of annual censuses. *Ecology*, **68**, 2046–55.

Pollard, S. (1988) Partial consumption of prey: the significance of prey water loss on estimates of biomass intake. *Oecologia*, **76**, 475–6.

Pollock, K.H., Nichols, J.D., Brownie, C. and Hines, J.E. (1990) Statistical inference for capture–recapture experiments. *Wildlife Monographs*, **107** 1–97.

Porres, M.A., McMurty, J.A. and March, R.B. (1975) Investigations of leaf sap feeding by three species of phytoseiid mites by labelling with radioactive phosphoric acid ($H_3{}^{32}PO_4$). *Annals of the Entomological Society of America*, **68**, 871–2.

Postek, M.T., Howard, K.S., Johnson, A.H. and McMichael, K.L. (1980) *Scanning Electron Microscopy – a Student's Handbook*, Ladd Research Industries, Louisiana.

Powell, W. (1980) *Toxares deltiger* (Haliday) (Hymenoptera: Aphidiidae) parasitising the cereal aphid, *Metopolophium dirhodum* (Walker) (Hemiptera: Aphididae), in southern England: a new host–parasitoid record. *Bulletin of Entomological Research*, **70**, 407–9.

Powell, W. (1982) The identification of hymenopterous parasitoids attacking cereal aphids in Britain. *Systematic Entomology*, **7**, 465–73.

Powell, W. (1986) Enhancing parasitoid activity in crops, in *Insect Parasitoids* (eds J. Waage and D. Greathead), Academic Press, London, pp. 319–40.

Powell, W. and Bardner, R. (1984) Effects of polyethylene barriers on the numbers of epigeal predators caught in pitfall traps in plots of winter wheat with and without soil-surface treatments of fonofos. *Bulletin IOBC/WPRS Working Group Integrated Control of Cereal Pests*, **8**(3), 136–8.

Powell, W. and Walton, M.P. (1989) The use of elactrophoresis in the study of hymenopteran parasitoids of agricultural pests, in *Electrophoretic Studies on Agricultural Pests* (eds H.D. Loxdale and J. den Hollander), Clarendon Press, Oxford, pp. 443–66.

Powell, W. and Wright, A.F. (1992) The influence of host food plants on host recognition by four aphidiine parasitoids (Hymenoptera: Braconidae). *Bulletin of Entomological Research*, **81**, 449–53.

Powell, W. and Zhang, Z.L. (1983) The reactions of two cereal aphid parasitoids, *Aphidius uzbeckistanicus* and *A. ervi*, to host aphids and their food plants. *Physiological Entomology*, **8**, 439–43.

Powell, W., Dean, G.J. and Bardner, R. (1985) Effects of pirimicarb, dimethoate and benomyl on natural enemies of cereal aphids in winter wheat. *Annals of Applied Biology*, **106**, 235–42.

Price, P.W. (1970) Trail odours: recognition by insects parasitic in cocoons. *Science*, **170**, 546–7.

Price, P.W. (1975) Reproductive strategies of parasitoids, in *Evolutionary Strategies of Parasitic Insects and Mites* (ed. P.W. Price), Plenum, New York, pp. 87–111.

Price, P.W. (1981) Semiochemicals in evolutionary time, in *Semiochemicals their Role in Pest Control* (eds D.A. Nordlund, R.L. Jones and W.J. Lewis), Wiley & Sons, New York, pp. 251–79.

Price, P.W. (1987) The role of natural enemies in insect populations, in *Insect Outbreaks* (eds P. Barbosa and J.C. Schultz), Academic Press, New York, pp. 287–312.

Price, P.W. (1988) Inversely density-dependent parasitism: the role of plant refuges for hosts. *Journal of Animal Ecology*, **57**, 89–96.

Primack, R.B. (1983) Insect pollination in the New Zealand mountain flora. *New Zealand Journal of Botany*, **21**, 317–33.

Principi, M.M. (1949) Contributi allo studio dei neurotteri italiani. 8. Morfologia, anafomia e funzionamento degli apparati genitali nel gen. *Chrysopa* Leach (*Chrysopa septempunctata* Wesm. e *Chrysopa formosa* Brauer). *Bolletino di Istituto Entomologia di Bologna*, **17**, 316–62.

Principi, M.M. and Canard, M. (1984) Feeding habits, in *Biology of the Chrysopidae* (eds M. Canard, Y. Semeria and T. New), W. Junk, The Hague, pp. 76–92.

Prokopy, R.J. (1981) Epideictic pheromones that influence spacing patterns of phytophagous insects, in *Semiochemicals: Their Role in Pest Control* (eds D.A. Nordlund, R.L. Jones and W.J. Lewis), Wiley & Sons, New York, pp. 181–213.

Pruess, K.P. (1983) Day–degree methods for pest management. *Environmental Entomology*, **12**, 613–19.

Pschorn-Walcher, H. (1971) Biological control of forest insects, *Annual Review of Entomology*, **22**, 1–22.

Putman, R.J. and Wratten, S.D. (1984) *Principles of Ecology*, Croom Helm, London.

Putman, W. (1963) Nectar of peach leaf glands as insect food. *Canadian Entomologist*, **95**, 108–9.

Putters, F.A. and van den Assem, J. (1988) The analysis of preference in a parasitic wasp. *Animal Behaviour*, **36**, 933–35.

Putters, F. and Vonk, M. (1991) The structure oriented approach in ethology: network models and sex ratio adjustments in parasitic wasps. *Behaviour*, **114**, 148–60.

Pyke, G.H. (1983) Animal movements: an optimal foraging approach, in *The Ecology of Animal*

Movement (eds I.R. Swingland and P.J. Greenwood), Clarendon Press, Oxford, pp. 7–31.

Pyke, G.H. (1984) Optimal foraging theory: a critical review. *Annual Review of Ecology and Systematics*, **15**, 523–75.

Pyke, G.H. (1991) What does it cost a plant to produce floral nectar? *Nature*, **350**, 58–9.

Quicke, D.L.J. (1993) *Principles and Techniques of Contemporary Taxonomy*, Chapman & Hall, London.

Quicke, D.L.J., Fitton, M.G. and Ingram S. (1992) Phylogenetic implications of the structure and distribution of ovipositor valvilli in the Hymenoptera (Insecta). *Journal of Natural History* **26**, 587–608.

Radwan, Z. and Lövei, G.L. (1982) Distribution and bionomics of ladybird beetles (Col., Coccinellidae) living in an apple orchard near Budapest, Hungary. *Zeitschrift für Angewandte Entomologie*, **94**, 169–75.

Ragsdale, D.W., Larson, A.D. and Newsom, L.D. (1981) Quantitative assessement of the predators of *Nezara viridula* eggs and nymphs within a soybean agroecosystem using ELISA. *Environmental Entomology*, **10**, 402–5.

Ragusa, S. (1974) Influence of temperature on the oviposition rate and longevity of *Opius concolor siculus* (Hymenoptera: Braconidae). *Entomophaga*, 19, 61–6.

Ratcliffe, N.A. (1982) Cellular defence reactions of insects. *Fortschritte Zoologie*. **27**, 223–44.

Ratcliffe, N.A. and King, P.E. (1969) Morphological, ultrastructural, histochemical and electrophoretic studies on the venom system of *Nasonia vitripennis* Walker (Hymenoptera: Pteromalidae). *Journal of Morphology*, **127**, 177–204.

Ratcliffe, N.A. and Rowley, A.F. (1987) Insect responses to parasites and other pathogens, in *Immune Responses in Parasitic Infections: Immunology, Immunopathology, and Immunoprophylaxis* (ed. E.J.L. Soulsby), CRC Press, Boca Raton, pp. 271–332.

Ratner, S. and Vinson, S.B. (1983) Encapsulation reactions in vitro by haemocytes of *Heliothis virescens*. *Journal of Insect Physiology*, **29**, 855–63.

Ravlin, F.W. and Haynes, D.L. (1987) Simulation of interactions and management of parasitoids in a multiple host system. *Environmental Entomology*, **16**, 1255–65.

Read, D.P., Feeny, P.P. and Root, R.B. (1970) Habitat selection by the aphid parasite *Diaeretiella rapae* (Hymenoptera, Braconidae) and hyperparasite *Charips brassicae* (Hymenoptera, Cynipidae). *Canadian Entomologist*, **102**, 1567–78.

Readshaw, J.L. (1973) The numerical response of predators to prey density. *Journal of Applied Ecology*, **10**, 342–51.

Redfern, M. and Askew, R.R. (1992) *Plant Galls*, Naturalists' Handbooks No 17, Richmond Publishing Co. Ltd, Slough.

Redfern, M., Jones, T.H. and Hassell, M.P. (1992) Heterogeneity and density dependence in a field study of a tephritid–parasitoid interaction. *Ecological Entomology*, **17**, 255–62.

Reeve, J.D. (1990) Stability, variability and persistence in host–parasitoid systems. *Ecology*, **71**, 422–6.

Reeve, J.D., Kerans, B.L. and Chesson, P.L. (1989) Combining different forms of parasitoid aggregation: effects on stability and patterns of parasitism. *Oikos*, **56**, 233–9.

Reitsma, G.. (1966) Pollen morphology of some European Rosaceae. *Acta Botanica Neerlandensis*, **15**, 290–307.

Reznick, D. (1985) Costs of reproduction: an evaluation of the empirical evidence. *Oikos*, **44**, 257–67.

Reznik S., Chernoguz, D.G. and Zinovjeva, K.B. (1992) Host searching, oviposition preferences and optimal synchronization in *Alysia manducator* (Hymenoptera: Braconidae), a parasitoid of the blowfly, *Calliphora vicina*. *Oikos*, **65**, 81–8.

Ricci, C. (1986) Seasonal food preferences and behaviour of *Rhyzobius litura*, in *Ecology of Aphidophaga* (ed. I. Hodek), Academia, Prague and W. Junk, Dordrecht, pp. 119–23.

Rice, R.E. and Jones, R.A. (1982) Collections of *Prospaltella perniciosi* Tower (Hymenoptera: Aphelinidae) on San Jose scale (Homoptera: Diaspididae) pheromone traps. *Environmental Entomology*, **11**, 876–80.

Richards, O.W. and Davies, R.G. (1977) *Imms' General Textbook of Entomology*, Vols 1 and 2, Chapman & Hall, London.

Richardson, B.J., Baverstock, P.R. and Adams, M. (1986) *Allozyme Electrophoresis*, Academic Press, London.

Richardson, P.M., Holmes, W.P. and Saul II, G.B. (1987) The effect of tetracycline on nonreciprocal cross incompatibility in *Mormoniella vitripennis*. *Journal of Invertebrate Pathology*, **50**, 176–83.

Ridley, M. (1988) Mating frequency and fecundity in insects, *Biological Reviews*, **63**, 509–49.

Rizki, R.M. and Rizki, T.M. (1984) Selective destruction of a host blood type by a parasitoid wasp. *Proceedings of the National Academy of Sciences of the United States of America*, **81**, 6154–8.

Robertson P.L. (1968) A morphological and functional study of the venom apparatus in representatives of some major groups of Hymenoptera. *Australian Journal of Zoology*, **16**, 133–66.

Roeder, K.D. (1935) An experimental analysis of the sexual behavior of the praying mantis (*Mantis religiosa* L.). *Biological Bulletin*, **69**, 203–20.

Rogers, D. (1972) Random search and insect population models. *Journal of Animal Ecology*, **41**, 369–83.

Rogers, D.J. and Hassell, M.P. (1974) General models for insect parasite and predator searching behaviour: interference. *Journal of Animal Ecology*, **43**, 239–53.

Roitberg B.D., Mangel, M., Lalonde, R.G. *et al.* (1992) Seasonal dynamic shifts in patch exploitation by parasitic wasps. *Behavioural Ecology*, **3**, 156–65.

Roitberg, B., Sircom, J., van Alphen, J.J.M. and Mangel, M. (1993) Life expectancy and reproduction. *Nature*, **364**, 108.

Roitt, I.M., Brostoff, J. and Male, D.K. (1989) *Immunology*, 2nd edn, Gower Medical Publishing, London.

Roland, J. (1990) Interaction of parasitism and predation in the decline of winter moth in Canada, in *Population Dynamics of Forest Insects* (eds A.D. Watt, S.R. Leather, M.D. Hunter and N.A.C. Kidd), Intercept, Andover, pp. 289–302.

Roland, J. (1994) After the decline: what maintains low winter moth density after successful biological control? *Journal of Animal Ecology*, **63**, 392–98.

Room, R.M. (1977) ^{32}P labelling of immature stages of *Heliothis armigera* (Hubner) and *H. punctigera* Wallengren (Lepidoptera: Noctuidae): relationships of doses to radioactivity, mortality and label half-life. *Journal of the Australian Entomological Society*, 16, 245–51.

Rose, N.R. and Friedman, H. (eds) (1980) *Manual of Clinical Immunology*, American Society for Microbiology, Washington DC.

Rosengren, R., Vepsaläinen, K. and Wuorenrinne. H. (1979) Distribution, nest densities and ecological significance of wood ants (the *Formica rufa* group) in Finland. *Bulletin SROP, 1979*, **2**(3), 181–213.

Rosenheim, J.A. and Rosen, D. (1992) Influence of egg load and host size on host-feeding behaviour of the parasitoid *Aphytis lingnanensis*. *Ecological Entomology*, **17**, 263–72.

Roskam, J.C. (1982) Larval characters of some eurytomid species (Hymenoptera, Chalcidoidea). *Proceedings Koninklijke Nederlandse Akademie van Wetenschappen*, **85**, 293–305.

Rotheram, S.M. (1967) Immune surface of eggs of a parasitic insect. *Nature*, **214**, 700.

Rotheram, S. (1973a) The surface of the egg of a parasitic insect. 1. The surface of the egg and first-instar larva of *Nemeritis*. *Proceedings of the Royal Society, London, B*, **183**, 179–94.

Rotheram, S. (1973b) The surface of the egg of a parasitic insect. 2. The ultrastructure of the particulate coat on the egg of *Nemeritis*. *Proceedings of the Royal Society, London, B*, **183**, 195–204.

Rotheray, G.E. (1979) The biology and host searching behaviour of a cynipoid parasite of aphidophagous Syrphidae. *Ecological Entomology*, **4**, 75–82.

Rotheray, G.E. (1981) Courtship, male swarming and a sex pheromone of *Diplazon pectoratorius* (Thunberg) (Hym., Ichneumonidae). *Entomologist's Gazette*, **32**, 193–6.

Rotheray, G.E. and Barbosa, P. (1984) Host related factors affecting oviposition behaviour in *Brachymeria intermedia*. *Entomologia Experimentalis et Applicata*, **35**, 141–5.

Rothschild, G.H.L. (1966) A study of a natural population of *Conomelus anceps* (Germar) (Homoptera: Delphacidae) including observations on predation using the precipitin test. *Journal of Animal Ecology*, **35**, 413–34.

Roush, R.T. (1989) Genetic variation in natural enemies: critical issues for colonization in biological control, in *Critical Issues in Biological Control* (eds M. Mackauer, L.E. Ehler and J. Rolands), Intercept, Andover, pp. 263–88.

Roush, R.T. (1990) Genetic considerations in the propagation of entomophagous species, in *New Directions in Biological Control: Alternatives for Suppressing Agricultural Pests and Diseases* (eds R.R. Baker and P.E. Dunn), Alan Royal Liss, pp. 373–87.

Royama, T. (1971) A comparative study of models for predation and parasitism. *Researches on Population Ecology, Supplement* **1**, 1–91.

Royama, T. (1977) Population persistence and density dependence. *Ecological Monographs*, **47**, 1–35.

Ruberson, J.R., Tauber, M.J. and Tauber, C.A. (1988) Reproductive biology of two biotypes of *Edovum puttleri*, a parasitoid of Colorado potato beetle eggs. *Entomologia Experimentalis et Applicata*, **46**, 211–19.

Ruberson, J.R., Tauber, C.A., and Tauber, M.J. (1989) Development and survival of *Telenomus lobatus*, a parasitoid of chrysopid eggs: effect of host species. *Entomologia Experimentalis et Applicata*, **51**, 101–6.

Ruberson, J.R., Tauber, M.J., Tauber, C.A. and Gollands, B. (1991) Parasitization by *Edovum puttleri* (Hymenopterm: Eulophidae) in relation to host density in the field. *Ecological Entomology*, **16**, 81–9.

Rubia, E.G. and Shepard, B.M. (1987) Biology of *Metioche vittaticollis* (Stål) (Orthoptera: Gryllidae), a predator of rice pests. *Bulletin of Entomological Research*, **77**, 669–76.

Rubia, E.G., Ferrer, E.R. and Shepard, B.M. (1990) Biology and predatory behaviour of *Conocephalus longipennis* (de Haan) (Orthoptera: Tettigoniidae). *Journal of Plant Protection in the Tropics*, **7**, 205–11.

Russel, R.J. (1970) The effectiveness of *Anthocoris nemorum* and *A. confusus* (Hemiptera: Anthocoridae) as predators of the sycamore aphid, *Drepanosiphum platanoides*. I. The number of aphids consumed during development. *Entomologia Experimentalis et Applicata*, **13**, 194–207.

Ruzicka, Z. (1975) The effects of various aphids as larval prey on the development of *Metasyrphus corollae* (Dipt.: Syrphidae). *Entomophaga*, **20**, 393–402.

Ryan, R.B. (1990) Evaluation of biological control: introduced parasites of larch casebearer (Lepidoptera: Coleophoridae) in Oregon. *Environmental Entomology*, **19**, 1873–81.

Ryoo, M.I., Hong, Y.S. and Yoo, C. K. (1991) Relationship between temperature and development of *Lariophagus distinguendus* (Hymenoptera: Pteromalidae), an ectoparasitoid of *Sitophilus oryzae* (Coleoptera: Curculionidae). *Journal of Economic Entomology*, **84**, 825–9.

Sabelis, M.W. (1990) How to analyse prey preference when prey density varies? A new method to discriminate between gut fullness and prey type composition. *Oecologia*, **82**, 289–98.

Sabelis, M.W. (1992) Predatory arthropods. In *Natural Enemies: The Population Biology of Predators, Parasites and Diseases* (ed. M.J. Crawley), Blackwell, Oxford, pp. 225–64.

Sabelis, M.W. and van de Baan, H.E. (1983) Location of distant spider-mite colonies by phytoseiid predators. Demonstration of specific kairomones emitted by *Tetranychus urticae* and *Panonychus ulmi* (Acari: Tetranychidae, Phytoseiidae). *Entomologia Experimentalis et Applicata*, **33**, 303–14.

Sabelis, M.W., Afman, B.P. and Slim, P.J. (1984) Location of distant spider mite colonies by *Phytoseiulus persimilis*: Localization and extraction of a kairomone. *Acarology*, **6**, 431–40.

Sahad, K.A. (1982) Biology and morphology of *Gonatocerus* sp. (Hymenoptera, Mymaridae), an egg parasitoid of the green rice leafhopper, *Nephotettix cincticeps* Uhler (Homoptera: Deltocephalidae). I. Biology. *Kontyû*, **50**, 246–60.

Sahad, K.A. (1984) Biology of *Anagrus optabilis* (Perkins) (Hymenoptera, Mymaridae), an egg parasitoid of delphacid leafhoppers. *Esakia*, **22**, 129–44.

Sahragard, A., Jervis, M.A. and Kidd, N.A.C. (1991) Influence of host availability on rates of oviposition and host-feeding, and on longevity in *Dicondylus indianus* Olmi (Hym., Dryinidae), a parasitoid of the Rice Brown Planthopper, *Nilaparvata lugens* Stål (Hem., Delphacidae). *Journal of Applied Entomology*, **112**, 153–62.

Salt, G. (1934) Experimental studies in insect parasitism, II. Superparasitism. *Proceedings of the Royal Society, London,B*, **114**, 455–76.

Salt, G. (1940) Experimental studies in insect parasitism. VII. The effects of different hosts on the parasite *Trichogramma evanescens* Westw. (Hym.; Chalcidoidea). *Proceedings of the Royal Entomological Society of London A*, **15**, 81–124.

Salt, G. (1941) The effects of hosts upon their insect parasites. *Biological Reviews*, **16**, 239–64.

Salt, G. (1958) Parasite behaviour and the control of insect pests. *Endeavour*, **17**, 145–8.

Salt, G. (1961) Competition among insect parasitoids. Mechanisms in biological competition. *Symposium of the Society for Experimental Biology*, **15**, 96–119.

Salt, G. (1968) The resistance of insect parasitoids to the defence reactions of their hosts. *Biological Reviews*, **43**, 200–232.

Salt, G. (1970) *The Cellular Defence Reactions of Insects*, Cambridge University Press, Cambridge.

Samson-Boshuizen, M., van Lenteren, J.C. and Bakker, K. (1974) Success of parasitization of *Pseudeucoila bochei* Weld (Hym.Cynip.): a matter of experience. *Netherlands Journal of Zoology*, **24**, 67–85.

Samways, M.J. (1986) Spatial and temporal population patterns of *Aonidiella aurantii* (Hemiptera; Diaspididae) parasitoids (Hymenoptera:

Aphelinidae and Encyrtidae) caught on yellow sticky traps in citrus. *Bulletin of Entomological Research*, **76**, 265–74.

Sandlan, K. (1979a) Sex ratio regulation in *Coccygomimus turionellae* Linnaeus (Hymenoptera: Ichneumonidae) and its ecological implications. *Ecological Entomology*, **4**, 365–78.

Sandlan, K. (1979b) Host feeding and its effects on the physiology and behaviour of the ichneumonid parasite *Coccygomimus turionellae*. *Physiological Entomology*, **4**, 383–92.

Sanger, C. and King, P.E. (1971) Structure and function of the male genitalia in *Nasonia vitripennis* (Walker) (Hym.: Pteromalidae). *Entomologist*, **104**, 137–49.

Sargent, J.R. and George, S.G. (1975) *Methods in Zone Electrophoresis*, 3rd edn, BDH Chemical Ltd., Poole, Dorset.

Sasaba, T. and Kiritani, K. (1975) A systems model and computer simulation of the green rice leafhopper populations in control programmes. *Researches on Population Ecology*, **16**, 231–44.

Sato, Y., Tagawa, J. and Hidaka, T. (1986) Effects of the gregarious parasitoids *Apanteles rufricus* and *A. kariyai* on host growth and development. *Journal of Insect Physiology*, **32**, 281–6.

Saugstad, E.S., Bram, R.A. and Nyquist, W.E. (1967) Factors influencing sweep net sampling of alfalfa. *Journal of Economic Entomology*, **60**, 421–6.

Saul G.B., Whiting, P.W., Saul, S.W. and Heidner, C.A. (1965) Wild-type and mutant stocks of *Mormoniella*. *Genetics*, **52**, 1317–27.

Sawoniewicz, J. (1973) Ichneumonidae (Hymenoptera) visiting the flowers of *Peucedanum oreoselinum* L. (Umbelliferae). *Folia Forestalia Polonica*, **21**, 43–78 (in Polish).

Sawyer, R. (1981) *Pollen Identification for Beekeepers*, University College, Cardiff Press, Cardiff.

Sawyer, R. (1988) *Honey Identification*, Cardiff Academic Press, Cardiff.

Schaupp, W.C. and Kulman, H.M. (1992) Attack behaviour and host utilization of *Coccygomimus disparis* (Hymenoptera: Ichneumonidae) in the laboratory. *Environmental Entomology*, **21**, 401–8.

Scheller, H.V. (1984) Pitfall trapping as the basis for studying ground beetle (Carabidae) predation in spring barley. *Tiddskrift voor Plantearl*, **88**, 317–24.

Schlinger, E.I. and Hall, J.C. (1960) The biology, behaviour and morphology of *Praon palitans* Muesebeck, an internal parasite of the the spotted alfalfa aphid, *Therioaphis maculata* (Buckton) (Hymenoptera; Braconidae, Aphidiinae). *Annals of the Entomological Society of America*, **53**, 144–60.

Schmidt, J.M. and Smith, J.J.B. (1985) Host volume measurement by the parasitoid wasp *Trichogramma minutum*: the roles of curvature and surface area. *Entomologia Experimentalis et Applicata*, **39**, 213–21.

Schmidt, J.M. and Smith, J.J.B. (1987a) The effect of host spacing on the clutch size and parasitization rate of *Trichogramma minutum*. *Entomologia Experimentalis et Applicata*, **43**, 125–31.

Schmidt, J.M. and Smith, J.J.B. (1987b) Short interval time measurement by a parasitoid wasp. *Science*, **237**, 903–5.

Schmidt, J.M. and Smith, J.J.B. (1989) Host examination walk and oviposition site selection of *Trichogramma minutum:* studies on spherical hosts. *Journal of Insect Behaviour*, **2**, 143–71.

Schneider, F. (1969) Bionomics and physiology of aphidophagous Syrphidae. *Annual Review of Entomology*, **14**, 103–24.

Schwartz, M. and Sofer, W. (1976) Diet induced alterations in distribution of multiple forms of alcohol dehydrogenase in *Drosophila*. *Nature*, **263**, 129–31.

Scott, J.A. (1973) Convergence of population biology and adult behaviour in two sympatric butterflies, *Neominois ridingsii* (Papilionoidea: Nymphalidae) and *Amblyscirtes simius* (Hesperoidea: Hesperiidae). *Journal of Animal Ecology*, 42, 663–72.

Scott, S.M. and Barlow, C.A. (1984) Effect of prey availability during development on the reproductive output of *Metasyrphus corollae* (Diptera: Syrphidae). *Environmental Entomology*, **13**, 669–674.

Scriber, J.M. and Slansky Jr, F. (1981) The nutritional ecology of immature insects. *Annual Review of Entomology*, **26**, 183–211.

Scudder, G.G.E. (1971) Comparative morphology of insect genitalia. *Annual Review of Entomology*, **16**, 379–406.

Seber, G.A.F. (1965) A note on the multiple-recapture census. *Biometrika*, **52**, 249.

Seber, G.A.F. (1973) *The Estimation of Animal Abundance and Related Parameters*, Griffin, London.

Sedivy, J. and Kocourek, F. (1988) Comparative studies on two methods for sampling insects in an alfalfa seed stand in consideration of a chemical control treatment. *Journal of Applied Entomology*, **106**, 312–8.

Sem'yanov, V.P. (1970) Biological properties of *Adalia bipunctata* L. (Coleoptera, Coccinellidae) in conditions of Leningrad region, *Zashchchita Rasteniĭ Vrediteleĭ i Bolezniĭ* **127**, 105–12.

Sequeira, R. and Mackauer, M. (1992a) Covariance of adult size and development time in the parasitoid wasp *Aphidius ervi* in relation to the size of its host, *Acyrthosiphon pisum*. *Evolutionary Ecology*, **6**, 34–44.

Sequeira, R. and Mackauer, M. (1992b) Nutritional ecology of an insect host–parasitoid association: the pea aphid–*Aphidius ervi* system. *Ecology*, **73**, 183–9.

Service, M.W. (1973) Study of the natural predators of *Aedes cantans* (Meigen) using the precipitin test. *Journal of Medical Entomology*, **10**, 503–10.

Service, M.W., Voller, A. and Bidwell, E. (1986) The enzyme-linked immunosorbent assay (ELISA) test for the identification of blood-meals of haematophagous insects. *Bulletin of Entomological Research*, **76**, 321–30.

Settle, W.H. and Wilson, L.T. (1990) Behavioural factors affecting differential parasitism by *Anagrus epos* (Hymenoptera: Mymaridae), of two species of erythroneuran leafhoppers (Homoptera: Cicadellidae). *Journal of Animal Ecology*, **59**, 877–91.

Shahjahan, M. (1974) *Erigeron* flowers as a food and attractive odour source for *Peristenus pseudopallipes*, a braconid parasitoid of the tarnished plant bug. *Environmental Entomology*, **3**, 69–72.

Shannon, R.E. (1975) *Systems Simulation: The Art and the Science*. Prentice-Hall, New Jersey.

Sharpe, P.J.H., Curry, G.L., DeMichele, D.W. and Cole, C.L. (1977) Distribution model of organism development times. *Journal of Theoretical Biology*, **66**, 21–38.

Shaw, C.R. and Prasad, R. (1970) Starch gel electrophoresis of enzymes – a compilation of recipes. *Biochemical Genetics*, **4**, 297–320.

Shaw, M.R. and Huddleston, T. (1991) Classification and biology of braconid wasps. *Handbooks for the Identification of British Insects,Royal Entomological Society of London* **7**(11), 1–126.

Sheehan, W., Wackers, F.L. and Lewis, W.J. (1993) Discrimination of previously searched, host-free sites by *Microplitis croceipes* (Hymenoptera: Braconidae). *Journal of Insect Behaviour*, **6**, 323–31.

Sheldon, J.K. and MacLeod, E.G. (1971) Studies on the biology of Chrysopidae. 2. The feeding behaviour of the adult *Chrysopa carnea* (Neuroptera). *Psyche*, **78**, 107–21.

Shepard, B.M. and Ooi, P.A.C. (1991) Techniques for evaluating predators and parasitoids in rice, in *Rice Insects: Management Strategies* (eds E.A. Heinrichs and T.A. Miller), Springer-Verlag, New York, pp. 197–214.

Shepard, M. and Waddill, V.H. (1976) Rubidium as a marker for Mexican bean beetles, *Epilachana varivestis* (Coleoptera: Coccinellidae). *Canadian Entomologist*, **108**, 337–9.

Sherratt, T.N. and Harvey, I.F. (1993) Frequency-dependent food selection by arthropods: a review. *Biological Journal of the Linnean Society*, **48**, 167–86.

Shimron, O., Hefetz, A. and Gerling, D. (1992) Arrestment responses of *Eretmocerus species* and *Encarsia deserti* (Hymenoptera, Aphelinidae) to *Bemisia tabaci* honeydew. *Journal of Insect Behaviour*, **5**, 517–26.

Short, J.R.T. (1952) The morphology of the head of larval Hymenoptera with special reference to the head of the Ichneumonoidea, including a classification of the final instar larvae of the Braconidae. *Transactions of the Royal Entomological Society, London*, **103**, 27–84.

Short, J.R.T. (1959) A description and classification of the final instar larvae of the Ichneumonidae (Insecta, Hymenoptera). *Proceedings of the United States National Museum*, **110**, 391–511.

Short, J.R.T. (1970) On the classification of the final instar larvae of the Ichneumonidae (Hymenoptera). *Transactions of the Royal Entomological Society, London, Supplement*, **122**, 185–210.

Short, J.R.T. (1978) The final larval instars of the Ichneumonidae. *Memoirs of the American Entomology Institute*, **25**, 1–508.

Siddiqui, W.H., Barlow, C.A. and Randolph, P.A. (1973) Effect of some constant and alternating temperature on population growth of the pea aphid *Acyrthosiphon pisum* (Hom: Aphididae). *Canadian Entomologist*, **105**, 145–56.

Sih, A. (1980) Optimal foraging: partial consumption of prey. *American Naturalist*, **116**, 281–90.

Sih, A. (1982) Foraging strategies and avoidance of predation by an aquatic insect, *Notonecta hoffmani*. *Ecology*, **63**, 786–96.

Sih, A. (1987) Nutritional ecology of aquatic insect predators, in *Nutritional Ecology of Insects, Mites, Spiders and Related Invertebrates* (eds F. Slansky Jr and J.G. Rodriguez), Wiley-Interscience, New York, pp. 579–607.

Simmonds, F.J. (1943) The occurrence of superparasitism in *Nemeritis canescens* Grav.. *Revue Canadienne de Biologie*, **2**, 15–58.

Singh, R. and Sinha, T.B. (1980) Bionomics of *Trioxys (Binodoxys) indicus* Sibba Rao and Sharma, an aphidiid parasitoid of *Aphis crac-*

civora Koch.. *Zeitschrift für Angewandte Entomologie*, **90**, 233–7.

Siva-Jothy, M.T. (1984) Sperm competition in the Libellulidae (Anisoptera) with special reference to *Crocothemis erythraea* (Brulle) and *Orthetrum cancellatum* (L.) *Advances in Odonatology*, **2**, 195–207.

Siva-Jothy, M.T. (1987) Variation in copulation duration and the resultant degree of sperm removal in *Orthetrum cancellatum* (L.). (Odon. Libellulidae). *Behavioural Ecology and Sociobiology*, **20**, 147–51.

Sivinski, J. and Webb, J.C. (1989) Acoustic signals produced during courtship in *Diachasmimorpha longicaudata* (Hym., Braconicae) and other Braconidae. *Annals of the Entomological Society of America*, **82**, 116–20.

Skinner, G.J. (1980) The feeding habits of the wood-ant, *Formica rufa* (Hymenoptera: Formicidae), in limestone woodland in north-west England. *Journal of Animal Ecology*, **49**, 417– 33.

Skinner, S.W. (1982) Maternally inherited sex ratio in the parasitoid wasp, *Nasonia vitripennis*. *Science*, **215**, 1133–4.

Skinner, S.W. (1985a) Clutch size as an optimal foraging problem for insects. *Behavioural Ecology and Sociobiology*, **17**, 231–38.

Skinner, S.W. (1985b) Son-killer: a third extra-chromosomal factor affecting sex-ratio in the parasitoid wasp, *Nasonia vitripennis*. *Genetics*, **109**, 745–59.

Skuhravý, V. (1970) Zur Anlockungsfahigkeit von Formalin für Carabiden in Bodenfallen. *Beitrage zur Entomologie*, **20**, 371–4.

Slansky Jr, F. (1986) Nutritional ecology of endoparastic insects and their hosts: an overview. *Journal of Insect Physiology*, **32**, 255–61.

Slansky Jr, F. and Rodriguez, J.G. (1987) Nutritional ecology of insects, mites, spiders and related invertebrates: an overview, in *Nutritional Ecology of Insects, Mites, Spiders and Related Invertebrates* (eds F. Slansky Jr and J.G. Rodriguez), Wiley-Interscience, New York, pp. 1–69.

Slansky Jr, F. and Scriber, M.J. (1982) Selected bibliography and summary of quantitative food utilization by immature insects. *Bulletin of the Entomological Society of America* , **28**, 43–55.

Slansky Jr, F. and Scriber, J.M. (1985) Food consumption and utilization, in *Comprehensive Insect Physiology, Biochemistry and Pharmacology* (eds G.A. Kerkut and L.I. Gilbert), Pergamon Press, Oxford, pp. 87–163.

Smilowitz, Z. and Iwantsch, G.F. (1973) Relationships between the parasitoid *Hyposoter exiguae* and the cabbage looper *Trichoplusia ni*: effects of host age on developmental rate of the parasitoid. *Environmental Entomology*, **2**, 759–63.

Smith, B.C. (1965) Growth and development of coccinellid larvae on dry foods (Coleoptera, Coccinellidae). *Canadian Entomologist*, **97**, 760–68.

Smith, B.C. (1961) Results of rearing some coccinellid (Coleoptera: Coccinellidae) larvae on various pollens. *Proceedings of the Entomological Society of Ontario*, **91**, 270–71.

Smith, H.S. and DeBach, P. (1942) The measurement of the effect of entomophagous insects on population densities of their hosts. *Journal of Economic Entomology*, **35**, 845–9.

Smith, J.G. (1976) Influence of crop background on natural enemies of aphids on Brussels sprouts. *Annals of Applied Biology*, **83**, 15–29.

Smith, J.W., Stadelbacher, E.A. and Gantt, C.W. (1976) A comparison of techniques for sampling beneficial arthropod populations associated with cotton. *Environmental Entomology*, **5**, 435–44.

Smith, K.G.V. (1974) Rearing the Hymenoptera Parasitica. *Leaflets of the Amateur Entomologist's Society*, **35**, 1–15.

Smith, L. and Rutz, D.A. (1987) Reproduction, adult survival and intrinsic rate of growth of *Urolepis rufipes* (Hymenoptera: Pteromalidae), a pupal parasitoid of house flies, *Musca domestica*. *Entomophaga*, **32**, 315–27.

Smith, S.M. and Hubbes, M. (1986) Isoenzyme patterns and biology of *Trichogramma minutum* as influenced by rearing temperature and host. *Entomologia Experimentalis et Applicata*, **42**, 249–58.

Snodgrass, R.E. (1935) *Principles of Insect Morphology*, McGraw-Hill, London.

Sokolowski, M.B. and Turlings, T.C.J. (1987) *Drosophila* parasitoid–host interactions: vibrotaxis and ovipositor search from the host's perspective. *Canadian Journal of Zoology*, **65**, 461–4.

Sol, R. (1961) Ueber den Eingriff von Insektiziden in das Wechselspiel von *Aphis fabae* Scop. und einigen ihrer Episiten. *Entomophaga*, **6**, 7–33

Sol, R. (1966) The occurrence of aphidivorous syrphids and their larvae on different crops, with the help of coloured water traps, in *Ecology of Aphidophagous Insects* (ed. I. Hodek), Academia, Prague and W. Junk, Dordrecht, pp. 181–4.

Solomon, M.E. (1949) The natural control of animal populations. *Journal of Animal Ecology*, **18**, 1–35.

Solomon, M.G., Fitzgerald, J.D. and Murray, R.A. (1995) Electrophoretic approaches to predator-prey interactions, in *The Ecology of Agricultural Pests: Biochemical Approaches* (eds W.O.C. Symondson and J.E. Liddell), Chapman and Hall, London. (in press).

Soper, R.S., Shewell, G.E. and Tyrrell, D. (1976) *Colcondamyia auditrix* nov. sp. (Diptera: Sarcophagidae), a parasite which is attracted by the mating song of its host, *Okanagana rimosa* (Homoptera: Cicadidae). *Canadian Entomologist*, **108**, 61–8.

Sopp, P.I. and Sunderland, K.D. (1989) Some factors affecting the detection period of aphid remains in predators using ELISA. *Entomologia Experimentalis et Applicata*, **51**, 11–20.

Sopp, P. and Wratten, S.D. (1986) Rates of consumption of cereal aphids by some polyphagous predators in the laboratory. *Entomologia Experimentalis et Applicata*, **41**, 69–73.

Sopp, P.I., Sunderland, K.D., Fenlon, J.S. and Wratten, S.D. (1992) An improved quantitative method for estimating invertebrate predation in the field using an enzyme-linked immunosorbent assay (ELISA). *Journal of Applied Ecology*, **29**, 295–302.

Sotherton, N.W. (1982) Predation of a chrysomelid beetle (*Gastrophysa polygoni*) in cereals by polyphagous predators. *Annals of Applied Biology*, **101**, 196–9.

Sotherton, N.W. (1984) The distribution and abundance of predatory arthropods overwintering on farmland. *Annals of Applied Biology*, **105**, 423–9.

Sotherton, N.W., Wratten, S.D. and Vickerman, G.P. (1985) The role of egg predation in the population dynamics of *Gastrophysa polygoni* (Coleoptera) in cereal fields. *Oikos*, **43**, 301–8.

Southwood, T.R.E. (1962) Migration of terrestrial arthropods in relation to habitat. *Biological Reviews*, **37**, 171–214.

Southwood, T.R.E. (1978) *Ecological Methods with Particular Reference to the Study of Insect Populations*, 2nd edn, Chapman & Hall, London.

Southwood, T.R.E. and Jepson, W.F. (1962) Studies on the populations of *Oscinella frit* L. (Diptera: Chloropidae) in the wheat crop. *Journal of Animal Ecology*, **31**, 481–95.

Southwood, T.R.E. and Pleasance, H.J. (1962) A hand-operated suction apparatus for the extraction of arthropods from grassland and similar habitats, with notes on other models. *Bulletin of Entomological Research*, **53**, 125–8.

Southwood, T.R.E., Hassell, M.P., Reader, P.M. and Rogers, D.J. (1989) Population dynamics of the viburnum whitefly (*Aleurotrachelus jelinekii*). *Journal of Animal Ecology*, **58**, 921–42.

Sparks, A.N., Chiang, H.C., Burkhardt, C.C., Fairchild, M.L. and Weekman, G.T. (1966) Evaluation of the influence of predation on corn borer populations. *Journal of Economic Entomology*, **59**, 104–7.

Spears, E.E. (1983) A direct measure of pollinator effectiveness. *Oecologia*, **57**, 196–9.

Speight, M.R. and Lawton, J.H. (1976) The influence of weed-cover on the mortality imposed on artificial prey by predatory ground beetles in cereal fields. *Oecologia*, **23**, 211–23.

Spence, J.R. (1986) Relative impacts of mortality factors in field populations of the waterstrider *Gerris buenoi* Kirkaldy (Heteroptera: Gerridae). *Oecologia*, **70**, 68–76.

Staak, C., Allmang, B., Kampe, U. and Mehlitz, D. (1981) The complement fixation test for the species identification of blood meals from tsetse flies. *Tropenmedizin und Parasitologie*, **32**, 97–8.

Starý, P. (1970) Biology of Aphid Parasites (Hymenoptera: Aphidiidae) with respect to integrated control. *Series Entomologica*, **6**, 1–643.

Stearns, S.C. (1992) *The Evolution of Life Histories*, Oxford University Press, Oxford.

Stelleman, P. and Meeuse, A.D.J. (1976) Anthecological relations between reputedly anemophilous flowers and syrphid flies. I. The possible role of syrphid flies as pollinators of *Plantago*. *Tijdschrift voor Entomologie*, **119**, 15–31.

Stephens, D.W. and Krebs, J.R. (1986) *Foraging Theory*, Princeton University Press, Princeton.

Sterling, W. (1989) Estimating the abundance and impact of predators and parasites on *Heliothis* populations, in *Proceedings of the Workshop on Biological Control of Heliothis: Increasing the Effectiveness of Natural Enemies* (eds E.G. King and R.D. Jackson), U.S. Department of Agriculture, New Delhi, pp. 37–56.

Steward, V.B., Smith, K.G. and Stephen, F.M. (1988) Predation by wasps on lepidopteran larvae in an Ozark forest canopy. *Ecological Entomology*, **13**, 81–6.

Stewart, L.A. and Dixon, A.F.G. (1989) Why big species of ladybird beetles are not melanic. *Functional Ecology*, **3**, 165–77.

Stewart, L.A., Hemptinne, J.-L. and Dixon, A.F.G. (1991) Reproductive tactics of ladybird beetles: relationships between egg size, ovariole number and development time. *Functional Ecology*, **5**, 380–85.

Stewart-Oaten, A. and Murdoch, W.W. (1990) Temporal consequences of spatial density dependence. *Journal of Animal Ecology*, **59**, 1027–45.

Steyskal, G.C. (1977) History and use of the McPhail trap. *The Florida Entomologist*, **60**, 11–16.

Steyskal, G.C. (1981) A bibliography of the Malaise trap. *Proceedings of the Entomological Society of Washington*, **83**, 225–9.

Stiling, P.D. (1987) The frequency of density dependence in insect host parasitoid systems. *Ecology*, **68**, 844–56.

Stiling, P. (1988) Density-dependent processes and key factors in insect populations. *Journal of Animal Ecology*, **57**, 581–94.

Stiling, P.D. (1990) Calculating the establishment rates of parasitoids in biological control. *American Entomologist*, **36**, 225–30.

Stinner, R.E., Gutierrez, A.P. and Butler, G.D. (1974) An algorithm for temperature dependent growth rate simulation. *Canadian Entomologist*, **106**, 519–24.

Stinner, R.E., Butler Jr, G.D., Bacheler, J.S. and Tuttle, C. (1975) Simulation of temperature-dependent development in population dynamics models. *Canadian Entomologist*, **107**, 1167–74.

Stoltz, D.B. (1981) A putative baculovirus in the ichneumonid parasitoid *Mesoleius tenthredinis*. *Canadian Journal of Microbiology*, **27**, 116–22.

Stoltz, D.B. (1986) Interactions between parasitoid-derived products and host insects: an overview. *Journal of Insect Physiology*, **32**, 347–50.

Stoltz, D.B. and Vinson, S.B. (1979) Viruses and parasitism in insects. *Advances in Virus Research*, **24**, 125–71.

Stoltz, D.B., Krell, P.J., Summers, M.D. and Vinson, S.B. (1984) Polydnaviridae – a proposed family of insect viruses with segmented, double-stranded, circular DNA genomes. *Intervirology*, **21**, 1–4.

Stone, N.D. and Gutierrez, A.P. (1986) Pink bollworm control in southwestern desert cotton. I. A field-oriented simulation model. *Hilgardia*, **54**, 1–24.

Storck-Weyhermüller, S. (1988) Einfluss naturlicher Feinde auf die Populationsdynamik der Getreideblattlause im Winterweizen Mittelhessens (Homoptera: Aphididae). *Entomologia Generalis*, **13**, 189–206

Stoutamire, W.P. (1974) Australian terrestrial orchids, thynnid wasps, and pseudocopulation. *American Orchid Society Bulletin*, **1974**, 13–18.

Stouthamer, R. (1993) The use of sexual versus asexual wasps in biological control. *Entomophaga*, **38**, 3–6.

Stouthamer, R., Luck, R.F. and Hamilton, W.D. (1990a) Antibiotics cause parthenogenetic *Trichogramma* (Hymenoptera: Trichogrammatidae) to revert to sex. *Proceedings of the National Academy of Sciences*, **87**, 2424–7.

Stouthamer, R., Luck, R.F. & Werren, J.H. (1992) Genetics of sex determination and the improvement of biological control using parasitoids. *Environmental Entomology*, **21**: 427–35.

Stouthamer, R., Pinto, J.D., Platner, G.R. and Luck, R.F. (1990b) Taxonomic status of thelytokous species of *Trichogramma* (Hymenoptera: Trichogrammatidae). *Annals of the Entomological Society of America*, **83**, 475–81.

Stradling, D.J. (1987) Nutritional ecology of ants, in *Nutritional Ecology of Insects, Mites, Spiders and Related Invertebrates* (eds F. Slansky and J.G. Rodriguez), John Wiley & Sons, New York, pp. 927–69.

Strand, M.R. (1986) The physiological interactions of parasitoids with their hosts and their influence on reproductive strategies, in *Insect Parasitoids* (eds J. Waage and D. Greathead), Academic press, London, pp. 97–136.

Strand, M.R. (1988) Variable sex ratio strategy of *Telenomus heliothidis* (Hymenoptera: Scelionidae): adaptation to host and conspecific density. *Oecologia*, **77**, 219–24.

Strand, M.R. (1989) Oviposition behavior and progeny allocation of the polyembryonic wasp *Copidosoma floridanum* (Hymenoptera: Encyrtidae). *Journal of Insect Behaviour*, **2**, 355–69.

Strand, M.R. and Godfray, H.C.J. (1989) Superparasitism and ovicide in parasitic Hymenoptera: theory and a case study of the ectoparasitoid *Bracon hebetor*. *Behavioural Ecology and Sociobiology*, **24**, 421–32.

Strand, M.R. and Vinson, S.B. (1982) Source and characterization of an egg recognition kairomone of *Telenomus heliothidis*, a parasitoid of *Heliothis virescens*. *Physiological Entomology*, **7**, 83–90.

Strand, M.R. and Vinson, S.B. (1983) Analyses of an egg recognition kairomone of *Telenomus heliothidis* (Hymenoptera: Scelionidae): isolation and function. *Journal of Chemical Ecology*, **9**, 423–32.

Strand, M.R. and Vinson, S.B. (1985) *In vitro* culture of *Trichogramma pretiosum*, on an artificial medium. *Entomologia Experimentalis et Applicata*, **39**, 203–9.

Strand, M.R., Johnson, J.A. and Culin, J.D. (1988) Developmental interactions between the parasitoid *Microplitis demolitor* (Hymenoptera: Braconidae) and its host *Heliothis virescens* (Lepidoptera: Noctuidae). *Annals of the Entomological Society of America*, **81**, 822–30.

Strand, M.R., Meola, S.M. and Vinson, S.B. (1986) Correlating pathological symptoms in *Heliothis virescens* eggs with development of the parasitoid *Telenomus heliothidis*. *Journal of Insect Physiology*, **32**, 389–402.

van Strien-van Liempt, W.T.F.H. (1983) The competition between *Asobara tabida* Nees von Esenbeck, 1834 and *Leptopilina heterotoma* (Thomson, 1862) in multiparasitized hosts. *Netherlands Journal of Zoology*, **33**, 125–63.

Stuart, M.K. and Burkholder, W.E. (1991) Monoclonal antibodies specific for *Laelius pedatus* (Bethylidae) and *Bracon hebetor* (Braconidae), two hymenopterous parasitoids of stored product pests. *Biological Control*, **1**, 302–8.

Stuart, M.K. and Greenstone, M.H. (1990) Beyond ELISA: a rapid, sensitive, specific immunodot assay for identification of predator stomach contents. *Annals of the Entomological Society of America*, **83**, 1101–7.

Stuart, M.K. and Greenstone, M.H. (1995) Serological diagnosis of parasitism: a monoclonal antibody-based immunodot assay for *Microplitis croceipes* (Hymenoptera: Braconidae), in *The Ecology of Agricultural Pests: Biochemical Approaches* (eds W.O.C. Symondson and J.E. Liddell), Chapman and Hall, London. (in press).

Stubbs, M. (1980) Another look at prey detection by coccinellids. *Ecological Entomology*, **5**, 179–82.

Sullivan, D.J. (1971) Comparative behaviour and competition between two aphid hyperparasites: *Alloxysta victrix* and *Asaphes californicus*. *Environmental Entomology*, **1**, 234–44.

Sunderland, K.D. (1975) The diet of some predatory arthropods in cereal crops. *Journal of Applied Ecology*, **12**, 507–15.

Sunderland, K.D. (1987) Spiders and cereal aphids in Europe. *Bulletin IOBC/WPRS Working Group Integrated Control of Cereal Pests*, **10**(1), 82–102.

Sunderland, K.D. (1988) Quantitative methods for detecting invertebrate predation occurring in the field. *Annals of Applied Biology*, **112**, 201–24.

Sunderland, K.D. (1995) Progress in quantifying predation using antibody techniques, in *The Ecology of Agricultural Pests: Biochemical Approaches* (eds W.O.C. Symondson and J.E. Liddell), Chapman and Hall, London. (in press).

Sunderland, K.D. and Sutton, S.L. (1980) A serological study of arthropod predation on woodlice in a dune grassland ecosystem. *Journal of Animal Ecology*, **49**, 987–1004.

Sunderland, K.D. and Vickerman, G.P. (1980) Aphid feeding by some polyphagous predators in relation to aphid density in cereal fields. *Journal of Applied Ecology*, **17**, 389–96.

Sunderland, K.D., Fraser, A.M. and Dixon, A.F.G. (1986a) Distribution of linyphiid spiders in relation to capture of prey in cereal fields. *Pedobiologia*, **29**, 367–75.

Sunderland, K.D., Fraser, A.M. and Dixon, A.F.G. (1986b) Field and laboratory studies on money spiders (Linyphiidae) as predators of cereal aphids. *Journal of Applied Ecology*, **23**, 433–47.

Sunderland, K.D., Crook, N.E., Stacey, D.L. and Fuller, B.T. (1987a) A study of feeding by polyphagous predators on cereal aphids using ELISA and gut dissection. *Journal of Applied Ecology*, **24**, 907–33.

Sunderland, K.D., Hawkes, C., Stevenson, J.H. McBride, J. Smart, L.E., Sopp, P., Powell, W., Chambers, R.J. and Carter, O.C.R. (1987b) Accurate estimation of invertebrate density in cereals. *Bulletin IOBC/WPRS Working Group Integrated Control of Cereal Pests*, **10**(1), 71–81.

Surber, E.W. (1936) Rainbow trout and bottom fauna production in one mile of stream. *Transactions of the American Fisheries Society*, **66**, 193–202.

Suzuki, Y., Tsuji, H. and Sasakawa, M. (1984) Sex allocation and the effects of superparasitism on secondary sex ratios in the gregarious parasitoid, *Trichogramma chilonis* (Hymenoptera: Trichogrammatidae). *Animal Behaviour*, 32, 478–84.

Svensson, B.G., Petersson, E. and Frisk, M. (1990) Nuptial gift size prolongs copulation in the dance fly *Empis borealis*. *Ecological Entomology*, **15**, 225–9.

Syme, P.D. (1975) Effects on the longevity and fecundity of two native parasites of the European Pine Shoot Moth in Ontario. *Environmental Entomology*, **4**, 337–46.

Symondson, W.O.C. and Liddell, J.E. (1993a) The development and characterisation of an anti-haemolymph antiserum for the detection of mollusc remains within carabid beetles. *Biocontrol Science and Technology*, **3**, 261–75.

Symondson, W.O.C. and Liddell, J.E. (1993b) The detection of predation by *Abax parallelepipedus* and *Pterostichus madidus* (Coleoptera: Carabidae) on Mollusca using a quantitative ELISA. *Bulletin of Entomological Research*, **83**, 641–47.

Symondson, W.O.C. and Liddell, J.E. (1993c) A monoclonal antibody for the detection of arionid slug remains in carabid predators. *Biological Control*, **3**, 207–14.

Symondson, W.O.C. and Liddell, J.E. (1993d) The detection of predator-mollusc interactions using advanced antibody technologies. *A.N.P.P. 3rd International Conference on Pests in Agriculture, Montpellier 7–9 December, 1993*, 417–24.

Symondson, W.O.C. and Liddell, J.E. (1993e) Differential antigen decay rates during digestion of molluscan prey by carabid predators. *Entomologia Experimentalis et Applicata*, **69**, 277–87.

Symondson, W.O.C. and Liddell, J.E. (1995) Polyclonal, monoclonal and engineered antibodies to investigate the role of predation in slug population dynamics, in *The Ecology of Agricultural Pests: Biochemical Approaches* (eds W.O.C. Symondson and J.E. Liddell), Chapman and Hall, London. (in press).

Syrett, P. and Penman, D.R. (1981) Developmental threshold temperatures for the brown lacewing, *Micromus tasmaniae* (Neuroptera: Hemerobiidae). *New Zealand Journal of Zoology*, **8**, 281–83.

Tagawa, J. (1977) Localization and histology of the female sex pheromone producing gland in the parasitic wasp *Apanteles glomeratus*. *Journal of Insect Physiology*, **23**, 49–56.

Tagawa, J. (1983) Female sex pheromone glands in the parasitic wasp, genus *Apanteles*. *Applied Entomology and Zoology*, **18**, 416–27.

Tagawa, J. (1987) Post-mating changes in the oviposition tactics of the parasitic wasp *Apanteles glomeratus* L. (Hym., Braconidae). *Applied Entomology and Zoology*, **22**, 537–42.

Tagawa, J. and Hidaka, T. (1982) Mating behaviour of the braconid wasp, *Apanteles glomeratus* (Hym., Braconidae): mating sequence and the factor for correct orientation of male to female. *Applied Entomology and Zoology*, **17**, 32–39.

Tagawa, J. and Kitano, H. (1981) Mating behaviour of the braconid wasp, *Apanteles glomeratus* (Hym., Braconidae) in the field. *Applied Entomology and Zoology*, **16**, 345–50.

Tagawa, J., Asano, S., Ohtsubo, T., Kamomae, M. and Gotoh, T. (1985) Influence of age on the mating behaviour of the braconid wasp, *Apanteles glomeratus*. *Applied Entomology and Zoology*, **20**, 227–30.

Takafuji, A. and Chant, D.A. (1976) Comparative studies on two a species of predacious phytoseiid mites (Acarina: Phytoseiidae) with special reference to their responses to the density of their prey. *Researches on Population Ecology*, **17**, 255–310.

Takagi, M. (1985) The reproductive strategy of the gregarious parasitoid, *Pteromalus puparum* (Hymenoptera: Pteromalidae). 1. Optimal number of eggs in a single host. *Oecologia*, **68**, 1–6.

Takahashi, S. and Sugai, T. (1982) Mating behavior of the parasitoid wasp *Tetrastichus hagenowii* (Hym., Eulophidae). *Entomologia Generalis*, **7**, 287–93.

Tauber, M.J., Tauber, C.A. and Masaki, S. (1986) *Seasonal Adaptations of Insects*, Oxford University Press, New York.

Taubert, S. and Hertl, F. (1985) Eine neue tragbare Insektensaugfalle mit Elektro-Batterie-Betreib. *Mitteilungen Deutsche Gesellschaft für Algemeine und Angewandte Entomologie*, **4**, 433–7.

Taylor, A.D. (1988a) Host effects on larval competition in the gregarious parasitoid *Bracon hebetor*. *Journal of Animal Ecology*, **57**, 163–72.

Taylor, A.D. (1988b) Parasitoid competition and the dynamics of host–parasitoid models. *American Naturalist*, **132**, 417–36.

Taylor, A.D. (1990) Metapopulations, dispersal, and predator–prey dynamics: an overview. *Ecology*, **71**, 429–33.

Taylor, A.D. (1991) Studying metapopulation effects in predator–prey systems. *Biological Journal of the Linnean Society*, **42**, 305–23.

Taylor, R.J. (1984) *Predation*, Chapman & Hall, London.

Thiede, U. (1981) Uber die Verwendung von Acrylglasrohrchen zur Untersuchung der Biologie und Ökologie solitarer aculeater Hymenopteren. *Deutsche Entomologische Zeitschrift*, **28**, 45–53.

Thiele, H.V. (1977) *Carabid Beetles in Their Environments*, Springer-Verlag, Berlin.

Thompson, D.J. (1978) Prey size selection by larvae of the damselfly *Ischnura elegans* (Odonata). *Journal of Animal Ecology*, **47**, 786–96.

Thompson, D.J. (1975) Towards a predator–prey model incorporating age-structure: the effects of predator and prey size on the predation of *Daphnia magna* by *Ischnura elegans*. *Journal of Animal Ecology*, **44**, 907–16.

Thornhill, E.W. (1978) A motorised insect sampler. *PANS*, **24**, 205–7.

Thornhill, R. (1980) Sexual selection in the black-tipped hangingfly. *Scientific American*, **242** (6), 138–45.

Thornhill, R. (1988) Mate choice in *Hylobittacus apicalis*. *Evolution*, **34**, 519–38.

Thornhill, R. (1992) Fluctuating asymmetry and the mating system of the Japanese scorpionfly, *Panorpa japonica*. *Animal Behaviour*, **44**, 867–79.

Thornhill R. and Alcock, J. (1983) *The Evolution of Insect Mating Systems*, Harvard University Press, Cambridge, Mass.

Thornhill, R. and Sauer, P. (1991) The notal organ of the scorpionfly *Panorpa vulgaris*: an adaptation to coerce mating duration. *Behavioural Ecology*, **2**, 156–64.

Thornhill, R. and Sauer, P. (1992) Genetic sire effects on the fighting ability of sons and daughters and mating success of sons in a scorpionfly. *Animal Behaviour*, **43**, 255–64.

Thorpe, W.H. (1939) Further experiments on olfactory conditioning in a parasitic insect. The nature of the conditioning process. *Proceedings of the Royal Society of London, B*, **126**, 370–97.

Thorpe, W.H. and Caudle, H.B. (1938) A study of the olfactory responses of insect parasites to the food plant of their host. *Parasitology*, **30**, 523–8.

Ticehurst, M. and Reardon, R. (1977) Malaise trap: a comparison of 2 models for collecting adult stage of gypsy moth parasites. *Melsheimer Entomological Series*, **23**, 17–19.

Tillman, P.G. and Powell, J.E. (1992) Intraspecific host discrimination and larval competition in *Microplitis croceipes, Microplitis demolitor, Cotesia kazak* (Hym.: Braconidae), and *Hyposoter didymator* (Hym.: Ichneumonidae), parasitoids of *Heliothis virescens* (Lep,: Noctuidae). *Entomophaga*, **37**, 429–37.

Tinbergen, L. (1960) The dynamics of insect and bird populations in pine woods. *Archives Néerlandaises de Zoologie*, **13**, 259–473.

Tinbergen, N. (1932) Ueber die Orientierung des Bienenwolfes (*Philanthus triangulum* Fabr.). *Zeitschrift für Vergleichende Physiologie*, **16**, 305–34.

Tinbergen, N. (1935) Ueber die Orientierung des Bienenwolfes (Philanthus triangulum Fabr.) II. Die Bienenjagd. *Zeitschrift für Vergleichende Physiologie*, **21**, 699–716.

Tinbergern, N. (1951) *The Study of Instinct*, Oxford University Press, Oxford.

Tinbergen, N. (1974) *Curious Naturalists*. Penguin Education, Harmondsworth.

Tingle, C.C.D. and Copland, M.J.W. (1988) Predicting development of the mealybug parasitoids *Anagyrus pseudococci, Leptomastix dactylopii,* and *Leptomastidea abnormis* under glasshouse conditions. *Entomologia Experimentalis et Applicata*, **46**, 19–28.

Tingle, C.C.D. and Copland, M.J.W. (1989) Progeny production and adult longevity of the mealybug parasitoids *Anagyrus pseudococci, Leptomastix dactylopii,* and *Leptomastidea abnormis* (Hym.: Encyrtidae) in relation to temperature. *Entomophaga*, **34**, 111–20.

Todd, F.E. and Vansell, G.H. (1942) Pollen grains in nectar and honey. *Journal of Economic Entomology*, **35**, 728–31.

Toft, C.A. (1983) Community patterns of nectivorous adult parasitoids (Diptera: Bombyliidae) on their resources. *Oecologia*, **57**, 200–215.

Toft, C.A. (1984a) Activity budgets in two species of bee flies (*Lordotus*: Bombyliidae, Diptera): a comparison of species and sexes. *Behavioural Ecology and Sociobiology*, **14**, 287–96.

Toft, C.A. (1984b) Resource shifts in bee flies (Bombyliidae): interactions among species determine choice of resources. *Oikos*, **43**, 104–12.

Tokeshi, M. (1985) Life-cycle and production of the burrowing mayfly, *Ephemera danica*: a new method for estimating degree-days required for growth. *Journal of Animal Ecology*, **54**, 919–30.

Tomiuk, J. and Wöhrmann, K. (1980) Population growth and population structure of natural populations of *Macrosiphum rosae* (L.) (Hemiptera, Aphididae). *Zeitschrift für Angewandte Entomologie*, **90**, 464–73.

Tonkyn, D.W. (1980) The formula for the volume sampled by a sweep net. *Annals of the Entomological Society of America*, **73**, 452–4.

Topham, M., and Beardsley, J.W. (1975) Influence of nectar source plants on the New Guinea sugarcane weevil parasite, *Lixophaga sphenophori* (Villeneuve). *Proceedings of the Hawaiian Entomological Society*, **22**, 145–54.

Toth, R.S. and Chew, R.M. (1972) Development and energetics of *Notonecta undulata* during predation on *Culex tarsalis*. *Annals of the Entomological Society of America*, **65**, 1270–79.

Townes, H.K. (1939) Protective odors among the Ichneumonidae (Hymenoptera). *Bulletin of the Brooklyn Entomological Society*, **34**, 29–30.

Townes, H. (1958) Some biological characteristics of the Ichneumonidae (Hymenoptera) in relation to biological control. *Journal of Economic Entomology*, **51**, 650–52.

Townes, H. (1962) Design for a Malaise trap. *Proceedings of the Entomological Society of Washington*, **64**, 253–62.

Traveset, A. (1990) Bruchid egg mortality on *Acacia farnesiana* caused by ants and abiotic factors. *Ecological Entomology*, **15**, 463–7.

Tretzel, E. (1955) Technik und Bedeutung des Fallenfanges für ökologische Untersuchungen. *Zoologische Anzeiger*, **155**, 276–87.

Trexler, J.C., McCulloch, C.E. and Travis, J. (1988) How can the functional response best be determined? *Oecologia*, **76**, 206–14.

Trimble, R.M. and Brach, E.J. (1985) Effect of color on sticky-trap catches of *Pholetesor ornigis* (Hymenoptera: Braconidae), a parasite of the spotted tentiform leafminer *Phyllonorycter blancardella* (Lepidoptera: Gracillariidae). *Canadian Entomologist*, **117**, 1559–64.

Tripathi, R.N. and Singh, R. (1991) Aspects of life-table studies and functional response of *Lysiphlebia mirzai*. *Entomologia Experimentalis et Applicata*, **59**, 279–87.

Tsuno, K., Aotsuka, N.T. and Ohba, S. (1984) Further genetic variation at the esterase loci of *Drosophila virilis*. *Biochemical Genetics*, **22**, 323–37.

Turnbow Jr, R.H., Franklin, R.T. and Nagel, W.P. (1978) Prey consumption and longevity of adult *Thanasimus dubius*. *Environmental Entomology*, **7**, 695–7.

Turnbull, A.L. (1962) Quantitative studies of the food of *Linyphia triangularis* Clerk (Araneae: Linyphiidae). *Canadian Entomologist*, **94**, 1233–49.

Ullyett, G.C. (1936) Host Selection by *Microplectron fuscipennis*, Zett. (Chalcididae, Hymenoptera). *Proceedings of the Royal Society of London, B*, **120**, 253–91.

Unruh, T.R., White, W., Gonzalez, D. and Luck, R.F. (1986) Electrophoretic studies of parasitic Hymenoptera and implications for biological control, in *Biological Control of Muscoid Flies* (eds R.S. Patterson and D.A. Rutz), Entomological Society of America, Miscellaneous Publication, pp. 150–63.

Unruh, T.R., White, W., Gonzalez, D. and Woolley, J.B. (1989) Genetic relationships among seventeen *Aphidius* (Hymenoptera: Aphidiidae) populations, including six species. *Annals of the Entomological Society of America*, **82**, 754–68.

Unwin, D.M. and Corbet, S.A. (1991) *Insects, Plants and Microclimate, Naturalists' Handbooks No 15* Richmond, Slough.

Van Handel, E. (1972) The detection of nectar in mosquitoes. *Mosquito News*, **32**, 458.

Varley, G.C. and Gradwell, G.R. (1960) Key factors in population studies. *Journal of Animal Ecology*, **29**, 399–401.

Varley, G.C. and Gradwell, G.R. (1965) Interpreting winter moth population changes. *Proceedings of the 12th International Congress of Entomology*, 377–8.

Varley, G.C. and Gradwell, G.R. (1968) Population models for the winter moth, in *Insect Abundance* (ed. T.R.E. Southwood), 4th Symposium of the Royal Entomological Society of London, **4**, 132–142.

Varley, G.C. and Gradwell, G.R. (1970) Recent advances in insect population dynamics. *Annual Review of Entomology*, **15**, 1–24.

Varley, G.C., Gradwell, G.R. and Hassell, M.P. (1973) *Insect Population Ecology: an Analytical Approach*, Blackwell, Oxford.

van Veen, J.C. (1981) The biology of *Poecilostictus cothurnatus* (Hymenoptera, Ichneumonidae) an endoparasite of *Bupalus pinarius* (Lepidoptera, Geometridae) *Annales Entomologici Fennici*, **47**, 77–93.

Veena, T. and Ganeshaiah, K.N. (1991) Non-random search pattern of ants foraging on honeydew of aphids on cashew inflorescence. *Animal Behaviour*, **41**, 7–15.

Vehrencamp, S.L. and Bradbury, J.W. (1984) Mating Systems and Ecology, in *Behavioural Ecology: An Evolutionary Approach* 2nd Edn. (eds J.R. Krebs and N.B. Davies), Blackwell, Oxford, pp. 251–78.

Vet, L.E.M. (1983) Host–habitat location through olfactory cues by *Leptopilina clavipes* (Hartig) (Hym.: Eucoilidae), a parasitoid of fungivorous *Drosophila*: the influence of conditioning. *Netherlands Journal of Zoology*, **33**, 225–48.

Vet, L.E.M. and van Alphen, J.J.M. (1985) A comparative functional approach to the host detection behaviour of parasitic wasps. I. A qualitative study on Eucoilidae and Alysiinae. *Oikos*, **44**, 478–86.

Vet, L.E.M. and Dicke, M. (1992) Ecology of infochemical use by natural enemies in a tritrophic context. *Annual Review of Entomology*, **37**, 141–172.

Vet, L.E.M. and Groenewold, A.W. (1990) Semiochemicals and learning in parasitoids. *Journal of Chemical Ecology*, **16**, 3119–35.

Vet, L.E.M, Janse, C., van Achterberg, C. and van Alphen, J.J.M. (1984) Microhabitat location and niche segregation in two sibling species of drosophilid parasitoids: *Asobara tabida* (Nees) and *A.rufescens* (Foerster) (Braconidae: Alysiinae). *Oecologia*, **61**, 182–8.

Vet, L.E.M., van Lenteren, J.C., Heymans, M. and Meelis, E. (1983) An airflow olfactometer for measuring olfactory responses of hymenopterous parasitoids and other small insects. *Physiological Entomology*, **8**, 97–106.

Vet, L.E.M., Lewis, W.J., Papaj, D.R. and van Lenteren, J.C. (1990) A variable-response model for parasitoid foraging behaviour. *Journal of Insect Behaviour*, **3**, 471–90.

van Vianen, A. and van Lenteren, J.C. (1986) The parasite–host relationship between *Encarsia formosa* Gahan (Hym., Aphelinidae) and *Trialeurodes vaporariorum* (Westwood) (Hom., Aleyrodidae), XIV. Genetic and environmental factors influencing body-size and number of ovarioles of *Encarsia formosa*. *Journal of Applied Entomology*, **101**, 321–31.

Vickerman, G.P. and Sunderland, K.D. (1975) Arthropods in cereal crops: nocturnal activity, vertical distribution and aphid predation. *Journal of Applied Ecology*, **12**, 755–66.

Viggiani, G. and Battaglia, D. (1983) Courtship and mating behaviour in a few Aphelinidae (Hym., Chalcidoidea). *Bolletino de Laboratoria Entomologia Agraria Filippo Silvestri, Portici*, **40**, 89–96.

Viktorov, G.A. (1968) The influence of population density on sex ratio in *Trissolcus grandis* Thoms. (Hymenoptera: Scelionidae). *Zoologichesky Zhurnal.*, **47**, 1035–19 (in Russian).

Viktorov, G.A. and Kochetova, N.I. (1971) Significance of population density to the control of sex ratio in *Trissolcus grandis* (Hymenoptera: Scelionidae). *Zoologichesky Zhurnal*, **50**, 1735–55 (in Russian).

Vinson, S.B. (1972) Effect of the parasitoid *Campoletis sonorenesis* on the growth of its host, *Heliothis virescens*. *Journal of Insect Physiology*, **18**, 1509–14.

Vinson, S.B. (1976) Host selection by insect parasitoids. *Annual Review of Entomology*, **21**, 109–33.

Vinson, S.B. (1985) The behaviour of parasitoids, in *Comprehensive Insect Physiology, Biochemistry and Pharmacology*, (eds G.A. Kerkut and L.I. Gilbert) Pergamon Press, New York, pp. 417–69.

Vinson, S.B. and Barras, D.J. (1970) Effects of the parasitoid *Cardiochiles nigriceps* on the growth, development and tissues of *Heliothis virescens*. *Journal of Insect Physiology*, **16**, 1329–38.

Vinson, S.B. and Iwantsch, G.F. (1980a) Host regulation by insect parasitoids. *Quarterly Review of Biology*, **55**, 143–65.

Vinson, S.B. and Iwantsch, G.F. (1980b) Host suitability for insect parasitoids. *Annual Review of Entomology*, **25**, 397–419.

Vinson, S.B. and Sroka, P. (1978) Effects of superparasitism by a solitary endoparasitoid on the host, parasitoid and field samplings. *Southwestern Entomologist*, **3**, 299–303.

Vinson, S.B. and Stolz, D.B. (1986) Cross protection experiments with two parasitoid (Hymenoptera: Ichneumonidae) viruses. *Annals of The Entomological Society of America*, **79**, 216–18.

Vinson, S.B., Harlan, D.P. and Hart, W.G. (1978) Response of the parasitoid *Microterys flavus* to the brown soft scale and its honeydew. *Environmental Entomology*, **7**, 874–8.

Visser, M.E. (1992) Adaptive self- and conspecific superparasitism in the solitary parasitoid *Leptopilina heterotoma*. *Behavioural Ecology*, **4**, 22–28.

Visser, M.E. and Driessen, G. (1991) Indirect mutual interference in parasitoids. *Netherlands Journal of Entomology*, **41**, 214–27.

Visser, M.E., van Alphen, J.J.M. and Nell, H.W. (1990) Adaptive superparasitism and time allocation in solitary parasitoids: the influence of the number of parasitoids depleting the patch. *Behaviour*, **114**, 21–36.

Visser, M.E., van Alphen, J.J.M. and Hemerik, L. (1992a) Adaptive superparasitism and patch time allocation in solitary parasitoids: an ESS model. *Journal of Animal Ecology*, **61**, 93–101.

Visser, M.E., van Alphen, J.J.M. and Nell, H.W. (1992b) Adaptive superparasitism and patch time allocation in solitary parasitoids: The influence of pre-patch experience. *Behavioural Ecology and Sociobiology*, **31**, 163–72.

Visser, M.E., Luyckx, B., Nell, H.W. and Boskamp, G.J.F. (1992c) Adaptive superparasitism in solitary parasitoids: marking of parasitised hosts in relation to the pay-off from superparasitism, *Ecological Entomology*, **17**, 76–82.

Völkl, W. and Starý P. (1988) Parasitisation of *Uroleucon* species (Hom., Aphididae) on thistles (Compositae, Cardueae). *Journal of Applied Entomology*, **106**, 500–506.

Voller, A., Bidwell, D.E. and Bartlett, A. (1979) *The Enzyme-Linked Immunosorbent Assay (ELISA)*, Dynatech Europe, Guernsey.

Volterra, V. (1931) Variations and fluctuations of the number of individuals in animal species living together (translation from 1928 version), in *Animal Ecology* (R.N. Chapman), Arno, New York.

Vonk, M., Putters, F. and Velthuis, B.J. (1991) The causal analysis of an adaptive system: sex-ratio decisions as observed in a parasitic wasp and simulated by a network model, in *From*

Animals to Animats (eds J.A. Meyer and S.W. Wilson), MIT Press, Cambridge, Massachusetts, pp. 485–91.

Vorley, W.T. (1986) The activity of parasitoids (Hymenoptera: Braconidae) of cereal aphids (Hemiptera: Aphididae) in winter and spring in southern England. *Bulletin of Entomological Research*, **76**, 491–504.

Vorley, W.T. and Wratten, S.D. (1985) A simulation model of the role of parasitoids in the population development of *Sitobion avenae* (Hemiptera: Aphididae) on cereals. *Journal of Animal Ecology*, **22**, 813–23.

Waage, Jeffrey K. (1978) Arrestment responses of a parasitoid, *Nemeritis canescens*, to a contact chemical produced by its host, *Plodia interpunctella*. *Physiological Entomology*, **3**, 135–46.

Waage, Jeffrey K. (1979) Foraging for patchily distributed hosts by the parasitoid *Nemeritis canescens*. *Journal of Animal Ecology*, **48**, 353–71.

Waage, Jeffrey K. (1982) Sib-mating and sex ratio strategies in scelionid wasps. *Ecological Entomology*, **7**, 102–12.

Waage, Jeffrey K. (1983) Aggregation in field parasitoid populations: foraging time allocation by a population of *Diadegma* (Hymenoptera, Ichneumonidae). *Ecological Entomology*, **8**, 447–53.

Waage, Jeffrey K. (1986) Family planning in parasitoids: adaptive patterns of progeny and sex allocation, in *Insect Parasitoids* (eds J.K. Waage and D.J. Greathead), Academic press, London, pp. 63–95.

Waage, Jeffrey K. (1990) Ecological theory and the selection of biological control agents, in *Critical Issues in Biological Control* (eds M. Mackauer, L.E. Ehler and J. Roland), Intercept, Andover, pp. 135–157.

Waage, Jeffrey K. (1992) Biological control in the year 2000, in *Pest Management and the Environment in 2000* (eds A.-A.S.A. Kadir and H.S. Barlow), CAB International, Wallingford, pp. 329–40.

Waage, Jeffrey K. and Godfray, H.C. (1985) Reproductive strategies and population ecology of insect parasitoids, in *Behavioural Ecology, Ecological Consequences of Adaptive Behaviour* (eds R.M. Sibly and R.H. Smith), Blackwell Scientific Publications, Oxford, pp 449–70.

Waage, Jeffrey K. and Greathead, D.J. (1988) Biological control: challenges and opportunities. *Philosophical Transactions of the Royal Society of London, B*, **318**, 111–28.

Waage, Jeffrey K. and Hassell, M.P. (1982) Parasitoids as biological control agents – a fundamental approach. *Parasitology*, **84**, 241–68.

Waage, Jeffrey K. and Lane, J.A. (1984) The reproductive strategy of a parasitic wasp. II. Sex allocation and local mate competition in *Trichogramma evanescens*. *Journal of Animal Ecology*, **53**, 417–26.

Waage, Jeffrey K. and Mills, N.J. (1992) Biological Control, in *Natural Enemies: The Population Biology of Predators, Parasites and Diseases* (ed. M.J. Crawley), Blackwell, Oxford, pp. 412–30.

Waage, Jeffrey K. and Ng. S.-M. (1984) The reproductive strategy of a parasitic wasp. I. Optimal progeny and sex allocation in *Trichogramma evanescens*. *Journal of Animal Ecology*, **53**, 401–16.

Waage, Jeffrey K., Carl, K.P., Mills, N.J. and Greathead, D.J. (1985) Rearing entomophagus insects, in *Handbook of Insect Rearing, Vol. 1* (eds P. Singh and R.F. Moore), Elsevier, Amsterdam, pp. 45–66.

Waage, Jeffrey K. (1973) Reproductive behavior and its relation to territoriality in *Calopteryx maculata* (Beauvois) (Odonata, Calopterygidae). *Behaviour*, **47**, 240–56.

Waage, Jeffrey K. (1979) Dual function of the damselfly penis: sperm removal and transfer. *Science*, **203**, 916–18.

Waage, Jeffrey K. (1984) Sperm competition and the evolution of odonate mating systems, in *Sperm Competition and the Evolution of Animal Mating Systems* (ed. R.L. Smith), Academic Press, New York, pp. 257–90.

Waddington, K.D. (1987) Nutritional ecology of bees, in *Nutritional Ecology of Insects, Mites, Spiders and Related Invertebrates* (eds F. Slansky and J.G. Rodriguez), Wiley International, New York, pp. 393–419.

Wagner, T.L., Wu, H.-I., Sharpe, P.J.H. and Coulson, R.N. (1984) Modeling distributions of insect development time: A literature review and application of the Weibull function. *Annals of the Entomological Society of America*, **77**, 475–87.

Wagner, T.L., Hsin-I, W., Feldman, R.M. Sharpe, P.J.H. and Coulson, R.N. (1985) Multiple-cohort approach for simulating development of insect populations under variable temperatures. *Annals of the Entomological Society of America*, **78**, 691–9.

Waldbauer, G.P. (1968) The consumption and utilization of food by insects. *Advances in Insect Physiology*, **5**, 229–88.

Walde, S.J. (1994) Immigration and the dynamics of predator-prey interaction in biological control. *Journal of Animal Ecology*, **63**, 337–46.

Walde, S.J. and Murdoch. W.W. (1988) Spatial density dependence in parasitoids. *Annual Review of Entomology*, **33**, 441–66.

Walker, T.J. (1993) Phonotaxis in female *Ormia ochracea* (Diptera: Tachinidae), a parasitoid of field crickets. *Journal of Insect Behaviour*, **6**, 389–410.

Wallin, H. (1985) Spatial and temporal distribution of some abundant carabid beetles (Coleoptera: Carabidae) in cereal fields and adjacent habitats. *Pedobiologia*, **28**, 19–34.

Wallin, H. (1986) Habitat choice of some field-inhabiting carabid beetles (Coleoptera: Carabidae) studied by recapture of marked individuals. *Ecological Entomology*, **11**, 457–66.

Wallin, H. and Ekbom, B.S. (1988) Movements of carabid beetles (Coleopt4era: Carabidae) inhabiting cereal fields: a field tracing study. *Oecologia*, **77**, 39–43.

Waloff, N. (1975) The parasitoids of the nymphal and adult stages of leafhoppers (Auchenorrhyncha, Homoptera). *Transactions of the Royal Entomological Society, London*, **126**, 637–86.

Waloff, N. and Jervis, M.A. (1987) Communities of parasitoids associated with leafhoppers and planthoppers in Europe. *Advances in Ecological Research*, **17**, 281–402.

Walsh, J.F and Garms, R. (1980) The detection of plant sugars in *Simulium damnosum s.l.* by means of the Cold Anthrone Test. *Transactions of the Royal Society of Tropical Medicine and Hygiene*, **74**, 811–13.

Walter, G.H. (1988) Activity patterns and egg production in *Coccophagus bartletti*, an aphelinid parasitoid of scale insects. *Ecological Entomology*, **13**, 95–105.

Walton, M.P. (1986) The Application of Gel-Electrophoresis to the Study of Cereal Aphid Parasitoids, Hatfield Polytechnic, Ph.D. Thesis.

Walton, M.P., Powell, W., Loxdale, H.D. and Allen-Williams, L. (1990) Electrophoresis as a tool for estimating levels of hymenopterous parasitism in field populations of the cereal aphid, *Sitobion avenae*. *Entomologia Experimentalis et Applicata*, **54**, 271–9.

Wardle, A.R. (1990) Learning of host microhabitat colour by *Exeristes roborator* (F.) (Hymenoptera: Ichneumonidae). *Animal Behavour*, **39**, 914–923.

Wardle, A.R. and Borden, J.H. (1990) Learning of host microhabitat form by *Exeristes roborator* (F.) (Hymenoptera: Ichneumonidae). *Journal of Insect Behaviour*, **3**, 251–63.

Waterhouse, D.F. (1991) *Guidelines for Biological Control Projects in the Pacific*, South Pacific Commission, Noumea.

Waters, T.F. (1962) Diurnal periodicity in the drift of stream invertebrates. *Ecology*, **43**, 316–20.

Way, M.J. and Banks, C.J. (1968) Population studies on the active stages of the black bean aphid, *Aphis fabae* Scop., on its winter host *Euonymus europaeus* L.. *Annals of Applied Biology*, **62**, 177–97.

Webb, J.C., Calkins, C.O., Chambers, D.L., Schweinbacher, W. and Russ, K. (1983) Acoustical aspects of behaviour of the Mediterranean fruitfly (*Ceratitis capitata*): analysis and identification of courtship sounds. *Entomologia Experimentalis et Applicata*, **33**, 1–8.

Weekman, G.T. and Ball, H.J. (1963) A portable electrically operated collecting device. *Journal of Economic Entomology*, **56**, 708–9.

Wei, Q. (1986) The foraging behaviour of *Cyrtorhinus lividipennis* (Reusen), as a predator of *Nilaparvata lugens* (Stål.). University of Wales M.Sc. Thesis.

Weisser, W.W. and Houston, A.I. (1993) Host discrimination in parasitic wasps: when is it advantageous? *Functional Ecology*, **7**, 27–39.

Wellings, P., Morton, R. and Hart, P.J. (1986) Primary sex-ratio and differential progeny survivorship in solitary haplo-diploid parasitoids. *Ecological Entomology*, **11**, 341–8.

Wellington, W.G. and Fitzpatrick, S.M. (1981) Territoriality in the drone fly, *Eristalis tenax* (Diptera: Syrphidae). *Canadian Entomologist*, **113**, 695–704.

Went, D.F. and Krause, G. (1973) Normal development of mechanically activated, unlaid eggs of an endoparasitic hymenopteran. *Nature, London*, **244**, 454–5.

Werren, J.H. (1980) Sex ratio adaptations to local mate competition in a parasitic wasp. *Science*, **208**, 1157–9.

Werren, J.H. (1983) Sex ratio evolution under local mate competition in a parasitic wasp. *Evolution*, **37**, 116–24.

Werren, J.H. and van den Assem, J. (1986) Experimental analysis of a paternally inherited extrachromosomal factor. *Genetics*, **114**, 217–33.

Werren, J.H., Nur, U. and Eickbush, D. (1987) An extrachromosomal factor causing loss of paternal chromosomes. *Nature*, **327**, 75–6.

Werren, J.H., Skinner, S.W. and Charnov, E.L. (1981) Paternal inheritance of a daughterless sex ratio factor. *Nature*, **293**, 467–8.

Weseloh, R.M. (1972) Sense organs of the hyperparasite *Cheiloneurus noxius* (Hymenoptera: Encyrtidae) important in host selection processes. *Annals of the Entomological Society of America*, **65**, 41–6.

Weseloh, R.M. (1981) Relationship between colored sticky panel catches and reproductive behavior of forest tachinid parasitoids. *Environmental Entomology*, **10**, 131–5.

Weseloh, R.M. (1986) Effect of photoperiod on progeny production and longevity of gypsy moth (Lepidoptera: Lymantriidae) egg parasite *Ooencyrtus kuvanae* (Hymenoptera: Encyrtidae). *Environmental Entomology*, 15, 1149–53.

Weseloh, R.M. (1989a) Temperature-based models of development for the Gypsy Moth (Lepidoptera: Lymantridae) predator, *Calosoma sycophanta* (Coleoptera: Carabidae). *Environmental Entomology*, **18**, 1105–11.

Weseloh, R.M. (1989b) Simulation of predation by ants based on direct observations of attacks on gypsy moth larvae. *Canadian Entomologist*, **121**, 1069–76.

Weseloh, R.M. (1990) Estimation of predation rates of gypsy moth larvae by exposure of tethered caterpillars. *Environmental Entomology*, **19**, 448–55.

Weseloh, R.M. (1993) Manipulation of forest ant (Hymenoptera: Formicidae) abundance and resulting impact on gypsy moth (Lepidoptera: Lymantriidae) populations. *Environmental Entomology*, **22**, 587–94.

Whalon, M.E. and Parker, B.L. (1978) Immunological identification of tarnished plant bug predators. *Annals of the Entomological Society of America*, **71**, 453–56.

Whalon, M.E. and Smilowitz, Z. (1979) The interaction of temperature and biotype on development of the green peach aphid, *Myzus persicae* (Sulz.). *American Potato Journal*, **56**, 591–6.

Whitcomb, W.H. (1980) Sampling spiders in soybean fields, in *Sampling Methods in Soybean Entomology* (eds M. Kogan and D.C. Herzog), Springer-Verlag, New York, pp. 544–58.

Whitcomb, W.H. and Bell, K. (1964) Predaceous insects, spiders, and mites of Arkansas cotton fields. *University of Arkansas, Agricultural Experiment Station Bulletin*, **690**, 1–84.

Whiting, A.R. (1967) The biology of the parasitic wasp *Mormoniella vitripennis* (Walker). *Quarterly Review of Biology*, **42**, 333–406.

Wigglesworth V.B. (1972) *The Principles of Insect Physiology*, Methuen, London.

Wilbert, H. (1977) Honeydew as a source of stimuli and energy for entomophagous insects. *Apidologie*, **8**, 393–400 (in German).

Wilbert, H. and G. Lauenstein (1974) Die eignung von *Megoura viciae* (Buckt.) (Aphid) für larven und erwachsene weibchen von *Aphelinus asychus* Walker (Aphelinidae). *Oecologia*, **16**, 311–22.

de Wilde, J. and de Loof, A. (1973) Reproduction, in *The Physiology of the Insecta, Vol. I* 2nd edn, (ed. M. Rockstein), Academic Press, New York, pp. 11–95.

Wilkes, A. (1963) Environmental causes of variation in the sex ratio of an arrhenotokous insect, *Dahlbominus fuliginosus* (Nees) (Hymenoptera: Eulophidae). *Canadian Entomologist*, **95**, 182–202.

Wilkinson, J.D., Schmidt, G.T. and Biever, K.D. (1980) Comparative efficiency of sticky and water traps for sampling beneficial arthropods in red clover and the attraction of clover head caterpillar adults to anisyl acetone. *Journal of the Georgia Entomological Society*, **15**, 124–31.

Williams, D.D and Feltmate, B.W. (1992) *Aquatic Insects*, CAB International, Wallingford.

Williams, F.M. and Juliano, S.A. (1985) Further difficulties in the analysis of functional-response experiments and a resolution. *Canadian Entomologist*, **117**, 631–40.

Williams, G. (1958) Mechanical time-sorting of pitfall captures. *Journal of Animal Ecology*, **27**, 27–35.

Williams, H.J., Elzen, G.W. and Vinson, S.B. (1988) Parasitoid–host plant interactions emphasizing cotton (*Gossypium*), in *Novel Aspects of Insect–Plant Allelochemicals and Host Specificity* (eds P. Barbosa and D. Letourneau), John Wiley, New York, pp. 171–200.

Williamson, M. (1972) *The Analysis of Biological Populations*, Edward Arnold.

Willis, J.C. and Burkill, I.H. (1895) Flowers and insects in Great Britain. I. *Annals of Botany*, **9**, 227–73.

Willmer, P.G. (1980) The effects of insect visitors on nectar constituents in temperate plants. *Oecologia*, **47**, 270–77.

Willmer, P.G. (1983) Thermal constraints on activity patterns in nectar-feeding insects. *Ecological Entomology*, **8**, 455–69.

Wilson, K. and Goulding, K.H. (1986) *A Biologist's Guide to the Principles and Techniques of Practical Biochemistry*, 3rd edn, Edward Arnold, London.

Wilson, L.T. and Gutierrez, A.P. (1980) Within-plant distribution of predators on cotton: comments on sampling and predator efficiences. *Hilgardia*, **48**, 1–11.

Wiskerke, J.S.C., Dicke, M. and Vet, L.E.M. (1993) Larval parasitoid uses aggregation pheromone of adult hosts in foraging behaviour: a solution to the reliability–detectability problem. *Oecologia*, **93**, 145–8.

Wolf, R. and Wolf, D. (1988) Activation by calcium ionophore injected into unfertilized ovarian eggs explanted from *Pimpla turionellae* (Hymenoptera). *Zoologische Jahrbucher, Abteilung für Allgemeine Zoologie und Physiologie der Tiere*, **92**, 501–12.

Wool, D., van Emden, H.F. and Bunting, S.D. (1978) Electrophoretic detection of the internal parasite *Aphidius matricariae in Myzus persicae*. *Annals of Applied Biology*, **90**, 21–26.

Wool, D., Gerling, D. and Cohen, I. (1984) Electrophoretic detection of two endoparasite species, *Encarsia lutea* and *Eretmocerus mundus* in the whitefly, *Bemesia tabaci* (Genn.) (Hom., Aleurodidae). *Zeitschrift für Angewandte Entomologie*, **98**, 276–9.

Wratten, S.D. (1973) The effectiveness of the coccinellid beetle, *Adalia bipunctata* (L.), as a predator of the lime aphid, *Eucallipterus tiliae* L.. *Journal of Animal Ecology*, **42**, 785–802.

Wratten, S.D. (1994) *Video Techniques in Animal Ecology and Behaviour*, Chapman & Hall, London.

Wratten, S.D. and Pearson, J. (1982) Predation of sugar beet aphids in New Zealand. *Annals of Applied Biology*, **101**, 178–81.

Wright, A.F. and Stewart, A.J.A. (1992) A study of the efficacy of a new inexpensive type of suction apparatus in quantitative sampling of grassland invertebrate populations. *Bulletin of the British Ecological Society*, **23**, 116–20.

Wright, D.W., Hughes, R.D. and Worrall, J. (1960) The effect of certain predators on the numbers of cabbage root fly (*Erioischia brassicae*) (Bouche) and on the subsequent damage caused by the pest. *Annals of Applied Biology*, **48**, 756–63.

Wylie, H.G. (1983) Delayed development of *Microctonus vittatae* (Hymenoptera: Braconidae) in superparasitised adults of *Phyllotreta cruciferae* (Coleoptera: Chrysomelidae). *Canadian Entomologist*, **115**, 441–42.

Wysoki, M., de Jong, M., Rene, S. (1988) *Trichogramma platneri* Nagarkatti (Hymenoptera: Trichogrammatidae), its biology and ability to search for eggs of two lepidopterous avocado pests, *Boarmia (Ascotis) selenaria* (Schiffermuller) (Geometridae) and *Cryptoblabes gnidiella* (Milliere) (Phycitidae) in Israel. *Colloques de l' INRA*, **43**, 295–301.

Yamamura, N. and Yano, E. (1988) A simple model of host–parasitoid interaction with host-feeding. *Researches on Population Ecology*, **30**, 353–69.

Yano, E. (1989a) A simulation study of population interaction between the greenhouse whitefly, *Trialeurodes vaporariorum* Westwood (Homoptera: Aleyrodidae) and the parasitoid *Encarsia formosa* Gahan (Hymenoptera: Aphelinidae). I. Description of the model. *Researches on Population Ecology*, **31**, 73–88.

Yano, E. (1989b) A simulation study of population interaction between the greenhouse whitefly, *Trialeurodes vaporariorum* Westwood (Homoptera: Aleyrodidae) and the parasitoid *Encarsia formosa* Gahan (Hymenoptera: Aphelinidae). II. Simulation analysis of population dynamics and strategy of biological control. *Researches on Population Ecology*, **31**, 89–104.

Yoshida, S. (1978) Behaviour of males in relation to the female sex pheromone in the parasitoid wasp, *Anisopteromalus calandrae* (Hym., Pteromalidae). *Entomologia Experimentalis et Applicata*, **23**, 152–62.

Yoshida, S. and Hidaka, T. (1979) Determination of the position of courtship display of the young unmated male *Anisopteromalus calandrae*. *Entomologia Experimentalis et Applicata*, **26**, 115–20.

Young, C.J., Turner, D.P., Killick-Kendrick, R., Rioux, J.A. and Leaney, A.J. (1980) Fructose in wild-caught *Phlebotomus ariasi* and the possible relevance of sugars taken by sandflies to the transmission of leishmaniasis. *Transactions of the Royal Society of Tropical Medicine and Hygiene*, **74**, 363–6.

Young, L.C. (1980) Field estimation of the functional response of *Itoplectis behrensi*, a parasite of the California oakworm, *Phryganidia californica*. *Environmental Entomology*, **9**, 49–500.

Yu, D.S. and Luck, R.F. (1988) Temperature dependent size and development of California Red Scale (Homoptera: Diaspididae) and its effect on host availability for the ecto-parasitoid *Aphytis melinus* De Bach (Hymenoptera: Aphelinidae). *Environmental Entomology*, **17**, 154–61.

Yu, D.S., Luck, R.F. and Murdoch, W.W. (1990) Competition, resource partitioning and coexistence of an endoparasitoid *Encarsia perniciosi* and an ectoparasitoid *Aphytis melinus* of the California red scale. *Ecological Entomology*, **15**, 469–80.

Yuval, B. and Schlein, Y. (1986) Leishmaniasis in the Jordan Valley. III. Nocturnal activity of *Phlebotomus papatasi* (Diptera: Psychodidae) in relation to nutrition and ovarian development. *Journal of Medical Entomology*, **23**, 411–15.

Zheng, Y., Hagen, K.S., Daane, K.M. and Mittler, T.E. (1993a) Influence of larval dietary supply on the food consumption, food utilisation efficiency, growth and development of the lacewing *Chrysoperla carnea*. *Entomologia Experimentalis et Applicata*, **67**, 1–7.

Zheng, Y., Hagen, K.S., Daane, K.M. and Mittler, T.E. (1993b) Influence of larval food consumption on the fecundity of the lacewing *Chrysoperla carnea*. *Entomologia Experimentalis et Applicata*, **67**, 9–14.

Zwölfer, H. (1971) The structure and effect of parasite complexes attacking phytophagous host insects, in *Dynamics of Populations* (eds P.J. den Boer and G.R.Gradwell), Proceedings of the Advanced Study Institute on 'Dynamics of Numbers in Populations', Oosterbeck, 1970, Centre for Agricultural Publishing and Documentation, Wageningen, pp. 405–18.

Zwölfer, H. (1979) Strategies and counterstrategies in insect population systems competing for space and food in flower heads and plant galls. *Fortschritte für Zoologie.*, **25**, 331–53.

AUTHOR INDEX

Note: page numbers in *italics* refer to tables, those in **bold** refer to figures.

GENUS AND SPECIES INDEX

Note: page numbers in *italics* refer to tables, those in **bold** refer to figures.

SUBJECT INDEX

Note: page numbers in *italics* refer to tables, those in **bold** refer to figures/figure captions.